KC's Problems and Solutions

for

Microelectronic Circuits **FOURTH EDITION**

Sedra / Smith

Kenneth C. Smith
University of Toronto
Hong Kong University of Science and Technology

New York Oxford
OXFORD UNIVERSITY PRESS
1998

Oxford University Press

Oxford New York
Athens Auckland Bangkok Bogota Bombay Buenos Aires
Calcutta Cape Town Dar es Salaam Delhi Florence Hong Kong
Istanbul Karachi Kuala Lumpur Madras Madrid Melbourne
Mexico City Nairobi Paris Singapore Taipei Tokyo Toronto Warsaw

and associated companies in
Berlin Ibadan

Published by Oxford University Press, Inc.,
198 Madison Avenue, New York, New York, 10016
http://www.oup-usa.org
1-800-334-4249

ISBN 0-19-511771-9

9 8 7 6 5 4

Printed in the United States of America
on acid-free paper

Cover Illustration: The chip shown is the ADXL-50 surface-micromachined accelerometer. For the first time, sensor and signal conditioning are combined on a single monolithic chip. In its earliest application, it was a key factor in the improved reliability and reduced cost of modern automotive airbag systems. Photo reprinted with permission of Analog Devices, Inc.

CONTENTS

PREFACE

I OVERVIEW

• THE MANUAL FORMAT

This manual, "KC's Problems and Solutions", is a collection of problems and solutions with compiled answers, designed to accompany the **Text** "*Microelectronics Circuits*", fourth edition, by Sedra and Smith, Oxford University Press, 1997.

The goal of this **Manual**, captured in its former subtitle "Trial and Success", is to motivate and assist in the dynamic process of active learning.

The mechanism provided here includes three parts: I: **Problems,** II: **Solutions,** III: **Answers. Specifically:**

> **Part I: Problems,** consists of a collection of problems keyed to the **Text** in a variety of ways: Most obviously, the problems are grouped according to the Sections of the **Text.** Possibly less apparent is their relationship both to segments of the **Text** and to the end-of-chapter problems contained there, about which more will be said shortly. As well, the problems are coded to indicate Complexity (C), Length (L), and Design content (D), with an appended asterisk notation to indicate the intensity of the associated attribute.
>
> **Part II: Solutions** provides solutions which are relatively detailed. While the presentation is usually in a somewhat compressed format, attention has been given to revealing intermediate analytical and computational steps. As well, additional comments on the interpretation of the Text, and the direction for additional work are relatively common.
>
> **Part III: Answers** allows readers to conveniently evaluate their success at problem solving without the inevitable hints that skimming the actual solution might provide.

• AN APOLOGY TO THE USER - THE LIKELYHOOD OF ERRORS

In a Manual such as this, intended as an aid to the student in a process of active learning, the issue of errors is a very critical one. Obviously, errors embodied in the problem solutions presented here can be very disconcerting to anyone who is less than secure in his or her knowledge of the subject matter. Thus the reduction of errors has been, and will continue to be, a high priority. It is in the latter sense that your indulgence and help are sought in the conjoined processes of error detection and error recovery. Certainly I will be most grateful for your help in reporting them!

In this process of error compensation, it is possibly useful to identify the types of errors you will inevitably find. In order of increasing subtlety and criticality, they are:

Typographical errors:

> There are many types of possible typographical errors which can be broadly characterized as *omission, exchange,* and *replacement,* either in word, number, symbol, phrase or sentence constructs. While unnecessarily confusing, they usually have the virtue of being easily detectable and correctable in context. To assist the detection process at its lowest level, solutions are relatively detailed with lots of intermediate calculations, relatively consistent variable naming, and relatively complete use of units for numerical results. Unfortunately, however, you may

possibly find missing solution lines, as well.

Arithmetic Errors:

These occur between steps in a computation as a result of calculator misuse or transcription error in the original work. They are distinguished from typographical errors by the fact that they propagate. They can be detected only by carefully checking and reproducing the preceeding substitutional and computational steps. Often the integrity of the following solution structure remains, but *not always*. One of the generic methods I use to help ensure structural integrity is an overall test for physical plausibility, or reasonableness, though this is often not documented. However, an explicit demonstration of the attempt to reveal such errors is in the use of frequent *Check* comments which typically employ a recent result in a somewhat-global verification process. Incidentally, this is a good approach *for you to use* in your solutions, as well!

Conceptual Errors:

These are of two kinds, either local or global. The former occur usually as a result of misinterpretation of a symbol, or of the scope of a question. Occasionally you may find a piece of a question that was not answered at all, or answered in a less than complete fashion. The only virtue of this sin is that it is normally detectable. On a far more serious scale will be the occasional occurrence of totally wrong solution methods. These are quite insidious and confusing to a novice, since they can easily be mistaken to be a valid alternative approach. While these are relatively unlikely, they are almost certainly present.

For all of these errors, please accept my apologies. While I have utilized many approaches to minimizing them, the limitations of available time and resources have produced the result you see before you. All that remains to be said, again, is that I beg your indulgence, and look forward to *your* help in improving the situation!

• SOLUTION-PRESENTATION FORMAT

As you will note, the solution format in **Part II: Solutions** in this **Manual** is often less-than-ideal, being basically a run-on string of what would ideally be separated lines. This choice was made in view of the need to reduce the overall size of the **Manual** while making the solution relatively complete, with lots of intermediate steps. Obviously fewer steps in a more structured format would be more readable, and certainly more beautiful, but probably less informative! To help in interpreting the string format, a somewhat-variable attempt at the use of bridging language, sentence structure, and punctuation has been made. For instructive variety, some solutions are presented more elegantly, including more explicit language, both with respect to physical arrangement and description, as well as mathematical structure.

II ADVICE TO THE STUDENT

• COPING WITH ERRORS

As noted earlier, I regret that you are likely to find errors in the solutions presented here. My regret concerns the fact that I am distressingly aware that an error of mine can be difficult to separate from a conceptual difficulty *you* may have. The only positive thing I can say is that learning to cope with imperfection is "good for the soul". Certainly a lot has been written about the positive effects of moderate stress on mental (and physical) development. Ask any reformed couch potato!

But what can *you* do? Certainly compare notes with your colleagues! Revel in the possibility that this **Manual** is an ideal candidate for leisure-time conversation, after a hard day in class or study hall! More seriously, it is certain that a minor degree of cross-checking with others can certainly avoid wasted time. Then, and even on your own, if your solution and mine differ, certainly be prepared for a quick check of obvious things – typos, arithmetic, etc. If you do not find the source of the discrepancy quickly, go on to another one, *as a way to test* **yourself**. If you have trouble there as well, suspect *your own need* for more reading and review of the **Text**. Otherwise a bit more work on checking the solutions is appropriate. Bear in mind, that it is regrettable, but true, that there are errors in these **Solutions**. *Feel good about yourself in finding them!* Feel sadness (and compassion) for my failure to do so! In any case, report them (through our WWW page). We will be grateful!

• THE ROLE OF CIRCUIT-RELATED SKETCHING IN ELECTRONICS-PROBLEM SOLUTION

The merits of sketching in the solution of problems in Electronics cannot be overemphasized! Properly organized, *sketching* constitutes a highly-efficient information-transmission mechanism, *a language* in which relatively complex issues in electronics design and analysis can be presented and communicated. As well, particularly for those broadly conversant with its idioms and dialects, circuit-related sketching can provide the basis for an enriching aesthetic experience, manifesting a kind of "poetry", or "music for the eyes", so to speak. This idea is a very important element in the graphic presentation style seen in the **Text** "*Microelectronics Circuits*", where a lot of use is made of schematic-circuit and waveform sketches. As well, the role of sketching in laboratory work is made quite explicit in the associated **Laboratory Manual** "*Laboratory Explorations*".

Regrettably, here in this **Manual**, "*Trial and Success*", it has not been possible to properly present anything like a complete view of the potential of sketching as language. There are two reasons, one economic, and one paedogogical.

The paedogogical issue appears first in problem presentations, in the use of circuit sketches in Part I: **Problems**. Thus, there, you see some problems posed almost exclusively in terms of circuit sketches. To better appreciate circuit sketches as language, pause for a moment to reflect on how to present problems like these, *without a sketch!* For large electronic assemblages, this can be a very daunting problem: For example, for those of you familiar with SPICE as a Circuit Simulator, contrast the sterility of the SPICE input file – the connection-specification list used in basic simulators (for example in Appendix D of the **Text**) – with the aesthetic elements of the circuit sketch it attempts to describe. It is for this reason that schematic-circuit input to circuit simulators is becoming more common, as you can see, for example, in the Electronic Workbench material, by Interactive Technologies, Inc., provided with the **Text**.

It is for exactly this reason that the graphical user interface provided in "Electronic Workbench" is recommended for practical work associated with the **Text**, particularly as a replacement for (or adjunct to) a "hands-on" laboratory.

On the other hand, to communicate situational detail using spoken and written language is also important! Certainly as a student of Electronics, or of engineering in general, you must be able to handle problems presented in spoken-language style. However *one of the best ways of dealing with such a word problem presented to you, is first to prepare a sketch* of the situation described. Incidentally, for a person proficient in the process of circuit sketching, such a sketch would normally be created incrementally as the text description is scanned, then augmented and checked later, as the text is reread.

In spite of all this, *economic issue* associated with the creation of well-formed drawings in a published work such as this is a very real one. Regrettably, because of the relatively-high cost of production and presentation, there are far fewer sketch-based problems provided to you in this **Manual** than good paedogogy would suggest. In particular, as well, there is a lot of reference to existing figures in the **Text**. Notice, however, that this is a good example of an important engineering principle, that reuse of a costly resource is a logical part of a good engineering solution to any (engineering) problem!

More critically, in terms of illustrating the best style for you to emulate, I must emphasize that there are *far too few sketches* used in the **Solutions** part of this Manual. The ones seen usually arise in response to a direct request for a sketch. While this is *paedogogically wrong*, it is *economically necessary*. More concretely, in *your* work in Electronics, normally without these constraints, the very best and most-effective style I would recommend is to **always** *try a sketch*. "*When in doubt, sketch*", would not be too strong a recommendation to follow. Notice that in the **Text**, an aspect of this idea is embedded in the recurring idea of "working on the diagram" that appears there, for example on pages 248 (numerically) and 267 (analytically). As is illustrated occasionally in the **Solutions** to follow, it is generally *a very good idea* to notate circuit sketches with small calculations or notations, whose role it is to present, memorably, in context, circuit-specific data. For example, a convenient way to notate event timing on digital or pseudo-digital circuits is illustrated on page 364 here in the **Solutions**. In a very broad sense, in general, but certainly in the solution of the relatively intricate problems which appear in this **Manual**, first try to capture the specified situation as a sketch. Then, at or near the appropriate node of the circuit, possibly connected by a pointer line or other reference notation, do the calculations that you can do easily, such as those, for example, relative to bias-point analysis, signal limits, etc. Use these (possibly approximate) results, then, to guide your more elegant and formal solution, and, as well, to provide a rough check on the plausibility of your final results.

• SOLVING A PROBLEM - SOME GENERAL ADVICE

Read the Problem carefully to see if you understand the general idea it attempts to present. As noted earlier, try to present the situation described in a labelled sketch. The preparation of this sketch may be somewhat iterative – first a rough idea with some labels (*to be left in place on your page), then a refined version added, with complete labelling. Note the idea of progression without erasure.* As a general rule, don't eliminate earlier work, either by erasure or abandonment, for it represents the path of your progress, the history of the process of your "learning to learn", the shoulders on which your final solution stands, the available evidence of the logical process you can use when reviewing your work, and so on. Perhaps, later, you may want to make your solution more beautiful for final presentation, but this is often not necessary in the engineering workplace, except for very formal reports required by top management. Notice also that in the phrase "*to be left in place*", I have attempted to suggest avoiding the scraps of paper, the legendary "back of the envelope", and so on, which are relatively inappropriate in a modern responsible decision-path-traceable engineering-design process. It is for these reasons that working engineers often use a bound "Engineering Workbook" to record their progress.

- *In general,* it is often a good idea to redraw the circuit presented in the original problem specification (or photocopy it with segmentation and enlargement, if complex), and then do your work while looking at it, and working *on it,* if that is convenient.
- Prepare an informal summary table of the symbolic and numeric values of specified variables and of the values which you must find in your calculations. It is often useful to organize the solution to your problem by first preparing a tabular format in which you might wish to present the results. Certainly from the point of view of real engineering problem solving, this is a very credible and effective way to both organize your thinking and to prepare for the ultimate presentation of your work to the "boss". Bear in mind, of course, that while all of this is a good idea (else I would not have written about it!), it is often difficult to do, and may be overkill in a simple situation. Whether you use the idea, or not, depends on your particular situation, in the same sense as does the use of refined sketches. **If it helps, do it!** Notice, in general, that most of life's problems are amenable to more than one solution style!
- As a generalization of the detailed comments above, always attempt to make the specifications of any problem you face, whether here, now, or later in real life, as *explicit* as you can. That is what the sketches and tables are all about! Set yourself up, as much as you can, for a multisensory input, for the possibility that a rapid review of the situation through, say, a quick glance at a circuit diagram can crystalize the issue before you, thereby avoiding the forgotten fact, the potential omission, the unnecessary rework, etc.

III GENERAL INFORMATION

• RELATIONSHIP OF THE PROBLEMS HERE TO THE EXERCISES AND PROBLEMS IN THE TEXT

The problems in this **Manual** are intentionally coupled in a variety of ways to the **Exercises** and **Problems** in the **Text**:

- First, you will see that a fraction of the **Problems** are direct variations of those in the **Text**. By and large, these can be seen to represent several situations: One is of the acknowledged existence of a set of relatively basic, classic problems that bear repeating. Another is where problem variety in some subject is somehow limited. Another is a concern for representing, by example, a general approach to creating numerically-different problems in an area where that is often not straightforward. Another is to provide, in conjunction with the Exercises or Problems in the **Text**, an opportunity to see the bigger picture as influenced by a particular set of circuit-design parameters, and thereby experience the issue of design variants, by viewing a few sample points in a related "design space".
- Second, a fraction of the **Problems** presented are coupled more subtley to those in the **Text** by being expansions, extensions, or decompositions of them. By *expansion*, I imply the more detailed examination of an interesting aspect of the **Text** problem. By *extension*, I imply the posing of questions which enlarge the domain of analysis, of design, or of application. By *decomposition*, I refer to the reuse of selected parts of a **Text** problem, often over a wider domain of device parameters, loads, frequencies, etc. The enlarged dimensionality implied by the words *expansion* and *extension* is indicative of the fact that the **Problems** presented are often relatively complex. The arguments, in support of the intended complexity, are many: that real life is complex, that complexity may reinforce in-depth and long-chain thinking, that complexity by added parts implies choice, and, finally, that the existence of **Solutions** as aids, all are intended to justify and support a complex situation that could otherwise be quite difficult.

• AIDS TO SIMULATION

You may notice that a large number of the circuit schematics used in this **Manual** have been prepared using software associated with "Electronics Workbench" by Interactive Image Technologies, Ltd. A major benefit of this approach is the availability of these circuits in a form-compatible with simulation using Electronics Workbench. In the near future, we proposed to make such material selectively available through our WWW site {sedrasmith.org} and in a CD-ROM.

• SOME FACTS OF INTEREST

This **Manual** contains 753 **Problems**, of which 202 involve direct design practice.

ACKNOWLEDGEMENTS

I would like to express my particular appreciation to some of those who made this work possible:

- To Laura Fujino, the love of my life, I am indebted for countless hours of discussion on the processes of problem creation and presentation, as well as for the final camera-ready production, both of this and the first edition.
- To Raymundo Tang Tang, who has prepared a majority of the circuit schematics you will see, both in the **Problems** and in the **Solutions** using Electronics Workbench.
- To Franky Leung, who has solved all of our continuing computer problems, both hardware and software.
- To the Computer Systems Research Institute at the University of Toronto, whose facilities and services were used so intensively in preparing the first edition of this work.
- To the Department of Electrical and Electronic Engineering at the Hong Kong University of Science and Technology, where this second edition was prepared.
-

To these and others more peripherally involved, I am most grateful.

But, for the errors and omissions, you will doubtless find here, I alone am responsible. For them, I must again apologize, and thank you in advance for your tolerance and forebearance in enduring and reporting them.

Kenneth Carless Smith, PhD, LFIEEE, PEng

Department of Electrical and Computer Engineering
University of Toronto
10 King's College Rd.
Toronto, Ontario, M5S 1A4
Canada

FAX: 416 971 2286

Email: lfujino@cs.toronto.edu

August 1997

PART I
PROBLEMS

pages 1 to 131

CHARACTERIZATION CODE

C Complex

D Design

L Long

Where suffixes * and ** indicate
indicate more and much more
of the preceeding attribute.

Chapter 1

INTRODUCTION TO ELECTRONICS

SECTION 1.1: SIGNALS

L

1.1 For the following circuits, identify the signal-source form, whether Thevenin or Norton, and provide, in an organized two-column table, sketches of both standard forms. Where appropriate, reduce the circuit to its single-source, single-impedance form. Be careful with the polarities of voltage and current generators.

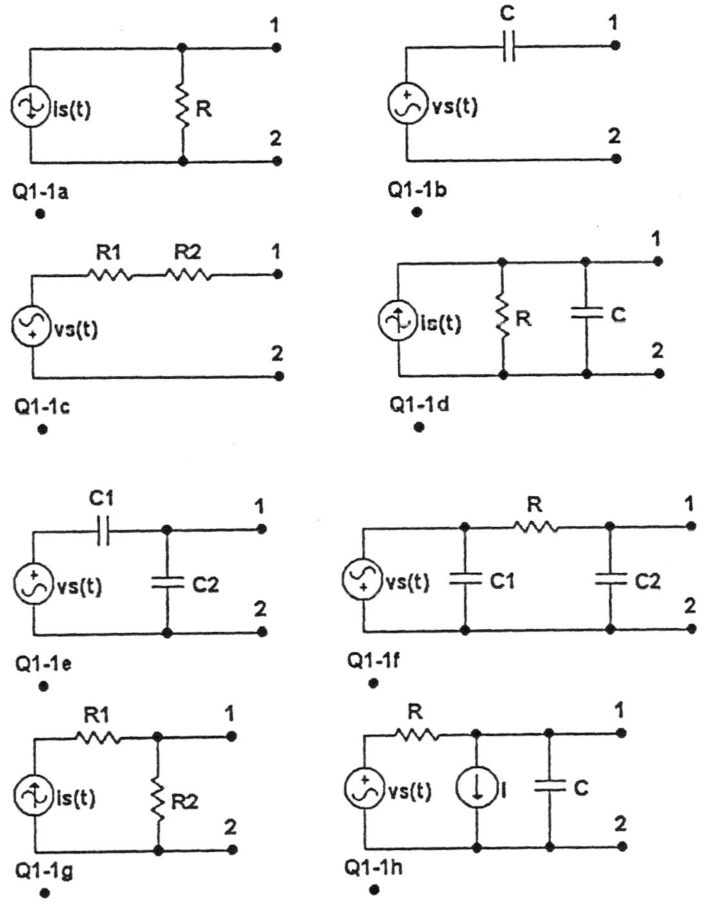

SECTION 1.2: FREQUENCY SPECTRUM OF SIGNALS

1.2 For the following signals whose frequency is expressed either in radians per second or Hertz, find the corresponding value in the alternate form. Provide your answers in a neat five-column format, a line label at the left, Hz next left, rad/s at middle right, and 2 blank columns at the far right.

(a) 60Hz, (b) 754 rad/s, (c) 2513.3 rad/s, (d) 1010 kHz, (e) 97.30 MHz, (f) 1 Hz, (g) 377 rad/s, (h) 1 rad/s, (i) 1 GHz, (j) 400 GHz.

L

1.3 For each part of the previous question find the period of the signal. Express it in seconds in two ways, using 3 significant digits:

a) with one left of the decimal point and with an appropriate power of 10, and

b) using the standard names for subdivisions (seconds(s), milliseconds(ms), microseconds(μs), nanoseconds (ns), picoseconds (ps), femptoseconds (fs)).

Create your answer in two ways:

i) directly from the specifications given in the previous question,

ii) the easiest way, using data from your table.

Use the 2 far-right columns in the answer table of P1.2 above for your answer (first using powers of 10, then names).

1.4 An oscillator, operating in an instrument at 10.7 MHz, is said to be stable within 3 parts-per-million per-degree-Celsius variation in temperature. What change of period would you expect from the moment it is first turned on in a room at 25°C, until it finally reaches its internal operating temperature at 50°C?

1.5 Three individuals, when asked to characterize different sine-wave signals presented to them, state:

a) 0.20 V peak-to-peak at 1000 Hz,

b) 2.12 V rms, with a 20μsec period,

c) 1.0 V peak amplitude, and a frequency of 12.57 rad/s.

Find the amplitude and frequency ratios which characterize the 3 signals using a) as the reference.

1.6 What fraction of the energy in a square wave of frequency f and 10 V amplitude is contained in harmonics above $9f$? at and above $3f$?

1.7 An ideal low-pass filter with cutoff frequency f passes all signal energy below f, and rejects all signal energy above. Find the cutoff frequency of a low-pass filter such that square waves at 1 khz and 2 kHz, with amplitudes of 1.1V and 1.2V respectively, provide nearly the same output-power levels.

SECTION 1.3: ANALOG AND DIGITAL SIGNALS

1.8 A square wave at frequency f can be considered to be the result of sampling a sine wave of frequency f twice per cycle (at a uniform rate of $2f$), and extending the measured value until the next sample. For this interpretation, characterize the result of sampling a 1V rms sine wave:

a) exactly at its peaks,

b) at 90° from a negative-going zero crossing,

c) at 45° from a positive-going zero crossing.

What waveform results for case a) if the sampling frequency is
i) doubled, ii) halved?

1.9 A designer wants to represent all decimal numbers from 0 to 33. How many bits are needed? What are the binary representations for 0, 7, 15, 31 and 33? What is the largest value that can be represented?

C

1.10 A second designer involved in creating a low-cost version of the application situation introduced in P1.9 above, realizes that only the even numbers from 0 to 30 must be represented. How many bits are needed? What are the binary representations she can use for 0, 8, 14, 28? What is the largest value that can be represented in this low-cost version?

1.11 Consider the 8-bit digital-signal representation shown in Figure 1.8 of the Text. If the most-significant bit (MSB) is sent first (at time 0), what value D is represented if a) all bits are positive, b) all but the MSB is positive; and the MSB has a negative weight (that is, b_o is negative, while b_1 through b_7, are positive). In each case, what is the value represented if the MSB is reversed (thus becoming logic 0)?

1.12 Reconsider the situation presented in P1.11 above, but with the MSB (b_o) appearing last in time. What is the value of D, the number represented? What value D is represented if a) all bits are positive, b) the MSB (alone) has a negative weight, c) the MSB is considered to be a sign bit with zero weight, 1 being the negative sign. What values are represented in each of these three interpretations, if the MSB is reversed (that is, to take on the logic value 1)?

1.13 For a 5-bit digital representation, what are the largest and smallest numbers that can be represented? What decimal value D corresponds to the 5-bit number 01101 written in *conventional* form. In a modern instrumentation system using a 3V supply, the digit voltages are 0V and 3V for logic 0 and logic 1 respectively. For an associated 5-bit DAC circuit, the most-significant digit (alone) produces an output of $3/2 =$ 1.5V. To what output voltage does the number 01101 correspond? What is the highest available voltage-output value? What is the smallest non-zero output value? What available output is closest to 1.00V? To what digital input to the DAC does this correspond?

SECTION 1.4: AMPLIFIERS

1.14 Measurements made on a set of amplifiers, labelled a) through e), provide the attributes tabulated below. Calculate those missing elements needed to characterize each. Each amplifier uses ± 10V supplies with no dc ground connection. Signal connections are with respect to ground, however. Signals are assumed to be sine waves whose peak values are given. Amplifier a) has been completely characterized by way of example.

	Supply			Input				Output				A_v		A_i		A_p		Eff.
#	I_+ mA	I_- mA	P mW	v_i mV	i_i μA	R_{in} kΩ	P_{in} μW	v_O V	i_O mA	R_{load} kΩ	P_{out} mW	ratio V/mV	dB	ratio mA/μA	dB	ratio mW/μW	dB	%
a	3	3	60	1	1	1	.0005	2	20	0.1	20	2	66	20	86	4×10^4	76	33
b		1		20		.01		1		1								
c					10^3	0.1			10		10							10
d			200			.01	10				40	0.2						
e						10		0.5		10						0.1		20

1.15 An amplifier operating from ± 10V supplies has a linear transfer characteristic passing through (0, 0), but with output saturation at +7V and −9V. If the amplifier gain is 50 V/V, what is the largest sine-wave input having no dc component, that can be applied without clipping?

1.16 For the situation described in P1.15 above, it is desired to have the largest possible unclipped output, and a dc component can be tolerated. What is the rms value of the largest possible sine wave at the output and at the input? What is the dc output component? To what dc value must the input be biassed?

1.17 An amplifier having a transfer characteristic

$$\upsilon_O = 8 - 4\,(\upsilon_I - 1)^2$$

with

$$1 \le \upsilon_I \le \upsilon_O + 1\ , \quad \upsilon_O \ge 0$$

is to operate with a dc output voltage of 4V. For an output signal of ≤1 volt peak amplitude at the input frequency ω, what % second-harmonic distortion results? (HINT: See Problem 1.15 on page 30 in the Text)

1.18 Repeat Example 1.2 on page 17 of the Text, for the situation in which

$$\upsilon_O = 5 - 10^{-10}\, e^{40\upsilon_I}$$

for $\upsilon_I > 0$ and $\upsilon_O \ge \upsilon_I$ with the output biassed at $V_O = +5/2$ volts. Find V_I, $L+$, $L-$, the peak magnitude V_i of the output sine wave allowed, and the voltage gain A_υ at the bias point.

SECTION 1.5: CIRCUIT MODELS FOR AMPLIFIERS

1.19 A voltage amplifier connected to a particular source υ_s has a no-load voltage gain of 100 V/V and a gain of 70 V/V with a 1 kΩ load. What is its output resistance? What is its gain with a 500 Ω load?

1.20 A voltage amplifier, when connected to a 10 kΩ source, has an overall gain (υ_o/υ_s) of 1667 V/V. When a second identical amplifier is connected in parallel to the same source, the corresponding gain for each is found to be 909 V/V. Estimate the input resistance of the amplifiers.

1.21 A voltage amplifier has an open-circuit voltage gain of $A_{\upsilon o}$, an input resistance R_i, and an output resistance R_o. Find the condition under which a cascade of n of these amplifiers has the same open-circuit gain as a single amplifier.

D

1.22 A design is required of a voltage amplifier to operate between a 1 MΩ source and a 100 Ω load. You have two amplifiers, each with a gain 10 V/V, but with the input and output resistances of A_1 being 1 MΩ and 10 kΩ, respectively, and of A_2 being 10 kΩ and 100 Ω, respectively. There are two possible ways to connect the two amplifiers between the source and load. Which is best? What is the highest overall gain? Contrast this with the gain using only one amplifier at a time? If a good fairy granted you one wish – to double (or halve) any one property of either amplifier – is there a best choice to be made? Why?

1.23 A voltage amplifier with a basic gain of 80 dB, has an output resistance of 10 kΩ. What is the voltage gain which results for loads of 1 MΩ, 10 kΩ, 10Ω? What is its equivalent transconductance when operating into a zero-ohm load?

DL*

1.24 This problem is intended to provide you with a basis for insight into Problem 1.21 on page 51 in the Text.

(a) Evaluate the gain υ_o/υ_s for each of the amplifier stages described there interposed individually between the stated source and load.

(b) From the process and results of (a), identify where the least loss occurs, whether at the source or load, for each amplifier. Use these observations to make 3 lists of amplifiers (in which amplifiers are put in descending order of merit), as input-stage coupler, output-stage coupler, and as provider of gain.

(c) Now consider a design with a pair of amplifiers, picking, as input, an amplifier high on list 1 and reasonable on list 3, and, as output, one high on list 2 and reasonable on list 3.

(d) What is the highest gain you can get from two stages?

(e) Reconsider the process outlined above, in an attempt to see if you could reach the same conclusion by simply thinking about it, rather than by making explicit lists.

DL*

1.25 You are required to design a two-stage current amplifier to operate between a current source having a 10 kΩ internal resistance and a load of 10 kΩ. Three types of amplifier stage are available:

(1) A low-input-resistance type, with R_i = 10 Ω, R_o = 10 kΩ and A_{is} = 100A/A

(2) A high-gain type, with R_i = 10 kΩ, R_o = 1 kΩ and A_{is} = 1000A/A

(3) A high-output resistance type, with R_i = 10 kΩ, R_o = 100 kΩ and A_{is} = 100A/A.

How many two-stage amplifier combinations are there? Rank them by available gain.

D

1.26 Reconsider Problem 1.25 above. Rank the 3 amplifiers on the basis of a *figure of merit* (for current amplifiers) which is $\left(\dfrac{A_{is} \times R_o}{R_i}\right)$. Select the two amplifiers of *lowest rank*, and use only those types to design a two-stage current amplifier of highest-possible gain between a 10 kΩ source and 10 kΩ load. What is the highest available gain?

D

1.27 Reconsider the three amplifiers introduced in Problem 1.25 above as transconductance amplifiers. Restate the specifications of each as a transconductance amplifier. Identify a figure of merit for a transconductance amplifier like that suggested in Problem 1.26 above for a current amplifier. Use this to rank the three as transconductance amplifiers.

1.28 Using the results of Example 1.4 (on page 25 of the Text) for a BJT, characterize its use with E grounded, B as input and C as output, both as a current amplifier and as a transconductance amplifier. Use $r_\pi = 5$ kΩ and $\beta = 200$. What are A_{is} and G_m respectively?

1.29 For the BJT circuit shown in Figure E1.14 on page 28 of the Text, find expressions for the voltage gain v_e/v_b and the resistance seen by resistor R_e connected between the emitter and ground. (Hint: to find the latter, use a test voltage as in Example1.4 in the Text)

1.30 For the BJT circuit shown in Figure E1.14 on page 28 of the Text, find expressions for the voltage gain v_c/v_b, and the resistance seen by R_L.

1.31 Use the results of Exercise 1.14 on page 28 of the Text and those from P1.29 above, to find an expression for the voltage gain v_e/v_s when a source v_s, whose source resistance is R_S, is connected to the base. What is the value of R_S for which v_e/v_s is half the value of v_e/v_b found in P1.29 above.

SECTION 1.6: FREQUENCY RESPONSE OF AMPLIFIERS

1.32 In passing through a particular amplifier, an input sine wave of 2 mV peak-to-peak amplitude at 1 kHz emerges with the same wave shape, an amplitude increased to 2V peak, and evidence that is has been delayed by 0.2 ms. For the amplifier transmission, what is the magnitude? What is the phase?

1.33 A direct-coupled (dc) amplifier (one whose response extends down to zero frequency) has an upper 3 dB frequency of 100 kHz. What is its bandwidth? When coupled to a signal source using a capacitor, its frequency response is found to deteriorate at low frequencies, the response being reduced by 3 dB at 20 kHz. What is the overall bandwidth of this arrangement?

1.34 Consider the circuits of Fig. 1.22 (on page 31 of the Text). In a particular system application, a new output $V_{out} = V_i - V_o$ is created in each case. What is the type of the corresponding output V_{out} for circuit a)? circuit b)?

1.35 An amplifier, considered to have a high-frequency response which can be characterized as STC, is measured at 3 frequencies, 1 kHz, 10 kHz and 20 kHz, at which the gain magnitude is found to be 11×10^3, 8×10^3, and 4×10^3V/V, respectively. Estimate the 3 dB frequency and the frequency at which the gain can be expected to drop to 1. At what frequency does a phase lag of 60° or so appear?

CDL

1.36 Consider one stage of the amplifier cascade in Fig. P1.37 (on page 54) of the Text. At what frequency is its response 3 dB down from the midband value? For 2 stages in cascade, what does the 3 dB frequency become? For a modified 2-stage cascade in which one of the resistors is decreased to kR (k≤1), find a process to calculate what the frequency becomes. For what value of k does $f_{3\,dB}$ of the modified 2-stage cascade have a value $\frac{0.95}{2\pi RC}$?

1.37 A voltage amplifier has the transfer function

$$T(f) = \frac{1000}{\left(1 + j\frac{f}{10^5}\right)\left(1 + \frac{10}{j\,f}\right)}$$

On a Bode magnitude plot, sketch asymtotes representing each of the terms shown. Then sketch the overall (sum) response. What do each of the three terms contribute (in dB) at f = 1, 10, 100, 10^4, 10^5 and 10^6 Hz. What is the overall response at the same frequencies? What is the 3 dB bandwidth of the amplifier? Over what frequency range is the phase 0 ± 6°?

1.38 A voltage amplifier has the transfer function

$$T(f) = \frac{10^7\, jf}{\left(jf + 10^5\right)\left(\frac{jf}{10} + 1\right)}$$

Note that this is not in the most useful standard form. Without converting it explicitly, what are the upper and lower 3 dB frequencies and what is the midband gain (i.e. the gain between the upper and lower cutoffs)? Now reduce $T(f)$ to standard form, and consider the same questions: Do you have a preference for one form over the other?

D

1.39 Consider the transconductance amplifier in Table 1.1 (on page 24) of the Text driving a load capacitance of C = 10pF and driven by a 10 kΩ source, R_s. Find expressions for the gain at low frequencies and the associated upper 3 dB frequency. For one particular amplifying device, namely a BJT, both R_i and

R_o are inversely proportional to bias current I, while G_m is directly proportion to it. Typically,

$$R_i = \frac{2.5}{I}, \quad R_o = \frac{200}{I} \quad \text{and } G_m = 40I$$

Design the circuit bias current so that the resulting upper 3 dB frequency is 1 MHz or more. What is the midband gain A_M that results? Using the expressions you have derived, find the product of gain and bandwidth. What is interesting about it? Use this result to state the gain of an amplifier whose bias is adjusted for a 3 dB frequency of 10 MHz. What current is needed?

DC

1.40 Consider the circuit of Figure 1.25 of the Text in which the output is augmented in two ways: capacitor C_2 couples R_L to another load resistor R_2, and C_i (a small capacitance) is shunted by a relatively large capacitor C_1. Here, R_s = 20kΩ, R_i = 100kΩ, R_o = 200Ω, R_L = 1kΩ, R_2 = 1kΩ and $\mu = 100V/V$. What is the nominal gain at midband frequencies, where the effects of C_1 and C_2 are ignored, that is, C_1 is considered to be very small, and C_2 is considered to be very large? Find values for C_1 and C_2 so that the amplifier has a relatively narrow midband region extending from 20kHz to 80kHz. What gain results at 40kHz? Over what frequency range is the gain within 1dB of the midband value. Here, the 3dB bandwidth is designed to be 80 – 20 = 60kHz. What is the 1dB bandwidth? (Hint: Follow the general appraoch implied in Equation 1.24 on page 34 of the Text and in Exercise 1.17 on page 38 there.

1.41 Find the transfer function of the circuit shown: Sketch its magnitude and phase.

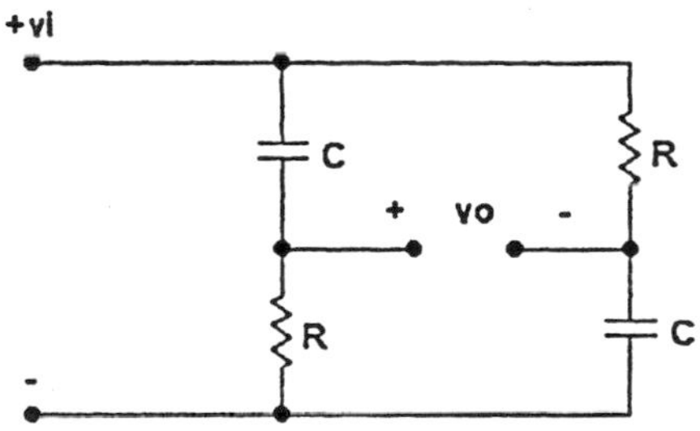

SECTION 1.7: THE DIGITAL LOGIC INVERTER

D

1.42 An amplifier, operating from a 5V supply limits 1.5V from the upper supply rail (at 5V) and 0.5V from the lower rail (at 0V). It has a relatively constant gain of –10V/V in the transition region which is centered at v_I = 2.5V. Using the three-segment-transfer-characteristic inverter model of Fig. 1.29 of the Text, find V_{OL}, V_{OH}, V_{IL}, V_{IH}, NM_L, NM_H. How wide is the transition region? If the transition region is doubled in width due to a manufacturing error, what do the noise margins become? By what factor do they change? If you, as a designer had a choice of relocating the center the transition region, what value would you chose in order to equalize the impact of lower gain on noise margins?

1.43 For a particular logic inverter modelled by the circuit of Fig.1.31c) of the Text, V_{DD} = 5V, R = 1kΩ, R_{on} = 50Ω, V_{offset} = 50mV. Find V_{OH} and V_{OL}. What static power is dissipated for input high? For input low? If the switch also has a 5kΩ leakage, what does V_{OH} become? What is the average static power loss of this "leaky inverter" for 50% duty cycle?

1.44 For a logic inverter whose operation is modelled by the complementary-switch circuits of Fig.1.32 of the Text, V_{DD} = 5V and R_{on} = 50Ω. Find V_{OL}, V_{OH}, and the average static power dissipation of the inverter. If each switch has a 5kΩ leakage, what do V_{OL}, V_{OH} and the average power become?

D

1.45 Consider the switched-current logic represented in Fig.1.33 of the Text. For I_{EE} = 4mA, what values of R_{C1} and R_{C2} are required to achieve a 1V logic swing? For V_{CC} = 0V, what values of V_{OL} and V_{OH} result? To achieve equal noise margins, at what value of υ_I should the switch be made to operate? (Note that the switch is usually modelled to have $V_{IL} = V_{IH}$.) If V_{EE} = 5V, what is the average static power dissipation in the circuit? Provided operation is otherwise OK, does switch resistance affect the total power dissipation of the gate?

1.46 Reconsider the situation described in P1.44 above in which the logic gate, loaded by a 10pF capacitor, operates at 100MHz. What is the dynamic power dissipation which results? Estimate the transition times and propagation delay for this inverter following the definition in Fig.1.35 of the Text and assuming that the switches operate instantaneously at $\upsilon_I = V_{DD}/2$.

1.47 Reconsider the situation presented in P1.44 above for V_{DD} reduced from 5V to 3V, with the switches still operating at $V_{DD}/2$.

1.48 A current-mode-logic gate modelled by the circuit in Fig.1.33 of the Text, uses I_{EE} = 4mA, V_{CC} = 0V, V_{EE} = 5V and $R_{C1} = R_{C2}$ = 250Ω. The logic load connected to each output can be modelled by a 3 pF capacitor. Sketch and label the output waveforms that result for a sequence of 2 switch reversals. Estimate values for V_{OL}, V_{OH}, t_{TLH}, t_{THL}, t_{PLH}, t_{PHL} for each output, assuming switch operation to occur instantaneously at its operating threshold. For this gate operating at 200 MHz with 50% duty cycle, what are the static, dynamic and total power consumptions?

Chapter 2

OPERATIONAL AMPLIFIERS

SECTION 2.1: THE OP AMP TERMINALS

2.1 What is the number of op amps that can be accomodated in an 8-pin IC package? In a 14-pin package? How many unused pins are there in each case?

SECTION 2.2: THE IDEAL OP AMP

2.2 An otherwise-ideal op amp, known to have a gain of 10^4 V/V, is measured in a circuit to have an output voltage of –3 V. While it would be difficult to measure, what would you expect the voltage from the negative input pin to the positive one to be? If the voltage at the positive pin is known to be +100 mV, what is the voltage you would expect at the negative one?

2.3 For the amplifier described in P2.2 above, connected in the circuit shown in Fig. P2.2 (on page 110 of the Text), what voltage v_I would be required at the input to produce v_O = 3.5 V?

SECTION 2.3: ANALYSIS OF CIRCUITS CONTAINING IDEAL OP AMPS – THE INVERTING CONFIGURATION

2.4 An inverting op-amp circuit with the topology of Fig. 2.4 on page 65 of the Text, has R_1 = 4.7 kΩ and R_2 = 47 kΩ. What closed-loop gain would you expect? In the laboratory, a student accidentally exchanges these two resistors. What gain would you expect him to find?

2.5 The circuit shown in Fig. P2.8 c), (on page 111 of the Text) using an op amp with a gain of 10^4 V/V, is found to have an output voltage of +10 V. What is the voltage required at the inverting input terminal of the op amp for this to occur? What is the current through the grounded 10 kΩ resistor? What is the precise input voltage, v_I, you would expect? (Hint: First, consider this question assuming that the gain (10^4) is very very high. Then, refine your answer with a calculation in which a very small error correction is made).

D

2.6 Design an op-amp circuit with a gain of –2 V/V, using three 100 kΩ resistors. How many solutions are there? What is the input resistance of each?

D

2.7 Design an inverting op-amp circuit with a gain whose magnitude is 10 V/V using one 220 kΩ resistor and another resistor no greater than 1 MΩ.

CD

2.8 Design an amplifier with a gain of –20 V/V, an input resistance of 100 kΩ, and no resistor greater than 1 MΩ. (Hint: you need more than 2 resistors! But not 4!)

2.9 An inverting op-amp circuit is designed to use one 10 kΩ and one 100 kΩ resistor. What are the two possible closed-loop gains you would expect with an ideal op amp? What gains do you get with an op amp whose open-loop gain is only 100 V/V?

2.10 An inverting op-amp circuit designed for a nominal gain of –100V/V uses a very high-frequency amplifier whose open-loop gain is relatively low. What must the amplifier gain be if the closed-loop gain is to lie within 10% of the nominal value? Within 1% of nominal?

2.11 For the inverting amplifier shown in Fig. 2.6 in the Text, find the input resistance R_i of the feedback circuit connected to the rightmost end of R_1 (namely the amplifier with gain –A and feedback resistor R_2). [Hint: Follow the general approach used in the analysis leading to Equation 2.1 with R_i being the ratio of the voltage at the negative input terminal and the current in R_2.] The mechanism, that causes R_i to be quite small is called the *Miller Effect*. Use R_i with R_1 and A to calculate G. Compare the result with Equation 2.1.

2.12 A relatively ideal op amp with open-loop gain A is connected in a circuit with its positive input grounded and an unmarked resistor R_f connected between its output and negative-input terminals. A 10μA test current is injected into the negative-input connection, where a voltage of 10.1 mV is measured. A corresponding measurement at the output shows v_O to be –978 mV. Estimate the value of the equivalent input resistance at the negative-input node, the amplifier open-loop gain A, and the actual value of the feedback resistor R_f. What is likely to be the nominal value of R_f? What is its corresponding tolerance? For what value of resistor joining a source v_S to the negative input terminal is v_O/v_S = – 10.00V/V.

D

2.13 Design an amplifier with a gain of +200 V/V and an input resistance of 100 kΩ using 2 op amps and resistors no larger than 1 MΩ. Share the gain as much as is convenient between the two amplifiers.

D

2.14 Reconsider P2.13 above if R_{in} must be 2 MΩ. Use a minimum number of resistors.

CD

2.15 Design the circuit of Fig. 2.8 on page 69 of the Text to have an input resistance of 1 MΩ and a gain of –22 V/V using resistors no larger than 1 MΩ. If resistors no smaller than 100 kΩ are available, what do you do?

C

2.16 Consider the circuit of Fig 2.8 of the Text with the grounded end of R_3 connected to input v_2, and v_1 connected to R_1. Use the approach in Example 2.2 (on page 69 of the Text) and superposition, to find an expression for v_O in terms of v_2 alone, and of v_1 and v_2 together.

SECTION 2.4: OTHER APPLICATIONS OF THE INVERTING CONFIGURATION

L

2.17 Find the transfer function of the following circuit:

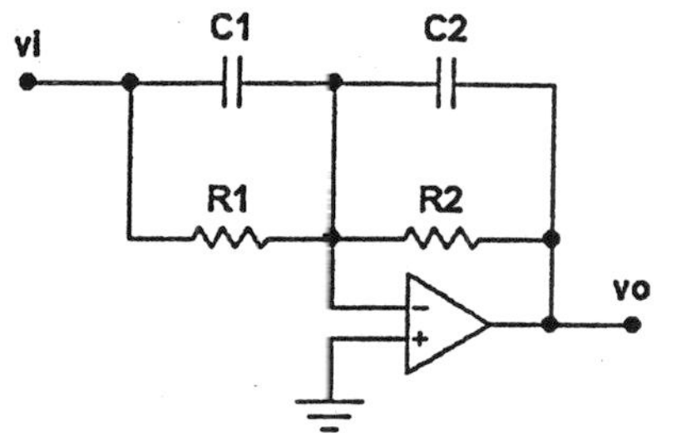

What is the condition for which the output is independent of frequency? Sketch Bode magnitude plots (in rad/s) for 3 cases:

a) $C_2 = 0.1C_1 = 0.1\mu F$, $R_2 = 10R_1 = 100k\Omega$;

b) R_2 is raised to 1MΩ;

c) R_2 is lowered to 10kΩ.

2.18 A Miller integrator for which the time constant is 1 ms is driven by a positive step of 1 volt amplitude. What does the output do? At what rate? If the initial output voltage is 10 V, how long does it take for the output to reach 0 V?

2.19 A Miller integrator with a time constant of 10 ms is driven by a 60 Hz sine wave of 0.1 V peak amplitude. Describe the resulting output waveform, in amplitude and phase. Is the output leading or lagging the input?

C

2.20 Consider a differentiator circuit such as that shown in Fig. 2.14 a), on page 79 of the Text, having a 5 ms time constant. For what rate of change of input signal is the output +1 V? An input signal begins to rise from zero volts at $t = 0$ at the rate of 1 V/ms, reaches a value of 20 V, then falls at the same rate to zero volts. Sketch and label the resulting output waveform over an interval of 50 ms.

L

2.21 The differentiator circuit of Fig 2.14 a) of the Text is augmented by a resistor $r = 100\ \Omega$ in series with $C = 1.0\ \mu F$. Resistor $R = 10\ \text{k}\Omega$. Sketch and label the output if the input is:

a) a positive pulse of 0.1 V amplitude and 10 μs duration,

b) a negative pulse of 50 mV amplitude and 0.1 s duration.

D

2.22 Design a circuit with 3 inputs to provide an output $\upsilon_O = -(\upsilon_1 + 2\ \upsilon_2 + 3\ \upsilon_3)$ using 10 kΩ as the smallest resistor.

D

2.23 Design a circuit to combine 3 inputs to form $\upsilon_O = \upsilon_1 + 2\ \upsilon_2 - 3\ \upsilon_3$. Use only inverting amplifiers, with 10 kΩ as the smallest resistor. There is more than one way! Find one which minimizes the total resistance used.

2.24 For the following circuit, find an expression for the output υ_O in terms of υ_1 and υ_2, assuming an ideal op amp.

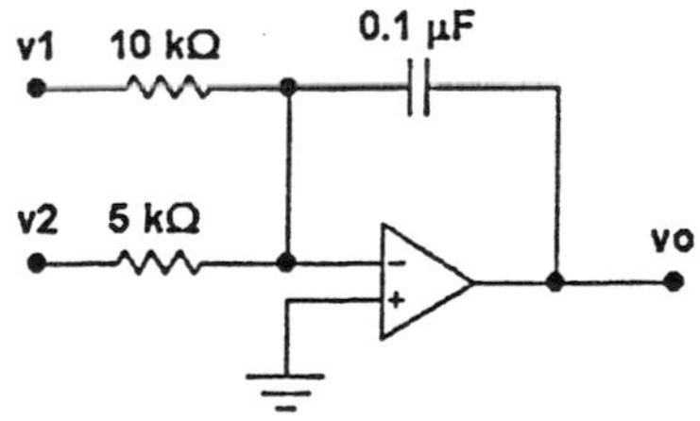

SECTION 2.5: THE NONINVERTING CONFIGURATION

2.25 A non-inverting op-amp circuit with the topology of Fig 2.16, on page 82 of the Text, has $R_1 = 4.7\ \text{k}\Omega$ and $R_2 = 47\ \text{k}\Omega$. What closed-loop gain would you expect? In the laboratory, a student accidentally exchanges these two resistors. What gain would you expect her to find?

D

2.26 Design a non-inverting amplifier with a gain of 1.5 V/V using three 1 kΩ resistors. Sketch two solutions.

CDL

2.27 Use the circuit idea shown in Fig P2.44 (on page 116 of the Text) to design a circuit whose output is $\upsilon_O = \upsilon_1 + 2\upsilon_2 - 3\upsilon_3$, with 10 kΩ as the smallest resistor used. There are several possible ways! Find one.

DL*

2.28 Use the general result outlined in P2.44 on page 116 of the Text for the arrangement shown there in Fig P2.44 to create a circuit to provide an output $v_O = 10\,(v_1 - v_2)$. (Hint: Use an additional positive input.) Have you seen this circuit before? What is it called? You may find the latter questions more straightforward after you have read the next Section of the Text.

D

2.29 A designer, needing to provide a unity-gain buffer, considers the use of the circuit topology shown in Fig. 2.19 on page 84 of the Text. However, the amplifier he has available has an open-loop gain of only 10. What closed-loop gain would the simple circuit produce? His boss suggests that he consider the circuit of Fig 2.16 on page 82 as a solution. As well, she requests that the smallest resistor used be 10 kΩ. What design would result?

SECTION 2.6: EXAMPLES OF OP-AMP CIRCUITS

CD

2.30 A designer wishes to use a simple modification of the circuit of Fig 2.20 on page 86 of the Text to implement a centre-zero voltmeter whose scale ends are ±1 volt. The meter movement provided is a 0 to 1 mA unit with a resistance of 50 Ω. Her boss suggests that a solution is possible using a single additional resistor and one of the ±10 V supplies from which the op amp is powered. What is the value of the additional resistor? To what supply is it connected? To what circuit node is the additional resistor connected? What is the required value of R?

CD

2.31 An analog-circuit designer requires a +5 V power source from which to run a small amount of digital logic requiring 20 mA at +5 V. The analog system uses ±15 V supplies which are quite well-regulated (that is stable over time and temperature and reasonably independent of load). Suggest a simple op-amp circuit, using a resistor network operating at 0.5 mA, to do the job. If the op amp requires a bias current of 2 mA from its supplies at no load, what is its total power dissipation when fully loaded at the maximum current required by the logic?

2.32 For a particular difference amplifier using the topology of Fig 2.21 on page 86 of the Text, $R_2 = R_4 = 100$ kΩ and $R_1 = R_3 = 10$ kΩ. What is the gain, $G = \dfrac{v_O}{v_1 - v_2}$, you would expect? (Be careful of what is asked!).

D

2.33 The difference amplifier described in P2.32 above is connected to two sources, v_{S1} and v_{S2}, each having a 10 kΩ internal resistance. What is the gain $\dfrac{v_O}{v_{S1} - v_{S2}}$ which results? What must you do to achieve a source-to-output gain of magnitude 10. As well, the source resistance of v_{S2} is found to be only 8 kΩ. What else must you do to achieve true difference action?

L

2.34 Reconsider the difference amplifier analyzed in Example 2.6 on page 86 of the Text, using Fig 2.21 and Fig 2.22, under the condition that resistor R_4 is connected to a 3rd input, v_3. Find the expression corresponding to Equation 2.13 for v_O. Simplify it for the case in which $\dfrac{R_2}{R_1} = \dfrac{R_4}{R_3}$.

L

2.35 Consider the circuit shown here which employs an ideal op amp.

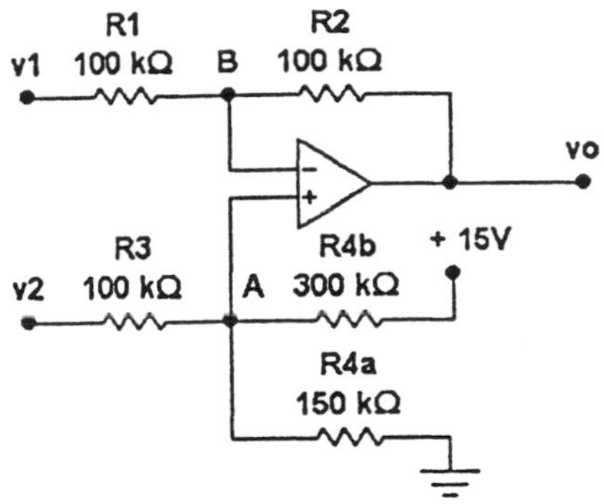

What is the value of v_O for

a) $v_1 = v_2 = 5$ V,

b) $v_1 = v_2 = 0$ V,

c) $v_1 = +3$ V, $v_2 = -2$ V?

Do your analysis from first principles. Afterward, consider using the answer to P2.34 above.

D

2.36 Using the circuit of Fig 2.25 on page 90 of the Text, design an instrumentation amplifier with a difference gain of 100 V/V shared equally between the input and output stages. Employ 10 kΩ as the smallest resistor. For your design, what voltages appear on the outputs of A_1 and A_2 for $v_1 = 5.0$ V and $v_2 = 4.9$ V?

2.37

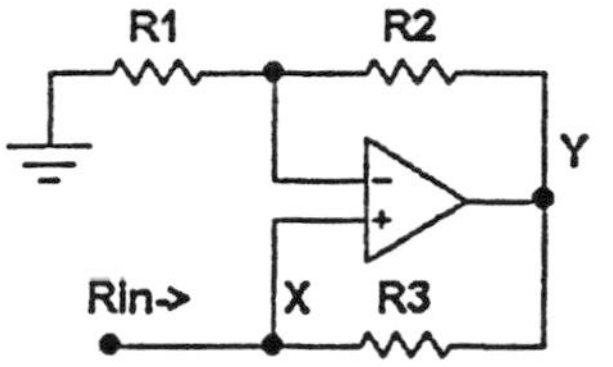

Show that the input resistance of the circuit shown is $R_{in} = -R_3\frac{R_1}{R_2}$, assuming the op amp to be ideal. (Hint: Use a test voltage v_X at node X and find the current it must supply.) To appreciate the significance of a negative resistance, connect it in series with a resistor R_4 to a signal source at node W. Sketch the circuit. Find expressions for the input resistance R_i seen by the signal source at W, and for the voltage ratios v_X/v_W and v_Y/v_W. What do these become for $R_1 = R_2$ and a) $R_4 = 2R_3$, b) $R_4 = R_3$, c) $R_4 = R_3/2$? For what value of R_4 is the voltage gain v_Y/v_W equal to +10V/V?

2.38

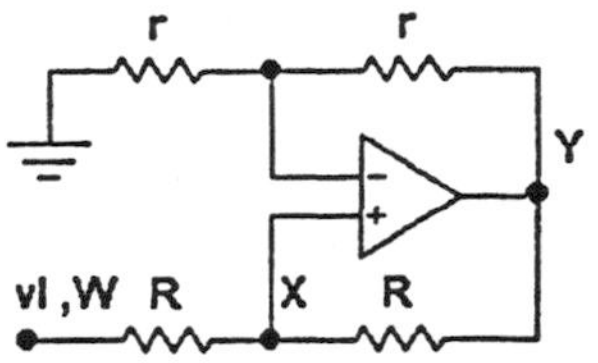

For the circuit shown with input v_I, find the Norton equivalent circuit at node X. Assume the op amp to be ideal. (Hint: Proceed as at the beginning of above.) What current will flow in an impedance Z connected to node X? Find the corresponding expression for the transmission from W to X in general, and when Z is a capacitor C, in terms of complex frequency s. Note that the latter circuit is actually a *noninverting integrator*. What is its integrator time constant? What is its unity-gain frequency?

2.39 A differential amplifier has a composite input signal consisting of 2 sine-wave components at different frequencies (60Hz and 1kHz) at each of its inputs: Both have a common component of 8 volts peak at 60Hz. At 1kHz, each has a component of 1mV amplitude, but of 180° relative phase. The output consists of a 0.6V peak component at 60Hz and a 60mV peak component at 1kHz. Find the difference-mode gain, and the common-mode gain. Using the definition of the Common-Mode-Rejection Ratio (CMRR) provided in Problem 2.60 on page 118 of the Text, calculate the CMRR in dB.

2.40 A differential amplifier is characterized by the first equation in Problem 2.60 on page 118 of the Text where CMRR is also defined. It is found to have a difference-mode gain of 200V/V and a CMRR of 100dB. For what amplitude of input common-mode signal is the unwanted output signal only 1% of the desired difference-mode output of 2Vpp?

SECTION 2.7: EFFECT OF FINITE OPEN-LOOP GAIN AND BANDWIDTH ON CIRCUIT PERFORMANCE

2.41 An internally-compensated op amp has f_t of 10 MHz and a dc gain of 10^6 V/V. What is the 3 dB frequency of its open-loop gain? If this amplifier is to be operated at 100 kHz, what open-loop gain is available?

2.42 The op amp in P2.41 above is to be used in a closed-loop amplifier having a gain of 20 dB. What corresponding break frequencies would you observe in the inverting and non-inverting versions? For what frequencies is the phase shift of the corresponding amplifier less than 6 degrees?

2.43 The op amp described in P2.41 above is to be used in a system for which low-frequency operation should extend (within 3 dB) to 10 kHz. What is the maximum closed-loop gain available from a single amplifier? From 2 identical amplifiers used in cascade? (See the result for 2 amplifiers in cascade developed in Problem 2.73 on page 119 of the Text).

2.44 A measurement of the closed-loop gain of an amplifier shows it to be −25 V/V at 120 kHz and −100 V/V at 5 kHz. Estimate the closed-loop gain at low frequencies and the corresponding 3 dB frequency. What is f_t for the op amp used? (Be careful!)

2.45 An amplifier intended for very-high-frequency operation, yet characterized by a single-pole rolloff, has f_t = 100 MHz and A_0 = 20 V/V. For a design in which the actual (rather than the nominal) closed-loop gain is −10 V/V, what 3 dB frequency results?

SECTION 2.8: LARGE-SIGNAL OPERATION OF OP AMPS

2.46 An op-amp circuit operating from ±10 V supplies has L+ and L− of +8 V and −8.5 V respectively, and a closed-loop gain of −10 V/V. What is the peak-to-peak value of the largest possible input sine wave having zero average, for which the output is not distorted?

2.47 An op amp has a slew rate of 10 V/μsec. What is the highest frequency at which it can reproduce a 6-V peak-to-peak triangle wave at its output?

2.48 Find an expression for the amplitude of the sine wave for which the small-signal and large-signal (SR-limited) bandwidths are the same. When the small-signal bandwidth is 0.5 MHz and the slew rate is 2 V/μsec, what is the amplitude for which equal bandwidths result?

SECTION 2.9: DC IMPERFECTIONS

D

2.49 For an amplifier operating with ±4 V saturation limits at a closed-loop gain of −100 V/V, what input offset voltage is required to assure less than 1% reduction in output swing capability due to offset?

D

2.50 An inverting amplifier with gain of −100 V/V and an input resistance of 100 kΩ, uses an op amp with 1 mV offset, a bias current of 30 nA and an offset current of 3 nA. What output offset results with a) a basic uncompensated design b) a bias-current-compensated design? In the latter case, what compensating resistor do you use? Which offset source dominates in each case? What is the net output offset if the dominant source is halved?

D

2.51 If the amplifier in P2.50 above is capacitor-coupled at the input, what output offset results in the basic and compensated designs? What compensation resistor should be used?

CD

2.52 Design a direct-coupled inverting op amp with a gain of –100 V/V, the highest possible input resistance, and an output offset ≤0.5 V, using an op amp with 2 mV offset, and bias currents of 1 μA equal to within ±10%. What is R_{in} of your design?

L

2.53 A basic integrator circuit such as that shown in Fig.2.11 on page 74 of the Text, operates from ±12V supplies. The op amp saturates at ±10V, has an input offset voltage of ±2mV, a bias current of 100nA (directed into the input terminals), and an offset current of ±10nA. For R = 10kΩ and C = 0.1μF, an input voltage of zero, and an initial charge of 0V on the capacitor, what is the minimum time it will take for the output to saturate, if imperfections lead to a) positive limiting, b) negative limiting. Consider the circuit shown as a means for improving operation.

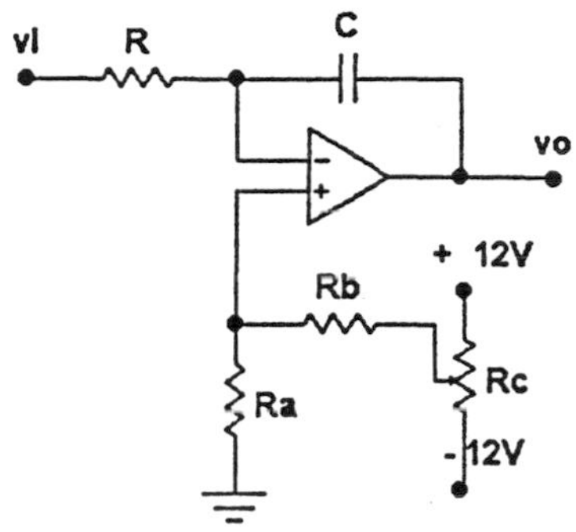

Assuming $R_b \gg R_a$, what value of R_a should be used? What do you expect the times to saturation to become now? If the bias current doubles, while the offset current remains the same, what (if anything) happens? For the offset voltage and bias current assumed to be stable at their most extreme values which cause positive saturation at the output, and with R_b = 10MΩ, to what voltage should the wiper on R_c be adjusted in order to reduce the rate of output-voltage change to essentially zero?

LD

2.54 A non-inverting amplifier using resistors of 10kΩ and 1MΩ to achieve a high gain is found to have an output offset voltage of +1.8V with input grounded. When a 10kΩ resistor is used in series with the positive input (and grounded), the output offset reduces to +0.6V. Estimate the nominal gain of the amplifier, and the input-bias current. What can you say about the input offset voltage and offset current? If the value of all 3 resistors is reduced by a factor of 10, the output offset reduces to 0.4V. What do you estimate the input offset voltage to be? Now, if the 10 kΩ resistor connected to the amplifier's negative input terminal is capacitor-coupled to ground, what does the output offset voltage become? What must you now do to compensate? What does the output offset voltage now become?

NOTES

Chapter 3

DIODES

SECTION 3.1: THE IDEAL DIODE

3.1 For the following circuits employing ideal diodes, find the labelled currents, I, and voltages, V, measured with respect to ground.

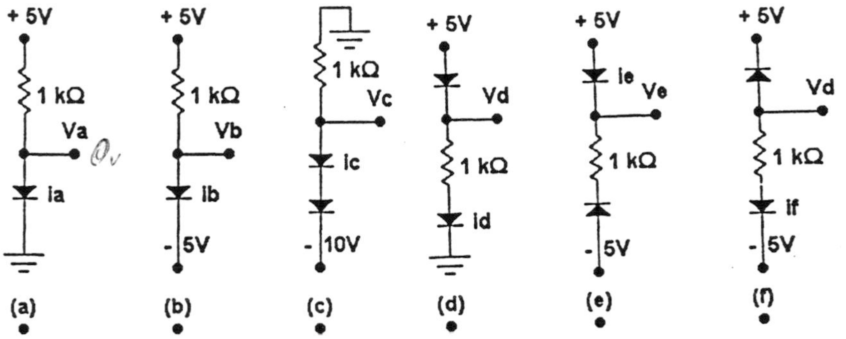

3.2 For the following logic gates using ideal diodes:

i) If $V_A = V_E = 5$ V, and $V_B = V_C = V_D = 0$ V, what is the value of V_Y produced?

ii) If logic '1' = 5 V and logic '0' = 0 V, identify the logic function performed.

iii) If logic '1' = 0 V and logic '0' = 5 V, identify the logic function performed.

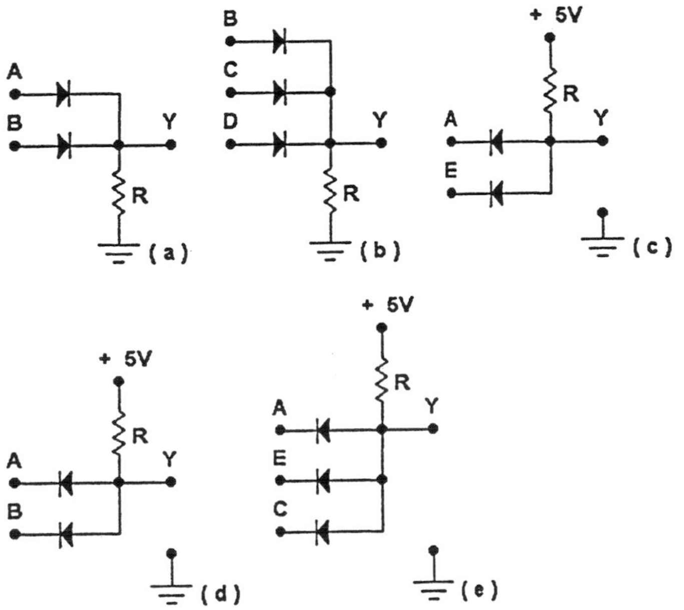

3.3

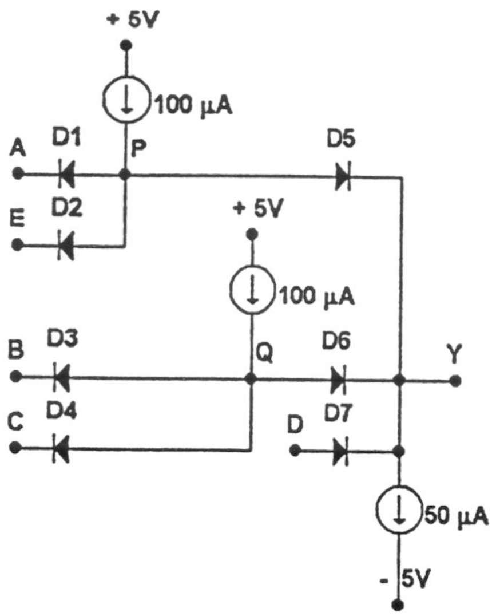

For the conditions stated in P3.2 ii) above, find an expression for the logic function $Y = f(A, B, C, D, E)$ of the circuit shown. In particular, for the input logic values stated, what is the logic output value?

3.4

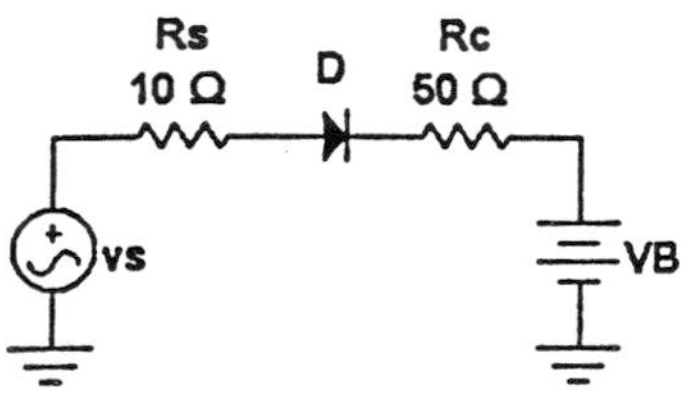

In the battery-charger circuit shown, the sinewave input v_S is 12 V rms, while the battery voltage varies from 12 V to 14 V from the discharged to fully-charged states. $R_S = 10\ \Omega$ is the charging-source resistance, D is an ideal diode and $R_C = 50\ \Omega$ is a current-controlling resistor established by the designer. Sketch and label the diode-current waveform for $V_B = 12$ V. What are its peak and average values? What do the peak and average diode currents become when V_B reaches 14 V?

3.5 Find the currents I_1, I_2, I_3, I_4 in each of the diodes D_1, D_2, D_3, D_4 of the circuit shown. What V_O results? The diodes are assumed to be ideal.

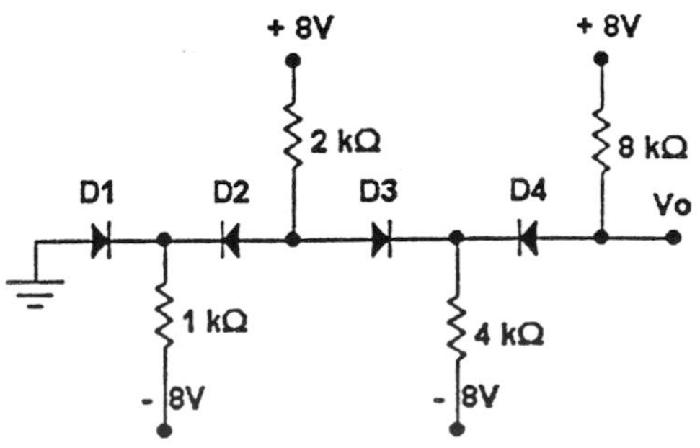

SECTION 3.2: TERMINAL CHARACTERISTICS OF JUNCTION DIODES

3.6 A very small discrete silicon diode (a "100 μA diode") is found to conduct 100 μA at 0.700 V and 1 mA at 0.815 V. Find the values of n and I_S which correspond.

3.7 A diode for which $n = 1$ conducts 0.1 mA at 0.7 V. Find its voltage drop at 1 mA. For what current is its voltage drop equal to 0.815 V?

3.8 A 10-A silicon diode for which $n = 2$ is known to have a forward voltage drop of 0.700 V at 10 A. What is the junction voltage at which it conducts 10 mA? 10 μA?

3.9 A particular "1 mA diode", which at 25° C conducts 1 mA at 0.7 V, is operated at 95° C in a circuit which provides it a constant 100 μA current. What does its junction voltage become if $n = 2$?

3.10 For the diode described in P3.9 above, the leakage current at 25° C is 1 nA. What does it become at 95° C? at 100° C?

SECTION 3.3: PHYSICAL OPERATION OF DIODES

NOTE: For a summary of important relationships and values of particular parameters or physical constants not stated explicitly in the following problems, please consult Table 3.1 on page 156 of the Text.

3.11 At a particular temperature, the fraction of ionized atoms in a piece of silicon is 10^{-n}. If the material is doped to a level of 1 in 10^m with acceptor atoms, what is the net concentration of holes and electrons in the resulting material?

3.12 Using Equations 3.6 and supporting data following it, find the intrinsic carrier density n_i at 200K, 300K and 400K, that is in the ± 100 °C range at and around room temperature. What is the % increase in concentration for the 100 °C rise above room temperature? At 127 °C, what fraction of the silicon atoms are ionized?

3.13 Find the resistivity of a) intrinsic silicon and b) n-type silicon with $N_D = 10^{16}/cm^3$. Use $n_i = 1.5 \times 10^{10}/cm^3$ with $\mu_n = 1350 cm^2/V_S$ and $\mu_p = 480 cm^2/V_s$ for intrinsic silicon, and mobility reduction to 80% for the doped material. To what values will the resistivity change in each case for a 100 °C rise in temperature of the material?

3.14 For a *pn* junction in which the *n* region is doped at ten times the concentration of the *p* region, in what region is the depletion region largest? By what factor?

3.15 For a junction in which the built-in voltage is 0.7V, what are the doping-concentrations in the two regions if: a) they are equal, b) they are in the ratio 10 to 1. [Hint: Use Eq.3.18 in the Text.] For each case, what is the width of the depletion region and the distance it extends each side of the junction? For a junction that is 30 μm by 50 μm in size, what is the magnitude of the uncovered charge on each side?

3.16 For a particular reverse-biased *pn* junction, the terminal current is 10 nA. If the drift current at the operating temperature is 15 nA, what must the voltage-dependent diffusion current be at this particular reverse voltage?

CL*

3.17 Find an expression for the charge q_J formed on either side of the junction in terms of the applied reverse voltage V_R, as represented in Fig. 3.14 of the Text. Calculate the value of q_J which applies to the junctions described in P3.15 above, for V_R = 0V, 10V and 11V. Use the latter pair of values to estimate the junction capacitance. Calculate this more directly at V_R = 10.5V, using Equations 3.25 and 3.26 on pages 148 and 149 of the Text. If the junction is not abrupt, but has a grading coefficient m = 1/3, what are the expected capacitances att V_R = 10.5 V? At V_R = 100V?

3.18 At a particular operating point of a reverse-biased *pn* junction, a change of 1 volt produces a transient current increase corresponding to a net charge flow of 0.1 pC. What is the corresponding depletion capacitance of the junction at this operating voltage?

3.19 For a particular junction for which m=1.6, a capacitance C_j of 1.8 pF is measured for a reverse junction voltage of 2 V, and 0.2 pF for a voltage of 10 V. What are the corresponding values of V_O, C_{j0} and C_j at 0 V?

3.20 A particular *pn* junction for which the breakdown voltage is 120 V, can dissipate 50 mW while maintaining its junction temperature at a value low enough to avoid permanent junction damage. What continuous reverse current flow appears likely to cause permanent failure? If reverse current flows only 10% of the time at the peaks of a cyclic applied voltage, what peak current can be tolerated?

3.21 In a diode intended for high-speed switching, the excess-minority-carrier lifetime for holes is 1 ns. Using the value of hole mobility in doped silicon from Ex. 3.12 on page 143 of the Text, and the Einstein relation (in Eq. 3.12 on page 141), find an estimate of the diffusion length in the forward-conducting diode. For this diffusion length, at what distance from the depletion-region edge will the excess hole density reach 10% of its value there?

3.22 For a $3\mu m \times 5\mu m$ junction, with $N_A = 10^{17}/cm^3$ and $N_D = 10^{16}/cm^3$, in which minority-carrier lifetimes are τ_p = 1 ns and τ_n = 2 ns, hole and electron mobilities are 400 and 1100 cm^{2}/Vs, respectively, find I_S.

3.23 Using Eq. 3.6 of the Text, evaluate the temperature dependence of I_S (as defined in Eq. 3.34 there) in %/°C at room temperature (say 300K).

3.24 For the diode in P3.22 above, conducting a 1 mA current, what fractions of the current are carried by holes and by electrons? Estimate both the hole and electron minority stored charges. What is the mean transit time τ_T of the diode? What is the associated small-signal diffusion capacitance?

3.25 For a junction conducting 1 mA at 700 mV, for which n = 2 and a diffusion capacitance of 1pF is associated, what is the value of τ_r which applies? For a junction 10 × larger what would τ_T be? In the original junction, what is the total stored charge at 1mA? At 10 mA?

3.26 Use the relationships given for charge Q in Eq. 3.38 on page 154 of the Text and thereafter, to calculate the diffusion capacitance of a junction characterized by n, υ, i in the diode equation.

SECTION 3.4: ANALYSIS OF DIODE CIRCUITS

3.27 A diode described by the exponential characteristic of Fig. 3.20 on page 159 of the Text is connected to a source whose Thevenin-equivalent voltage is V_T and resistance is R_T. {Note that V_T is a Thevenin voltage, not a thermal voltage!} Draw load lines and find operating points (V_D, I_D) for:

(a) $V_T = 1$ V, $R_T = 100\ \Omega$,

(b) $V_T = 0.9$ V, $R_T = 100\ \Omega$,

(c) $V_T = 0.9$ V, $R_T = 90\ \Omega$.

Note that the graphical process, while tedious for a single analysis, can be quite effective if a variety of related or similar situations are to be evaluated.

3.28 Determine the diode current I_D and voltage V_D for the circuit in Fig. 3.18 on page 157 of the Text, with $V_{DD} = 1.0$ V and $R = 100\ \Omega$. Consider the diode to be much like the one sketched in Fig. 3.20, having a current of 1 mA at a voltage of 0.7 V, exhibiting a voltage change of 0.1 V per decade of current change. Use an iterative solution of the diode logarithmic voltage-current relationship.

3.29 Repeat problem P3.28 above utilizing a piecewise-linear diode model whose parameters are $V_{DO} = 0.65$ V and $r_D = 20\ \Omega$.

3.30 Repeat problem P3.29 above utilizing a lower-resistance piecewise-linear model whose parameters are $V_{DO} = 0.70$ V and $r_D = 10\ \Omega$. What do I_D and V_D become if a simple 0.75 V battery model (for which $r_D = 0\ \Omega$) is used?

L

3.31 In the context of the sequence of problems P3.28, 3.29 and 3.30 preceding, note that the degree of adequacy of a simple model depends on the choice of its parameters in the particular context. To illustrate this dependence, consider the circuit Figure 3.18 of the Text, with V_{DD} reduced to 0.8 V while R remains at 100 Ω. Find the operating point (V_D, I_D) for:

(a) a diode characterized by Fig. 3.20 on page 159 of the Text, by plotting the load line,

(b) a piecewise-linear diode for which $V_{DO} = 0.65$ V and $r_D = 20\ \Omega$,

(c) a piecewise-linear diode for which $V_{DO} = 0.70$ V and $r_D = 10\ \Omega$,

(d) a constant-voltage model with $V_D = 0.75$ V.

D

3.32 A series string of 5 diodes is connected through a resistor R to a 10 V supply. For diodes having 0.7 V drop at 1 mA and a 0.1 V/decade characteristic, find R required to establish a total diode-string voltage of 4.0 V.

3.33 In problem P3.32 above, if R is reduced to 500 Ω, what does the voltage across the string of 5 diodes become?

3.34 A 1-mA diode having a 0.1 V/decade characteristic operates from a constant-current supply with $V_D = 0.8$ V. If it is shunted by two more identical diodes, what does the voltage drop become?

SECTION 3.5: THE SMALL-SIGNAL MODEL AND ITS APPLICATION

D

3.35 A junction diode for which n=2 operates in a particular circuit with a current that varies over the range 0.1 mA to 10 mA. What is the diode incremental resistance at the extreme values of current? If you were asked to state an "average"resistance, which would be "best" – an arithmetic mean $(r_1 + r_2)/2$, or a geometric mean $(r_1 r_2)^{1/2}$? Calculate both, as well as the resistances at 5.05 mA and at 1 mA (the mean currents of each kind). What do you conclude?

3.36 A diode for which $n = 2$ operates in a circuit for which the current is essentially a constant value of 2 mA. Find the corresponding diode incremental resistance. A second identical diode is used to shunt the first. What does the current in each diode become? What is the incremental resistance of each? What is their parallel combination? What can you conclude about the relation of diode incremented resistance to junction size?

3.37

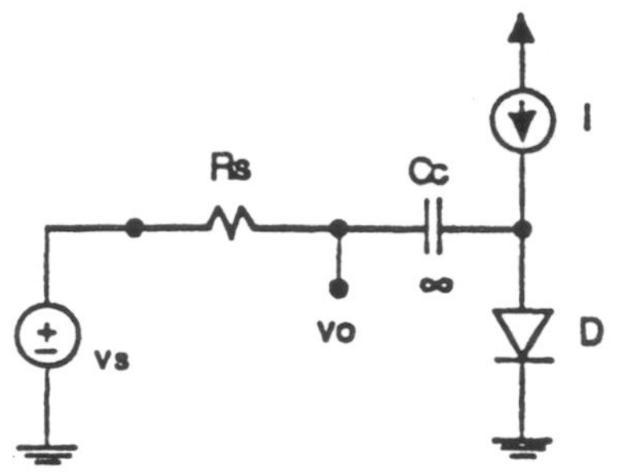

In the circuit shown, v_S is a sine wave of amplitude V and C_C is a large capacitor which blocks direct-current, and allows all of I to flow in D. For $R_S = 1\ k\Omega$ and $v_S \leq 10$ mV, find v_O/v_S for $I = 10$, 1, 0.1 and 0.01 mA. Use $n = 2$.

3.38

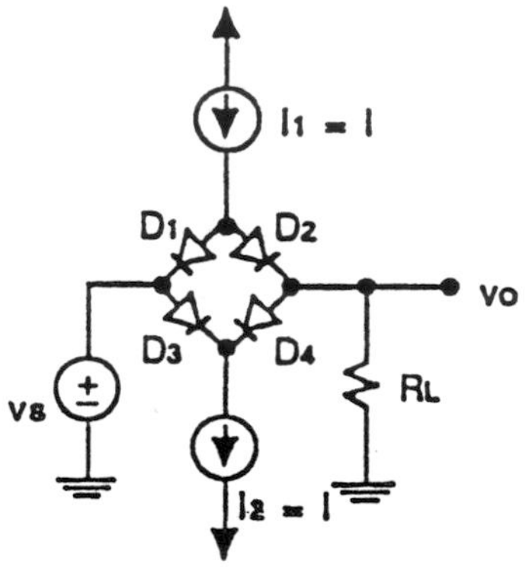

In the circuit shown, v_S is a small signal having a relatively low source resistance. $I_1 = I_2 = I$ is variable. All diodes are identical, with $n = 2$. For $v_O = v_S = 0$V, how does the current I split among the diodes? In general, as the output voltage varies, what is the relationship amongst the diode currents? Find an expression for the Thevenin-equivalent source resistance seen by R_L as a function of I for v_S around zero volts. For $R_L = 10\ k\Omega$, find v_O/v_S for $I = 1$ mA and 1 μA. For which current is the input signal size most critical? What is the peak signal for which reasonably linear operation is possible at $I = 1\mu A$? Note that this circuit, when compared to that in P3.37 above, gives you a preview of a general principal that you will see much more of in your Text. It is that an increased use of semiconductors (such as in the diodes and the current sources, here) reduces the need for other (often much larger) components (such as the capacitor C_C in P3.37).

3.39 For the design described in problem P3.32 above, of a 4.0 V regulator using five diodes and a 600 Ω resistor with a 10 V supply, the supply voltage is found to vary by ±10%. What output-voltage variation results? Use $n = 2$. If, separately, a load of 2 mA is applied to the output, what drop in output voltage would you expect? For both low input voltage and maximum load, what is the lowest output voltage you would find? Express the resulting changes in output voltage (for supply variation alone, load variation alone, and both together) both in absolute terms and as % changes.

CL*

3.40 For a particular pn junction, the following measurements are taken:

$C_j = 0.8$ pF at $v_r = 1$ V reverse bias,

$C_j = 0.2$ pF at $v_r = 5$ V reverse bias,

$C_j = 0.1$ pF at $v_r = 10$ V reverse bias,

$C_T = C_j + C_d = 10$ pF at $v_f = 0.70$ V where $i_f = 1$ mA,

$r_d = 50\ \Omega$ at $i_f = 1$ mA.

Find C_{j0}, V_0, m, n, τ_T. What is $C_T = C_j + C_d$ at $i_f = 10$ mA?

3.41 For a diode 10 times the junction area of that in P3.40 above, find r_d, C_d, C_j, C_T at $i_f = 5$ mA forward bias and C_j at 10 V reverse bias.

SECTION 3.6: OPERATION IN THE REVERSE BREAKDOWN REGION – ZENER DIODES

3.42 A 6.8 V Zener diode specified at 5 mA to have $V_Z = 6.8$ V and $r_Z = 20\ \Omega$ with $I_{Zk} = 0.2$ mA, is operated in a regulator circuit using a 200 Ω resistor and a 9 V supply. Estimate the knee voltage of the Zener. For no load, what is the lowest supply voltage for which the Zener remains in breakdown operation? For the nominal supply voltage, what is the maximum load current for which the Zener remains in breakdown operation? For half this load current, what is the lowest supply voltage for breakdown operation?

3.43 For the situation described in P3.42 above, what are the line regulation and load regulation (as defined in Equations 3.60 and 3.61 in the Text).

D

3.44 For the situation described in P3.42 above, a modified design is required for the situation in which the supply variation is ±5%, the Zener-diode nominal voltage variation is ±3%, and the load variation is from 2 to 10 mA. Find the value of R for which the minimum Zener current is $\geq 2\ I_{Zk}$. For this situation, what are the limits on the output voltage produced? Assume $r_Z = 20\ \Omega$ and $I_{Zk} = 0.2$ mA for all available zeners.

D

3.45 A designer needing a well-regulated 15 V supply in an application where a poorly-regulated 24 V source is available, considers the use of a shunt regulator string consisting of a series string of two 6.8 V Zeners and two junction diodes. Available Zener diodes are specified to have $V_Z = 6.8$V at 20 mA with $r_Z = 5\ \Omega$. Available junction diodes are 10-mA types modelled at 10 mA by a 0.7 V drop and a 2.5 Ω series resistance. Design a suitable regulator for desired operation with a 15 mA nominal load. What does the output of your design become if the supply is 10% high, the series resistor is 5% low, and the load is accidentally removed? What is the power dissipated in each 6.8 V Zener under the worst combination of these conditions?

SECTION 3.7: RECTIFIER CIRCUITS

3.46 A half-wave rectifier using diodes for which $V_D = 0.7$ V, is supplied by an 8 V rms sine wave at 60 Hz. What is the peak value of the output voltage for very light loads? For what fraction of a cycle does the diode conduct (first approximately, and then more exactly)? What is the average value of the output voltage? What is the peak-inverse voltage across the diode? Now, if the diode resistance is 10 Ω, the source resistance is 50 Ω, and the load resistance is 1 kΩ, what do the peak and average output voltages become? [Hint: The average value of half-sine wave of peak amplitude V_P is V_P/π.]

3.47 In a half-wave rectifier employing an 8 V rms sine-wave supply and driving a 1 kΩ load, a 6.8 V Zener diode connected with its cathode at the output is accidentally substituted for the rectifier diode. Using 6.8 V and 0.7 V drops for diode conduction in the two directions, sketch the output voltage. What average value of the output voltage results? [Hint: The average value of a half-sinewave of peak amplitude V_P is V_P/π.]

C

3.48 In a full-wave rectifier, using a centre-tapped transformer winding with full-output voltage of 16 V rms, and having a 1 kΩ load, 6.8 V Zener diodes are accidentally installed in place of high-breakdown diodes, but with the same cathode polarities. Sketch and label the output voltage waveform in the event that the total-winding equivalent resistance is 100 Ω. What peak diode currents flow?

3.49 A transformer secondary winding whose output is a 12 V rms sinusoid at 60 Hz is used to drive a bridge rectifier whose diodes' conduction can be modelled as 0.7 V voltage drops. The load is a 1 kΩ resistor. Sketch the load waveform. What is its peak value? Over what time interval is it zero? What is its average value? What is the *PIV* for each diode? [Hint: The average value of a full-wave rectified sinewave of peak amplitude V_P is $2V_P/\pi$.]

3.50 A half-wave rectifier employing a 12-V-rms 60-Hz sine-wave source and no *dc* load is filtered using a polarized electrolytic capacitor having a small leakage current. For diodes assumed to have a 0.7 V drop independent of current, characterize the resulting output. What is the *PIV* required of the diode?

D

3.51 To the circuit in P3.50 above, a load which can be modelled as a 1 mA constant current is connected. If an output ripple of 0.4 V pp results, what is the value of the filter capacitor used? For half this ripple, and double the load current, what capacitor is necessary? In each case, what average current flows during the diode's conduction interval?

D

3.52 For both situations described in P3.51 above, but with full-wave rectification, what capacitor values are necessary? What average diode currents flow? What diode *PIV* is required?

D

3.53 A design is required of a full-wave rectifier with capacitor filter to supply 12 volts dc to a 100 Ω load. A ripple voltage of less than or equal 0.4 V pp is necessary. Diodes are assumed to conduct with 0.7 V drop. Characterize the required 60 Hz transformer secondary, the capacitor and the diodes. For the diodes, provide the required *PIV* and the peak-current ratings.

C*L*

3.54 Consider a full-wave bridge rectifier operating at 60 Hz from a single transformer-secondary winding having a 1.0 Ω equivalent internal resistance and 20 V pp open-circuit output. The load consists of a 1000 μF capacitor and 200 Ω resistor in parallel. Consider the diodes to have a constant 0.7 V drop during conduction. [Hint: To characterize this situation, first consider the ideal case of a zero-impedance source, finding the usual parameters, including the average diode current during conduction.] Assuming the diode current to be a triangular pulse, limited by some combination of the sinewave slope and the charging-circuit time constant, find the corresponding average voltage drop in the transformer resistance.

SECTION 3.8: LIMITING AND CLAMPING CIRCUITS

3.55

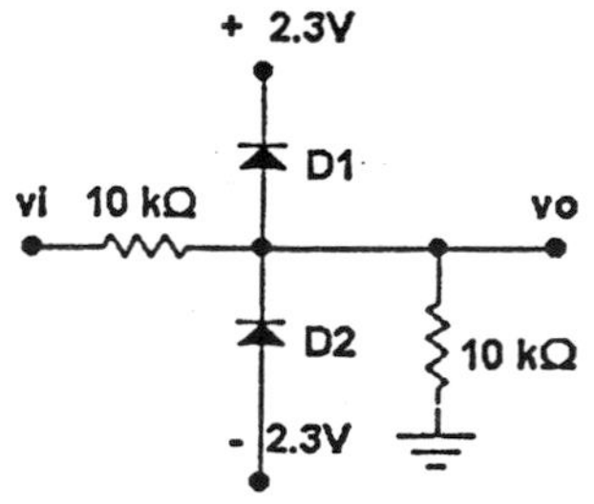

For the passive symmetric hard limiter shown, find the upper and lower limiting levels (including a 0.7 V diode drop), the gain K, and the upper and lower input threshold levels. What is the input current required from an input which is twice the upper threshold value?

CD*

3.56 Convert the circuit in P3.55 above to a soft limiter with $K = 1/4$ after limiting begins, by adding two additional components. Add two more components (and additional power supplies) for hard limiting at ±5.0 V. Using the new supplies and two more components, create an equivalent circuit having the same hard-

and soft-limiting characteristics, but only two power supplies.) [Hint: Your final circuit should employ six resistors, four diodes, and two power supplies!]

D

3.57 Design a passive symmetric high-gain pseudo-hard-limiting circuit using four diodes for ±1.4 V limits. Use only a connection to ground (that is, **use no supplies**). This limiter is to be used to create an approximation of a square wave from a sine-wave input. For what peak-to-peak amplitude of the sine wave is the rise and fall time of the pseudo-square wave ≤5% of the wave period. What resistor is required for a peak diode current of about 10mA?

3.58 A simple clamped-capacitor circuit such as that shown in Fig. 3.56 of the Text, utilizing a capacitor C and grounded-cathode diode, has a square-wave input with upper and lower levels at 100 V and 10 V respectively. Describe the resulting output for a very high-resistance load to ground and a 0.5 V diode drop at very low currents. What happens as the load resistance R reduces? Describe the output waveform when $RC = 2T$ where T is the period of the input square wave.

3.59 A voltage doubler, consisting of a clamped capacitor and a half-wave rectifier, operates at 20 kHz using two 0.1 μF capacitors. For a sine-wave input signal of 100 V peak, what output voltage would you expect at no load? For a peak-to-peak ripple voltage of 5% of the peak voltage, what average output voltage would you expect? What is the load current for which this situation applies?

CI

3.60 Consider the detailed operation of an unloaded half-wave doubler circuit with positive output using equal capacitors C and driven by a 100 V pp square wave. In particular, follow the cycle-by-cycle operation immediately upon turn-on, with the output voltage initially equal to zero and the input low. What is the output voltage after the first half cycle (after the input has risen by 100 V and fallen again)? After the first cycle (just after the input rises again)? After the first two cycles? After four cycles? After eight cycles? Sketch the output voltage against time as measured by the number of cycles.

3.61 Continue to think about the operation of the half-wave doubler as suggested in P3.60 above. In particular, if the load is a current which discharges a capacitor of value C by 5% in ½ cycle, find the steady-state average output voltage as a function of V_0, the peak-to-peak input (V_0 = 100 V here). (Hint: Note a) that the total effective capacitance is 2C for half the cycle and b) that two equal capacitors when joined, share their charge difference equally).

SECTION 3.9: SPECIAL DIODE TYPES

3.62 A particular silicon Schottky-barrier power rectifier conducts 0.1 A at 0.30 V, and 1 A at 0.37 V. Evaluate n and I_S for this diode assuming that the diode equation applies. At 20 A, the rectifier diode has a voltage drop of 0.8 V. Estimate its series resistance.

3.63 A small-signal silicon Schottky diode conducts 10 mA at 0.42 V with n = 1.5. If it has a series resistance of 10Ω, what is its voltage drop at 1 mA and at 50 mA?

3.64 A small GaAs SBD has n = 1.1 and $I_S = 10^{-15}A$. Find the forward voltage drop at 0.1 mA and at 10 mA.

3.65 A Zetex silicon ion-implanted hyperabrupt-junction varactor, for which the grading coefficient is m = 0.9 and V_0 = 2.2V, has a capacitance of 33 pF at a reverse bias of 2 V. Find its capacitance at 0 V, 1 V and 10 V. What is the value of C_{j0}?

3.66 A particular photodiode capable of operating over a range of wavelengths from 350 nm to 1100 nm, has a peak sensitivity of 0.7 μA/mW/cm^2 at 750 mm, and a dark current of 1.5 nA at 20 V reverse bias. Incidentally, intended for high-speed applications, it has a junction capacitance of 12 pF and current response time of 4 ns. For each 10°C rise in temperature, the dark current doubles, and the photocurrent increases by about 3.5%. For illumination in bright sunlight with intensity estimated at 1000 W/m^2, what photocurrent flows in a suitably-reverse-biased diode at 25°C? At 125°C? What is the dark current in each case? Note the relative insensitivity of photocurrent to temperature.

CD

3.67 Consider the operation of the photodiode introduced in P3.66 above, in a circuit in which the cathode is connected to a +10-V supply and the anode to ground through a 100 kΩ resistor. The photodiode is illuminated either directly by a light beam at 10 mW/cm^2 or indirectly by reflected light at 0.5 mW/cm^2. What are the two output-signal levels available? Modify the circuit using a ±5 V supply, a second diode, a single resistor, and a fixed-intensity light source of your choice, in order to produce a ± 1 V output signal.

3.68

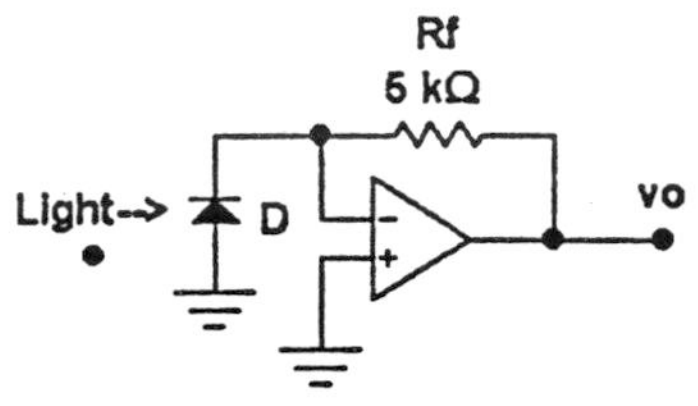

The photodiode introduced in P3.66 above, can be used for detecting relatively high-speed light pulses using the circuit shown, in which the op amp gain and bandwidth are large. If D is an ordinary diode (in a light-sealed package), what value of v_O would result? If D is a photodiode (or even a conventional diode whose junction is exposed to light), what happens? For the diode described in P3.66 above exposed to light of intensity of 20 mW/cm^2, what does the output become?

CL*

3.69 A silicon solar panel intended to operate with a 12 V automobile battery, consists of a large number of large-area diodes, or cells,in series. Its commercial specifications, an optimistic combination of average, expected, uncoordinated-limiting-case and optimized behaviours, include the following: In bright sunlight (at 1000 W/m^2) the panel has an open-circuit voltage at 24 V, and a short-circuit output current of 110 mA. The panel generates a maximum load power at 17.5 V (where at higher voltages the forward conduction loss reduces available load current significantly). Nominal operation (in the battery environment) is at 14.5 V with nominal 100 mA output current. At what power level is the panel normally operating? If each diode has a forward-conduction voltage of 0.67 V at around 10 mA, how many series diodes are used in the panel? Estimate the value of n for these diodes and the corresponding current lost to diode forward conduction at the normal operating voltage of 14.5 V.

D

3.70 A GaAs LED producing red light at 635 nm operates at 10 mA with a junction voltage of 1.9 V. What is the corresponding input-power level? This diode has a 60 mW power rating. If the junction is characterized by $n =$ 1.2, what current and voltage correspond to operation at half the maximum rated power level? The LED is to be driven by a logic gate for which V_{OL} = 0.5 V, using a series resistor connected to a + 5 V supply. What resistors are needed for operation in the low- and high-current modes suggested above?

CD

3.71 A particular opto-isolator, the Siemens IL 300-X016, available in a wide-body 8-pin dual-in-line package, includes three optically-coupled diodes – one emitter diode and two detector diodes – with physical separation between diodes capable of withstanding 7500 V. The two detectors are closely matched to allow feedback from one of them to the emitter driver circuit to ensure a linear relationship between the input current and the output current from the second isolated detector diode. For the AlGaAs emitter

LED, the operating point is typically at 1.25 V and 7 mA. For the detector, the corresponding open-circuit voltage is about 500 mV, and the short-circuit current is about 70 μA. Overall, the current-transfer gain is specified to be within the range from 0.6% to 1.6%. For what mode of operation are the detector diodes specified? For an emitter current of 5 mA, what is the range of short-circuit diode currents you would expect? Sketch an isolating driver and receiver system using two op amps and a small number of resistors, operating from two isolated sets of ± 5 V supplies.

NOTES

Chapter 4

BIPOLAR JUNCTION TRANSISTORS (BJTs)

SECTION 4.1: PHYSICAL STRUCTURE AND MODES OF OPERATION

4.1 Various transistors, either *npn* or *pnp* are measured with the following voltages on their terminals labelled E, B, C. Identify the likely transistor type and its operating mode:

#	E	B	C	Type	Mode
1	2.1	2.8	4.9		
2	1.0	1.2	10.0		
3	2.1	2.4	-1.1		
4	2.2	1.4	1.9		
5	1.8	1.4	-8.9		
6	0.6	1.4	0.9		

4.2 The BJT transistor, whose simplified structure is shown Fig. 4.2 of the Text, has two junctions, each of which can be either forward-biased or reverse-biased. How many different modes of operation are possible? What is missing from Table 4.1? It is called the inverse or inverted or reverse mode of which something is said on pages 234 and 308 of the Text. This mode is relatively rarely used directly, except in one form of BJT digital logic (called TTL), but can occur in the dynamic operation of other more-conventional circuits.

SECTION 4.2: OPERATION OF THE NPN TRANSISTOR IN THE ACTIVE MODE

4.3 A particular *npn* transistor operating at about 25°C conducts a collector current of 2.0 mA at a base-emitter voltage of 0.70 V. For an IC process in which $n = 1$, what is the value of the saturation current I_S? For this device, $N_A = 10^{18}/cm^3$ where $\mu_n = 1100\ cm^2/Vs$. Find the value of D_n which applies, and estimate the emitter-base junction area A_E in terms of the effective base width W. What does it become for $W = 2\mu m$? [Hint: Use Equations 4.4 and 3.12 with $n_i = 1.5 \times 10^{10}/cm^3$, $q = 1.6 \times 10^{-19}C$.]

4.4 For a transistor which is 100 times larger in emitter-base junction area than that described in P4.3 above, find the value of I_S, the current at $V_{BE} = 0.70$ V, and the voltage at 1.0 mA, which apply. What do I_S and V_{BE} (at 1 mA) become if the temperature is raised by 100°C? [Hint: Use Equations 4.3, 4.4 and 3.6.]

4.5 A particular *npn* transistor has an emitter area of $20\mu m \times 20\mu m$. The doping concentrations are: $N_D = 10^{19}/cm^3$ in the emitter, $N_A = 10^{17}/cm^3$ in the base, and $N_D = 10^{15}/cm^3$ in the collector. The transistor operates at $T = 300K$, where $n_i = 1.5 \times 10^{10}/cm^3$. For electrons diffusing in the base: L_n 19µm and $D_n = 21.3 cm^2/s$. For holes diffusing in the emitter, $L_p = 0.6$ µm and $D_p = 1.7 cm^2/s$ and $q = 1.60 \times 10^{-19}$ C. For base widths $W = 1\mu m$ and 0.1µm, and $v_{BE} = 700$ mV, calculate n_{p0}, $n_p(0)$, I_n, I_S, i_C, β, α. [Hint: Use Equations 4.1, 4.2, 4.3, 4.4, 4.12, 3.30.]

4.6 Using the information provided in P4.2 in the Text, find the value of base width W for which β is a) 1000, b) 2000.

4.7 A particular BJT for which $n = 1$ has a base-emitter voltage of 0.650V at $i_C = 10\mu A$. What value of I_S applies? By what factor must the base-emitter junction area be increased to provide a 0.500 V drop at $10\mu A$? For what current does this large junction have a base-emitter voltage of 0.650 V? 0.700 V?

4.8 A particular *npn* BJT operates in the active mode with $i_C = 10$ mA, $i_B = 75\mu A$, and $V_{BE} = 0.69$ V. With a view to providing values for all the parameters in the large-signal T models presented in Fig. 4.5 of the Text, find the corresponding diode scale currents, as well as α and β.

4.9 Measurements made on the emitter and collector currents of a particular BJT show values of 0.753 mA and 0.749 mA. Assuming these results to be accurate to within ±1 in the third decimal place, find the range of α and β which may apply.

4.10 For the transistor and situation described in P4.3 above, find the stored base charge at $i_C = 2$ mA. If the common-emitter current gain β is found to be 120, and assuming recombination to be the dominant source of base current, estimate the minority-carrier lifetime τ_b.

4.11 A particular npn BJT operates with the base-emitter junction forward-biased, with the base-to-emitter voltage being 700 mV. For active-mode operation, in what range must the collector-to-base and collector-to-emitter voltages lie? When the transistor is appropriately biased in the active mode, the collector current is found to be 10 mA. What is the corresponding value of I_S for this transistor if n is assumed to be 1? Under the same conditions, the base current is found to be 100 μA. What is the value of β for this transistor? If measured, what would the emitter current be found to be?

4.12 For the devices and situations described in the following table, provide the missing entries. Line a) is given completely by way of example.

Device #	I_C mA	I_B mA	I_E mA	α	β
a	10	0.1	10.1	0.99	100
b	1				50
c			2	0.98	
d		0.01		0.995	
e			110		10
f		0.001			1000

Note that transistors like device f) are constructed with very thin bases in order to achieve high β, but suffer accordingly from reduced breakdown-voltage ratings.

4.13 Of the first-order large-signal equivalent-circuit models of an npn BJT shown in Fig. 4.5 of the Text, consider the two which employ α and β explicitly. Draw these side-by-side to emphasize current flow directly from collector to emitter with base current entering from the left (the final shape can be called a "tilted T", or, in the spirit of livestock branding in the far west of North America, a "Lazy T" or, simply, a T (model)). Label all currents and v_{BE} in each case. What two labels can be applied to each of the two controlled current sources?

4.14 A particular BJT operates in its usual current range with $v_{BE} = 0.7$ V, and $i_C = 1$ mA. What would be its v_{BE} at $i_C = 0.1$ μA for $n = 1$? Repeat for $n = 2$.

4.15 For a "1 mA transistor", that is, one for which i_C = 1 mA for v_{BE} = 700 mA, the collector-base reverse current, I_{CBO}, is 0.1 nA at 25°C. Device β is nominally 100. For a transistor operating with its base open-circuited (in which case I_{CBO} constitutes the only source of base current), what collector current flows, at 25°C? At 95°C?

4.16 For a particular BJT fabricated in the style shown in Fig. 4.6 of the Text, the collector-base junction is 100 times larger in area than the emitter-base junction. If, for this device, normal β = 150, what would you expect it to become if the roles of emitter and collector are reversed? That is, estimate β_R.

SECTION 4.3: THE PNP TRANSISTOR

4.17

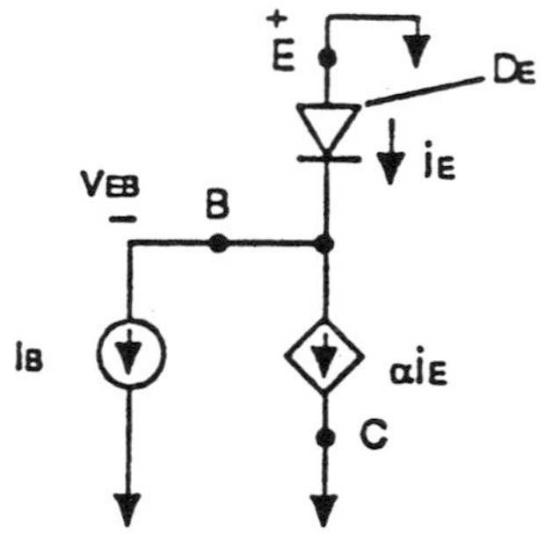

For the accompanying equivalent circuit of a pnp transistor (drawn to emphasize the direction of current flow), α = 0.975. For an external current extracted from the base, I_B = 10 μA, what collector and emitter currents would you expect? If for this device, v_{EB} = 0.70 V at I_C = 1 mA, what value of v_{EB} would you expect? (Assume that n=1)

4.18 For a particular pnp transistor for which D_B has a scale current of 10^{-13} A, and D_E has a scale current of 10^{-11} A, calculate β and i_C for v_{BE} = 0.643 V, for n = 1.

SECTION 4.4: CIRCUIT SYMBOLS AND CONVENTIONS

4.19

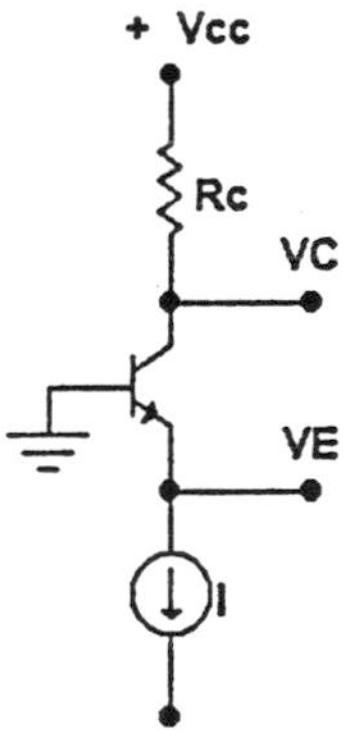

The operation of the BJT in this circuit can be shown to be conveniently independent of device parameters. Of course, it requires the complexity of the current sink I. However, this will be shown later (in Sections 4.10 and 6.4 of the Text) to be relatively simply constructed in an integrated-circuit environment. Specifically, for operation of this circuit, consider various different devices for which v_{BE} varies from 0.6 to 0.8 V at 1 mA with β variation from 10 to 300.

(a) For I = 1 mA, what is the range of values expected for emitter current i_E, emitter voltage V_E and collector current i_C?

(b) For V_{CC} = 10 V, R_C = 5 kΩ, what is the corresponding expected range of values of V_C? Does variation of V_E from device to device make any difference?

(c) To ensure operation in the active mode, $v_{CB} \geq 0$. What is the largest value of R_C which maintains active-mode operation?

4.20 Reconsider the situation described in P4.19 above, modified to include an additional (signal) current source i_E connected to the emitter. For our purposes here, i_E can be considered to be a sine-wave current with peak amplitude of 0.1 mA, that is 1/10 of the emitter-bias current I = 1 mA. For this situation, express the collector current i_C in terms of I, i_E and α. What is the largest value of i_C for a) very high β, b) β=10? Under these conditions, what is the largest value you can use for R_C to ensure active-mode operation? [Recall that V_{CC} = +10 V.] For this value of R_C, what is the peak-to-peak value of the signal voltage v_C at the collector that is produced for a) very high β and b) β = 10?

4.21 Reconsider the situation described in P4.19 above in which I is implemented using a resistor R_E and a negative voltage supply. With a −10 V supply what value of R_E ensures I_E = 1 mA for a transistor for which V_{BE} = 0.7 V. For R_E, select a "pseudo-standard value", one specified to two significant digits, chosen to produce a current on the high side of your calculated value. Now, for V_{BE} varying from 0.6 V to 0.8 V, β from 10 to ∞, and R_E by ±1%, what is the largest available collector current? Now chose R_C as large as possible while ensuring active-mode operation. For R_C, select a pseudo-standard value (as specified above) on the low side. What is the lowest possible value of υ_C for your chosen R_C varying by ±1%?

SECTION 4.5: GRAPHICAL REPRESENTATION OF TRANSISTOR CHARACTERISTICS

4.22 A BJT which conducts i_C = 10 mA at V_{BE} = 0.7 V and 25°C is operated with V_{BE} fixed at 0.62 V. What is the collector current at 0°C, 25°C and 50°C? Assume n=1.

4.23 A BJT operating at a fixed V_{BE} is found to have i_C = 2.1 mA at V_{CE} = 2 V, and i_C = 2.19 mA at V_{CE} = 9 V. What is its output resistance r_o at this current level? What value of V_A corresponds? What would its output resistances be at 0.1 mA and 10 mA (approximately)?

4.24 A BJT for which V_A = 200 V operates at V_{CE} = 5 V at a current of 100 μA. What would its current become (provided breakdown does not occur) if V_{CE} is raised to 50 V?

SECTION 4.6: ANALYSIS OF TRANSISTOR CIRCUITS AT DC

L

4.25 For the following circuits, find node voltages, V_E, V_C, and branch currents I_E, I_C, I_B. Use V_{BE} = | V_{EB} | = 0.7 V and β = 50.

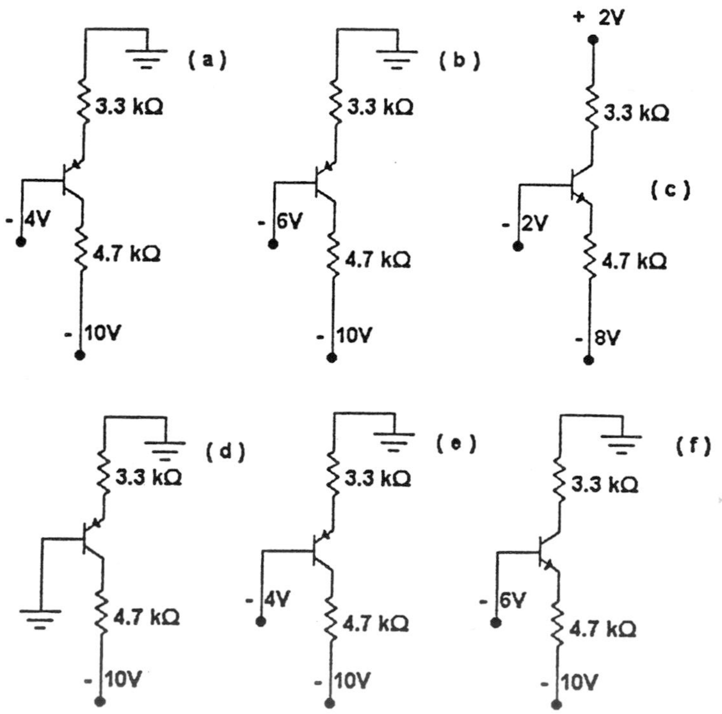

D

4.26 For the circuits shown in P4.25 a), b) above, find emitter and collector resistors (to replace the present ones) such that I_E = 0.5 mA and V_{BC} = 0 for $\alpha \simeq 1$.

4.27 For the following circuits in which $|V_{BE}|$ = 0.7 V and $\beta = 10$, find the collector, emitter and base currents and voltages.

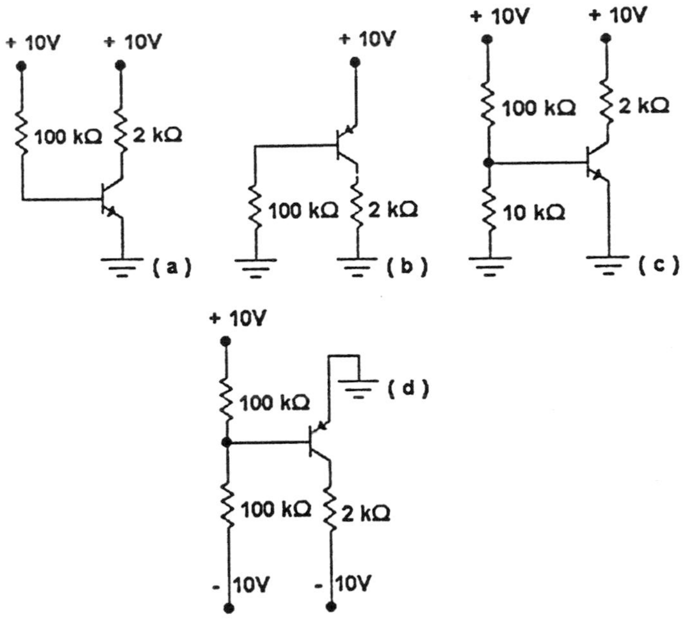

4.28 For the following circuits in which $|V_{BE}|$ = 0.7 V and $\beta = 20$, find the collector, base and emitter voltages and currents.

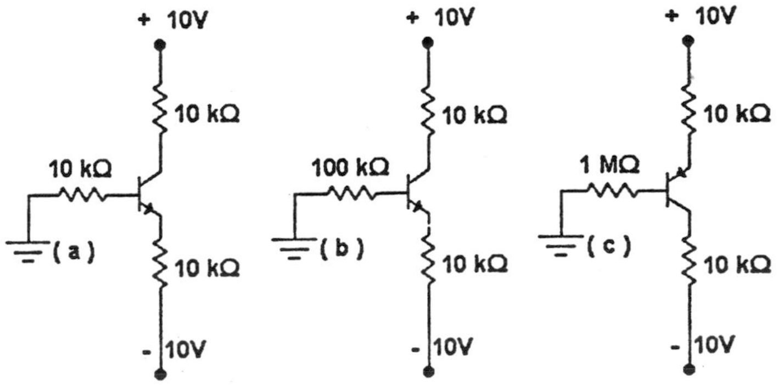

4.29

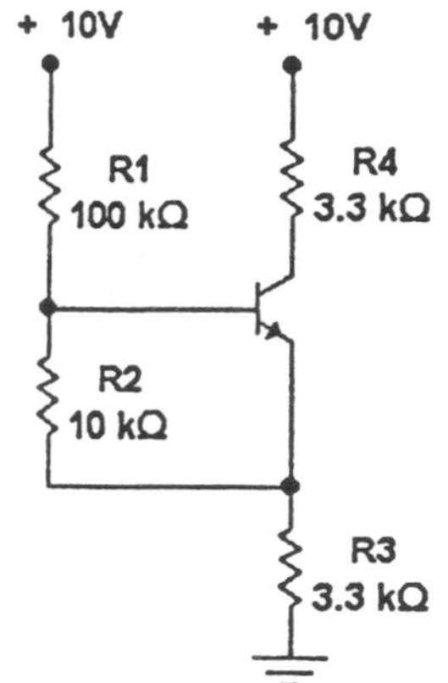

For the circuit shown, find the voltages at the base, emitter and collector for $\beta = \infty$, 100 and 10. Assume $V_{BE} = 0.7$ V.

4.30 For the circuit of P4.29 above, for what value of β does the emitter current reduce to 80% of that for $\beta = \infty$?

4.31

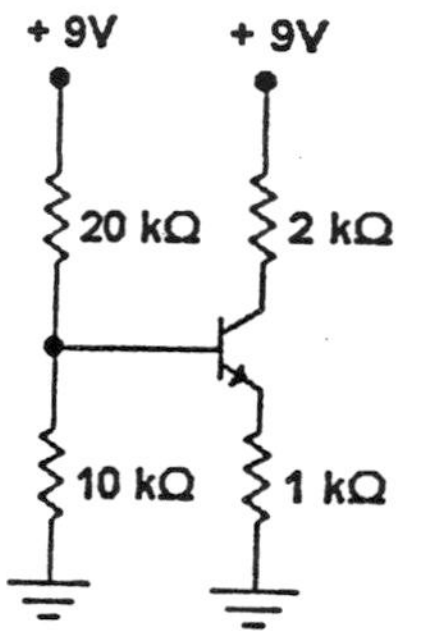

For the circuit shown, find I_E and V_{CE} for $V_{BE} = 0.7$ V and

a) $\beta = \infty$
b) $\beta = 100$
c) $\beta = 10$

4.32 For the following circuits, find the currents I_C and the voltages V_{CE}. Use $\beta = 50$ and $V_{BE} = 0.7$ V.

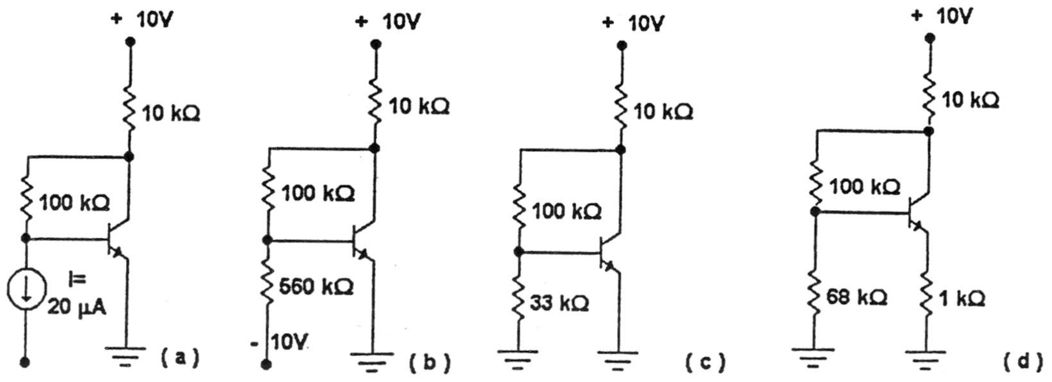

4.33

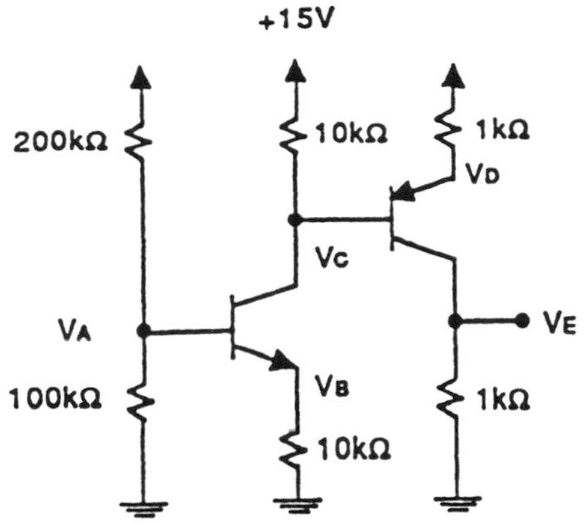

For the circuit shown, find the labelled node voltages when β is a) ∞, b) 100.

SECTION 4.7: THE TRANSISTOR AS AN AMPLIFIER

4.34 What values of transconductance apply to BJTs biased at 1 μA, 100 μA, 1 mA, and 100 mA?

4.35 For the current levels listed in P4.34 above, what equivalent small-signal input resistances model operation as seen at the emitter? At the base, for $\beta = 100$?

D

4.36 In the design of a particular amplifier, a young engineer considers the use of bias currents I_E, from 0.1 to 10 mA. Unfortunately, the application requires that the dc voltage across the load resistor be held constant to provide correct biassing of a connected amplifier stage. Find the range of gains she can expect from this gain stage.

4.37 A particular amplifier utilizes a BJT biased at $I_E = 100$ μA and having $\beta = 150$ to drive a load of 10 kΩ. For the emitter grounded for signals, what is the input resistance at the base, and the voltage gain from base to collector?

4.38

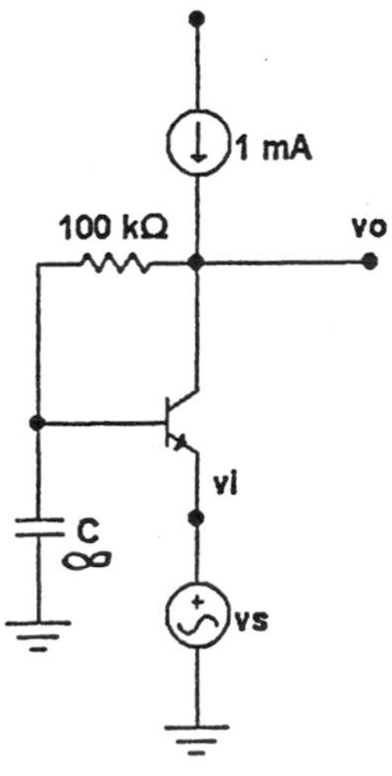

What is the voltage gain v_o/v_i of the amplifier shown? Note that capacitor C grounds the base of the amplifier for *ac* signals. Note that the gain is essentially independent of β (although the *dc* voltage V_O is not). What is the input resistance "seen" by the source v_s. What does the gain v_o/v_s become if the source resistance is 75 Ω.

SECTION 4.8: SMALL-SIGNAL EQUIVALENT-CIRCUIT MODELS

4.39 A BJT having a particular β and bias current, has a resistor r_E added in series with the emitter. Use the T model shown in Fig. 4.27 a) of the Text to create a simplified hybrid–π model for the overall amplifier (including r_E). For this model find g_m', r_π' in terms of r_E, g_m and r_π of the basic BJT. What is the equivalent input resistance (r_π') and transconductance (g_m') of the modified amplifier, for $\beta = 100$, $I_C = 1$ mA and $r_E = 3r_e$?

L

4.40 An appropriate choice of one of the BJT models of Figs. 4.26 and 4.27 of the Text, often makes the solution of a particular problem somewhat easier. To illustrate, find the gain v_o/v_s for each of the circuits below using the model(s) suggested as Π_{gm}, Π_β, T_{gm}, T_α, corresponding to Fig. 4.26 a) 4.26 b), 4.27 a), 4.27 b) respectively. In each case, assume (for simplicity) that $I_E = 1$ mA, $r_e = 25\ \Omega$, $r_\pi = 2.5$ kΩ, $\beta \approx 100$, $\alpha = 0.99$ and $g_m = 40$ mA/V. (Note that biassing is generally not shown in detail).

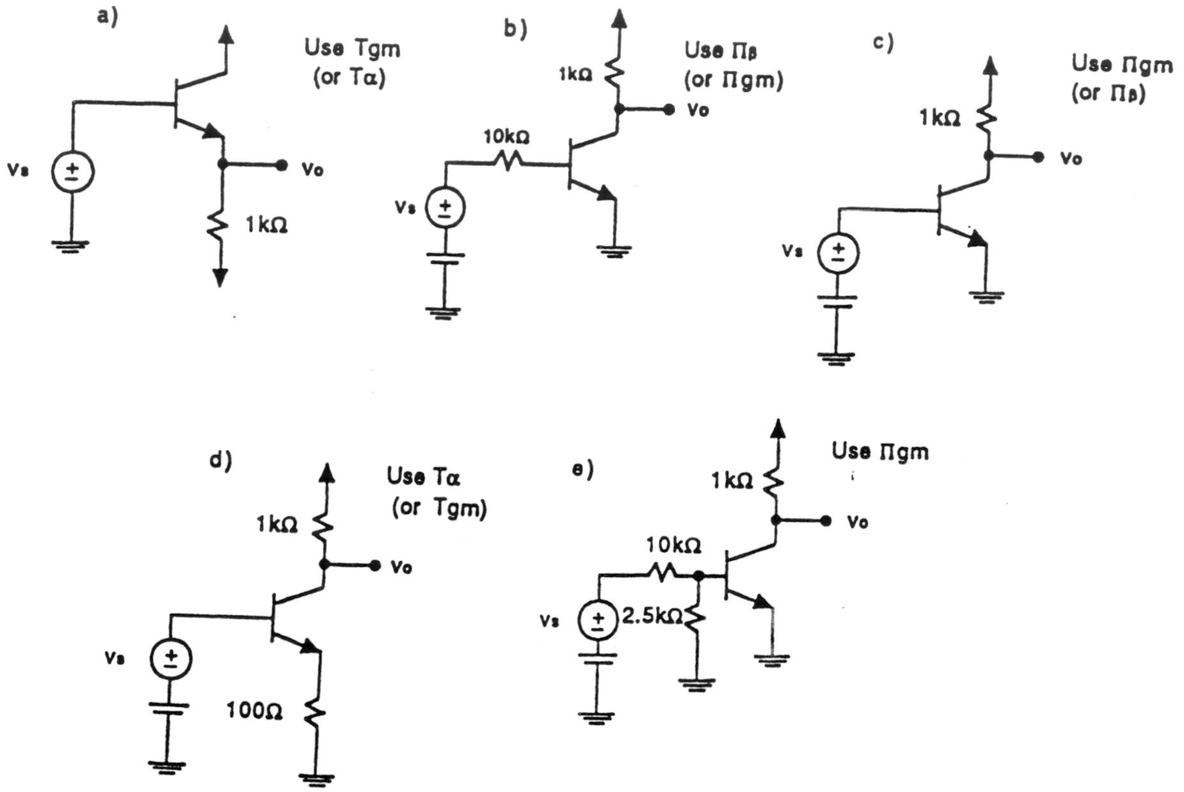

4.41

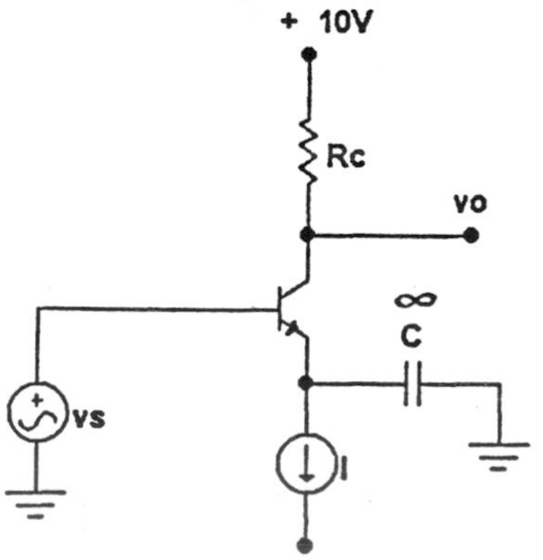

In the circuit shown, $I = 1$ mA, $R_C = 7.5$ kΩ and $\alpha = 0.99$. What is the voltage gain v_o/v_s? What is the largest sine-wave output for which the transistor remains in the active region? What is the peak value of the corresponding input?

4.42 For reasonably-linear operation of a transistor amplifier, it is customary to limit the base-emitter voltage swing to ±10 mV around the operating point. For a transistor for which $n=1$, to what fraction of the emitter bias current does this voltage range correspond? For the circuit shown in P4.41 above, find the value of R_C which provides the largest-possible reasonably-linear output while operation remains in the linear region.

4.43

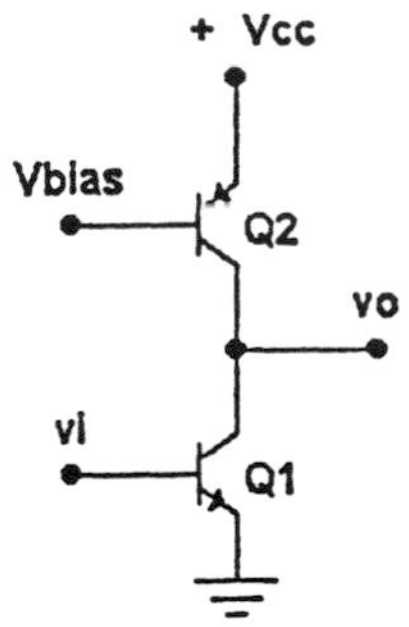

An amplifier employs the components shown, together with others that maintain $i_{C1} = i_{C2} = 100\ \mu A$ and $V_O \approx V_{CC}/2$. What is the gain v_o/v_i for $V_A = 200$ V?

C

4.44

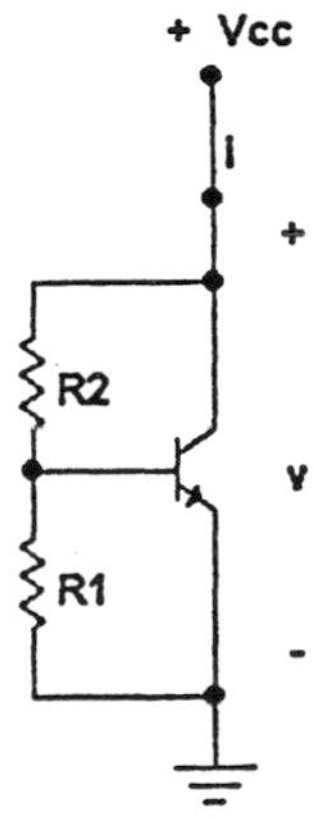

Find the equivalent resistance of the circuit shown, as a two-terminal device, (i.e., find $r = v/i$) in terms of β, r_e, R_1, R_2. Assume that the transistor remains biassed at current I. What does r become when:

a) $R_2 = 0$, $R_1 = \infty$
b) $R_1 = \infty$, $R_2 = r_\pi$
c) $R_1 = R_2 = r_\pi$?

4.45

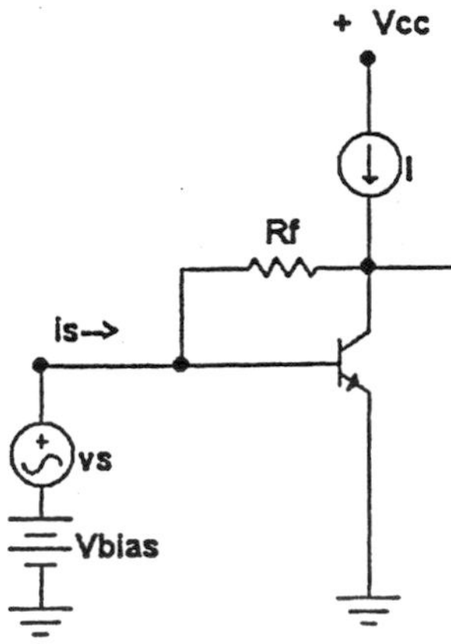

Use the hybrid-π model with g_m and r_o to find the voltage gain, $A_v = v_o/v_s$, and the input resistance, $R_i = v_s/i_s$, of this circuit, assumed biassed at $i_C \approx I$. What do these parameters reduce to when $R_f = r_o$. (Hint: Use the fact that the signal at the base must be relatively small).

4.46

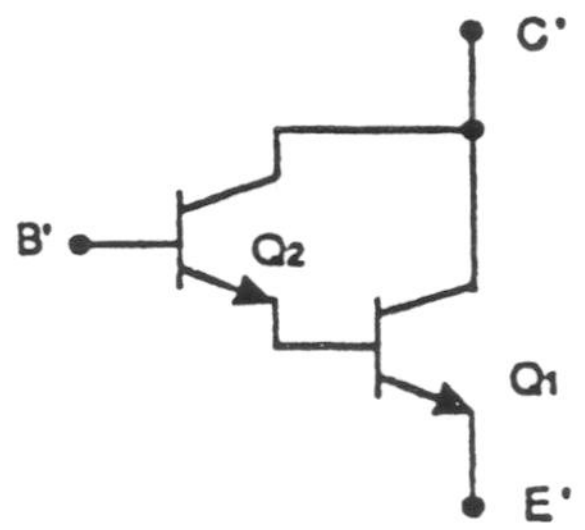

Find the equivalent hybrid-π model of the following circuit (called a Darlington connection) expressing overall values r_π' and g_m' in terms of $r_{\pi 1}$, g_{m1}, $r_{\pi 2}$ and g_{m2}. Now, realizing that the total collector current will flow predominantly in Q_1 and, accordingly, that the parameters of Q_2 will differ from those of Q_1 by a factor of approximately β_1 (for $\beta_1 >> 0$), find a corresponding approximate model.

SECTION 4.9: GRAPHICAL ANALYSIS

4.47 Sketch the $i_C - \upsilon_{EC}$ characteristics for a pnp transistor having $\beta = 200$ and $V_A = 100$ V. Sketch the characteristic curves for $i_B = 1, 2, 5, 8, 10$ μA. Assume for this sketch that $i_C = \beta i_B$ at $\upsilon_{EC} = 0$. Sketch the load line for $V_{CC} = 10$ V and $R = 5$ kΩ. Operation is defined by a dc bias current of $I_B = 5$μA. Identify the operating point, and estimate its coordinates. For a triangular signal of 3 μA peak superimposed on I_B, find the corresponding signal component of i_C and υ_{EC}. For operation in the active region, defined for convenience as $\upsilon_{EC} \geq 0$, estimate the maximum peak of an output triangle wave and the corresponding peak signal current. Sketch the output which corresponds to a superimposed base-current triangle wave of 10 μA peak amplitude. Assume for this purpose that υ_{EC} can reduce to zero. For what fraction of a cycle is its output clipped?

SECTION 4.10: BIASING THE BJT FOR DISCRETE-CIRCUIT DESIGN

D

4.48 Consider the one-supply bias scheme in Fig. 4.39 of the Text. For $R_B = R_E$, above what value of β is I_E constant to within 1%?

D

4.49 Using the rule ($V_{BB} = V_{CB} = V_{CC}/3$) and $R_B = \beta R_E/10$, provide a design for the circuit of Fig. 4.39 in the Text, in which $V_{CC} = 12$ V and $I_E = 100$ mA. For the BJT, $\beta = 50$ and $V_{BE} = 0.7V$. Find R_E, R_1, R_2, R_C to the nearest single significant digit. What values of I_E and V_{CE} does your design provide?

D

4.50 For the bias arrangement shown in Fig. 4.40 in the Text, using ±5 V supplies, a design is required for which I_E is fixed to within 5% and a ±1 V signal output range is available, for $\beta \geq 20$ and $R_C = 1$ kΩ.

D

4.51 A design is required of the feedback-bias scheme shown in Fig. 4.41 in the Text which will maintain $V_{CB} \geq 0.5$ V for $\beta \leq 200$, $V_{CC} = 5$ V and $R_C = 3.6$ kΩ, with $V_{BE} = 0.7$ V. For $\beta \geq 50$, what is the range of I_E and V_{CB} you achieve?

D

4.52 In the situation described in P4.51 above, a designer, faced with the possibility of β being uncontrollably high, choses to shunt the base-emitter junction with resistor R_β. What is its value for $\beta_{eq} \leq 200$? Find R_B to meet the other specifications. What ranges of I_E and V_{CB} result for $\beta \geq 50$?

D

4.53 Repeat P4.52 above for $\beta_{eq} \leq 100$.

4.54 For a BJT operating with the constant-current-source bias shown in Fig. 4.42a) in the Text, the manufacturer specifies β to lie in a range from 40 to 200. The bias-current source operates at 1 mA for voltages at its upper end in the range ±5V. For $V_{BE} =$ 0.70V, what is the largest value of R_B that can be tolerated? For this value of R_B, what is the range of dc voltages to be found at the base? For $R_B = 100\ r_\pi$ at the lowest value of β, what range of base voltages results?

4.55 A current source using the current-mirror circuit shown in Fig. 4.42b) of the Text operates from ±5V supplies. Select a value of R for $I = 1$ mA. Over what range of voltages V, does the current remain essentially constant? Use $V_{BE} = 0.7$ V and assume that linear operation is possible until V_{BC} reaches the edge of conduction at 0.5 V.

SECTION 4.11: BASIC SINGLE-STAGE BJT AMPLIFIER CONFIGURATIONS

4.56 For the circuit in Fig. 4.43 in the Text, $R_s = 0.5$ kΩ, $R_C = 0.5$ kΩ, $I = 10$ mA $V_{CC} = V_{EE} = 10$ V. For the BJT, $\beta = 100$ and $V_A = 100$ V. Find the corresponding values of V_B, V_E, I_C and V_C. Find g_m, r_e, r_π and r_o which correspond.

4.57 Consider the common-emitter amplifier, whose bias design was analyzed in P4.56 above. Find the values of R_i, R_o, A_v, A_i. What does A_v become if the collector is coupled appropriately to a 500 Ω load? Compare your results with those in Ex. 4.31 (page 285 of the Text). What conclusions can you reach related to resistance-scaled (or current-scaled) designs?

CD

4.58

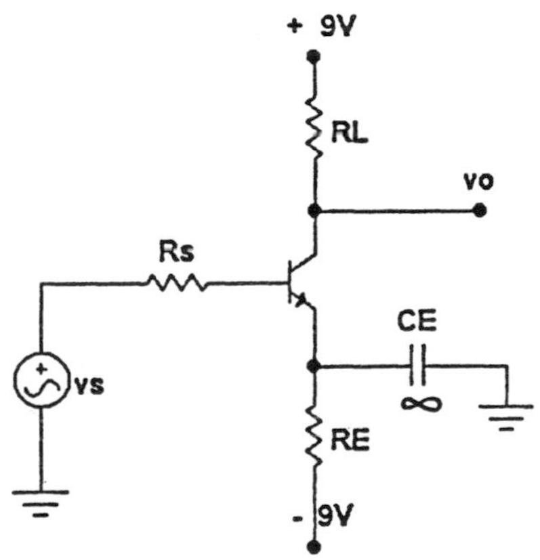

In an attempt to reduce the number of components in a space-critical design, a designer employs the circuit shown, which incorporates the source resistance R_S and load resistor R_L as part of the bias design. In this particular situation, $R_L = 10$ kΩ, $R_S = 100$ kΩ, $\beta \geq 90$ and $V_A = 100$ V. Design for the highest possible gain and an output-signal swing of 1 Vpp under all bias conditions. What are the extreme values of V_E, V_C and I_C which your design produces? What is the range of voltage gains v_o/v_s you expect? What is the range of voltages v_b, and v_s which correspond to ±1V output signals?

4.59 A common-emitter amplifier operating between a 10 kΩ source and a 10 kΩ capacitor-coupled load, with ±10 V supplies, employs $R_E = R_C = 10$ kΩ. For $V_A = 200$ V and β ranging from 50 to 150, what range of voltage gains v_o/v_s results?

4.60 An alternative to the *CE* amplifier described in P4.59 above is considered in which a 100 Ω part of R_E is left unbypassed. What range of voltage gains v_o/v_s results?

4.61

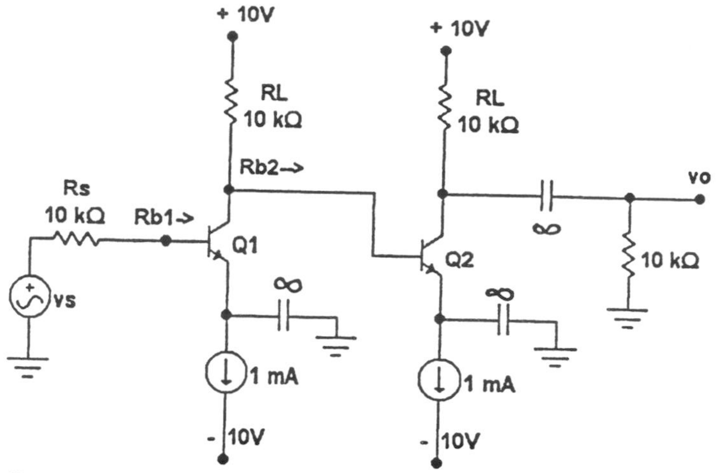

For $\beta = 150$, find $\frac{v_o}{v_{b2}}$, R_{b2}, $\frac{v_{b2}}{v_{b1}}$, R_{b1}, $\frac{v_{b1}}{v_s}$, $\frac{v_o}{v_s}$.

D

4.62 Provide a design using the basic circuit in Fig. P4.84 on page 344 of the Text in which R_E (shown as 125 Ω) is chosen so R_{in} is 10 kΩ for $\beta =$ 50. what is its voltage gain from v_s, for the load reduced (from 10 kΩ) to 1 kΩ?

4.63 A common-base amplifier, biased at an emitter current of 3 mA, employs an unbypassed base resistor $R_B =$ 2 kΩ, with $R_C = 3$ kΩ, $R_E = 3$ kΩ and $R_L = 1$ kΩ. For $\beta \geq 150$, what range of input resistances result? What range of voltage gains result from a 100 Ω source? What does the input resistance and gain become if $R_B =$ 0 Ω? Note how much simplier the design now becomes!

D

4.64

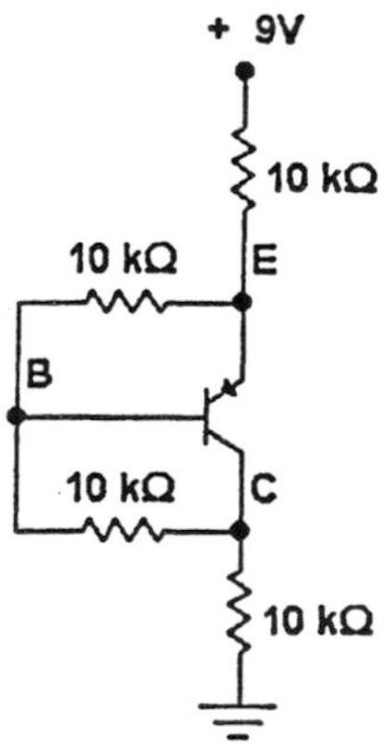

For the circuit shown, evaluate V_E, V_B, V_C and I_C for $\beta = 100$. Show capacitor-coupled connections to a 0 Ω source, a 10 kΩ load, and ground to achieve voltage gains of:

a) ≈ + 1 V/V

b) − 1 V/V (Hint: Use an extra resistor)

c) − K, where K is large

d) + K, where K is large

4.65 For each of the designs created in P4.64 above, calculate the exact gains assuming $\beta = 100$, $V_A = \infty$.

4.66 An emitter follower biassed at 0.1 mA employs a 100 kΩ base resistor and a 50 kΩ emitter resistor. The BJT has $\beta = 50$ and $V_A = 100$ V. When driven by a capacitor-coupled 20 kΩ source and driving a 2 kΩ capacitor-coupled load, what is the voltage gain which results?

CL

4.67

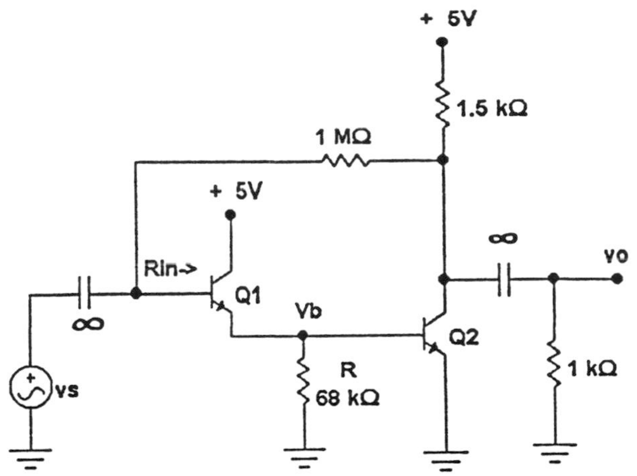

The circuit shown is a combination of a follower and common-emitter amplifier (It is called a $CC - CE$ cascade.), which has the advantage of a simple biassing structure and relatively high input resistance. For $\beta = 100$ and $V_A = 100$ V, find υ_o/υ_s and R_{in} for the circuit a) as shown, and b) with R removed. [Hint: To calculate R_{in}, realize that the 1MΩ resistor R_f has an important effect since at its right-hand-side, the voltage is υ_o/υ_s times that at its left; Thus for a test-voltage input υ_x at the left, the input current is $i_x = \upsilon_x(1-\upsilon_o/\upsilon_s)/R_f$. This is an example of the Miller-Effect idea introduced earlier in P2.11 in this book.

SECTION 4.12: THE TRANSISTOR AS A SWITCH – CUTOFF AND SATURATION

4.68 In the circuit shown in Fig. P4.97 in the Text, the transistor operates with $V_{CE} = 0.2$ V, $V_{BE} = 0.7$ V, and forced β of three. What must the value of R_B be? For $\beta_{forced} \le \beta/2$, what is the largest value R_B can have?

4.69

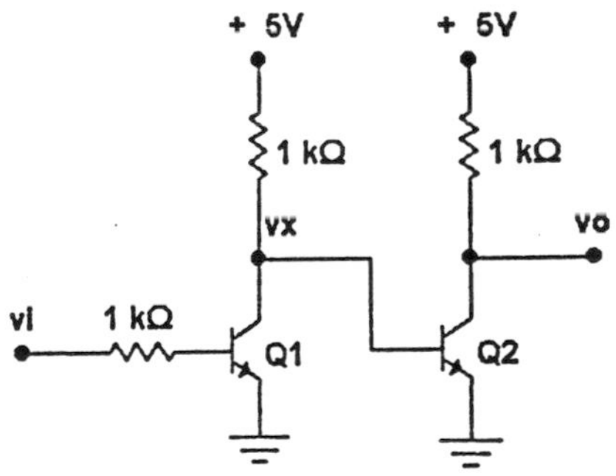

For the circuit shown, characterize the mode of operation of each transistor and the voltages υ_X and υ_O for υ_I equal to:

a) 0 V

b) +5 V

Assume that $V_{CE\,sat} = 0.2$ V, $V_{BE} = 0.7$ V. At what value of forced β do Q_1 and Q_2 operate, when saturated?

4.70 In the following circuits, $\beta = 100$, $V_{BE} = 0.7$ V, and $V_{CE\,sat} = 0.2$ V. Find V_E, V_B, V_C and the value of β at which each BJT operates. At what value of I does each transistor just leave saturation?

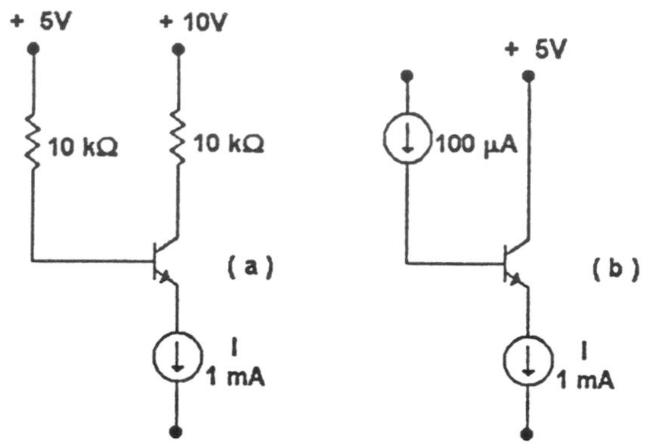

SECTION 4.13: A GENERAL LARGE-SIGNAL MODEL FOR THE BJT: THE EBERS-MOLL (EM) MODEL

4.71 A particular large *npn* BJT is known to have a base-emitter diode whose scale curent is 2×10^{-13}A, and a base-collector junction which is 40 times larger. Current gain β_F is measured to be 150. For this transistor, what are the values of I_{SE}, I_{SC}, α_F, α_R, and β_R?

4.72 The transistor in P4.71 above is operated as a diode. If operation is in the forward active mode, what is the resulting diode drop at a diode current of 100 mA?

4.73 In an application of the transistor in P4.72 above, the base connection is changed, being wired to the emitter rather than to the collector. Since the base and emitter voltages are the same, the base-emitter junction is cut off. What happens to the direction of current flow? What is the mode of operation called? What is the voltage drop between emitter and collector for a 100 mA current flow?

4.74 A particular BJT for which β_F is 200 is known to have CBJ 50 times larger than EBJ. What is its value of β_R? This transistor is to be operated as a saturated switch with I_B = 1 mA and I_C = 0. What value of $V_{CE_{sat}}$ results if the transistor is used in the normal saturation mode?

4.75 Consider the possibility that a BJT operating with collector open, and a somewhat-variable base current, can be used as a very-low-voltage regulator. What value of β_{forced} applies? For base currents varying from 1 to 4 mA, a junction-area ratio of 10 to 1, and β_F ranging from 70 to 280, what range of collector-to-emitter voltages result? Note that for a range of transistors, this voltage is relatively constant, even for varying base currents!

4.76

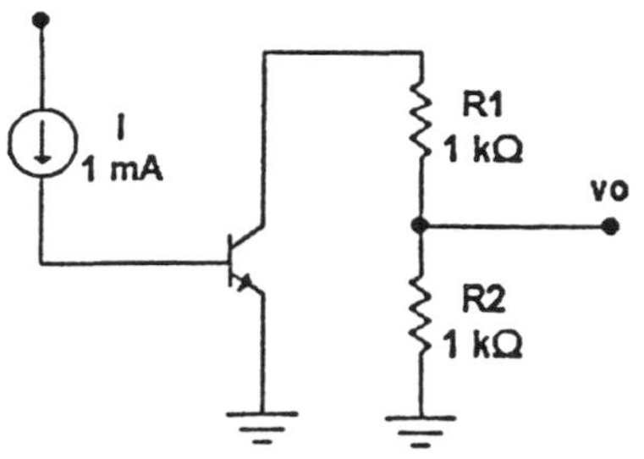

A particular BJT for which $\beta_F =$ 100, when operated as suggested in P4.75 above, with open collector and 1 mA base current, has a collector-to-emitter voltage of 100 mV. What is the output voltage, v_O, of the circuit shown? For $I =$ 1 mA and $R_1 = R_2 =$ 500Ω, what does v_O become? What is the equivalent $R_{CE_{sat}}$ of the device itself?

4.77 Considering Table 4.4 on page 307 of the Text, prepare a table for the same transistor but with emitter and collector roles interchanged. What is the limiting value for forced β? Find table entries at 90%, 50%, 20%, 10% and 1% of this value, as well as for $\beta_{forced} =$ 0.

4.78 Prepare a table of saturation voltages versus β_{forced} for an npn transistor for which $\alpha_F =$ 0.995 and the collector-junction area is 5 times that of the emitter. What is the limiting value of forced β? Tabulate $V_{CE_{sat}}$ for β_{forced} values as suggested in P4.77 above. For $I_B =$ 10 mA and $I_C =$ 1 mA, what voltage exists between collector and emitter with the transistor operated in the normal mode? inverted mode?

4.79 For a grounded-emitter pnp transistor for which $\beta_F =$ 200 and $\beta_R =$ 2, operated in a circuit for which $I_B =$ 1 mA and $\beta_{forced} =$ 10:

a) Calculate and label all currents in the branches of the *EM* model (with diodes reversed from Fig.4.55b).

b) For $I_S = 10^{-14}$ A, find the voltages across the two junctions and $V_{EC_{sat}}$.

c) Verify $V_{EC_{sat}}$ using Equation 4.114.

4.80

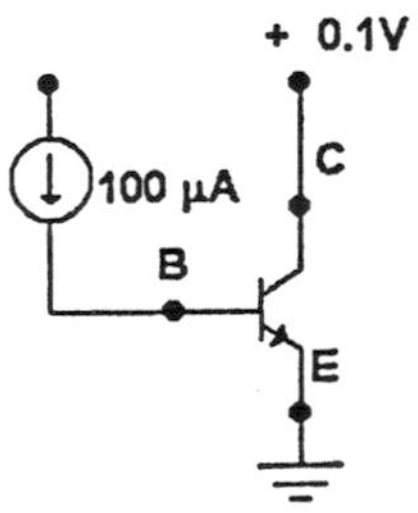

Use the transport model to find the current flowing from 0.1 V to ground in the circuit shown using a transistor for which $\beta_F =$ 50, $\beta_R =$ 0.1, and $V_{BE} =$ 0.70 V for $i_C =$ 10 mA in normal active mode.

SECTION 4.14: THE BASIC LOGIC INVERTER

4.81 A particular version of the logic inverter circuit of Fig. 4.60 of the Text (called Resistor-Transistor Logic), popular in the early days of integrated circuits, used $V_{CC} =$ 3V, $R_B =$ 450Ω and $R_C =$ 640Ω. For a transistor for which $V_{BE} =$ 0.70 V in saturation with conduction beginning at about 0.5 V, and $V_{CE_{sat}} =$ 0.3 V, with $\beta =$ 30, find V_{OH}, V_{OL}, V_{IH} and V_{IL} and the noise margins for 2 cases of fanout to similar circuits: a) 10, b) 1. What is the voltage gain in each case for $v_O =$ 0.7 V?

CL

4.82 Reconsider P4.81 above for fanout of 1, using a transistor for which v_{BE} = 0.70 V for i_C = 1 mA, n = 1, β_F = 30, and the collector junction is 5 times the size of the emitter. Using a detailed analysis, find V_{OH}, V_{OL}, V_{IH}, V_{IL}, NM_H, NM_L, the overall voltage gain, and the small-signal gain at v_O = 0.7 V.

SECTION 4.15: COMPLETE STATIC CHARACTERISTICS, INTERNAL CAPACITANCES, AND SECOND-ORDER EFFECTS

4.83 For a particular transistor, for which V_A = 200 V, β = 120, operating in the grounded-base configuration, the collector current is found to increase by 50 nA from its former value of 0.1 mA when the collector voltage is raised by 10 V. Estimate r_o and r_μ. (Hint: Use the results of P4.119 in the Text)

4.84 A particular transistor for which BV_{CBO} = 50 V, BV_{EBO} = 7 V, BV_{CEO} = 30 V is used in the following circuits: In each case, find V_O. Note that X represents (the shears causing) an open circuit.

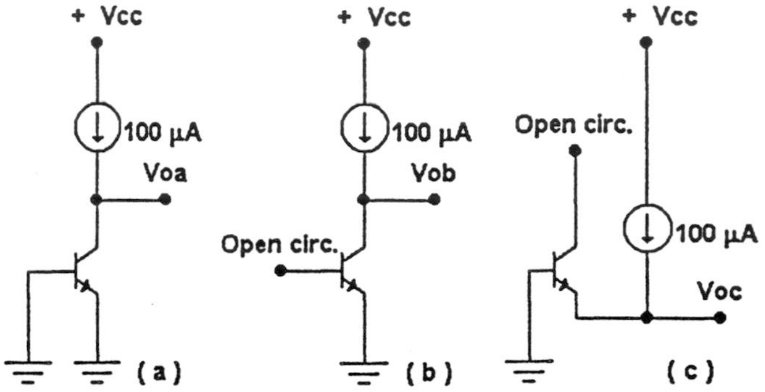

4.85 A particular BJT with grounded emitter and a constant base current of 0.1 mA, is found to have $V_{CE\ sat}$ of 0.2 V at I_C = 3 mA and $V_{CE\ sat}$ of 0.1 V at I_C = 1 mA. What are the corresponding values of $R_{CE\ sat}$ and $V_{CE\ off}$?

4.86 A BJT operating at a constant collector-to-emitter voltage of 10V is found to have I_C = 1.20 mA with I_B = 11 μA. When I_B is increased to 12 μA, I_C becomes 1.29 mA. What are the values of h_{FE} and h_{fe} for this transistor in this situation? Estimate V_A.

4.87

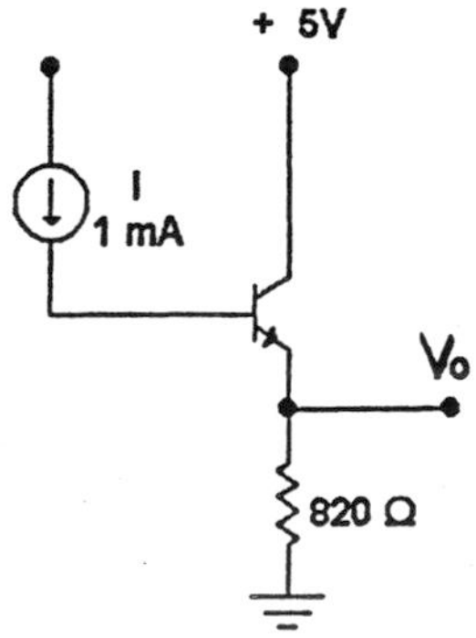

For $V_{CE\ off}$ = 50 mV and $R_{CE\ sat}$ = 50 Ω for a transistor operating at a base current of 1 mA in the circuit shown, what output voltage results? If the base drive were quadrupled, what would you expect V_O to become?

4.88 Consider the relationships shown in Fig. 4.68 of the Text between β, I_C and temperature T. Estimate an average value for the temperature coefficient of β in %/°C for I_C = 1 μA and 1 mA.

4.89 Consider the npn transistor whose detailed physical parameters are as specified in P4.2 of the Text, when operating at $I_C = 1$ mA. For $W = 1$ μm and 5 μm, calculate: the stored base charge, the forward base transit time and the emitter diffusion capacitance.

4.90 For the BJT specified in P4.2 of the Text having a base width of 1 μm, calculate C_{je} and C_μ using Equations 4.123 and 4.124 with Equations 3.26 and 3.18. Use a grading coefficient of 0.4 for the CBJ, which is 10 times the area of the EBJ and reverse biased by 2.0 V. Note that the permittivity of silicon is $\varepsilon_S = 1.04 \times 10^{-12}$ F/cm. Using the result for the emitter diffusion capacitance at 1 mA found in P4.89 above, calculate C_π and f_T.

4.91 A particular BJT for which f_T is 10 GHz at $I_C = 10$ mA, has f_T reduce to 7 GHz at $I_C = 1$ mA. Estimate values for $C_\mu + C_{je}$, and C_{de} at 10 mA and at 10 μA. What is f_T at 10 μA?

4.92 A need arises to adapt the BJT described in Exercise 4.44 of the Text to a new application in which operation is desired at $I_C = 4$ mA, but with the base-emitter voltage unchanged. A decision is made to use the same process but to double the perimeter of the square base area. What do the values of τ_F, C_{je0}, $C_{\mu0}$, V_{0e}, m_{CBJ}, C_{de}, C_{je}, C_π, C_μ and f_T become for operation at 4 mA? What f_T results at $I_C =$ 1 mA?

4.93 The 500 MHz transistor in Exercise 4.45 of the Text is being considered for operation at $i_C = 10$μA and even $i_C = 1$ μA. What unity gain frequencies would likely apply? If it is possible to reduce each of the sides of the square base used in this device structure by a factor of 10, what values of f_T would you expect at 10 μA and 1 μA?

NOTES

Chapter 5

FIELD-EFFECT TRANSISTORS (FETs)

SECTION 5.1: STRUCTURE AND PHYSICAL OPERATION OF THE ENHANCEMENT – TYPE MOSFET

5.1 An n-channel enhancement MOS transistor for which $V_t = 1.5$ V is operated with a source voltage of 0 V. For what range of values of v_{GS} is a channel induced? For $v_{GS} = 3.0$ V, for what range of values of v_{DS} is the channel pinched off at the drain end? For what range of values of v_D does the drain current saturate? For what range of values of v_D does the transistor operate in the triode mode?

5.2 Complete the following table for devices (a) through (g)

#	Channel Type	V_t V	v_S V	v_G V	v_D V	Mode (region)
a	n	1	0	3	2.1	
b	n	2	−2	2	−0.1	
c	p	−2	0	−1	−3	
d	p	−1	2	0	−1	
e		2	−3	0		saturated
f		−2	3	0	−1	
g		−2	3		−3	cutoff

SECTION 5.2: CURRENT-VOLTAGE CHARACTERISTICS OF THE ENHANCEMENT MOSFET

5.3 The n-channel enhancement MOSFET characteristics shown in Fig. 5.11b of the Text represent the relationship between i_D, v_{DS} and v_{GS} for a range of devices for which $k = k'(W/L) = 0.5\ \text{mA/V}^2$ with V_t as a parameter. For a particular device for which $V_t = 1$ V, use the data in the following table to locate the point (or points) of operation and, thereby, the missing attribute. [Hint: It may help to mark and label the points on Fig. 5.11b in the Text, or, preferably, a photocopy of it.

Reference	v_{GS} V	v_{DS} V	i_D mA
a	4	4	
b		4	2.25
c	4		2.25
d	3	3	
e	3	2	
f	3	1	
g		5	4
h	5		3

L

5.4 The characteristic curves in Fig. 5.11b of the Text are even more useful than you have perhaps realized. On a photocopy or other facsimile of the curves, relabel the axes and v_{GS} values to correspond to the following situations:

a) $k'(W/L) = 0.50 mA/V^2$, $V_t = 1\ V$

b) $k'(W/L) = 0.50 mA/V^2$, $V_t = 0.5\ V$

c) $k'(W/L) = 0.25\ \mu A/V^2$, $V_t = 1\ V$

d) $k'(W/L) = 1.00\ mA/V^2$, $V_t = 1\ V$

*e) $k'(W/L) = 2.0\ mA/V^2$, with the five v_{GS} lines labelled from (and including) the lower axis as 1.0, 1.5, 2.0, 2.5 and 3.0 volts. [Hint: The i_D axis remains unchanged.]

**f) $k'(W/L) = 0.50\ mA/V^2$, with the five v_{GS} lines labelled from (and including) the lower axis as 0.5, 1.0, 1.5, 2.0, 2.5 volts.

5.5 An n-channel enhancement MOSFET having $\mu_n C_{ox} = 20\ \mu A/V^2$, $W/L = 10$, and $V_t = +1$ V is operated with $v_S = 0$ V and $v_G = 3$ V. For what range of voltages, v_D, on the drain is operation in the triode region? What current flows for $v_{DS} = 2$ V? 1 V? 0.5 V? What is the value of r_{DS} for v_{DS} relatively small? At what value of v_{DS} does r_{DS} increase beyond its very low-voltage value by 1%? 10%?

DL

5.6 An n-channel enhancement MOSFET having $\mu_n C_{ox} = 20\ \mu A/V^2$, $W = 20\mu m$, $L = 2\mu m$, and $V_t = +1$ V is to be used for small signals as a linear resistor in the range 1 kΩ to 1 MΩ. What is the corresponding range of values of v_{GS} required? What are the corresponding ranges of operating current (i_D) and voltage (v_{DS}) for which the resistance provided is within 10% of its desired value?

5.7 Using Equations 5.5 and 5.6 of the Text, find a relationship for v_{DS} for which i_D is 100%, 99%, 90% and 50% of its saturated value. For a device for which $k'W/L = 0.50$ mA/V^2 and $V_t = 2$ V, find the values of v_{DS} for $v_{GS} = V_t + |V_t| = 2|V_t|$.

5.8 For a particular MOSFET operating in saturation at $i_D = 2.10$ mA and $v_{DS} = 3.0$ V, the drain current is found to increase to 2.20 mA when v_{DS} is raised by 5 V. Find the corresponding output resistance and estimates for the channel-length-modulation factor λ, and the equivalent Early voltage V_A.

5.9 A p-channel MOSFET for which $V_t = -2$ V has a channel width of 100 μm and length of 3 μm. If it is fabricated in a process for which $\mu_n C_{ox} = 20\ \mu A/V^2$ and $\lambda = -0.01\ V^{-1}$, estimate the drain current for saturation operation with $v_{GS} = v_{DS} = -5$ V. [Hint: Note μ_n above, not μ_p; Use $\mu_n = 2.5\ \mu_p$]

5.10 A p-channel MOSFET for which the nominal threshold is – 1 V (when $v_{SB} = 0$) operates in a circuit for which operation is such that the source voltage takes on any value between 0 V and +5 V. In a process for which $\gamma = 0.6\ V^{1/2}$ and $\phi_f = 0.3$ V, what is the threshold voltage which applies at 5 V? at 0 V? [Hint: Note that the substrate must be connected to a voltage which prevents the substrate-to-channel junction from becoming forward-biassed; here, + 5 V would be the normal choice.]

5.11 For the following circuits employing enhancement MOSFETS, for which $|V_t| = 2$ V and $\mu_n C_{ox} = 20$ $\mu A/V^2$, $W = 20\mu m$, and $L = 2\mu m$, find the labelled voltages and currents.

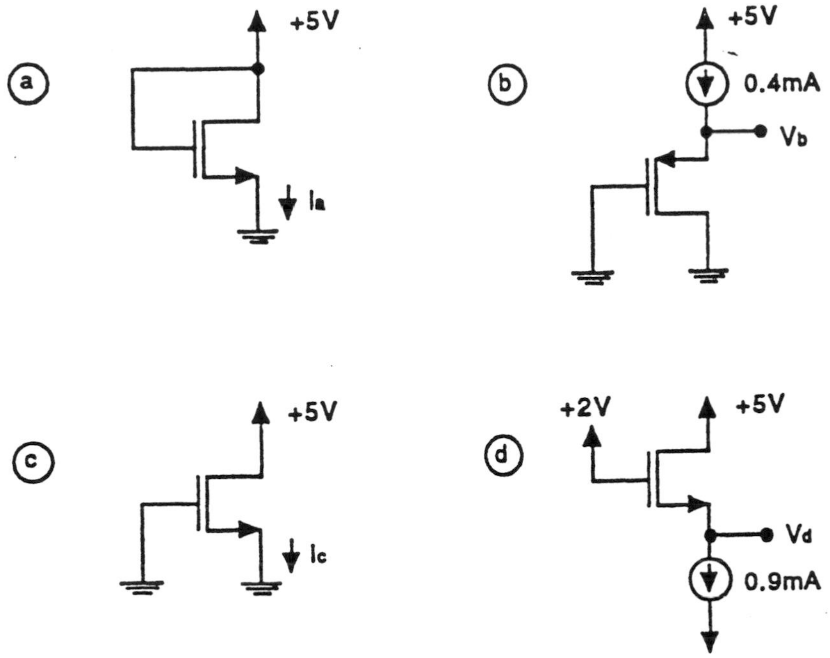

SECTION 5.3: THE DEPLETION-TYPE MOSFET

5.12 A depletion-type n-channel MOSFET for which $\mu_n C_{ox} = 20\mu A/V^2$, $W = 200\mu m$, $L = 2\mu m$ and $V_t = -4$ V is operated under a variety of conditions as stated partially in the following table: Complete the table by providing the missing entries:

#	v_S V	v_G V	v_{GS} V	v_D V	v_{DS} V	Operation Mode	i_D mA
a	0	–4		5			
b	0		–2		3	saturation	
c	0	0		5			
d	0		0		2		
e		0	+1		5		25
f	0	+2		5		triode	
g	0		+2	0			
h		0	+2	0		triode	

5.13 For the following circuits employing depletion MOSFETS, for which $|V_t| = 2$ V and $\mu_n C_{ox} = 20\mu A/V^2$ and $W/L = 10$, find the labelled voltages and currents.

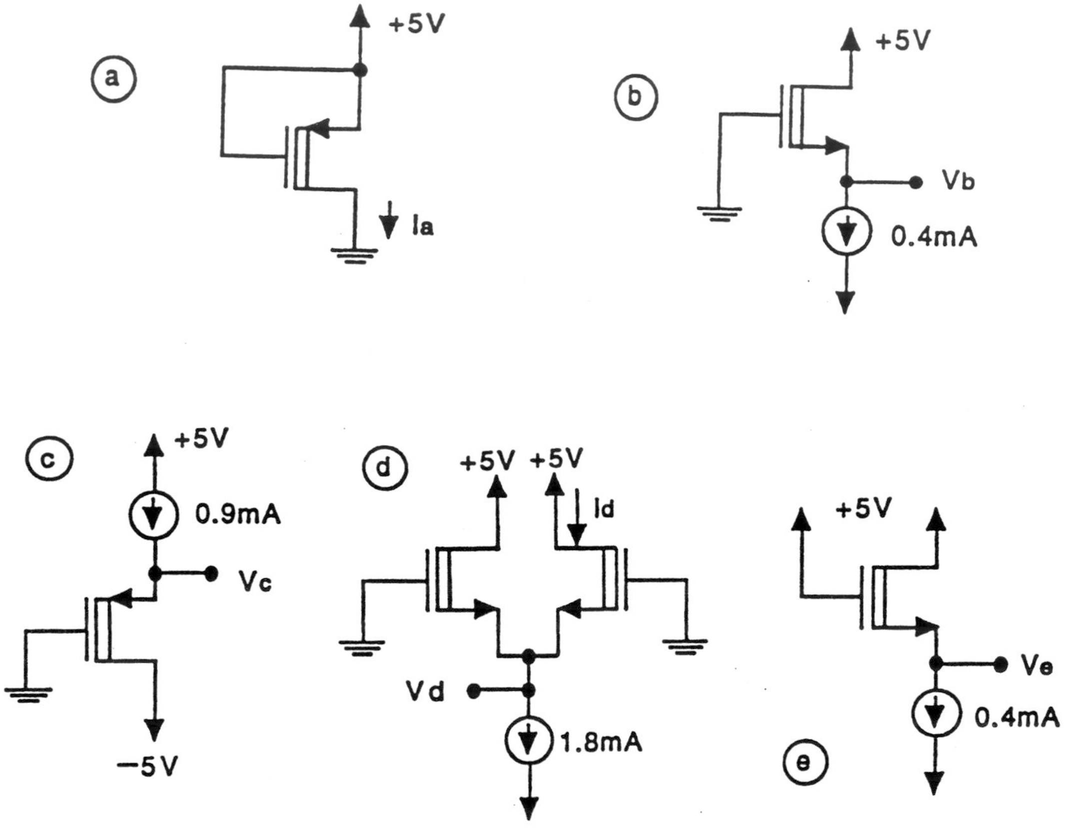

5.14

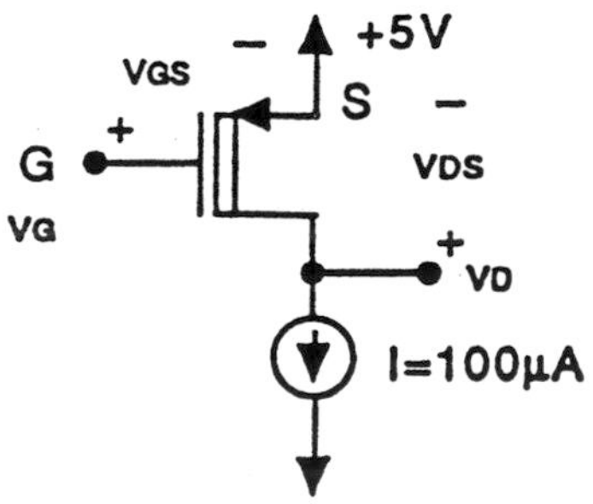

A depletion-type PMOS transistor operates in the circuit shown with $v_D = 4.8$ V when $v_G = 5$V, and $v_D = 4.95$ V when $v_G = 0$ V. Find its I_{DSS} and V_t.

SECTION 5.4: MOSFET CIRCUITS AT DC

D

5.15 For a circuit whose topology is as shown in Fig. 5.24 of the Text, employing an n-channel enhancement transistor for which $V_t = 1$ V, $\mu_n C_{ox} = 20\ \mu A/V^2$ and $W = 40\ L$, and with $R_D = 7.5$ kΩ and ±5 V supplies, V_D is measured to be +2 V. What is the current I_D? What source voltage would you expect? What is the value of R_S used? Assume $\lambda = 0$.

5.16 A circuit using the topology shown in Fig. 5.24 of the Text with $R_D = R_S = 7.5$ kΩ is found to have $V_D = +2$ V. If V_t is known to be 1 V, what value of $K = 1/2 k'(W/L)$ applies? For K of half this value, what V_D results? By what factor has it changed to compensate for a 50% change in K?

5.17 Design the circuit of Fig. 5.25 of the Text, to obtain $I_D = 0.4$ mA with the transistor specified in P.5.15 above.

D

5.18

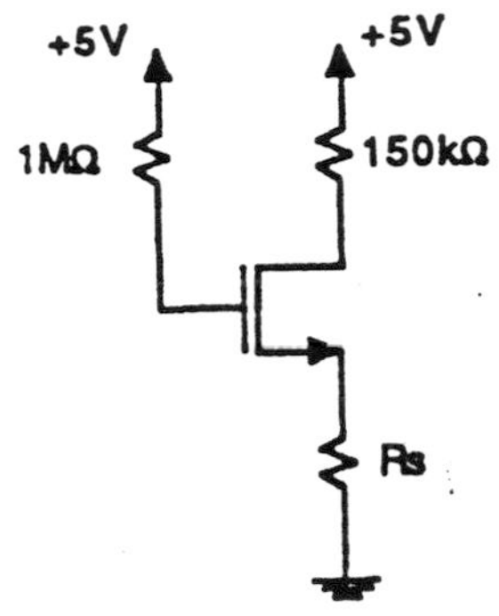

For a transistor for which $V_t = 1$ V and $k'(W/L) = 1\ mA/V^2$ used in the circuit shown, find the value of R_S for which $V_D = +2$ V. Specify R_S to 1 significant digit. For this value of R_S, what would a more precise measurement of V_D show it to be?

C

5.19 In the following circuit, the transistor employed has nominal values of $K = 1/2 k'(W/L)$ and $|V_t|$ of 0.5 mA/V^2 and 1 V respectively.

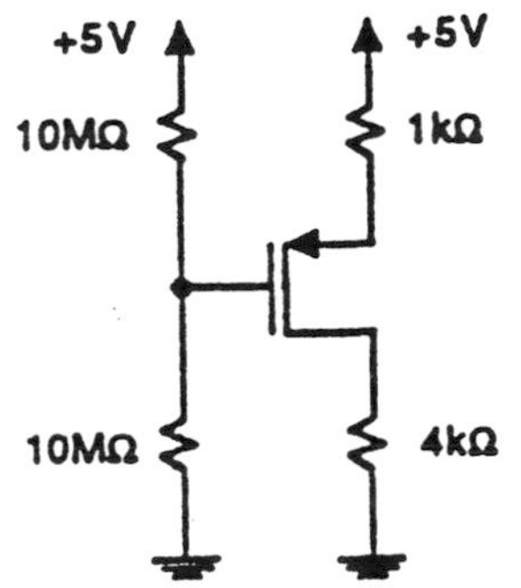

What voltage would you expect at V_D? If the voltage actually measured at V_D is 90% of that expected, what % change from the value specified for in V_t alone or of K alone would account for the result?

5.20 In the circuit shown, the depletion PMOS has $\mu_p C_{ox} = 8\mu A/V^2$, $W =$ 500 μm, $L =$ 2μm and $V_t = 2$ V. What is its value of I_{DSS}?

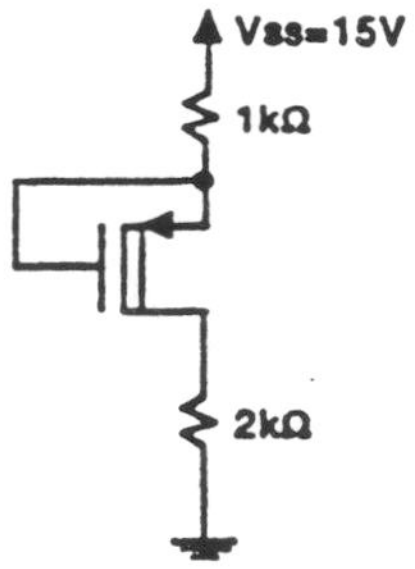

What are the voltages at the source and drain? What is the lowest value of V_{SS} for which the device remains in the saturated mode?

5.21

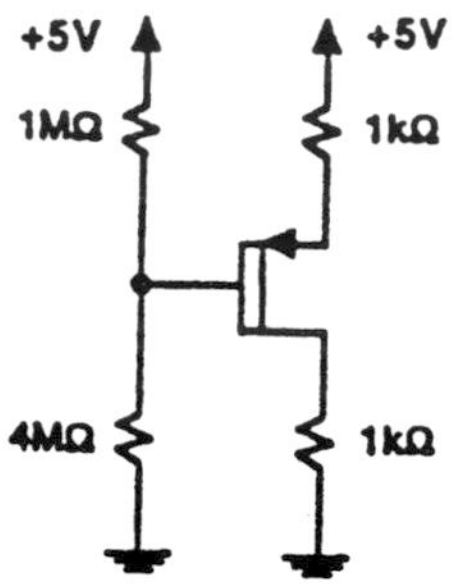

In this circuit, the depletion PMOS has $V_t = 2$ V, $\mu_p C_{ox} = 8\mu A/V^2$, and $W/L =$ 250. What is the corresponding value of V_{DS}?

5.22

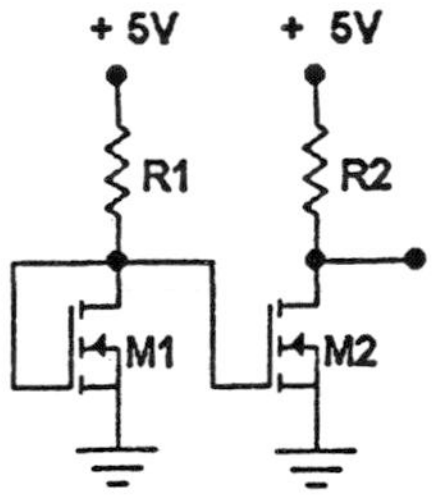

In this circuit, the enhancement NMOS have $\mu_n C_{ox} = 20\mu A/V^2$, $V_t =$ 1V, $L = 2\mu m$ and $W = 30\mu m$. Find R_1 so the current i_{D1} is 150 μA. If $R_1 = R_2$, what voltage results at the drain of M_2. Suggest a connection for a third transistor M_3 which will potentially double the current in R_2. What value must R_2 now be to make $\upsilon_{DS2} = \upsilon_{DS1}$?

SECTION 5.5: THE MOSFET AS AN AMPLIFIER

5.23 Consider the generalized amplifier circuit of Fig. 5.31 of the Text, in which is installed a transistor with $V_t =$ 2V, and $k'(W/L) = 2mA/V^2$. For $V_{DD} =$ 12V, $R_D =$ 0.5kΩ and $V_{GS} =$ 5V, what i_D and υ_D result? For $\upsilon_{gs} = \pm$ 0.5V peak, what is the total variation in drain current? [Hint: Perform 2 more bias calculations for this result.] What is the peak-to-peak value of υ_d.

5.24 For the situation described in P5.23 above, what is the largest load resistance for which operation remains in the saturation mode for $V_{DD} =$ 12 V and $\upsilon_{gs} = \pm 0.5$ V. For the load resistance increased to 1 kΩ, what is the ratio of the peak voltages of the output signal produced by a ±0.5 V input signal?

5.25 For the situation described in P5.23 above. Use Eq. 5.38 of the Text to find g_m. What is the voltage gain you expect (using Eq. 5.40). Using this result, with a ±0.5 V input, what output signal should result? Using Eq. 5.35, what peak output signal voltages would you expect? [Hint: Use the last term in Eq. 5.35 extended which represents a dc output shift.] Compare generally with the results of P5.23 above.

D

5.26 A p-channel MOSFET, for which $\mu_p C_{ox} = 10\ \mu A/V^2$, $W = 300\ \mu m$, and $L = 3\ \mu m$ is operated at $I_D = 4$ mA. What is the corresponding value of g_m. For what value of R_L is the gain of a simple amplifier equal to −10 V/V? For what peak input signal value is operation reasonably linear. [Hint: use Eq. 5.36 with Eq. 5.32.]

5.27 A MOS device operating at a *dc* bias current of 1 mA with a 10 kΩ load has a gain of −9.091 V/V for small signals. When the current is reduced to ¼ mA, the gain reduces to −4.808 V/V. What values of $K = 1/2k'(W/L)$ and V_A apparently prevail? At $I_D = 1$ mA, as a sine wave input signal is raised in amplitude, the output signal peaks are found to change by 10% from their expected value for input peaks of ±0.5 V. What is the value of $(\upsilon_{GS} - V_t)$ which apparently applies?

5.28

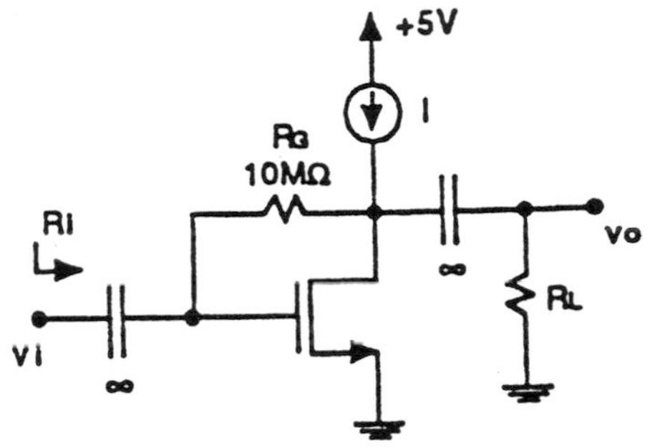

For the transistor shown, $k'(W/L) = 2\ mA/V^2$, $V_t = 1$ V, and $V_A = 50$ V. Find V_D, I_D, g_m, r_o, υ_o/υ_i and R_i for $R_L = R$. For $I = 1$ mA, what are R_i, υ_o/υ_i for $R_L = R_G$? r_o?, R_i?. Note that the latter gain is the one to be used for each stage in a cascade of n identical stages.

5.29

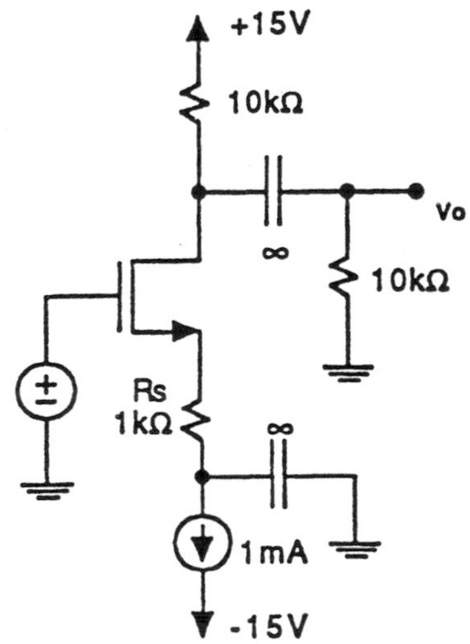

For the transistor shown, $g_m = 1$ mA/V and $r_o = 100$ kΩ. Find υ_o/υ_i. What does the gain become for $R_S = 0$? for $R_S = 3.76$ kΩ?.

5.30 An NMOS source follower operates with a constant bias current I_S for which $g_m = 0.725$ mA/V and $r_o = 47$kΩ. Ignoring the body effect, use the T model to find the output resistance of the follower. What is the no-load voltage gain G_0? For what range of load resistances is the follower gain greater than 0.95? 0.90?

5.31 Consider the impact of body effect on the situation described in P5.30 above. Generally speaking, there are two effects of substrate bias on V_t: The first is the average change in V_t from V_{t0} as the average value of υ_S changes. The second is the instantaneous signal-induced change in V_t due to the changing channel-to-substrate voltage and modelled by χ. For this situation (where I_S is fixed), does the first effect matter? [Hint: Consider what controls g_m.] For $\chi = 0.2$, find a better estimate of the follower output resistance. What is the follower no-load gain? For what load resistor is the follower gain greater than 50%?

SECTION 5.6: BIASING IN MOS AMPLIFIER CIRCUITS

5.32 A particular n-channel enhancement MOS device for which $V_t = 2$ V and $k'(W/L) = 1.0\ mA/V^2$ is to be biassed using the circuit of Fig. 5.39a of the Text with a 9 V supply. For $R_{G1} = R_{G2} = 10\text{M}\Omega$ and $R_S = R_D = 10$ kΩ, find I_D and V_{DS}. For what peak amplitude of output signal will operation remain in the saturation region? What are all the corresponding values when a device with $V_t = 1$ V is substituted in the same circuit?

CD

5.33 A design of a bias circuit using the scheme shown in Fig. 5.39a is required for a family of MOSFETs, for which V_t ranges from 1 to 2 V and $K = 1/2k'(W/L)$ ranges from 0.3 to 0.5 mA/V^2. The design should provide the largest possible gain using a drain current limited to the range 0.5 to 1 mA. For the situation in which the largest resistor available is 10 MΩ, what are the values of R_{G1}, R_{G2}, R_S and R_D to be used with a 9 V supply? Arrange that the largest output signal for which operation in saturation is assured, is 0.5V peak.

5.34 A particular n-channel enhancement MOS device for which $V_t = 2$ V and $k'(W/L) = 1.0\ mA/V^2$ is to be biased using the circuit of Fig. 5.39d of the Text with $R_G = 10$ MΩ, $R_D = 20$ kΩ and $V_{DD} = 9$V. Find I_D and V_{DS}. For what peak amplitude of the output signal will operation remain in the saturation region? What are the corresponding values when a device with $V_t = 1$ V is substituted?

5.35 The circuit and situation described in P5.34 above, is modified by a second resistor, $R_{G2} = 10$ MΩ shunted from gate to source. Repeat the computations requested there.

D

5.36 Using the circuit of Fig. 5.39d of the Text, prepare a design for the situation described in P5.33 above, but with a 5 V supply.

D

5.37 Reconsider the situation presented in P5.36 above using the topology of Fig. 5.39d of the Text, but with a resistor R_{G2} added from gate to source to increase the output swing by a factor of 1.5, all other conditions being the same.

D

5.38 For a depletion MOS device for which $V_p = -4$ V and $I_{DSS} = 32$ mA, design a bias circuit of the type shown in Fig. 5.39a of the Text for a drain current of 8 mA, using a 9 V supply, and the largest possible value of R_D that allows for a drain-signal swing of ±2 V. Find values for R_S, R_D, R_{G1}, R_{G2} using 10 MΩ as the largest available resistor value.

D

5.39 A basic current mirror circuit resembling that in Fig. 5.40 of the Text, operates with two transistors for which $V_t = 1$V, $\mu_n C_{ox} = 20\mu\text{A}/V^2$, and $W = 4$ μm and $L = 2$ μm, from a 5 V supply, with $\upsilon_{GS} = 2.5$V. What is the output current flowing into Q_2? What value of R is to be used? For what range of voltages on Q_2 does it operate in saturation? For what value of output voltage does the output current reduce by a factor of 2?

D

5.40 Using transistors for which $V_t = 1$ V, $V_A = 10$ V/μm of channel length, and $k_n' = 20\ \mu\text{A}/V^2$, design a mirror circuit, using a reference current of 25μA, to produce a nominal output current of 100μA with an output resistance of 1MΩ. Arrange that the output transistor remains in saturation mode for voltages to within 0.5 V of the negative supply. What values of L, W_1 and W_2 should be used? At what output voltage (measured from the negative supply) will the output be exactly 100μA? What does the output current become for an output that is 5V above the negative supply?

CD

5.41 Sketch the topology of a multiple-output current-steering circuit using a 10μA reference current source, to produce current sinks of 5μA, 40μA and 100μA, and sources of 20μA and 40μA. Use transistors all of the same length and having a minium width of 2μm. How many individual transistors are needed? What is the total width of all the NMOS? of all the PMOS?

SECTION 5.7: BASIC CONFIGURATIONS OF SINGLE-STAGE IC MOS AMPLIFIERS

5.42 For each of the circuits of Fig. 5.44 of the Text operating from a ±5V supply with transistors for which $V_t = 1V$ and operation is at $V_{GS} = 2V$ using current/sources/sinks whose minimum operating voltage is 0.5V, find the nominal operating voltage or voltage range for each of the input and output terminals.

5.43 A CMOS amplifier using the topology of Fig. 5.45 of the Text employs devices for all of which $k = k'(W/L) = 20μA/V^2$ and $V_A = 100$ V. What is the small-signal voltage gain which results for $I_{REF} = 25$ μA, and for 0.25 μA?

5.44 A CMOS amplifier using the topology of Fig. 5.45 of the Text is fabricated using a process for which 0.5 $μ_n C_{ox} = μ_p C_{ox} = 20$ μA/V², $|V_t| = 1$ V, $|V_A| = 50$ V, and $L = 10$ μm, to have $W_n = W_p = 100$ μm. I_{REF} is created using a diode-connected NMOS device half the width of Q_1. For the amplifier biased to have $V_{GS1} = V_{DS1}$ using a 5 V supply, find the *total* supply current, the *dc* output voltage V_O, the voltage gain $υ_o/υ_i$, and the output votlage signal range for which all devices operate in saturation.

5.45 Consider the CMOS common-gate configuration shown in Fig. 5.47 of the Text, using transistors for which $|V_t| = 1V$, $μ_n C_{ox} = 2μ_p C_{ox} = 20μA/V^2$, $|V_A| = 50$ V, $L = 10$ μm and $W_n = W_p = 100$ μm, $I_{REF} = 50$ μA. For the input signal source having an average voltage of 0 V, what must V_{BIAS} be? For χ found to be 0.2, what are the values of the voltage gain $υ_o/υ_i$, and the input resistance at the source R_i, which result?

C

5.46 Consider the common-drain circuit shown in Fig. 5.48 of the Text, using transistors for which $μ_n C_{ox} = 20$ μA/V², $V_t = 1$ V, $V_A = 50$ V, $L = 10 μm$, $W = 100$ μm, $\chi = 0.2$, $I_{REF} = 50$ μA, and $V_{DD} = V_{SS} = 5$ V. What are the limits on $υ_I$ and $υ_O$ for saturation-mode operation? Find the no-load voltage gain $υ_o/υ_i$, and the output resistance R_o. For what value of load resistor is the gain reduced by a factor of 2?

5.47 A source follower employing a constant-current bias supply is measured to have an output resistance of 952 ohms. When its bias current is quadrupled, its output resistance reduces to 455 ohms. Find values for g_m and r_o in the original situation. If this bias current was 1 mA at first, what is V_A for this transistor? At 1 mA bias, for what range of loads is the follower gain ≥ 0.900 V/V?

DL

5.48 Two p-channel transistors, one enhancement and one depletion, are operated as 2-terminal devices. For each, how many connections are there in which current can flow between the two terminals? (Be careful!) In total, how many configurations allow the device to operate in saturation? Sketch them and identify for each the minimum terminal voltage at which current flows in the saturation mode.

L

5.49 For the following circuits, for which $k'(W/L) = 2\ mA/V^2$ and $|V_t| = 2$ V, find the labelled currents and voltages.

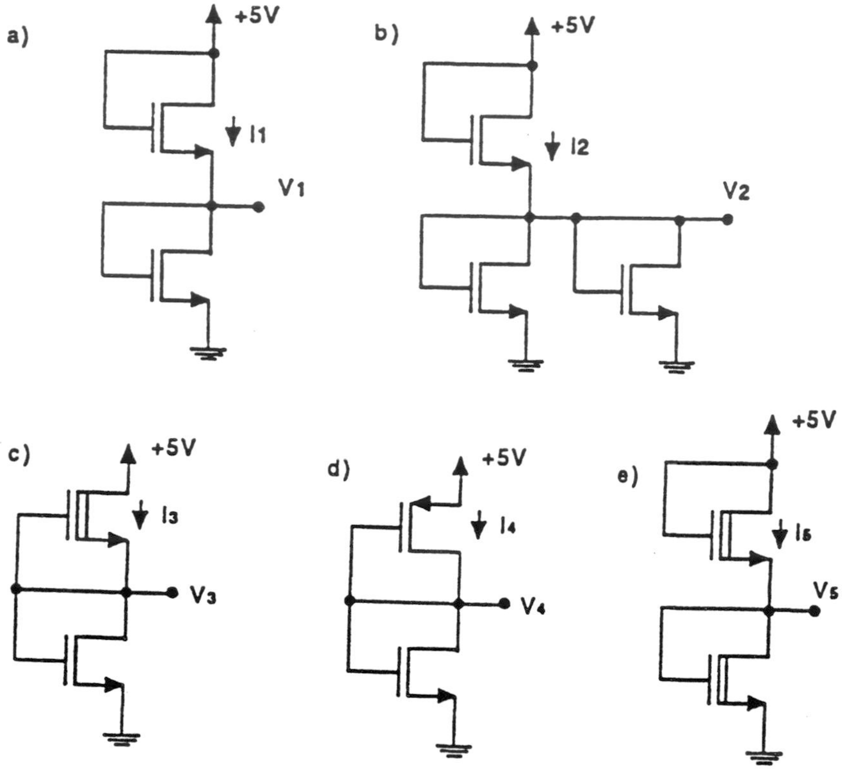

5.50 A particular amplifier employs two enhancement p-channel transistors for which $V_t = -1$ V. For the driver, $k_D = k_D'(W/L) = 180\mu A/V^2$. For the load, $k_L = k_L'(W/L) = 20\mu A/V^2$. The power supply is +5 V. Ignoring the body effect, when this amplifier is operated in its linear range, what is the gain, υ_o/υ_i? What is the value of υ_I for which $\upsilon_O = V_{DD}/2$? What are the upper and lower values of output voltage, and corresponding input voltage, for which Equation 5.81 applies?

5.51 For an NMOS enhancement-load transistor for which $V_{to} = 0.9$V, $2\Phi_f = 0.6$V, $\gamma = 0.5V^{1/2}$, $k'(W/L) = 20\mu A/V^2$, connected to a +5 V supply, what is the upper limit of the output-voltage range? [Hint: Use Eq. 5.30 on page 374 of the Text]. What are the values of V_t, χ, and g_m for outputs near the upper limit, and near 0 V? [Hint: Use Eq. 5.51 on page 399 of the Text.]

5.52 For the load situation described in P5.51 above, employing a driver for which $k_{nD}'(W/L) = 180\ \mu A/V^2$, what voltage gain results for $V_O = 2.5$ V?

5.53 An NMOS amplifier employs a driver for which $V_t = 1$ V and $k_D' = 180\mu A/V^2$ with a depletion load for which $V_t = -2$ V and $k_L'(W/L) = 45\mu A/V^2$. For both devices $V_A = 50$ V. The power supply is +5 V and $\chi = 0.2$. Find the gain for V_O around 2.5 V. What is the range of outputs for which this gain value applies?

SECTION 5.8: THE CMOS DIGITAL LOGIC INVERTER

5.54 A CMOS logic inverter such as that in Fig. 5.55 of the Text employs matched transistors in a 0.5μm process in which $\mu_n C_{ox} = 100\mu A/V^2$, $(W/L)_n = (2\mu m/1\mu m)$, $|V_A| = 40$ V and $|V_t| = 0.8$ V, with a 3.3 V supply. Find values for V_{OH}, V_{OL}, V_{IH}, V_{IL}, NM_H, NM_L and $V_{th} = V_M$ (where $\upsilon_I = \upsilon_O$), [Hint:

Use Equations 5.93 to 5.96.]. What is the current I_{peak} which flows from the supply when $\upsilon_I = \upsilon_O = V_M = V_{th} = V_{DD}/2$? At what input voltages is the current half that value? one-tenth that value? What are the output resistances of this gate in the high and low output states? Estimate the output voltage levels for load-current levels equal to I_{peak}.

5.55 For the CMOS inverter described in P5.54 above, loaded by a capacitance of 50fF, estimate the average propagation delay $t_P = (t_{PLH} + t_{PHL})/2$. For operation at the frequency $1/(4t_P)$, estimate the average power dissipation for transition times which are a) ideal (that is, zero), and b) equal to $2t_P$. [Hint: For the latter case, consider the triangular current pulses conducted through the two devices.] Estimate the delay-power product on the latter basis. How does the result compare with CV_{DD}^2 suggested on page 435 of the Text. What is the reason for the difference?

SECTION 5.9: THE MOSFET AS AN ANALOG SWITCH

D

5.56 A MOSFET switch is to be used to ground an internal node of a network whose open-circuit voltage ranges from 1.1 to 3.3 V and source resistance is 21 kΩ. The available FET control voltage available switches between 0 and 5 V. For the technology used, $V_t = 1$ V, $\mu_n C_{ox} = 20\ \mu A/V^2$ and $L = 10\ \mu m$. What switch width is required to guarantee that the node can be brought to within 10 mV of ground?

5.57 A CMOS transmission gate uses devices for which $W_p = 2W_n = 100\ L$, $|V_t| = 2$ V and $\mu_n C_{ox} = 20\ \mu A/V^2$. For control signals of ±5 V and a load of 5 kΩ to ground, what is the fraction of the *ac* input signal lost in the switch, for an input $\upsilon_I = \upsilon_i + V_I$, for $V_I = -5$ V, 0 V or +5 V?

SECTION 5.10: THE MOSFET INTERNAL CAPACITANCES AND HIGH-FREQUENCY MODEL

L

5.58 The gate-to-channel capacitance of a MOS transistor is often used as an explicit capacitor in MOS circuits, in which case the gate and source are joined to form the second electrode. Use the data provided in Table 5.1 on page 364 of the Text to calculate the dimensions of a square capacitor of 1 pF for the range of technologies cited, where oxide thickness ranges from 20 nm to 100 nm. In a 0.8 μm feature-size technology, in which the minimum-size NMOS digital device has $L = 1.2\mu m$ and $W = 2.4\mu m$, to how many such transistors do these capacitor areas correspond?

5.59 For the 1.2 μm technology whose parameters are provided in Ex. 5.41 on page 444 of the Text, calculate values of C_{ov}, C_{gs}, G_{gd}, C_{sb}, C_{db}, for transistors operating in saturation at $|V_{SB}| = |V_{DB}| = 2$ V for which: a) $L = 2.4$ μm, $W = 100$ μm; b) $L = 24$ μm, $W = 10$ μm. [Hint: Recall in calculating C_{sb} and C_{db} that the values at C_{sb0} and C_{db0} provided are approximately proportional to source and drain areas, respectively and correspondingly to device widths.]

5.60 For each of the transistors in P5.59 above, operating at $I_D = 100$ μA with $k_n' = 100\ \mu/V^2$, calculate f_T. What does f_T become in each case if the operating current is reduced to 10μA?

5.61 For the transistor evaluated in Ex. 5.42 and Ex. 5.41 on pages 447 and 444, respectively, of the Text, what is the gate input impedance of the NMOS device with output shorted when operating at f_T. What does the input impedance become at $f_T/10$ if the FET operates there with a voltage gain of –2 V/V?

SECTION 5.11: THE JUNCTION FIELD-EFFECT TRANSISTOR (JFET)

D

5.62 An n-channel JFET with I_{DSS} = 10 mA and $V_p = -2$ V operates with gate and source grounded and drain connected to a positive voltage V_+. What current flows when $V_+ = +4$ V? +2 V? +1 V? At what value of V_+ does i_D become half its saturation value?

D

5.63 An n-channel JFET for which I_{DSS} = 10 mA and $V_P = -2$ V operates with source grounded and drain at +1 V . For what value of gate voltage is the drain current 5 mA? 1 mA?

5.64 An n-channel JFET for which I_{DSS} = 10 mA and $V_P = -$ 2 V operates as a switch with small v_{DS}. What is the series switch resistance for v_{GS} = 0 V? −1 V? − 2V?

5.65 An n-channel JFET for which I_{DSS} = 10 mA, and $V_P = -$ 2 V operates at v_{DS} = 2 V with i_D = 5 mA, and at v_{DS} = 7 V with i_D = 5.1 mA. If v_{GS} is the same in each case, what is its value? What values of r_o, λ and V_A corespond?

L

5.66 For the following JFET circuits, using devices for which I_{DSS} = 4 mA and $|V_p|$ = 2 V, find the labelled voltages and currents.

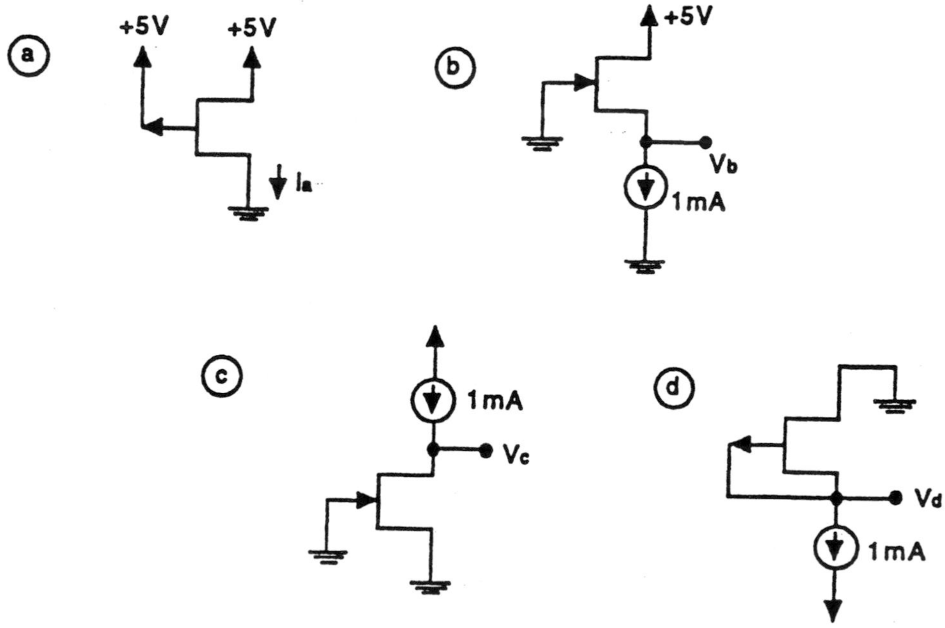

5.67

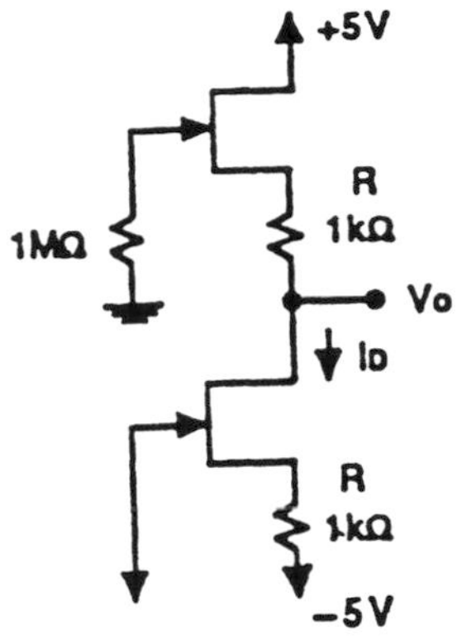

For the FETs shown, I_{DSS} = 4 mA, V_P = − 2 V. What values of I_D and V_O result? What do they become if both resistors are accidentally replaced by ones of value 2 kΩ?

5.68

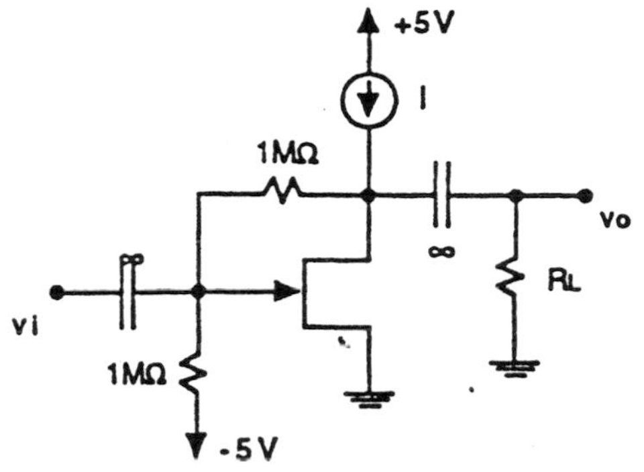

For the FET in the circuit shown, V_P = − 2 V, I_{DSS} = 10 mA, V_A = 100 V. For I = 10 mA, what are I_D, V_D, r_o, g_m, and υ_o/υ_i and R_i for $R_L = \infty$? for $R_L = r_o$?

SECTION 5.12: GALLIUM ARSENIDE (GaAS) DEVICES – THE MESFET

5.69 A GaAs MESFET for which $\beta = 10^{-4} A/V^2$ for each μm of gate width, $\lambda = 0.2\ V^{-1}$ and $V_t = -1.0$ V, and having a width of 100 μm, is operated at $\upsilon_{GS} = 0 \pm 0.2$ V, with υ_{DS} of about 3 V. Find the range of g_m, r_o and the highest available voltage gain you can expect for such operation.

5.70 The transistor described in P5.69 above is operated with a 3 V supply and a 100 Ω load. What values of υ_{DS} result for the inputs stated? What is the corresponding "voltage gain" for a ± 0.2 V input signal?

5.71 The amplifier in Example 5.11 and Fig. 5.73 on page 457 of the Text is modified by increasing the width of Q_2 to equal that of Q_1. For the output stabilized at 5 V by some external means, what values of i_{D1}, V_{GS1}, g_{m1}, and small-signal voltage gain result?

5.72 For the situation described in P5.71 above, the output is stabilized at +3 V. What is the value of V_{GS1} required? What are g_{m1} and the gain υ_o/υ_i?

NOTES

Chapter 6

DIFFERENTIAL AND MULTISTAGE AMPLIFIERS

SECTION 6.1: THE BJT DIFFERENTIAL PAIR

6.1 For the BJT differential-pair configuration, find the differential signal ($\upsilon_d = \upsilon_{B1} - \upsilon_{B2}$) sufficient to cause

a) $i_{C2} = 99\% \; I$

b) $i_{C1} = 95\% \; I$

c) $i_{C1} = 9.0 \; i_{C2}$

L

6.2 For situations related to those shown in Fig. 6.2 of the Text, some measurements are taken as tabulated below: For all cases, $V_{CC} = +10$ V, $R_C = 4\ k\Omega$, and $I = 2$ mA. For the BJTs, assume β is high, $V_{BE} = 0.7$ V, V_{CE} sat = 0.2 V, all essentially independent of the detail of junction-current magnitude. Find the missing values.

Case	υ_{B1} V	υ_{B2} V	$\upsilon_{E1,2}$ V	υ_{C1} V	υ_{C2} V
a	0		−0.7		6
b	2	2		6	
c		1	1.3		
d	−2		0.3	10	
e	1		2.8		3
f	−4	−4			8
g		0	+3.3		
h	1	3.5			

L

6.3 For an npn BJT differential pair using a +10 V supply and collector resistors of 4 kΩ, partial measurements provide results as follows. Find the missing entries, assuming $\alpha = 1$ and $n = 1$.

Case	I mA	υ_{B1} V	υ_{B2} V	υ_E V	υ_{C1} V	υ_{C2} V
a	0.2	0.00		−.700		9.60
b	0.2	0.01	0.00			
c	0.2	0.00	0.05			
d	0.2		0.00	−.675		
e	0.2	−1.00				9.90
f	2.0	−1.00	−1.00			
g	2.0	0.01	0.00			
h	2.0	0.00	0.05			
i	2.0	1.00			3.00	

SECTION 6.2: SMALL-SIGNAL OPERATION OF THE BJT DIFFERENTIAL AMPLIFIER

C

6.4 Explore the nature of the small-signal assumption made following Eq. 6.11 on page 493 of the Text, in the creation of Eq. 6.12, by including one additional term of the exponential series ($e^x = 1 + x + x^2/2$) in the creation of a higher-order alternative. What is the error made in using Eq. 6.12 for $\upsilon_d/2 \le 10\ mV$? For what value of υ_d is the error made by the linear approximation equal to 10%, 5%, 1%?

6.5 A particular differential amplifier resembling that in Fig. 6.5 of the Text, uses $I = 200\ \mu A$, $R_C = 10k\Omega$ and $V_{CC} = +3$ V. What is the differential gain achieved for outputs taken differentially? If taken from one or the other collectors separately? What is the upper limit of common-mode input voltage, for which operation maintains $\upsilon_{CB} \ge -0.4$ V?

D

6.6 A differential amplifier resembling that in Fig. 6.5 of the Text, employs collector resistors of 100 kΩ and a bias source of 200 μA. What is its differential voltage gain for outputs taken differentially? What is its differential input resistance? Transistor $\beta \ge 150$. Emitter resistors are added to double the input resistance. What are their values? What does the differential voltage gain become?

6.7 A differential amplifier employing 10 kΩ collector resistors, and for which the emitter bias current is 400 μA, uses BJTs for which $n = 1$ and $\beta = 200$. It is driven differentially by signal sources whose output resistances are 10 kΩ. The emitter-current source has an output resistance of 0.5 MΩ. For outputs taken both differentially and single-endedly, find the differential input resistance, the differential-mode gain from the source, the common-mode input resistance, the common-mode gain, and the CMRR as a ratio and in dB.

6.8 For the situation described in P6.7 above, the collector resistors are mismatched. For outputs taken differentially, find the common-mode gain and CMRR (as a ratio and in dB) for load resistors specified to be ±1%, and to be ±10%.

C

6.9 For the situation described in P6.7 above, the source resistors are mismatched by 10% and device betas vary by ±10% from their nominal value. For outputs taken differentially, find the nominal differential gain, the worst-case common-mode gain, and the corresponding CMRR in dB. Hint: Note that the half-circuit idea does not work directly here; rather current division in a Y-shaped resistor network must be considered.

6.10 For the situation described in P6.9 and P6.7 above, fixed emitter resistors, each of value $R_E = 9r_e$ (where r_e is the incremental emitter resistance) are added. What do A_d, A_{cm} and CMRR become for the output taken differentially from matched collector resistors?

SECTION 6.3: OTHER NON-IDEAL CHARACTERISTICS OF THE DIFFERENTIAL AMPLIFIER

6.11 A BJT differential amplifier operating at a total bias current of 200 μA employs collector resistors that have a ±5% tolerance. What is the worst-case input offset voltage you would expect? If emitter resistors are added, with $R_E = 9r_e$ what input-offset voltage results?

6.12 If in P6.11 above, the added emitter resistors each have a ±5% tolerance, what might the most extreme input offset become? What might be a more realistic estimate of its expected value [Hint: Use the idea of uncorrelated variations presented in Eq. 6.55 on page 506 of the Test.] If the collector resistors are now trimmed to have exactly equal values, what does the input offset become?

6.13 Four uncorrelated sources of input offset to which a differential amplifier is subject, produce essentially equal individual contributions of 2 mV. Estimate the total offset resulting. If closer examination reveals that the offsets are 0.5, 1, 2, and 4 mV individually, what overall offset might be expected?

6.14 For a BJT differential pair biased at current I, both a β mismatch of 10% and a source-resistance mismatch of 10% are present. For nominal values of I, β, and R_S of 100 μA, 100, and 100 kΩ, respectively, what worst-case input-voltage offset is possible?

6.15 An npn BJT differential amplifier for which the bias current is 300 μA employs a 15 V supply and 60kΩ resistors. For peak signals of 10 mV across the junctions of each input transistor and V_{CEsat} limited to 0.4 V with V_{BE} = 0.7 V, what is the most positive usable common-mode input signal?

SECTION 6.4: BIASING IN BJT INTEGRATED CIRCUITS

6.16 A diode-connected transistor is operated at a bias current of 100 μA. What is the resistance between its two terminals? If two such transistors are connected, a) in parallel, b) in series, to the same biassing source, what do the resistances across the combinations become?

6.17 For what value of β would a simple current mirror have a gain error of 1%? 0.1%?

DL*

6.18 A simple mirror operating at a current of 1 mA is augmented by resistors in series with each emitter across which the nominal voltage drop is 1/10 V_{BE}. For transistors for which V_{BE} = 0.700 V at 10 mA and n = 1, what resistors would be used (specify to 1 significant digit only). Now one of these resistors is to be laser-trimmed to raise its value to compensate for a nominal value of β equal to 90. Which resistor must be adjusted? What is its required value? What is the current error at 0.5 mA and at 2 mA with nominal β? At each of the 3 currents with β = 70?

6.19 A simple current mirror operating at 100 μA employs devices for which β = 150 and V_A = 150 V. For what value of output voltage do the two imperfections cancel? Over what output-voltage range is the net error less than 1%?

D

6.20 In the design of a simple current mirror for a particular application, there is a concern for the effect of temperature change on the output current. Chose a value of R and V_{CC} to provide a nominal current of 100 μA at 25°C, at which V_{BE} = 0.700V, and for which the change at 75°C limited to 5% (Use a junction temperature coefficient of –2 mV/°C).

D

6.21 Given several identical npn transistors and a reference current of 1 mA, sketch the circuit of a multiple-output mirror whose nominal current values are 0.5 mA, 1 mA, and 2 mA. How many transistors do you need? If you also are provided with some matched pnp transistors, is it possible to save transistors? Sketch an alternative circuit topology assuming only one end of I_{REF} is available. How many transistors do you need? What is the number needed if both ends of I_{REF} are available (as in Fig. 6.18 of the Text).

6.22 Repeat the analysis of the circuit of Fig. 6.16 of the Text, by starting with $i_{B1} = i_{B2} = i$. What does the current gain become with two outputs using two identical transistors Q_{2a} and Q_{2b} with separate collectors?

CDL

6.23 For the compensated circuit of Fig. 6.19 of the Text, repeat the analysis leading to Eq. 6.67, but maintain all 3 β values separately. For 3 transistors having current gains of β and $\beta(1 \pm k)$, select an optimal placement of each in the circuit. Is there a particular value of k for which your design is particularly good?

DC

6.24 For the Wilson Mirror in Fig. 6.20 of the Text, follow through the process with separate betas as suggested in P6.23 above for the base-current-compensated mirror.

D

6.25 Design a two-output Widlar current source using a 100 μA reference to provide outputs of both 1 μA and 10 μA using 1mA transistors for which $V_{BE} = 0.700$V at 1mA with $n = 1$.

SECTION 6.5: THE BJT DIFFERENTIAL AMPLIFIER WITH ACTIVE LOAD

6.26 The amplifier shown in Fig. 6.25 of the Text uses a bias source of 100 μA with devices for which V_A = 150 V and $\beta = 75$. What is its overall transconductance, its open-circuit voltage gain, its output resistance, and its differential input resistance? What does the voltage gain become when feeding a load equal to the input resistance?

6.27 The differential amplifier in Fig. 6.25 of the Text, using a 100 μA bias source, is augmented with 500 Ω resistors in the emitters of each of Q_1 through Q_4. For $\beta = 75$ and V_A = 150 V, what is the overall transconductance, the output resistance, and the open-circuit voltage gain. (Hint: Note that the common connection between emitter resistors is virtual ground for differential inputs).

6.28

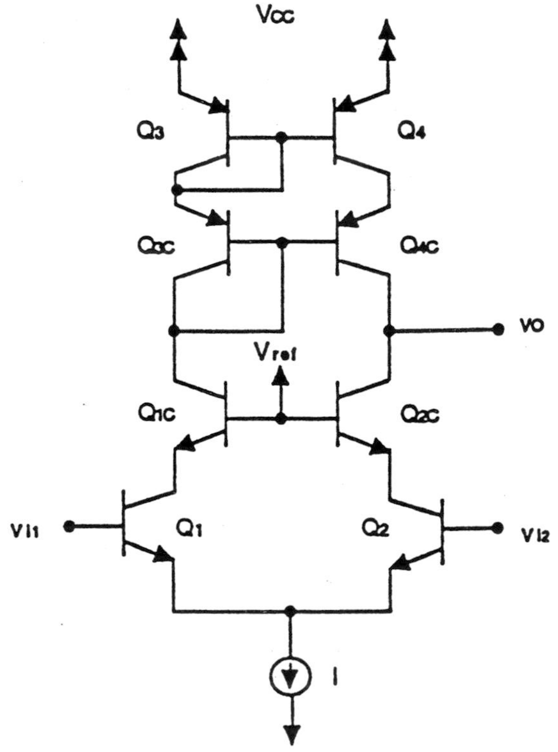

For the circuit shown, with $\beta = 75$, V_A = 75 V and I=100 μA, find the overall g_m, the output resistance and the open-circuit gain. Incorporate the fact that $r_\mu \simeq 10\,\beta r_o$.

SECTION 6.6: MOS DIFFERENTIAL AMPLIFIERS

6.29 Consider Fig. 6.30 in the Text, where the slope of the curves represents g_m. For what values of $\frac{v_{id}}{V_{GS}-V_t}$ does g_m of each device deviate from its value at $v_{id} = 0$ by 10%? 5%? 1%? [Hint: Consider Eq. 6.101.]

6.30 A PMOS differential amplifier utilizing a bias current of $I = 25$ µA uses devices for which $V_t = 1$ V, $W = 120$ µm, $L = 6$ µm, $\mu_p C_{ox} = 10$ µA/V^2 and $V_A = 50$ V. Find V_{GS}, g_m and the maximum possible voltage gain using current-source loads which are a) ideal, b) have $V_A = 50$ V.

6.31 For the amplifier in P6.30 above, the current-source loads are unbalanced, one being 10% higher and the other 10% lower than the nominal value of $I/2$. What input offset voltage is required to compensate?

6.32 Reconsider Ex. 6.16 on page 533 of the Text, for the situation in which the R_D and W/L tolerances are ±1%, and the V_t tolerance is ±0.6 mV. Find the separate offsets corresponding. What is the worst-case total offset? What is a likely value for the offset, assuming the offset sources are independent? [Hint: Use a root-sum-of-squares estimate.]

6.33 Consider the cascode mirror circuit of Fig. 6.32b of the Text, for the situation in which $V_t = 1$ V, $k'(W/L) = 200$ µA/V^2, $I_{REF} = 100$ µA, and $V_A = 20$ V, for all transistors. Include the effect of λ in your bias calculations. What is V_{GS1}? What is I_O for $V_{D3} = V_{D4}$? For $V_{D3} = 12$V? What is the output resistance?

6.34 Repeat P6.33 above for the (simple) Wilson mirror of Fig. 6.32c.

D

6.35 Consider the cascode mirror as described in P6.33 above, augmented by another transistor Q_5 having gate and source connections common with those of Q_3 and providing a second output, I_{02}. What values of I_{01} and I_{02} result? (Be careful!) What is the output resistance of the outputs, when joined? When operated independently? What change can be made to provide two high-resistance outputs of $I_{REF}/2$ each? Compare the total width of the transistors used in the two cases with that of the original single-output mirror.

6.36 For the CMOS amplifier of Fig. 6.34 in the Text, all transistors have $|V_A| = 20$ V, $|V_t| = 1$ V, and $k'(W/L) = 200$ µA/V^2. For $I = 200$ µA, find the voltage gain. For what external load does the gain reduce by a factor of 2?

SECTION 6.7: BiCMOS AMPLIFIERS

6.37 For $I = 10$ µA, find g_m, R_i, r_o and the voltage gain of the CE and CS amplifiers shown in Figs. 6.35a, b. For the BJT, $V_A = 100$ V, $\beta = 100$. For the MOSFET, $V_A = 20$ V, $\mu_n C_{ox} = 20$ µA/V^2, $L = 2$ µm and $W = 20$ µm.

6.38 For the BiCMOS cascode shown in Fig. 6.35c of the Text, using the parameters provided in P6.37 above, find the overall voltage gain v_o/v_i, for $I = 10$ µA.

6.39 For the BiCMOS cascode shown in Fig. 6.35d of Text, using the parameters provided in P6.37 above, find the voltage gain v_o/v_i, for $I = 100$ µA.

6.40 Consider the BiCMOS double-cascode mirror in Fig. 6.37 of the Text using devices described in P6.37 above, and operating at 10 µA. What does the output resistance become if Q_3 and Q_6 are not used?

Recall that $r_\mu \approx 10\ \beta r_o$. What does the output resistance become if Q_6 and Q_3 are retained, but Q_5, Q_2 are eliminated?

SECTION 6.8: GaAs AMPLIFIERS

6.41

For the following GaAs circuits, using devices characterized by the normalized data given in Table 5.2 (see page 456 of the Text) with width in μm as noted near each, find labelled values of I and V.

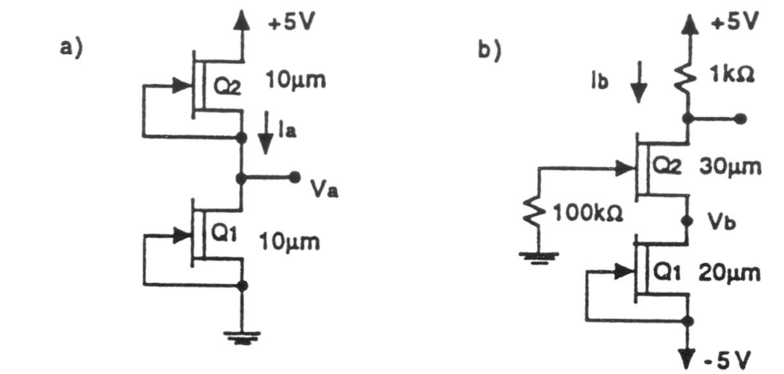

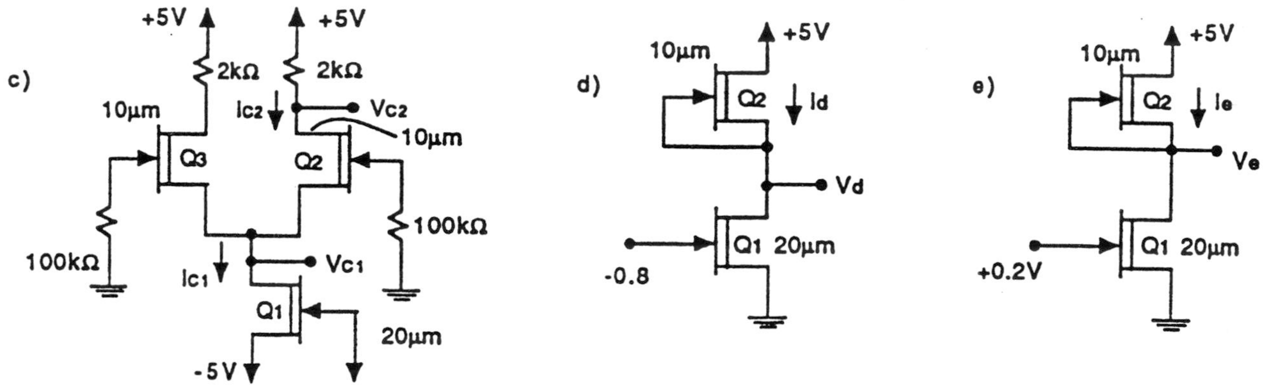

6.42 Repeat Ex. 6.24 on page 544 of the Text for the cascode current source, for conditions as stated, except that W_1=5 μm.

L

6.43 The circuit of Fig. 6.40 of the Text is extended to a double cascode by adding a transistor Q_3 of width 20 μm whose gate is connected to $V_{BIAS\ 2}$, with a change of V_{SS} to 6 V, and V_{BIAS} to $V_{BIAS\ 1} = -5.3$ V. For this design provide the data requested in Exercise 6.24 on page 544 of the Text.

D

6.44 Consider the circuit of Fig. 6.41. Using the data from Table 5.2 (on page 456 of the Text), select appropriate values for FET channel and diode widths for operation at I = 5 mA, with V_{DD}=5 V and $V_A \approx 2$ V. For this design, calculate $\alpha = \dfrac{v_b}{v_a}$ from Eq. 6.132 on page 546 of the Text, and R_o from both Eq. 6.133 and Eq. 6.134.

CDL

6.45 Using Table 5.2 values, design a composite MESFET for nominal operation with I_{DSSeq} = 0.5 mA, υ_{DS} = 3 V and υ_{DS1} = 0.7 V. What are W_1 and W_2 required? Using basic relationships for device currents and voltages, what does I_{DSS} become when υ_{DS} is increased to 6 V? Using formal small-signal analysis, what is R_{oeq} nominally? How well does this account for the change in I_{DSS} already found?

D

6.46 Use two of the composite devices created in P6.45 above to implement the amplifier in Fig. 6.43c. For V_{DD} = 6 V, what nominal value of bias V_I is required for $V_O \simeq V_{DD}/2$? Using the results of P6.45 above, what is the output resistance of the amplifier? What is the overall gain?

CDL

6.47 For the gain-enhanced MESFET differential amplifier in Fig. 6.45 of the Text, prepare a design for operation of as many as possible of the devices at a nominal i_D of 0.5 mA and υ_{GS} = −0.5 V. What current sources are needed for υ_{CM} around 0 V and V_{DD} = 5 V? What is V_O? What is the resulting gain υ_o/υ_i?

SECTION 6.9: MULTISTAGE AMPLIFIERS

6.48 For the multistage amplifier of Fig. 6.46 of the Text, evaluate the input resistance, output resistance and overall gain for the situations in which:

a) $\beta = \infty$

b) $\beta = 50$

DL

6.49 Consider the amplifier in Fig. 6.46 of the Text,, with the third and fourth stages modified to operate at a higher current to allow more output-drive capability. For i_{C7} and i_{C8} increased by a factor of 2 and 4 respectively, what resistor values must be changed? What do A_3, A_4, A and R_0 become for β = 100? For what load does the gain reduce to 0.8 of the value calculated? For this load, what output signal swing is possible?

D

6.50 Consider the multistage amplifier in Fig. 6.46 of the Text with the supplies reduced to ± 10 V: Find the values of all resistors to accommodate the charge in supply while maintaining υ_O = 0 V, I_{C3} = 0.5 mA, I_{C7} = 1 mA, I_{C8} = 5 mA, 3 V across the collector resistors of Q_1, Q_2, and 2 V across the collector resistor of Q_5. Use $|V_{BE}|$ = 0.7 V and β = 50. Now, find the overall gain using the method described on page 557 of the Text. [Hint: First find critical input resistances, namely R_{i1}, R_{i2}, R_{i3}, R_{i4}; then evaluate the local current-transmission factors.] What happens to this method when $\beta = \infty$? (Essentially, when $\beta = \infty$, each amplifier stage becomes exclusively voltage-controlled, and the usual voltage-factor method must be used.)

NOTES

Chapter 7

FREQUENCY RESPONSE

SECTION 7.1: S-DOMAIN ANALYSIS: POLES, ZEROS, AND BODE PLOTS

7.1 Find the transfer function $T(s) = V_o(s)/V_i(s)$ of the circuit shown: Is this an STC circuit? If so, of what type? For $10C_1 = C_2 = 0.5\mu F$ and $R_1 = 10$ kΩ, find the location of the pole(s) and zero(s) and sketch Bode plots for the magnitude and phase responses.

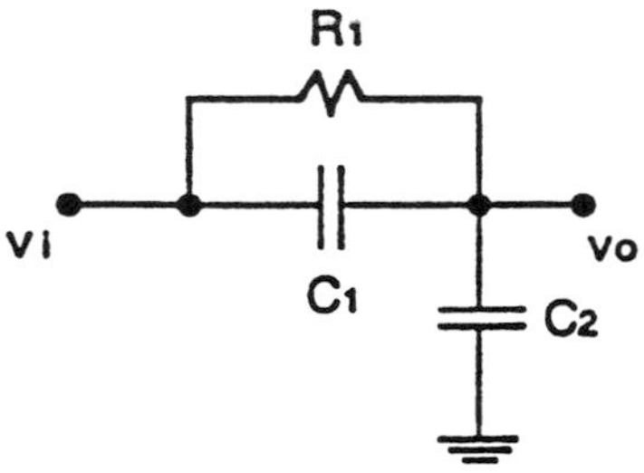

L*

7.2 For the transfer function:

$$T(s) = \frac{10^8\,(s)\,(s + 10)}{(s + 1)\,(s + 100)\,(s + 10^5)\,(s + 10^6)}$$

find

a) The equivalent expression in which the factors are of the form $(1 + \frac{s}{a})$;

b) The gain and phase at very low frequencies and at very high frequencies;

c) The poles and zeros;

d) The greatest-available gain and corresponding phase;

e) The gain at 10^3 rad/s, 10^5 rad/s.

Sketch Bode plots for gain and phase. Note that you have been asked to prepare Bode plots at the very end. Is this appropriate? At what stage in your overall answer would the Bode sketches have been easiest? most useful?

7.3 For the situation presented in P7.2 above, use exact analysis to calculate the amplitude and phase of $T(s)$ at 100 rad/s and 2×10^5 rad/s.

SECTION 7.2: THE AMPLIFIER TRANSFER FUNCTION

7.4 For an amplifier whose overall response is characterized as in P7.2 above, find (expressions for) A_M, $F_L(s)$, $F_H(s)$, $A_L(s)$, $A_H(s)$.

7.5 From a dominant-pole point of view, find approximate transfer functions for $F_L(s)$, $F_H(s)$ and $A(s)$ for an amplifier characterized by the complete transfer function in P7.2 above.

7.6 For the transfer function in P7.2 above, find a value for the lower 3-dB frequency a) from a dominant-pole viewpoint, b) using the root-squares approach, and c) exactly.

7.7 Proceed as in P7.6 above, but for the upper 3-dB frequency.

C

7.8 An amplifier having the transfer function described in P7.2 above, is augmented by circuitry which causes the addition of a zero and pole at 10^5 and 2×10^6 rad/s respectively. What is the new transfer function? Estimate the new upper 3-dB frequency. Calculate it exactly. Note that the technique demonstrated is called *pole-zero cancellation.*

CL*

7.9 For the circuit shown, find the upper 3-dB frequency using the method of open-circuit time constants, and exactly, for the conditions that:

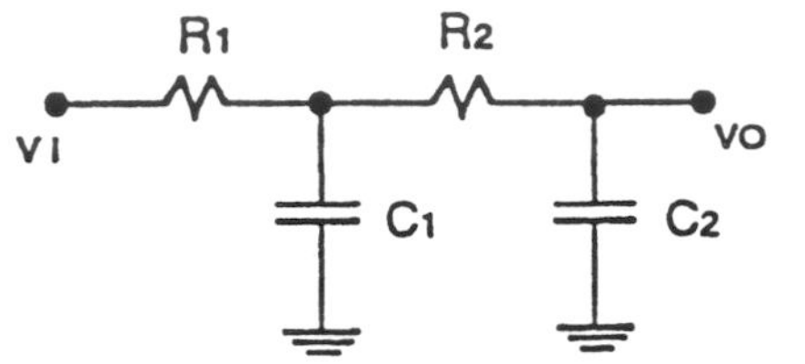

a) $R_1 = R_2 = 10$ kΩ,
$C_1 = C_2 = 100$ pF.

b) $R_1 = 10$ kΩ, $R_2 = 100$ kΩ,
$C_1 = 100$ pF, $C_2 = 10$ pF.

c) As in b), but with $C_1 = 10$ pF.

L

7.10 For the circuit shown, find the lower 3-dB frequency using the method of short-circuit time constants, for the conditions that:

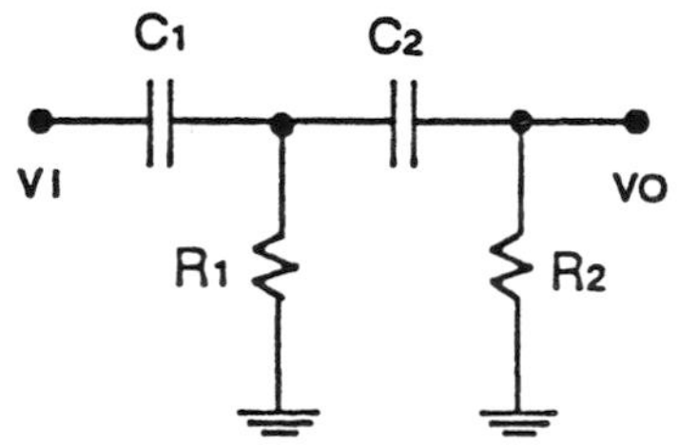

a) $R_1 = R_2 = 10$ kΩ
$C_1 = C_2 = 1$ μF

b) $R_1 = 10$kΩ, $R_2 = 100$kΩ
$C_1 = 1$ μF, $C_2 = 0.1$ μF

c) As in b) but with $C_1 = 0.1$ μF

SECTION 7.3: LOW-FREQUENCY RESPONSE OF THE COMMON-SOURCE AND COMMON-EMITTER AMPLIFIERS

7.11 A MOSFET amplifier using the topology of Fig. 7.10 fo the Text, employing $R = 100$kΩ, $R_{G2} = 10\ M\Omega$, $R_{G1} = 22M\Omega$, $R_S = R_D = 10$ kΩ, $R_L = 20$ kΩ, $C_{C1} = 0.01$ μF, $C_{C2} = 0.1$ μF and $C_S = 1$ μF, operates with $g_m = 2$ mA/V. Find the midband gain, and 3 poles and a zero at low frequencies.

D

7.12 For the situation described above in P7.11 above, it is desired to have a single dominant pole at 10 Hz or less, and two coincident ones at about 1 Hz. What values of coupling and bypass capacitors should be used which minimize the total capacitance? Specify the capacitors to 1 significant digit. What poles and zero actually result?

CDL

7.13 The circuit and situation described in P7.11 above is modified by the addition of a resistor r_S in series with C_S. Find expressions for the associated zero and pole, and the gain which correspond to Equations 7.37, 7.38 and 7.42, respectively. For gain reduced from its maximum value by a factor of 2 using V_S, what do the new pole and zero associated with C_S become?

7.14 For a particular BJT CE amplifier using the circuit of Fig. 7.13 of the Text, R_S = 10 kΩ, $R_1 \parallel R_2$ = 40 kΩ, R_E = 8.2 kΩ, R_C = 9.1 kΩ, R_L = 10 kΩ, and V_{CC} = 5 V. Under these conditions, I_E is 0.15 mA, at which β = 150 and r_o = 500 kΩ. Coupling capacitors of value $C_{C1} = C_{C2}$ = 1 μF and a bypass capacitor C_E = 10 μF are used. Calculate the three associated pole and zero frequencies, and estimate the gain and lower 3dB frequency.

D

7.15 For the situation described in P7.14 above, find suitable values for C_{C1}, C_{C2} and C_E so that the dominant low-frequency pole is at 20 Hz, another pole is at 2 Hz, and the third pole and zero coincide.

7.16 For the situation described in P7.14 above, an additional resistor of 350 Ω is included in series with C_E. Calculate the midband gain and the associated pole and zero frequencies and estimate the lower 3dB frequency.

SECTION 7.4: HIGH-FREQUENCY RESPONSE OF THE COMMON-SOURCE AND COMMON-EMITTER AMPLIFERS

L

7.17 Find values of the FET unity-gain frequency f_T (for operation in the grounded-source (CS) configuration) for:

a) A JFET for which I_{DSS} = 4 mA, V_p = –2V, C_{gs} = 2 pF and C_{gd} = 0.2 pF, operating at 1 mA.

b) A MOSFET with gate-to-channel capacitance of 0.15 pF, overlap capacitance of 20 fF, gate-to-substrate capacitance of 0.1 pF, having V_t = 1 V, and $k'(W/L)$ = 200 $\mu A/V^2$, operating at 200 μA.

c) A GaAs MESFET for which g_m = 10 mA/V at relatively high bias currents with C_{gs} = 0.15 pF and C_{gd} = 15 fF.

[Hint: See the development associated with Eq. 5.115 on page 446 of the Text.]

7.18 An n-channel enhancement MOSFET, for which $C_{ox} = 1.0\ fF/\mu m^2$, $\mu_n = 0.05 m^2/Vs$, $L = 3\ \mu m$, $W = 27\ \mu m$, and V_t = 0.5 V, operates with $v_{GS} = v_{DS}$ = 2.5 V, and the source grounded. The gate overlap is about 0.3 μm. Estimate C_{gs}, C_{gd} and the unity-gain frequency f_T which corresponds.

7.19 A particular FET transistor is to be operated in one of two grounded-source topologies for which the gain from gate to drain is either –1 or –100 V/V. Its C_{gs} = 200 fF, C_{db} = 100 fF, C_{gd} = 20 fF. Find the equivalent capacitances to ground at the gate and at the drain of each circuit.

D

7.20 A particular FET operates in a common-source circuit environment in which g_m = 1 mA/V, r_o = 50 kΩ, C_{gs} = 1 pF, C_{gd} = 0.5 pF, R_s = 100 kΩ, $R_{in} = 1 M\Omega$, R_D = 10 kΩ and R_L = 30 kΩ. Find the equivalent input capacitance at the gate, output capacitance at the drain, two poles, and an estimate of the upper 3-dB frequency. What is the highest frequency to which f_H can be raised by lowering R_s? What is the value of R_s which reduces f_H to 90% of that frequency?

7.21 For the situation described in P7.20 above, find exact values of the associated poles and zero, and an estimate of f_H. [Hint: Use the results associated with Eq.7.61 of the Text.] To what do all these frequencies change if the signal-source resistance R_s is reduced to 1 kΩ?

C

7.22 A high-performance n-channel MOS device for which V_t = 1 V and f_T = 1 GHz [See Eq. 5.115 of the Text.] operates at 1 mA and V_{GS} = 2 V in a common-source amplifier stage for which the gain is –3 V/V. If C_{gd} is known to be $\le 0.2 C_{gs} \approx C_{db}$, what is the equivalent input capacitance? If driven from a similar amplifier whose output impedance is approximately $(3/g_m) \parallel 4C_{db}$, what 3-dB frequency would you estimate? Consider this situation in the context of that described in P7.50 of the Text.

D

7.23 A MOS amplifier resembling that in Fig. P7.50 of the Text uses a resistor R_f connected from the drain to the gate of Q_1 for biasing. If the stage gain is 3, and the output resistance of the source is 10 kΩ, what is the minimum value of R_f to ensure at most a 5% loss in signal at the input?

7.24 For a particular BJT transistor, for which $C_\mu = 0.5$ pF, operating at 2 mA, with $\beta = 200$ and $f_\beta = 12.7$ MHz, find the corresponding values of unity-gain frequency f_T and C_π. [Hint: See Eq. 4.131 on page 321 of the Text.] If the bias current is increased to 10 mA, what values of f_β and C_π apply? For the usual situation described, for what range of currents is f_T maintained at the value estimated, as defined by the situation in which $C_\pi \geq C_\mu$?

7.25 For a particular BJT *CE* amplifier using the circuit of Fig. 7.13 of the Text, $R_s = 10$ kΩ, $R_1 \parallel R_2 = 40$ kΩ, $R_E = 8.2$ kΩ, $R_C = 9.1$ kΩ, $R_L = 10$ kΩ, and $V_{CC} = 5$ V. Under these conditions, I_E is 0.15 mA, at which $\beta = 150$, $r_o = 500$ kΩ, $r_x = 50$ Ω, $f_T = 1$ GHz and $C_\mu = 0.3$ pF. Estimate the mid-band gain and the upper 3-dB cutoff frequency assuming large bypass and coupling capacitors.

7.26 For the situation described in P7.25 above, with all coupling and bypass capacitors appropriately large, an additional resistor $R = 350$ Ω is included in series with C_E. What new values of midband gain and upper 3 dB frequency result?

7.27 An amplifier having a gain of −50 V/V and dominant poles at 50 Hz and 50 MHz is supplied by a negative pulse of 50 mV amplitude and 50 μs duration. Completely characterize the output pulse produced. [Hint: Consult Appendix F, pages F-14 to F-17 of the Text.]

SECTION 7.5: THE COMMON-BASE, COMMON-GATE AND CASCODE CONFIGURATIONS

7.28 For a particular BJT common-base amplifier using the topology of Fig. 7.21 of the Text, $R_s = 100$ Ω, $R_C = 9.1$ kΩ, with an external load $R_L = 10$ kΩ. The bias current is $I = 0.2$ mA, at which $\beta = 150$, $r_o = 400$ kΩ, $r_x = 50$ Ω, $f_T = 1$ GHz and $C_\mu = 0.3$ pF. Estimate the midband gain and the upper 3-dB frequency.

7.29 For a particular BJT cascode amplifier using the circuit of Fig. 7.33 of the Text, $R_s = 10$ kΩ, $R_C = 9.1$ kΩ and an external load $R_L = 10$ kΩ is connected. The bias current I is 0.15 mA, at which $\beta = 150$, $r_o = 500$ kΩ, $r_x = 50$ Ω, $f_T = 1$ GHz and $C_\mu = 0.3$ pF. Estimate the midband gain and the upper 3-dB frequency.

SECTION 7.6: FREQUENCY RESPONSE OF THE EMITTER AND SOURCE FOLLOWERS

7.30 For a particular BJT emitter-follower circuit resembling that in Fig. 7.25 of the Text, $R_s = 10$ kΩ, $R_E = 8.2$ kΩ and $R_L = 10$ kΩ is coupled through a capacitor $C_C = 1$ μF. Under these conditions $I_E = 0.15$ mA, at which $\beta = 150$, $r_o = 500$ kΩ, $r_x = 50$Ω, $f_T = 1$ GHz and $C_\mu = 0.3$ pF. Estimate the midband gain and the upper and lower 3-dB frequencies.

7.31 For a particular FET source-follower circuit operating at a 1 mA bias current, $R_s = 1$ MΩ, $R_L = 10$ kΩ, $g_m = 1$ mA/V, $C_{gs} = C_{gd} = 1$ pF. Estimate the midband gain and upper 3dB frequency resulting.

SECTION 7.7: THE COMMON-COLLECTOR COMMON-EMITTER CASCADE

CL

7.32

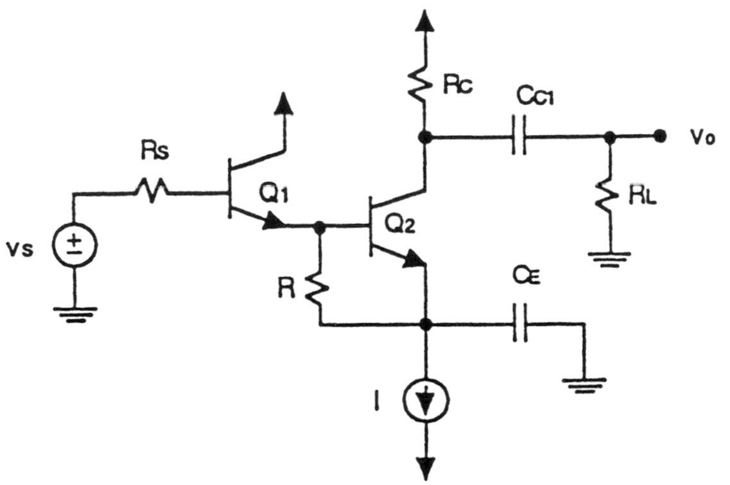

For the circuit shown, R_s = 100 kΩ, R_L = 10kΩ, R_C = 9.1 kΩ, R_E = 10 kΩ, I = 160 μA, R = 70 kΩ, C_{C1} = 1 μF, C_E = 10 μF. Under these conditions, β = 150, V_A = 100 V, r_x = 50 Ω, f_T = 1 GHz and C_μ = 0.3 pF. Estimate the midband gain and the upper and lower 3-dB frequencies. Use V_{BE} = 0.7 V. Note that for low-current operation, f_T may reduce, since $C_{\pi\,min} \approx C_\mu$.

L*

7.33 Repeat P7.32 above for the situation in which a) R = 14 kΩ, b) ∞.

SECTION 7.8: FREQUENCY RESPONSE OF THE DIFFERENTIAL AMPLIFIER

7.34 A BJT differential amplifier operates with a 300 μA emitter-current source and collector resistors of 4 kΩ. For differential input from a source having a total resistance of 10 kΩ, and output to a 10 kΩ load connected differentially, what are values of the gain v_o/v_s and the upper 3-dB frequency? For the BJTs used, β = 150, f_T = 1 GHz and C_μ = 0.3 pF.

7.35 The situation described in P7.34 above, is modified so that one BJT base is grounded, with the other connected to a 10 kΩ source. Both collector resistors remain, but the 10 kΩ load is capacitively-coupled to only one of them. There are two possible topologies. Find the midband gain and upper cutoff frequency for each.

7.36 The situation described in P7.35 is further modified so that the 10 kΩ load is connected to the collector resistor of the grounded-base BJT, while the collector of the other BJT is connected directly to the supply. What midband gain and upper cutoff frequency result?

7.37 The situation described in P7.34 above is modified by the addition of resistors equal in value to r_e connected in series with the emitter of each of the BJTs of the differential pair. What midband gain and cutoff frequency result?

C

7.38 For the situation described in P7.34 above, the emitter-current source uses a single transistor for which C_μ = 0.3 pF and V_A = 200 V. Wiring at the common-emitter connection accounts for an additional 0.5 pF. For a common-mode input signal of 5 V peak, what is the corresponding peak voltage on the ends of the load resistor at low frequencies? At what frequency does it reach 1V peak? At what frequency of the common-mode signal, do the input transistors saturate, for V_{CC} = 10V?

D

7.39 A wideband amplifier using the topology shown in Fig. 7.34 of the Text, employs a 300 μA emitter-current source, a source resistance of 10 kΩ, and an equivalent collector resistance of 2.7 kΩ. For the BJTs used, $\beta = 150$, $f_T = 1$ GHz and $C_\mu = 0.3$ pF. What midband gain and high cutoff frequency result? [Ignore the effect of bias imbalance due to the asymmetric drive.]

D

7.40 For the situation described in P7.39 above, emitter resistors are used to double the input resistance presented to the 10 kΩ source. Find the modified midband gain and cutoff frequency.

7.41 In the CD-CG amplifier shown in Fig. 7.35 of the Text, the MOS devices each operate with g_m = 1mA/V, C_{gs} = 200 fF, C_{db} = 100 fF, and C_{gd} = 20 fF. The drain resistor is 5 kΩ. Find the value of f_T which characterizes these transistors. Find the midband gain and upper 3dB frequency.

7.42 In the common-base differential circuit shown in Fig. 7.36a) of the Text, assume all transistors are relatively well-matched with specifications as in P7.39 above, namely $\beta = 150$, $f_T = 1$ GHz and $C_\mu = 0.3$ pF. Feedback biasing assures that each transistor operates at 150 μA emitter current. The collector load resistors are 2.7 kΩ. The source resistances are each 10 kΩ. Find the midband gain and upper 3dB frequency.

7.43 Repeat P7.42 above, for the situation in which the pnp transistors, while matched to each other, have $\beta = 50$, $f_T = 300$ MHz and $C_\mu = 1$ pF.

Chapter 8

FEEDBACK

SECTION 8.1: THE GENERAL FEEDBACK STRUCTURE

8.1 A feedback amplifier having the structure shown in Fig. 8.1 of the Text, is found to have $x_o = 3.0$ V and $x_f = 0.99$ V when a signal of $x_s = 1.00$ V is applied. What value of x_i results? What must the values of A and β be? What is the open-loop gain? What is the amount of feedback? What is the closed-loop gain? If by accident the connection to the β network were removed, what value would the load voltage tend toward? Is that value likely to be measured? What would happen instead?

8.2 For the circuit in Fig. E8.1 on page 670 of the Text, with $A = 10^2$ V/V, find R_2/R_1 for a closed-loop voltage gain of 8 V/V. What is the corresponding value of β? What is the amount of feedback in decibels? For $V_s = 0.125$ V, find V_o, V_f and V_i. If A increases by 100%, what is the % change in A_f?

SECTION 8.2: SOME PROPERTIES OF NEGATIVE FEEDBACK

8.3 An amplifier for which design was done with $A = 10^3$ and $\beta = 10^{-2}$ is manufactured using an amplifier with half the intended gain. What is the desensitivity factor? What is the sensitivity of closed-loop to open-loop gain $\left(\dfrac{\partial A_f/A_f}{\partial A/A}\right)$ in dB? What closed-loop gain results?

8.4 An amplifier for which the loop gain is designed to be 89 V/V is suspected to be deteriorating over an extended period of operation at high temperature. The closed-loop gain is found to be 98 V/V rather than the value of 99 V/V measured shortly after original installation. On the assumption that the change has occurred in the active components within the basic amplifier itself, what % deterioration would you expect to find in the open-loop gain if it were possible to measure it directly?

8.5 An amplifier whose open-loop response is characterized by a *dc* gain of 10^4 V/V and a 3-dB rolloff at 10^4 Hz, is connected in a feedback loop for which the overall low-frequency gain is 10^2 V/V. What is the 3 dB rolloff with feedback? What are the values of the Gain-Bandwidth product of the basic amplifier and of the feedback arrangement?

8.6 If in P8.5 above, a manufacturing error reduces the upper 3-dB cutoff to 2×10^3 Hz, what does the closed-loop upper cutoff become? Is this consistent with the sensitivity idea? To appreciate this situation better, find the new gain at 10^4 Hz using a relatively basic calculation. How does the desensitivity factor manifest itself?

8.7 Performance of a basic power amplifier having a signal-to-noise ratio of −3 dB (at the output) is to be improved, using a low-noise preamplifier and feedback, by 40 dB. What is the gain of the preamplifier required? What does the *S/N* ratio at the output become?

8.8 An amplifier exhibiting a non-linear transfer characteristic with a gain $\geq 10^3$V/V for $v_O \leq 0.1$V, a gain $\geq 10^2$V/V for $0.1\text{V} \leq v_O \leq 1$V, but which hard-limits at $v_O = 1$ V, is connected in a feedback loop with $\beta = 0.01$ V/V. Characterize the transfer characteristic of the closed-loop circuit by finding values of closed-loop gain for the 3 regions of operation, and the values of *input* and *output* voltage which bound them.

SECTION 8.3: THE FOUR BASIC FEEDBACK TOPOLOGIES

8.9 Characterize each of the following amplifiers by feedback type. As well, for each, find β in terms of the labelled components. In all cases, assume that the op amp is ideal.

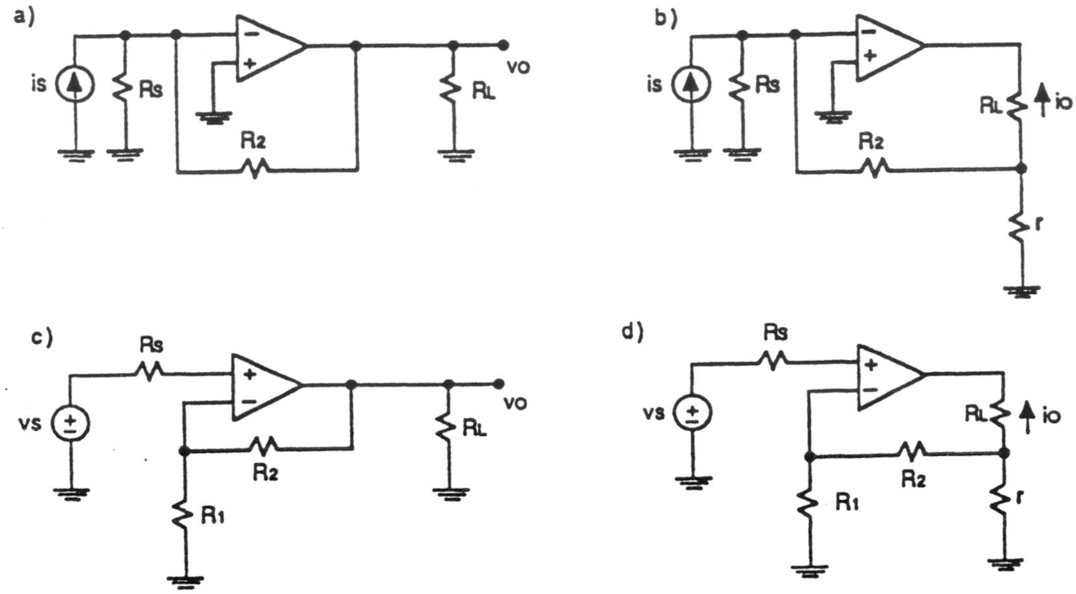

8.10 For each of the circuits below, identify the feedback type and find an expression for β. Assume for the present purposes that g_m for each FET is very high.

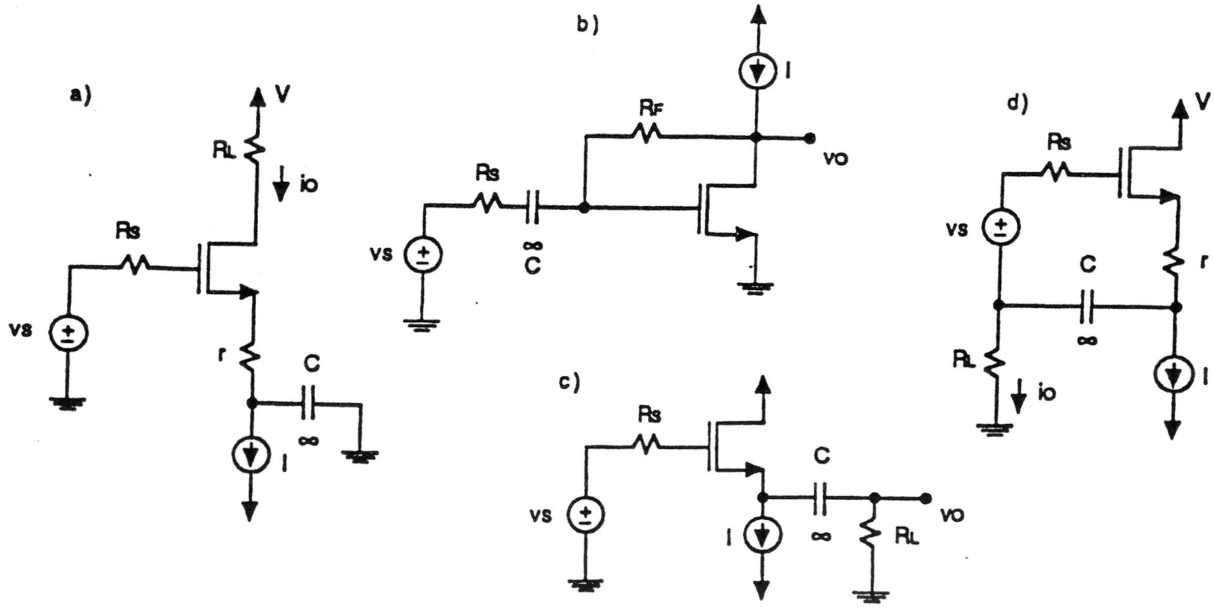

8.11 A series-series feedback circuit representable by Fig. 8.4c in the Text, and which uses an ideal transconductance power amplifier, operates with V_s = 1 V, V_f = 0.1 V and I_o = 2 A. What values of A and β correspond (with correct units noted)?

SECTION 8.4: THE SERIES-SHUNT FEEDBACK AMPLIFIER

8.12 A series-shunt feedback amplifier, has an A circuit for which A = 100 V/V, R_i = 10 kΩ and R_o = 10 Ω, and a β circuit with β = 0.1 V/V, R_1 = 2 kΩ and R_2 = 18 kΩ. When operating from a zero-impedance source and with no load, what is the overall gain and input and output resistances that result with feedback? If this feedback amplifier is connected between a 0.1 V rms source whose resistance is 10 kΩ and a load of 100 Ω, what does the output voltage become?

8.13 A feedback amplifier connected in the series-shunt topology employs an amplifier having a gain of 900 V/V, an input resistance of 20 kΩ, and an output resistance of 1 kΩ, with a feedback network employing two resistors of 10 kΩ and 190 kΩ at its output and input respectively. The amplifier operates between a 10 kΩ source and a 1 kΩ load. Find A, β, R_{11}, R_{22} and A_f as well as the overall gain and input and output resistances seen by the source and load respectively.

CL

8.14

For the circuit shown, the transistors have BJT $\beta = h_{fe}$ = 120. Find v_o/v_s, R_{in} and R_{out}. For what values of R_S and R_L (considered separately) does v_o/v_s drop to ½ the value just found. (Hint: be careful!)

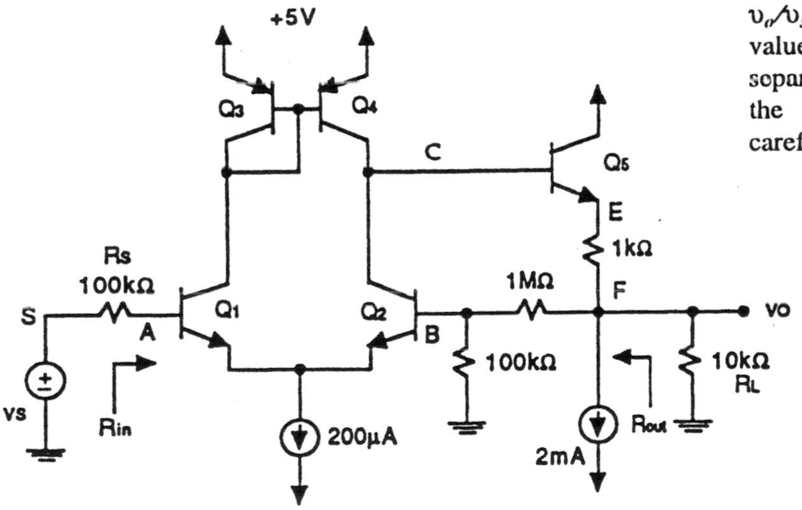

L

8.15

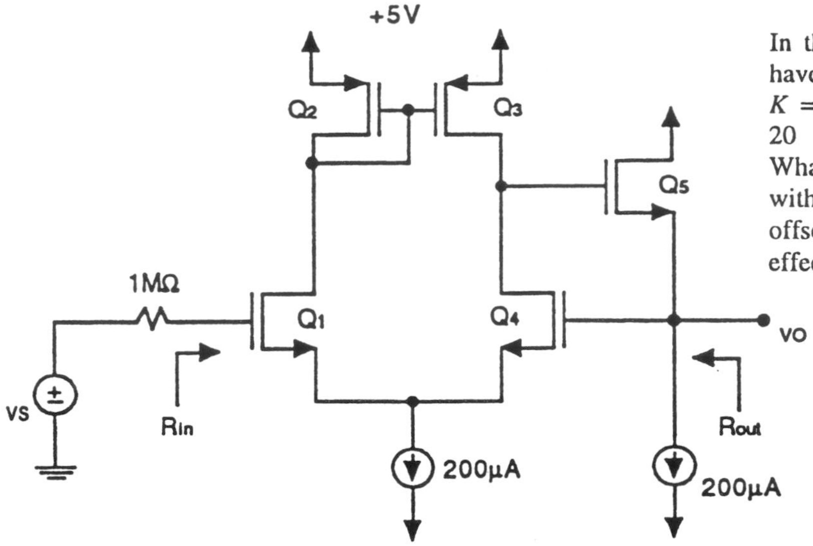

In the circuit shown, all transistors have $V_t = 1$ V, $K = \frac{1}{2}k'(W/L) = 100\mu A/V^2$, $V_A = 20$ V. Find, v_o/v_s, R_{in}, R_{out}. What does the overall gain become with a 1 kΩ load? Estimate the offset voltage. [Hint: Consider the effects of r_{01}, r_{02}, r_{03} and r_{04}.]

L*

8.16

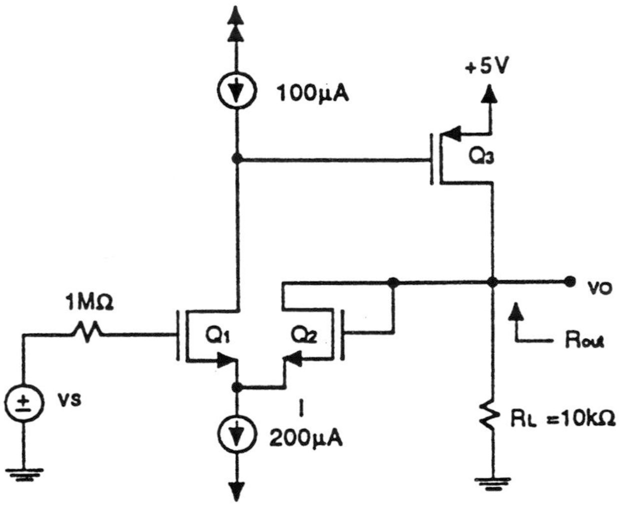

In the circuit shown, all transistors have $V_t = 1$ V, $K = \frac{1}{2}k'(W/L) = 0.1$ mA/V^2, $V_A = 20$ V. Find v_o/v_s and R_{out} using a feedback approach. Formulate the process in two slightly different ways involving interpretations of whether Q_2 is part of A or part of β. What happens if Q_2 and Q_3 are increased by a factor of 10 in width, and I is increased to 1.1 mA?

8.17

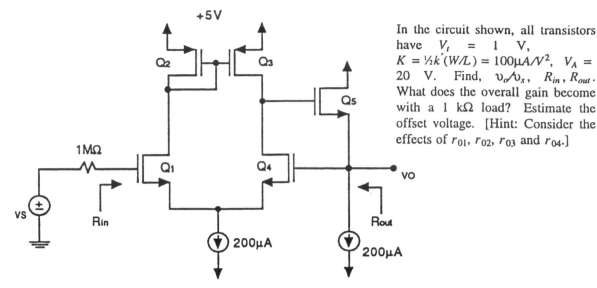

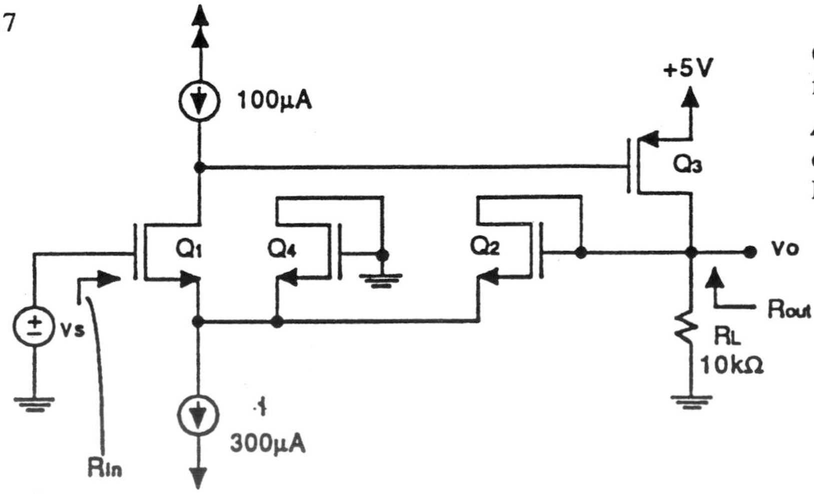

Considering Q_2 and Q_4 to be the feedback network, find

A, R_{11}, R_{22}, A_f and R_{out}. Use the device specifications provided in P8.16 above.

SECTION 8.5: THE SERIES-SERIES FEEDBACK AMPLIFIER

8.18 A feedback amplifier connected in the series-series topology uses a basic amplifier having a gain of 900 V/V, an input resistance of 20 kΩ and an output resistance of 1 kΩ, with a feedback network for which β = 50 V/A, R_{11} = 10 kΩ and R_{22} = 200Ω. The amplifier operates between a 10 kΩ source and a 1 kΩ load. Find A, A_f as well as the resistance R_{in} and R_{out} seen by the source and the load. [Hint: Note the specification of A is as a voltage amplifier: You must transform it suitably!]

8.19 Reconsider the situation described in Example 8.2 and Fig. 8.17 on pages 692 through 696 of the Text, modified to place R_L = 600 Ω between the emitter of Q_3 and the connection to R_{E2} and R_F. The output current is that measured in R_L. Find A, β, A_f and the input and output resistances seen by the source and load respectively.

8.20

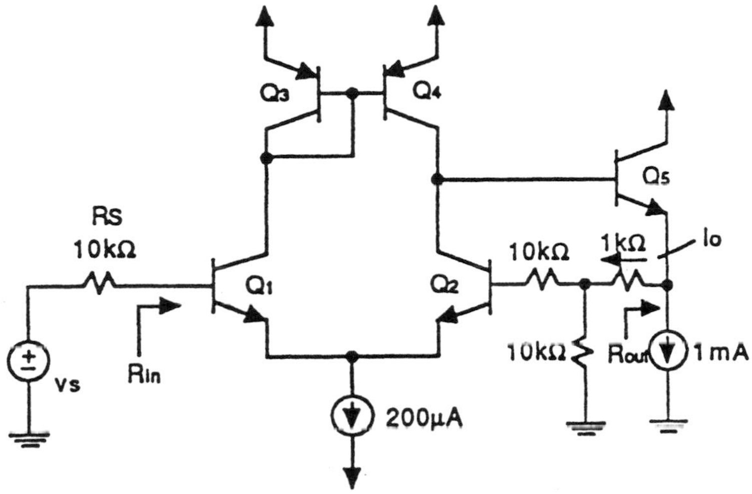

For the circuit shown, $h_{fe} = h_{FE}$ = 100, V_A = 200V. Find β, R_{11}, R_{22}, A, A_f, and the resistances R_{in} and R_{out}.

SECTION 8.6: THE SHUNT-SHUNT AND SHUNT-SERIES FEEDBACK AMPLIFIERS

CDL

8.21 A shunt-shunt feedback circuit uses a basic amplifier whose voltage gain is 900 V/V, input resistance is 20 kΩ, and output resistance is 1 kΩ, with a feedback network consisting solely of a 100 kΩ resistor. The amplifier operates between a 10 kΩ source and 1 kΩ load. What is the transresistance of the basic amplifier? What are R_{11}, R_{22}, A, A_f and the input and output resistances presented to the source and load? What does the gain become if the load resistance is halved? What change in R_S (nominally 10 kΩ) is needed to compensate?

L*

8.22

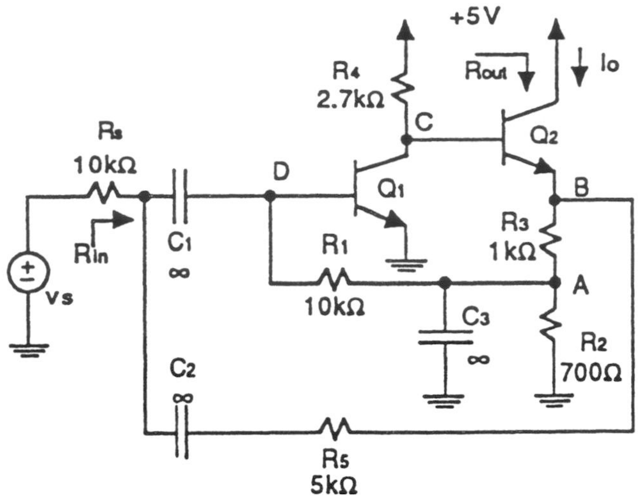

The circuit shown combines two feedback loops: One, involving R_1, R_2 and R_3 operates at dc, but because of C_3, not at high frequencies. The other, involving R_3 and R_5 operates at high frequencies, but because of C_2, not at low frequencies. At low frequencies, the feedback is intended to establish the dc voltage at A at a value which is nearly the voltage V_{BE1}, and thereby the entire bias-current situation at nodes B and C. What is the feedback type? At high frequencies the object is to create an output current i_o. What is the feedback type? For the latter feedback loop, find β (with its characteristic resistances R_{11}, R_{22}), A, $A_f{}' = i_o/v_s$, R_{in}, R_{out}. Assume $h_{FE} = \infty$, and $h_{fe} = 100$.

8.23 For the shunt-shunt loop described in P8.22 above, which operates at low frequencies, what is β, the corresponding R_{11} and R_{22}, and A_f? What is the corresponding resistance seen by C_3? What is the upper cutoff frequency for a 100 μF capacitor? What would you have judged it to be before considering the feedback situation?

C

8.24 Again consider the circuit in P8.22 above: As noted in P8.23 above, capacitor C_3 must be very large (and therefore costly). What occurs at high frequencies if it is removed? What are β, R_{11}, R_{22}, A and $A_f = i_o/v_s$ corresponding?

8.25 Using the results of P8.23, and those of P8.24 above with C_3 removed, find C_1 and C_2 for poles at 1 Hz and 10Hz respectively.

CDL

8.26 For the circuit of P8.10d above, with I such that g_m = 2 mA/V, r_o = 10 kΩ, R_S = 100 kΩ, and r = 1 kΩ, find i_o/v_s for a reasonable range of choices of R_L. What is the source resistance associated with i_o? What constrains the values of R_L that can be used?

CDL

8.27 Reconsider the situation in P8.23 above with respect to C_3. Consider the effect of moving C_3 to some tap on R_1: That is, split R_1 into two parts R_{1a}, R_{1b} where $R_1 = R_{1a} + R_{1b}$. For what ratio of R_{1a}/R_1 is the resistance seen at the tap, a maximum? What is the input resistance seen at the tap at this setting when the low-frequency feedback loop is closed? By what factor can the capacitor C_3 be reduced to maintain the same pole frequency as found in P8.23?

SECTION 8.7: DETERMINING THE LOOP GAIN

C*

8.28 A particular feedback loop intended for relatively high-frequency operation, when opened and terminated, returns a signal of 1.27 V rms when a signal of 20 mV rms at 1 kHz is injected, and 3.1 V rms when a 2 mV rms signal at 10 Hz is injected. What are the loop gains at the two frequencies? Assuming a single capacitor is associated with low-frequency bias stabilization of a direct-coupled amplifier, what do you imagine to be happening? Estimate the lower 3-dB point associated with the loop-gain response. What would you estimate the lower 3-dB response of the closed loop to be? If a third measurement indicates the loop gain to be only about twice as great at 1 Hz as at 10 Hz, estimate a lower bound for the open-loop gain of the basic amplifier. What might the closed-loop gain at 1 kHz be? If the capacitor which stabilizes the bias loop has been identified to be 1 μF, what is the equivalent (open-loop) resistance at the node to which it is attached? (Hint: see P8.29 following).

DL

8.29 A useful circuit, which is also a possible model of the situation alluded to in P8.28 above, is shown here. For an ideal amplifier with $A = 1550$ V/V and $R_1 = 2$ kΩ, $R_2 = 47$ kΩ, find the loop gain at high frequencies and at very low frequencies. For what value of C does the loop gain have a zero at 2.45 Hz? What is the associated pole of the loop gain? What is the corresponding 3-dB frequency of the closed-loop gain? If the capacitor used must be 10 μF, what values of R_1 and R_2 are needed to keep the same frequency response?

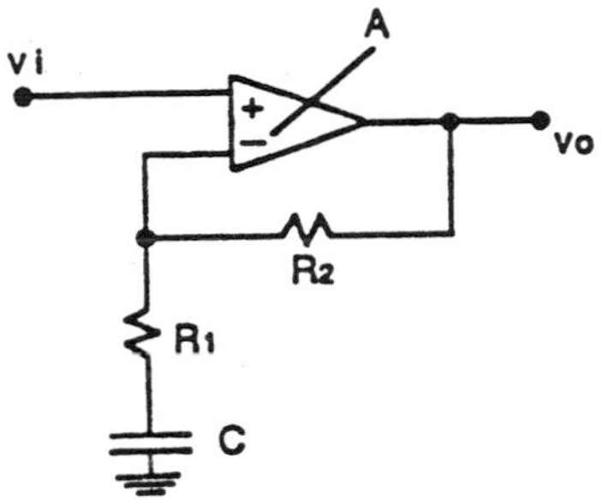

8.30 A non-inverting op-amp circuit for which the two resistors in the β network are 100 Ω and 10 kΩ is measured for loop gain by disconnecting the larger resistor from the output, injecting a 10 mV signal and measuring the returned signal to be 1.2 V. What is the loop gain found? What is the basic op-amp open-loop gain?

C

8.31 For a particular situation involving a non-inverting series-shunt feedback amplifier in which a complex unlabelled feedback network is used, measurements at a particular frequency show both the loop gain magnitudes and the closed-loop gain to be 10. Estimate the (gain) of the basic amplifier and β at this frequency.

8.32 For the circuit shown in P8.15 above, evaluate the loop gain for the conditions stated with a) no load, and b) a load of 1 kΩ. Using the fact that $\beta = 1$, what closed-loop gain results in each case? Check your results with those obtained in P8.15 using the direct method.

SECTION 8.8: THE STABILITY PROBLEM

8.33 An amplifier with a midband gain of 10^3 and dominant pole at 10^3 rad/s has two other poles coincident at 10^5 rad/s. At what frequency does its total phase shift become 180°? At that frequency, what is its gain magnitude? When incorporated in a feedback loop for which the feedback factor is independent of frequency, for what range of β is the resulting amplifier stable?

8.34 For the situation described in P8.33 above, sketch Nyquist plots for the loop gain $|A\beta|$, for three values of β: equal to the critical value, and 20 dB more and less than that.

C

8.35 Through a manufacturing error, an op amp whose *dc* gain is 10^6 V/V and dominant pole is at 10^3 rad/s, acquires a second non-dominant pole at 10^8 rad/s. When used with a frequency-independent feedback network, what total phase shift can result when $|A\beta| = 1$ for $\beta \le 1$. Can oscillation occur? Unfortunately, in a particular application, a stray capacitance of up to 5 pF to ground is associated with the output of the feedback network. For nominal $\beta = 0.5$, what is the maximum tolerable equivalent output resistance of the β network for which oscillation will **not** begin?

SECTION 8.9: EFFECT OF FEEDBACK ON THE AMPLIFIER POLES

8.36 A *dc* amplifier having a single-pole response with pole frequency at 5×10^3 Hz and unity-gain frequency of 20 MHz is operated in a loop whose feedback factor is 0.125, independent of frequency. Find the low-frequency gain, the 3-dB frequency and the unity-gain frequency of the closed-loop amplifier. By what factor did the pole shift?

D

8.37 Using the dc amplifier described in P8.36, above, a design is required of a closed-loop amplifier whose 3 dB frequency is at least 1 MHz. What is the corresponding amount of feedback required? The loop gain? The feedback factor? What low-frequency gain results?

8.38 A two-stage *dc* amplifier having a low-frequency gain of $10^4\,K$, one pole at 10^5 Hz and a second pole at $10^6/K$, where K is a factor depending on the choice of a resistor in one of the stages. For a particular feedback application using a frequency-independent feedback factor, coincident closed-loop poles at 5×10^5 Hz are acceptable. What value of K can be used? What is the *dc* open loop gain of the amplifier? the frequency of its second pole? the value of β for which the poles are coincident? and the corresponding low-frequency closed-loop gain?

DL

8.39 A two-pole amplifier with *dc* gain of 10^3 and poles at 10^6 Hz and 2×10^7 Hz is available to a designer interested in exploring bandwidth extension using feedback. For a maximally flat design, what pole frequencies result? What is the corresponding value of Q? of ω_o? of the 3-dB frequency? What values of β and closed loop gain are achieved with this design?

CL

8.40 Consider the amplifier of P8.33 above in a feedback loop with frequency-independant β. Find the closed-loop poles as a function of β. Sketch a root-locus diagram. At what frequency and for what β is the complex-pole-pair $Q = 0.707$? What is the corresponding phase margin?

SECTION 8.10: STABILITY USING BODE PLOTS

8.41 Use equation 8.48 on page 726 of the Text to explore response peaking as a function of phase margin: For what phase margin is there no peak? For what margin is the peaking factor equal to 2? to 10?

8.42 A particular amplifier with a *dc* gain of 10^4 has one pole at 10^6 Hz and two coincident poles at 10^8 Hz. Provide a sketch of the Bode magnitude and phase plots. Use them to estimate the following: What is the value of frequency-independent β for which the margins are zero? For what value of β is the phase margin 78°? 45°? What are the corresponding closed loop gains? What is the phase margin for $\beta = 3 \times 10^{-2}$?

L

8.43 For the situation described in P8.42 above, express the amplifier transfer function in a form corresponding to Eq. 8.50 on page 727 of the Text. Then use the approach exemplified in Eq. 8.51 to find more precise values for the results requested in P8.42 above.

DL*

8.44 For the parameterized amplifier design described in P8.38 above, and frequency-independent feedback, use the rate-of-closure rule to design an amplifier with 20 dB of feedback at low frequencies and the greatest available bandwidth. What is the available bandwidth? For a closed-loop gain of 10, what is the bandwidth? For each design, what are the corresponding values of K, *dc* gain, and the second-pole frequency?

SECTION 8.11: FREQUENCY COMPENSATION

D

8.45 For the amplifier described in P8.42 above, consideration is being given to various frequency-compensation ideas. In the case of an *added* dominant pole, what must its location be for a closed-loop gain of 10? of 1? What are the corresponding closed-loop cutoff frequencies?

D

8.46 Continue the compensation evaluation begun in P8.45 above, by considering the possibility of lowering the existing dominant pole. To what frequency must it be lowered for a closed-loop gain of 10? of 1? What closed-loop cutoff frequencies correspond? (Hint: Concerning the double pole, find the frequency at which each pole contributes 22.5° as the frequency to use in the design to give a net margin of 45° or so.)

DL**

8.47 For a particular amplifier in which one pole is at 10^5 Hz and two at 10^7 Hz, one approach to compensation involves lowering the dominant pole to 10^4 Hz. It is realized that two of the poles are controlled by one amplifier stage, for which input and output capacitances and resistances are $C_1 = C_2 = C$, $R_1 = 100\ R_2 = 1\ \text{M}\Omega$, and the gain is $-g_m R_2 = -100$, and across which a Miller capacitor can be installed. What is the required value of C_f? What is the overall effect on the other poles? What is the resulting cutoff frequency? In view of the pole shift, by how much can the dominant pole be raised to maintain the same margins? What new value of C_f is needed? What poles result? What is the resulting cutoff? For comparison, what was the original frequency at which the phase margin was 45°?

NOTES

Chapter 9

OUTPUT STAGES AND POWER AMPLIFIERS

SECTION 9.1: CLASSIFICATION OF OUTPUT STAGES

9.1 A particular amplifier for which the output-stage bias current is 50 mA is intended to produce single-sine-wave output signals of 1, 10, 100 V rms across a load R_L. In each case, classify the mode of operation that prevails for $R_L = 1$ kΩ, and 0.25 kΩ. If a manufacturing error causes the bias current to reduce to zero, yet the output appears nearly normal for large signals, what mode of operation is apparently possible?

SECTION 9.2: CLASS A OUTPUT STAGE

9.2 The emitter-follower shown in Fig. 9.2 of the Text operates from ± 3 V supplies with $R = 1.5$ kΩ, using three identical transistors. For $V_{CE\ sat} = 0.3$ V, what is the largest undistorted sine wave with zero average that can be produced across a 1 kΩ load? a 10 kΩ load? For what range of load resistances is the output symmetrically clipped? For a circuit modification involving connecting a second device in parallel with Q_2, what change occurs?

9.3

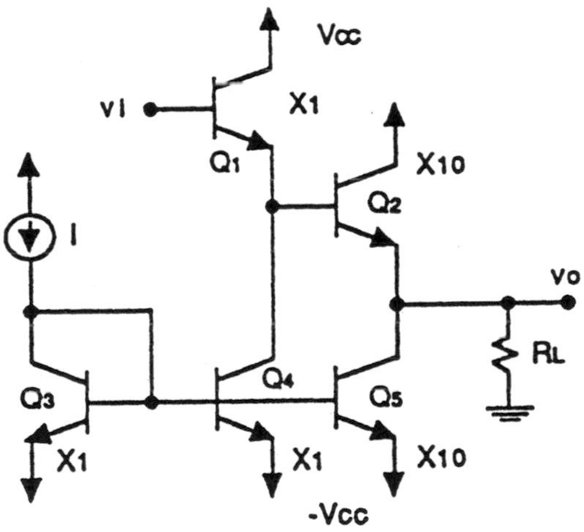

The follower design shown is intended to provide a relatively high input resistance. For $V_{CC} = 5$ V, $V_{EB} = 0.7$ V, $V_{CE\ sat} = 0.2$ V, what is the largest possible unclipped zero-average sine-wave output? For a minimum load resistance of 100 Ω, what is the minimum value of I which maintains $I_{E2} \geq I_{E1}$ for the largest possible undistorted output? Note the transistor sizing indicated by the notation $\times\ n$.

9.4 The class-A follower shown in Fig. 9.2 of the Text operates from ±9 V supplies with $I = 10$ mA using three identical transistors. Ignoring the power loss in R and Q_3, find the load power, the supply power and the conversion efficiency for:

a) The largest-possible sine-wave output and the smallest-possible load resistance (assuming $V_{CE\ sat} = 0.3$ V).

b) A sine wave of half the amplitude in a) across a load which is half the resistance of that in a)

c) Repeat a) including the loss in Q_3 and R (note that it is connected to ground!)

d) Repeat b) including the loss in Q_3 and R.

9.5

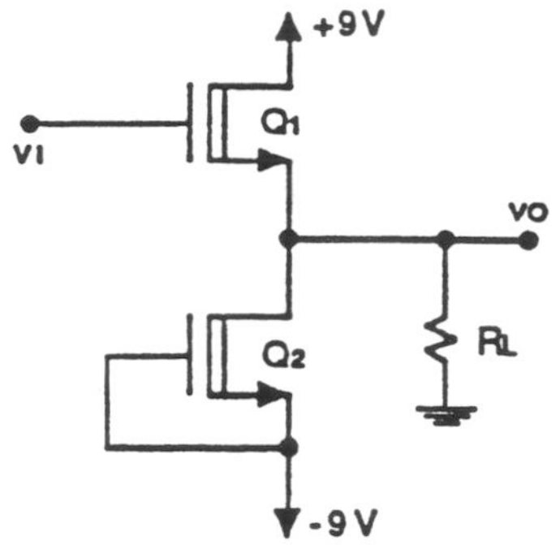

For the *FET* follower shown, V_t = –2 V, I_{DSS} = 10 mA. For υ_O = 0 V, what value of υ_I is measured? For R_L = 1 kΩ, what are the most positive and most negative outputs for which both transistors remain in saturation? What are the corresponding inputs? What are the corresponding values of load power, supply power, and conversion efficiency? What is the largest possible relatively undistorted output sine wave, and the corresponding conversion efficiency?

SECTION 9.3: CLASS B OUTPUT STAGE

C

9.6 For the circuit shown, sketch the transfer characteristics for $R_L = \infty$ and R_L = 10 kΩ. For both devices, $|V_t|$ = 1 V and $K = \frac{1}{2}k'(W/L)$ = 1 mA/V². What is the amplitude of the maximum possible output sine wave for which Q_1, Q_2 remain in saturation? What are values for the corresponding input voltage equivalent gain, supply power, load power, and conversion efficiency?

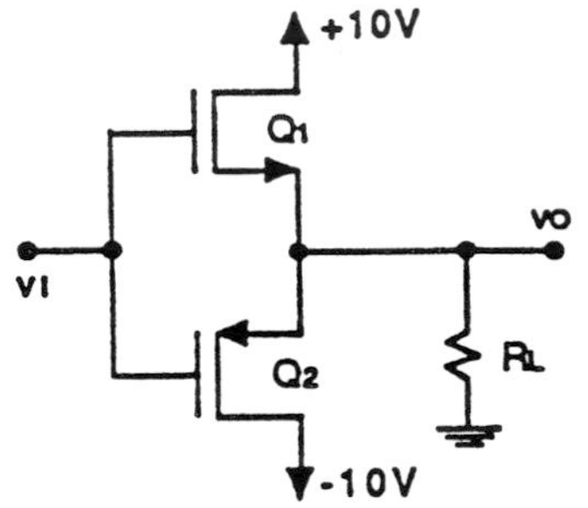

9.7 Consider the circuit of Fig. 9.9 of the Text, in which the supplies for A_o and the output stage are ±3 V, and A_o is a CMOS amplifier with rail-to-rail (i.e. ±3 V) outputs, having a transconductance of 10 mA/V. For R_L = 100Ω, $|V_{BE}|$ = 0.7 V, β = 50, find the input voltages required for outputs of +10 mV, +100 mV and 1 V.

9.8 Consider a particular automotive application of the circuit of Fig. 9.10 of the Text, for which R_L = 16 Ω and $2V_{CC}$ = 12 V nominally, and the base input is biassed at V_{CC}. What is the power level of the largest possible undistorted output signal? What is the corresponding supply power? What is the power dissipation in the two transistors? What is the corresponding efficiency? What are the values of output power, supply power, total device dissipation, and efficiency for output sine-wave signals of 4-V peak amplitudes. For the signal level maintained at the maximum undistorted level found above, but with the supply raised to 14.5 V, what do the total device dissipation and efficiency become?

SECTION 9.4: CLASS AB OUTPUT STAGE

9.9 A BJT class-AB output stage using the structure shown in Fig. 9.11 of the Text employs transistors for which V_{BE} = 0.690 V at 10 mA and n = 1. For a small-signal output resistance at light loads of ≤5 Ω, what quiescent current is needed? What value of V_{BB} should be used? For a 50 Ω load and peak output voltage of 5 V, what is the peak input voltage required? What large-signal gain results? For R_L = 50 Ω, what is the corresponding small-signal gain?

D

9.10 Design the quiescent current of a class-AB MOS output stage so that the incremental voltage gain for v_I near zero volts, is in excess of 0.99 V/V, for loads larger than 100 Ω. The MOS devices have $|V_t| = 1$ V and $K = ½k'(W/L) = 200$ mA/V^2. What value of V_{BB} is required?

SECTION 9.5: BIASING THE CLASS AB CIRCUIT

D

9.11 For the situation described in P9.9 above, in which the output transistors have $\beta \geq 30$, design a 2-junction biassing scheme, such as that shown in Fig. 9.14 of the Text, with devices having one-fourth the junction area of those at the output. What is the maximum output current available into a short circuit? For a 50 Ω load, what is the value of the peak output signal for which the current in the bias junctions reduces to 1/10 of the no-load value?

DL

9.12 For the situation described in P9.9 above, in which the output transistors have $\beta \geq 30$, design a V_{BE} multiplier, such as that shown in Fig. 9.15 of the Text, with a device for which the junction area and β are the same as those of the output transistors. For a 1-volt positive output across a 50 Ω load, arrange that the bias network current reduces to no less than 20% of its normal value, and that the current in the biassing transistor is no less than half of that. What is the incremental gain for your design with the peak positive output described, and $\beta = 30$?

9.13 For the V_{BE} multiplier embodied in Fig. 9.15 of the Text and operating at total current I, find an expression for the incremental resistance between its terminals in terms of I, R_1, V_{BE} and β, and a multiplication factor $k = 1 + R_2/R_1$, $k \geq 1$. For $k = 2$, $I = 1$ mA, and $\beta \geq 50$, what is the available value of incremental resistance when $R_1 = r_\pi$?

SECTION 9.6: POWER BJTs

9.14 The base-emitter junction voltage of a heat-sink-mounted power BJT, measured at a particular test current immediatedly following the application of power, is found to be 630 mV at what is assumed to be $T_J = 30°$. Subsequently, it is found to display a junction voltage of 500 mV when operated for some time at ten times the initial test current and with a larger supply voltage at a power level of 45W. Estimate the new junction temperature. If the ambient temperature remains at 30°C, what is the thermal resistance of this device? To maintain operation with a junction temperature of 180°, at the same collector voltage, what collector current would be needed? What voltage drop across the base-to-emitter junction would you expect under these conditions? Use a junction TC of –2 mV/°C.

D

9.15 A power transistor, for which the maximum safe junction temperature is believed to be 150°, has a thermal resistance junction-to-case of ≤1.1°C/W. What is the maximum power it can dissipate for a case temperature ≤55°C? For half that power level, to what temperature can the case be allowed to rise? For an ambient temperature of 30°C, what is the thermal resistance of the heat sink needed in each of these situations? For a modular heat sink design for which the rating is 3°C/W for each cm of length, how long a heat sink is needed in each case? If all the thermal resistance measurements cited can be in error by as much as 20%, to what length must the large heat sink be increased to guarantee safe operation?

9.16 A BJT for which $T_{J\,max} = 150°$, and $\Theta_{JC} = 2$°C/W, operates with an electrically-isolating bond to a heat sink for which the resistances are $\Theta_{CS} = 0.5$°C/W, and $\Theta_{SA} = 1$°C/W for each cm of heat-sink length, respectively. For a 10 cm heat sink, what is the maximum power this device can dissipate for ambient temperatures less than 40°C? What is the effect of doubling the heat-sink length? What do you conclude?

9.17 A power BJT operating at a high current density at $I_E = 5$ A is found to have a base current of 0.2 A and a base-input resistance of 0.72 Ω. Estimate a value for the base spreading resistance. Estimate a value for the base-input resistance at $I_E = 3$ A.

SECTION 9.7: VARIATIONS ON THE CLASS AB CONFIGURATION

D

9.18 A version of the circuit in Fig. 9.24 of the Text, uses *equal* current sources, I, in place of R_1 and R_2. What must their value be for the following situation: the maximum load current is 100 mA, $\beta_{npn} \geq 100$, $\beta_{pnp} \geq 80$; the minimum current in Q_1 and Q_2 must be no less than the maximum base currents of Q_3 or Q_4, nor smaller than 1.5 mA. For the situation in which Q_3 and Q_4 have junction areas 5 times those of Q_1 and Q_2 and conduct 50 mA at 0.7 V, (with $n = 1$ for all), find $R_3 = R_4$ so that the quiescent output current is equal to I calculated above. For what value of R_L is the small-signal gain ≥0.90 for output voltages a) near zero volts b) near +10 volts, c) near −10 volts.

D

9.19 For the output stage using compound devices as shown in Fig. 9.27 of the Text, the output npns are 100 mA devices while the other transistors are 1mA devices. For the npns, $\beta \geq 100$, and for the pnp, $\beta \geq 20$. Find the voltage needed across the V_{BE} multiplier for a standing current, I_Q, of 10 mA in the output, when a) the circuit is as shown, b) the base-emitter junctions of Q_2 and Q_4 are shunted to provide 1 mA currents in each of Q_1 and Q_3. For each situation described, with calculated voltages applied between the bases of Q_1 and Q_3, estimate the total effect of all transistors having a β which is ten times their minimum specified value.

D

9.20 For the circuit of Fig. 9.28 of the Text, in which transistors conduct 1 mA at 0.7 V, with $n = 1$ and $\beta \geq 100$, and $I_{bias} = 1$ mA, find $R_{E1} = R_{E2}$ so that the outgoing short-circuit current is limited to 25 mA. What is the current available without Q_5?

9.21

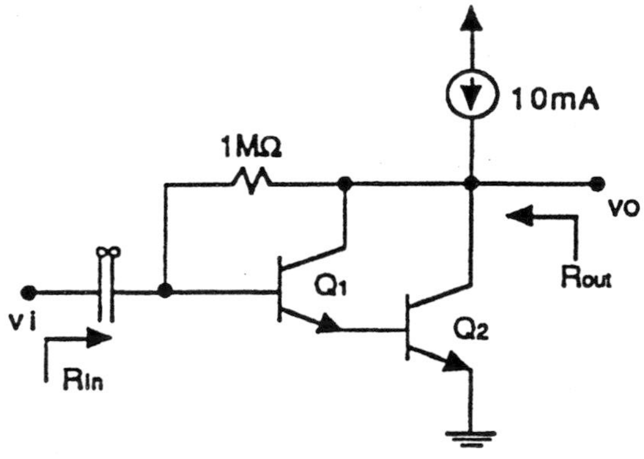

For the circuit shown, with β ranging from 50 to 150, $V_A = 100$ V, and $r_\mu = 10\,\beta r_o$, find the limits of the ranges of values of $g_{m\,eq}$, v_o/v_i, R_{in} and R_{out}.

D

9.22

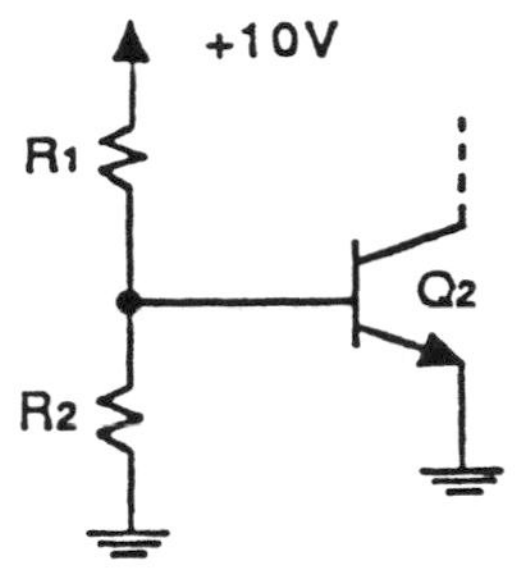

Consider the circuit shown as a simple candidate for use in a thermal-shut-down mechanism. The transistor has V_{BE} of 0.7 V at 100 μA, β = 100 and a TC of –2 mV/°C at 25°C. Design the circuit so that at 125°C the current in R_1 and Q_2 are each 100 μA. What is the current in Q_2 at 25°C? For what supply voltage (nominally 10 V) would the current at 25°C be double the value found? For operation at 100°C, at what value of supply voltage does the current in Q_2 become 50 μA?

SECTION 9.8: IC POWER AMPLIFIERS

D

9.23 Consider the circuit of Fig. 9.30 in the Text. For operation with a supply voltage of 25 V, it is desired to reduce the input bias current to 0.5 μA while maintaining nearly the original gain. Find new values for all the important resistors that must be changed. As a result of your change, what happens to the steady currents in Q_{12} and Q_9? Assume β_{npn} = 100, β_{pnp} = 20, $|V_{BE}|$ = 0.7V.

9.24 For the circuit shown in Fig. 9.30 in the Text, operating from a 27 V supply with all transistors (but Q_7 and Q_9) of the same junction size, and V_A = 100 V, and with β_{npn} = 100 and β_{pnp} = 20, estimate the gain A (as shown in Fig. 9.31) with no external load, and the corresponding 3 dB rolloff. Note that the current level in Q_9 is about 10 times that in Q_{11}.

D

9.25 For the circuit of Fig. 9.33 of the Text, find the value of R_3 for which a load current of 50 mA is shared equally by Q_3 and Q_5. By what factor does the current in Q_3 increase for a total output current of 1A? Q_5 and Q_6 both have an emitter-base voltage of 1.0 V at 1A, while Q_3 and Q_4 are 10 mA units, and Q_1, Q_2 are 1mA units. All show a 0.1 V/decade junction-voltage variation. Use β = 30 for Q_5, Q_6 and 100 for the other devices. What values of R_5 and R_6 ensure a quiescent current of 2mA in Q_3 and Q_4?

D

9.26 Design the bridge amplifier shown in Fig. 9.34 of the Text, to provide the largest possible output sine wave from an input sine of 0.1 V peak. The input resistance should be 10 kΩ or more, with the largest available resistor limited to 10 MΩ. The op amps saturate at (at most) 2 V from the supply rails. Use ±12 V supplies.

D

9.27 Modify the connections shown in Fig. 9.34 of the Text to create a bridge amplifier with infinite input resistance. Choose values to provide a 20-V peak output from a 1-volt peak input signal.

SECTION 9.9: MOS POWER TRANSISTORS

9.28 A power MOSFET for which $\mu_n C_{ox} = 30\mu A/V^2$ has $W = 10^5\mu m$, $L = 5\mu m$. For electrons in silicon, $U_{sat} = 5\times10^4$ m/s and $\mu_n = 5\times10^{-2} m^2/Vs$. For V_t = 2 V, find the value of v_{GS} for which velocity saturation begins. What is the corresponding current? What are the currents at twice and half this value? Find g_m at all three values of current.

9.29 For the class-AB amplifier shown in Fig. 9.38 of the Text, operating in the non-saturated-velocity region, $K = \frac{1}{2}k'(W/L) =$ 200 mA/V², $|V_t|$ = 2 V, $|V_{BE}|$ = 0.7V, β is high, and the quiescent currents $I_P = I_N = I_2 = I_3$ = 5 mA. Find the required voltage V_{14} between the bases of Q_1 and Q_4. What is the value of resistor R? If the TC of V_t (at low currents) is –3 mV/°C and of V_{BE} is –2 mV/°C, what fraction of the voltage V_{14} must appear across Q_6? Estimate the voltage gain of the resulting stage for a load of 100 Ω for a) outputs around 0 volts b) around +20 V.

D

9.30

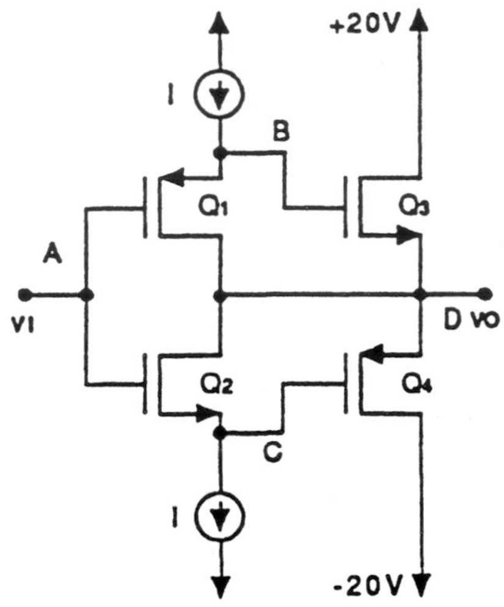

In the circuit shown, where $K = \frac{1}{2}k'(W/L)$, $K_3 = K_4 = 100\ K_1 = 100K_2 = 100\ \text{mA/V}^2$, and $|V_t| = 1$ volt for all devices. Find I for a quiescent current in Q_3 and Q_4 of 10 mA. What is the gain of the amplifier with a 100 Ω load, for an output voltage of a) 0 V b) +10 V. Find voltages at A, B, C, D for each case. Comment on the mode of operation of each transistor in each case.

Chapter 10

ANALOG INTEGRATED CIRCUITS

SECTION 10.1: THE 741 OP-AMP CIRCUIT

D

10.1 In the 741 op-amp circuit of Fig. 10.1 of the Text, transistors Q_{11} and Q_{12} with R_5 establish I_{REF} for the rest of the circuit. For ±15 V supplies and $|V_{BE}| = 0.7$, what value of I_{REF} results? What is I_{REF} for ±5 V supplies? To what value must R_5 be changed to restore I_{REF} to its high-supply value? Specify it to 2 significant digits.

D

10.2 For the situation described in P10.1 above, consider the operation of the Widlar current source involving Q_{10} and Q_{11} with R_4. For both devices having $I_S = 10^{-14}$A and $n=1$, calculate the Widlar output current for ±15 V and ±5 V supplies. Choose a new value for R_4 which reestablishes the ±15 V current level in Q_{10} for the ±5 V situation.

CD*

10.3 A designer, attempting to reduce the number of high-value resistors in the 741 design shown in Fig. 10.1 of the Text, notes the possibility that R_9 can be replaced by another transistor Q_{25} and a small resistor R_{12}. Suggest two possible connections and corresponding resistor values to establish a suitable current. Which is best? Why?

C

10.4 Reflect upon the ability of the extended input stage of the 741 Op Amp of Fig. 10.1 (consisting of Q_1, Q_2, Q_3 and Q_4, with Q_5, Q_6, Q_8, Q_9, Q_{10}) to withstand the application of voltages outside the normal operating range. For this purpose, approximate the breakdown voltages as follows: 7 V between the npn base and emitter, and 50 V for both pnp junctions and the npn base-to-collector junction. Specifically, consider a) In+ and In- connected in all possible ways to the ±15 V rails, b) In+ and In- having voltages somewhat outside the supply range. What are the limitations you see?

SECTION 10.2: DC ANALYSIS OF THE 741

10.5 For the Widlar current source in Fig. 10.2, find the effect on Equation 10.1 of making the base-emitter junction of Q_{10} k times that of Q_{11}. For $I_{REF} = 730$ μA, $R_4 = 5000\Omega$, $k = 0.5$, what does I_{C10} become? For what value of R_4 does it become 19 μA again?

10.6 For $I_{C10} = 19$ μA, and the inputs to the bases of Q_1 and Q_2 both at zero, what voltage on the bases of Q_3 and Q_4 results?

10.7 For the mirror circuit consisting of Q_5, Q_6, Q_7, and assuming all devices conduct approximately the same current with $R_1 = R_2 = 1\text{k}\Omega$ and $R_3 = 50\text{k}\Omega$, find the ratio I_{C6}/I_{C3} in terms of β, assumed equal for all devices. What is its value for $\beta = 200$?

10.8 For the mirror consisting of Q_5, Q_6, Q_7, find the ratios I_{C6}/I_{C5} for R_1 and R_2 shorted separately. Assume β is high, and operation is at $I_{C5} + I_{C6} = 19\mu\text{A}$.

10.9 For the extreme situations described in P10.8 above, what are the corresponding input-offset voltages (between the bases of Q_1, Q_2) which result? Note in general that offsets originating anywhere in the input stage can be compensated by varying R_1 and R_2.

D

10.10 Consider the second stage of the 741 op amp shown in Fig. 10.1 of the Text modified slightly by connecting the lower end of R_9 to the emitter of Q_{17}. What should its value be to maintain the current in Q_{16} at its present value? What must (the new) R_9 be to reduce its current to 4 times that in the base of Q_{17}, for which $\beta = 200$ is assumed?

D

10.11 For the output-stage-biasing scheme of the 741 op amp shown in Fig. 10.1 of the Text, what value of R_{10} would be needed to increase the quiescent current in Q_{14}, Q_{20} by 50%? First assume $R_6 = R_7 = 0$. Then consider $R_6 = R_7 = 27\Omega$.

SECTION 10.3: SMALL-SIGNAL ANALYSIS OF THE 741 INPUT STAGE

10.12 For a particular application, it is desired to raise the input resistance of a 741-like input stage to 3.6 MΩ for $\beta_n \geq 180$. If a change of input-stage bias current is to be used for this purpose, to what value must the current be changed? What is the corresponding value of G_{m1} which results?

10.13 Reconsider the output resistances of Q_4 and Q_6 as calculated in association with Fig. 10.8a, b of the Text. What is the value of R_2 for which $R_{06} = R_{04} = 10.5\text{M}\Omega$? What is the corresponding value of R_{01}? Contrast the open-circuit voltage gains for the original and R_2-modified versions of the input stage.

10.14 In an input-stage design for which the new value of R_1 and R_2, as established for R_2 in P10.13 above, is used, the actual resistors installed are different from each other by 2%. Calculate the corresponding input-offset voltage which results. Contrast that with the result for the original design as calculated in Example 10.1 on page 825 of the Text.

D

10.15 Use the result of Exercise 10.9 on page 826 of the Text to evaluate the required degree of match between R_1 and R_2 to ensure that the CMRR without feedback (that is the ratio of G_{m1} to G_{mcm}) is ≥80 dB.

D

10.16 In the calculation of R_0 performed in Exercise 10.10 on page 827 of the Text, its value is seen to be dominated by that of R_{09}. Augment Q_8 and Q_9 with additional resistors which will raise R_{09} to be equal to R_{010}. What resistors are needed? What value of R_0 results? For this situation, find a new value for G_{mcm} as requested in Exercise 10.11 on page 827 of the Text, and CMRR as requested in Exercise 10.12, there.

SECTION 10.4: SMALL-SIGNAL ANALYSIS OF THE 741 SECOND STAGE

10.17 Find the input resistance of the 741 second stage in the event that R_9 is replaced by a transistor Q_{25} and 1 kΩ resistor R_9', all connected suitably to the emitter of Q_7.

D

10.18 A designer wishes to capitalize on the change suggested in P10.17 above by lowering R_8 to a value which causes R_{i2} to return to the value of about 4 MΩ. What value of R_8 is appropriate? What value of G_{m2} results? What does R_{017} become? What is the new value of R_{02}? What is the new value of the no-load voltage gain of stage 2? Contrast this with the original value (having the same input resistance).

SECTION 10.5: ANALYSIS OF THE 741 OUTPUT STAGE

10.19 Assuming the existence of some degenerative process which causes the β of Q_{23} to reduce, what is the value for which R_{i3} reduces to 4 times the value of R_{02}? What does the second-stage gain become with $R_L = 2\ \text{k}\Omega$ and the conditions assumed just following Equation 10.22 on page 832 of the Text? What does the corresponding output resistance become?

L

10.20 Evaluate the sensitivity of the output-current-limiting scheme involving R_6 to the β of Q_{14} and Q_{15}. In particular, calculate the output current levels at which the excess current available from Q_{13A} (180 μA) is absorbed by Q_5, for the cases in which transistor β is 400, 200 or 100. Recall that $I_{S15} = 10^{-14}$A.

SECTION 10.6: GAIN AND FREQUENCY RESPONSE OF THE 741

10.21 For the situation described in P10.18 above, calculate the overall gain which results when the amplifier is loaded with $R_L = 2\ \text{k}\Omega$. Find the frequency of the dominant pole associated with $C_C = 30$ pF. Find the unity-gain frequency f_t. Provide a corresponding Bode plot for this situation. What do you notice about f_t in this and the original cases?

D

10.22 Consider the possibility of a 741-like amplifier in which C_C is available for the customization of frequency response for high-gain applications. What values of C_C would be necessary to maintain phase margins of 45° and of 60° for closed-loop gains of 1000 and of 10^4. In each of the four cases, what upper 3 dB frequencies are possible?

10.23 For the two modified versions of the 741 amplifier alluded to in P10.22 above, for which the phase margin is 45°, calculate the corresponding values of slew rate SR and full-power bandwidth f_M, for outputs of ±10 V.

10.24 Reconsider the amplifier shown in Fig. P10.44 on page 880 of the Text, modified to have the bias currents reduced by a factor of 2, to 50 μA and 500 μA respectively, and the junction sizes of Q_6 and Q_7 made 2 times that of Q_5. Assume $\beta = 120$ and $V_A = 200$ V for all devices. What is the classification of the output stage type? What is its standing current? What is the gain with a load of 10kΩ? Calculate the capacitor C required for an open-loop 3dB frequency of 1kHz?

SECTION 10.7: CMOS OP AMPS

L*

10.25 Consider a CMOS amplifier which uses the topology of Fig. 10.23 of the Text, but with somewhat different device sizing than that in Example 10.2 on page 842 of the Text. In particular, all devices have $L = 10\mu\text{m}$, the p devices have $W = 200\mu\text{m}$, and the n devices are such that $W_6 = 2W_4 = 2W_3 = 200\mu\text{m}$. Generally, $V_A = 25$ V, $V_{DD} = V_{SS} = 5$ V, and $I_{REF} = 25\mu\text{A}$. For all devices, evaluate I_D, $|V_{GS}|$, g_m, and r_o, with V_A ignored in the bias calculations. Present these results in tabular form. Also find A_1, A_2, the dc open-loop gain, the input common-mode range, and the output voltage range. Use $\mu_n C_{ox} = 20\mu\text{A/V}^2 = 2\mu_p C_{ox}$, and $|V_t| = 1\text{V}$.

L

10.26 Repeat Example 10.2 on page 842 of the Text for $I_{REF} = 12\mu\text{A}$.

D

10.27 A young designer having forgotten the issue revealed in Eq.10.46 on page 844 of the Text, uses $(W/L)_6 = (W/L)_4 = 50/10$ in a design otherwise the same as considered in Example 10.2. What new value of V_{GS6} is required? Recalculate the value of A_2 and the overall gain which result. What is the input offset voltage produced?

D

10.28 For the modified amplifier described in P10.25 above, find the value of C_C that will result in $f_t = 1$ MHz. Find the value of R that places the transfer-function zero at infinity. Find the frequency of the second pole under the condition that the total capacitance at the output (C_2) is 10 pF. Find the excess phase it introduces at $\omega = \omega_t$. For a 10 pF load, what value of C_C is required to reduce the excess phase shift to 6° at a modified f_t frequency (f_t'). What is the maximum available slew rate for the two cases of excess phase?

10.29 In the CMOS bias circuit in Fig. 10.25 of the Text, all transistors have the same W/L ratio, $W/L = 2$, except for Q_{12} which is m times larger. For all devices operating at a bias current of 5 μA in a process in which $\mu_n C_{ox} = 2.5\mu_p C_{ox} = 20\ \mu A/V^2$, with V_A very large, find R_B, g_{m12}, and g_{m9}, for m selected to be a) 2, and b) 5. Note that the circuit incorporates a positive feedback loop involving all the transistors and R_B. Breaking the loop at the gate of Q_9, evaluate the loop gain in terms of m. Are there any restrictions on the value of m which can be used? What factors determine a good choice for the value of m? Assume, for simplicity, that operation is at a fixed bas current I_B.

SECTION 10.8: ALTERNATIVE CONFIGURATIONS FOR CMOS AND BICMOS OP AMPS

D

10.30 Consider the mirror used for differential-to-single-ended signal conversion in the cascode first-stage configuration in Fig. 10.26 of the Text. In this context, consider the relative merit of configuring Q_3, Q_{3C}, Q_4, Q_{4C} as a simple cascode mirror, or as a Wilson mirror. Assuming all transistors identical and characterized by $|V_t|$, K, V_{GS}, and r_o, find the (effect of the) output resistance of each (as seen at the drain of Q_{4C}). Also, find the minimum voltage between $V_{BIAS\ 2}$ and $-V_{SS}$ to ensure that all transistors operate in saturation.

10.31 Repeat Exercise 10.29 on page 852 of the Text for the current 2I reduced to 10 μA, but all other conditions the same. What values of R_o and A_1 result?

D

10.32 A designer inspired by the additional gain provided by the cascode connection as exemplified in Fig. 10.26 of the Text, contemplates the use of a 3-layer cascode using Q_{3CC}, Q_{4CC} and $V_{BIAS\ 3}$ with $Q_{3\ CC}$ and $Q_{4\ CC}$ in a simple cascode connection. Find an expression for the output resistance R_o now available. For conditions corresponding directly to those in Ex. 10.29 on page 852 of the Text, find the new values of R_o and A_1. Assuming the values of $V_{BIAS\ 2}$ and $V_{BIAS\ 3}$ to be optimized, what is the total voltage required for saturation operation, as measured from the gates of Q_1, Q_2 to the supply, $-V_{SS}$?

D

10.33 For the circuit in Fig. 10.27 of the Text, and conditions as stated in Exercise 10.30, on page 853, but with $2I = I_B = 10\mu A$, find values of $V_{BIAS\ 1}$, $V_{BIAS\ 2}$, $V_{BIAS\ 3}$ so that Q_6 and Q_7 operate at the edge of saturation. As well, find the output resistance and overall gain in the event that $|V_A| = 25$ V for all devices.

10.34 A folded cascode such as that shown in Fig. 10.27 of the Text, operates with a bias current of 10 μA flowing in each of the devices Q_1, Q_2, Q_{1C}, Q_{2C} and Q_{3C}, Q_{4C} all having the same value of $K = \frac{1}{2}k'(W/L)$ and $V_A = 25$ V. For a load capacitance, C_L, of 10 pF, a unity-gain frequency of 1 MHz is found. What is the value of g_m for the critical devices? What is the low-frequency voltage gain of the amplifier? What is the dominant-pole frequency? The slew rate?

10.35 Repeat Problem P10.59 on page 882 of the Text for the situation in which $I_B = 800\mu A$ and $(W/L)_{1,2} = 600/10$, with $\mu_p C_{ox} = 10\mu A/V^2$, and $C_L = 2$ pF. What output-pole frequency f_t results? For all BJTs of the same size, what is the minimum value of unity-gain frequency of the BJT which ensures that the parasitic pole is at least a factor of 10 higher than the overall f_t.

SECTION 10.9: DATA CONVERTERS – AN INTRODUCTION

10.36 A sample-and-hold circuit operating at a 100 kHz rate uses a sampling switch which closes for a 10-nsec interval during each cycle. What is the frequency, f, of the highest-frequency square-wave input singal for which the output of the S/H provides an adequate representation? Sketch the output of the S/H for input square waves having a frequency f, $0.5f$, $1.1f$ and $2f$. Note in this context that Shannon's sampling theorem states that sampling must be performed at a frequency at least twice that of the highest frequency component to be represented. For a sampling capacitor of 100 pF, what is the maximum value of total source and switch resistance that will normally provide samples which are accurate to within 1%? (Hint: think again of the square-wave situation).

D

10.37 A signal, to be sampled and converted to digital format, has a dynamic range of ±5 V. If it is important to resolve signals and signal changes as small as 0.1 V, what is the number of binary digits (bits) required in the converter? What is the greatest possible resolution that a 10-bit converter would provide?

SECTION 10.10: D/A CONVERTER CIRCUITS

10.38 For the DAC circuit shown in Fig. 10.32 of the Text, using resistors no smaller than 1 kΩ, what is the largest resistor required to implement an 8-bit converter? (Be careful!). What is the largest value of switch resistance for which the associated error is at most ±½ LSB? Assuming that resistors are adjusted to compensate for the average value of switch resistance, what variability in switch resistance is acceptable if the corresponding error is limited to ±½ LSB? If the two sources of error identified are switch-resistance variation and resistor tolerance, and each is allowed to contribute equally to the overall error, what switch-resistance variation and resistor tolerance are needed for the 8-bit converter described?

10.39 For an R-2R ladder of 8 bits using a 10 V reference, what is the value of R which ensures that the total reference-supply current is 1 mA? What switch resistance would cause an error of ½ LSB if left uncompensated? If compensated by an appropriate change in the 2R resistor, what value of switch resistance, when doubled, produces an error ≤ ½ LSB?

10.40 A D/A circuit modelled after that shown in Fig. 10.34 of the Text operates with $I_{ref} = 1$ mA and $R = 0.5$ kΩ. If device EB Junction area can be maintained only to within 1%, while other components and parameters are ideal, what is the greatest number of bits possible for which the corresponding error is ± ½ LSB?

SECTION 10.11: A/D CONVERTER CIRCUITS

D

10.41 Sketch the design of a 2-bit flash converter for signals in the range ±1 V using uncompensated op amps as comparators. How many comparators do you need? What reference voltages would you use? Assuming op amps whose output limiting levels are precisely ±10 V, sketch a converter circuit using 2 op amps, one of which detects the polarity of the input, and, thereby, controls the reference voltage for the second. For the analog input connected to positive input terminals of each op amp, find, for the both the 3-op-amp and 2-op-amp designs, the output codes corresponding to inputs of +0.75 V, +0.25 V, −0.25 V and −0.75 V. Let +10 V correspond to logic '1', and −10 V correspond to logic '0' in both cases.

10.42 The following circuit, whose operation is related to that of the charge-redistribution converter, can be used as a comparator component in a flash converter. With **X** indicating an NMOS switch operated by non-overlapping positive pulses Φ_A or Φ_B, describe v_O: a) during Φ_A, b) following Φ_A, before Φ_B, c) during Φ_B, d) following Φ_B, for i) $V_A > V_{REF}$, ii) $V_A < V_{REF}$. Assume that the op amp saturates at ±10 V.

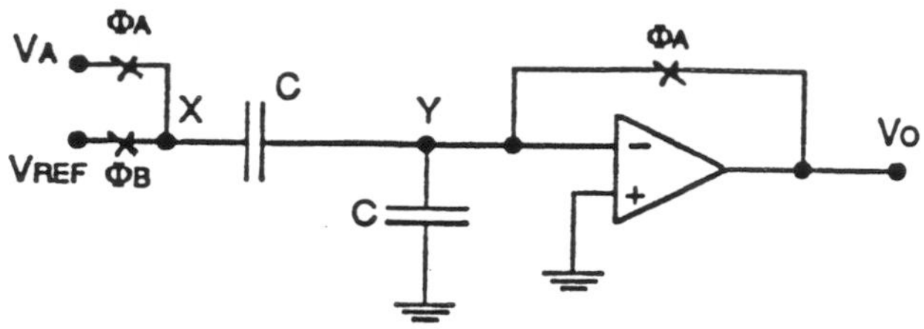

Chapter 11

FILTERS AND TUNED AMPLIFIERS

SECTION 11.1: FILTER TRANSMISSION, TYPES, AND SPECIFICATION

L

11.1 The transfer function of a first-order high-pass filter (such as that realized by an RC circuit) can be expressed as $T(s) = \dfrac{s}{s + \omega_o}$, where ω_o is the 3 dB frequency of the filter. Provide a table of values of $|T|$, Φ, G and A at $\omega = \infty$, $2\omega_o$, ω_o, $\omega_o/2$, $\omega_o/5$, $\omega_o/10$, $\omega_o/100$, $\omega_o/1000$.

11.2 A high-pass filter has an equiripple magnitude response and specifications resembling those in Fig. 11.3 of the Text, but with the passband and stopband exchanged on the frequency axis. Provide a sketch corresponding to this situation. If $A_{max} = 1$ dB and $A_{min} = 50$ dB, find $|T|$ at $\omega = \infty$, $\omega = \omega_p$ and $\omega = \omega_s$

11.3 A high-pass filter is required to pass all signals within its passband extending from 4 kHz to ∞, with a transmission variation of at most ±5%. The transmission in the stopband, which extends below 3.2 kHz, should not exceed 0.05% of the maximum passband transmission. Find corresponding values of A_{max}, A_{min}, and the selectivity factor.

11.4 A high-pass filter is specified to have $A_{max} = 0.5$ dB and $A_{min} = 20$ dB. It is found that its specifications can be just met with a single-time-constant RC circuit with a time constant τ of 10^{-3} s and a high-frequency transmission of unity. What are the values of ω_p, ω_s and the selectivity factor which correspond? In a modified design for which A_{min}, A_{max} and ω_p/ω_s are the same, but $f_p = 10^3$ Hz, what value of τ is required? What is the 3-dB frequency of the STC network? What is the resulting attenuation at 100 Hz?

SECTION 11.2: THE FILTER TRANSFER FUNCTION

11.5 A third-order high-pass filter has transmission zeros at $\omega = 0$ and $\omega = 0.1$ rad/s. Its natural modes are at $s = -1$ and $s = -0.5 \pm 0.8$. The high-frequency gain is unity. Find $T(s)$.

11.6 Find the order N and the form $T(s)$ of a bandpass filter having transmission zeros as follows: one at $\omega = 0$, one at $\omega = 1 \times 10^3$ rad/s, one at 2×10^3 rad/s, one at 6×10^3 rad/s, one at 12×10^3 rad/s and one at ∞. If this filter has a monotonically decreasing passband transmission with a peak at the center frequency of 4×10^3 rad/s. and equiripple response in the stop bands, sketch the shape of its $|T|$.

SECTION 11.3: BUTTERWORTH AND CHEBYSHEV FILTERS

11.7 Determine the order N of the Butterworth filter for which $A_{max} = 0.5$ dB, $A_{min} \geq 40$ dB, and the selectivity ratio $\omega_s/\omega_p = 1.6$. What is the actual value of minimum stopband attenuation realized? If A_{min} is to be exactly 40 dB, to what value can A_{max} be reduced? For A_{max} raised by 0.1 dB, what change in A_{min} results for the same filter order?

D

11.8 Design a Butterworth filter having the following low-pass specifications: $f_p = 20$ kHz, $A_{max} = 1$ dB, $f_s = 30$ kHz and $A_{min} = 20$ dB. Find N, the natural modes, and $T(s)$. What attenuation is provided at 25 kHz? 40 kHz?

11.9 Contrast the attenuation provided by a 3rd-order low-pass Chebyshev filter at $\omega_s = 2\omega_p$ to that provided by a Butterworth filter of the same order. For both, $A_{max} = 1$ dB. Sketch $|T|$ of both filters on the same axes.

11.10 Repeat P11.7 above for a Chebyshev filter.

L

11.11 For P11.7 and P11.10 above, with $\omega_p = 10^3$ rad/s, and a dc gain of 1, find the required natural modes.

SECTION 11.4: FIRST-ORDER AND SECOND-ORDER FILTER FUNCTIONS

11.12 Consider the op-amp circuit of Fig. 11.13a) on page 902 of the Text driven so as to have a infinite input resistance, a dc voltage gain of 11, and a 3 dB frequency of 10 kHz using 10 kΩ as the lowest resistor value. What is the frequency of the associated zero? What is the gain at high frequencies? For the pole and zero separated by a factor of 100, what must its dc gain become?

D

11.13 Use a combination of the table entries in Fig. 11.13 on page 902 of the Text to design a first-order bandpass filter using a single op amp with 3 dB frequencies at 100 and 1000 Hz. Arrange for a mid-band gain of −1 V/V and midband input resistance around 10 kΩ. Available capacitors can be specified only to one significant digit.

DL

11.14 Find the transfer function $T(s)$ of the circuit shown: Use it to create a spectrum-shaping network, having a midband gain of −10 V/V, with gain at high and low frequencies of −1 V/V. The 3-dB limits of midband gain should be at 100 and 1000 Hz. The midband input resistance should be around 10 kΩ. Available capacitors can be specified to only one significant digit. Sketch and label a pole-zero plot for this design.

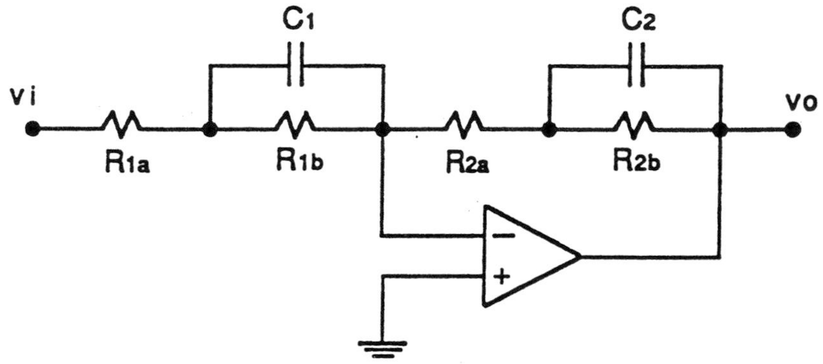

DL

11.15 Use the circuit in P11.14 above, and any insight you may have derived about it, to design a spectrum-shaping network, for which the gain between (3-dB points at) 100 Hz and 1000 Hz is −1 V/V, but the gain at high and low frequencies is −10 V/V. Arrange that the lowest resistance presented to the input source is around 10 kΩ, with the same capacitor-availability restriction as stated in P11.14 above. Provide a labelled pole-zero plot.

11.16 For the op-amp phase-shifter circuit derived from that shown in Fig. 11.14 of the Text by the exchange of R and C, find the transfer function $T(s)$ and the corresponding pole and zero. What phase shift results at $\omega_o = 1/CR$? For an input frequency of 10^4 Hz, and $C = 1.59$ nF, what values of R are required for phase-shift magnitudes of 6°, 12°, 30°, 60°, 90°, 120°, 150°, 168°, and 174°?

11.17 Use the information in Fig. 11.16c) on page 907 of the Text, to obtain the transfer function of a second-order bandpass filter with $\omega_o = 10^3$ rad/s, a maximum gain of 1, and a 3-dB bandwidth of 200 rad/s. What are the frequencies at the stopband edge as characterized by $A_{min} = 20$ dB?

L

11.18 Using the information provided in Fig. 11.16b) on page 907 of the Text, find the transfer function for 2 second-order high-pass filters for which A_{max} = 3-dB, $\omega_p = 1$ rad/s, and the maximum gain is 1. For one version, the maximum gain occurs at very high frequencies; for the other, the maximum is at relatively low frequencies. Each has the maximum possible Q. For each, what is the lower 3 dB frequency, and A_{min} at $\omega_s = ½$ rad/s?

DL

11.19 Use the result of Exercise 11.15 on page 905 of the Text to find the transfer function of a notch filter that is required to eliminate 60 Hz power-supply hum in an instrumentation system. To accommodate some variability in the power-supply frequency component values, design the filter to provide 20 dB or more attenuation over a band of 1 Hz total width, nominally centred at 60 Hz. The high-frequency transmission is to be unity. What is the 3-dB bandwidth of the notch? At what frequencies is the notch transmission down by 1 dB? by 1%?

SECTION 11.5: THE SECOND-ORDER LCR RESONATOR

D

11.20 Using the data in Fig. 11.16c and Equation 11.29 on pages 907 and 905, respectively, of the Text, design the LCR resonator of Fig. 11.17b) (on page 910 of the Text) for operation at 1 MHz with a 20 kHz bandwidth, for $R = 10$ kΩ. What is the rms output when driven by a 1 MHz 1 mA-rms current?

D

11.21 The user of an FM radio receiver, discovering that reception is bad due to overload from a nearby broadcast transmitter radiating at 99.9 MHz, considers the use of a "wave trap" employing the notch circuit of Fig. 11.18e) on page 912 of the Text. The input resistance of his receiver is 75Ω. Design the trap's LC components so that the loss presented to stations 2 MHz from the interfering one is only 3 dB. If the initial adjustment and stability of the trap components is such that the trap can be off-tuned by as much as 100 kHz, what attenuation would you expect of the offending station's signal?

D

11.22 Using the circuit of Fig. 11.18c) on page 912 of the Text, design an RLC high-pass filter having a maximally flat response with a 3-dB frequency of 100 kHz, to operate as part of a circuit for which the loading resistance is 10 kΩ. If coil Q of 50 is available, find suitable values of L and C.

SECTION 11.6: SECOND-ORDER ACTIVE FILTERS BASED ON INDUCTOR REPLACEMENT

D

11.23 Design the inductance-simulator circuit of Fig. 11.20 of the Text to realize inductances of 10 H and 0.1 H for operation around 1 kHz. Choose a majority of resistors of value 10 kΩ and a capacitor whose impedance is about 10 kΩ at the operating frequency. Available capacitors can be specified to only one significant digit.

DL

11.24 Use the designs prepared in P11.23 above to create a circuit which, when connected to a source having an output resistance of 20 kΩ, provides a bandpass function at 1 kHz. with a) the highest possible Q, b) the least-possible capacitor spread c) equal-valued capacitors (specified to 1 significant digit) and a change of a single resistor but whose value is not less than 1kΩ. What are the Q values which result in each case?

D

11.25 Design a fifth-order Butterworth LP filter having a 3-dB bandwidth of 10^4 Hz, and a dc gain of 10. Use a cascade of the circuits shown in Figs. 11.22a) and 11.13a) of the Text. Assume that capacitors of any required two-digit value are available as composed of 2 paralleled parts, each specified to 1 significant digit. Use as many resistors of 33 kΩ as possible. [Your boss has a surplus of these, since all the other employees seem to prefer to use 10 kΩ in their designs! As well, she likes the colour code!]

CD

11.26 Rework the transfer function of the BP filter given in Table 11.1 on page 922 of the Text, to employ equal capacitors C with two resistors chosen to control Q and ω_o as conveniently as possible, with all the other resistors of some fixed value. For Q ranging from 0.5 to 50, and capacitors specifiable using single digits 1, 2, 3 or 5, what is the range of values, specified by the ratio of maximum to minimum value, required for the resistors selected to tune Q and ω_o.

SECTION 11.7: SECOND-ORDER ACTIVE FILTERS BASED ON THE TWO-INTEGRATOR-LOOP TOPOLOGY

CDL

11.27 Design the KHN circuit of Fig. 11.24a) of the Text to realize a bandpass filter with center frequency of 10 kHz and 3 dB bandwidth of 500 Hz. Use 1 nF capacitors. Arrange that the input resistance is 100 kΩ, that 100 kΩ resistors predominate, and that no resistor larger than 1 MΩ is used. [Hint: In general, you may have to replace R_3 by a T network.] Sketch the complete circuit and specify all components. What center-frequency gain results?

D

11.28 Use the KHN circuit with an output summing amplifier, to design a low-pass notch filter with f_o = 5 kHz, f_n = 7.5 kHz, Q = 10 and a dc gain magnitude of 3. Where possible, use the values found in Exercise 11.22 on page 929 of the Text.

D

11.29 Consider the bandpass filter described in P11.27 above, implemented using the Tow - Thomas biquad. Maintain the same center-frequency gain. Find suitable capacitor and resistor values, (assuming no value restriction).

SECTION 11.8: SINGLE-AMPLIFIER BIQUADRATIC ACTIVE FILTERS

DL

11.30 Design the circuit of Fig. 11.29 of the Text to realize a pair of poles for which $\omega_o = 10^5$ rad/s and $Q = 1/\sqrt{2}$. Use the largest possible values of resistance, but with 1 MΩ as the largest available (precisely specifiable) value. Precision capacitors available are specified to one significant digit, either 1, 2 or 5.

D

11.31 Design the Bridged-T-based circuit of Fig. 11.30 of the Text, to realize a bandpass filter with center frequency of 10 kHz, center-frequency gain magnitude of 1, and 3-dB bandwidth of 500 Hz. Use 1 nF capacitors. What is its input resistance at very-low and very-high frequencies?

CL

11.32 For the bandpass circuit of Fig. 11.30 of the Text, find an expression for the input impedance as a function of component values and α, and then as a function of center-frequency gain, ω_o and Q. What is the value of input impedance at very low frequencies, very high frequencies, and at the center frequency?

DL

11.33 Design a 7th-order Butterworth low-pass filter with a 3-dB bandwidth of 5 kHz and a dc gain of unity, using the cascade connection of 3 Sallen-and-Key biquads [Fig. 11.34c)], and a first-order section [Fig. 11.13a)]. Use a 3.3 nF value for all capacitors.

SECTION 11.9: SENSITIVITY

11.34 In a particular implementation of the feedback loop of Fig. 11.34a) of the Text, for which the values of ω_o and Q are given by Equations 11.77 and 11.78 respectively, each of the capacitors C_3 and C_4 is created using the parallel combination of 2 others. In particular, $C_3 = C_{3a} + C_{3b}$ where $C_{3b} = k_3\, C_{3a}$ and $C_4 = C_{4a} + C_{4b}$ where $C_{4b} = k_4\, C_{4a}$ for $0 \le k_3,\ k_4 \le 1$. Find the passive sensitivities of ω_o and Q relative to C_{3a}, C_{3b}, C_{4a}, C_{4b} in terms of k_3, k_4, and when $k_3 = k_4 = 1$.

L

11.35 The feedback loop of Fig. 11.34 of the Text, for which ω_o and Q are described by Equations 11.77 and 11.78 there, operates over a range of temperatures for which the resistors, nominally equal, have a resistance which is not exact and, as well, a function of temperature, T. Thus $R_1 = k_1\, R$ and $R_2 = k_2\, R$ where $k_1,\ k_2 = 1 \pm k,\ k << 1$, and $R = R_o\ (1 + a(T_o - T))$. Find the sensitivities of ω_o and Q to both k and T. Note that although this is a special situation in which the resistors vary equally around a fixed mean value, the results can be usefully combined with the average resistor sensitivity of –1/2, as needed.

SECTION 11.10: SWITCHED-CAPACITOR FILTERS

11.36 For a dc voltage of 1 V applied to the input of the circuit of Fig. 11.35b) of the Text in which the capacitance C_1 is 0.1 pF, what charge is transferred for each cycle of a 1 MHz clock. What is the average current drawn from the input source? What is the equivalent resistance seen by the input source? For a 2 pF feedback capacitance, what change in output would you expect for each cycle of the input clock? In what direction? For an amplifier saturating at ±10 V, how many cycles does it take for the amplifier output to go from one limit to the other? What is the average slope of the output? What does the slope become for an input of –0.1 V?

D

11.37 Design the circuit of Fig. 11.37b) of the Text to realize, at the second integrator output, a Butterworth (maximally-flat) low-pass function with $f_{3dB} = 10^5$ Hz, and unity dc gain. Use a clock frequency of $f_c = 1/T_c = 1$ MHz with $C_1 = C_2 = 2$ pF. Find values for C_3, C_4, C_5, C_6 (Note that for a 2nd-order maximally-flat response, $Q = 1/\sqrt{2}$ and $\omega_{3dB} = \omega_o$). Characterize the output of the first integrator: What is its centre frequency? Its 3-dB bandwidth? Its associated maximum gain?

SECTION 11.11: TUNED AMPLIFIERS

11.38 A signal source with internal resistance $R_S = 10\text{k}\Omega$ is connected to the input of a common-emitter BJT amplifier having an emitter resistor R_E. From base to ground, a tuned circuit having $L = 1\mu H$ and $C = 200$ pF is connected. The transistor is biased at 1 mA, with $\beta = 200$, $C_\pi = 10$ pF and $C_\mu = 1$ pF. The transistor has a load resistance of 5 kΩ. Find ω_o, Q, the 3-dB bandwidth and the center-frequency gain which result for: a) $R_E = r_e$, b) $R_E = 9r_e$. (Hint: Use a Miller calculation for the effect of both C_μ and C_π).

L

11.39 Repeat P11.72 on page 971 of the Text for the situation as described, except that the base is connected to a coil tap located at a fraction k from the grounded end (the one to which the emitter is connected.). Find ω_o, Q, the bandwidth and overall center-frequency gain for: a) $k = 0.5$ and b) $k = 0.1$. Contrast these results with those found in P11.38 above.

D

11.40 A coil having an inductance of 2 μH and intended for operation at 10 MHz, has a Q specified to be 200. What is its equivalent series resistance, its equivalent parallel resistance, the value of the resonating capacitor, and the parallel resistance, which, when added, provides a bandwidth of 200 kHz?

D

11.41 A particular LC single-tuned amplifier has a center frequency of 1 MHz and a 3-dB bandwidth of 100 kHz. What is the corresponding Q? How many such amplifiers synchronously-tuned would result in a bandwidth reduced to 50 kHz? For the basic amplifier, what is the 30-dB bandwidth and skirt selectivity? What do these become for the synchronously-tuned cascade with a 50 kHz bandwidth?

D

11.42 A two-stage 4th-order stagger-tuned Butterworth design is required with center frequency of 10.7 MHz and an overall bandwidth, B = 250 kHz. Using 3 μH inductors, find C and R for each of the two stages. What is the relative peak gain of each of the two stages?

Chapter 12

SIGNAL GENERATORS AND WAVEFORM - SHAPING CIRCUITS

SECTION 12.1: BASIC PRINCIPLES OF SINUSOIDAL OSCILLATORS

12.1 Consider a sinusoidal oscillator consisting of a non-inverting transconductance amplifier and an RLC tank circuit. The amplifier is characterized by transconductance G_m, input resistance R_{in}, and output resistance R_o. The tank circuit employs a tapped inductor L for which the turns ratio is n (see Fig. 11.42 on page 951 of the Text), and whose total series resistance is R_{ls}, with a capacitor C whose (small) loss is represented by a parallel resistance R_{cp}. For the amplifier input connected across the small part of the coil, and the amplifier output connected so as to drive it all, find an expression for the conditions for oscillation, the oscillation frequency, and the required relationship amongst parameters.

CD

12.2 For the situation described in P12.1 above, sketch the corresponding circuit topology. Prepare a second sketch to show what must be done if the available transconductance is of an inverting kind. What are the new conditions for oscillation?

12.3 For a situation related to that described in P12.1 above, the connections between the coil and transconductor are exchanged, such that its output drives the tap, while its input is connected across the entire coil. What are the conditions which prevail at the threshold of oscillation?

12.4 For a particular oscillator application, the limiting-amplifier topology shown in Fig. 12.3 of the Text is being considered. The requirement is a design with a maximum gain of 5 which reduces to 0.5 at ±2.5 V, and has an input resistance of 10 kΩ. The op amp to be used has an open-loop gain of 1000V/V or more. Supply voltages of ±10 V are available. Assume a diode drop of 0.6 V at the levels of conduction implied. Find appropriate component values. If the op amp, assumed to have a single-pole rolloff, has an f_t of 1 MHz, what is the maximum frequency of operation for which an amplifier phase shift of 2° can be tolerated.

12.5 Reconsider the circuit of Fig. P12.8b) on page 1032 of the Text with R_2 = 10 kΩ, R_1 = 7.5 kΩ and a resistor R_3 = 10 kΩ connected in series with Z_1 and Z_2. Sketch the resulting transfer characteristic under the conditions that V_Z = 6.8 V, V_f = 0.7 V and I_{ZK} is 100 μA. Show what happens if, for a better zener, I_{ZK} reduces to 10 μA.

SECTION 12.2: OP-AMP-RC OSCILLATOR CIRCUITS

DL

12.6 Design a basic Wien-Bridge oscillator circuit, using the topology of Fig. 12.4 augmented by a gain-control mechanism consisting of two paralleled anode-to-cathode-connected diodes and a series resistor equal to R_2. The diodes are 1 mA units for which n = 2, and the voltage drop at 1 mA is 0.7 V. Select component values for an output voltage of 2 V peak-to-peak at 10 kHz using 10 nF capacitors. Assume the op amp to have A_o and f_t high enough to be ignored.

C*

12.7 A designer, wishing to use the idea described in P12.6 above at a higher frequency, is concerned with amplifier phase shift. She considers an otherwise-suitable amplifier for which f_t = 1 MHz. What is the highest nominal frequency of operation for which the actual frequency of oscillation is different by 10%, because of amplifier phase shift. What must the amplifier closed-loop gain be for operation in this mode?

CL

12.8

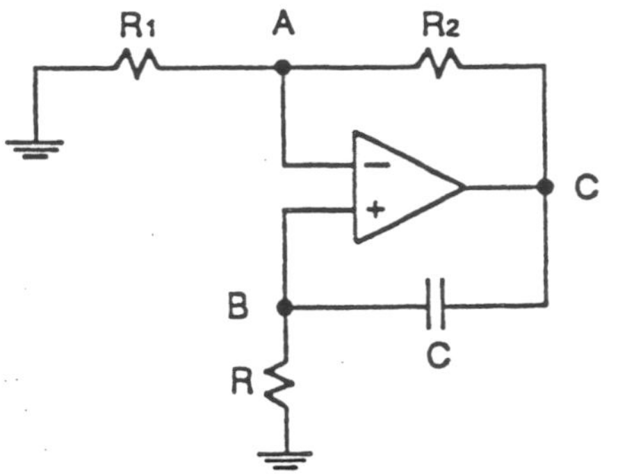

A designer, wishing to employ the phaseshift inherent in a compensated op amp, operated in a closed loop in the vicinity of its 3dB frequency, considers the circuit shown: For an op amp having a relatively large open-loop gain and high-order poles remote from f_t, find expressions for the loop gain and potential conditions for oscillation. For what value of R_2 is the frequency of oscillation $4/RC$? Now for excess phase shift at f_t, what occurs? How could you use this idea to evaluate f_t, and phase margins under various conditions?

CL

12.9 The circuit of Fig. P12.18 on page 1033 of the Text is modified by removing the rightmost resistor R and connecting the lower node of the rightmost capacitor C to the op-amp negative input. Find an expression for the loop gain, the frequency of possible oscillation and the conditions for which it occurs. What is the sensitivity of the frequency of oscillation, and of the critical value of R_f to a deviation of value of *any one of* the resistors R.

12.10 Another possible variant of the phase-shift oscillator of Fig. P12.18 of the Text can be considered in which the right-most RC circuit is removed, and a capacitor C is shunted across R_f. Find an expression for the corresponding loop gain, and the frequency and gain conditions for possible oscillation.

CD

12.11 A phase-shift oscillator is considered which resembles that of Fig. 12.7 of the Text, but uses a positive-gain amplifier. What is the minimum number of RC sections required? What is the corresponding frequency of oscillation and critical value of gain, K?

C*

12.12 For the circuit of Fig. 12.8 of the Text, consider the effect of adding an additional RC section. What are the conditions of oscillation? For the modified and original circuits, find the effect on frequency and critical gain of a simultaneous change in value of all its resistors R. [Hint: Economize your effort by using the technique described in Exercise 12.5 on page 963 of the Text. Correspondingly, note the solution of Exercise 12.5 which emerges in your work.]

C*D

12.13 Design the active-filter tuned oscillator shown in Fig. 12.11 of the Text for operation at 10 kHz using 10 nF capacitors, and diodes for which the voltage drop is a 0.7 V at 1 mA. Establish Q to reduce the 3rd-harmonic distortion to less than 1%. What is the output voltage which results?

SECTION 12.3: LC AND CRYSTAL OSCILLATORS

C*L**

12.14 For the Colpitts oscilator shown in Fig. P12.21b) on page 1034 of the Text, in which both FETs are matched with I_{DSS} = 4 mA, V_p = –2 V and V_A = 100 V, the load resistor is 10 kΩ, and the inductor has an inductance L = 10 μH with a Q of 100 at the operating frequency of 1 MHz. Find the values of C_1 and C_2 to ensure oscillation at 1 MHz. With C_2 set 5% lower than the value calculated, and the gate-to-source diode of Q_1 characterized as having a 0.7 V drop at 1 mA and n = 1, find the output signal amplitude, assuming the supply voltages to Q_1 and Q_2 are high enough. For supplies of ±3 V, estimate the amplitude of the output signal. (Hint: Use the incremental slope of the characteristic in the upper end of the triode region for a value of v_{DS} slightly less than $|V_p|$.

C*DL*

12.15 A particular quartz crystal is measured to have a series resonance at 2.015 MHz, a parallel resonance at 2.018 MHz, a parallel plate capacitance of 4pF, and a Q of 50×10^3. Find corresponding values of L, C_s, and r. When used in the circuit of Fig. 12.16 of the Text, C_2 = 10 pF, and C_1 can be as low as 1 pF. For device sizing and V_{DD} such that $g_{mn} = g_{mp}$ = 1 mA/V and V_A is large, with R_f = 1 MΩ, choose values of C_1 and R_1, for which the loop gain can be assumed to be >1. (Hint: Consider the CMOS amplifier as an op amp with finite gain loaded by $R_1 \parallel R_f$, with gain established by a network consisting of Miller-multiplied R_f reflected back through the C_1/C_2 capacitor ratio, driven by R_1. Using this idea, find a maximum value for R_1).

SECTION 12.4: BISTABLE MULTIVIBRATORS

CDL**

12.16 For the CMOS circuit shown, in which all transistors are matched with $K = ½k'(W/L) =$ 1 mA/V², and $|V_t|$ = 1 V, V_A = 30 V, design a version whose output voltage range extends beyond +4.9V and +0.1V for inputs in the range 0 to 5V, with V_{TL} = 2.0V, and V_{TH} = 3.0V. What is the consequence on the input thresholds of K and V_t varying by ±20%? For amplifier capacitance characterized as 10 pF at each inverter input and 1 pF at each inverter output, estimate the output rise and fall times. For an input rising and falling at a rate of 1 V/μsec, estimate delay time from the time the source begins to change until regeneration begins.

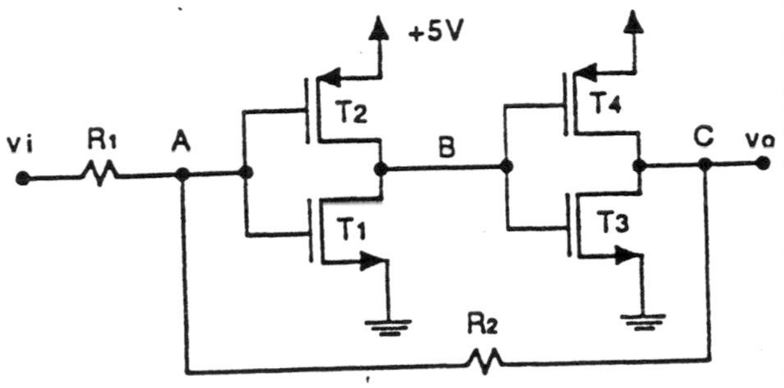

DL

12.17 In the circuit shown, using a high-gain op amp, the output signal is ±13 V or so with ± 15V supplies, and diode drops are 0.7 V for reasonable currents. Design R_1, R_2, and R_3 so that each diode, when conducting, has a current in the range 1 to 4 mA, with as low a current level as possible preferred. What are the input thresholds V_{TL}, V_{TH} for this circuit?

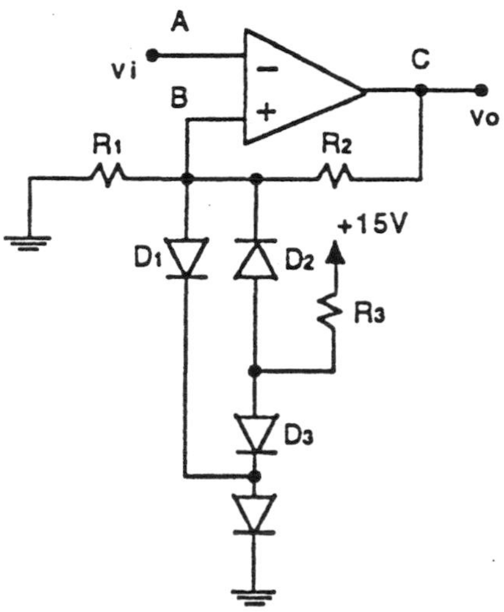

12.18 In the circuit shown, all transistors have $|V_t|$ = 1 V and, with $K = ½k'(W/L)$, $K_1 = K_2 = K_3 = K_4 = 2K_5 = 2K_6 = 100\mu A/V^2$.

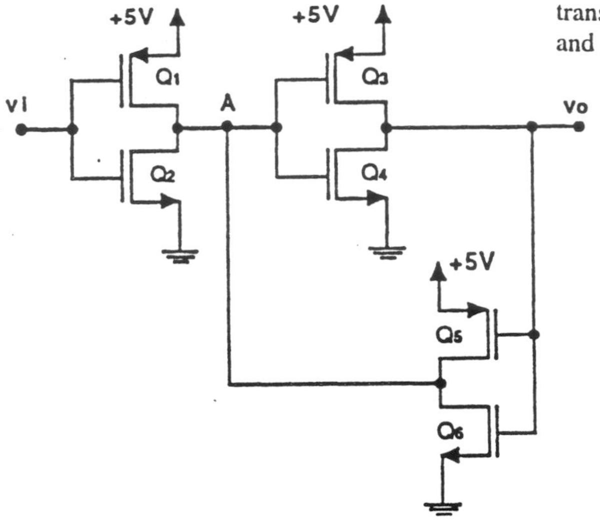

Sketch and label the transfer characteristic of the $Q_3 - Q_4$ inverter. Now sketch and label the transfer characteristic υ_o versus υ_i. What are V_{IH} and V_{IL}?

SECTION 12.5: GENERATION OF SQUARE AND TRIANGULAR WAVEFORMS USING ASTABLE MULTIVIBRATORS

DL

12.19

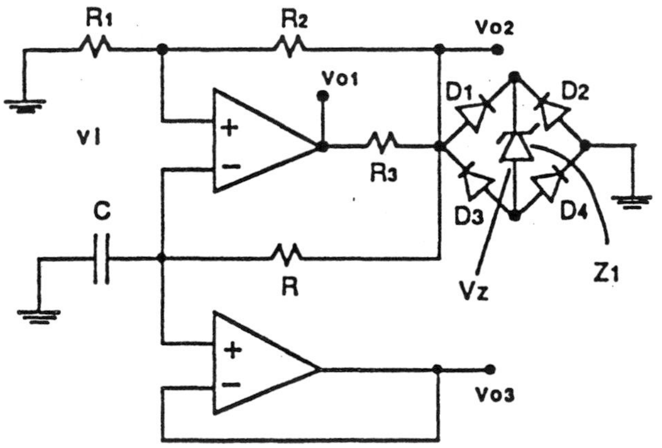

With high-gain op amps for which υ_o of ±13 V is guaranteed, design the circuit shown, using a 6.8V zener diode, to provide a triangle-wave output at υ_3 of ±1 V amplitude at 10 kHz, using a 1000 pF capacitor. Arrange a current of at least 1.0 mA in the zener diode and approximately equal current flow in R_2 and R. For this design, find both the average and extreme slopes of the output triangle wave. If it is important to reduce the slope variation by a factor of 2, show a simple circuit modification involving the addition of 2 resistors and suitable changes in values, which does the job.

CD

12.20 Use the topology of Fig. 12.25 of the Text, implemented using the output-voltage-regulator scheme shown in P12.19 above, to provide square and triangular waves at 10 kHz, each of ±1 V peak amplitude, the square wave being obtained from an appropriate voltage divider using a 1 mA current. Notice that the required bistable is of the non-inverting type. What must its upper and lower thresholds be? Sketch the complete circuit, and select values consistent with specifications provided in P12.19 above.

SECTION 12.6: GENERATION OF A STANDARDIZED PULSE – THE MONOSTABLE MULTIVIBRATOR

12.21 The circuit shown uses an op amp for which the limiting output voltages are ±10 V.

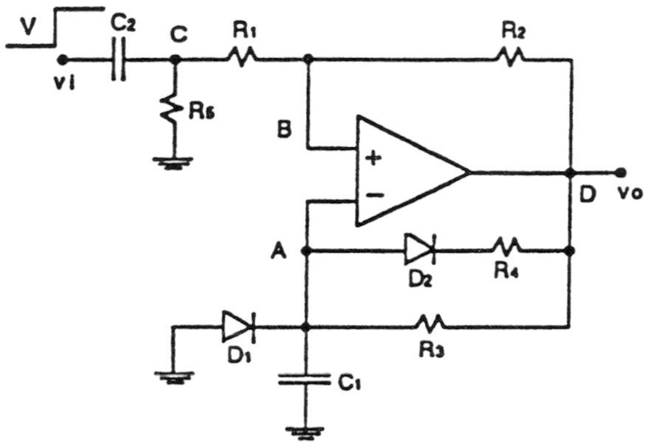

For this design, $R_2 = R_3 = 100\text{k}\Omega$, $R_1 = R_5 = R_4 = 10\text{k}\Omega$, $C_1 = 10$ nF, $C_2 = 1000$ pF. What are the resting voltages at A, B, C, D? What is the minimum amplitude of a positive step (V) at the input that will trigger the circuit? Sketch waveforms at all nodes for the case in which V is a 5 V step. What is the length of the output pulse produced? What is the purpose of R_4? For V, a rate-limited rising input of 5 V, what is the slowest rate of rise to ensure triggering? How soon after the output falls can the circuit be retriggered with the expectation of an output pulse of the usual duration?

12.22 For the circuit shown, Q_1 through Q_4 have $|V_t| = 1$ V and equal values of $K = \frac{1}{2}k'(W/L)$. Correspondingly, their active (amplifying) region is around $V_{DD}/2 = 2.5$ V. Normally A is held at 0 V by R_2, B is high at 5 V and C is low at 0V. When input υ_i is applied, D rises and A rises with it, lowering B and causing C to rise, replacing υ_i through D_2. For $R_1 = R_2 = 10$ kΩ, $C_1 = 10$ nF, sketch the waveforms at all nodes following the rise of υ_i. What is the expected maximum duration of υ_i? What happens if it stays high longer? In this event, the circuit operates open-loop for the fall of υ_o (and does *not* regenerate upon turnoff). If υ_i is too long and the gain of each stage is 40 V/V, what output fall time results?

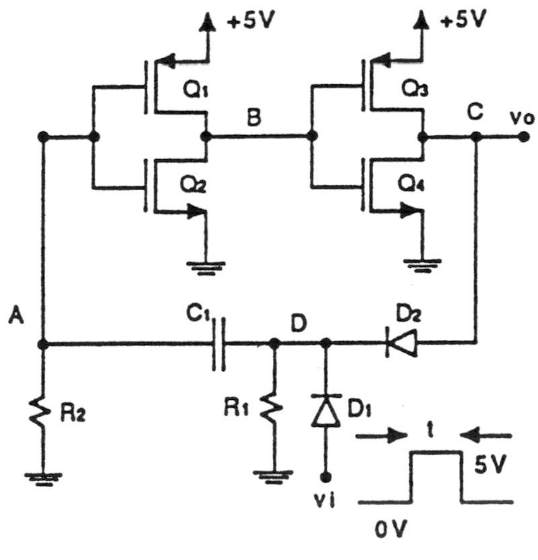

SECTION 12.7: INTEGRATED-CIRCUIT TIMERS

12.23 With reference to Fig. 12.28 in the Text, showing the **555 timer** operating as a monostable multivibrator with $R = 10\ k\Omega$ and $C = 10\ nF$, what is the length of output pulse produced by a negative trigger pulse? What is the maximum length of input pulse which might logically be used? What happens to the flip-flop if the input is longer than this? It is usual in the case of such conflict that the flip flop is arranged to be so-called "set-dominant", such that υ_o (that is, Q) remains high (with $\overline{Q}$ low) while S is high (held by V trigger being low).

12.24 With reference to the situation described in P12.23 above, and for normal operation with a 5 V supply, what is the % change in output pulse length for each 100 mV increase in the saturation voltage of Q_1?

D

12.25 For the astable connection of the **555 timer** depicted in Fig. 12.29 of the Text, using $C = 10$ nF and $R_A = R_B = 10\ k\Omega$, what is the frequency of the output at υ_o? What is the duty cycle? What do these become if R_B is reduced to 1 kΩ? What change must be made to R_A to return to the same frequency? What combination of resistors will provide an output at 10 kHz with a duty cycle of 10%?

SECTION 12.8: NONLINEAR WAVEFORM-SHAPING CIRCUITS

DL*

12.26 Design a sine-wave shaper using a series resistance R and two shunting diodes connected to ground, one with anode grounded, and the other with cathode grounded. The input is a triangle wave whose amplitude is such that its zero-crossing slope equals that of the desired output sinewave. The diodes are characterized as conducting 1 mA at 0.700 volts, with $n = 2$. Find the triangle-wave peak voltage, and a suitable value for R. Then find the angles Θ ($\Theta = 90°$ at the wave peak) where the output of the circuit is 0.7, 0.65, 0.6, 0.55, 0.5, 0.4, 0.3, 0.2, 0.1 and 0 V. Use these angles to find the values of the prototype sine wave (i.e., $\upsilon_o = 0.7 \sin \Theta$) and the corresponding errors. Present your results in tabular form.

DL*

12.27 Use the results obtained in Exercise 12.22 on page 1017 of the Text, for the square-law shaper shown in Fig. E12.22 there, using ideal diodes, as the basis of a design with real diodes. First consider pseudo-1-mA diodes modelled by an ideal diode, a 600 mV offset and a 100 Ω resistor. Chose new values for R_1, R_2, R_3 and associated voltages V_2 and V_3. Second, using these values of V_2 and V_3, create a revised design using diodes for which the drop is 0.7 V at 1 mA with $n = 2$. In each case, in approximating $i = 0.1\upsilon^2$, arrange that the approximation is perfect at 2, 4, and 8 volts, and calculate the error at 3, 5, 7 and 10 volts.

CD

12.28 Using the idea embodied in Fig. P12.44 on pages 1037 and 1038 of the Text and similar component values, prepare a design of a circuit for which the output is $\upsilon_o = \upsilon_i\ \upsilon_2/\upsilon_3$ for $\upsilon_1, \upsilon_2, \upsilon_3 > 0$. Check circuit performance at various combinations of input voltages, say, 0.5, 1, 2, 3 volts. Assume that all diodes are identical with 700 mV drop at 1 mA with $n = 2$.

SECTION 12.9: PRECISION RECTIFIER CIRCUITS

D

12.29 Combine the circuits of Fig. 12.33 and Fig. 12.34 of the Text into a single one connected both to a common source υ_i and a common load R_o as in P12.30 below. Sketch the composite circuit and its composite transfer characteristic. Since its output is always positive, yet proportioned to the size of the input, what is such a circuit called?

D

12.30 Note that the circuit created in answer to P12.29 above is not symmetric, in the sense that one amplifier is allowed to saturate. Add an additional resistor R_3 and diode D_4 to correct for this, using the idea embodied in the use of D_2 in the more complete sub-circuit. How should the value of R_3 be chosen in relation to R_1, R_2, and R_0?

D

12.31 A designer wishes to create an expanded-scale ac voltmeter in which voltages in the range 100V rms to 140 V rms are displayed. Note that for voltages ≤ 100 V rms, the meter reads 100 V at the left end of the moving-coil-meter scale. Use a combination of the circuit in Fig. 12.35 of the Text, the idea embodied in Fig. E12.28 on page 1021 there, and a 1-mA 50-Ω meter connected in series with R_4 = 10 kΩ of the filter. Within the circuit itself, design for signals whose largest value in the normal operating range is ±10 V. Chose R_1 to absorb the very large input voltage involved. Note that ±15 V supplies are available.

12.32

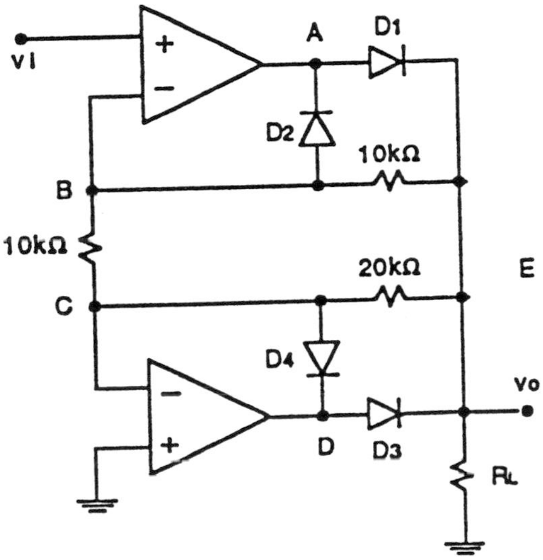

Consider the operation of the circuit shown: What is the output voltage for v_I = +5 V, 0 V, −5 V. What is the input resistance of the circuit? What might the circuit be called? Notice the relationship to the bridge amplifier in Fig. P9.52 on pages 808 and 809 of the Text.

12.33

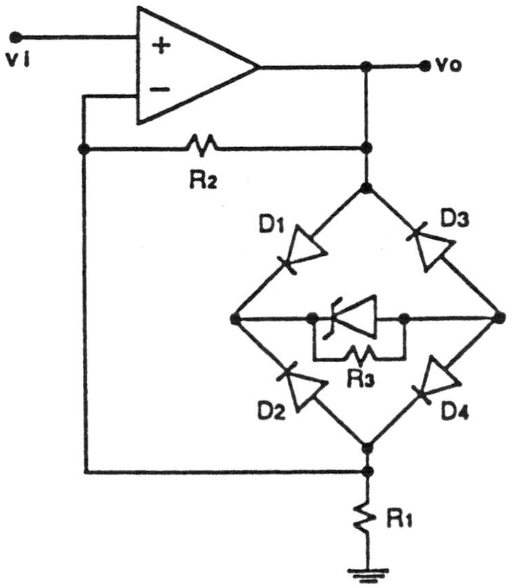

Consider the transfer characteristic of the circuit shown. Sketch and label it for R_1 = 1kΩ, R_2 = 100 kΩ, R_3 = 100 kΩ, with $r_Z = r_D \approx 0$, but V_Z = 6.8 V, V_D = 0.7 V.

DL

12.34

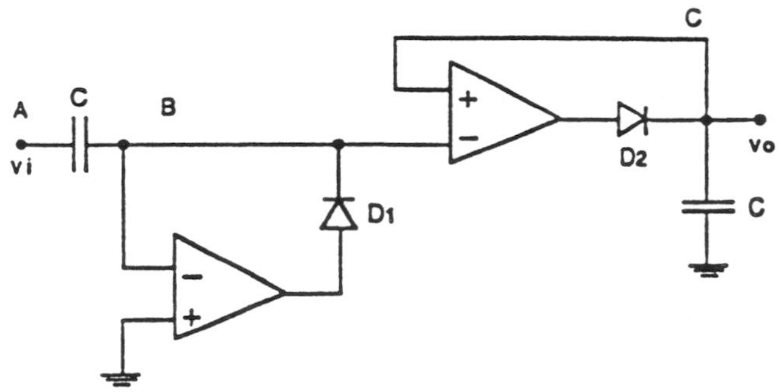

For the circuit shown using ideal op amps, what output results for the application of a 100 mV peak sine wave of frequency f at the input? What happens to the output when the input signal is removed? Add a component which will return the output 95% of the way to zero in a time $10/f$. What happens if the average value of the input drifts around by a volt or so at a rate $f/100$ Hz? Add a component to correct for drift at the stated rate or less. Provide a labelled sketch of the output with these two components added.

Chapter 13

MOS DIGITAL CIRCUITS

NOTE CONCERNING STANDARD DEVICE SPECIFICATIONS: In some of the problems to follow, we will consider the design of a 3.3 V CMOS inverter in a *generic* 0.8 μm process in which: $V_{tn} = -V_{tp} = 0.6$ V, $\mu_n C_{ox} = 100\ \mu A/V^2$, $\mu_p C_{ox} = 40\ \mu A/V^2$, and the minimum size digital device has $W/L = 1.2\ \mu m/0.8\ \mu m = 1.5$. *For capacitance calculations*, use $C_{ox} = 1.8\ fF/\mu m^2$, gate-drain overlap capacitance as 0.5 fF/μm of gate width, and drain-to-body capacitance as $C_{db} = 2.5$ fF/μm of gate width. *For gain calculations*, use $|V_A| = 20$ V for all 0.8 μm channels. *For body-effect calculations*, $|V_{t0}| = 0.6V$, $\gamma = 0.5\ V^{1/2}$, $2\Phi_f = 0.6$ V.

SECTION 13.1: DIGITAL CIRCUIT DESIGN: AN OVERVIEW

13.1 The voltage-transfer characteristic (VTC) of a particular logic inverter is observed to have the following salient features:

a) The upper output voltage level is 3.3 V.

b) The lower output voltage level is 0 V.

c) The slope of the characteristic is −1 V/V for inputs of 1.8 V and 1.2 V.

d) The maximum slope is about − 40 V/V occuring where the input and output voltages each equal 1.5 V.

Sketch and label the transfer characteristics. Find values for V_{OL}, V_{OH}, V_{IL}, V_{IH}, V_{th}, V_M, NM_L, NM_H.

13.2 Five inverters of the type described in P13.1 above are connected in a ring (each driving the next; each with a fanout of 1). Convince yourself that this circuit will oscillate by sketching waveforms at each successive gate input. Though your sketch can be quite rough, make sure it includes identifiable propagation delays. The circuit is called a *ring oscillator*. How many transitions does each inverter make in one cycle of the overall oscillation? How many gate-propagation delays t_{PLH} are there in one cycle? How many t_{PHL}? How many transitions in total? If the ring of 5 oscillates at 100 MHz, what is the average propagation delay, t_P? If the asymmetry of the inverter makes t_{PLH} 20% greater than t_{PHL}, estimate each value.

13.3 During oscillation at 100 MHz, the ring-of-five described in P13.2 above is found to use 300 μA from the 3.3 V supply. When the loop is opened, and the input to the string grounded, the supply current is essentially zero. What is the dynamic power P_D of each inverter (whose fanout is 1)? What is the capacitance associated with each? What is the delay-power product DP for this logic?

13.4 A particular two-input NOR logic gate has $t_{PLH} = 30$ ns and $t_{PHL} = 10$ ns. Five such gates, each with one input low, are connected in a ring. What is the frequency of the oscillation which results? Two additional NOR gates, A and B, are connected to the ring, gate A to the outputs of gates 1 and 3, and gate B to the outputs of gates 1 and 4. Sketch the resulting waveforms at the NOR-gate outputs, paying attention to their relative timing.

SECTION 13.2: DESIGN AND PERFORMANCE ANALYSIS OF THE CMOS INVERTER

13.5 For the generic process described in the introductory NOTE, above, what is the W/L ratio for the PMOS device in a matched minimum-size inverter (for which the output drive currents are equal). For this design, what are the values of V_{th}, V_{IL}, V_{IH}, NM_H, NM_L? [Hint: Use Equations 5.93, 5.94, 5.95, 5.96 and 13.8 in the Text.]

13.6 For a general CMOS inverter characterized by V_{tn}, V_{tp}, k_n and k_p, and V_{DD} where $k = k'(W/L)$, derive the expression for $V_M = V_{th}$ given in Equation 13.8 in the Text. As an extra challenge, you might consider deriving corresponding expressions for V_{IH}, V_{IL}, NM_H and NM_L. If you try this, you can test your results for the special case of a matched inverter as expressed in Equations 5.93, 5.94, 5.95 and 5.96 in the Text.

13.7 For the the general CMOS inverter alluded to in P.13.6 above, for which Early voltages are V_{An}, V_{Ap}, find an expression for the slope of the transfer characteristic at V_M. For the general minimum-size matched inverter (as described in the introductory NOTE above), what is the voltage gain for a linear amplifier biased at $v_O = v_I = V_M$.

13.8 Consider a CMOS inverter for which $\mu_n C_{ox} = 2\mu_p C_{ox} = 20\mu A/V^2$, $(W/L)_n = 8\mu m/2\mu m$, $(W/L)_p = 16\mu m/2\mu m$, $V_{tn} = -V_{tp} = 1$ V, and $V_{DD} = 5$ V. What are the resistances from the output node to the supply rails for inputs high (+5 V) and low (0 V).

13.9 For the inverter described in P13.8 above, with standard inputs of 0 V or V_{DD}, find the maximum currents that can be sourced or absorbed with the output joined to ground or V_{DD}, respectively. What do these currents become for an output voltage of $V_{DD}/2$? for an output voltage of 0.1 V_{DD} from either V_{DD} or 0V?

13.10 A CMOS inverter for which the device thresholds and k' factors are matched, with $|V_t|$ nominally 1 V, operates from a nominal supply voltage of 5 V. For possible variation of both V_t (of both devices) and V_{DD} by ±25%, what ranges of V_{IL}, V_{IH}, V_{OL}, V_{OH} result? [Hint: Use Equations 5.93 and 5.94 in the Text.] What ranges of NM_H and NM_L result between various pairs of these inverters?

13.11 Consider the propagation delay associated with the generic minimum-size matched inverter described in the introductory NOTE above, in the context of the description of the associated capacitances identified in Fig. 13.6 of the Text. In particular, evaluate the total load capacitance C seen by the test inverter using Equation 13.12. Assume the wiring capacitance C_w to be about equivalent to the n-channel gate capacitance. Now, estimate t_{PHL}, t_{PLH} and t_P from Equations 13.18 and 13.19.

13.12 In the computation for propagation delay leading to Equation 13.17 of the Text, both saturation and triode operation of the driving transistor are considered. What propagation delay would result if saturation-mode current was available for the entire half-signal transition? Put your result in the standard form represented by Equation 13.18. For the standard minimum-size matched inverter for which $t_{PHL} = t_{PLH} = t_P$, what are the 3 available estimates? Which is easiest to calculate from first principles? (But see the effect of decreased $|V_t|$ in P13.13 below.)

13.13 Repeat the analysis leading to Equation 13.18 of the Text, for the situation in which $V_t = 0.1\ V_{DD}$. For the approximation in which the initial saturation current is presumed to continue at the level provided by Equation 13.14, what relationship results? Finally, substitute for this value of V_t in Equation 5.101 to find the presumably most accurate result. What do you prefer for rapid analysis?

13.14 The CMOS inverter specified in P13.8 above, and having a 0.5 pF load, is switched at a rate of 20 MHz by an input wave whose rise and fall times are each ¼ of the wave period. What is the peak current due to gate self-conduction? Assuming the gate self-current to be triangular in pulse form, flowing for input voltages $v_I = |V_t|$ to $v_I = V_{DD} - |V_t|$, find the average supply current due both to self-conduction and capacitor charging. What is the gate power dissipation for this situation? What is it with the load capacitor removed?

13.15 For the inverter specified in P13.8 above, loaded with a 0.5 pF capacitor, estimate t_{PHL} and t_{PLH} using Equations 13.18 and 13.19 of the Text. What is the oscillation frequency of a five-stage oscillator? What is the corresponding delay-power product?

13.16 A CMOS inverter for which $k_n = k_p = k = 20\ \mu A/V^2$ and $|V_t| = 0.8$ V, is considered for operation in an electronic-watch environment with $V_{DD} = 1.3$ V.

(a) For what range of input voltages does Q_n conduct? Q_p conduct? What happens for $\upsilon_I = V_{DD}/2$?

(b) What range of output voltages is possible? What values of V_{OH} and V_{OL} can be identified?

(c) What are the values of V_{IL} and V_{IH} which apply? What currents flow in the transistors for voltages between V_{IL} and V_{IH}?

(d) Sketch the transfer characteristic, under the assumption that the input is a triangle wave extending from V_{OL} to V_{OH}, and that the output is loaded with a very small capacitor to ground.

(e) For the inverter loaded by a 1 pF capacitor, and input driven by an ideal square wave, sketch the output waveform. Estimate the times taken for the output to rise and fall by the amount V_t following a change in state of the input? Estimate the propagation delay, as defined as the time taken for the output to move to the switching threshold of a succeeding gate, following the crossing of its own threshold.

(f) Estimate the frequency of oscillation of a ring of 5 such gates, once an oscillation is initiated. Note that this oscillator is *not* ordinarily self-starting, but must be "kick-started" by an external signal. Otherwise, the ring has a stable state with all voltages in a range around $V_{DD}/2$ (in fact from 0.5V to 0.8V).

13.17 Consider an inverter for which $\mu_n C_{ox} = 2\mu_p C_{ox} = 20\mu A/V^2$, $(W/L)_n = 18\mu m/2\mu m$, $(W/L)_p = 4\mu m/2\mu m$, $V_{tn} = -V_{tp} = 1$ V, and $V_{DD} = 5$ V. What values of V_{OH}, V_{OL}, V_{IL}, V_{IH} and V_{th} apply? For the latter threshold voltage used in defining propagation delay, find t_{PLH} and t_{PHL} with the gate loaded by a 0.5 pF capacitor.

SECTION 13.3: CMOS LOGIC-GATE CIRCUITS

D

13.18 For the Boolean function $\overline{Y} = A(B + C)$, synthesize the PDN and PUN networks for a CMOS implementation using the direct method. Now, use the PDN to obtain a PUN design. Is it identical, or are the details of some transistor connections different? Did you have other choices? Now, perform the complementary process, obtaining the PDN from the PUN. Note in your designs that some transistors are connected directly to a power supply while others are more remote. Correspondingly, several different arrangements are usually possible. How many different PDN are there? How many PUN? How many different gate topologies exist? We will see in other contexts that the position of inputs can effect gate dynamic performance, and that, correspondingly, gate performance can be optimized depending on the statistics on input activity.

D

13.19 As noted in Fig. 13.15 of the Text, the PUN and the PDN there, each synthesized directly, are not dual networks. Sketch the PDN which is dual to the PUN shown, and the PUN which is dual to the PDN shown. How many possible XOR circuits of this general kind are possible using these four networks? How many are there if the proximity of an input to a power-supply connection is of importance?

D

13.20 Design a minimum-size 4-input CMOS NOR gate for ideal performance similar to that of the minimum-size matched inverter suggested in the introductory NOTE. What is its area? Compare that with the area of the basic inverter.

D

13.21 Repeat P13.20 above for a 4-input CMOS NAND.

D

13.22 Repeat 13.20 for the circuit whose topology was considered in P13.18 above.

D

13.23 For the 2-input NOR circuit of Fig. 13.12 of the Text, in which $\mu_n = 2\mu_p$, $k_n = k_n'(W/L)_n = k$, $V_{tn} = -V_{tp} = 1$ V, and $V_{DD} = 5$ V, find the value of k_p such that the threshold for a single input active is 2.50 V. What is the corresponding threshold for both inputs tied together?

D

13.24 Create a CMOS implementation of the logic function $\overline{Y} = AB + ACD$ both directly and in factored form (ie $\overline{Y} = A(B + CD)$). Size all devices for worst-case speed performance equivalent to that of the minimum-size matched inverter described in the introductory NOTE above. What is the total device width needed in each case? Clearly, Boolean factoring and minimization are important elements in digital logic-circuit design!

D

13.25 In a so-called *buffered-logic family* for which minimum overall size and input capacitance is important, a minimum-size input stage is buffered by two additional larger inverters. For example, a buffered-inverter input stage is matched, using the smallest possible n-channel device of unit area with a p-channel device of twice its width. In the 2nd and 3rd stages, all corresponding devices are each 3× wider. What is the total area of the buffered inverter? What is the total area of a buffered 4-input NOR whose first stage is made basic-inverter-compatible? What is the input capacitance relative to the inverter? A second design is considered with the same stage-size ratio (ie 3 ×) but using minimum-size matched input inverters and an intermediate-stage NAND. What is its total area? What is the relative input-capacitance presented?

SECTION 13.4: PSEUDO-NMOS LOGIC CIRCUITS

D

13.26 Using the parameters introduced in the NOTE preceding Section 13.1, design a pseudo-NMOS inverter to operate with a 3.3 V supply using a minimum-size NMOS device and a PMOS of minimum area chosen to provide equal positive and negative current drive to a load capacitor at $v_O = V_{DD}/2$. What value of r is needed? What are V_{OH}, V_{OL}, V_{IH}, V_{IL}, and V_M for this design? What do you think of the design?

D

13.27 A designer, wanting to reproduce the dynamic behaviour of a complementary matched CMOS inverter, considers a pseudo-NMOS inverter using a minimum-size NMOS with a PMOS that provides the same maximum load-driving current as is available from a complementary gate. What is the problem with this idea? What value of V_{OL} and V_M result? Use $V_{DD} = 3$ V.

13.28 As noted on page 1072 of the Text, pseudo-NMOS is a ratioed logic for which $r = k_n/k_p = k_n'(W/L)_n/k_p'(W/L)_p$ has a value from 4 to 10. Using a technology in which $\mu_n = 2.0\ \mu_p$ and the minimum gate length and width are each 1 unit, consider two types of designs for each of the extreme values of r:

a) With the minimum-size NMOS and $W_p = 1$ unit, find L_p and the combined areas of the n-channel and p-channel devices.

b) With the minimum-size PMOS, and $L_n = 1$ unit, find W_n and the combined areas of the n-channel and p-channel devices.

By what factor does the output-current drive improve in design b) over that in a)? Now consider an intermediate design c) in which the extreme dimensions of devices are reduced by "sharing" the ratio between the two devices, making each more "square" and reducing the total device area. Find such a design for which as many dimensions as possible are of miniumum value. What is the total area required? By what factor does the output-current drive exceed that in a)?

C*L

13.29 Use Equations 13.41 and 13.42 to find the value of r for which the noise margins of a pseudo-NMOS inverter are equal. What values of V_{OH}, V_{OL}, V_{IH}, V_{IL}, V_M, NM_H, NM_L correspond, when V_{DD} = 3.3 V and V_t = 0.6 V? For the NMOS specified as the minimum-size device in the NOTE preceeding Section 13.1 above, what is I_{stat}?

13.30 For V_{OL} of a pseudo-NMOS gate expressed as a fraction α of $V_t = V_{tn} = -V_{tp}$, find r. For V_{DD} = 3.3 V and V_t = 0.6 V, what is the value of r for which α = 0.5?

13.31 For a pseudo-NMOS inverter for which V_{DD} = 3.3 V and $|V_t|$ = 0.6 V, find r for: a) $V_M = V_{DD}/2$; b) $V_{OL} = V_t$; c) V_{OL} = 0.1 V; d) V_{OL} = 0.01 V; e) $V_{IL} = 2V_t$. For case a), find V_{OL}; For cases c), d), find V_{IL}.

L

13.32 For a pseudo-NMOS inverter, use Equations 13.43 and 13.44 and the fact that $r = k_n/k_p$ to find expressions for t_{PHL} and t_{PLH} in terms of a) k_p, b) k_n. Further, find the ratio t_{PLH}/t_{PHL}. For $t_{PLH} = t_{PHL} = t_P$, what value of r is needed? For this value, what are V_M, V_{OL} and the noise margins? By what factor is t_{PHL} longer than that of a matched CMOS inverter using the same devices, supply voltage, and capacitive loading. Is such a design useful? Recall that the major advantage of pseudo-NMOS occurs with complex logic functions where a large saving in gate area can be achieved by eliminating the PUN. In some such circumstances a high-threshold complementary load inverter may be used to accommodate the high V_{OL} level of a design like this one.

L

13.33 An 8-input pseudo-NMOS NOR gate operating at 3.3 V uses the standard minimum-size NMOS devices described in the introductory NOTE preceding Section 1.1. It is to be loaded with an inverter of similar design. For r = 4 and r = 10, evaluate t_{PLH} and t_{PHL} for a single input active and for two inputs operating together.

13.34 Compare the total device areas of pseudo-NMOS and matched complementary CMOS implementations of an 8-input NOR. Use all minimum-size NMOS, with $k_n' = 2.5k_p'$, and $r = 2$.

SECTION 13.5: PASS-TRANSISTOR LOGIC CIRCUITS

L

13.35 For the standard minimum-size NMOS transistor (specified in the introductory NOTE) used as a pass transistor in a 3.3-V system, find V_{OL} and V_{OH} assuming both the logic and gate inputs to be full-swing signals, 0 to 3.3 V. For a second pass gate driven by this one, what does V_{OH} become? For this value of V_{OH}, what current will flow in a connected minimum-size matched complementary CMOS inverter? What will the inverter output voltage be? For a single pass gate driving the basic inverter, with a second passgate connected to ground at the inverter input, estimate the total capacitance there, and then t_{PLH} and t_{PHL}, as the logic input to the pass transistor rises and falls.

13.36 Consider the pass-gate situation described in P13.35 above agumented by the circuit of Fig. 13.28 of the Text, using a PMOS of minimum width with $(W/L)_p$ = 0.1. Using the notation of that figure, at what value of υ_{O1} rising from 0 V does Q_R begin to conduct? What is the effect on t_{PLH}? What is the general effect on the upper end of the rising transition? What does V_{OH} become? As measured at υ_{O1}, estimate what t_{PHL} becomes.

13.37 Consider the operation of a single CMOS transmission gate in a pass-transistor logic configuration connected to a standard matched inverter and a single NMOS grounding switch (as additional capacitance). At the inverter input, what do V_{OL} and V_{OH} become? Estimate the capacitance at the inverter input node. Estimate t_{PLH} and t_{PHL}. Assume all the pass transistors have $(W/L) = 1.2/0.8$ **to the inverter input.**

D*

13.38 Following the pattern suggested in Fig. 13.30 of the Text, sketch a PTL version of a 4-to-1 multiplexor with automatic decoding of two control bits (C_1 and C_2). Label inputs X_1, X_2, X_3, X_4 and output Y. [Hint: Consider 2 inputs at a time.]

D*

13.39 Consider the Exclusive-OR function PTL implementation for two variables shown in Fig. 13.31 of the Text. Extend the Boolean expression to a third input variable C. [Hint: A with B produce X and $\overline{X}$ with C produces Y.] Sketch the corresponding PTL realization.

D*

13.40 The circuit shown illustrates the existence of yet other possibilities for the design of flexible logic functions using CMOS devices:

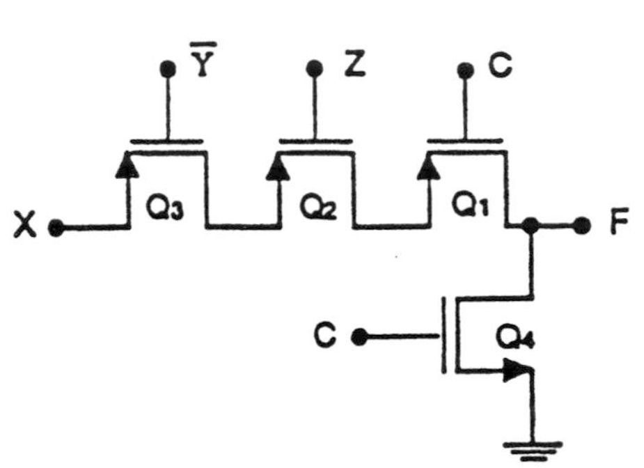

It combines aspects of transmission-gate logic and conventional CMOS logic, with neither in its complete form. In usual applications, the signal C acts as a clock, being normally held high while X, Y and Z change. Finally, C falls while X, Y, C are stable. For signals having 0 V and 5 V levels and using the positive logic convention (where logic '1' is high), prepare a truth table for F. Express F as a function of C, X, Y, Z. In an application in which variable $\overline{X}$ is available, but X is not, suggest a means to provide the required function without adding a (complete) additional inverter.

For dynamic operation of the original circuit equivalent to that of a basic symmetric inverter with $(W/L)_n = 10\mu m/5\mu m$, chose appropriate device widths on the assumption that all device lengths are $5\mu m$.

CD

13.41 Following the idea suggested in Exercise 13.9b) in conjunction with Fig. 13.32 of the Text, sketch the circuit of a CPL XOR-XNOR of 3 variables A, B, C. It is known that A is the most active signal changing twice as often as B and four times as often as C. The dynamic response of your design can be improved using this fact. How?

SECTION 13.6: DYNAMIC LOGIC CIRCUITS

13.42 A dynamic logic 3-input NAND gate uses 5 minimum-size devices with V_{DD} = 3.3 V as specified in the introductory NOTE at the beginning of the problems of this Chapter. Assuming that it is reasonable to consider all stray capacitance to be represented by C = 100 fF at the gate output, find t_{PLH} and t_{PHL} in response to Φ with inputs A, B, C assumed high. What does t_{PHL} become for a corresponding 3-input NOR?

13.43 For a 3-input dynamic-logic NAND implemented with 5 minimum-size standard devices (as specified in the NOTE early in this Chapter), estimate the capacitances at each of the circuit nodes. If all of these are accumulated at the output node, what does the equivalent capacitance become? If this circuit drives the equivalent of two minimum-size matched inverters, what does the output load capacitance become? What does the equivalent output load capacitance become? Now, consider the inputs as labelled A, B,

C, from the top, with Φ as the clock signal, and the nodes labelled 1, 2, 3, 4 from the top, with 1 as the output node. Now with Φ recently gone high at time t_0 with C, B high and A low, first C goes low at t_1, then A goes high at t_2. Prepare a table which records the voltages at all nodes at times t_{0+}, t_{1+}, and t_{2+}, the + sign indicating a time just after the change. Pay particular attention to the entries for nodes 1 and 2 at t_{2+}. They show the impact of *charge sharing.*

13.44 Consider the impact of charge leakage in defining the functional threshold of a dynamic n-input NOR gate. Initially, all inputs are low, and Φ has just gone high for a short evaluation period t_E. Estimate the voltage at an input which will cause the output capacitor voltage to fall by an amount V_t in t_e. Assume that all transistors are minimum size as specified in the introductory NOTE using a 3.3 V supply, and that the equivalent output capacitor is (50 + 10n) fF. For n = 5, and t_E = 10 ns, find the threshold for a single input, and for all n inputs acting together. Note that the latter situation can occur readily due to noise at the grounded source of the evaluation transistor. Note as well that subthreshold conduction can make the effective noise threshold even lower.

13.45 Consider the simple dynamic inverter string shown in Fig. 13.35 of the Text operating from 3.3 V with standard minimum-size transistors (as specified in the introductory NOTE) and C_L = 50 fF (arbitrarily). Estimate how long Y_1 remains above V_{tn} with A high following the rise of Φ. In this interval, what is the average current in Q_2? What voltage change would you expect on C_{L2}?

13.46 For the Domino CMOS logic scheme shown in Fig. 13.37 of the Text using minimum size devices, estimate the capacitances at nodes X_1 and Y_1 and the low-to-high propagation delay from A to Y_1. Include the effect of the low threshold voltage of the simple inverter. How long must Φ be high to evaluate a string of 10 such logic gates. In practice, for worst-case device tolerances, wiring capacitances, etc, a much longer time would be needed. One approach is to control the evaluation interval pseudo-asynchronously by detecting the propagation delay of a model Domino chain.

SECTION 13.7: LATCHES AND FLIP FLOPS

13.47 Consider two standard minimum-size matched CMOS static inverters in a positive-feedback loop. For each latch, find V_{IL}, V_{IH}, V_M and the voltage gain at $v_I = v_O = V_M$. For the pair, what is the maximum loop gain?

13.48 Two CMOS inverters operating from a 5 V supply with V_{IL} and V_{IH} of 2.40 V and 2.90 V, respectively for which corresponding outputs are 4.7 V and 0.4 V, are connected as a latch. Approximating the corresponding transfer characteristic of each gate by a straight line between threshold points, sketch the latch open-loop transfer characteristic. At what points are the open-loop input and output voltages equal? What is the loop gain at each?

13.49 Consider the CMOS SR flipflop implimentation shown in Fig. 13.40 of the Text using only standard minimum-size devices with V_{DD} = 3.3 V. What is V_M for the regenerating inverters, say at node Q. For R high, what voltage is required at Φ to lower Q from 3.3 V to V_M?

13.50 Sketch the circuit of a NAND *SR* flip-flop using CMOS, and prepare a truth table whose entries are in terms of stable output voltages available with a 3 V supply, and devices for which $|V_t|$ < 3/2 V.

13.51 For the SR flipflop shown in Fig. 13.40 of the Text, implimented with minimum-size devices of the standard kind (as specified in the introductory NOTE), estimate the total capacitance of the Q and $\overline{Q}$ nodes, each with a matched inverter load. Find the propagation delay from Φ rising (with R high) to Q falling, and then to $\overline{Q}$ rising.

D

13.52 Consider the design of the clocked SR flipflop shown in Fig. 13.42 of the Text driven by the equivalent of matched CMOS inverters of *non-minimum* size. Assume that all NMOS, Q_1, Q_3, Q_5, Q_6 are of minimum size and Q_2 and Q_4 wide enough to make $V_M = V_{DD}/2$ for each internal inverter. Use the standard process described in the introductory NOTE at the beginning of this Chapter. Note, for example, with Q initially high and $\overline{Q}$ low, that Q_6 will be able to provide more current than Q_5, due primarily to the source follower configuration of Q_5 and its body effect as the driven node rises. Correspondingly, we will concentrate on the process of pulling Q low via Q_6. Now, assume for ease of calculation, that during the gating process, Q_4 will conduct the full saturation current for $|v_{GS}| = V_{DD}$, and that the current flowing through Q_6 will cause S to rise, say to V_t as a worst case. Find the width W_6 of Q_6 to allow Q to be brought to $V_{DD}/2$ where regeneration is assured. What minimum width W_D of the driver at the S input is required?

DL*

13.53 Design a D flipflop resembling that in Fig. 13.44 of the Text whose PMOS and NMOS are all of minimum-size. Assume that the D input is driven by a minimum-size matched CMOS inverter and that the internal switches are simple NMOS. To what level V_{th} must the input of the internal inverter be brought to assume full switching at Q? Assume each of the internal nodes has an equivalent capacitance of 50 fF. Calculate the propagation delay from the rise of Φ until the change of Q for D high, and then low. How long must Φ be high to ensure correct data flow? Is there any restriction on how long $\overline{\Phi}$ must be high? Note that the clock phases are said to be "non-overlapping", meaning that Φ and $\overline{\Phi}$ must not be simultaneously high. Does overlap matter at the falling edge of Φ? Why? Does overlap matter at the falling edge of $\overline{\Phi}$? Why? What is your view of a highest possible frequency for Φ if the non-overlap interval must be at least 5% of the longest clock phase.

L*

13.54 Reconsider P13.53 above for the situation in which each switch is a full CMOS transmission gate using minimun-size PMOS and NMOS. Estimate the input propogation delays, the minimum phase durations, and the maximum clock frequency. What advantages do the added PMOS bring to speed and to power-supply current reduction?

13.55 Consider the master-slave D flipflop in Fig. 13.45 of the Text, using full CMOS transmission gates and internal inverters. How many transistors are needed? If every device is minimum size, what is the total device width needed? If the current-driving capabilities of the NMOS and PMOS are matched, what total width is needed? What is the perecentage cost increase of matching? Suggest locations where matching and/or even larger devices would provide an advantage, with reasons for your choice, emphasizing reliability, speed, output capability, etc.

13.56

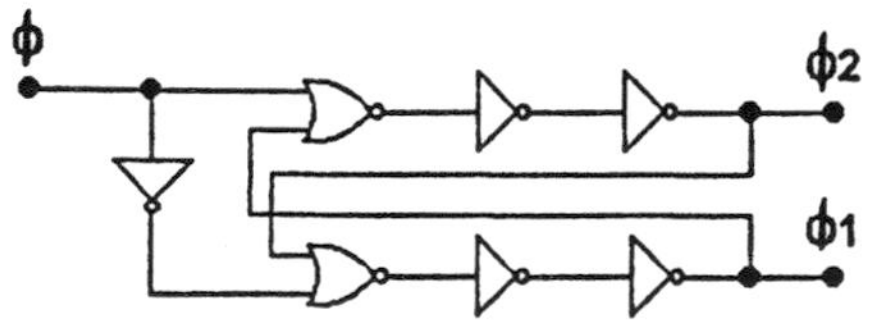

The logic circuit shown is driven by a square wave Φ of period 20t where t is the propagation delay of each inverter/gate. Sketch a timing chart with relative timing of Φ, Φ_1, Φ_2 carefully shown. What are the non-overlap periods for Φ_1, Φ_2? Sketch the effect of eliminating, successively, the inverters feeding Φ_1, Φ_2, both Φ_1 and Φ_2. What do the non-overlap intervals become in each case? To increase the gap between Φ_1 falling and Φ_2 rising, which inverter string should be increased (the top or the bottom one)? By how many inverters should the string be increased to approximately double the gap?

13.57 Sketch a CMOS implementation of the circuit using P13.56 above, using minimum-size unmatched complementary CMOS everywhere except for the outputs which should provide 10 times the normal matched inverter output. What total device width is needed?

SECTION 13.8: MULTIVIBRATOR CIRCUITS

13.58 Consider the monostable circuit of Fig. 13.47 of the Text, implemented with devices from an SSI CMOS package operating at V_{DD} = 5 V with devices for which $V_{th} = V_{DD}/2$, t_P = 15 ns, R_{on} = 200 Ω and the protection diodes begin to conduct at 0.5 V and are fully conducting at 0.7 V. Using a 20 pF capacitor, find R required for a positive 200 ns pulse at v_{O2}. What is the minimum length of the positive input triggering pulse? What happens if the input pulse is positive for 1μsec? If the large-signal voltage gain of G_2 is about 20 V/V, what is the fall time of v_{O2} for a long input pulse? What is the maximum length of input pulse for optimal behaviour? What would you estimate the output pulse transition times to be [Hint: t_T is in the range of $t_P/2$ to 3 t_P.]

13.59 For the monostable circuit of Fig. 13.47 of the Text, whose waveforms are given in Fig. 13.50 and an expression for T given in Exercise 13.15 (on page 1110), let V_{DD} = 5 V, V_{th} = 0.6 V_{DD}, R = 22kΩ, C = 1500 pF, and R_{on} = 180Ω. Find the values of T, ΔV_1, and ΔV_2. By how much does v_{O1} change during the interval T? What are the peak sink and source currents of G_1? Use v_D = 0.7 V.

D

13.60 In a particular CMOS implementation of Fig. 13.47 of the Text, G_2 is a simple inverter and G_1 a NOR, both of which use all minimum-sized devices for which (W/L) = 2. For this process, $|V_t|$ = 1 V, $\mu_n C_{ox} = 2\mu_p C_{ox} = 20\mu A/V^2$, and V_{DD} = 5 V. The function of R is implemented using a simple current mirror employing two minimum-sized p-channel devices and a grounded source diode-connected minimum-width n-channel device of 10 times the minimum length. Find the value of C for a 10 μs output pulse, accounting for the non-zero value of V_{OL} of G_1 and the actual value of V_{th} of G_2.

CL

13.61 The circuit shown is a One-Shot intended for operation using the standard technology introduced in the introductory NOTE of this Chapter, with V_{DD} = 3.3V. All NMOS have minimum L and 10× minimum width. Q_6 is minimum size. Q_4 is matched in current to Q_3. Q_5 has minimum width but 10× minimum length.

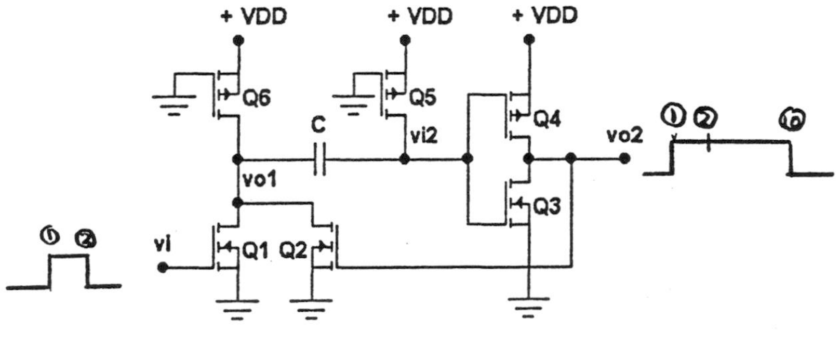

For C = 10 pF, estimate the output pulse length. Estimate the parasitic capacitances at the output and internal node (for which C is a short-circuit), and then the propagation and transition times for each of the gates. What is the minimum-length trigger pulse? What is the maximum trigger pulse which still allows regenerative turnoff. Between pulse inputs, what is the required power-supply current? What is the supply current immediately after triggering? What is the voltage across C during the period between inputs? How long does it take C to recover to zero volts following the end of a pulse at the output (or after a long input)? [Hint: Q_5 operates for a time with drain and source functions interchanged.]

13.62 For the astable multivibrator modified as suggested in P13.77 on page 1154 of the Text, provide a design for operation at 1MHz, using a 100 pF capacitor, V_{DD} = 5 V, and V_{th} = 0.44 V_{DD}. What values of resistors would you use?

D

13.63 Describe the operation of the circuit shown. It is implemented with a 3.3 V supply using the standardized technology introduced earlier in the introducotry NOTE. Q_1, Q_2 is a regular minimum-size matched CMOS inverter. Q_3, Q_4 is matched but 10 times wider than minimum. Q_5 has minimum width and a length which is 100× the minimum. Capacitor C = 10 pF. For inverter thresholds at $V_{DD}/2$, how long does the output stay positive? How long negative? Suggest a way to arrange for a 50% duty cycle.

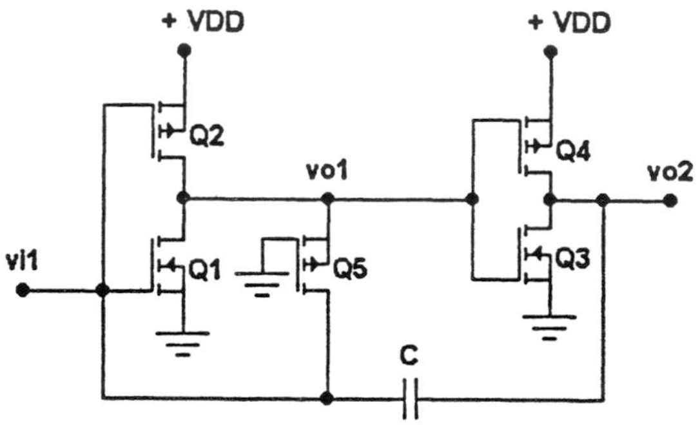

13.64 (For other ring-oscillator problems, see also P13.2 and P13.3 above.)

A ring of 5 inverters is constructed with the basic matched CMOS inverter described in the NOTE at the beginning of this Chapter. Operation in this technology is at 3.3 V. Estimate the equivalent load capacitance, the propagation delay and oscillation frequency. What would you expect the frequency to become if: a) an additional inverter loads each stage, b) the supply is reduced to 2.0 V?

SECTION 13.9: SEMICONDUCTOR MEMORIES: TYPES AND ARCHITECTURES

13.65 For various Random-Access Memories listed, complete the missing Table entries.

	Address Bits				Structures						
#	Block	Row	Col	Total	Blocks	Rows	Columns	Words	Bits/Word	Bits/Block	Total Bits
a		10	7	22	32	1024	128	4M	1		4M
b	0		8			256			16M		
c		11		25						16	256M
d	4	10		24				16M		16M	
e	3	12	11						1		
f							1024		4	8M	64M

13.66 A 1 M-bit memory chip is organized as 4 square blocks, each of which uses simple NOR decoders for row and column selection. How many inputs would each decoder need? For a bit address consisting (from its left) of block-number bits, row bits, and column bits, what are the column-address digits of bit 102,476 on the chip?

SECTION 13.10: RANDOM-ACCESS MEMORY (RAM) CELLS

13.67 Consider an SRAM cell of the type shown in Fig. 13.55 of the Text using the standard 3.3 V, 0.8 μm technology described in the NOTE at the beginng of this Chapter's problems. For each of the inverters, use a minimum-sized matched design. For the access transistors, use NMOS of 3× the minimum width. What is the total area of the gates of all the devices in the cell? Assuming that the connection overhead in the cell causes the cell area to be twice as large as the gate area, what would the dimensions of a square cell be, approximately? For the Read operation, let us examine the situation in which the word line is activated to select a cell for which $\upsilon_B = \upsilon_{\overline{B}} = V_{DD}/2$ initially. For this situation, evaluate the available current to charge/discharge the bit line, assuming that υ_Q and $\upsilon_{\overline{Q}}$ do not change, but including the body effect for one of the gating devices. Now consider the bit-line capacitance, assuming 128 cells on the line and capacitance about the same as a 1 μm gate stripe 128 cells long. Using this value and the current data, how long will it take to establish a differential voltage of about 0.2 V between the bit lines?

13.68 For the situation described in P13.67 above, consider the Write operation in which the bit lines have complementary values, either 0 or V_{DD} as the word line is raised from 0 to V_{DD}. For this analysis, consider the cell to consist of 2 separated inverters with fixed inputs, either 0 or V_{DD}, and to be concrete, with Q high and $\overline{Q}$ low. The write operation will be successful if either input is moved beyond the regeneration point at $V_{th} = V_{DD}/2$. The excess driving current at $\upsilon_Q = V_{DD}/2$ or $\upsilon_{\overline{Q}} = V_{DD}/2$ will be a measure of operating speed. For $\upsilon_Q = \upsilon_{\overline{Q}} = V_{DD}/2$, evaluate the direction and magnitude of the net current flow in the access transistors with the $\overline{bit}$ line, $\overline{B}$, low and bit line, B, high. Is switching possible? Now estimate the total capacitance at Q (or $\overline{Q}$). For the successful case (or cases), evaluate the excess drive current initially (at $V_{DD}/2$), and on average. Use the average current(s) to estimate the time for regeneration to begin.

D*L

13.69 A proposed CMOS static RAM uses cells such as that shown in Fig. 13.55 of the Text, having a cell supply voltage of 5V and word-line selection voltages of 0 at rest, 5 V for reading, and 5 V for writing. Digit-line voltages are precharged to 2.5 V for reading, and 0 and 5 V for writing. The cell itself uses minimum-size devices for which $(W/L) = 2\mu m/3\mu m$ and $|V_t| = 1V$ with $\mu_n C_{ox} = 2.5\mu_p C_{ox} = 25\mu A/V^2$. What is the threshold voltage at the drain of Q_1 or Q_2 at which the cell will change state? What currents supplied to or from the cell, move the cell output voltage half way from its stable state to the threshold? What must the width of Q_5 (and Q_6) be to ensure that readout is nondestructive? What currents can be supplied during the writing process by such a device? Is the design viable? Why or why not?

13.70 Consider the one-transistor dynamic RAM cell in Fig. 13.58 of the Text, using a minimum-size NMOS transistor in the standard 0.8 µm technology with a storage capacitor of 40 fF. The cell pitch (bit-to-bit spacing) in the bit-line direction is 2.5 µm. A 1-µm-wide bit line is used to couple the 256 cells in a column. Bit-line capacitance per unit area is about the same as gate capacitance. Sense amplifiers and drivers add another 70 fF to the bit-line capacitance. What is the total bit-line capacitance? For maximum cell signals υ_{CS} of $(V_{DD} - V_t)$ and 0 V, subject to a possible charge deterioration due to leakage in one direction *or* the other of 20% of full signal, and bit-line pre-charge to $(V_{DD}/2 - V_t)$, what bit-line signals will result? [Hint: Recall that V_t is subject to body effect.]

13.71 For the situation described in P13.70 above, where the voltage $(V_{DD} - V_t)$ on each 40 fF cell capacitor can deteriorate by as much as 20% in the 10-ms interval between guaranteed refresh cycles, estimate the corresponding leakage resistance.

13.72 In a particular dynamic-RAM technology, cell leakage currents can be reduced to 10 fA (1 fempto-ampere = $10^{-15}A$). What is the minimum allowable capacitor for a 4 ms refresh interval with a recoverable cell-voltage loss of 1.5 V?

13.73 For a particular DRAM having 1024 rows and 1024 columns in each of 16 blocks, with a read-write cycle time of 30 ns and a refresh cycle of 10 ms, what fraction of the available cycles is spent on refresh if one word in each block is refreshed in parallel in one cycle. How many sense amplifiers does such a design require? If all blocks share a single set of 1024 sense amplifiers, what does the refresh overhead become?

SECTION 13.11: SENSE AMPLIFIERS AND ADDRESS DECODERS

D

13.74 Consider the bit-line-voltage equalization process involving transistor Q_7 in Fig. 13.60 of the Text. Assume bit-line capacitances of 1 pF. Using a minimum-size NMOS in the standard technology introduced in the introductory NOTE of this Chapter, how long must Φ_P be activated to reduce the voltage difference on the bit lines to 1% of its regular value? [Hint: Recall that the bit-line average voltage is to be $V_{DD}/2$, that Q_7 is subject to body effect, and that $V_{DD} = 3.3$ V.] If precharge must be complete in 1 ns, how wide must Q_7 (and the cooperating path through Q_8 and Q_9) be made?

D

13.75 Consider the regenerative process occuring in the sense amplifier in Fig. 13.60 of the Text, immediately following the rise of the Φ_S signal. Each of the constituent inverters in the sense amp is matched, but with $(W/L)_n$ to be determined. The sense-line capacitances are each 2 pF. Size the sense-amplifier inverter to achieve a time constant of 2 ns. Estimate the time for the sense line to rise from 0.5 V_{DD} to 0.9 V_{DD} initiated by a 50 mV signal, and for the line to fall from 0.5 V_{DD} to 0.1 V_{DD} initiated by a 25 mV signal.

13.76 A particular 1 Mb DRAM uses a square cell array with 4-bit readout. How large a word-line (row) decoder is needed? How large a bit-line (column) decoder is needed? For a 1024-row decoder, how many address bits are used? Using the design in Fig. 13.63 of the Text, how many decoder-array NMOS are needed? How many dynamic-load PMOS? How many input address-bit inverters are needed?

D

13.77 For a tree column decoder such as that shown in Fig. 13.65 of the Text, how many input layers are needed for 256 lines? 1024 lines? How many transistors are used in each case? If each transistor is the minimum-size standard NMOS (as specified in the introductory NOTE) , what series resistance is acquired with each switch? For a 1-pF bit line directly connected, what is the greatest number of layers that can be used while ensuring that a logic zero settles to $V_{DD}/10 =$ 0.33 V within 7 ns. What can you do to improve this situation?

13.78 For the NOR address decoder, part of which is shown in Fig. 13.63 on page 1132 of the Text, draw row 13, indicating the connection of its transistors to the first 4 address lines. How many transistors, including the load, are connected to each row line of a 256 K-bit square array?

SECTION 13.12: READ-ONLY MEMORY (ROM)

13.79 A CMOS ROM of the general type shown in Fig. 13.66 of the Text uses a gated load structure in which the PMOS loads are turned on only at evaluation time. Minimum-size standard 0.8 μm NMOS (see the introductory NOTE) are used in the array. Connections to each cell require 30% overhead in each device dimension. Approximately, what would the typical cell dimensions be? If the array has a 15% overall overhead for decode, sensing buffering and connection, how many bits of ROM can be installed in a chip $1 mm^2$ in area. If the ROM is configured for a 32-bit-word output, how many words (expressed as a power of 2) can be accommodated in a chip of about this size?

D

13.80 Design a bit pattern to be stored in a (14 × 5) ROM which provides the results of division of one two-bit number, X, by another, Y. The 4-bit word address is to be (x_1, x_0, y_1, y_0). The output is to be (f, q_1, q_0, r_1, r_0) where F is the 1-bit overflow (divide-by-0) flag. Q is the 2-bit quotient, and R is the 2-bit remainder. Give a circuit implementation resembling that in Fig. 13.66, but in which an installed transistor represents a logic 1 internally. (Hint: While this saves ROM transistors, it requires additional inverters). Excluding the input decoder, how many transistors do you need in the heart of the ROM? How many are used in inverters? How many transistors would be required in total, without using the extra inverters? How many transistors would be required for the 4-bit input decoder using the circuit of Fig. 13.63?

13.81 Sketch the decoder and array parts of a 16-word 8-bit MOS ROM combining Fig. 13.63 with Fig. 13.66 of the Text, as modified in P13.79 above. For simplicity, represent the NMOS devices by circles at the intersection of access and output lines, and the required inverters by triangle symbols; but show the PMOS explicitly. Provide a sketch of the relative timing of the precharge and access signals.

D

13.82 A ROM used to record the presence or absence of unusual properties of diverse materials in an inventory-control system. A designer has a choice of representing this data as either high or low signals at the output of a transistor array such as that shown in Fig. 13.66 of the Text. What choice of data representation is best if

a) Transistors exist at array nodes and must be selectively "removed" to allow the digit line to go high.

b) Transistors must be created and connected at array nodes as needed to lower the digit line.

c) Fuses must be blown to disconnect an npn emitter from a digit line to allow it to remain low.

d) High-voltage programming pulses must be applied to raise the threshold of a floating-gate transistor allowing the bit line to be always high.

[Hint: In each case, identify the relative cost of the production or programming technique for representing a logic zero or logic one. Obviously, it will be best ot chose a logic representation which is cheapest for the statistics of the data being represented, which in this case has mostly zeros with only a few ones.]

Chapter 14

BIPOLAR AND ADVANCED - TECHNOLOGY DIGITAL CIRCUITS

SECTION 14.1: DYNAMIC OPERATION OF THE BJT SWITCH

L

14.1 A particular BJT inverter of the type shown in Fig. 14.1a) of the Text, uses V_{CC} = 5 V, R_C = 2 kΩ, and R_B = 10 kΩ, with a 0 to 4.3 V input signal. For the transistor, β_F = 100, β_R = 0.25, V_{BE} = 700 mV at I_C = 1 mA, r_x = 50 Ω and n = 1. Find α_F, α_R, I_{SE}, I_{SC}, I_S, V_{BEsat}, V_{CEsat}, I_{Csat}, I_{Bsat}, β_{forced}. [Hint: Use Equations 4.100, 4.102, 4.109, 4.110, 4.113, 4.114.]

C**L**

14.2 Consider the dynamic operation of the circuit described in P14.1 above in response to a 0 to 4.3 V input pulse. For the BJT, f_T = 1 GHz, C_{je} = C_{jc} = C_μ = 0.5 pF and τ_s = 1.5 ns. Estimate t_d, t_r, t_{on}, t_s, t_f, t_{off} and the time the base voltage remains at essentially 0.7 V following the fall of the input. [Hint: Think a lot about the details of Fig. 14.1.]

14.3 A BJT for which the storage time constant is 20 ns, and β = 200, is operated in a circuit for which $I_{C\ sat}$ = 10 mA and the base turn-on current, I_{B2}, is 1 mA. Calculate the storage delay under the conditions that the base turn-off current, I_{B1}, is a) 0 mA, b) 1 mA, c) 10 mA.

D

14.4

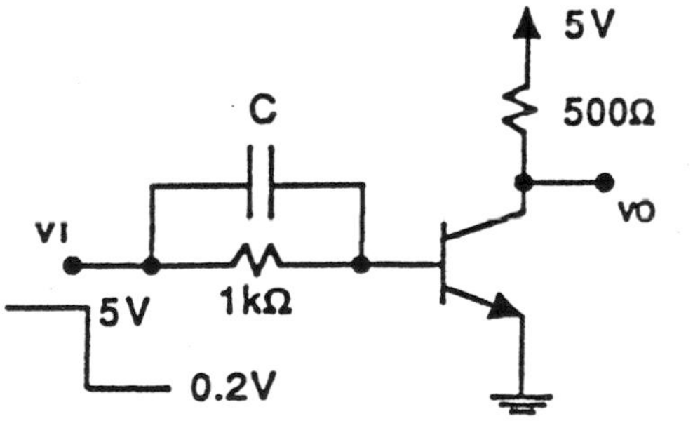

In the following circuit, t_s is measured for various values of C: For C = 0, t_s = 80 ns; for C = 8 pF, t_s = 30 ns. Estimate values of τ_s and β if V_{BE} = 0.7 V and $V_{CE\ sat}$ = 0.2V. What value of C would you chose to reduce t_s to zero? For β twice the present value, what would t_s be with the capacitor you have chosen? What would it have been with C = 0?

SECTION 14.2: EARLY FORMS OF BJT DIGITAL CIRCUITS

14.5 With reference to Fig. 14.4 on page 1164 and P14.5 on page 1232 of the Text, provide a better estimate for V_{IL}, as that voltage v_I for which the gain is −1 V/V, for a gate with a single fanout, assumed to consist of 450Ω connected to the low-impedance base of a saturated transistor. [Note that a fanout of zero is not normal in logic applications, and 2 or more lowers the gain, implying a need for higher V_{IL}]. Assume that v_{BE} = 0.700 V at i_C = 1 mA and that β = 50. [Hint: Proceed by finding the required r_π and the corresponding currents and voltages.] For two inputs operating simultaneously (for example joined), what does V_{IL} become?

14.6 For the RTL NAND logic gate shown in Fig. P14.7 on page 1232 of the Text, evaluate V_{IH} for each of the inputs under the condition that for V_{IH}, β_{forced} = $\beta_F/2$ and β_F = 50, β_R = 0.1 with V_{BE} = 0.700 V at i_E = 1 mA. [Hint: Since α_R is quite small, V_{BE} in saturation can be calculated approximately using the external emitter current.]

DL

14.7 Your boss is considering the possibility of raising the input threshold of the DTL circuit of Fig. 14.6. She asks you to help, by calculating, some important parameters, for operation with input B high and input A controlling:

a) The base current in Q required to lower the output to $V_{DD}/2$. Use $\beta = 30$.

b) The input threshold voltage V_{th} (at A) corresponding. All junctions have a 0.7 V drop at 1 mA.

c) The maximum base current available to Q when A is high.

d) The input current flowing from D_1 when A is at 0 V.

Now, she suggests that you add a diode and change a resistor so that V_{th} is raised by about 0.7 V, while the maximum base drive remains the same. What change do you make? What is the input current needed now when $V_A = 0$? What is the fanout available with this redesign, for which $\beta_{forced} \leq \beta/2$ is maintained?

D

14.8 Following the general direction suggested by the two-input NAND DTL gate in Fig. 14.6 of the Text, sketch a circuit using only a single transistor, but many diodes, to provide the function $Y = \overline{ABC + D + EF}$ using voltage and resistor values the same as those in Fig. 14.6. What is the base current which results when D alone is high? When all inputs are high?

SECTION 14.3: TRANSISTOR-TRANSISTOR LOGIC (TTL OR T^2L)

D

14.9 Modify the design of the IC-form DTL gate in Fig. 14.7 to double the turnoff current of Q_3. Using the results of Ex. 14.4 on page 1168 of the Text, what does the turnon base current of Q_3 become? What is the absolutely greatest fanout N that ensures that Q_3 is barely saturated? What does V_{OL} become for ¼ that value of fanout? Use $\beta_R = \beta_F/100$, $\beta_F = 50$.

14.10 For the modification and current levels suggested in P14.9 above, and $\tau_s = 10$ ns, what does the storage delay become for $N = 0$? for N equal to ¼ of the maximum fanout? Recalculate these values in the event that β is doubled (from 50 to 100), but with the N values kept the same.

14.11 Consider the output stage of a DTL gate, consisting of a transistor with $\beta_F = 50$, operating at forced β of 10, with a load resistor of 2 kΩ connected to a +5 V supply. Assume $V_{CEsat} = 0.2$ V, what is the minimum value of external load resistance connected to +5 V, that still ensures saturation? For a resistor of twice that value, calculate the 10% to 90% rise and fall times of the output with a 10 pF load capacitor.

D*L

14.12 Consider the circuit of Fig. 14.9 of the Text as the basis of a very-low-voltage BJT logic structure. For $V_{BE} = 0.7$ V nominally, but 0.6 V at turnon, and $V_{CE} = 0.2$ V in saturation, with $\beta_F = 40$ and $\beta_R = 0.1$, prepare a design meeting the following specifications: a) the total collector current that can be sustained by Q_3 with $\beta_{forced} = \beta_F/2$ is ≥ 20 mA, b) resistors of a single value, R, are used, c) V_{CC} = V is chosen as small as possible, d) $NM_H \geq 1.5\ NM_L$ with fanout $N = 10$ for which $\beta_{forced} \leq 20$. For what value of N does Q_3 reach the edge of saturation with v_I high?

D

14.13 Extend the structure of the logic gate of Fig. 14.9 of the Text to provide a logic gate to perform the following function: $Y = \overline{A\ B + C\ D}$ while maintaining the same input thresholds. Use 6 transistors and 3 resistors in total.

14.14 A manufacturing-process deviation in the production of T^2L gates using the circuit of Fig. 14.19 of the Text, reduces current gain such that $\beta_F = 9$ and $\beta_R = 0.05$. For input high, estimate all node voltages and branch currents, for $V_{BE} = 0.7$ V, and a load of 1kΩ connected to the 5 V supply. What is the largest possible fanout (excluding the 1kΩ load), for which saturation of Q_3 is still possible?

14.15 Repeat the analysis suggested in P14.14 above, with input low (at 0.3 V) and a resistor of 1 kΩ connected from the output to ground.

L*

14.16 For the situation described in P14.14 above, and with input and output joined by a 200Ω resistor, estimate all the node voltages and branch currents. Calculate $V_{CE\,sat}$ for Q_1 relatively precisely using a negative value for β_{forced}.

L*

14.17 Modify your response to P14.16 above for the situations in which a) a load resistor of 200Ω is connected from the output to: a) ground, b) +5 V.

SECTION 14.4: CHARACTERISTICS OF STANDARD TTL

14.18 Using a similar analysis style to that found following Eq. 14.6 on page 1180 of the Text, find the value of R_2 which raises point C of Fig. 14.23b) to 3.0 V. What does the slope of the BC segment become? If R_2 is to be kept at 1kΩ by a desire to maintain the turnoff current level that R_2 provides, what change in R_1 would be needed to raise C to 3.0 V? What change in turnon current to Q_3 does this produce (in absolute value and as a percentage)? What is the effect on gate storage delay? And so you see, once again, the nature of compromise in real design!

14.19 Using the data provided in the answers to Exercise 14.11 on page 1182 of the Text, find the noise margins that apply at the interface between two sets of T^2L logic gates, one operating at –55°C and the other at 125°C. Note that there are two pairs of margins depending on the relative temperatures of the driving and driven gates.

14.20 For the circuit of Fig. 14.26 of the Text extended to as have 4 *OR* inputs, what is the maximum base current provided to Q_3? In a particular circuit with values shown, and no load, with 3 of the 4 *OR* inputs already low, the storage delay is 10 ns. What delay would you expect when all 4 inputs are brought low simultaneously?

14.21 For the tristate gate shown in Fig. E14.16 on page 1187 of the Text, find the voltage at the tristate input at which the collector current in Q_6 just reaches 1 mA. For all junctions, the voltage drop is 0.700 V at 1 mA, $\beta_F = 50$ and $\beta_R = 0.1$.

14.22 For the tristate gate shown in Fig. E14.16 of the Text, estimate the possible current flow in the base of Q_3 if the connection from the tristate input is low, but the corresponding link to Q_1 is broken. Use $\beta_F = 50$, $V_{BE} = 0.7$ and $V_{CE\,sat} = 0.2$ V.

SECTION 14.5: TTL FAMILIES WITH IMPROVED PERFORMANCE

14.23 A Schottky npn transistor consists of 2 elements: a BJT for which $I_E = 1$ mA at $V_{BE} = 0.75$ V with $n = 1$ and $\beta = 50$, and a Schottky diode for which $I = 1$ mA at 0.5 V with $n = 1$. For the emitter grounded, find the base and collector voltages and transistor base and diode currents, for input and load currents, respectively, of

a) 1 mA, 0 mA,

b) 1 mA, 1 mA,

c) 1 mA, 10 mA,

d) 10 mA, 10 mA,

e) 10 mA, 1 mA.

L

14.24 Consider the Schottky TTL circuit of Fig. 14.28 in the Text, with both inputs A and B high. For devices as specified in P14.23 above, find the voltages at the base and collector of Q_3, Q_2, Q_1 and Q_6. Assume that the circuit has a fanout of two similar gates.

D

14.25 Consider the active-pulldown circuit shown in Fig. 14.28 of the Text. Find the value of R_5 for which half the emitter current from Q_2 flows in the base input of Q_3, but only 1 mA is required to cause the SBD in Q_3 to conduct.

C

14.26 Consider the output stage of the Schottky TTL gate shown in Fig. 14.28 of the Text. For both inputs high and a current I flowing into the output, find the incremental output resistance for a) I = 1 mA, b) I = 10 mA. Use the device data provided in P14.23 above.

L

14.27 For the low-power Schottky TTL gate of Fig. 14.31, find the current that flows in the power supply with a) inputs both high, or b) inputs both low, both for the output i) open-circuited or ii) short-circuited to ground. What is the power dissipated in the gate under all 4 conditions? For a propagation delay of 10 ns average, what is the delay-power product for operation with a 10 pF load at 30 MHz?

SECTION 14.6: EMITTER-COUPLED LOGIC (ECL)

L

14.28 Reconsider P14.38 on page 1235 of the Text, for the situation in which V_{IL} and V_{IH} are based on currents ranging from 0.95I to 0.05I.

14.29 Consider Fig. P14.38 on page 1235 of the Text, for the situation in which V_{BE} = 0.75 at current I = 4 mA. Find R so that V_{th} = −1.32 V. What are the values of V_{OH} and V_{OL} that result? Find V_{IH} and V_{IL} for a current split in Q_R and the input transistor in the ratio of 1000 to 1. What are the corresponding noise margins?

D

14.30 Modify the circuit of Fig. P14.38 on page 1235 of the Text to create an ECL-to-T^2L converter by connecting the upper supply connections (now grounded) to +5 V, and adding a number of 0.75 V diodes in series with the emitters of Q_2 and Q_3 (to lower the output voltage). Maintain the input threshold at −1.32 V by a suitable choice of $R/2$ (still connected to ground) with I = 4 mA, and corresponding V_{BE} = 0.75V. Select the resistors (called R_1) connected to the bases of Q_1 and Q_2, and the number of diodes, N, to meet T^2L worst-case output specifications, namely V_{OL} = 0.5V and V_{OH} = 2.7 V, while keeping R_1 as small as possible.

14.31

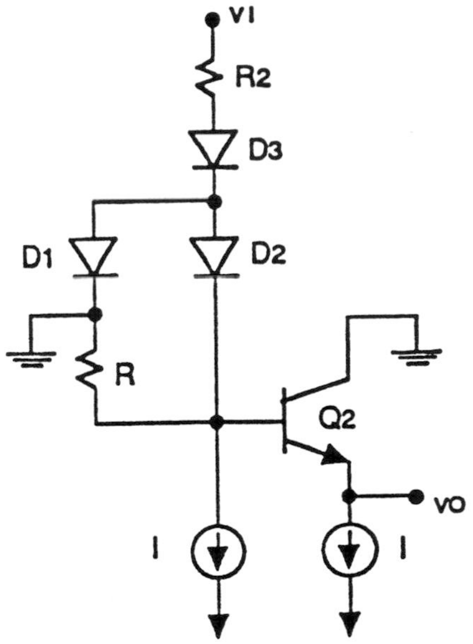

Consider the circuit shown as a T^2L-to-ECL converter: Use values of I and R as specified and calculated in P14.29 above. Arrange that the current from the T^2L gate is 8 mA when its output is at the minimum specified high value of output, ie V_{OH} = 2.7 V. What is the T^2L output current at $V_{OL} \leq 0.5$ V? What are V_{OL} and V_{OH} of the converter circuit? Use 0.75 V for all junctions, when operated at 4 mA.

14.32 For the circuit of Fig. 14.37 of the Text, a manufacturing error reduces the junction size of Q_2 by a factor of 2 and its β to 30. What is the corresponding effect on NM_H?

14.33 For the circuit of Fig. 14.35 of the Text, calculate the small-signal voltage gain from input to OR output for v_I biased at V_R.

14.34 Estimate the propagation delays expected to the OR output of the ECL gate of Fig. 14.35 of the Text, loaded with a single fanout for which C_{EQ} = 3 pF. Assume that the capacitance at the output of an unloaded gate is 2 pF, and that for the transistors, f_T = 5 GHz and C_μ = 0.1 pF. What do the delays become for a fanout of 10?

14.35 For signals whose rise and fall times are 1.2 ns, what length of unterminated gate-to-gate interconnect can be used if a ratio of rise time to return time of 6-to-1 is required. Assume that the signal propagates at 2/3 the speed of light (which is 0.3 mm/ps or 300 μm/ps).

D

14.36 Consider a version of the ECL gate in Fig. 14.33 of the Text, in which resistors R_E, R_2, R_3 are replaced by current sources. What values would you use in each case? One other resistor must be changed. Which one? To what value? Using the techniques in Example 14.2 on page 1208 of the Text, evaluate the temperature-related changes associated with V_{OH}, V_{OL} and V_R for this new circuit. Are there any other changes you would suggest?

SECTION 14.7: BICMOS DIGITAL CIRCUITS

14.37 For the circuit of Fig. 14.44c) of the Text using $V_{DD} = 5V$, with $k_p = \frac{1}{2}k_n = k$, $|V_t| = 1$ V, and with $R_1 = R_2 = r_{SDp}$ at small $|V_{SDp}|$ and $|V_{SGp}| = V_{DD}$, find V_{th}, V_{OH} and V_{OL} with a) no load, b) a 5 kΩ load to 2.5 V. Assume $\beta = 100$, $V_{BE} = 0.7$ V, and $k = 400\mu A/V^2$.

D

14.38 The circuit of Fig. 14.44d) of the Text has the apparent advantage over the circuits of Figs. 14.44c) and d) of using no resistors, which generally occupy far greater area than a MOS device. However the R-circuit in Fig. 14.44e) has the additional advantage of eventual full-swing outputs. Show how the circuit in Fig. 14.44d) of the Text can be augmented by an additional CMOS inverter connected between input and output to achieve full output swing. For devices as specified in P14.37 above, what are the average equivalent resistances provided at the output for the last 0.7 V of output swing?

CD*L**

14.39 A set of 3.3 V CMOS matched-logic circuits whose device specifications are given in the introductory NOTE to the the problems of Chapter 13 in this book, are to be augmented by BJT output devices. Design such an inverter circuit patterned after that in Fig. 14.44d) of the Text, using predominantly minimum-size matched-inverter-scaled MOS devices and npn transistors for which $V_{BE} =$ 0.7 V at full conduction, $V_{BE} =$ 0.5 V at the edge of conduction, and $\beta =$ 50. For the resulting circuit, estimate V_{OL}, V_{OH}, (both for immediate and longer-term responses), V_{IL} and V_{IH}. For a load capacitor of 10 pF, what are the peak charging currents in each direction? What are the available currents at $\upsilon_O = V_{DD}/2$? Estimate the propagation delay, assuming that it is dominated by the output circuit. If the circuit is augmented by a matched inverter connected between υ_I and υ_O, what do V_{OH} and V_{OL} become on a short-term and a long-term basis? What is the time needed for a full-swing signal to be established [to, within, say, 0.2 V of the power rails]?

14.40

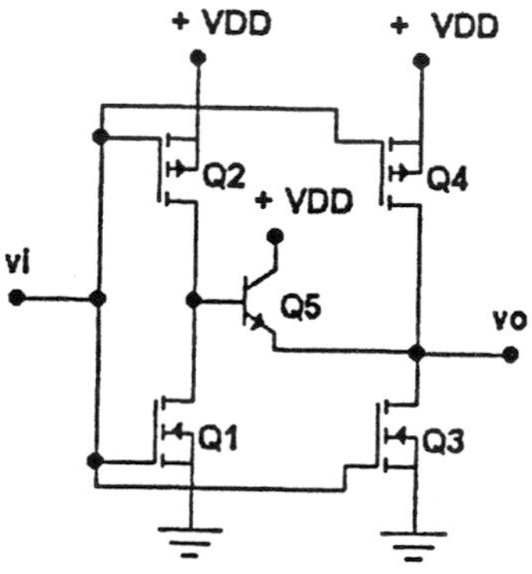

In the 3.3 V circuit shown, MOS devices are the same as in P14.39 above, with $|V_t| =$ 0.6 V, $\mu_n C_{ox} = 2.5\mu_p C_{ox} =$ $100\mu A/V^2$, with $(W/L)_1 =$ 1.2 μm/0.8 μm, $(W/L)_2 = (W/L)_3 = 2.5(W/L)_1$, and $\beta =$ 50. Find V_{OL}, V_{OH}, t_{PLH}, and t_{PHL} for a 10 pF load. What supply current flows if $\upsilon_O = V_{DD}/2$?

D

14.41 Following the general direction indicated in Fig. 14.46 of the Text, provide a circuit sketch of a 2-input BiCMOS NOR gate. For Q_P and Q_N of the basic inverter having the same W/L ratio, what are the device ratios in an equivalent NOR gate? For one input low, what is the threshold voltage for the other input, namely that voltage at which two of the 3 active FETs operate in saturation? Assume $\mu_n/mu_p =$ 2.0, and $|V_t| =$ 1 V.

SECTION 14.8: GALLIUM ARSENIDE DIGITAL CIRCUITS

14.42 Repeat the *dc* analysis of a CDFL NOR gate as sketched in Example 14.3 of the Text, for the conditions stated there, except with a reduction of the load MESFET width to 5μm. Find V_{OH}, V_{OL}, V_{IL}, V_{IH}, NM_H and NM_L.

14.43 For the situation described in P14.42 above, find the average supply current, the average static power dissipation, the average propagation delay for a 30 fF equivalent load, the dynamic power loss at 2GHz, and the corresponding delay-power product.

DL*

14.44 For the *FL* gate produced in Fig. 14.57 of the Text, whose transfer characteristic is shown in Fig. 14.52 on page 1222 there, evaluate the sensitivity of V_{OH}, V_{OL}, V_{IL}, V_{IH}, NH_L, and NM_H to some of the various device parameters. For the nominal design, all $L =$ 1μm, $W_S = W_L =$ 20μm, $W_{PD} =$ 10μm, β (per μm) $= 10^{-4}A/V^2$, $V_{tD} = -$ 0.9 V, $\lambda =$ 0, $V_{DD} = +$ 3 V, $V_{SS} =$ 2 V. In particular, conisder (separate) changes, as follows: a) all V_{tD} to –0.8 V, b) all V_{tD} to – 1.0 V, C) $W_L = 2W_{PD}$ to 5 μm. Note that the latter design choice will also make propagation delays quite asymmetric.

D

14.45 Using the FET-sizing ideas associated with the *FL* design of Fig. 14.51 of the Text, provide a directly-corresponding design for the SDFL NOR in Fig. 14.53b) of the Text. Using the results for the *FL* design, find V_{OL}, V_{OH}, V_{IL}, V_{IH} and noise margins for your SDFL gate. Note that the design requested, for which the device-size ratios correspond directly to those used for Fig. 14.51, will have a problem with fanout. What can be done to increase the fanout from 1 to 4?

PART II
SOLUTIONS

pages 133 to 425

Chapter 1

INTRODUCTION TO ELECTRONICS

SECTION 1.1: SIGNALS

1.1 **Results**: (See "Rough Work/Notes" at the end).

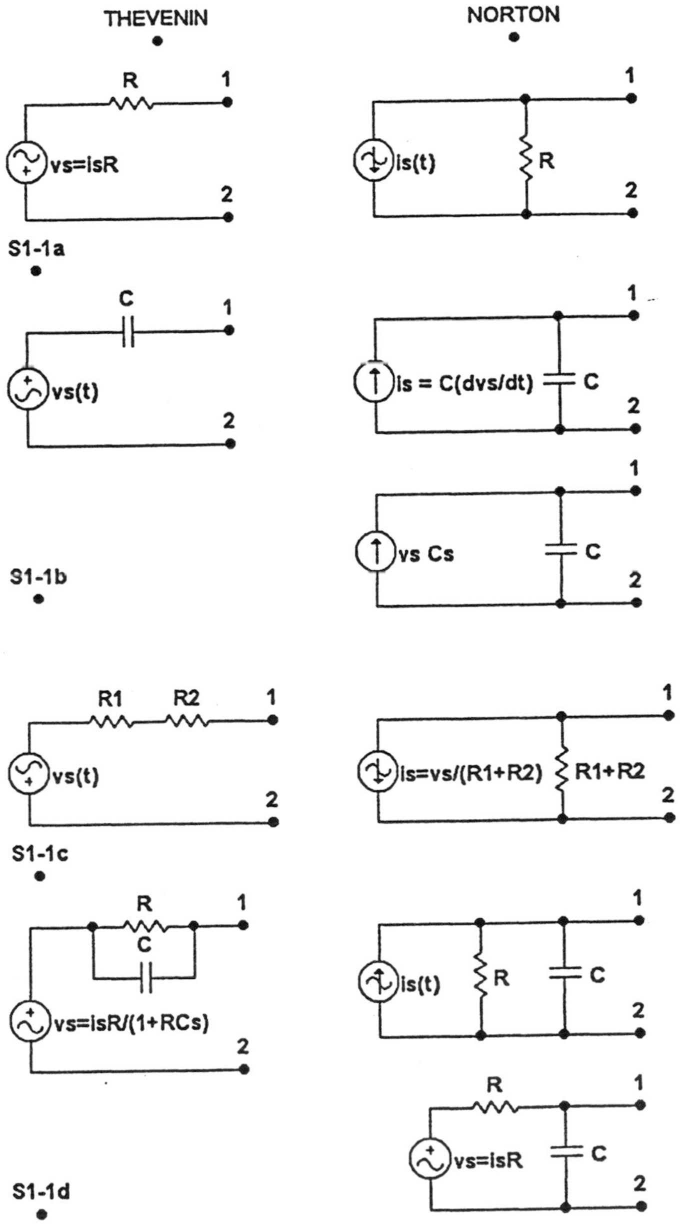

1.1 (continued)

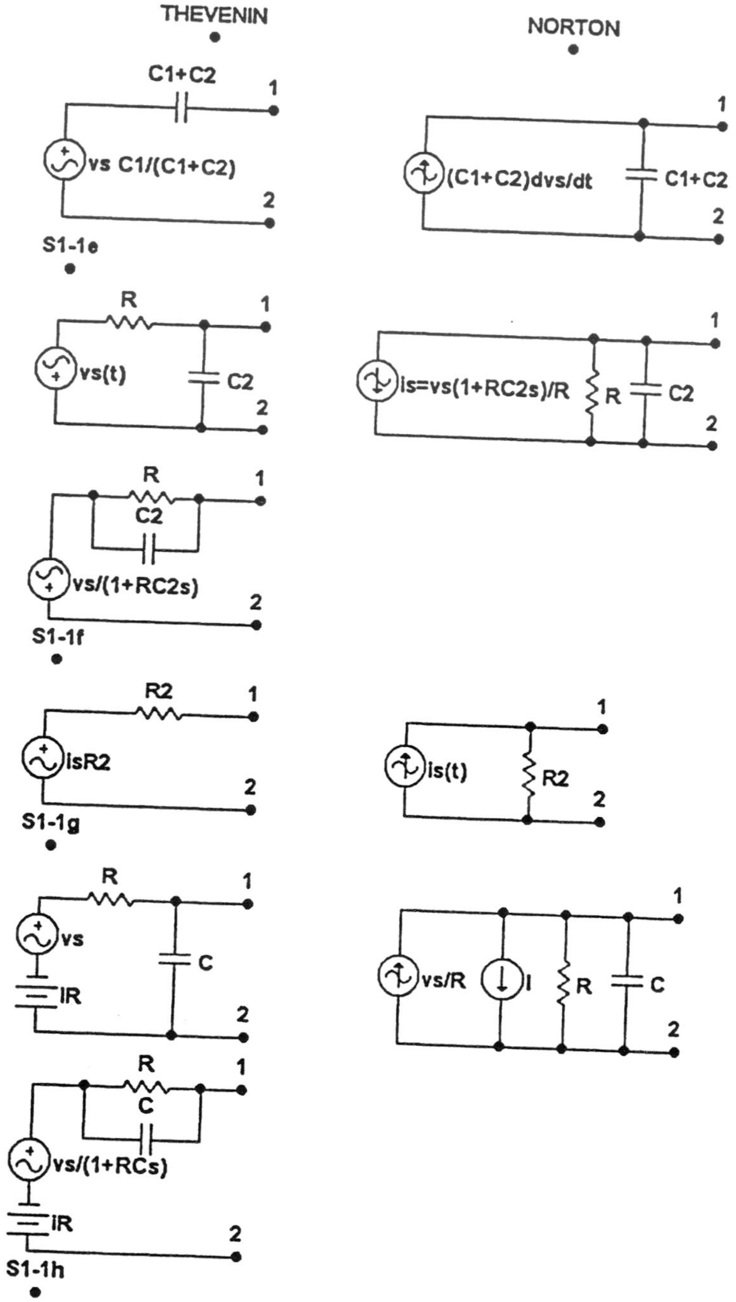

Rough Work/Notes:

a) Simply recall that the Thevenin generator is the open-circuit voltage, while the Norton generator is the short-circuit current:

b) The Norton equivalent is presented in the time-domain by the derivative, and in the complex frequency domain using Laplace-transform notation.

d) Converting only the resistive part to the Thevenin form or Norton form avoids the use of complex frequency notation.

SECTION 1.2: Frequency Spectrum of Signals

1.2 **Results:** In general, $\omega = 2\pi f$, 2π rad/s = 1 Hz = 6.283 rad/s, and 1 rad/s = 0.159 Hz. Note that the rightmost columns are the results from P1.3.

Line Label	Frequency (Hz)	Frequency (rad/s)	**Period (s)**	**Period**
a	60	**377**	**1.67×10^{-2}**	**16.7 ms**
b	**120.0**	754	**8.33×10^{-3}**	**8.33 ms**
c	**400.0**	2513.3	**2.50×10^{-3}**	**2.50 ms**
d	1.01×10^{6}	**6.346×10^{6}**	9.90×10^{-7}	990 ns, **0.99 μs**
e	97.3×10^{6}	**611.4×10^{6}**	1.03×10^{-8}	**10.3 ns**
f	1	**6.28**	**1.00**	**1.00 s**
g	**60.0**	377	**1.67×10^{-2}**	**16.7 ms**
h	**0.159**	1	**6.29**	**6.29 s**
i	10^{9}	**6.28×10^{9}**	**1.00×10^{-9}**	**1.00 ns**
j	4×10^{11}	**25.1×10^{11}**	**2.50×10^{-12}**	**2.50 ps**

1.3 **Results:** These are tabulated in the two rightmost columns of the table above.

Examples:

a) For 60 Hz, period $= \frac{1}{60} =$ **$0.016\dot{6}$** seconds $= 1.66 \times 10^{-2}$ s $= 16.\dot{6}$ or **16.7 ms.**

g) For 377 rad/s, corresponding to $\frac{1}{377} = .00265$ sec/rad, period $= 2\pi$ rad $= 2\pi \times .00265 =$ **0.01665 s** $\equiv$ **16.7 ms.**

j) For 400 GHz, period $= 1/(400 \times 10^{9}) = .0025 \times 10^{-9} = 2.50 \times 10^{-12} =$ **2.50 ps.**

Conclusion: Clearly, dealing with frequency is easier, either directly from the specification in Hz, or from the tabulated calculation derived from rad/s.

1.4 $\Delta T = 50 - 25 = 25C°$.

Period of a 10.7 MHz wave = $1/(10.7 \times 10^6) = 93.46$ ns.

Total variation in the period would be $93.46 \times 10^{-9} \times 3 \times 10^{-6} \times 25 = 7.009 \times 10^{-12} \equiv$ **7.01 ps**.

1.5 a) 0.2 V peak-to-peak ≡ 0.1 V peak ≡ $0.1/\sqrt{2} = 0.0707$ Vrms.

1000 Hz ≡ $2\pi \times 10^3$ rad/s = 6.28×10^3 rad/s, with a period of 1/1000 = 1 ms. Since this is the reference: Amplitude Ratio is 1 times; Frequency Ratio is 1 times.

b) 2.12 Vrms, 20μsec period.

Amplitude ratio, $\dfrac{V_b}{V_a} = \dfrac{2.12}{.0707} = 29.98$, or about **30 times.**

Frequency ratio, $\dfrac{f_b}{f_a} = \dfrac{1}{\textit{Period ratio}} = \dfrac{T_a}{T_b} = \dfrac{1\times10^{-3}}{20\times10^{-6}} =$ **50 times.**

c) 1.0 V peak amplitude, 12.57 rad/s frequency.

Amplitude ratio, $\dfrac{V_c}{V_a} = \dfrac{1.00}{0.1} =$ **10 times.**

Frequency ratio, $\dfrac{f_c}{f_a} = \dfrac{12.57}{2\pi \times 10^3} = 2\times10^{-3} = \dfrac{2}{1000} =$ **1/500 times.**

1.6 The Fourier series for a square wave of frequency f and 10 V peak amplitude is (from Eq. 1.2 of the Text):

$\upsilon(t) = 4(10)/\pi\ (\sin 2\pi ft + 1/3 \sin 3(2\pi ft) + 1/5 \sin 5(2\pi ft) + 1/7 \sin 7(2\pi ft) + 1/9 \sin 9(2\pi ft)$
$+ 1/11 \sin 11\,(2\pi ft) + \ldots)$.

Energy per unit time in a voltage wave $\upsilon(t)$ of duration τ, across a unit load = $\dfrac{1}{\tau}\displaystyle\int_o^{\tau} \upsilon^2(t)dt$.

In one cycle of the square wave (period $T = 1/f$), associated energies are proportional to:

a) For the square wave: $(10^2)\ T = 100T$ Ws.

b) For the fundamental (first harmonic): $\left[\dfrac{40}{\pi}\dfrac{1}{\sqrt{2}}\right]^2 T = \left[\dfrac{40}{\pi}\right]^2 T/2,$

third harmonic: $\left[\dfrac{1}{3}\right]^2 \left[\dfrac{40}{\pi}\right]^2 T/2,$

fifth harmonic: $\left[\dfrac{1}{5}\right]^2 \left[\dfrac{40}{\pi}\right]^2 T/2,$

seventh harmonic: $\left[\dfrac{1}{7}\right]^2 \left[\dfrac{40}{\pi}\right]^2 T/2,$

ninth harmonic: $\left[\dfrac{1}{9}\right]^2 \left[\dfrac{40}{\pi}\right]^2 T/2.$

For the first 9 harmonics: $\left[\dfrac{40}{\pi}\right]^2 \dfrac{T}{2}\left[1 + \dfrac{1}{9} + \dfrac{1}{25} + \dfrac{1}{49} + \dfrac{1}{81}\right]$

$$= 1.184 \times \left[\frac{40}{\pi}\right]^2 \frac{T}{2} = 95.95\ T \text{ Ws.}$$

c) Above the ninth harmonic, total energy is proportional to $100\ T - 95.95\ T = 4.05\ T$, corresponding to $\frac{4.05T}{100T} \times 100 =$ **4.05%** of the total. (See page 200 following Chapter 3, Solutions, for a graphic view of this).

d) At and above the 3rd harmonic, the total energy is proportional to $100T - \left[\frac{40}{\pi}\right]^2 T/2$.

Of the total, this is: $$\frac{100T - (\frac{40}{\pi})^2 \frac{T}{2}}{100\ T} \times 100 = \left[100 - \left[\frac{40}{\pi}\right]^2 \frac{1}{2}\right] = \mathbf{18.9\%}.$$

1.7 For a square wave of amplitude V, $v = \frac{4V}{\pi}(\sin \omega t + 1/3 \sin 3\omega t + 1/5 \sin 5\omega t + \cdots)$.

Assume the pass band includes both fundamentals, i.e., $f > 2$ kHz, and *totally* excludes energy outside the band. For cutoff at $f = 4$ kHz, power levels for unit loads are:

$$P_1 = \left[\frac{4\times 1.1}{\pi}\right]^2 \left[1^2 + ()^2\right] = 1.96 \times 1.111 = 2.18 \text{ W, and}$$

$$P_2 = \left[\frac{4\times 1.2}{\pi}\right]^2 \left[1^2\right] = 2.33 \text{ W, where } P_2 - P_1 = 0.15 \text{ W.}$$

For cutoff at $f = 5+$, between 5 and 6 kHz,

$$P_1 = 1.96\left[1^2 + 1/3^2 + 1/5^2\right] = 2.26 \text{ W,}$$

$P_2 = 2.33$ W, where $P_2 - P_1 = 0.07$ W.

For cutoff at $f = 8$ kHz,

$$P_1 = 1.96\left[1 + (1/3)^2 + (1/5)^2 + (1/7)^2\right] = 1.96\ (1.172) = 2.29 \text{ W,}$$

$$P_2 = 2.33\left[1 + (1/3)^2\right] = 2.33\ (1.11) = 2.59 \text{ W, where } P_2 - P_1 = 0.29 \text{ W.}$$

See that the closest one can get to equal power is for filtering at a frequency **between 5 kHz and 6 kHz,** where the difference is 0.07 W, or $\frac{0.07}{(2.26 + 2.33)/2} \times 100$, or 3% of the average energy level.

SECTION 1.3: ANALOG AND DIGITAL SIGNALS

1.8

1 Vrms corresponds to $\sqrt{2}$ V or 1.414 V peak.

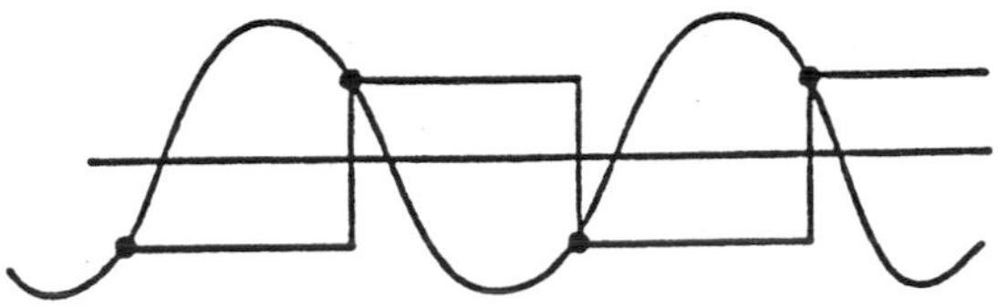

a) For sampling at the peaks, the square wave amplitude would be **1.414 V** peak or 2.818 Vpp.

b) 90° from a negative-going zero crossing is at a negative peak. The next sample is at the positive peak. The result is the same as a).

c) 45° from a positive-going zero crossing, the amplitude is $\sqrt{2}$ sin 45° = 1.00 V. The result would be a square wave of amplitude **1V** and frequency f.

For case a) and sampling at other frequencies:

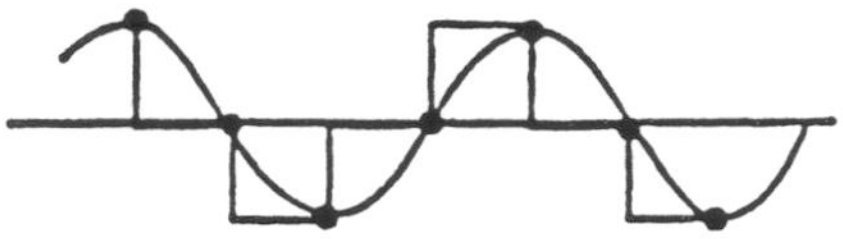

i) For sampling at $2(2f) = 4\ f$ Hz, the result is a sequence of positive and negative pulses of amplitude $\sqrt{2}$ V, width **1/4f**, spaced **1/4f** apart.

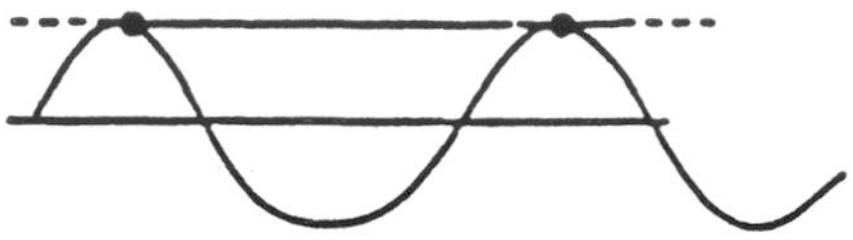

ii) For sampling at $\frac{1}{2}(2f) = f$ Hz, the result is a dc level of $\sqrt{2}$ V = **1.414 V**.

1.9 Five bits correspond to $2^5 =$ 32 values. **Six bits** are needed with values ranging from 000000 $= 0_{10}$ to 111111 $= 63_{10}$ where the subscript 10 indicats a radix 10 number, and the least-significant digit is at the right. Thus **63** is the largest value. Generally, for particular cases: b_5, b_4, b_3, b_2, b_1, b_0 represent $b_0 2^0 + b_1 2^2 r + \cdots + b_5 2^5$. Thus, $0_{10} = 000000_2$; $7_{10} \equiv 000111_2$, $15_{10} \equiv 001111_2$; $31_{10} = 011111_2$; and $33_{10} =$ **100001.**

1.10 The even numbers from 0 to 30 are: 0, 2, 4, 6, 8, 10, 12, 14, 16, 18, 20, 22, 24, 26, 28, 30, a total of 16 values in all. Thus, one could use only **4 bits** (for 16 values) with a recoding. For particular values: 0 ≡ **0000**; 8 ≡ (4) ≡ **0100**; 14 ≡ (7) ≡ **0111**; 28 ≡ (14) = **1110**. The biggest value is 1111 ≡ (15) ≡ **30.**

1.11

	← Most to Least Significant →							
Time	0	1	2	3	4	5	6	7
Bit #	7	6	5	4	3	2	1	0
Value	128	64	32	16	8	4	2	1

In order of presentation (MSB first, at time 0, at the left), the digits have weight magnitude 128, 64, 32, 16, 8, 4, 2, 1. The 8-digit binary number in Fig. 1.8 is 101110100: a) For all bits positive, its value is 128 + 32 + 16 + 4 = **180.** b) For MSB negative, its value is − 128 + 32 + 16 + 4 = **− 76.**

For the MSB digit reversed (becoming 0, at the left), the value is a) **52,** b) **52.**

1.12 *In order of presentation* (MSB last, at time 7, at the right), the digits have weights 1, 2, 4, 8, 16, 32, 64, 128. The number in Fig. 1.8 in time order is 10110100 with value. a) All digits positive: 1 + 4 + 8 + 32 = **45.** b) MSB negative: same, **45.** c) MSB as a sign (1 negative): same, **45.**

For the MSB digit reversed (becoming 1, at the right), the digits are 10110101 with values as follows: a) All digits positive: 45 + 128 = **173.** b) MSB negative: 45 = 128 = − **83.** c) MSB a sign: 45 × (− 1) = **− 45.**

1.13 For a 5 bit representation, the largest number is **11111** of value $2^5 = 1 = 32 - 1 =$ **31**, and the smallest is **00000** = **0**. In conventional form (MSB, at the left),

01101 = 0(16) + 1(8) + 1(4) + 0(2) + 1(1) = **13**. For a system where 10000 ≡ 3/2 V, 00001 ≡ (3/2)/16, and 01101 ≡ 13 × (3/2)/16 = **1.219 V**. The highest available output is 11111 ≡ 31/(3/32) = **2.906 V**. The smallest (non-zero) output is 1(3/32) = **0.09375 V**. *Check:* 31(0.09375) = 2.906 V, OK. *For closest to 1.000 V:* See 1.000/0.9875 = 10.67, for which the nearest integer is 11 and the nearest representation is 11(3/32) = **1.03125**. This corresponds to $11_{10} \equiv 01011_2$.

SECTION 1.4: AMPLIFIERS

1.14 For each amplifier in turn, beginning with b):

b) No dc connection to ground implies that $I+ = I- =$ **1 mA**:

$P = 2\ (1\text{ mA} \times 10\text{ V}) = \mathbf{20\ mW}$.

$$i_i = \frac{\upsilon_i}{R_{in}} = \frac{20\times10^{-3}}{.01\times10^{3}} = 2000\mu\text{A} = 2\times10^{3}\mu\text{A} = \mathbf{2\ mA}.$$

$$P_{in} = \frac{20}{\sqrt{2}} \times \frac{2\times10^{3}}{\sqrt{2}}\ \text{nW} = \mathbf{20\mu W}.$$

$$i_o = \frac{\upsilon_o}{R_L} = 1/1 = \mathbf{1\ mA}.$$

$$P_{out} = \frac{1}{\sqrt{2}} \times \frac{1}{\sqrt{2}} \times 10^{-3} = \mathbf{0.5\ mW}.$$

$$A_\upsilon = \frac{1}{20\times10^{-3}} = 50\text{ V/V} = .05\text{ V/mV} \equiv \mathbf{34\ dB}.$$

$$A_i = \frac{i_o}{i_i} = \frac{1\times10^{-3}}{2\times10^{3}\times10^{-6}} = 0.5\text{ A/A} = 0.5\times10^{-3}\text{ mA/}\mu\text{A} \equiv \mathbf{-6\ dB}.$$

$$A_p = \frac{P_{out}}{P_{in}} = \frac{0.5\times10^{-3}}{20\times10^{-6}} = 25\text{ W/W} = .025\text{ mW/}\mu\text{W} \equiv 10\log 25 = 14\text{ dB, i.e., } \frac{34-6}{2} = \mathbf{14\ dB}$$

$$Eff = \frac{P_{out}}{P} = \frac{0.5\times10^{-3}}{20\times10^{-3}} = 0.025 \equiv \mathbf{2.5\%}.$$

	Supply			Input				Output				A_v		A_i		A_p		Eff.
#	I_+ mA	I_- mA	P mW	υ_i mV	i_i μA	R_{in} kΩ	P_{in} μW	υ_O V	i_O mA	R_{load} kΩ	P_{out} mW	ratio V/mV	dB	ratio mA/μA	dB	ratio mW/μW	dB	%
a	3	3	60	1	1	1	0.0005	2	20	0.1	20	2	66	20	86	40×10^3	76	33
b	1	1	20	20	2×10^3	0.01	20	1	1	1	0.5	0.05	34	0.5×10^{-3}	−6	0.025	14	2.5
c	0.05	0.05	1	100	10^3	0.1	50	2	10	0.2	10	0.02	26	0.01	20	0.2	23	10
d	10	10	200	14.1	1.41×10^3	0.01	10	2.82	28.2	0.1	40	0.2	46	0.02	26	4	36	20
e	3.1×10^{-3}	3.1×10^{-3}	0.063	50	5	10	0.125	0.5	0.05	10	0.013	0.01	20	0.01	20	0.1	20	20

c) $P_{out} = 0.1$ mW, $Eff = 10\% = \dfrac{P_{out}}{P} \times 100.$

$P = 10\,P_{out} = 10\,(0.1) = \mathbf{1\ mW}.$

$I = \dfrac{P}{2(10)} = \dfrac{1\times10^{-3}}{2(10)} = \mathbf{0.05\ mA}.$

$\upsilon_i = i_i\,R_{in} = 10^3\times10^{-6}\times0.1\times10^3 = 0.1\text{ V} = \mathbf{100\ mV}.$

$P_{in} = \dfrac{100}{\sqrt{2}}\times10^{-3}\times\dfrac{10^3\times10^{-6}}{\sqrt{2}} = 5\times10^{-5}\text{ W} = \mathbf{50\ \mu W}.$

$p\upsilon_o = \dfrac{10\times10^{-3}}{10\sqrt{2}\times10^{-3}} = \sqrt{2}\text{ Vrms} = \mathbf{2\ V\ peak}.$

$R_{load} = \dfrac{2}{10\times10^{-3}} = 200\Omega = \mathbf{0.2\ k\Omega}.$

$A_\upsilon = \dfrac{2}{100\times10^{-3}} = \dfrac{2000}{100} = 20\text{ V/V} = .02\text{ V/mV} \equiv \mathbf{26\ dB}.$

$A_i = \dfrac{i_o}{i_i} = \dfrac{10\times10^{-3}}{10^3\times10^{-6}} = 10\text{ A/A} = .01\text{ mA}/\mu\text{A} \equiv \mathbf{20\ dB}.$

$A_p = \dfrac{10\times10^{-3}}{50\times10^{-6}} = 200\text{ W/W} = 0.2\text{ mW}/\mu\text{W} \equiv 10\log 200 = \mathbf{23\ dB}.$

d) $I = \dfrac{P}{V_+ + V_-} = \dfrac{200}{2(10)} = \mathbf{10\ mA}.$

$\upsilon_i = (P_{in}\,R_{in})^{1/2} = (10\times10^{-6}\times.01\times10^3)^{1/2} = 10^{-2}\text{ V} = 10\text{ mV rms} = \mathbf{14.1\ mV\ peak}.$

$$i_i = \frac{\upsilon_i}{R_{in}} = \frac{14.1 \times 10^{-3}}{.01 \times 10^3} = 1.41 \text{ mA} = 1.41 \times 10^3 \mu\text{A}.$$

$\upsilon_o = A_\upsilon \, \upsilon_i = 0.2 \times 10^3 \times 14.1 \times 10^{-3} = \mathbf{2.82\ V.}$

$$R_L = \frac{\upsilon_o^2}{P_{out}}$$

$$i_o = \frac{\upsilon_o}{R_L} = \frac{2.82}{0.1} = \mathbf{28.2\ mA.}$$

$A_\upsilon = 0.2 \times 10^3 \equiv 20\log_{10}(200) = \mathbf{46\ dB.}$

$$A_i = \frac{i_o}{i_i} = \frac{28.2 \times 10^{-3}}{1.41 \times 10^3 \times 10^{-6}} = 20 \text{ A/A} = 0.02 \text{ mA}/\mu\text{A} \equiv \mathbf{26\ dB.}$$

$$A_p = \frac{P_{out}}{P_{in}} = \frac{40 \times 10^{-3}}{10 \times 10^{-6}} = 4 \times 10^3 \text{ W/W} = 4 \text{ mW}/\mu\text{W} \equiv \mathbf{36\ dB.}$$

$$Eff = \frac{P_{out}}{P} = \frac{40}{200} \times 100 = \mathbf{20\%.}$$

e) $$i_o = \frac{\upsilon_o}{R_L} = \frac{0.5}{10 \times 10^3} = \mathbf{0.05\ mA\ peak.}$$

$P_{out} = \upsilon_o i_o = (0.5/\sqrt{2}) \times (.05 \times 10^{-3}/\sqrt{2}) = \mathbf{0.0125\ mW.}$

$$P_{in} = \frac{P_{out}}{A_p} = \frac{.0125 \times 10^{-3}}{0.1 \times 10^3} = 0.125\mu\text{W}.$$

$A_p = 0.1 \times 10^3 \equiv 10\log_{10}(100) = \mathbf{20\ dB.}$

$\upsilon_i = (R_{in} P_{in})^{1/2} = (10 \times 10^3 \times .125 \times 10^{-6})^{1/2} = (1.25 \times 10^{-3})^{1/2} = .0354$ V = 35.4 mV rms = 35.4 × 1.414 = **50 mV peak.**

$$i_i = \frac{\upsilon_i}{R_{in}} = \frac{50\text{mV}}{10 \times 10^3} = 5\mu\text{A}.$$

$$A_\upsilon = \frac{\upsilon_o}{\upsilon_i} = \frac{0.5}{50 \times 10^{-3}} = 10 \text{ V/V} = 0.01 \text{ V/mV} \equiv \mathbf{20\ dB.}$$

$$A_i = \frac{i_o}{i_i} = \frac{.05 \times 10^{-3}}{5 \times 10^{-6}} = 10 \text{ A/A} = 0.01 \text{ mA}/\mu\text{A} \equiv \mathbf{20\ dB.}$$

$$P = \frac{P_o}{Eff} = \frac{.0125 \times 10^{-3}}{20/100} = \mathbf{0.0625\ mW.}$$

$$I = \frac{P}{V_+ + V_-} = \frac{0.0625 \times 10^{-3}}{2(10)} = 3.125\mu\text{A} = 0.003125 \text{ mA} = 3.1 \times 10^{-3} \text{ mA.}$$

1.15 Largest undistorted positive output signal is 7V peak.

Largest undistorted output sine wave can be 7 volt peak.

Corresponding input $= \frac{7.0}{50} = 140$ mV peak.

Largest sine wave input (having no dc component) is **140 mV peak.**

1.16 For the largest possible unclipped output, center the output between +7 V and −9 V. The corresponding sine wave has a peak voltage of (7 − −9)/2 = 8 V and an rms of $8/\sqrt{2}$ = **5.66 V**, with an offset of +7 −8 = **−1 V.**

Required dc input offset = −1/50 = **−20 mV.**

Required ac input signal = 8/50 = +160 mV peak, or **113 mV rms.**

1.17 For $V_O = 4$ V, $V_O = 8 - 4(V_I - 1)^2 = 4$, $-4 = -4(V_I - 1)^2$, $(V_I - 1)^2 = 1$, $V_I - 1 = \pm 1$, and $V_I = 0$ or 2 V, with 0 V forbidden.

Now, for a sine wave input $\upsilon_i = V_i \cos\omega t$ correctly biassed, $\upsilon_I = 2 + V_i \cos \omega t$, where,

$$\upsilon_O = 8 - 4(2 + V_i \cos \omega t - 1)^2 = 8 - 4(1 + V_i \cos\omega t)^2 = 8 - 4(1 + 2V_i \cos\omega t + V_i^2 \cos^2 \omega t)$$
$$= 8 - 4(1 + 2V_i \cos \omega t + V_i^2 (1 + \cos 2\omega t)/2)$$
$$= 4 - 8V_i \cos \omega t - 2V_i^2 - 2V_i^2 \cos 2\omega t$$
$$= 4 - 2V_i^2 - 8V_i \cos \omega t - 2V_i^2 \cos 2\omega t.$$

Now, for an output signal (at the input frequency ω) ≤ 1 V peak, $8V_i \leq 1$, $V_i \leq 0.125$ V.

Now, % 2nd harmonic distortion $= \dfrac{2V_i^2}{8V_i} \times 100 = \dfrac{2(.125)^2}{8(.125)} \times 100 = \dfrac{.125}{4} \times 100 =$ **3.125%.**

1.18 $\upsilon_O = 5 - 10^{-10} e^{40\upsilon_I}$ for $\upsilon_I > 0V$, $\upsilon_O \geq \upsilon_I$.

υ_O is largest when $\upsilon_I = 0$, at which point $\upsilon_O = L+ = 5 - 10^{-10} e^{40(0)} =$ **5 V.**

For bias at $V_O = 5/2$, $5/2 = 5 - 10^{-10} e^{40 V_I}$, for which

$e^{40 V_I} = 2.5 \times 10^{10}$, and $V_I = \dfrac{\ln 2.5 \times 10^{10}}{40} =$ **0.598 V.**

Now for $L-$: $\upsilon_O = \upsilon_I = 5 - 10^{-10} e^{40\upsilon_I}$.

Solve Iteratively: $\upsilon_I = \dfrac{\ln (5 - \upsilon_I)10^{10}}{40}$, with $\upsilon_{I0} = 0.6$V. Thus $\upsilon_{I1} = \dfrac{\ln (5 - 0.6)10^{10}}{40} = 0.612$V, and

$\upsilon_{I2} = \dfrac{\ln (5 - 0.612)\ 10^{10}}{40} = 0.613$ V. See convergence: $L- =$ **0.613 V.**

Peak sine-wave allowed is limited by $L-$ to $2.5 - 0.613 =$ **1.89 V** (peak).

Gain, $A_\upsilon = \dfrac{d\,\upsilon_O}{d\,\upsilon_I}\Big|_{V_O = 2.5} = \dfrac{d}{d\,\upsilon_I}\left[5 - 10^{-10} e^{40\upsilon_I}\right]\Big|_{V_I = 0.598} = -10^{-10}(40)\, e^{40\upsilon_I}\Big|_{V_I = 0.598}$.

i.e., $A_\upsilon = -10^{-10}\left[40\, e^{40(0.598)}\right] =$ **−97.8 V/V.**

SECTION 1.5: CIRCUIT MODELS FOR AMPLIFIERS

1.19 $A_{\upsilon L} = A_{\upsilon o} \times \dfrac{R_L}{R_L + R_o}$. Thus $\dfrac{R_L + R_o}{R_L} = 1 + \dfrac{R_o}{R_L} = \dfrac{A_{\upsilon o}}{A_{\upsilon L}}$,

and $R_o = R_L\left[\dfrac{A_{\upsilon o}}{A_{\upsilon L}} - 1\right] = 1k\left[\dfrac{100}{70} - 1\right] = 0.429$ kΩ. Use $R_o \approx$ **430 ohms.**

For a 500Ω load, gain $= 100 \times \dfrac{500}{500 + 429} =$ **53.8 V/V.**

1.20

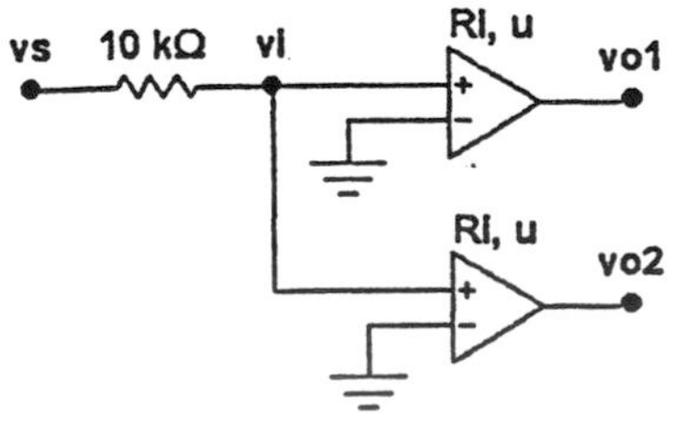

Originally, $\upsilon_o = 1667\,\upsilon_s = \dfrac{R_i}{R_i + 10k}\,\mu\upsilon_s$,

whence $\mu\,\dfrac{R_i}{R_i + 10} = 1667$ V/V − − − (1).

Now, with a second amplifier connected,

$\mu\,\dfrac{R_i/2}{R_i/2 + 10} = 909$V/V − − − (2).

$$(1)/(2) \rightarrow \frac{\dfrac{R_i}{R_i + 10}}{\dfrac{R_i/2}{R_i/2 + 10}} = \frac{1667}{909} = \frac{R_i + 20}{R_i + 10}.$$

Thus, $1667\, R_i + 16670 = 909R_i + 18180$, and $R_i = \dfrac{18180 - 16670}{1667 - 909} = 1.99\ \text{k}\Omega \approx \mathbf{2.0\ k\Omega}$.

1.21

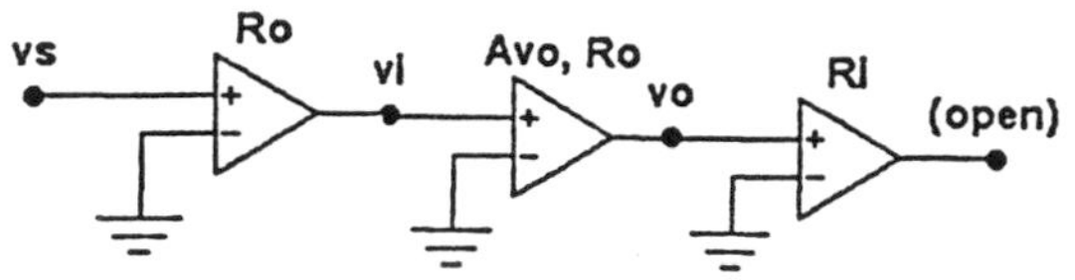

Gain of an internal element in the cascade $= \dfrac{R_i}{R_i + R_o} \times A_{vo}$. For the condition stated, this must be 1.

Thus, $\dfrac{R_i}{R_i + R_o} \times A_{vo} = 1$, or $\mathbf{A_{vo} = 1 + \dfrac{R_o}{R_i}}$.

Particular Cases:

See, for $R_o = 0$, $A_{vo} = 1$;

for $R_o = R_i$, $A_{vo} = 2$;

for $R_i = \infty$, $A_{vo} = 1$;

for $A_{vo} = 11$, $R_o = 10\, R_i$.

1.22

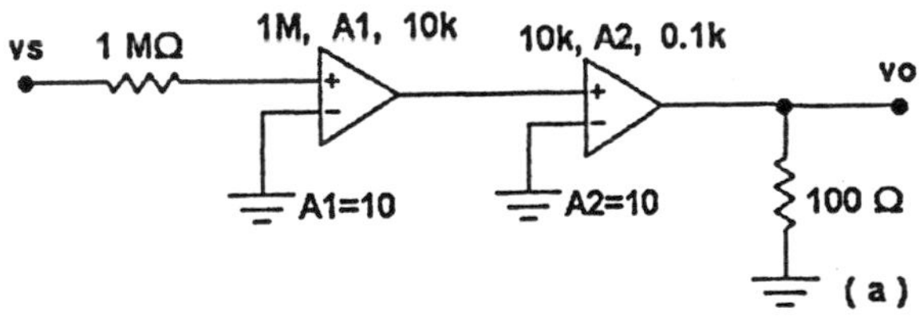

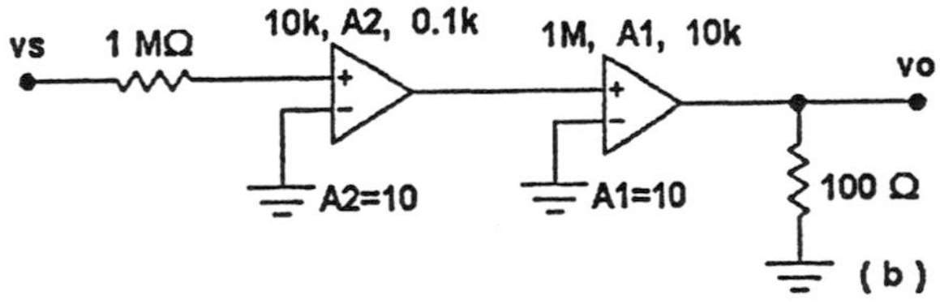

$$\text{Gain } (A_1 A_2) = \frac{100}{100 + 100} \times 10 \times \frac{10k}{10k + 10k} \times 10 \times \frac{1M}{1M + 1M} = 10^2 \left(\frac{1}{2^3}\right) = \mathbf{12.5\ V/V}, \text{ the best,}$$

where:

$$\text{Gain } (A_2 A_1) = \frac{100}{100 + 10^4} \times 10 \times \frac{10^6}{100 + 10^6} \times 10 \times \frac{10^4}{10^4 + 10^6} = 10^2 \left(\frac{1}{101}\right)^2 \times (1) \approx \mathbf{0.01\ V/V}.$$

$$\text{Gain } (A_1) = \frac{10 \times 100}{100 + 10^4} \times \frac{10^6}{10^6 + 10^6} = 10\, \frac{1}{101} \times \frac{1}{2} = \mathbf{0.05V/V}.$$

$$\text{Gain } (A_2) = 10 \times \frac{100}{100 + 100} \times \frac{10^4}{10^4 + 10^6} = 10 \left(\frac{1}{2}\right) \left(\frac{1}{101}\right) = \mathbf{0.5V/V}.$$

$$\text{Gain (wire)} = 1 \times \frac{100}{100 + 10^6} \approx \frac{1}{10^4} = \mathbf{0.0001\ V/V.}$$

Note how important source-to-load matching can be: Only one of 4 possible one-or-two-amplifier combinations does any good; *and* one two-amplifier arrangement is worse than either one-amplifier arrangement!

Choices: **double gain, double input resistance, halve output resistance.**

In general, a change in an input or output resistance by a particular factor provides less than that factor of improvement, due to the comparison process inherent in a voltage divider. Thus **change of gain has the greatest affect** for a given factor. Next choice would be output resistance (reduction by 2) since this implies reduced power loss in many applications.

1.23 80 dB $\equiv 10^{80/20} = 10^4$ V/V:

For 1MΩ load, $A_v = 10^4 \frac{10^6}{10^4 + 10^6} = 0.99 \times 10^4$ **V/V.**

For 10 kΩ load, $A_v = 10^4 \frac{10^4}{10^4 + 10^4} = 0.5 \times 10^4$ **V/V.**

For 10 Ω load, $A_v = 10^4 \frac{10}{10^4 + 10} = \frac{1}{1001} 10^4 = 9.99$ **V/V.**

For a 0 Ω load, $i_o = \frac{v_s \times 10^4}{10^4} = v_s$, for which $g_m = \frac{i_o}{v_s} = 1$ **A/V.**

1.24 **Individual Amplifiers:**

Amp	R_i	A_{vo}	R_o
1	10^6	10	10^4
2	10^4	100	10^3
3	10^4	1	20

With a $0.5 \times 10^6 \Omega$ source and a 100 Ω load:

a) $$\left.\frac{v_s}{v_o}\right|_1 = \frac{10^6}{0.5 \times 10^6 + 10^6} \times 10 \times \frac{10^2}{10^2 + 10^4} = \frac{1}{1.5} \times 10 \times \frac{1}{101} = \mathbf{0.066\ V/V.}$$

$$\left.\frac{v_s}{v_o}\right|_2 = \frac{10^4}{0.5 \times 10^6 + 10^4} \times 100 \times \frac{10^2}{10^2 + 10^3} = \frac{1}{51} \times 100 \times \frac{1}{11} = \mathbf{0.178\ V/V.}$$

$$\left.\frac{v_s}{v_o}\right|_3 = \frac{10^4}{0.5 \times 10^6 + 10^4} \times 1 \times \frac{10^2}{20 + 10^2} = \frac{1}{51} \times 1 \times \frac{10}{12} = \mathbf{0.016\ V/V.}$$

b) **One-Stage designs:**

	Loss		
Amp	Input	Output	Least Loss
1	low	high	source
2	med	med	load
3	med	low	load

Ranking (best first): As Input Coupler: **1, 2/3, 3/2**; As Output Coupler: **3, 2, 1**; As Provider of Gain: **2, 1, 3.**

c) **Two-Stage Designs:**

Input: clearly A_1 is #1 on list 1, and reasonable on list 3;

Output: clearly A_3 is #1 on list 2, but worst on list 3;

But, A_2 is #2 on list 2, *but* #1 on list 3.

Conclude: Try (A_1 A_2) and (A_1 A_3).

d) **Highest Gain for a Two-Stage Design:**

For (A_1 A_2):

$$\text{Gain} = \frac{10^6}{10^6 + 0.5 \times 10^6} \times 10 \times \frac{10^4}{10^4 + 10^4} \times 100 \times \frac{100}{10^2 + 10^3} = \frac{1}{1.5} \times 10 \times \frac{1}{2} \times 100 \times \frac{1}{11}$$

$$= 30.3 \text{ V/V}.$$

For (A_1 A_3):

$$\text{Gain} = \frac{10^6}{10^6 + 0.5 \times 10^6} \times 10 \times \frac{10^4}{10^4 + 10^4} \times 1 \times \frac{100}{100 + 20} = \frac{1}{1.5} \times 10 \times \frac{1}{2} \times 1 \times \frac{10}{12}$$

$$= 2.78 \text{ V/V}$$

Certainly (A_1 A_2) seems **best** with an overall gain of **30.3 V/V.**

e) **Reconsidering:**

Certainly, maximizing the gain is a good idea, since coupling is never perfect, (i.e. there is always a loss). Of the highest gain choices, pick the highest input resistance for the input stage and the lowest output resistance for the output stage, i.e., A_1 and A_2 respectively.
chose (A_1 A_2).

Try also:

For (A_2 A_3):

$$\text{Gain} = \frac{10^4}{10^4 + 0.5 \times 10^6} \times 100 \times \frac{10^4}{10^4 + 10^3} \times 1 \times \frac{100}{100 + 20} = \frac{1}{51} \times 100 \times \frac{10}{11} \times 1 \times \frac{10}{12}$$

$= 1.49$ V/V.

Now if two amps of the same kind can be used:

For (A_2 A_2):

$$\text{Gain} = \frac{10^4}{0.5 \times 10^6 + 10^4} \times 100 \times \frac{10^4}{10^3 + 10^4} \times 100 \times \frac{100}{10^2 + 10^3} = \frac{1}{51} \times 100 \times \frac{10}{11} \times 100 \times \frac{1}{11}$$

$= 16.2$ V/V.

Note in retrospect, that the only way to possibly better the value of 30.3 V/V is to use $(A_1 A_1)$, $(A_2 A_2)$, $(A_2 A_1)$. See that the loss at the output is too great in the $(A_1 A_1)$ and $(A_2 A_1)$ cases.

1.25

Amp #	R_i Ω	A_{is}	R_o A/A
1	10	100	10^4
2	10^4	1000	10^3
3	10^4	100	10^5

With 10 kΩ source, and 10 kΩ load:

a) **There are 9 possible amplifier pairs:**

$$\text{Gain } 1,1 = \frac{10^4}{10+10^4} \times 100 \times \frac{10^4}{10+10^4} \times 100 \times \frac{10^4}{10^4+10^4}$$

$$= \frac{1000}{1001} \times 100 \times \frac{1000}{1001} \times 100 \times \frac{1}{2} = 4901 \text{ A/A.}$$

$$\text{Gain } 1,2 = \frac{10^4}{10+10^4} \times 100 \times \frac{10^4}{10^4+10^4} \times 100 \times \frac{10^3}{10^3+10^4}$$

$$= \frac{1000}{1001} \times 100 \times \frac{1}{2} \times 1000 \times \frac{1}{11} = 4541 \text{ A/A.}$$

$$\text{Gain } 1,3 = \frac{10^4}{10+10^4} \times 100 \times \frac{10^4}{10^4+10^4} \times 100 \times \frac{10^5}{10^5+10^4}$$

$$= \frac{1000}{1001} \times 100 \times \frac{1}{2} \times 100 \times \frac{10}{11} = 4541 \text{ A/A.}$$

$$\text{Gain } 2,1 = \frac{10^4}{10^4+10^4} \times 10^3 \times \frac{10^3}{10+10^3} \times 100 \times \frac{10^4}{10^4+10^4}$$

$$= \frac{1}{2} \times 10^3 \times \frac{100}{101} \times 100 \times \frac{1}{2} = 24752 \text{ A/A.}$$

$$\text{Gain } 2,2 = \frac{10^4}{10^4+10^4} \times 10^3 \times \frac{10^3}{10^3+10^4} \times 10^3 \times \frac{10^3}{10^3+10^4}$$

$$= \frac{1}{2} \times 10^3 \times \frac{1}{11} \times 10^3 \times \frac{1}{11} = 4132 \text{ A/A.}$$

$$\text{Gain } 2,3 = \frac{10^4}{10^4+10^4} \times 10^3 \times \frac{10^3}{10^3+10^4} \times 10^2 \times \frac{10^5}{10^5+10^4}$$

$$= \frac{1}{2} \times 10^3 \times \frac{1}{11} \times 10^2 \times \frac{10}{11} = 4132. \text{ A/A.}$$

$$\text{Gain } 3,1 = \frac{10^4}{10^4+10^4} \times 100 \times \frac{10^5}{10+10^5} \times 100 \times \frac{10^4}{10^4+10^4}$$

$$= \frac{1}{2} \times 100 \times 1 \times 100 \times \frac{1}{2} = 2500 \text{ A/A.}$$

$$\text{Gain } 3,2 = \frac{10^4}{10^4+10^4} \times 100 \times \frac{10^5}{10^5+10^4} \times 10^3 \times \frac{10^3}{10^3+10^4}$$

$$= \frac{1}{2} \times 100 \times \frac{10}{11} \times 10^3 \times \frac{1}{11} = 4132 \text{ A/A}.$$

$$\text{Gain } 3,\, 3 = \frac{10^4}{10^4 + 10^4} \times 100 \times \frac{10^5}{10^4 + 10^5} \times 10^2 \times \frac{10^5}{10^4 + 10^5}$$

$$= \frac{1}{2} \times 100 \times \frac{10}{11} \times 10^2 \times \frac{10}{11} = 4132 \text{ A/A}.$$

Summary:

Gain	*Combination*
24752	(2,1)
4901	(1,1)
4541	(1,2), (1,3)
4132	(2,2), (2,3), (3,2), (3,3)
2500	(3,1)

Note the relative superiority of (1,1) is due essentially to the R_i of amplifier 1 being 10Ω.

1.26 Figure of merit: $A_{is} R_o / R_i$

Amp	R_i	A_{is}	R_o	$\frac{A_{is} \times R_o}{R_i}$	Rank
1	10	10^2	10^4	10^5	1
2	10^4	10^3	10^3	10^2	3
3	10^4	10^2	10^5	10^3	2

Lowest ranked are A_2, A_3:

Consider: $G_{2,2}$, $G_{2,3}$, $G_{3,2}$, $G_{3,3}$ with values (from P1.25 above) of 4132, 4132, 4132, 4132, respectively. Thus the highest available gain with the lowest-ranked amplifiers is **4132 A/A**.

1.27

Amp	R_i	A_{is}	g_m	R_o	Fig. of Merit (1) $g_m R_o R_i = A_{is} R_o$	Rank (1)	Fig. of Merit (2) $g_m R_o R_i^2 = A_{is} R_o R_i$	Rank (2)
1	10	10^2	10	10^4	10^6	2	10^7	3
2	10^4	10^3	10^{-1}	10^3	10^6	2	10^{10}	2
3	10^4	10^2	10^{-2}	10^5	10^7	1	10^{11}	1

Now, $g_m = \frac{i_o}{v_i} = \frac{i_o}{R_i\ i_i} = \frac{A_{is}}{R_i}$:

For A_1, $g_m = \dfrac{10^2}{10} = 10$ A/V.

For A_2, $g_m = \dfrac{10^3}{10^4} = 10^{-1}$ A/V.

For A_3, $g_m = \dfrac{10^2}{10^4} = 10^{-2}$ A/V.

Figure of merit (FM) for a transconductance amplifier is $g_m R_o R_i$. But $g_m R_o R_i = \dfrac{A_{is}}{R_i} \times R_o \times R_i$ $= \mathbf{A_{is} R_o}$. Use as FM1. However, high R_i is obviously very important for the g_m generator. Consider $A_{is} R_o R_i = \mathbf{g_m R_o R_i^2}$, as FM2.

1.28 $A_{is} = \dfrac{i_c}{i_b} \Big|_{\upsilon_c=0} = \beta.$

$R_i = \dfrac{\upsilon_{be}}{i_b} \Big|_{\upsilon_c=0} = \dfrac{\upsilon_{be}}{\upsilon_{be}/r_\pi} = r_\pi.$

$R_o = \dfrac{\upsilon_c}{i_c} \Big|_{i_b=0} = \infty$, as described.

$G_m = \dfrac{i_c}{\upsilon_{be}}$; but $i_c = \beta\, i_b$, and $i_b = \dfrac{\upsilon_{be}}{r_\pi}$.

$$G_m = \frac{\beta\, i_b}{\upsilon_{be}} = \frac{\beta\, \upsilon_{be}/r_\pi}{\upsilon_{be}} = \frac{\beta}{r_\pi}.$$

Numerically:

$A_{is} = \beta =$ **200 mA/mA, and** $G_m = \dfrac{\beta}{r_\pi} = \dfrac{200}{5 \times 10^3} =$ **40 mA/V.**

1.29 **For the gain υ_e/υ_b:**

From Fig. E1.14, see $i_b = i_x$, $\upsilon_b = \upsilon_x$, and $i_e = i_b + i_c = i_b(1 + \beta) = (\beta + 1)i_x$. Thus $\upsilon_e = i_e R_e = (\beta + 1)R_e i_x$, and $\upsilon_x = i_x(r_\pi) + \upsilon_e = i_x(r_\pi + (\beta + 1)R_e)$. Correspondingly, $\upsilon_e/\upsilon_b = \upsilon_e/\upsilon_x = [(\beta + 1)R_e i_x]/[(r_\pi + (\beta + 1)R_e)i_x]$, or $\upsilon_e/\upsilon_b = (\beta + 1)R_e/[(r_\pi + (\beta + 1)R_e]$, or $\upsilon_e/\upsilon_b = 1/[1 + \mathbf{r_\pi/[(\beta + 1)R_e]}] = 1/(1 + r_e/R_e) = R_e/(r_e + R_e)$.

For the resistance $\mathbf{R_{in}}$ seen at the emitter: apply a test voltage υ_x at E, with the base input grounded, and R_e removed. See $i_b = -\upsilon_x/r_\pi$; $i_C = \beta ib = -\beta\upsilon_x/r_\pi$. Thus total current from υ_x is $i_x = \upsilon_x/r_\pi — \beta\upsilon_x/r_\pi = \upsilon_x(\beta + 1)/r_\pi$. Thus $R_{in} = \upsilon_x/i_x = 1/[(\beta + 1)/r_{pi}] = r_\pi/(\beta + 1) = r_e$. Thus the resistance seen by R_e is $\mathbf{R_{in} = r_\pi/(\beta + 1) = r_e}$

1.30 For the gain υ_c/υ_b: See (from P1.29 above) that $i_b = (upsilon_b - \upsilon_e)/r_\pi$, that $\upsilon_e = (\beta + 1)i_b R_e$, and that $upilon_c = r_\pi i_b + (\beta + 1)i_b R_e = i_b(r_\pi + (\beta + 1)R_e)$. Thus $\upsilon_c/\upsilon_b = -\beta R_L/[r_\pi + (\beta + 1)R_e]$, or dividing by $(\beta + 1)$, $\upsilon_c/\upsilon_b = -[\beta R_L/(\beta + 1)]/[r_\pi/(\beta + 1) + R_e] = -\alpha\mathbf{R_L/(r_e + R_e)}$. For the resistance seen by R_L, note that the output at C is a current source β_{ib} whose current is independent of υ_c. Thus the output resistance seen by R_L is **infinite.**

1.31 From Exercise 1.14, $R_{in} = r_\pi + (\beta' + 1)R_e$ at the base. Now with a resistor R_S connected from a source υ_s to the base, $\upsilon_b = [R_{in}/(R_S + R_{in})]\upsilon_s$, or $\upsilon_b = \upsilon_s[r_\pi + (\beta + 1)R_e]/[R_S + r_\pi + (\beta + 1)R_e]$. Now $\upsilon_e/\upsilon_s = [\upsilon_e/\upsilon_b][\upsilon_b/\upsilon_s]$, where from P1.29 above, $\upsilon_e/\upsilon_b = (\beta + 1)R_e/[r_\pi + (\beta + 1)R_e]$. Thus $\upsilon_e/\upsilon_s = [(\mathbf{r_\pi + (\beta + 1)R_e})/(\mathbf{R_S + r_\pi + (\beta + 1)R_e})] \times [(\beta + 1)\mathbf{R_e/[r_\pi + (\beta + 1)R_e]}]$.

This reduces to half the value without R_S when the leftmost factor becomes ½, that is **when** $R_S = [r_\pi + (\beta + 1)R_e]$.

This is a very logical result which can be seen directly, since when $R_S = R_{in}$, half the input signal is lost in the resulting voltage divider.

SECTION 1.6: FREQUENCY RESPONSE OF AMPLIFIERS

1.32 $\left| A \right| = \frac{v_o}{v_i} = \frac{2(2V)}{2mV} =$ **2000 V/V.** At 1 kHz, the period is $\frac{1}{10^3} = 1$ ms.

Delay of 0.2 ms corresponds to $\frac{0.2}{1} \times 360 = 72°$. Thus, the corresponding phase shift is **72°, lagging.**

1.33 The 3 dB bandwidth = 100 kHz − 0 kHz = **100 kHz.** For capacitor coupling, the bandwidth is 100 kHz − 20 kHz = **80 kHz.**

1.34 See for circuit a) that $v_{out} = v_i - v_o$ is the voltage across R, fed by C, i.e. a **high-pass output.** Correspondingly, for circuit b), $v_{out} = v_i - v_o$ is the voltage across C, a **low-pass output.** In fact, the circuits are really the same, with both output types available: high-pass across R, and low-pass across C.

1.35

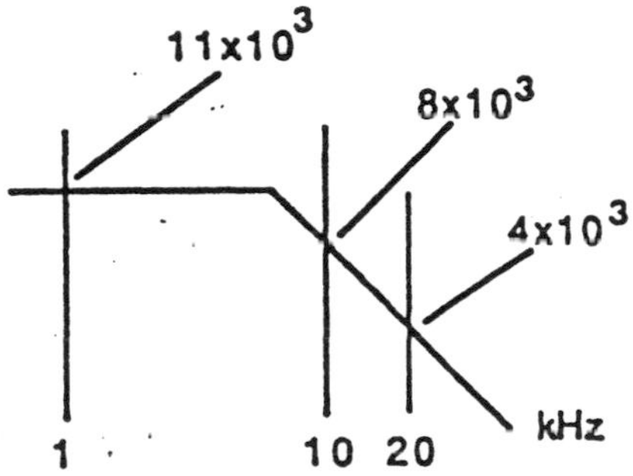

See immediately, that as frequency goes from 10 kHz to 20 kHz, gain drops by a factor of 2 from 8×10^3 to 4×10^3 V/V. Conclude that 10 kHz and 20 kHz are on the − 20 dB/decade rolloff, such that

$Af = f_t = 20 \times 10^3 \times 4 \times 10^3 = 80 \times 10^6$Hz. Thus f_t, the unity-gain frequency, is **80 MHz.** Now, since at 1 kHz, the gain is only 11/8 larger than at 10 × the frequency, one can conclude that the midband gain is likely to be 11×10^3 V/V. Thus the 3 dB frequency is $f_H = \frac{80 \times 10^6}{11 \times 10^3} =$ **7.27 kHz.** At frequency f, the phase shift is $\tan^{-1} f/f_H$. For $\tan^{-1} f/f_H = 60°$, $f/f_H = \tan 60° = 1.73$.

Thus, $f = 1.73$ (7.27 kHz) = **12.6 kHz** is the frequency where the phase lag is 60°.

1.36 For each stage of the amplifier, the 3 dB frequency is at $\omega_H =$ **1/RC rad/s.** For 2 stages, the output reaches $1/\sqrt{2}$ of midband amplitude at the frequency where each stage contributes $(1/2)^{1/4}$, that is when

$$\left[\frac{1}{[1^2 + (\omega/\omega_H)^2]^{1/2}}\right]^2 = \left[\frac{1}{2}\right]^{1/2}, \text{ or } 1 + \left[\frac{\omega}{\omega_H}\right]^2 = 2^{1/2} = 1.4142, \text{ or}$$

$\omega = \omega_H \, (1.4142 - 1)^{1/2} = 0.644 \, \omega_H$.

Thus, for 2 stages, the 3 dB frequency becomes **0.644/(RC) rad/s.**

Now, for a modified cascade, where one stage has $\omega_H = 1/RC$ and the other has $\omega_{H1} = 1/(kRC) = \omega_H/k$, the response will be:

$$T(s) = \frac{K}{(1 + j\,\omega/\omega_H)\,(1 + j\,k\omega/\omega_H)} = \frac{K}{1 + j\,(k+1)\,\frac{\omega}{\omega_H} - \frac{k\omega^2}{\omega_H^2}}$$

The response is 3 dB down when $\left[1 - \frac{k\omega^2}{\omega_H^2}\right]^2 + \left[(k+1)\frac{\omega}{\omega_H}\right]^2 = 2$,

$$\text{or, } 1 - \frac{2k\omega^2}{\omega_H^2} + k^2\frac{\omega^4}{\omega_H^4} + (k+1)^2\frac{\omega^2}{\omega_H^2} = 2, \text{ or } k^2\frac{\omega^4}{\omega_H^4} + (k^2+1)\frac{\omega^2}{\omega_H^2} - 1 = 0,$$

whence $\left[\dfrac{\omega}{\omega_H}\right]^2 = -\dfrac{(k^2+1)\pm((k^2+1)^2+4k^2)^{\frac12}}{2k^2} = \dfrac{-(k^2+1)\pm(k^4+6k^2+1)^{\frac12}}{2k^2}$.

Thus $\omega = \omega_H\left[\dfrac{-k^2-1\pm(k^4+6k^2+1)^{\frac12}}{2k^2}\right]^{\frac12}$ $---$ (1).

Check: *for* k $=$ *1:*

See $\omega = \omega_H\left[\dfrac{-1-1\pm(1+6+1)^{\frac12}}{2}\right]^{\frac12} = \omega_H\left[\dfrac{-2\pm\sqrt8}{2}\right]^{\frac12} = \omega_H\left[-1\pm\sqrt2\right]^{\frac12}$

$= \omega_H(0.414)^{\frac12} = 0.644\omega_H$: OK.

To find K for which $W/W_H = 0.95$ from (1), solve by *Trial and Success:*

Now, for the convenience of smaller k, from (1): $\omega = \omega_H\left[-\dfrac12-\dfrac{1}{2k^2}\pm\left[\dfrac14+\dfrac{1.5}{k^2}+\dfrac{1}{4k^4}\right]^{\frac12}\right]^{\frac12}$,

in general.

For k $= 0.1$: $\omega = \omega_H\left[-0.5\pm50\pm(.25+150+2500)^{\frac12}\right]^{\frac12} = \omega_H\left[-50.5\pm51.48\right]^{\frac12} = 0.99\omega_H$.

For k $= 0.2$: $\omega = \omega_H\left[-0.5-12.5\pm(.25+37.5+156.25)^{\frac12}\right]^{\frac12} = 0.964\omega_H$.

For k $= 0.25$: $\omega = \omega_H\left[-0.5-8\pm(.25+24+64)^{\frac12}\right]^{\frac12} = 0.946\omega_H$.

Thus, the required value of k is about **0.25**.

1.37

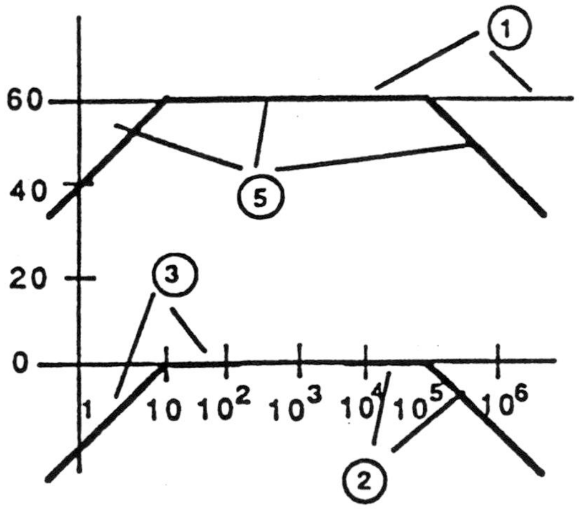

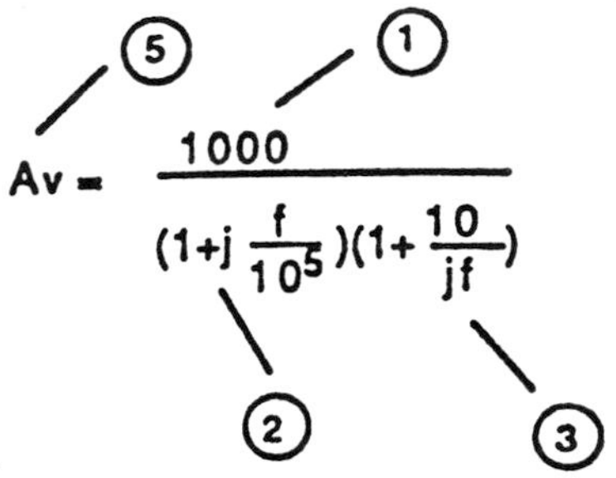

$Av = \dfrac{1000}{(1+j\frac{f}{10^5})(1+\frac{10}{jf})}$

Overall, $1000 \equiv 20\log 10^3 = 60$ dB, At $_\_ = 1$:

$$\left|\dfrac{1}{1+j\frac{1}{10^5}}\right| \approx \dfrac11 \equiv 0\text{ dB, and }\left|\dfrac{1}{1+\frac{10}{j1}}\right| \approx \dfrac{1}{10} \equiv -20\text{ dB, totalling }60+0-20 = \textbf{40 dB}.$$

At $f = 10$:

$$\left|\dfrac{1}{1+j\frac{10}{10^5}}\right| \approx 1 \equiv 0\text{ dB, and }\left|\dfrac{1}{1+\frac{10}{j10}}\right| = \dfrac{1}{\sqrt2} \equiv -3\text{ dB, totalling }60+0-3 = \textbf{57 dB}.$$

At $f = 100$: $1000 \times \left| \dfrac{1}{1 + j\dfrac{100}{10^5}} \right| \times \left| \dfrac{1}{1 + \dfrac{10}{j100}} \right| \approx 1000 \times 1 \times 1 \equiv 60 + 0 + 0 = \mathbf{60\ dB}$.

At $f = 10^4$: $1000 \times \left| \dfrac{1}{1 + j\dfrac{10^4}{10^5}} \right| \times \left| \dfrac{1}{1 + \dfrac{10}{j\,10^4}} \right| \approx 1000 \times 1 \times 1 \equiv 60 + 0 + 0 = \mathbf{60\ dB}$.

At $f = 10^5$: $1000 \times \left| \dfrac{1}{1 + j\dfrac{10^5}{10^5}} \right| \times \left| \dfrac{1}{1 + \dfrac{10}{j\,10^5}} \right| \approx 1000 \times 1/2 \times 1 \equiv 60 - 3 + 0 = \mathbf{57\ dB}$.

At $f = 10^6$: $1000 \times \left| \dfrac{1}{1 + j\dfrac{10^6}{10^5}} \right| \times \left| \dfrac{1}{1 + \dfrac{10}{j\,10^6}} \right| \approx 1000 \times \dfrac{1}{10} \times 1 \equiv 60 - 20 + 0 = \mathbf{40\ dB}$.

See, 3 dB bandwidth = 10^5 Hz – 10 Hz = 10^5 Hz.

Now, phase is 6° at a frequency which is a factor of 10 on the midband side of f_L and f_H.

Thus, theregion for which phase extends from +6° through 0 to –6° is from 10 (10 Hz) to $\dfrac{10^5\ \text{Hz}}{10}$, or from **100 Hz to 10^4 Hz**.

1.38

a) We see in general that: $A_v = \dfrac{10^7\, jf}{(jf + 10^5)\,(\frac{jf}{10} + 1)}$

See at very high frequencies, that $\left| A(f) \right| \rightarrow \dfrac{10^7 f}{f \times f/10} = \dfrac{10^8}{f}$.

See at very low frequencies, $\left| A(f) \right| \rightarrow \dfrac{10^7 f}{(10^5)\,(1)} = 10^2 f$.

See at midband frequencies, $\left| A(f) \right| = \dfrac{10^7 f}{10^5\,(f/10)} = 10^3$, where the midband extends from $\dfrac{f}{10} = 1$, or f = **10 Hz**, to f = **10^5 Hz**.

Check:

See, at f = 10 Hz, $A(f) = \dfrac{10^7\, j\,10}{(j\,10 + 10^5)\,(\frac{j10}{10} + 1)}$, and $\left| A(f) \right| = \dfrac{10^8}{10^5\sqrt{2}} = \dfrac{10^3}{\sqrt{2}}$.

See, at f = 10^5 Hz, $A(f) \doteq \dfrac{10^7\, j\,10^5}{(j\,10^5 + 10^5)\,(j\,\frac{10^5}{10} + 1)} \approx \dfrac{10^{12}}{10^5(j{+}1)\,(10^4)}$, and $\left| A(f) \right| \approx \dfrac{10^3}{\sqrt{2}}$

Thus the midband gain is verified to be **10^3V/V**.

b) Now, $A_v = \dfrac{10^7\, jf}{(jf + 10^5)\,(jf/10 + 1)} = \dfrac{10^7/10^5}{\left[\dfrac{jf}{10^5} + 1\right]\left[\dfrac{1}{10} + \dfrac{1}{jf}\right]} = \dfrac{10^3}{\left[1 + \dfrac{jf}{10^5}\right]\left[1 + \dfrac{10}{jf}\right]}$

From this form, we see that the critical frequencies occur when:

$\dfrac{f}{10^5} = 1 \rightarrow f = 10^5$ **Hz**, and $\dfrac{10}{f} = 1 \rightarrow f =$ **10 Hz**,

such that between these frequencies, $A = \dfrac{10^3}{(1 + j\varepsilon_1)\,(1 - j\varepsilon_2)}$ and $\left| A \right| = 10^3$ **V/V**.

Note that the latter approach is more straightforward.

1.39

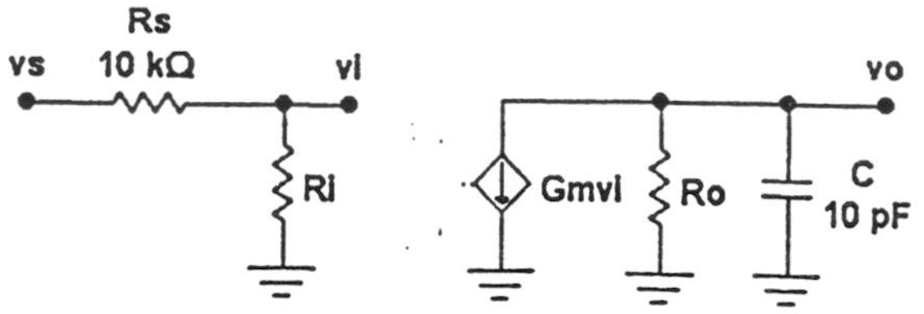

At low frequencies, $A_v = \dfrac{v_o}{v_s} = \dfrac{\mathbf{R_i}}{\mathbf{R_s + R_i}} \times \mathbf{G_m\, R_o = A_M}$. The upper 3 dB frequency, $\omega_H = 1/(\mathbf{R_o\, C})$.

Now $R_i = 2.5/I$, $R_o = 200/I$, $G_m = 40\, I$. For $\dfrac{1}{2\pi\, R_o\, C} > 10^6$, i.e. $\dfrac{1}{2\,\pi\, 200/I \times 10 \times 10^{-12}} \geq 10^6$,

$I = 10^6 \times 2\,\pi \times 200 \times 10^{-11} = 0.0126\, A = 12.6$ mA.

Now, $A_M = G_m\, R_o\, \dfrac{R_i}{R_s + R_i} = 40 \times 12.6 \times \dfrac{200}{12.6} \times \dfrac{2.5/12.6}{2.5/12.6 + 10}$

$= 40 \times 200 \times 0.198/(0.198 + 10) =$ **155.3 V/V.**

Gain-Bandwidth

$(GB) = \dfrac{R_i}{R_s + R_i} \times G_m\, R_o \times \dfrac{1}{R_o\, C} = \dfrac{\mathbf{G_m}}{\mathbf{C}} \times \dfrac{\mathbf{R_i}}{\mathbf{R_s + R_i}} = \dfrac{40I}{C}\; \dfrac{(2.5/I)}{R_s + \dfrac{2.5}{I}} = \dfrac{100/C}{R_s + \dfrac{2.5}{I}}$.

See that for large I, **GB** $= \dfrac{100}{\mathbf{R_s\, C}}$, *independent of I!*

Now, Gain $\times\, 2\pi \times 10^7 = \dfrac{100}{C\, R_s} = \dfrac{100}{10 \times 10^{-12} \times 10^4} = 10^9$, whence Gain $\approx \dfrac{10^9}{2\pi \times 10^7} =$ **15.9 V/V.**

See directly (but approximately), that the previous design has a gain of 155 and a bandwidth of 1 MHz. Thus this design, with a bandwidth $10\times$ greater, should have a gain $10\times$ less. Required current $I = 10^7 \times 2\pi \times 200 \times 10^{-11} =$ **126 mA.**

1.40 The circuit described is as shown,

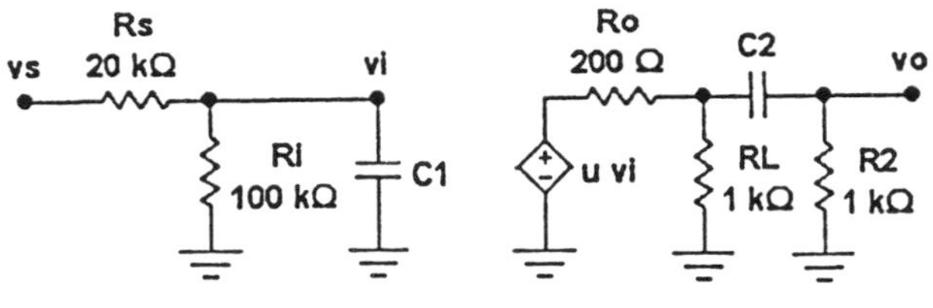

In this circuit, C_i is small and included within C_1 (or ignorcd). The resistance seen by C_1 is $R_{C1} = R_s \parallel R_i = 20k\Omega \parallel 100k\Omega = 100k\Omega = 100k\Omega/(5+1) = 16.67k\Omega$. The resistance seen by C_2 is $R_{C2} = (200\Omega \parallel 1k\Omega + 1k\Omega) = (1k\Omega/(5+1) + 1k\Omega) = 1167\Omega$. For a low cutoff frequency at 20 kHz, $2\pi(20)10^3 = 1/(R_{C2}C_2)$, where $C_2 = 1/[2\pi(20)10^3(1167)] =$ **6.8 nF**. For a high cutoff frequency at 80 kHz, $2\pi(80)10^3 = 1/(R_{C1}C_1)$, whence $C_1 = 1/[2\pi(80)10^3(16.67 \times 10^3) =$ **119 pF**. The midband gain {for whose calcultion $C_1 = 0$ and $C_2 = \infty$} is $v_o/v_s = \dfrac{100}{100+20} \times 100 \times \dfrac{1111}{1111+0.2} = 100/120 \times 500/700 =$ **59.5 V/V.**

Now for the transfer function (in terms of $p = s/2\pi$, a complex Hertz value, and (using Table 1.2 on page 32 of the Text): $v_o/v_s(p) = \dfrac{59.5p}{(1+p/80)(p+20)} = \dfrac{595}{(1+p/80)(1+20/p)}$. {*Check:* For $p = 0$, $v_o/v_s(p) = 0$; For $p = \infty$, $v_o/v_s(p) = 0$; OK.}

Now at midband (40 kHz), $p = jf = j40$, and $v_o/v_s(jf) = \dfrac{59.5}{(1+j40/80)(1+20/j40)} = \dfrac{59.5}{(1+j0.5)(1-j0.5)}$. Now at 40 kHz $|v_o/v_s| = \dfrac{59.5}{(1+0.5^2)^{1/2}(1+0.5^2)^{1/2}} =$ **47.6 V/V and** $\angle(v_o/v_s(40)) = -\tan^{-1}0.5 - \tan^{-1}(-0.50) = 0^0$.

For the 1 dB bandwidth where: -1 dB $\rightarrow 20\log_{10}K = -1$; $K = \log_{10}^{-1}(-1/20) = 0.89125 = 0.891$.

Now $\left|\dfrac{1}{1+jf/80}\right| = 0.891$, when $1 + f^2/80^2 = (1/0.891)^2 = 1.259$, and $f = (0.259)^{1/2}80 = 0.509(80) =$ **40.708 kHz.**

Likewise $\left|\dfrac{1}{1-j20/f}\right| = 0.891$, when $1 + 20^2/f^2 = 1.259$ and $f = 20/0.259^{1/2} =$ **39.304 kHz.**

Thus the 1 dB bandwidth is $40.708 - 39.304 =$ **1.404 kHz.**

1.41

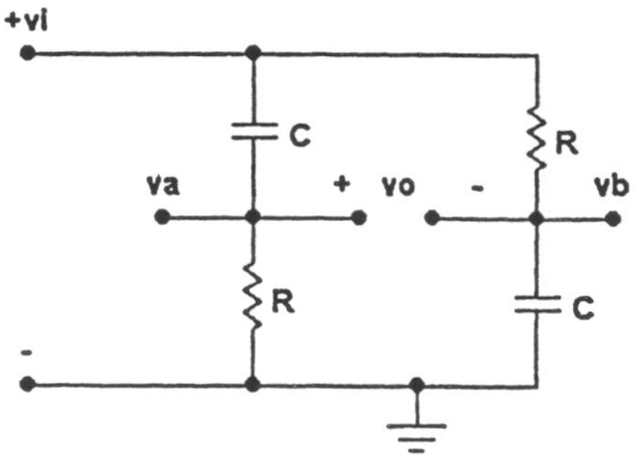

$$\upsilon_a = \frac{R}{R + \frac{1}{Cs}} \times \upsilon_i .$$

$$\upsilon_b = \frac{\frac{1}{Cs}}{R + \frac{1}{Cs}} \upsilon_i .$$

$$\upsilon_o = \upsilon_a - \upsilon_b .$$

$$\frac{\upsilon_o}{\upsilon_s} = \frac{R - \frac{1}{Cs}}{R + \frac{1}{Cs}} = \frac{\mathbf{RCs - 1}}{\mathbf{RCs + 1}} .$$

At low frequencies , $\frac{\upsilon_o}{\upsilon_s} \to -1$; Magnitude = 1, $\Phi = \pm 180°$.

At high frequencies : $\frac{\upsilon_o}{\upsilon_s} \to \frac{RC}{RC} = 1$; Magnitude = 1, $\Phi = 0°$.

At $\omega = \frac{1}{RC}$: $\frac{\upsilon_o}{\upsilon_s} = \frac{j-1}{j+1}$, with magnitude $\frac{\sqrt{2}}{\sqrt{2}} = 1$, and $\Phi = -45° -45° = -90°$.

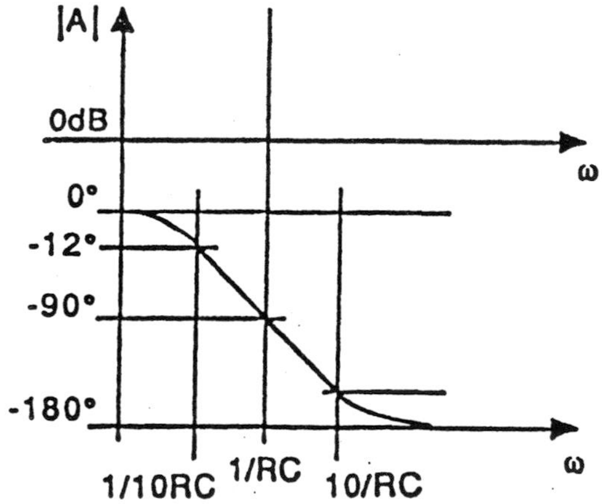

SECTION 1.7: THE DIGITAL LOGIC INVERTER

1.42

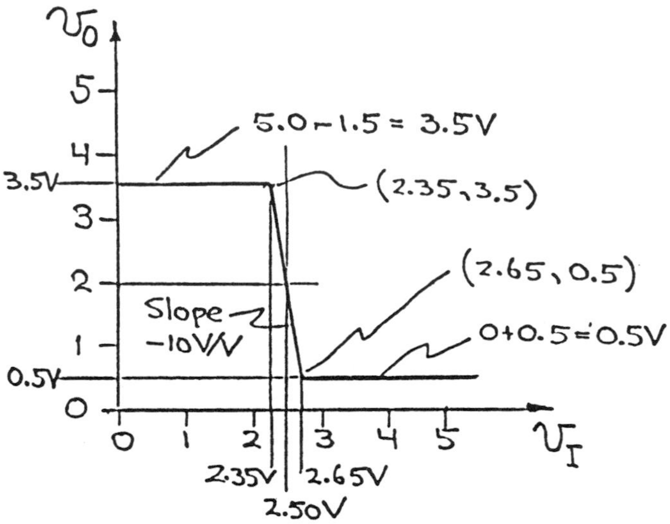

From the description (and diagram), $V_{OL} = 0 + 0.5 =$ **0.5 V**; $V_{OH} = 5 - 1.5 =$ **3.5 V**. The transition-region gain is essentially constant at [− 10 V/V]. The transition-region width $= (V_{OH} - V_{OL})/|\,gain\,| = (3.5 - 0.5)/10 =$ **0.3 V**. The transition region is centred at $V_M =$ 2.5 V. Correspondingl;y, $V_{IH} = 2.5 + 0.3/2 =$ **2.65** V, and $V_{IL} = 2.5 - 0.3/2 =$ **2.35** V. Thus $NM_L = V_{OH} - V_{IH} = 3.5 - 2.65 =$ **0.85 V** and $NM_L = V_{IL} - V_{OL} = 2.35 - 0.50 =$ **1.85 V**. If the transition region doubles to 0.6 V, the margins change by $0.6/2 = 0.30V$ to $NM_H = 0.85 - 0.30 =$ **0.55 V**, and $NM_L = 1.85 - 0.30 =$ **1.55 V**. To equalize the noise margins for lower gains, the transition centre must be moved by $(1.55 - 0.55)/2 =$ 0.50 V (lower) to $V_M = 2.5 - 0.5 =$ **2.0 V**.

1.43 See with the switch open (as in Fig. 1.31(b), $V_{OH} = V_{DD} =$ **5 V**. With switch closed (as in Fig. 1.31(c), $V_{OL} = 0.050 + [50/(50 + 1000)](5.000 - 0.50) = 0.050 + 0.0476(4.95) =$ **0.286 V**.

For input high, $P_{S_1} = 5[4.95/1050] =$ **23.6 mW**.

For input low, $P_{S_2} = 5(0) =$ **0 mW**.

For a 5 kΩ "leak" and assuming that the lower end of the switch is essentially grounded, $V_{OH} = 5(5/(1 + 5)) =$ **4.17 V**, and the static power is essentially $P_{S_3} = V^2/R = 5^2/(1 + 5) =$ 4.17 mW.

Now for 50% duty cycle, average dissipation is $P_{S_{av}} = 0.5(23.6) + 0.5(4.17) =$ **13.9 mW**.

1.44 For ideal switches, $V_{OH} = V_{DD} =$ **5.0 V**, and $V_{OL} =$ **0 V**. Since there are no static current paths, static power is $P_S =$ **0 mW**. For each switch with a leakage resistance of 5 kΩ, $V_{OL} = 5V(50/5050) =$ **49.5 mV** and $V_{OH} = 5 - 0.50 =$ **4.95 V**. In each state, the static power loss is $5^2/5050 =$ 4.95 mW. Thus, the average power loss is **4.95 mW**.

1.45 For a 1 V logic swing, $R_{C1} = R_{C1} = R = 1V/4mA =$ **250 Ω**. For a 0 V upper supply, $V_{OH} =$ **0.0 V** and $V_{OL} = 0 - 1V =$ **− 1.0 V**. For equal noise margins, the switch should operate at $(0 + 1)/2 =$ **0.5V.**. For $V_{EE} =$ 5 V, and $I_{EE} =$ 4 mA, the power loss is constant at $P_S = 5V(4mA) =$ **20 mW**. Since the circuit switches currents, gate operation is unaffected by reasonable switch resistances ($<R_C$). Switch resistance changes only the voltage at the switch common point. **No**, the power dissipated (assuming no load capacitances) is unaffected by switching.

1.46 For V_{DD} = 5 V, R_{on} = 50 Ω (and ignoring leakage) with a 10 pF capacitor and operation at 100 MHz, dynamic power is $P_D = fCV_{DD}^2 = 100 \times 10^6 \times 10 \times 10^{-12} \times 5^2 =$ **25 mW**. Transition times are governed by $\tau = R_{on}C = 50 \times 10 \times 10^{-12} =$ 0.5 ns. For the transition times $e^{-t/0.5} =$ 0.1 for $t = -0.5\ln 0.1 =$ 1.15 ns. Thus the 10% to 90% rise and fall times are essentially **1.15 ns** (ignoring the very small time for the initial 10% swing). The output reaches $V_{DD}/2$ when $t = -0.5\ln 0.5 =$ 0.35 ns. Thus the propagation delay is **0.35 ns.**

1.47 Here, the supply is 3V, switches operate at 1.5 V, R_{on} = 50 Ω, and R_{leak} = 5 kΩ. For no leakage, V_{OL} = **0 V** and V_{OH} = **3.0 V**. For leakage, V_{OL} = 3(50/5050) = **29.7 mV** = 0.03 V and $V_{OH} = 3V - 30mV =$ **2.97 V**. For no leakage, the static power is **0 mW**. For leakage, the static power is $3^2/5050 =$ **1.78 mW**.

1.48

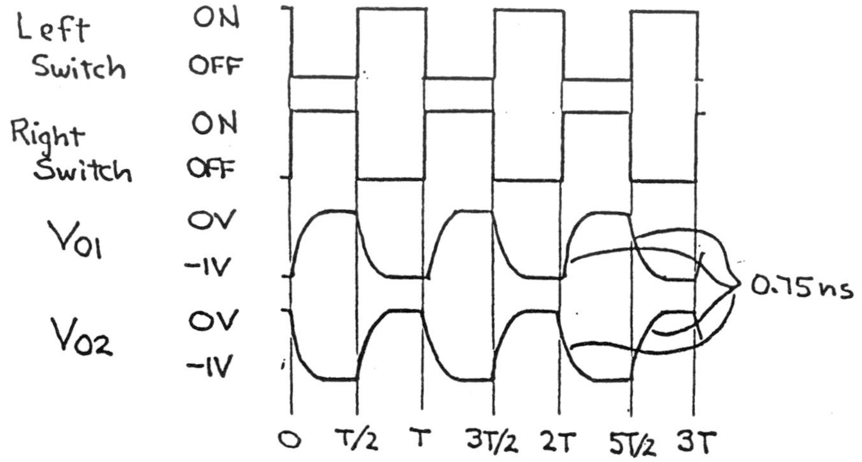

The time constant for each transition is $\tau = R_C C = 250 \times 3 \times 10^{12} =$ 0.75 ns. See V_{OH} = **0 V** and $V_{OL} = 0 - 4mA\,(250\Omega) =$ **− 1 V.** Now $t_{TLH} = t_{THL} = -\tau\ln 0.1 = 2.3\tau = 2.3(0.75) =$ **1.73 ns,** and $t_{PLH} = t_{PHL} = -\tau\ln 0.5 = 0.69\tau = 0.69(0.75) =$ **0.52 ns.** For 200 MHz operation (independent of duty cycle) the dynamic power at each output is $P_{D1} = fC(\Delta V)^2 = 200 \times 10^6 \times 3 \times 10^{-12} \times 1^2 =$ 0.6 mW. For 2 outputs, the dynamic dissipation at 200 MHz (independent of duty cycle) 2(0.6) = **1.2 mW**. The static dissipation is $4mA\,(5V) =$ **20 mW**. The total power dissipation is **21.2 mW**. Note that if the power-supply voltage is reduced, the dominance of static power also reduces, to give 13.2 mW total for a 3V supply.

Chapter 2

OPERATIONAL AMPLIFIERS

SECTION 2.1: OP AMP TERMINALS

2.1 Each op amp has pins for

input : 2- unique
output: 1- unique
power : 2- sharable

Thus an 8-pin package can accommodate: **2 op amps**, using 2(3) + 2 = 8 pins, with **none unused**

Thus a 14-pin package can accommodate: **4 op amps**, using 4(3) + 2 = 14 pins, with **none unused.**

SECTION 2.2: THE IDEAL OP AMP

2.2 Voltage between input pins, $\upsilon = \upsilon_+ - \upsilon_- = -3V/10^4$ or **$-300\ \mu$V**.

In particular, from the negative to the positive input, one would expect -300μV or -0.3 mV.

If the positive pin is at +100 mV, the negative would be at 100 − − 0.3 = **100.3 mV.**

2.3 For $\upsilon_O = 3.5$ V: $\upsilon_+ = 3.5\ V/10^4 = 0.35$ mV, across 1 kΩ.

$$\text{required } \upsilon_I = \upsilon_+ + \upsilon_{IM} = 0.35 \times 10^{-3} + \left(\frac{0.35 \times 10^{-3}}{1 \times 10^3} \right) 10^6 = 0.35035 \text{ V} \approx \mathbf{0.35\ V.}$$

Check: Overall gain is $\dfrac{10^3}{10^3 + 10^6} \times 10^4 \approx \dfrac{10^3}{10^6} \times 10^4 = 10$ V/V.

0.35 V at the input produces 3.5 V at the output.

SECTION 2.3: ANALYSIS OF CIRCUITS CONTAINING IDEAL OP AMPS – THE INVERTING CONFIGURATION

2.4

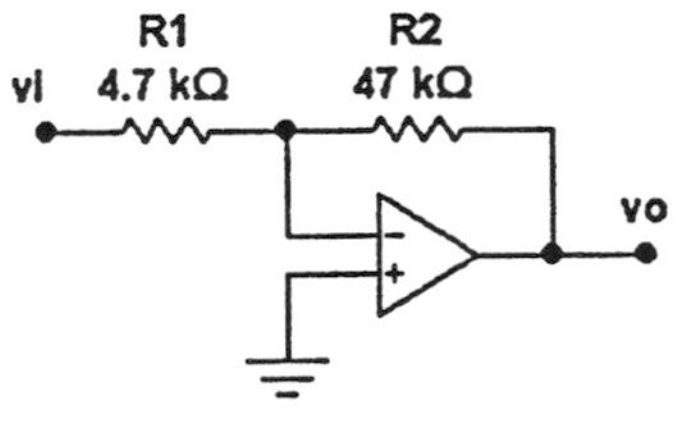

For the desired connection:

$$\frac{\upsilon_O}{\upsilon_I} = -\frac{R_2}{R_1} = -\frac{47}{4.7} = \mathbf{-10\ V/V.}$$

For R_1 and R_2 exchanged:

$$\frac{\upsilon_O}{\upsilon_I} = -\frac{R_2'}{R_1'} = -\frac{4.7}{47} = \mathbf{-0.1\ V/V.}$$

2.5

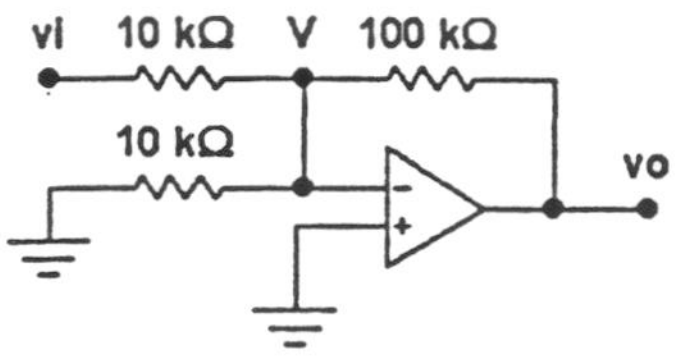

For an ideal op amp: to obtain $\upsilon_O = +10$ V, $\upsilon_- = 0$ V, and the current in the grounded 10kΩ resistor is zero. Thus, $\upsilon_I = -\dfrac{10}{100\text{k}\Omega}\,10\text{k}\Omega =$ **−1 V**.

For gain $= 10^4$: For $\upsilon_O = +10$ V, $\upsilon_- = -\,10/10^4 = -10^{-3}$ V.

Whence current in 100 kΩ is $(10 - -\,10^{-3})/10^5 = 10^{-4}$ A, to the υ_- node;

current in grounded 10 kΩ is $10^{-3}/10^4 = 10^{-7}$ A, to the υ_- node;

current in input 10 kΩ is $(10^{-4} + 10^{-7})$ A $= 10^{-4}$ A, to the input.

$\upsilon_I = -\,10^{-3} - 10^4\,(10^{-4} + 10^{-7}) = -10^{-3} - 1 - 10^{-3} = -\,(1 + 2 \times 10^{-3}) =$ **−1.002 V.**

2.6 Want Gain of −2 V/V with three 100 kΩ resistors:

There are 2 **solutions**:

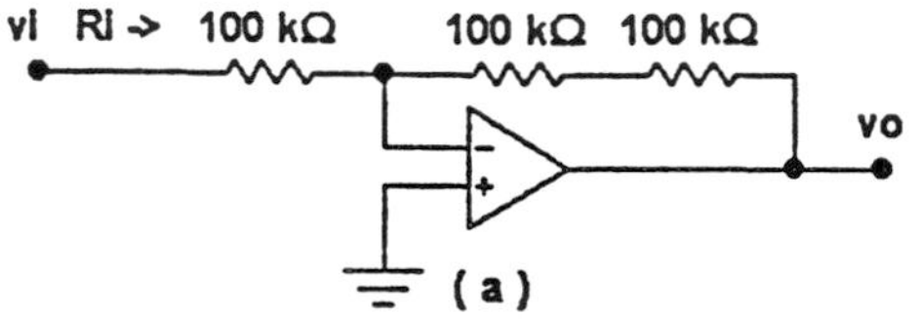

(a)

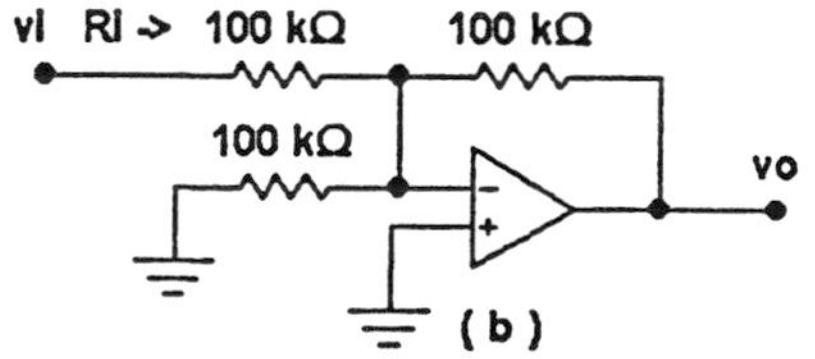

(b)

For (a), $\dfrac{\upsilon_o}{\upsilon_i} = -\dfrac{200\text{k}\Omega}{100\text{k}\Omega} = -2$ V/V, and $R_{in} = $ **100 kΩ.**

For (b), $\dfrac{\upsilon_o}{\upsilon_i} = -\dfrac{100\text{k}\Omega}{100\text{k}\Omega/2} = -\,2$ V/V, and $R_{in} = \dfrac{100\text{k}\Omega}{2} =$ **50 kΩ.**

2.7 There are 2 potential solutions:

a) $R_1 = 220$ kΩ, and $R_2 = 10(220\text{k}\Omega) = 2.2$ MΩ > 1 MΩ; no good.

b) $\mathbf{R_2 = 220}$ **kΩ, and** $\mathbf{R_1} = \dfrac{220k\Omega}{10} =$ **22 kΩ** << 1 MΩ; OK.

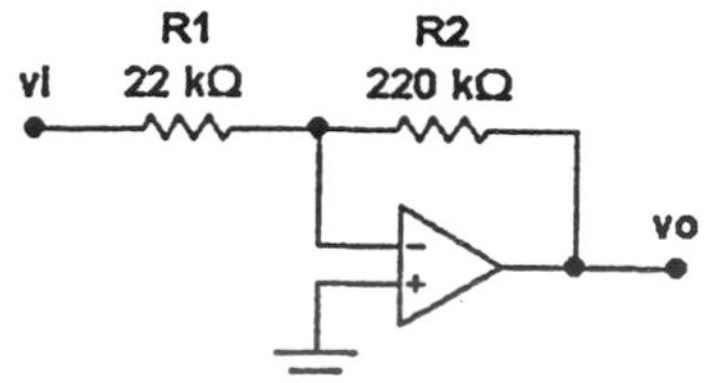

2.8 For an inverting op amp, with R_{in} = 100 kΩ, use R_1 = 100 kΩ.

(a) For a direct design of gain = −20 V/V, $R_2 = 20(R_1)$ = 2 MΩ; not allowed directly, but $R_2 = 1M\Omega + 1M\Omega$ in series is OK. If very large resistors are to be avoided completely, and even 1 MΩ is too large, consider a network for R_2:

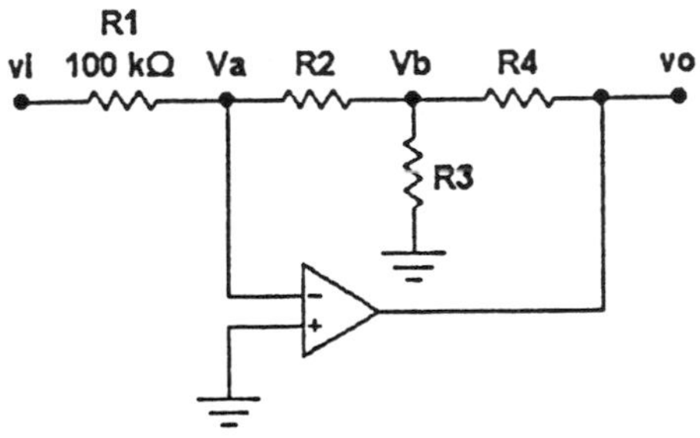

There are many possible designs, but only three which use 3 100Ω resistors to meet the specifications, all having R_1 = 100 kΩ.

(b) Make $R_1 = R_2 = R_3 =$ **100 kΩ**:

$$\upsilon_b = -\frac{R_2}{R_1}\upsilon_i = -\upsilon_i$$

and $\upsilon_o = \upsilon_b \left[1 + \frac{R_4}{R_2 \parallel R_3}\right]$.

$$\upsilon_o = -\upsilon_i \left[1 + \frac{R_4}{50k}\right] = -20\upsilon_i.$$

$1 + \frac{R_4}{50k} = 20 \rightarrow R_4 = (20–1)\ 50 =$ **950 kΩ**, but this may be too large for some applications.

(c) Make $R_1 = R_2 = R_4 =$ **100 kΩ**:

See $\upsilon_b = -\upsilon_i$ and, summing currents at υ_b, $\frac{\upsilon_o - \upsilon_b}{R_4} = \frac{\upsilon_b}{R_3} + \frac{\upsilon_b - \upsilon_a}{R_2}$, where υ_a = 0.

Thus $\frac{\upsilon_o}{R_4} = \upsilon_b \left[\frac{1}{R_3} + \frac{1}{R_2} + \frac{1}{R_4}\right]$, and thus $\frac{\upsilon_o}{\upsilon_i} = -R_4 \left[\frac{1}{R_3} + \frac{1}{R_2} + \frac{1}{R_4}\right] = -20$ V/V.

$\frac{100}{R_3} + 1 + 1 = 20$, or $R_3 = \frac{100}{20–2} = \frac{100}{18} =$ **5.55 kΩ**, a very good solution!

(d) Make $R_1 = R_3 = R_4 =$ **100kΩ**:

See Thevenin equivalent at υ_b (to the right) is a source $\upsilon_o/2$ with source resistance $R_3 \parallel R_4$ = 50 kΩ. Now, gain to equivalent source ($\upsilon_o/2$) must be 10 V/V:

$R_2 = 10\ R_1 - 50 = 10(100) - 50 =$ **950 kΩ**, which, again may be too large.

2.9 Two possible gains:

a) With an ideal op amp:

Gains are $\frac{-100}{10}$ = **−10 V/V**, and $\frac{-10}{100}$ = **−0.1 V/V**.

b) With an amplifier with gain of A = 100 V/V, and $G = \frac{-R_2/R_1}{1 + (1 + R_2/R_1)/A}$,

Gains are $G = -\frac{100/10}{1 + (1 + 100/10)/100} = \frac{-10}{1 + 11/100} =$ **−9.009 V/V**,

and $G = -\frac{10/100}{1 + (1 + 10/100)/100} = \frac{-0.1}{1 + 1.1/100} =$ **−0.0989 V/V**.

See that the error in the high-gain case ≈ 10%, where G/A ≈ 1/10, and ≈ 1% in the low-gain case, where G/A ≈ 1/1000 (We will see why in Chapter 8).

2.10 From Equation 2.1: $G = \dfrac{-R_2/R_1}{1 + (1 + R_2/R_1)/A}$. For $\dfrac{R_2}{R_1} = 100 \rightarrow |G| = \dfrac{100}{1 + 101/A}$.

Now, $|G| \gtrsim 0.9\ (100)$ when $\dfrac{100}{1 + 101/A} \geq 0.9\ (100)$, or $1 + 101/A \leq \dfrac{1}{0.9} = 1.11$ or $101/A \leq 0.11$.

Whence $A \geq \dfrac{101}{0.111} =$ **909 V/V**.

Now, $|G| \gtrsim 0.99\ (100)$ when $101/A \leq \dfrac{1}{0.99} - 1 = 0.01010$.

Whence $A \geq \dfrac{101}{.0101} =$ **10^4 V/V**.

Check: $G = \dfrac{-100}{1 + (1 + 100)/10^4} = \dfrac{100}{1 + 101/10^4} = 99.0001$.

2.11 For a small test voltage v applied at the negative op-amp input terminal (where R_2 returns), the op amp output voltage $v_O = -Av$. The current flowing into R_2 is $i = (v - -Av)/R_2$. The input resistance $R_i = v/i = v/[v(1 + A)/R_2]$ $R_i = R_2/(1 + A)$. Now, for an input resistor R_1, and input voltage v_I, the voltage at the op amp negative input is $v = v_I R_i/(R_i + R_1)$ and $v_O = -Av$. Thus, combining, $G = v_O/v_I = \dfrac{-AR_i}{R_i + R_1} = \dfrac{-AR_2/(1 + A)}{R_2/(1 + A) + R_1}$, $G = \dfrac{-AR_2}{R_2 + R_1 + R_1 A} = -\dfrac{\mathbf{R_2/R_1}}{1 + (1 + \mathbf{R_2/R_1})/\mathbf{A}}$, as expected.

2.12

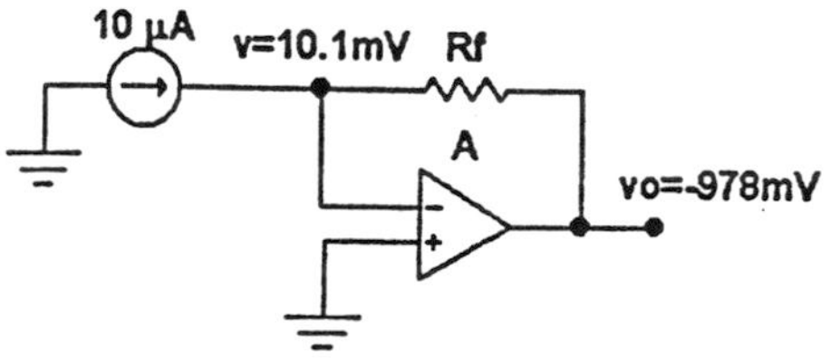

Since the op amp is conventionally ideal with zero input current, 10 μA flows in R_f. Thus $R_f = (10.1 - -978)10^{-3}/(10 \times 10^{-6}) =$ **98.8 kΩ**. R_f is likely to be a nominal **100 kΩ** resistor, with a tolerance of 98.8 – 100) **100 – 98.8**/100 × 100 = – **1.2%**. [Even more likely, this is a ±5% resistor near the middle of its value distribution.] The open-loop gain $A = -978/10.1 =$ **– 96.8 V/V**. The input resistance at the negative input node is $R_i = 10.1mV/10\mu A =$ **1010 Ω**. *Check:* Using the result in P2.11 above $R_i = R_f/(1 - A) = 98.8/(1 - -96.8) =$ 1.01 kΩ, as above. OK. For input resistor

R, $v_O/v_S = -[R_i/(R_S + R_i)] \times 96.8$. Now, $v_O/v_S = -10.00$ for $R_S + R_i = 96.8R_i/10.00 = 9.68R_i$, $R_S = 8.68R_i = 8.68(1.01) =$ **8.77 kΩ**.

Check: From Eq. 2.1, $G = \dfrac{-R_2/R_1}{1 + (1 + R_2/R_1)/|A|} = \dfrac{-98.8/8.77}{1 + (1 + 98.8/8.77)/96.8} = -10.00V/V$, as required.

Note how much smaller R_S is than the nominal value of $100k\Omega/10V/V = 10k\Omega$!

2.13

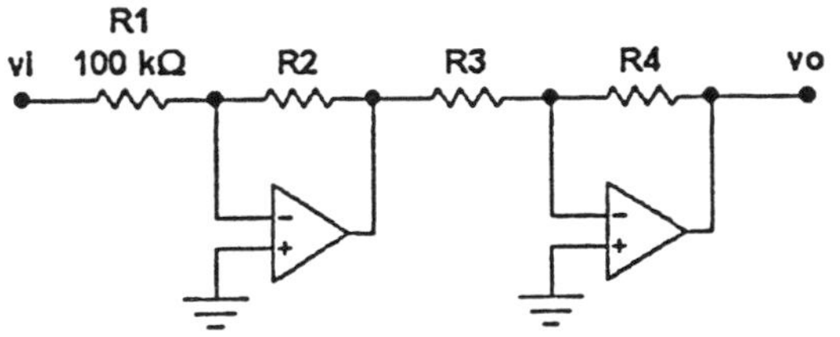

Since $R_{in} = 100\text{k}\Omega$, $R_1 = 100\text{k}\Omega$.
For equal gain/stage,
$G_1, G_2 = \pm\sqrt{200} = -14.14\text{V/V}$,
and $R_2 = 14.14\ (100\text{k}\Omega) = 1.414\text{M}\Omega$, which is too large.

Use $R_2 = 1\text{M}\Omega$, for which $G_1 = -10^6/10^5 = -10$ V/V.

$$G_2 = -\frac{200}{10} = -20\text{V/V}.$$

Use $R_4 = 1$ MΩ, $R_3 = R_4/{-G_2} = 10^6/20 = 50$ kΩ.

Thus, use **R_1 = 100 kΩ, R_2 = 1 MΩ, R_3 = 50 kΩ, R_4 = 1 MΩ.**

2.14 Now for the circuit above, with $R_{in} = 2\text{M}\Omega$, make $R_1 = R_{1a} + R_{1b}$, each 1MΩ, since larger resistors are not available, and $R_2 = 1$ MΩ. Correspondingly, $G_1 = -1/2$ V/V, and G_2 must be $+200/(-1/2) = -400$ V/V. Use $R_4 = 1$ MΩ, and $R_3 = 10^6/400 = 2.5$ kΩ.

In summary, use **R_1 = 1 MΩ in series with 1MΩ, R_2 = 1 MΩ, R_3 = 2.5 kΩ, R_4 = 1MΩ.**

2.15

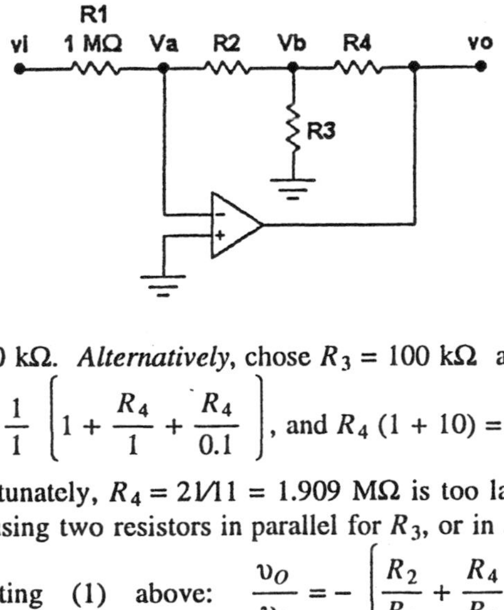

Since $R_{in} = 1$ MΩ, use $R_1 = 1\ \text{M}\Omega = R_2 = R_4$. From page 70 of the Text,

$$\frac{v_O}{v_I} = -\frac{R_2}{R_1}\left(1 + \frac{R_4}{R_2} + \frac{R_4}{R_3}\right) \quad \text{--- (1)}$$

$$\text{Thus} \quad -22 = -\frac{1}{1}\left(1 + \frac{1}{1} + \frac{1}{R_3}\right), \text{ and}$$

$R_3 = 1/20$ MΩ = 50 kΩ. If resistors ≥ 100 kΩ *only* are available, one could make $R_3 = R_{3a} \parallel R_{3b}$, each of 100 kΩ. *Alternatively*, chose $R_3 = 100$ kΩ and select a suitable value for R_4:

$$22 = \frac{1}{1}\left(1 + \frac{R_4}{1} + \frac{R_4}{0.1}\right), \text{ and } R_4\ (1 + 10) = 22 - 1 = 21\text{k}\Omega.$$

Unfortunately, $R_4 = 21/11 = 1.909$ MΩ is too large! One can see that there are in fact no other choices than using two resistors in parallel for R_3, or in series for R_2 or R_4.

Rewriting (1) above: $\dfrac{v_O}{v_I} = -\left(\dfrac{R_2}{R_1} + \dfrac{R_4}{R_1} + \dfrac{R_2 R_4}{R_1 R_3}\right)$, which, for $R_1 = 1$ MΩ, becomes

$$\frac{v_o}{v_i} = -\left(R_2 + R_4 + \frac{R_2 R_4}{R_3}\right), \text{ and for } R_3 = 0.1\ \text{M}\Omega, \text{ then } 22 = R_2 + R_4 + 10\ R_2 R_4.$$

Now, if either R_2 or R_4 are 1MΩ, say $R_2 = 1$ MΩ, then $22 = 1 + R_4 + 10R_4$, and $R_4 = 21/11 > 1$ MΩ.

Obviously, using two resistors in series for R_4 is possible also, but not as nice from a practical point of view, since a new circuit node (for connection of the two resistors) must be found. Thus, the first solution (where $R_3 = R_{3a} \parallel R_{3b}$) is the preferred one:

In summary, use $R_1 = R_2 = R_4 = 1\ \text{M}\Omega$, and $R_3 = 100\ \text{k}\Omega\ \|\ 100\ \text{k}\Omega$.

2.16

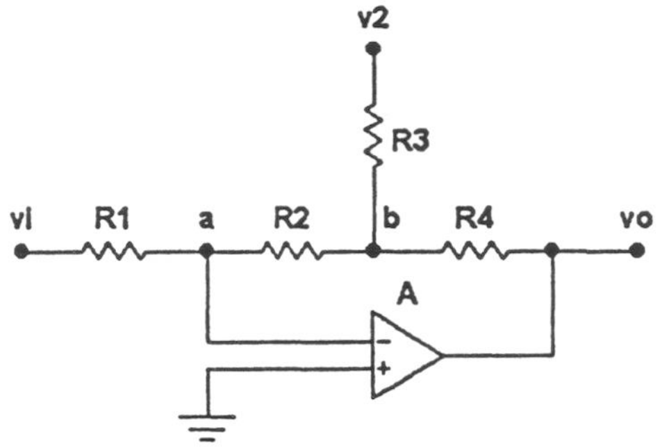

For $v_I = 0$: For $A = \infty$, and the feedback working, $v_a = 0$. Since $i_1 = i_2$, then $v_b = 0$. Now, for $v_b = 0$, $i_3 = i_4$, and $v_O = -(R_4/R_3)\,v_2$. *For* $v_2 = 0$: $v_O = -v_I \left[\dfrac{R_2}{R_1}\right]\left[1 + \dfrac{R_4}{R_2} + \dfrac{R_4}{R_3}\right]$, as derived in Example 2.2 on page 69 of the Text. Using superposition:

$$v_O = -\frac{R_2}{R_1}\left[1 + \frac{R_4}{R_2} + \frac{R_4}{R_3}\right] v_I - \frac{R_4}{R_3} v_2.$$

SECTION 2.4: OTHER APPLICATIONS OF THE INVERTING CONFIGURATION

2.17

$$\frac{v_o(s)}{v_i(s)} = -\frac{Z_2(s)}{Z_1(s)} = -\frac{R_2 \parallel \dfrac{1}{C_2 s}}{R_1 \parallel \dfrac{1}{C_1 s}} = -\frac{\dfrac{R_2/C_2 s}{R_2 + 1/C_2 s}}{\dfrac{R_1/C_1 s}{R_1 + 1/C_1 s}} = -\frac{\dfrac{R_2}{1 + R_2 C_2 s}}{\dfrac{R_1}{1 + R_1 C_1 s}} .$$

Thus $\dfrac{v_o(s)}{v_i(s)} = -\left[\dfrac{R_2}{R_1}\right] \times \left[\dfrac{1 + R_1 C_1 s}{1 + R_2 C_2 s}\right]$,

which is independent of frequency if $R_1 C_1 = R_2 C_2$.

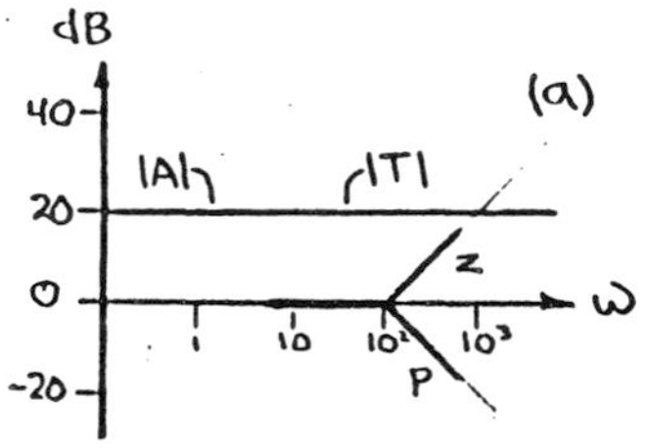

(a) For $C_2 = 0.1\ C_1 = 0.1\mu\text{F}$, and $R_2 = 10\ R_1 = 10^5\Omega$:

See $\dfrac{v_o}{v_i} = -10\left[\dfrac{1 + 10^4 \times 10^{-6}\, s}{1 + 10^5 \times 10^{-7}\, s}\right] = -10\left[\dfrac{1 + 10^{-2} s}{1 + 10^{-2} s}\right] =$

– 10V/V, independent of frequency.

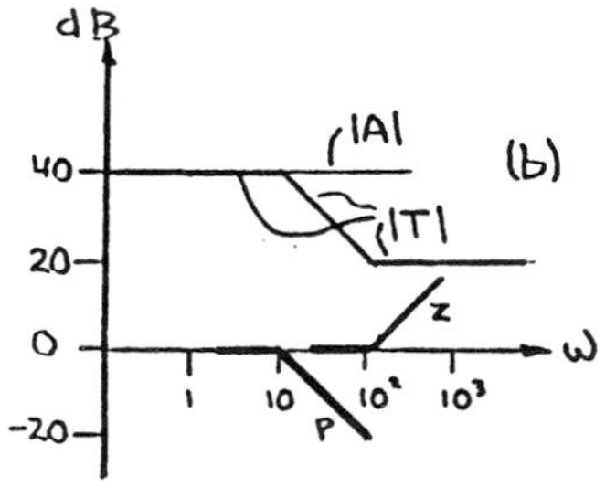

(b) For R_2 raised to $1\text{M}\Omega$:

See $\dfrac{v_o}{v_i} = -\dfrac{10^6}{10^4}\left[\dfrac{1 + 10^{-2} s}{1 + 10^{-1} s}\right] = -100\left[\dfrac{1 + s/100}{1 + s/10}\right].$

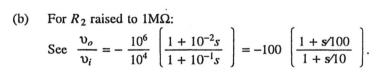

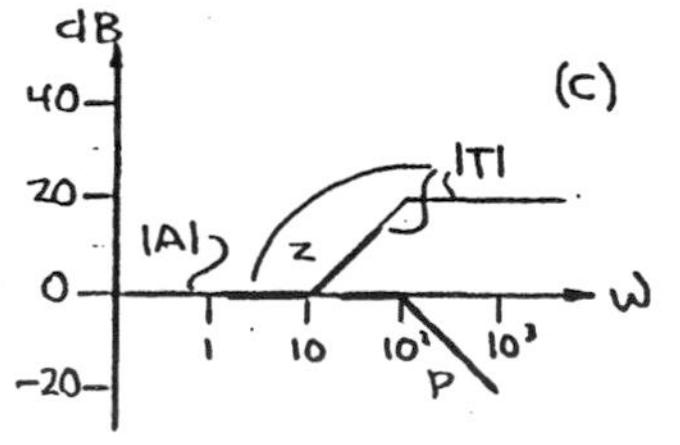

(c) For R_2 lowered to $10\ \text{k}\Omega$:

See $\dfrac{v_o}{v_i} = -1\left[\dfrac{1 + s/100}{1 + s/1000}\right].$

2.18 *Using* $v_O = V_C - \dfrac{1}{CR}\displaystyle\int_0^t v_I(t)dt$, see:

For $v_I = +1$ V, the output $v_O = V_C - \dfrac{1}{1 \times 10^{-3}}\displaystyle\int_0^t 1\text{dt} = V_C - 10^3\, t$.

That is, the output is a *negative-going ramp* with slope of **1 V/ms** or 1000 V/s, proceeding from V_C = 10 V downward, reaching zero in **10 ms**.

Directly: Following the 1 V step, the input current is $\frac{1}{R}$, charging C, causing v_O to fall at a rate $\frac{V}{T} = \frac{I}{C} = \frac{1}{RC} = \frac{1}{10^{-3}} = 10^3$ V/s, moving 10 V in $\frac{10}{10^3}$ = 10 ms.

2.19 Assuming V_C = 0 V, $v_O = \frac{-1}{CR}\int v_I\,dt = \frac{-1}{CR}\int 0.1 \sin 2\pi\, 60\, t = -\frac{1}{10 \times 10^{-3}}\, 0.1 \left(\frac{-1}{2\pi\, 60}\right) \times \cos 2\pi\, 60\, t$ $= 26.5 \times 10^{-3} \cos 2\pi\, 60\, t = 26.5 \times 10^{-3} \sin(2\pi 60t + 90°) = -26.5 \times 10^{-3} \sin(2\pi 60t - 90°)$
indicating that the output is a sine wave of 26.5 mV peak, **leading the input by 90°**, or, alternatively, is an inverted sine wave, lagging by 90°. Note that the latter idea, that of a lagging inverted output is the most consistent with the STC low-pass view of the Miller integrator.

2.20

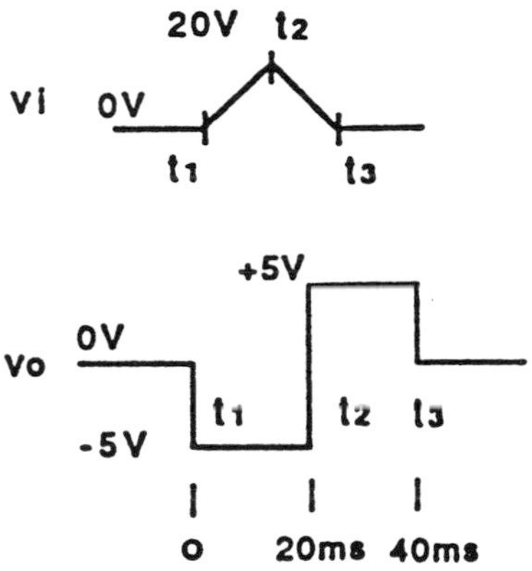

$v_O = -CR\frac{dv_I}{dt} = -5 \times 10^{-3}\frac{dv_I}{dt} = 1$ V.

$\frac{dv_I}{dt} = -200$ V/s. That is, for an output of +1V, the input must **fall** at a rate of **200 V/s**.

See $t_1 = 0$,

$t_2 = \frac{20-0}{1/10^{-3}} = 20$ ms,

$t_3 = 20\times10^{-3} - \frac{20}{-1/10^{-3}} = 40$ ms.

For the rise, $v_O = -5$ ms × 1 V/ms = −5 V.

For the fall, $v_O = -5$ ms × −1 V/ms = +5 V.

2.21

(a)

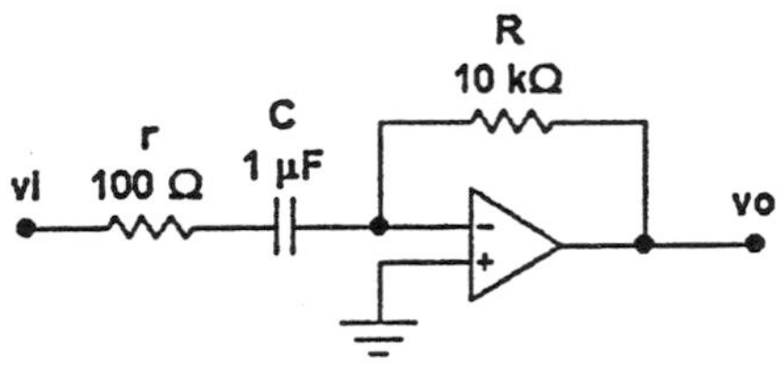

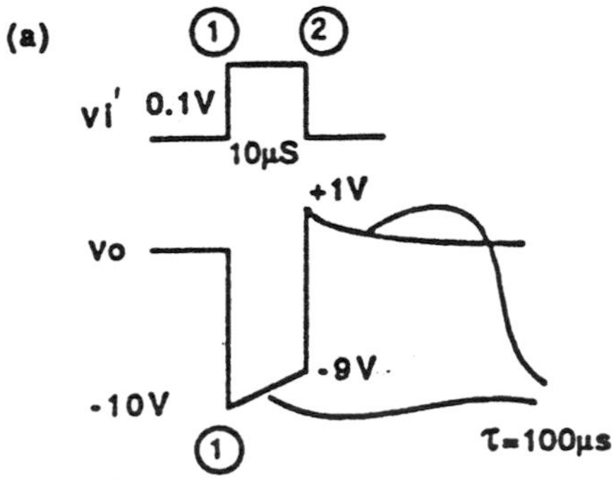

Immediately, upon the input rise, $v_O = -\left(\frac{10k\Omega}{100\Omega}\right) = -10$ V. Time constant: $\tau = 100\Omega \times 1 \times 10^{-6} = 10^{-4}$ s = 100 μs. In 10 μs: the output rises (almost linearly) to $-10 + \frac{10 \times 10^{-6}}{100 \times 10^{-6}} \times 10 = -9$ V. Then, as the input falls, the output rises by 10 V to +1 V.

(b)

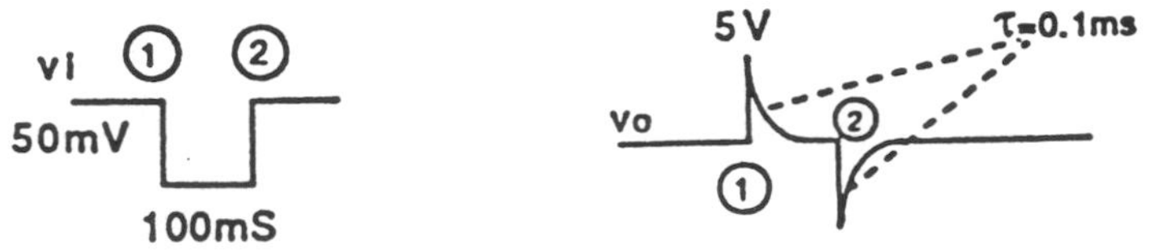

At the falling edge of the input (time t_1) the output rises by 50 mV × 10kΩ/100Ω = 5 V, then begins to fall with a time constant, $\tau = RC = 100 \times 10^{-6} = 10^{-4}$s = 0.1 ms. By the rising edge of the input (100/0.1 = 1000 time constants later), the output has reached 0 V.

2.22

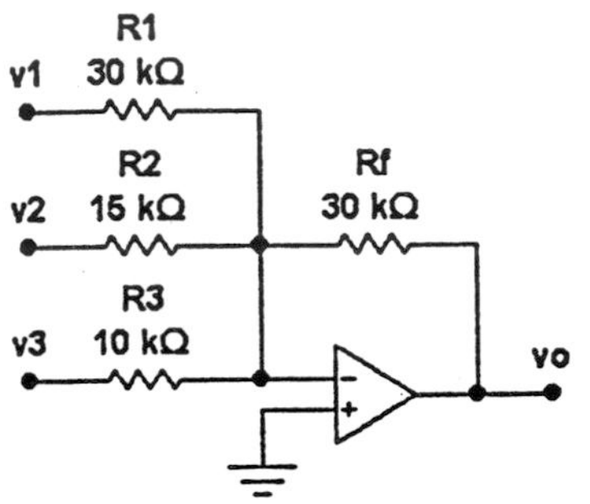

Want, $\upsilon_O = -(\upsilon_1 + 2\upsilon_2 + 3\upsilon_3)$

$$= -3\left[\frac{\upsilon_1}{3} + \frac{\upsilon_2}{3/2} + \frac{\upsilon_3}{1}\right].$$

But, $\upsilon_O = -R_f\left[\frac{\upsilon_1}{R_1} + \frac{\upsilon_2}{R_2} + \frac{\upsilon_3}{R_3}\right]$.

Thus, make R_3 = 10 kΩ, R_f = 30 kΩ,

R_1 = 30 kΩ, and $R_2 = \frac{30}{2}$ = 15 kΩ.

2.23 Want $\upsilon_O = \upsilon_1 + 2\upsilon_2 - 3\upsilon_3$.

See that there are several decompositions:

a) $\upsilon_O = -\left[-\upsilon_1 - 2\upsilon_2 + 3\upsilon_3\right]$,

b) $\upsilon_O = -\left[-(\upsilon_1 + 2\upsilon_2) + 3\upsilon_3\right]$,

c) $\upsilon_O = -\left[-(\upsilon_1 + 2\upsilon_2 - 3\upsilon_3)\right]$, with corresponding circuits:

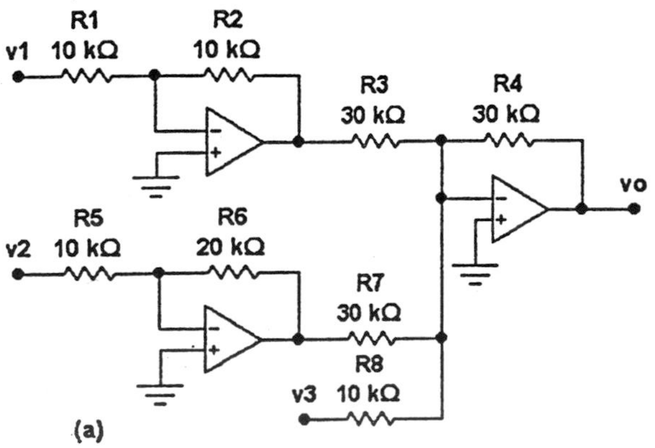

(a)

2.23 (continued)

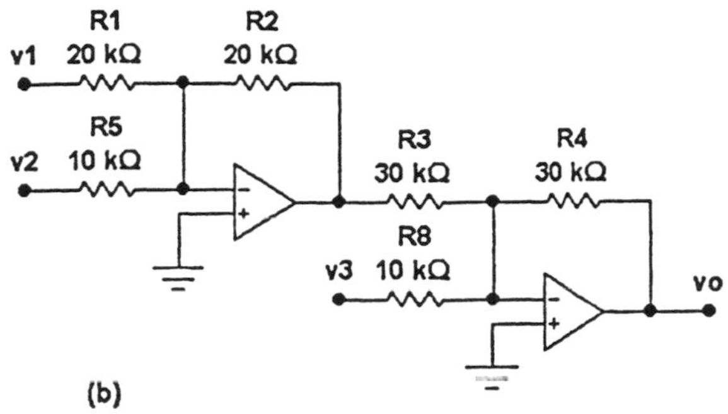

(b)

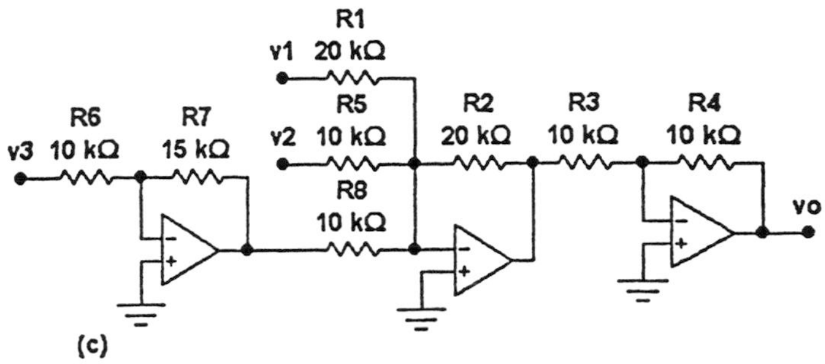

(c)

For each circuit, there are many variants of which some are:

Version	R_1	R_2	R_3	R_4	R_5	R_6	R_7	R_8	ΣR	# Amp.
a_1	10	10	30	30	10	20	30	10	150	3
a_2	10	10	30	30	10	10	15	10	125	3
a_3	10	10	10	10	10	20	10	10‖10‖10	110	3
b_1	20	20	30	30	10			10	120	2
b_2	20	10	15	30	10			10	95	2
b_3	10	10	15	30	10‖10			10	95	2
c_1	20	20	10	10	10	10	15	10	105	3

Conclusion:

Note that b) is obviously the simplest, using the fewest op amps, derivable directly from c) which is the brute-force approach. Note that a) is clearly not a good choice, using an extra op amp to separate (unnecessarily) the v_1 and v_2 inversions. Conclude b_2 is best, with the fewest op amps, the fewest resistors, the lowest total resistance, and on input resistance ≥10 kΩ.

2.24 For an ideal op amp and virtual ground at υ_-, $i_C = \dfrac{\upsilon_1}{10k} + \dfrac{\upsilon_2}{5k}$, and

$$\upsilon_O = 0 - \frac{i_C}{Cs} = -\frac{1}{10k}\left[\upsilon_1 + 2\upsilon_2\right]\frac{1}{10^{-7}s} = \upsilon_O$$

$$\upsilon_O = V_o - 1000\int_0^t \left[\upsilon_1(t) + 2\upsilon_2(t)\right] dt.$$

SECTION 2.5: THE NON-INVERTING CONFIGURATIO

2.25 Normally: $\dfrac{\upsilon_O}{\upsilon_I} = 1 + \dfrac{R_2}{R_1} = 1 + \dfrac{47}{4.7} = \mathbf{11\ V/V}$.

With resistor exchange: $\dfrac{\upsilon_o}{\upsilon_i} = 1 + \dfrac{R_2}{R_1} = 1 + \dfrac{4.7}{47} = \mathbf{1.10\ V/V}$.

2.26 Want $\dfrac{\upsilon_O}{\upsilon_I} = 1.5$ V/V, with three 1 kΩ resistors:

Solutions:

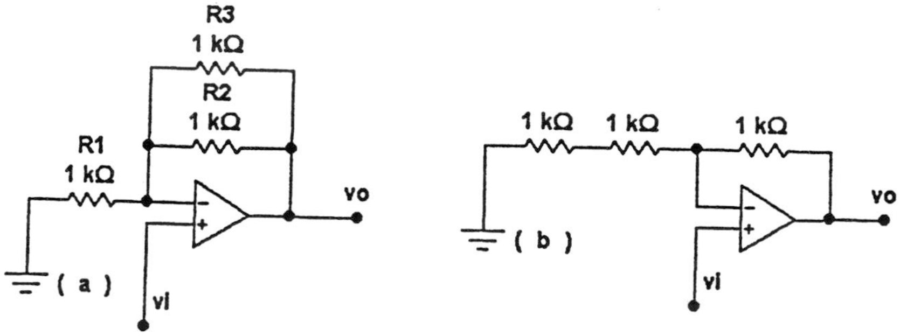

2.27 Want $\upsilon_O = \upsilon_1 + 2\upsilon_2 - 3\upsilon_3$:

Simple approach:

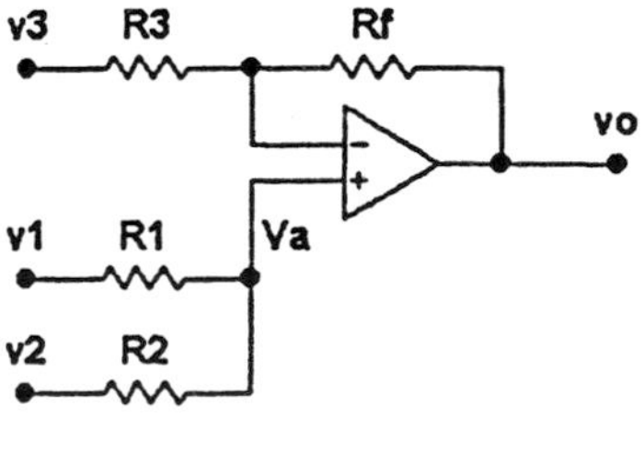

Make $R_2 = R_1/2 \rightarrow R_1 = 2R_2$.

$$\upsilon_a = \frac{R_2}{R_1 + R_2}\upsilon_1 + \frac{R_1}{R_1 + R_2}\upsilon_2 = \frac{R_2}{3R_2}\upsilon_1 + \frac{2R_2}{3R_2}\upsilon_2 = \frac{1}{3}(\upsilon_1 + 2\upsilon_2).$$

Now if $R_f = 2R_3$, and $\upsilon_3 = 0$:

$$\upsilon_O = \left[1 + \frac{2R_3}{R_3}\right] \frac{1}{3}(\upsilon_1 + 2\upsilon_2) = (\upsilon_1 + 2\upsilon_2).$$

In general, $\upsilon_O = \upsilon_1 + 2\upsilon_2 - 2\upsilon_3$. Unfortunately, neat and simple, but **not quite right**! We see the need for more gain for υ_3.

Second approach: (The idea is to reduce the gain given to $(\upsilon_1 + 2\upsilon_2)$). Let $R_f = 3R_3$, such that for υ_3 alone, $\upsilon_O = -3\upsilon_3$. But, now, $\upsilon_O/\upsilon_a = 4$. Introduce an additional input $\upsilon_4 = 0$ V connected to υ_a by R_4. Now by superposition,

$$\upsilon_a = \frac{R_2 \parallel R_4}{R_1 + R_2 \parallel R_4}\upsilon_1 + \frac{R_1 \parallel R_4}{R_2 + R_1 \parallel R_4}\upsilon_2 = \frac{\dfrac{R_2R_4}{R_2+R_4}}{R_1+\dfrac{R_2R_4}{R_2+R_4}}\upsilon_1 + \frac{\dfrac{R_1R_4}{R_1+R_4}}{R_2+\dfrac{R_1R_4}{R_1+R_4}}\upsilon_2,$$

or

$$\upsilon_a = \frac{R_2 R_4}{R_1 R_2 + R_1 R_4 + R_2 R_4}\upsilon_1 + \frac{R_1 R_4}{R_1 R_2 + R_2 R_4 + R_1 R_4}\upsilon_2.$$

Now, if it is required (for $\upsilon_3 = 0$), that $\upsilon_1 + 2\upsilon_2 = \upsilon_o$, need $R_1 = 2R_2$, for which

$$\upsilon_a = \frac{1}{\dfrac{R_1}{R_4} + \dfrac{R_1}{R_2} + 1}(\upsilon_1 + 2\upsilon_2).$$

Now for the desired output, $\upsilon_O = 4\upsilon_a$, and $\dfrac{1}{\dfrac{R_1}{R_4} + 2 + 1} = \dfrac{1}{4}$, for which $\dfrac{R_1}{R_4} = 4 - 2 - 1 = 1$, or $R_4 = R_1 = 2R_2$. Thus use $R_2 =$ **10 kΩ** $\rightarrow$ $R_1 = R_4 =$ **20 kΩ**, and use $R_3 =$ **10 kΩ** $\rightarrow$ $R_f = 3R_3 =$ **30 kΩ**.

2.28

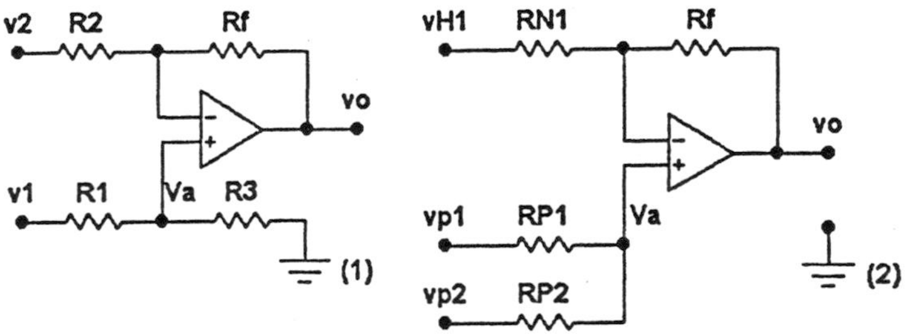

Want $\upsilon_O = 10(\upsilon_1 - \upsilon_2)$. Try the circuit in 1) using R_1, R_2, R_f: For υ_2 alone, $\upsilon_O = -10\upsilon_2 = -R_f/R_2\,\upsilon_2$. $R_f = 10R_2$

For $R_2 = 10$ kΩ $\rightarrow R_f = 100$ kΩ, for which $\upsilon_O = -10\,\upsilon_2$

Now for υ_1 alone, $(R_3 = \infty)$, $\upsilon_O = \left[1 + R_f/R_2\right]\upsilon_1 = (1 + 10)\,\upsilon_1 = 11\upsilon_1$

and, together, $v_O = 11v_1 - 10v_2$

We see that there is too much of v_1: It needs to be attenuated:

Introduce an additional resistor R_3 from v_a to ground:

$$v_a = \frac{R_3}{R_1 + R_3} v_1 = \frac{10}{11} v_1$$

Now for $R_1 = 10\text{k}\Omega$, $R_3/(10 + R_3) = 10/11$, and $11\ R_3 = 100 + 10R_3$,

whence $R_3 = 100\text{k}\Omega$, with the result that $v_O = 11\ v_1 \times 10/11 - 10\ v_2 = 10(v_1 - v_2)$, as required.

In summary: $R_1 = R_2 = 10$ **kΩ**, $R_f = R_3 = 100$ **kΩ**.

This configuration, with $R_f/R_2 = R_3/R_1$, is referred to as a **difference amplifier**.

Alternatively: From the equations supplied in P2.44 on page 1116 of the Text:

$$v_O = -\left(\frac{R_F}{R_{N1}}\right) v_{N1} + \left(1 + \frac{R_F}{R_{N1}}\right)\left(v_{P1}\left(\frac{R_{P1} \parallel R_{P2}}{R_{P1}}\right) + v_{P2}\left(\frac{R_{P1} \parallel R_{P2}}{R_{P2}}\right)\right)$$

We want $v_O = 10\ (v_1 - v_2)$. Now, for $v_{N1} = v_2$, see $\frac{R_F}{R_{N1}} = 10$, and for $R_{N1} = 10\ \text{k}\Omega$, $R_F = 100\ \text{k}\Omega$.

Now for $v_{P1} = v_1$, see $\left(1 + \frac{R_F}{R_{N1}}\right)\left(\frac{R_{P1} \parallel R_{P2}}{R_{P1}}\right) = 10 = 11\left(\frac{R_{P2}}{R_{P1} + R_{P2}}\right)$, and for $R_{P1} = 10\ \text{k}\Omega$, R_{P2} $= 100\ \text{k}\Omega$. Clearly, if $v_{P2} = 0$, then $v_O = 10(v_1 - v_2)$.

2.29

(a) For the circuit of Fig. 2.19, using the result in Equation 2.11 on page 83 of the Text,
$G = \frac{1 + R_2/R_1}{1 + (1 + R_2/R_1)/A}$. Now with $R_2 = 0$, $R_1 = \infty$, $A = 10$ V/V, see $G = \frac{1}{1 + (1/10)} =$ **0.909 V/V.**

(b) Now for the circuit of Fig. 2.16, with R_1, R_2 finite, $G = \frac{1 + R_2/R_1}{1 + (1 + R_2/R_1)/10}$. Now for $G = 1.00$, $1 + R_2/R_1 = 1 + (1 + R_2/R_1)/10 = 1 + 0.1 + 0.1\ R_2/R_1$, and $0.9\ R_2/R_1 = 0.1$, $R_2/R_1 = 0.1/0.9$, and $R_1 = 9R_2$. Thus, $R_2 = 10\ \text{k}\Omega$, $R_1 =$ **90 kΩ**. Thus

Gain (nominal) = 1 + 10/90 = 1.11 V/V.

Gain (actual) = $\frac{1.111}{1 + 1.11/10} = \frac{1.111}{1.111} =$ **1.00 V/V**, as required.

SECTION 2.6: EXAMPLES OF OP-AMP CIRCUITS

2.30

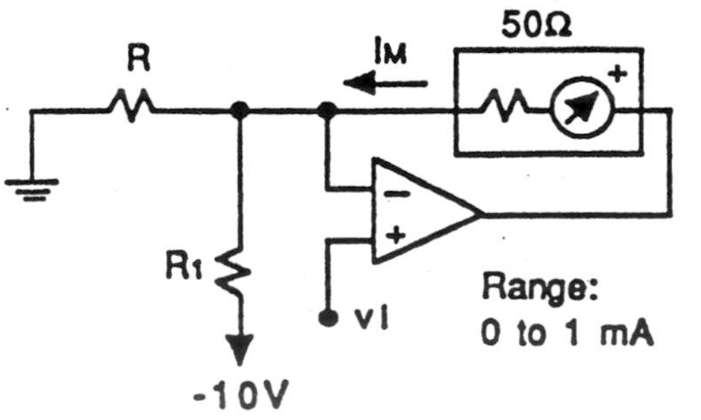

For $v_I = 0$, $v_R = 0$, and the meter is required to read mid-scale. That is, $I_M = 0.5\text{mA}$. Thus $R_1 = (0 - -10)/0.5\text{mA} =$ **20kΩ** with the −10V supply. Now, for full-scale reading with $v_I = +1\text{V}$, I_M must be 1mA. $v_I/R + (v_I - -10)/R_1 = 1\text{mA}$, and $1/R + (1 + 10)/R_1 = 1$, whence $1/R = 1 - 11/20 = 9/20 = 0.45$, and $R = 1/0.45 =$ **2.22kΩ**. *Check*: for $v_I = -1$ V: $I_M = \frac{-1 - -10}{R_1} - \frac{1}{R} = \frac{9}{20} - \frac{1}{2.22} = 0.45 - 0.45$ = 0 V: OK.

2.31

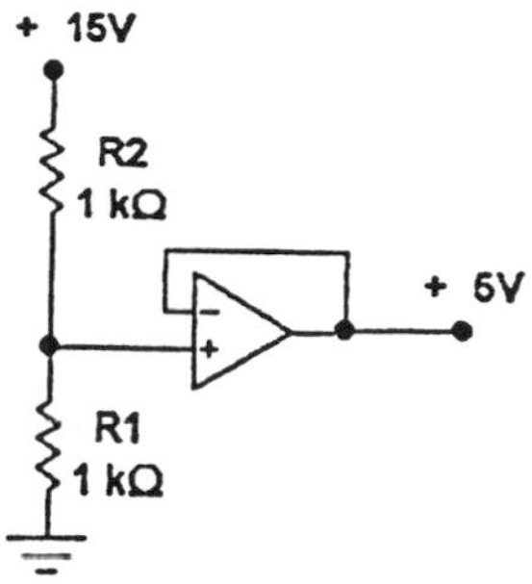

R_1 = +5V/0.5mA = 10 kΩ; R_2 = (15 − 5)/0.5mA = 20 kΩ. Op amp standby dissipation is 2(15V) × 2mA = 60 mW. Op amp dissipation when loaded is 20 mA (15 − 5) = 200 mW. Total op amp dissipation is 200 + 60 = 260 mW.

Note also: That the bias network dissipation is 15 × 0.5 mA = 7.5 mW, and the load power consumption is 5 × 20 mA = 100 mW.

Thus the total supply power is 60 + 200 + 7.5 + 100 = **367.5 mW**.

2.32 From Equation 2.13 (or directly by superposition):

$$\upsilon_O = -\frac{R_2}{R_1}\upsilon_1 + \left[1 + \frac{R_2}{R_1}\right]\left[\frac{R_4}{R_3 + R_4}\right]\upsilon_2 = -\frac{R_2}{R_1}\upsilon_1 + \frac{1 + R_2/R_1}{1 + R_3/R_4}\upsilon_2.$$

Hence $\upsilon_O = -\frac{100}{10}\upsilon_i + \frac{1 + 100/10}{1 + 10/100}\upsilon_2 = -\frac{100}{10}\upsilon_1 + \frac{100 + 1000}{100 + 10}\upsilon_2 = 10\,(\upsilon_2 - \upsilon_1)$.

gain $A = \frac{\upsilon_O}{\upsilon_1 - \upsilon_2}$ = **−10 V/V**: (Note the sign!)

2.33 For this situation, gain $\frac{\upsilon_O}{\upsilon_{s1} - \upsilon_{s2}} = -\frac{100}{10 + 10}$ = **−5 V/V.**

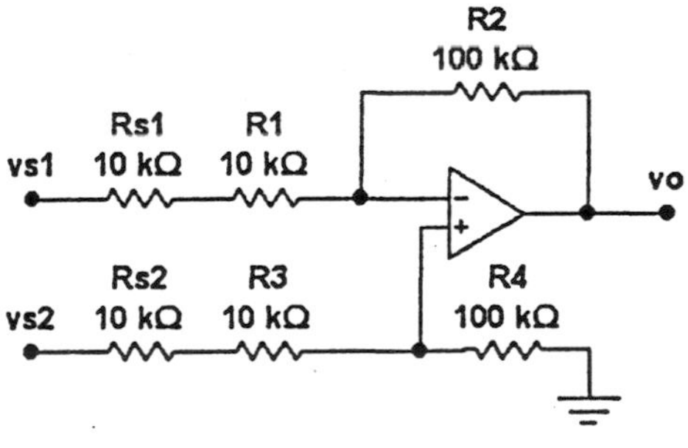

To recover the gain, two possibilities exist:

a) Remove R_1 and R_2 and connect sources directly.

b) Make $R_2 = R_4 = 200$kΩ.

The latter is probably best for secondary reasons. For example, it reduces (by a factor of 2), the effect of minor variations in R_{S1} and R_{S2}.

For the case in which R_{S2} is 8 kΩ, there are two basic approaches: a) "Pad out" R_3, i.e., add an additional 2kΩ resistor in series with R_{S2} and R_3, or, b) change R_4 to 180kΩ to compensate. (Making $\frac{R_2}{R_{1eq}} = \frac{R_4}{R_{3eq}}$)

2.34 Using superposition, υ_O consists of 3 parts. Thus $\upsilon_O = -\upsilon_1\frac{R_2}{R_1} + \upsilon_2\frac{R_4}{R_3 + R_4}\left[1 + \frac{R_2}{R_1}\right] +$

$$\upsilon_3\left[\frac{R_3}{R_3 + R_4}\right]\left[1 + \frac{R_2}{R_1}\right], \text{ or } \upsilon_O = -\upsilon_1\frac{R_2}{R_1} + \upsilon_2\frac{\left[1 + \frac{R_2}{R_1}\right]}{\left[1 + \frac{R_3}{R_4}\right]} + \upsilon_3\frac{\left[1 + \frac{R_2}{R_1}\right]}{\left[1 + \frac{R_4}{R_3}\right]}.$$

Now, $\dfrac{R_2}{R_1} = \dfrac{R_4}{R_3}$, and

$$v_O = -v_1 \frac{R_2}{R_1} + v_2 \frac{\left[1 + \dfrac{R_2}{R_1}\right]}{\left[1 + \dfrac{R_1}{R_2}\right]} + v_3 \frac{\left[1 + \dfrac{R_2}{R_1}\right]}{\left[1 + \dfrac{R_2}{R_1}\right]} = -v_1 \frac{R_2}{R_1} + v_2 \frac{R_2}{R_1} \frac{(R_1 + R_2)}{(R_1 + R_2)} + v_3.$$

whence $v_O = \dfrac{R_2}{R_1}(v_2 - v_1) + v_3$. Note that the operation is the same, but with the output established by v_3 for $(v_2 - v_1) = 0$. This is an interesting and important result!

2.35 *Check first* that it is a balanced difference amplifier for v_1 and v_2: see that $R_{4b} \parallel R_{4a} = 300k\Omega \parallel 150k\Omega = 300/3 = 100$ k$\Omega = R_2$, (OK).

(a) For $v_1 = v_2 = 5$ V: See (by superposition), $v_A = 5 \dfrac{100}{100 + 100} + 15 \dfrac{100 \parallel 150}{300 + 100 \parallel 150} + 0$, or
$v_A = 5\,(1/2) + 15\,(60/360) = 2.5 + 15/6 = 2.5 + 2.5 = 5\text{V}$. Now, since $v_A = 5\text{V}$, and $v_1 = 5\text{V}$, current in $R_1 = 0$, and $v_O = \mathbf{5V}$.

(b) For $v_1 = v_2 = 0\text{V}$, $v_A = 15 \dfrac{100 \parallel 150}{300 + 100 \parallel 150} = 15\left[\dfrac{60}{360}\right] = \dfrac{15}{6} = 2.5\text{V}$, and $v_O = 2.5\,(2) = \mathbf{5V}$.

(c) For $v_1 = 3\text{V}$, $v_2 = -2\text{V}$, see $v_A = -2\left[\dfrac{1}{2}\right] + 15\left[\dfrac{1}{6}\right] = -1 + 2.5 = 1.5\text{V}$. Thus $v_B = 1.5\text{V}$, and

$$v_O = v_B - \frac{v_1 - v_B}{R_1} \times R_2 = v_B - v_1 + v_B = 2\,(1.5) - 3 = \mathbf{0V}.$$

Alternatively, to use the extended result from P2.34 above, realize that R_{4b} with R_{4a} and the connected supplies have a Thevenin equivalent of $v_3 = \dfrac{150}{150 + 300} \times 15 = 5\text{V}$, and $150k \parallel 300k = 100k\Omega$. Thus, since $R_1 = R_2 = R_3 = R_4$, $v_O = v_2 - v_1 + v_3$. Therefore the output is the conventional difference plus the extra (reference) input. Thus (a) $v_O = \mathbf{5V}$, (b) $v_O = \mathbf{5V}$, (c) $v_O = -2 - 3 + 5 = \mathbf{0V}$.

2.36 For an overall gain of 100 V/V, with 10 V/V from the input stage, and 10 V/V from the output stage, see that $1 + 2R_2/R_1 = 10$, and $R_4/R_3 = 10$. For $R_1 = 10k\Omega$, $R_2 = (10 - 1)\,R_1/2 = 45k\Omega$, and for $R_3 = 10k\Omega$, $R_4 = 100k\Omega$. Now, for $v_1 = 5.0\text{V}$, $v_2 = 4.9\text{V}$, $v_{O1} = 5.0 + \dfrac{5.0 - 4.9}{10k\Omega}\,(45k\Omega) = \mathbf{5.45}$ V, and
$v_{O2} = 4.9 - \dfrac{5.0 - 4.9}{10k\Omega}\,(45k\Omega) = \mathbf{4.45}$ V. *Check:* $v_{O1} - v_{O2} = 5.45 - 4.45 = 1.00\text{V} = 10\,(0.1)$, as expected!

2.37 See that $v_Y = (1 + R_2/R_1)v_X$ and that the current into X is $i_X = (v_X - v_Y)/R_3$.

Thus $R_{in} = v_X/i_X = \dfrac{v_X}{[v_X - (1 + R_2/R_1)v_X]/R_3}$, or $R_{in} = R_3(1 - 1 - R_2/R_1) = -R_1 R_3/R_2$, as required.

See, from the diagram that $R_i = R_4 + R_{in} = \mathbf{R_4 - R_1 R_3/R_2}$.

Also, using a voltage-divider ratio, $\dfrac{v_X}{v_W} = \dfrac{R_{in}}{R_4 + R_{in}} = \dfrac{-R_1 R_3/R_2}{R_4 - R_1 R_3/R_2}$, or $v_X/v_W = \dfrac{1}{1 - (R_2/R_1)(R_4/R_3)}$,

and $v_Y/v_W = (1 + R_2/R_1)(v_X/v_W) = \dfrac{1 + R_2/R_1}{1 - (R_2/R_1)(R_4/R_3)}$. For $R_1 = R_2$, these become: $R_i = R_4 - R_3$,

$\upsilon_X/\upsilon_W = 1/(1 - R_4/R_3)$ and $\upsilon_Y/\upsilon_W = 2/(1 - R_4/R_3)$

a) For $R_4 = 2R_3$, $R_i = 2R_3 - R_3 = \mathbf{R_3}$, $\upsilon_X/\upsilon_W = 1/(1-2) =$ **– 1V/V** and $\upsilon_Y/\upsilon_W =$ **– 2V/V.**

b) For $R_4 = R_3$, $R_i = R_4 - R_3 = 0\Omega$ $\upsilon_X/\upsilon_W = 1/(1 - R_4/R_3) = \infty$, and $\upsilon_Y/\upsilon_W = \infty$, also.

c) For $R_4 = R_3/2$, $R_i = -R_4 - R_3 = -\mathbf{R_3/2}$, $\upsilon_X/\upsilon_W = 1/(1 - R_4/R_3) = 1/(1 - 1/2) =$ **2 V/V**, and $\upsilon_Y\upsilon_W =$ **4 V/V.**

For $\upsilon_Y/\upsilon_W =$ 10 V/V, $10 = 2/(1 - R_4/R_3)$. Thus $(1 - R_4/R_3) = 2/10 =$ 0.2, and

$R_4/R_3 =$ 0.8, whence $R_4 = 0.8\mathbf{R_3}$.

2.38 Use the results from P2.37 above: Input resistance at X to the right is $(-R)$, and to the left is R. Overall, at X, the resistance is $R \parallel (-R) = \dfrac{R(-R)}{R - R} = \infty$! Now the short circuit current at X to ground, with υ_I applied is υ_I/R (from W) and 0 from Y for a total of υ_I/R. Thus the Norton equivalent at X is a current source $I_N = \upsilon_I/\mathbf{R}$ with a shunt resistance $R_N = \infty$. For an impedance Z connected at X, the current flowing will be $\upsilon_I/\mathbf{R}$ and the voltage will be $\upsilon_I Z/R$. In general, the transmission $\upsilon_X/\upsilon_W =$ **Z/R** and $\upsilon_Y/\upsilon_W =$ 2Z/R. For a capacitor for which $Z = 1/(Cs)$, $\upsilon_X/\upsilon_W = 1/sCR$. For this non-inverting integrator, the time constant is **RC**. The unity-gain frequency is $\omega_o = 1/(\mathbf{RC})$

2.39 The common-mode input (at 60Hz) is 8 V peak, and the output is 0.6 V peak. The difference-mode input (at 1 kHz) is 1+1 = 2 mV peak, and the output is 60 mV peak. Thus, the common-mode gain = 0.6V/8V = **0.075 V/V**, and difference-mode gain = 60mV/2mV = **30 V/V**. Thus the CMRR = 30/.075 = 400 ≡ $20\log_{10} 400 =$ **52 dB**.

2.40 CMRR = $10^{100/20} = 10^5$. For a difference-mode output of 2 Vpp, the required common-mode signal output = 1/100 × 2 = 20 mVpp. Thus $\dfrac{200V/V}{20mV/\upsilon_{icm}} = 10^5$, and $\upsilon_{icm} = \dfrac{10^5}{200} \times 20 \times 10^{-3} =$ **10 Vpp.**

SECTION 2.7: EFFECT OF FINITE OPEN-LOOPS GAIN AND BANDWIDTH ON CIRCUIT PERFORMANCE

2.41 At and above the cutoff frequency f_{3dB}, $A \times f = f_t = 10^7$. Thus $f_{3dB} = f_t/A_o = 10^7/10^6 =$ **10Hz.** At f = 100 kHz, the available gain, $A = f_t/f = 10^7/10^5 =$ **100V/V.**

2.42 Closed-loop gain of 20 dB corresponds to a gain ratio of $10^{20/20}$ = 10 V/V. In general, $f_{dB} = f_t/(1 + R_2/R_1)$. For a gain of –10 V/V, $R_2/R_1 = 10$, and $f_{3dB} = 10^7/(1 + 10) =$ **0.909 MHz.** For a gain of +10 V/V, $1 + R_2/R_1 = 10$, $R_2/R_1 = 9$, and $f_{3dB} = 10^7/(1 + 9) =$ **1.00 MHz.** The phase shift at the 3 dB frequency is 45°, and 6° at 1/10 the 3 dB frequency. Thus 6° shift is reached at **90.9 kHz** for the inverting amplifier, and at **100 kHz** for the non-inverting, and is less than 6° for all lower frequencies.

2.43 Amplifiers have $A_o = 10^6$ V/V and $f_t = 10^7$ Hz: For maximum bandwidth, use the noninverting form with gain $(1 + R_2/R_1)$. *For a single amplifier*, with f_{3dB} = 10 kHz: $10^4 = 10^7/(1 + R_2/R_1)$. Thus $1 + R_2/R_1 = 10^7/10^4 = 10^3$, and the available gain is 10^3 V/V. *For 2 amplifiers in cascade*, each with a 3dB frequency at f_1, $f_{3dB} = (\sqrt{2} - 1)^{1/2} f_1 = 0.644 f_1$. Now, for f_{3dB} = 10kHz, f_1 = 10kHz/.644 = 15.54 kHz, with corresponding gain per stage of $\left(1 + \dfrac{R_2}{R_1}\right) = \dfrac{10^7}{15.54 \times 10^3}$ = 643.6 V/V, and, for two stages in cascade, a gain of $(643.6)^2 = 4.14 \times 10^5$ V/V.

2.44 Gain with feedback at low frequencies is likely to be: $-R_2/R_1 =$ **–100 V/V.** Assume $|A_0| >> 100$: Now, for a 3dB frequency f_L < 120 kHz, 100 f_L = 120(25), and $f_L = \dfrac{120}{100}(25) =$ **30 kHz.** For the closed-loop amplifier, the unity-gain frequency (where $R_2/R_1 = 1$) is 120(25) kHz = 3 MHz. But the unity-gain frequency of the negative-gain amplifier is $\frac{f_t}{(1 + R_2/R_1)}$, which is $\frac{f_t}{2}$ for $\frac{R_2}{R_1} = 1$. Thus $f_t/2 = 3$ MHz, and for the op amp itself, f_t = **6 MHz.**

2.45 f_t = 100 MHz, A_o = 20 V/V: Now, for $\frac{v_o}{v_i} = \frac{-R_2/R_1}{1 + (1 + R_2/R_1)/A_o} = -10$ V/V, $\frac{R_2}{R_1} = 10 + \frac{10}{20}\left[1 + \frac{R_2}{R_1}\right] = 10 + \frac{1}{2} + \frac{1}{2}\frac{R_2}{R_1}$, $\frac{1}{2}\frac{R_2}{R_1} = 10.5$, and $\frac{R_2}{R_1} = 21$. *Check:* $\frac{v_o}{v_i} = \frac{-21}{1 + (22/20)} = 10$ V/V, OK. Correspondingly, $f_{3dB} = \frac{f_t}{1 + R_2/R_1} = \frac{100 \times 10^6}{1 + 21}$ = **4.55 MHz.**

SECTION 2.8: LARGE-SIGNAL OPERATION OF OP AMPS

2.46 The largest possible peak output with zero average is **8 V**. The corresponding input has a peak-to-peak value of 2(8)/10 = **1.6 V.**

2.47 A 6-V pp triangle wave at f Hz moves 6 V in $1/2 \times 1/f$ seconds with a slope of $6/(1/2f) = 12f$ V/s. Now, if this just matches the slew rate: $12f$ V/s = 10V/μs, and $f = 10/12 \times 10^6$ Hz = **0.833 MHz.**

2.48 The slew-rate-limited bandwidth of an amplifier with a sinewave peak output V_o is $f_R = \frac{SR}{2\pi V_o}$. Now $f_R = f_b$, when $V_o = SR/(2\pi f_b)$. Now, for SR = 2 V/μs, and $f_b = 0.5 \times 10^6$, $V_o = 2 \times 10^6/(2\pi \times 0.5 \times 10^6)$ = **0.64 V peak.**

SECTION 2.9: DC IMPERFECTIONS

2.49 Nominal 4V peak swing is reduced by less than 4V/100 = 40mV. Since gain = −100 V/V, $R_2/R_1 = 100$ and the gain for the offset voltage is $1 + R_2/R_1 = 101$. Thus, the required offset < $\frac{40}{101}$ ≈ **0.40 mV.**

2.50 $A_f = -100$; $R_{in} = 100\text{ k}\Omega = R_1 \rightarrow R_2 = 10\text{ M}\Omega$.

a) For no compensation: $v_O = (1 + 100)(1\text{mV} + 30 \times 10^{-9} \times 100k \parallel 10M) = 101\ (1\text{ mV} + 3\text{ mV}) = 404$ mV ≈ **0.40 V.**

b) For compensation, using $R_3 = R_1 \parallel R_2 \simeq$ **100 kΩ**, only the offset current and offset voltage apply. Thus $v_O = (101)(1\text{mV} + 3 \times 10^{-9} \times 100\text{ k}) = 131.3$ mV ≈ **0.13 V.** For case (a), **bias current dominates;** For case (b), **offset voltage dominates.** For each dominant effect halved, the output offset becomes: (a) 101 (1 + 3/2) = **0.25 V**; (b) 101 (1/2 + 0.3) = **0.08 V.**

2.51

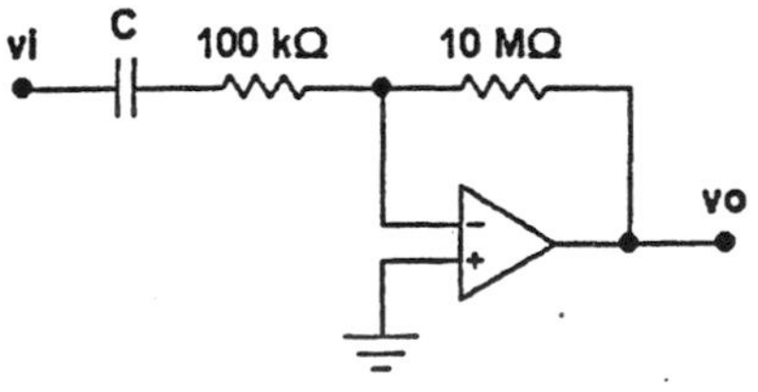

Note that the offset gain = 1V/V:

(a) For no compensation, $v_O = 1 \times 10^{-3} + 30 \times 10^{-9} \times 10 \times 10^6 = 301$ mV ≈ **0.3 V.**

(b) For compensation with R_3 = 10 MΩ from the negative input to ground, $v_O = 1\text{ mV} + 3 \times 10^{-9} \times 10 \times 10^6 = 31$ mV = **0.03 V.** Thus, use 10 MΩ to compensate.

2.52

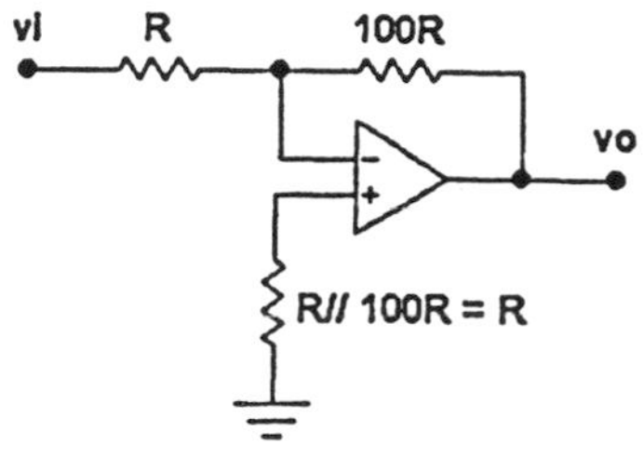

For no offset compensation,

$v_O = 101\ [2\times10^{-3} + (R \parallel (100\ R))\ 1.1\times10^{-6}] \le 0.5V$. Thus $2\times10^{-3} + 1.1\times10^{-6}R \le 4.95\times10^{-3}$, and $R = \dfrac{2.95\times10^{-3}}{1.1\times10^{-6}} = 2.68\ k\Omega$.

For compensation, with $R_3 = R$, $v_O \approx 101\ [2\times10^{-3} + R(2\times\frac{1}{10}\times10^{-6})] \le 0.5V$.

Thus $R = \dfrac{2.95\times10^{-3}}{0.2\times10^{-6}} = 14.75\ k\Omega$, (use 15 kΩ).

For this design, $R_{in} \approx 14.75k\Omega \approx$ **15 kΩ.**

2.53

a) Positive limiting is caused by current flowing out of the capacitor at the op amp negative-input node. Contributions include: bias current of 100 nA, offset current (worst case polarity) of 10 nA, current from the input resistor when the offset voltage is + 2mV is $2mV/10k\Omega = 200nA$. Total current is $100 + 10 + 200 = 310nA$. The ouput voltage will reach positive saturation in $T = CV/I = 0.1\times10^{-6}(10)/(310\times10^{-9}) =$ **3.22s.**

b) For negative limiting, current must flow into the input end of the capacitor. Maximum current from input resistor (with offset – 2mV) is 200 nA. Maximum offset current is 10 nA. Minimum bias current is – 100 nA (flowing into the top amp). Thus the total current is $200 + 10 - 100 = 110nA$ for which negative saturation occurs in $T = 0.1\times10^{-6}(10)/110\times10^{-9} =$ **9.09s.**

For Compensation: Make $R_a = R_1 = 10k\Omega$, in which case the bias current flows through R_1 (and R_a) to produce voltages at each input which are equal and cancel. Adjusting R_c with R_b provides an additional ± 2mV on the positive op-amp input to cancel the voltage offset, which though unknown may be relatively stable (certainly if the temperature is fixed). Even the offset current can be compensated on the short term. Typically a factor of 10x improvement is possible (at least) with saturation time increased to perhaps one minute. Because $R_a \approx R_1$, doubling the bias current does not matter much. (Again cancellation is usually good to within a 10% difference error.)

For compensated positive saturation, the choice of $R_a = 10k\Omega$ compensates bias current, while the current from R_b compensates for the offset voltage. Here, $v_C = \dfrac{10k\Omega}{10k\Omega + 10M\Omega}\times V_C = 2mV$ or $1/1001V_C = 2/1000$ and $V_C \approx 2V$. Note that the offset current flowing in 10 kΩ produces a voltage of $10\times10^{-9}\times10\times10^3 = 0.1$ mV. For this included, $V_C \approx (2.0 + 0.1)1001/1000 =$ **2.1V**

2.54 The nominal closed-loop gain is $G = 1 + R_2/R_1 = 1 + 10^6/10^4 =$ **101 V/V.** The bias current flowing in $10k\Omega(\parallel 10M\Omega)$ is responsible for an offset of $\dfrac{1.8 - 0.6}{101} = 11.88$ mV. Thus the bias current is $\dfrac{11.88\times10^{-3}}{10\times10^3} = 1.19\mu A$. The input offset voltage is **0.6/101 = 5.94 mV.** Note that a fraction of the 5.94 mV voltage offset is due to the offset current flow in 10 kΩ. Now, with all resistors reduced by a factor of 10, the change in input offset voltage is $\dfrac{0.6 - 0.4}{101} = 1.98$ mV, due to the difference between I_{offset} flowing in 10 kΩ and in 1 kΩ. That is, $I_{off}(10k - 1k) = 1.98$. $I_{offset} = 1.98/9k\Omega =$ **0.22 μA** . Now the original offset voltage 5.94 mV at the input includes the $V_{OS} + I_{offset}\times10k\Omega$, or $V_{OS} = 5.94mV - 220\times10^{-9}\times10^4 =$ **3.74 mV.**

To summarize, for the basic amplifier, $V_{OS} =$ **3.74 mV**, $I_{bias} =$ **1.19 μA**, $I_{offset} =$ **0.22 μA**.

Now if the 10 kΩ resistor at the negative input is capacitor-coupled, the offset voltage is multiplied only by 1, *but* the bias and offset currents flow in 1 MΩ. If they add, the output offset becomes $1M\Omega(1.19 + 0.22)\times10^{-6} =$ **1.41V**!. Two possible compensations are possible:

a) If the resistances R_1 and R_2 are reduced by a factor of 10, the output offset reduces to **0.14V.**

b) If a resistor $R_2 = 1M\Omega$ is connected from the positive op-amp input terminal to the actual input, which is grounded, bias currents are compensated, and the output offset becomes $1M\Omega(0.22 \times 10^{-6}) =$ **0.22V.**

c) If both techniques are used (ie using a 1 kΩ resistor and two 100 kΩ resistors), the offset due to offset current reduces to about 22 mV. With the voltage offset remaining, about 22 + 3.74 = **25.7 mV** would be found at the output.

Chapter 3

DIODES

SECTION 3.1: THE IDEAL DIODE

3.1 Diodes are ideal: Thus the forward voltage drop is 0V and reverse current is 0 mA:

(a) Diode is polarized to conduct by the +5V and 0V connections: Thus V_a = **0V**, and I_a = (5–0)/1kΩ = **5mA.**

(b) As in (a), but the most negative supply is –5V. Thus V_b = –5+0 = **–5V**, and I_b = (+5 – 0 – (–5))/1kΩ = **10mA.**

(c) Diodes both conduct: Thus V_c = –10 + 0 + 0 = **–10V**, and I_c = (0 –0 –0 –(–10))/1kΩ = **10mA.**

(d) Both diodes are polarized to conduct. Thus V_d = +5 –0 = **5V**, and I_d = (+5 –0–0–0)/1kΩ = **5mA.**

(e) Upper diode is polarized to conduct, but lower to cut off. Thus current I_e = **0mA**, and V_e = 5V – 0V = **5V** (with the upper diode conducting the meter current).

(f) Upper diode polarized to cut off, and lower to conduct. Thus current, I_f = **0mA**, and V_f = –5V –0 – 0 = **–5V** (with the lower diode conducting the meter current (assumed very small) through 1kΩ).

3.2 (a)

i) V_A = +5V; upper diode conducts; V_Y = **+5V.**

ii) The output is high if either input is high. For high (+5V) defined as logic '1', the output is 1, if $A = 1$ or $B = 1$; ie, $Y = A + B$. Thus, the function is **logic OR** (in positive logic).

iii) For high (+5V) defined as logic '0', the output is high (logic '0') if either A or B is high (logic '0'). That is $\overline{Y} = \overline{A} + \overline{B}$, or $Y = \overline{\overline{Y}} = \overline{\overline{A} + \overline{B}} = A \bullet B = AB$ (for simplicity). Thus the function is a **logic AND** in negative logic. The AND idea can be verified by noting that the output is low (logic '1') only if A *and* B are both low.

(b)

i) V_B, V_C and V_D are all 0V. V_Y follows the highest input. Thus V_Y = **0V.**

ii) $Y = A + B + C$, an **OR** (in positive logic).

iii) $Y = A \bullet B \bullet C = ABC$, an **AND** (in negative logic).

(c)

i) See that the output follows the lower of A or E (just as the output followed the upper of A or B in (a)). That is, $V_A = V_E$ = 5V → V_Y = **5V.**

ii) From above, $\overline{Y} = \overline{A} + \overline{B}$, ie $Y = \overline{\overline{Y}} = \overline{\overline{A} + \overline{B}} = A \bullet B = AB$ That is, the function is **AND** or, directly, $Y = A \bullet B = AB$ (since both inputs must be high for the output to be high).

iii) Directly, from logic first principles, or in analogy to the circuit in (a), $Y = A + B$ (in negative logic). That is, the function is an **OR.**

(d)

i) V_A = 5V, V_B = 0V → V_Y = **0V.**

ii) **AND** (positive logic).

iii) **OR** (negative logic).

(e)

i) $V_A = V_E = 5V$, $V_C = 0V \rightarrow V_Y =$ **0V**.

ii) **AND**; $Y = A \bullet E \bullet C = AEC$.

iii) **OR** ; $Y = A + E + C$.

3.3 Use positive logic, that is, '1' = 5V, '0' = 0V.

For D_5 open, $P = A \bullet E = AE$.

For D_6 open, $Q = B \bullet C = BC$.

Now, note that the current available from node P or Q (100μA) exceeds that drawn from node Y: Thus if P or Q goes high, Y is also pulled high. Thus $Y = P + Q + D$ and, **Y = AE + BC + D** (in positive logic).

Now, for $V_A = V_E = 5V$; $V_B = V_C = V_D = 0V$; that is $A = E =$ '1' and $B = C = D =$ '0', for which $Y = 1 \bullet 1 + 0 \bullet 0 + 0 = 1$. Thus the output is **logic '1'**, or **5V** (for ideal diodes)

3.4 12V rms = 12(1.414) = 16.97V peak

For a 12V battery and an ideal diode, the peak diode current is (16.97 −12)/(10 + 50) = **82.8mA**

The diode begins to conduct (and ceases to conduct) when 16.97 sinωt = 12V or sinωt = 12/16.97 = 0.707 or $\omega t = \pi/4 \equiv 45°$.

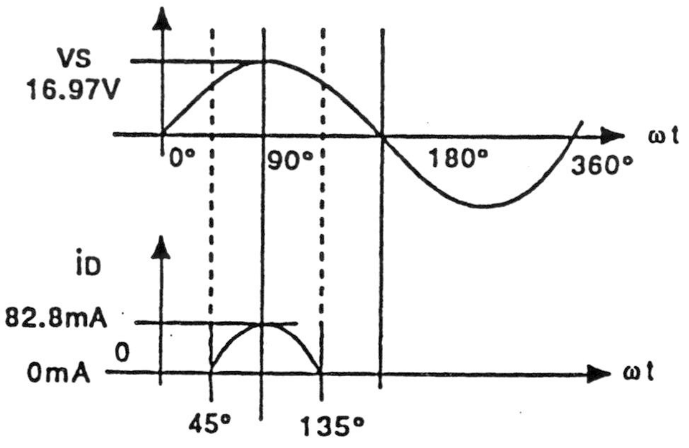

Average value of $i_D = \frac{1}{2\pi}\int_{\pi/4}^{3\pi/4}$ (16.97 sin Θ - 12) $d\Theta$)/60 =

$$\frac{1}{2\pi}\left[-\frac{16.97}{60}\cos\Theta - \frac{12}{60}\Theta\right]_{\pi/4}^{3\pi/4}$$

$$= \frac{0.282}{2\pi}(0.707 + 0.707)$$

$$-0.2\,\frac{3\pi/4 - \pi/4}{2\pi} = (.0635 - .0500)$$

A = **13.5mA.** Now for V_B = 14V: Peak current = (16.97 − 14)/60 = 0.0495A or **49.5mA.** Conduction begins when 16.97 sinωt = 14, or sinωt = 14/16.97 = 0.825, or $\omega t \equiv$ 55.6°.

$$\text{Average current} = \frac{1}{2\pi}\left[-\frac{16.97}{60}\cos\Theta - \frac{14}{60}\Theta\right]_{55.6^\circ}^{180^\circ-55.6^\circ} = -\frac{.282}{2\pi}(-.565 - .565) - 0.233$$

$$\left[\frac{\frac{124.4 - 55.6}{360}\times 2\pi}{2\pi}\right] = 0.0507 - 0.0445 = 0.0062\text{A or } \mathbf{6.2mA}.$$

3.5

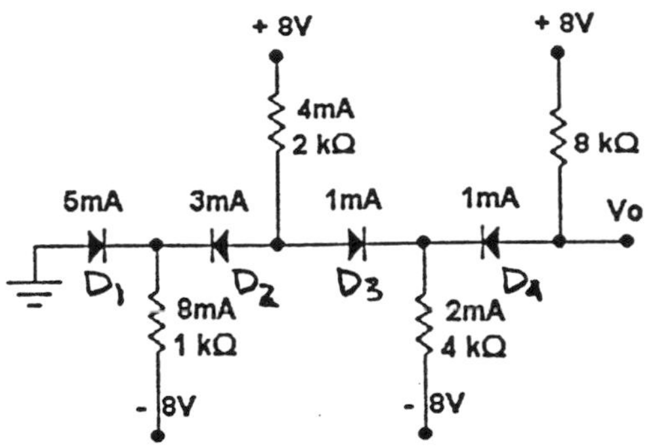

Consider the conducting state of each of the diodes: Assume D_1 conducts; thus its cathode is at 0V. Correspondingly, the current in the 1kΩ resistor, I_{1K} = (0 – – 8)/1k = 8mA. Assume D_2 conducts; thus its cathode is at 0V, and I_{2K} = (8 – 0)/2k = 4mA.

Now, since $I_{1K} > I_{2K}$, then D_1 and D_2 both conduct as assumed. Now, noting that succeeding resistors are progressively larger, assume that both D_3 and D_4 also conduct, and that their anode and cathode voltages are all zero volts. Thus, I_{4K} = (0 – – 8)/4k = 2mA, and I_{8K} = (8 – 0)/8k = 1mA.

Thus, overall: $I_{D4} = I_{8K}$ = **1mA**; $I_{D3} = I_{4K} - I_{D4}$ = 2 – 1 = **1mA**; $I_{D2} = I_{2K} - I_{D3}$ = 4 – 1 = **3mA**; $I_{D1} = I_{1K} - I_{D2}$ = 8 – 3 = **5mA**; Thus diodes are all conducting, and V_o = **0V**.

SECTION 3.2: TERMINAL CHARACTERISTICS OF JUNCTION DIODES

3.6 Given: V_1 = 0.700V @ I_1 = 100μA, and V_2 = 0.815V @ I_2 = 1mA.

Now $i_D = I_s\, e^{v_D/nV_T} \rightarrow v_D = nV_T \ln i_D/I_s = nV_T \ln i_D - nV_T \ln I_S$.

$v_{D1} - v_{D2} = nV_T (\ln i_{D1}/i_{D2})$.

Here, 0.700 – 0.815 = $n\ V_T$ ln 0.1/1 = 0.025n (– 2.3026).

n = 0.115/(.025(2.303)) = 1.997 or **2.00**.

Now $I_S = i_D\, e^{-v_D/nV_T} = 100 \times 10^{-6}\, e^{-700/(2(25))} = 10^{-4}\, e^{-14} = 8.32 \times 10^{-17}$**A**.

3.7 $i = I_S\, e^{v/nV_T}$, and $v_1 - v_2 = nV_T \ln i_1/i_2$ – – – (1).

Thus, $v_1 = v_2 + nV_T \ln i_1/i_2$ = 700 + 1(25) ln 1/0.1 = 700 + 25 ln 10 = 757.6mV, or **0.758V**.

Now from (1), $i_1 = i_2\, e^{\frac{v_1 - v_2}{nV_T}} = 0.1\, e^{(815 - 700)/1(25)}$ = **9.95mA**.

3.8 $i = I_S\, e^{v/nV_T} \rightarrow v = nV_T \ln i/I_S$, or $v = v_o + nV_T \ln i/i_o$.

At 10mA, v = 700 + 2(25) ln (10 × 10^{-3}/10) = 355mV, or **0.355V**.

At 10μA, v = 700 + 2(25) ln (10 × 10^{-6}/10) = **9.2mV**.

3.9 At 100μA and 25°C, $v = v_o + nV_T \ln i/i_o$ = 700 + 2(25) ln 0.1/1 = 585mV.

At 100μA and 95°C, v = 585 + 2.0 (25 – 95) = 445mV, or **0.445V**.

3.10 The leakage doubles for each 10°C rise in temperature.

Now at 95°C, it is $2^{(95-25)/10} = 2^7$ = 128 times its value at 25°C, or **128nA,** or 0.128μA.

Now at 100°C, it is $2^{(100-25)/10} = 2^{7.5}$ = 181 times its 25°C value, or **181nA**.

Note that it is $2^{(100-95)/10} = 2^{0.5}$, or 1.41 times as large as at 95°C, ie 1.41 × 128 = 181nA, as already noted.

SECTION 3.3: PHYSICAL OPERATION OF DIODES

3.11 Acceptor concentration is $N_A = 10^{-m}$. Thus the hole concentration is $p_{p0} \approx 10^{-m}$, while the electron concentration is $n_{p0} = 10^{-n}$, $n >> m$. *More precisely,* in 10^n atoms, one is ionized at a particular temperature to produce one hole and one electron. As well, the number of acceptor - produced holes is $10^n \times 10^{-m}$. Thus the total number of holes is $1 + 10^n \times 10^{-m}$ in 10^n atoms, where $\mathbf{p_{p0}} = \dfrac{1 + 10^n \times 10^{-m}}{10^n} \approx 10^{-m}$, while $\mathbf{n_{p0}} = 10^{-n}$ directly.

3.12 From Eq. 3.6, $n_i^2 = BT^3 e^{-E_G/kT}$, where for silicon, $B = 5.4 \times 10^{31}$, $E_G = 1.12eV$, $k = 8.62 \times 10^{-5}$.

At 200 K, $n_i^2 = 5.4 \times 10^{31} \times 200^3 e^{-1.12/(8.62 \times 10^{-5} \times 200)} = 2.639 \times 10^{10}$,

and $n_i =$ **1.62 × 10⁵ carriers/ cm³**.

At 300 K. $n_i^2 = 5.4 \times 10^{31} \times 300^3 e{-1.12/(8.62 \times 10^{-5} \times 300)} = 2.2616 \times 10^{20}$,

and n_i = 1.50×bold {10 sup 10}carriers/bold {cm sup 3}..*IPAt*400*K*,n sub i sup 2 ~=~ 5.4 ~times~ 10 sup 31 ~times~ 400 sup 3 e {-1.12/(8.62 ~times~ 10 sup -5 ~times~ 400)} ~=~ 2.706 ~times~ 10 sup 25,.*IP*andn sub i ~=~ **5.20 × 10¹² carriers/ cm³**.

For the 100 °C rise above room temperature, the increase is $\dfrac{5.2 \times 10^{12} - 1.5 \times 10^{10}}{1.5 \times 10^{10}} \times = \mathbf{3.46 \times 10^4\%!}$

At this temperature (127 °C), the fraction of ionized atoms is $\dfrac{5.2 \times 10^{12}}{5 \times 10^{22}} \approx 10^{-10}$, or **one in every 10 (US) billion!** Note that the reference on page 139 of the Text is to British billion.

3.13 From Eq. 3.10C, resistivity is $\rho = 1/[q(p\mu_p + n\mu_n)]\Omega cm$.

(a) For intrinsic silicon at room temperature: $n = p = n_i = 1.5 \times 10^{10}/cm^3$.

Thus $\rho = 1/[1.6 \times 10^{-19}(1.5 \times 10^{10})(1350 + 480)]$ or $\rho =$ **2.28 × 10⁵Ωcm**

(b) For n-type silicon with $N_D = 10^{16}/cm^3$: $n_{no} = 10^{16}$, and $p_{no} = n_i^2/N_D = (1.5 \times 10^{10})^2/10^{16}$, or $p_{no} = 2.25 \times 10^4$.

Thus (including mobility reduction):

$\rho = 1/[1.6 \times 10^{-19}(2.25 \times 10^4 \times 0.8 \times 480 + 10^{16} \times 0.8 \times 1350)]$ or $\rho =$ **1.73 Ωcm.**

As noted in the solution to P3.12 above, for a 100 °C rise n_i will increase by a factor of 3.46×10^4. Thus for the intrinsic material ρ will decrease to $2.28 \times 10^5/3.46 \times 10^4 =$ **6.59Ωcm,** while for the doped material, it will stay the same, at **1.73Ωcm**

3.14 The depletion region will be **larger in the lighter-doped p region.** In fact, it will be **10 × larger** there than in the *n* region.

3.15 Eq. 3.18 in the Text states that the built-in voltage is $V_0 = V_T \ln[N_A N_D/n_i^2]$, where $n_i = 1.5 \times 10^{10}/cm^3$ for intrinsic silicon at room temperature. Here $700 = 25\ln[N_A N_D/(1.5 \times 10^{10})^2]$.

Thus $N_A N_D = (1.5 \times 10^{10})^2 e^{700/25} = 3.254 \times 10^{32}/cm^6$ – – – (1)

(a) *For Equal Doping:*

$N_A N_D = N^2 = 3.254 \times 10^{32}$, whence $N =$ **1.804 × 10¹⁶/cm³** $= 1.80 \times 10^{22}/m^3$.

From Eq. 3.20 in the Text, $W_{dep} = x_n + x_p = 2x = [(2\varepsilon_s/q)(1/N_D + 1/N_A)V_0]^{1/2}$ – – – (2)

where (from Table 3.1 on page 157 of the Text), $q = 1.60 \times 10^{-19}C$, and $\varepsilon_s = 11\varepsilon_o = 11(8.85 \times 10^{-14})F/cm = 9.74 \times 10^{-13}F/cm = 9.74 \times 10^{-11}F/cm$.

Thus $W_{dep} = [2(9.74 \times 10^{-13})/(1.60 \times 10^{-19})(2/1.804 \times 10^{16})0.7]^{½} = 3.07 \times 10^{-6} cm =$ **0.307μm.**

The distance depletion extends into each region is $W/2 =$ **154 nm**. From the preamble to Eq. 3.21, the uncovered charge on each side is

$q_J = qN_D x_n = 1.6 \times 10^{-19} \times 1.80 \times 10^{32} \times 154 \times 10^{-9} \times 30 \times 10^{-6} \times 50 \times 10^{-6} =$ **0.665 pC.**

(b) *For 10-to-1 doping:*

From (1), $N_A N_D = 3.254 \times 10^{32}/cm^6$, and

for $N_D = 10N_A$, say, see $10N_A^2 = 3.254 \times 10^{32}$, where $N_A =$ **5.70 × 10^15/cm^3** $= 5.70 \times 10^{21}/m^3$, and $N_D =$ **5.70 × 10^16/cm^3** $= 5.70 \times 10^{22}/m^3$.

From (2), $W_{dep} = [2(9.74 \times 10^{-13})/(1.60 \times 10^{-19})(1/(5.70 \times 10^{16})(1 + 10)0.7]^{½} 4.046 \times 10^{-5} cm =$ **0.405μm.**

Now $W_{dep} = x_n + x_p = x_p(N_A/N_D + 1) = x_p(0.1 + 1)$.

Thus $x_p = 0.405/1.1 =$ **0.368μm**

$x_n = x_p/10 =$ **0.037μm.**

The charges uncovered on each side are equal, and of value

$q_J = qN_D x_n A = 1.6 \times 10^{-19} \times 5.70 \times 10^{22} \times 0.368 \times 10^{-7} \times 30 \times 10^{-6} \times 50 \times 10^{-6} =$ **0.504 pC**

3.16 $I_S - I_D = I \rightarrow 15 - I_D = 10$ and $I_D = 15 - 10 = 5$. Thus the diffusion current is $I_D =$ **5nA.**

3.17 Combining Eq. 3.21 and Eq. 3.22, see

$$q_J = q\frac{N_A N_D}{N_A + N_D} A W_{dep} = \frac{qN_A N_D}{N_A + N_D} A \left\{ \frac{2\varepsilon_s}{q} \left[\frac{1}{N_A} + \frac{1}{N_D} \right] (V_O + V_R) \right\}^{½}, \text{ whence}$$

$$\mathbf{q_J = A \left[2\varepsilon_s q \frac{N_A N_D}{N_A + N_D} (V_O + V_R) \right]^{½}}.$$

For the $(30 \times 50)\mu m^2$ junctions in P3.15 above, for which $V_0 = 0.7V$:

a) $N_A = N_D = 1.80 \times 10^{22}/m^3$.

b) $N_A = 5.70 \times 10^{21}/m^3$, $N_D = 5.70 \times 10^{22}/m^3$.

(a) For $N_A = N_D$:

$q_J = 30 \times 50 \times 10^{-12}[2 \times 9.74 \times 10^{11} \times 1.6 \times^{-19} \times (1.80 \times 10^{22}/2)(0.7 + V_R)]^{½} = 7.94 \times 10^{-13}[0.7 + V_R]^{½}$

For $V_R = 0.0V$, $q_J = 7.94 \times 10^{-13}(0.7)^{½} = 6.65 \times 10^{-13} C$ or **0.665 pC**, as in P3.15.

For $V_R = 10V$, $q_J = 7.94 \times 10^{-13}(0.7 + 10)^{½} =$ **2.597 pC.**

For $V_R = 11V$, $q_J = 7.94 \times 10^{-13}(0.7 + 11)^{½} =$ **2.716 pC.**

From $\Delta Q = C\Delta V$, the junction capacitance is $C = \Delta Q/\Delta V = (2.716 - 2.597)/(11 - 10) =$ **119 fF.**

Now, from Eq. 3.26,

$$C_{j0} = A \left\{ \left[\frac{\varepsilon_s q}{2} \right] \left[\frac{N_A N_D}{N_A + N_D} \right] \left[\frac{1}{V_O} \right] \right\}^{½} \quad --- (1)$$

Thus, $C_{j0} = 30 \times 50 \times 10^{-12}[9.74 \times 10^{-11} \times (1.6 \times 10^{-19}/2)(1.6 \times 10^{-19}/2)(1.80 \times 10^{22}/2)/0.7]^{½} = 0.475 fF$.

At 10.5 V, $C_j = C_{j0}/[1 + V_R/V_O]^{½} = 0.475/[1 + 10.5/0.7]^{½} = 0.475/4 = 0.119 pF =$ **119 fF** same as found above.

Now for a graded junction with $m = 1/3$:

At $V_R = 10.5V$, $C_j = 0.475/(1 + 10.5/0.7)^{1/3} =$ **189 fF.**

At $V_R = 100V$, $C_j = 0.475/(1 + 100/0.7)^{1/3} =$ **90.6 fF**

(b) For $N_A = N_D/10 = 5.70 \times 10^{21}/m^3$

$q_J = A[2\varepsilon_s q N_A N_D/(N_A + N_D)(V_O + V_R)]^{1/2} = 30 \times 50 \times 10^{-12}[2(9.74 \times 10^{-11})(1.6 \times 10^{-19}) \times 5.70 \times 10^{21} \times 10/11]^{1/2}(0.7 + V_R)^{1/2} = 6.028 \times 10^{-13}(0.7 + V_R)^{1/2}$.

For $V_R = 0$, $q_J = 6.028 \times 10^{-13}(0.7)^{1/2} =$ **0.5043 pC.**

For $V_R = 10V$: $q_J = 6.028 \times 10^{-13}(0.7 + 10)^{1/2} =$ **1.9718 pC.**

For $V_R = 11V$, $q_J = 6.028 \times 10^{-13}(0.7 = 11)^{1/2} =$ **2.0619 pC.**

Now, the junction capacitance is $C = \Delta Q/\Delta V = \dfrac{2.0619 - 1.9718}{1} =$ **90.1 fF.**

Now, from Eq. 3.26 ((1) above),

$C_{jo} = 30 \times 50 \times 10^{-12}[9.74 \times 10^{-11} \times (1.6 \times 10^{-19}/2)(5.7 \times 10^{21})(10/11)(1/0.7)]^{1/2} = 0.360pF$.

and $C_j = C_{jo}/[1 + V_R/V_O]^{1/2}$.

At $V_R = 10.5V$, $C_j = 0.360/(1 + 10.5/0.7)^{1/2} = 0.360/4 = 0.090 =$ **90 fF**, very much like the earlier estimate. Now for a graded junction with $m = 1/3$.

At $V_R = 10.5V$, $C_j = 0.360/(1 + 10.5/0.7) = 0.360/(16) =$ **143 fF.**

At $V_R = 100V$, $C_j = 0.360/(1 + 100/0.7) =$ **68.7 fF**

3.18 $Q = CV \rightarrow C = \dfrac{0.1pC}{1V} = 0.1\text{pF}$. That is, the depletion capacitance is **0.1pF.**

3.19 From Eq. 3.27, $C_j = C_{j0}(1 + V_R/V_0)^{-m}$.

For $V_R = 2V$, $1.8 = C_{j0}(1 + 2/V_0)^{-1.6}$.

For $V_R = 10V$, $0.2 = C_{j0}(1 + 10/V_0)^{-1.6}$.

Divide these to get $9 = [(1 + 2/V_0)/(1 + 10/V_0)]^{-1.6}$

Thus $1 + 2/V_0)/(1 = 10/V_0) = 9^{-1/1.6} = 0.253$, and $1 + 2/V_0 = 0.253 = 2.53/V_0$, and

$0.53/V_0 = 1 - 0.253 = 0.747$, whence $V_0 = 0.53/0.747 =$ **0.709V.**

Now, $0.2 = C_{j0}(1 + 10/0.706)^{1.6} = 0.2(15.1)^{1.6} =$ Thus $C_{j0} =$ **15.4 pF.**

For $V_R = 0V$, $C_j = 15.4(1 + 0/0.71)^{-1.6} =$ **15.4 pF,** as expected, the same as C_{j0}, by definition.

3.20 $P_D = V_B I_R$. Thus, $I_R = \dfrac{P_D}{V_B} = \dfrac{50 \times 10^{-3}}{120} =$ **0.42mA.**

For breakdown only 10% of the time, a peak current of 0.42/0.1 = **4.2mA** can be tolerated.

3.21 From Exercise 3.12, in doped silicon, $\mu_p = 400cm^2/Vs$.

From Eq. 3.12, $D_P = V_T\mu_p = 25 \times 10^{-3} \times 400 = 10cm^2/s$.

From Eq. 3.30, the diffusion length, $L_p = \sqrt{D_P\tau_P}$ where $\tau_P = 1ns$.

Thus $L_P = \sqrt{10(1)10^{-9}} = 10^{-4}cm = 10^{-6}m =$ **1μm.** For the diffusion profile, the excess hole concentration is $p_n(x) = p_n(0)e^{-x/L}$ where a 10% level is reached when $e^{-x/L} = 0.1$ or $x/L = 2.3$.

Thus the conentration reaches 10% at $2.3L =$ **2.3μm** from the depletion-region edge.

3.22 From Eq. 3.34, $I_S = Aqn_i^2\left(\dfrac{D_p}{L_pN_D} + \dfrac{D_n}{L_nN_A}\right)$

where from Table 3.1 on page 156 $q = 1.6 \times 10^{-19}C$, $n_i = 1.5 \times 10^{10}/cm^3$.

Now, from Eq. 3.12, $D_n/\mu_n = D_p/\mu_p = V_T$ and from Eq. 3.30, $L = \sqrt{D \times \tau}$
whence $D/L = \mu V_T/(\mu V_T \tau)^{1/2} = (\mu V_T/\tau)^{1/2}$.
Thus $D_P/L_P = (400(25 \times 10^{-3})/10^{-9})^{1/2} = 10^5 cm/Vs = 10^3 m/Vs$
$D_n/L_n = (1100(25 \times 10^{-3}/(2 \times 10^{-9}))^{1/2} 1.17 \times 10^5 cm/Vs = 1.17 \times 10^3 m/Vs$.
Thus
$I_S = 3 \times 10^{-6} \times 5 \times 10^{-6} \times 1.6 \times 10^{-19} \times (1.5 \times 10^{10} \times 10^6)^2[10^3/(10^{16} \times 10^6) + 1.17 \times 10^3/(10^{17} \times 10^6)]$
$= 5.4 \times 10^2[10^{-19} + 1.17 \times 10^{-20}] =$ **6.03×10^{-17}A**

3.23 From Eq. 3.34,

$$I_S = Aqn_i^2\left(\frac{D_p}{L_p N_D} + \frac{D_n}{L_n N_A}\right) \text{ where } n_i^2 = BT^3 e^{-E_G/kT}, \text{ and}$$

$B = 5.4 \times 10^{31}$, $E_G = 1.12eV$, $k = 8.62 \times 10^{-5} eV/K$.
We see that the temperature dependence originates in the n_i^2 term. Evaluate this at two temperatures and find the coefficient:
At 300 K, $n_i^2 = 5.4 \times 10^{31}(300)^3 e^{-1.12/(8.62 \times 10^{-5}(300))} = 1.458(10^{39})e^{-43.3}$
$= 1.458(10)^{39} 1.551 \times 10^{-19} = 2.2616 \times 10^{20}$.
At 303K, $n_i^2 = 5.4 \times 10^{31}(303)^3 e^{-1.12/(8.62 \times 10^{-5}(303))} = 1.502(10)^{39} 2.382 \times 10^{-19} = 3.5782 \times 10^{20}$.

Thus the temperature coefficient is $\dfrac{(3.5782 - 2.2616)}{(2.2616)(303 - 300)} \times 100 = 19.4\%/^\circ C$

3.24 From the solution to P3.22 above, in which (from Eq. 3.33),

$$I = Aqn_i^2\left(\frac{D_p}{L_p N_D} + \frac{D_n}{L_n N_A}\right)\left(e^{V/V_T} - 1\right)$$

the ratio of the hole to electron components $I_p/I_n = 10^{-19}/1.17 \times 10^{-20} = 8.55$
Of the total, $I = I_p + I_n$,

$I_p/I = \dfrac{I_p}{I_p + I_n} = \dfrac{1}{1 + I_n/I_p} =$ **89.5%** (holes), and $I_n/I = \dfrac{I_n}{I_p - I_n} = \dfrac{1}{1 = I_p/I_n} =$ **10.5%** (electrons).

For a total current of 1 mA flowing in the junction, $I_p = 895\mu A$ flows as hole current and $I_n = 105\mu A$, as electron current.
Now, in general, the stored charge is $Q = \tau I$ for each current component.
For holes, $Q_p = \tau_p I_p = 1 \times 10^{-9} \times 895 \times 10^{-6} =$ **0.895 pC.**
For electrons, $Q_n = \tau_n I_n = 2 \times 10^{-9} \times 105 \times 10^{-6} =$ **0.210 pC.**
Total stored charge is $Q = 0.895 + 0.210 = 1.105 pC$, and the mean transit time $\tau_T = Q/I = 1.105 \times 10^{-12}/1 \times 10^{-3} =$ **1.1 ns.**
Here, from Eq. 3.39, $C_d = (\tau_T/V_T)I = \tau_T I/V_T = 1.1 \times 10^{-9} \times 1 \times 10^{-3}(25 \times 10^{-3})$, or $C_d =$ **4 pF**

3.25 From Eq. 3.39, generalized, $C_d = (\tau_T/nV_T)I$. Thus $\tau_T = nV_T C_d/I$. For this junction, $\tau_T = 2(25 \times 10^{-3})1 \times 10^{-12}/1 \times 10^{-3} =$ **50 ps.**
For a junction 10× larger, τ_T is the same, namely **50 ps**, simply because it relates stored charge to current level at an essentially-constant junction-voltage. For a larger junction, the current density changes, but the total charge does not.
Total stored charge at 1 mA is $Q = \tau_T I = 50 \times 10^{-12} \times 1 \times 10^{-3} =$ **50 fC.**
For 10 mA, $Q_{10} = 10(50) = 500 fC =$ **0.5 pC.**

3.26 From Eq. 3.38, $q = \tau_T i$ and $i = I_S e^{\upsilon/(nV_T)}$ Thus $q = \tau_T I_S e^{\upsilon/nV_T}$.

Now, $C_d = \partial q/\partial \upsilon = \tau_T I_S/(nV_T) e^{\upsilon/nV_T}$.

Thus $\mathbf{C_d = \tau_T I/(nV_T)}$, as Eq. 3.39 indicates (I being substituted for i and n added for generality).

SECTION 3.4: ANALYSIS OF DIODE CIRCUITS

3.27

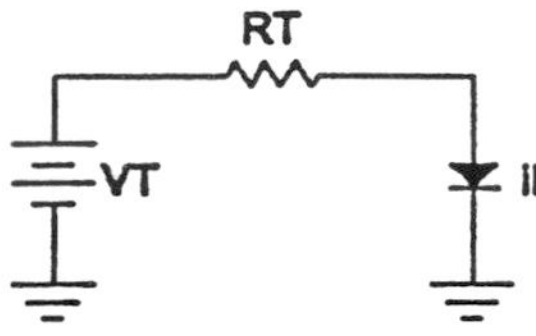

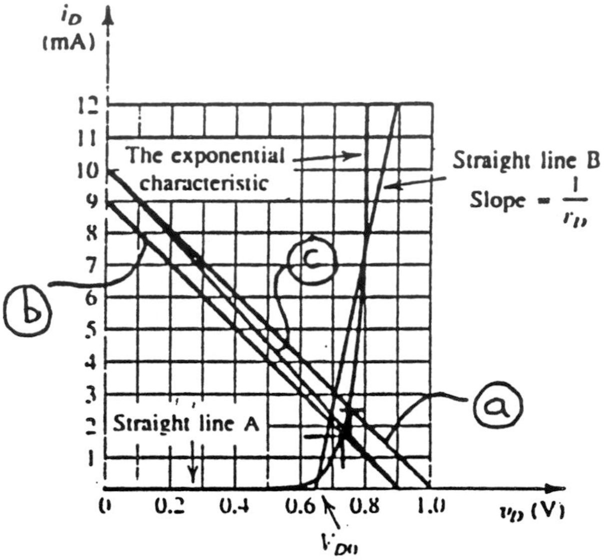

For the Load Lines:

(a) $V_T = 1$V, $R_T = 100\Omega$, $I_T = 1/100 = 10$mA.

(b) $V_T = 0.9$V, $R_T = 100\Omega$, $I_T = 0.9/100 = 9$mA.

(c) $V_T = 0.9$V, $R_T = 90\Omega$, $I_T = 0.9/90 = 10$mA.

From the Graph:

(a) $V_D = \mathbf{0.75V}$, $I_D = \mathbf{2.5mA}$.

Check: $I_D = 10 - V_D/0.1 = 10 - 10V_D = 10 - 10\,(.75) = 2.5$ mA.

(b) $V_D = \mathbf{0.73V}$, $I_D = \mathbf{1.7mA}$.

Check: $I_D = 9 - V_D/0.1 = 9 - 10V_D$, or $I_D = 9 - 10\,(.73) = 1.7$ mA.

(c) $V_D = \mathbf{0.74V}$, $I_D = \mathbf{1.8mA}$

Check: $I_D = 10 - V_D/.09 = 10 - 11.11\ V_D = 10 - 11.11\,(.74) = 1.77$ mA.

3.28 Though the diode is "like the one sketched in Fig. 3.20", use its analytical description to provide a comparison with the results in P3.27a) above.

For the diode: $\upsilon = 0.7 + 0.1 \log i/1$.

For the circuit: i, $V_{DD} = 1.00\text{V} = (100\Omega)\ i + 0.7\text{V} + 0.1 \log i/1$,

or $i = 0.3\text{V}/0.1\text{k} - (0.1/0.1) \log i$, or $i = 3 - \log i$, in mA.

Solve iteratively: Initially with $i = 1$:

$i = 1 \rightarrow i = 3 - \log 1 = 3 - 0 = 3$mA,

$i = 3 \rightarrow i = 3 - \log 3 = 3 - .48 = 2.52$mA,

$i = 2.52 \rightarrow i = 3 - \log 2.52 = 2.60\text{mA}$, whence

$i = 2.60 \rightarrow i \times 30 - \log 2.60 = 2.59mA$, whence

$I_D =$ **2.59mA**, and $V_D = 0.7 + 0.1 \log 2.59 =$ **0.741V**.

3.29 *For the diode,* $\upsilon - V_{Do} + r_{Di}$ Here, $V_{DD} = 1.00 = (100\Omega)\, i + 0.65\text{V} + (20\Omega)\, i$,

or for i in mA, $I_D = i = \dfrac{1.00 - 0.65}{0.1 + 0.02} = \dfrac{0.35V}{0.12k\Omega} =$ **2.92mA**, and $V_D = 0.65 + 2.92\ (.02) =$ **0.708V**.

3.30 For $V_{DO} = 0.70\text{V}$, and $r_D = 10\Omega$, $1.00 = 0.1i + 0.70 + 0.01i$, and

$I_D = i = \dfrac{1.00 - 0.70}{0.1 + 0.01} = \dfrac{0.30}{0.11} =$ **2.73mA**, with $V_D = 0.70 + 2.73\ (0.01) =$ **0.727V**.

For $V_{DO} = 0.75\text{V}$ and $r_D = 0\Omega$, $I_D = i = \dfrac{1 - 0.75}{0.1 + 0} =$ **2.50mA**, with $V_D =$ **0.750V**.

3.31

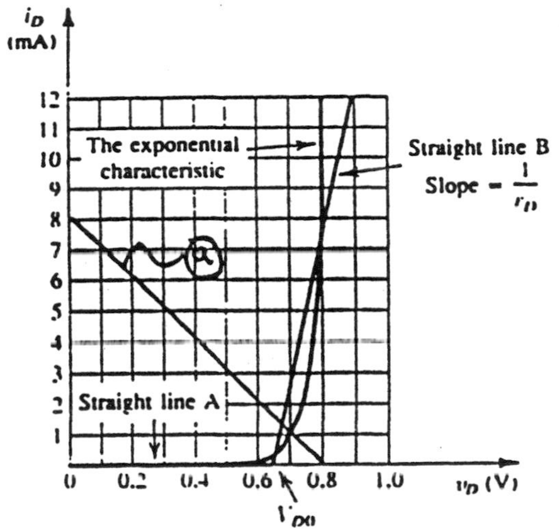

(a) For $V_T = 0.8\text{V}$ and $R_T = 100\Omega$, $I_T = 0.8/100 = 8\text{mA}$. From the Figure, see $\mathbf{V_D = 0.7V}$, $I_D =$ **1mA**. *Check:* $I_D = 8 - V_D/0.1 = 8 - 10V_D = 8 - 10\ (0.7) = 1$ mA.

(b) $I_D = \dfrac{0.8 - 0.65}{0.1 + 0.02} =$ **1.25mA**, and $V_D = 0.65 + 0.22\ (1.25) =$ **0.675V**.

(c) $I_D = \dfrac{0.8 - 0.70}{0.1 + 0.01} =$ **0.91mA**, and $V_D = 0.70 + 0.01\ (0.91) =$ **0.701V**.

(d) $I_D = \dfrac{0.8 - 0.75}{0.1} =$ **0.50mA**, and $V_D =$ **0.750V**.

Check: Overall, using the logarithmic model, the supply voltage is $\upsilon = 0.1(I_D) + 0.7 + 0.1\log(I_D/1)$ which for $I_D = 1mA$, $\upsilon = 0.1(1) + 0.7 + 0.1\log(1) = 0.1 + 0.7 + 0 = 0.8$, as proviced, OK.

3.32 Drop across each diode $= 4.0/5 = 0.8\text{V}$. Thus $0.8\text{V} = 0.7\text{V} + 0.1\ \log\ (I_D/1mA)$, or $\log I_D = (0.8 - 0.7)/0.1 = 1$, whence $I_D = 10^{\log I_D} = 10^1 = 10$, or $I_D = 10\text{mA}$ (as can be seen directly). Now $R = (10 - 4.0)/10\text{mA} = 0.6\text{k}\Omega$, or **600Ω**.

3.33 For $R = 500\Omega$ used in P3.32 above, with each diode having a drop of υ volts:

See, $i = \dfrac{10 - 5\upsilon}{0.5} = 20 - 10\upsilon$, also $\upsilon = 0.7 + 0.1 \log i/1$.

Solve iteratively:

$\upsilon = 0.8\text{V} \rightarrow i = 20 - 8 = 12$, $\upsilon = 0.7 + 0.1 \log 12 = 0.7 + 0.1079 = 0.8079\text{V}$;

$\upsilon = 0.8079\text{V} \rightarrow i = 20 - 8.079 = 11.92$, $\upsilon = 0.7 + 0.1 \log 11.92 = 0.7 + 0.1076 = 0.8076\text{V}$;

$\upsilon = 0.8076\text{V} \rightarrow i = 20 - 8.076 = 11.92$ mA, $\upsilon = 0.7 + 0.1 \log 11.92 = 0.7 + 0.1076 = 0.8076\text{V}$.

Thus the diode-string voltage becomes $5(0.8076) = 4.038\text{V}$, or **4.04V**.

3.34 After shunting, three identical diodes in parallel share the current equally.
Original current is $i = (1)\ 10^{(0.8 - 0.7)/0.1} = 10mA$ (as can be seen directly).
Final current in each diode = 10/3 = 3.33mA. Final voltage drop = 0.7 + 0.1 log 3.33/1 = **0.752V.**

SECTION 3.5: THE SMALL-SIGNAL MODEL AND ITS APPLICATION

3.35 In general, $r = nV_T/I_D$. At 0.1mA, $r = \frac{2(25mV)}{0.1mA} = 500\Omega$. At 10mA, $r = \frac{2(25)}{10mA} = 5\Omega$.
Now the "average" resistance would more likely be considered at $(0.1 \times 10)^{1/2} = 1mA$, than at (0.1 + 10)/2 = 5.05mA. Thus the geometric mean is likely to be most relevant, and therefore "best". But, let us check:

See $(r_1 + r_2)/2 = (500 + 5)/2 =$ **252.5Ω**, and $(r_1\, r_2)^{1/2} = \sqrt{500 \times 5} =$ **50Ω.** Now, at 1mA, $r = \frac{2(25)}{1mA} = 50\Omega$, and at 5.05mA, $r = \frac{2(25)}{5.05mA} =$ **9.9Ω.** *Note* the correspondence of the value at 1mA with the geometric mean, as expected. Note that the arithmetic-mean results appear to be less easily interpreted.

3.36 At 2mA: $r = \frac{nV_T}{I_D} = \frac{2(25mV)}{2mA} =$ **25 Ω.** With a second identical diode shunting the first, the current is shared equally: **1mA** in each. At 1mA: $r = \frac{2(25)}{1} =$ **50 Ω**, and for two in parallel, $r_{eq} = 50 \| 50 =$ **25 Ω.** This demonstrates that diode incremental **resistance is independent of diode junction size.**

3.37 For small v_S, use the small-signal slope resistance, r. In general, $r = \frac{nV_T}{I}$ and $\frac{v_O}{v_S} = \frac{r}{r + R_S}$.

For $I = 10mA$: $r = \frac{2(25)}{10} = 5\Omega$; $v_O/v_S = \frac{5}{5 + 1000} =$ **0.00498V/V** ≈ 0.005V/V.

For $I = 1mA$: $r = \frac{2(25)}{1} = 50\Omega$; $\frac{50}{50 + 1000} =$ **0.0476V/V** ≈ 0.05V/V.

For $I = 0.1mA$: $r = \frac{2(25)}{0.1} = 500\Omega$; $\frac{500}{500 + 1000} =$ **0.333V/V** ≈ 0.33V/V.

For $I = 0.01mA$: $r = \frac{2(25)}{.01} = 5000\Omega$; $\frac{5000}{5000 + 1000} =$ **0.833V/V** ≈ 0.83V/V.

3.38 For $v_O = v_S = 0$, see that I_1 and I_2 both **split equally** between D_1, D_2 and D_3, D_4 respectively. Thus all diode **currents are equal.**

In general, for $v_S \neq 0$ (say $+\varepsilon$), the currents in D_1, D_2 redistribute with their sum = I_1. Extra current in D_2 needed to drive R_L high implies that there is less current in D_4 than in the balanced case, the extra part of I_2 flowing in D_3. In general, current needed by R_L for $v_O \neq 0$ is provided half by that in D_2 increasing and half by that in D_4 decreasing, with corresponding changes in D_1 and D_3, such that load current originates ultimately in v_S, as Kirchoff's current law would require.

For small v_S (around zero volts), with all diode currents equal to $i/2$,

$$\mathbf{R_T} = \left[r_1 + r_2\right] \| \left[r_3 + r_4\right] = \left[\frac{50}{I/2} + \frac{50}{I/2}\right] \| \left[\frac{50}{I/2} + \frac{50}{I/2}\right] = \frac{200}{I} \| \frac{200}{I} = \frac{100}{\mathbf{I}}.$$

For $I = 10mA$, $R_T = \frac{100mV}{10mA} = 10\Omega$, and with $R_L = 10k\Omega$, $v_O/v_S = \frac{10k}{10 + 10k} =$ **0.999V/V.**

For $I = 1\mu A$, $R_T = \frac{100mV}{1\mu A} = 100k\Omega$, and $v_O/v_S = \frac{10k}{100k + 10k} =$ **0.0909V/V.**

Concerning Linearity: Signal size (for $R_L = 10\text{k}\Omega$) is of little concern at 1mA, since the load current is likely to be a small part of I. But, at low I, the load current may cause the diode current to vary over a wide range. Operation is linear for diode-voltage variation of ±10mV or so: Note that the corresponding current variation is Δi, where $10 = 2(25)\ \ln \frac{\Delta i}{i}$, or $\ln \frac{\Delta i}{i} = \frac{10}{50} = .20$, or $\Delta i/i = 1.2$, corresponding to a variation of about 20%. Now for a positive output signal, i_{D2} can increase by 20% of its normal current $I/2$, while i_{D4} decreases by the same amount, the output current being the sum of the two changes. Thus for $R_L = 10\text{k}\Omega$ and $I = 1\mu\text{A}$, υ_O is limited to $2 \times 0.20\ \frac{1\mu A}{2} \times 10\text{k}\Omega = 2\text{mV}$ peak, in which case υ_S is restricted to $\frac{2mV}{.0909} = $ **22mV** peak.

An Alternative View: The largest (positive) output occurs when the drop in D_2 increases by 10mV, and that in D_1 decreases by 10mV. Thus, at the limit, $\upsilon_S - \upsilon_O = 2(10) = 20\text{mV}$. But for 1µA and a 10kΩ load, $\upsilon_O/\upsilon_S = 0.09$, or $\upsilon_O = 0.09\ \upsilon_S$. Thus, the limit is at $\upsilon_S - .09\ \upsilon_S = 20\text{mV}$, or $\upsilon_S =$ **22mV**.

3.39 For the regulator, $I_D = \frac{10-4}{600} = 10\text{mA}$. For each of 5 diodes, $r = \frac{nV_T}{I_D} = \frac{2(25)}{10} = 5\Omega$; Resistance of the total diode string = 5(5) = 25Ω. For ±10% supply variation, expect an output variation of $\pm \frac{25}{25+600} \times (0.1 \times 10) = \pm$ **40mV**, equivalent to $\pm \frac{40mV}{4V} \times 100 =$ **±1%**. For a 2mA load increase, the diode current reduces from 10mA to 8mA (by a small amount) and the output drops by 5(5Ω) (2mA) = **50mV**, or $\frac{50}{4V} \times 100 =$ **−1.25%**. For both effects, assuming approximately linear operation, the combined drop would be [−40mV −50mV] = **−90mV** or **−2.25%**. The lowest output voltage would be 4.00 − .09 = **3.91V**.

From First Principles: (for a 0.1V/decade current change (which is *not* $n = 2$ precisely!)), $V_S = 90\%$ of 10 = 9V. For diode current i and voltage υ, see $\frac{9 - 5\upsilon}{0.6} = i + 2$, whence $i = 13 - 8.33\ \upsilon$ − − − (1), and $\upsilon = 0.7 + 0.1 \log i/1$ − − − (2). Thus $\upsilon = 0.7 + 0.1 \log (13 - 8.33\ \upsilon)$.

Iterate: with $\upsilon = 0.8 - .09/5 = .782$ initially. Thus $\upsilon = 0.7 + 0.1 \log (13 - 8.33\ (.782)) = 0.7812\text{V}$, and $\upsilon = 0.7 + 0.1 \log (13 - 8.33\ (.7812)) = 0.78123\text{V}$.

Thus the output drop is 5 (0.8000 − .78123) = **93.9mV**.

3.40 $r_d = nV_T/I_A \rightarrow n = (50\Omega \times 1mA)/25mV =$ **2**. Also $C_j = \frac{C_{j0}}{(1 + V_R/V_0)^m}$. Thus, at $V_R =$, $0.75 = \frac{C_{j0}}{(1 + 1/V_0)^m}$ − − − (1). Thus, at 5V, $0.2 = \frac{C_{j0}}{(1 + 5/V_0)^m}$ − − − (2). Thus, at −10V, $0.1 = \frac{C_{j0}}{(1 + 10/V_0)^m}$ − − − (3). Now (1)/(2) → $0.75/0.2 = \left[\frac{1 + 5/V_0}{1 + 1/V_0}\right]^m = \left[\frac{V_0 + 5}{V_0 + 1}\right]^m = 3.75$ − − − (4), and (2)/(3) → $0.2/0.1 = \left[\frac{V_0 + 10}{V_0 + 5}\right]^m = 2.0$ − − − (5).

Explore trial solutions of (4), (5).

For $V_0 = 0$ V: $\left[\frac{0+5}{0+1}\right]^m = 3.75 \rightarrow m = 0.82$,

and $\left[\frac{0+10}{0+5}\right]^m = 2.0 \rightarrow m = 1$, $\Delta = 0.18$.

For $V_0 = 1.0$: $\left[\frac{1+5}{1+1}\right]^m = 3.75 \rightarrow m = 1.21$,

and $\left[\frac{1+10}{1+5}\right]^m = 2.0 \rightarrow m = 1.15$, $\Delta = 0.06$.

For $V_0 = 2.0$: $\left[\dfrac{2+5}{2+1}\right]^m = 3.75 \rightarrow m = 1.56,$

and $\left[\dfrac{2+10}{2+5}\right]^m = 2.0 \rightarrow m = 1.27 \quad , \Delta = 0.29.$

See result is nearer $V_0 = 1 \rightarrow$ Try:

$V_0 = 0.5$: $\left[\dfrac{0.5+5}{0.5+1}\right]^m = 3.75 = 3.666^m \rightarrow m = 1.02,$

and $\left[\dfrac{0.5+10}{0.5+5}\right]^m = 2.0 = 1.909^m \rightarrow m = 1.07 \quad , \Delta = 0.05.$

See between 0.5 and 1.0 $\rightarrow$ Try:

$V_0 = 0.7$: $\left[\dfrac{0.7+5}{0.7+1}\right]^m = 3.75 = 3.353^m \rightarrow m = 1.095,$

and $\left[\dfrac{0.7+10}{0.7+5}\right]^m = 2.00 = 1.877^m \rightarrow m = 1.10 \quad , \Delta = 0.005 \simeq 0.$

Overall, conclude that n = **2**, m = **1.10**, V_0 = **0.7V**, and $C_{j0} = 0.75\ (1 + 1/0.7)^{1.1}$ = **1.99 pF**.

Now, at $V_0 = 0V$, $C_j(0) = \dfrac{1.99}{(1 - 0/0.7)^{1.2}} = 1.99$ pF.

For forward conduction: As noted on page 155 of the Text, Eq. 3.27 does not properly represent C_j for forward conduction. [For example, check here, where if the diode is forwarded biased at $V_D = V_0$, and that $V_R = -V_0$ and C_j becomes infinite!] Rather, for forward conduction, one uses from experience $C_j = 2C_{j0} = 2(2) = 4pF$, here.

Now, at 1 mA, as $C_T = C_j + C_d = 10$pF, then $C_d = 10 - 4 = 6$pF.

Generally, $C_d = (\tau_T/nV_T)I_D$. Thus $\tau_T = nV_T C_d/I_D = 2(25 \times 10^{-3}) \times 6 \times 10^{-12}/1 \times 10^{-3} = 300 \times 10^{-12}$ = **300 ps.**

Correspondingly, at 10mA, $C_T \simeq 10(6) + 4$ = **64pF.**

3.41 For a diode 10× the area of that in P3.40, but otherwise using the same technology, both n and τ_T will be the same, 2 and 300 ps, respectively. Thus, at $I_D = 5mA$, $r = nV_T/I_D = 2(25)/5 = 10\ \Omega$, and $C_d = (\tau_T/nV_T)I_D = [300 \times 10^{-12}/(2 \times 25 \times 10^{-3})] \times 5 \times 10^{-3} = 30pF$. [*Aside:* This could be seen more directly, since C_d must be 5× larger at 5 mA then at 1 mA. Correspondingly, $C_d = 5 \times 6 = 30pF$.

Since the junction is 10× larger, so must be C_j. Thus, $C_j = 10 \times 4pF$ = **40 pF.**

Thus at 5 mA, $C_T = C_j + C_d = 40 + 30$ = **70 pF.**

For 10V reverse bias, C_j will be 10× that value given in P3.40 (which was 0.1 pF). Thus, here, $C_j(10) = 10(0.1)$ = **1 pF**

SECTION 3.6: OPERATION IN THE REVERSE BREAKDOWN REGION – ZENER DIODES

3.42 Knee voltage $\simeq 6.8V - \dfrac{20}{1000}(5 - 0.2)$ = **6.70V**. For no load, breakdown is sustained for supply voltages down to about **6.7V**. For a 9V supply, and bare breakdown, the load can increase to $\dfrac{9 - 6.7}{200\Omega} - 0.2$ $= 11.5 - 0.2$ = **11.3mA**. For $\dfrac{11.3}{2} = 5.65$mA load, the lowest supply voltage for regulation is V, where $\dfrac{V - 6.7}{0.2k} - 5.65 = 0.2$, from which V = 6.7 + 0.2 (5.85) = 7.87V. Thus the lowest supply for regulation with half-maximum load ≈ **7.9V.**

3.43 Line regulation (Eq. 3.60) is $\frac{r_z}{R + r_z} = \frac{20}{200 + 20}$ = 0.0909V/V = **90.9mV/V.**

Load regulation (Eq. 3.61) is $-(r_z \parallel R) = -(20 \parallel 200) = -18.2\Omega$ or, **−18.2mV/mA.**

3.44 *Worst case*: V_S low, I_L high, V_Z high, where $\frac{9(.95) - 6.8\,(1.03)}{R}$ = 10mA + 2 (0.2)mA, whence $R \le \frac{9\,(.95) - 6.8\,(1.03)}{10 + 0.4} = \frac{8.55 - 7.004}{10.4}$ = 0.148kΩ. Use **150Ω.**

The lowest output occurs for 10mA load, V_Z low, V_S low, where $I_Z \approx \frac{9\,(.95) - 6.81\,(.97)}{0.148} - 10.0 = \frac{8.55 - 6.596}{0.148} - 10.0$ = 3.20mA, for which $V_Z \approx 6.8\,(.97) - (5 - 3.2)\,20\Omega = 6.596 - .036 = 6.56$V.

More precisely: $I_Z = \frac{8.55 - 6.56}{0.148} - 10$ = 3.45mA, for which $V_Z \approx 6.8\,(.97) - (5 - 3.45)\,20 = 6.596 - .031 = 6.565\text{V} \approx$ **6.57V.**

The highest output occurs for 2mA load, V_Z high, V_S high, where $I_Z \approx \frac{9(1.05) - 6.8\,(1.03)}{0.148} - 2 = \frac{9.45 - 7.004}{0.148} - 2$ = 14.52mA, for which $V_Z = 6.8\,(1.03) + (14.52 - 5)\,20 = 7.004 + .1904 = 7.194\text{V} \approx$ **7.19V.**

More precisely, $I_Z = \frac{9.45 - 7.194}{0.148}$ = 15.24mA, for which $V_Z = 6.8\,(1.03) + (15.24 - 5.0)\,20 = 7.004 + 0.205 = 7.209$, or **7.21V.**

3.45

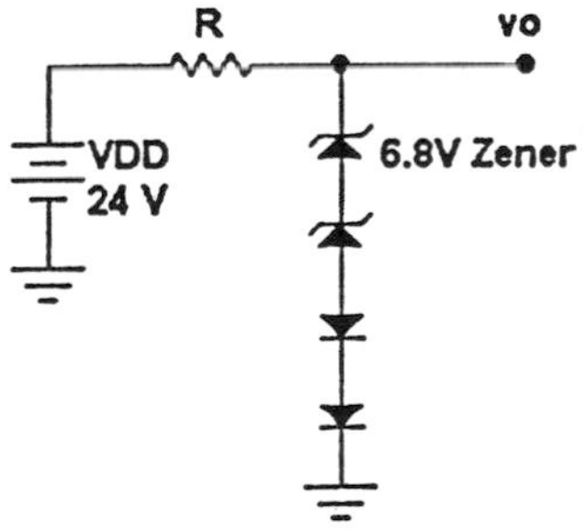

$\Sigma V = 2\,(6.8) + 2\,(0.7) = 15$V, but since V_Z is specified at 20mA and V_D at 10mA, use an I_Z in between, where $2\,(20 - I_Z)\,(5\Omega) = 2\,(I_Z - 10)\,(2.5\Omega)$, or $2\,(20 - I_Z) = I_Z - 10$, $40 - 2I_Z = I_Z - 10$, $3I_Z = 50$, and $I_Z = 16.7$mA. For a nominal load of 15mA, and nominal 24V supply, $R = \frac{24 - 15}{16.67 + 15}$ = 0.284kΩ. Use a **270Ω** resistor, as a standard value. For supply 10% high, resistor 5% low, and no load, $I_Z = \frac{24\,(1.1) - V_Z}{(0.270)\,(0.95)} = 102.9 - 3.899V_Z$. Also $V_Z = 15\text{V} + (I_Z - 16.67)\,(2\,(5) + 2\,(2.5)) \times 10^{-3} = 15 + .015\,I_Z - 0.250 = 14.75 + .015\,I_Z$, whence $I_Z = 102.9 - 3.899\,(14.75 + 0.015\,I_Z) = 102.9 - 57.51 - .0585\,I_Z$, and $I_Z = \frac{45.39}{1.0585}$ = 42.88mA, for which $V_Z = 14.75 + .015\,(42.88)$ = **15.39V.** For each zener at 42.88mA, $V_Z = 6.8\text{V} + (42.88 - 20)\,(.005) = 6.911$V, and $P_D = 6.911 \times 42.88$ = **296mW.**

SECTION 3.7: RECTIFIER CIRCUITS

3.46 The peak output is $8\sqrt{2} - 0.7\text{V} = 11.31 - 0.70$ = **10.6V**: The diode conducts for about ½ **cycle**, but more precisely, between the points where $11.31 \sin\Theta = 0.7$, or $\sin\Theta = 0.0619$, for which $\Theta = 3.55°$ and $180 - 3.55°$, that is, for $\frac{180 - 2\,(3.55)}{360}$ = 0.48 or, **48%** of a cycle. Ignoring the diode drop, the average output = $8\sqrt{2}/\pi$ = **3.60V**. With a constant 0.7V drop for 0.48 of a cycle, the average output is 3.6 − 0.48 (0.7) = **3.26V**. Alternatively, and perhaps better, including the diode, the peak voltage is 10.6 V, and assuming the waveform to be approximately a half-sine, the average output is $10.6/\pi$ = **3.37 V** The peak inverse voltage across the diode is approximately the peak input = $8\sqrt{2}$ =

11.3V. For $R_S = 50\Omega$. $r_D = 10\Omega$ and a load of 1kΩ, $v_O \approx (v_S - 0.7)\dfrac{1k}{(1+.01+.05)k}$ for a half cycle, or $v_O \approx 0.94\, v_S - 0.66$. Correspondingly, at the peak, the peak output is 0.94 (11.3) − 0.66 = **9.97V.** Overall, the average output is 0.94 (3.26) = **3.06V.**

3.47 Peak inputs are $\pm 8\sqrt{2} = \pm 11.31$V. Peak outputs are 11.31 − 0.70 = 10.61 V, and −11.31 + 6.80 = −4.51V.

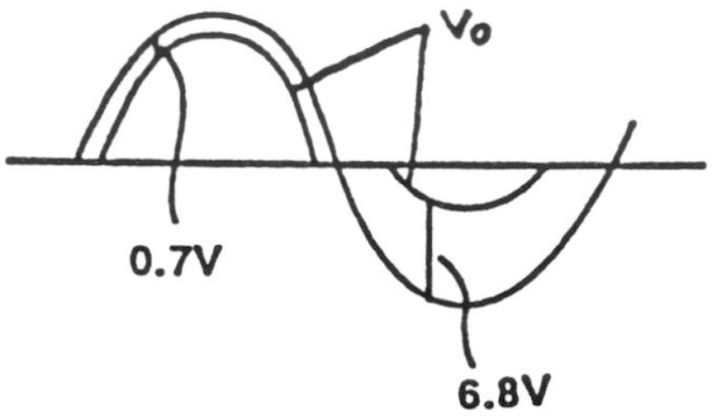

For conventional forward conduction: (As noted in P3.46), conduction is from 3.6° to 176.4°, or about 48% of the cycle, *For reverse conduction:* current flows for the parts of the cycle between which 11.31 sin Θ = 6.8V, or $\sin\Theta = \dfrac{6.8}{11.31} = 0.601$, for which Θ = 36.96° or 37°, and 180° − 37° = 143°, that is for (143 − 37)/360 = 29.4% of a cycle. Assuming that the rectified half-waves are sinusoidal-like, for the conduction period, the forward current average is about (10.61/π)(48/50) = 3.24*V*, and the reverse current average is about (4.51/π)(29.4/50) = 0.844*V*. Overall, the average value of the output is 3.24 − 0.84 = **2.40V.**

3.48 Full-winding peak transformer output voltage = $16\sqrt{2}$ = 22.62V.

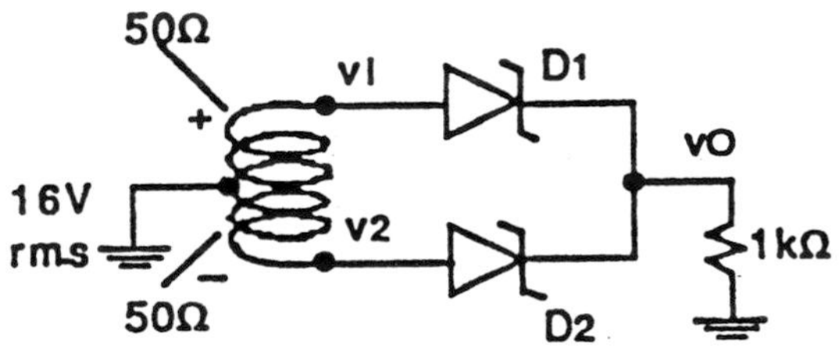

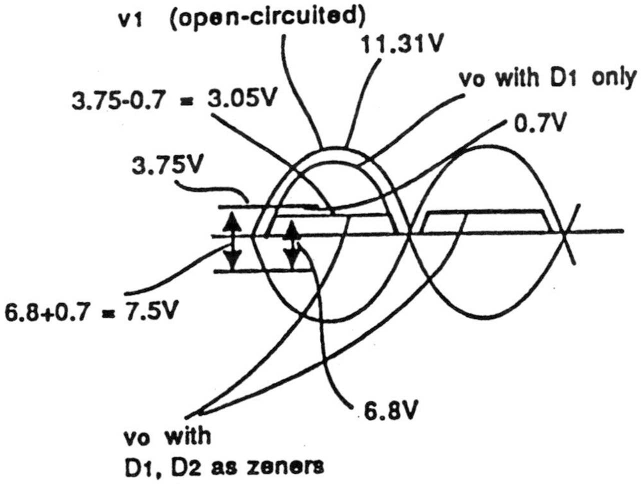

For full-winding voltages ≥ (6.8 + 0.7) = 7.5V, D_1 conducts in the forward direction (0.7V), while D_2 breaks down (6.8V). At the peak, the shorted-diode current flow is (22.62 − 7.5)/100Ω = 151.2mA with v_O = 22.62/2 − 0.7 − (100Ω/2) × 151.2mA = (11.31 − 0.7 − 7.56V) = 3.05V, {or, see this as 7.5/2 −0.7 = 3.05V}. But current also flows in $R_L, i_{RL} \approx \dfrac{3.05}{1k}$ = 3mA from an equivalent source of $\dfrac{100}{2} \parallel \dfrac{100}{2}$ = 25Ω, to produce an additional drop of 3mA × 25Ω = 0.075V. Correspondingly, $v_O \approx$ 3.05 −.075 ≈ 2.975V.

Thus, the peak value of the output is 2.975V, relatively constant while the sum of the open-circuit winding voltages exceeds 7.5V. The peak diode current flow is 151.2mA + 3*mA*/2 = **152.7mA.**

3.49

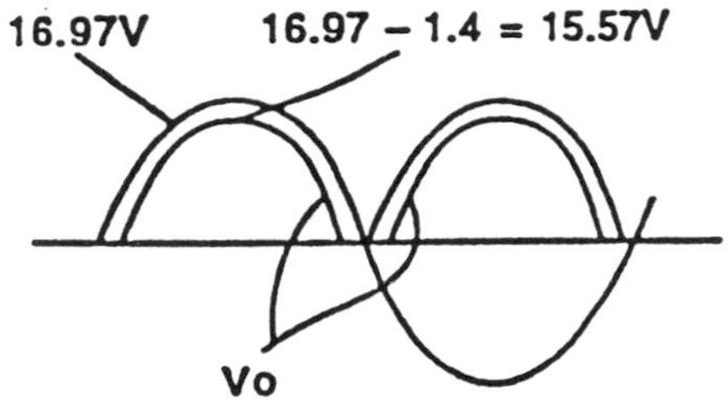

Transformer Peak Voltage = $12\sqrt{2}$ = 16.97V. Load Peak Voltage = 16.97 − 2 (0.7) = **15.57V**. Now, 16.97 sin Θ = 0.7 + 0.7 = 1.4, for which sin Θ = $\frac{1.4}{16.97}$ = 0.0825, and Θ = 4.73°. That is the output is zero for 4(4.73°) = 18.92° per cycle or $\frac{18.92}{360} \times \frac{1}{60}$ = **0.876msec**. Thus, the average output ≃ $\frac{2}{\pi}$ 16.97 − 1.4 = 10.80 − 1.4 = **9.40V**. Peak Inverse Voltage for each diode is 16.97 − 0.7 = **16.3V**.

3.50 For a 12V sinusoid, the peak is $12\sqrt{2}$ = 16.97V. Capacitor charges to the peak voltage less one diode drop (due to the flow of small capacitor leakage currents), that is, to 16.97 − 0.7 = 16.27V. Thus the output is a dc voltage of **16.27V**. The PIV required of the diode is 16.27 + 16.97 = **33.2V**.

3.51 For a constant current of 1mA flowing for one cycle, the voltage drop is $V = \frac{IT}{C}$ = 0.4V = $\frac{1mA \times 1/60}{C}$, whence $C = \frac{1 \times 10^{-3}}{0.4 \times 60}$ = **41.7μF**. For ½ the ripple and 2x the load, need 4x the capacitance or 4(41.7) = **167μF**. For 0.4V ripple with 16.97V peak input, diode conduction occurs for inputs from 16.97 − 0.4 = 16.57V to 16.97V, or for an angle from $\sin^{-1}$ (16.57/16.97) to 90°, that is, from 77.54° to 90°, or 12.5°. Thus for 12.5/360 = 0.0347 of a cycle, diode current flows to replace charge lost through the 1mA load. The average diode current during conduction is 1/0.0347 = **28.8mA**. For 0.2V ripple, the interval is from $\sin^{-1}$ 16.77/16.97 = 81.2° to 90°, or 8.8°, corresponding to 8.8/360 = 0.0244 of a cycle. Thus, the average diode current for a 2mA load is 1/.0244 × 2 = **81.8mA**!

3.52 For full-wave rectification, the discharge interval is essentially halved. Using the results of P3.51 above, for the same ripple, half the capacitance is needed, namely 41.7/2 = **20.9μF** for 1mA and 0.4V ripple, and 167/2 = **83.4μF**, for 2mA and 0.2V ripple. Also diodes conduct twice per cycle. Thus the diode average conduction currents are ½ of the originals, ie 28.8/2 = **14.4mA**, and 81.8/2 = **40.9mA**, respectively. In each case, diode PIV = 16.27 + 16.97 = **33.24V**.

3.53 12V applied to a 100Ω load implies a 120mA load current. Now $CV = IT$.

Thus $C = (120 \times 10^{-3} \times ½ \times 1/60)/0.4$ = **2500μF**. Assume negligible transformer resistance: Thus the peak sine wave required is 12 + 0.4/2 + 0.7 = 12.9V and the transformers RMS voltage = $12.9/\sqrt{2}$ = 9.12V per side. Thus the transformer should have an **18.24V rms centre-tapped secondary**.

For diodes: PIV = 12 + 0.2 + 9.12 $\sqrt{2}$ = **25.1V**. Diode current flows from $\sin^{-1}$ (12.9 − 0.4)/12.9 = 75.69° to 90°, or (90 − 75.69)/360 = 0.0397 of a cycle. Average diode current for each diode = 120/0.0397 × ½ = **1.509A**, with the peak current being about twice as high, ie (1.509 − 0.12) 2 = **2.78A**!!

3.54

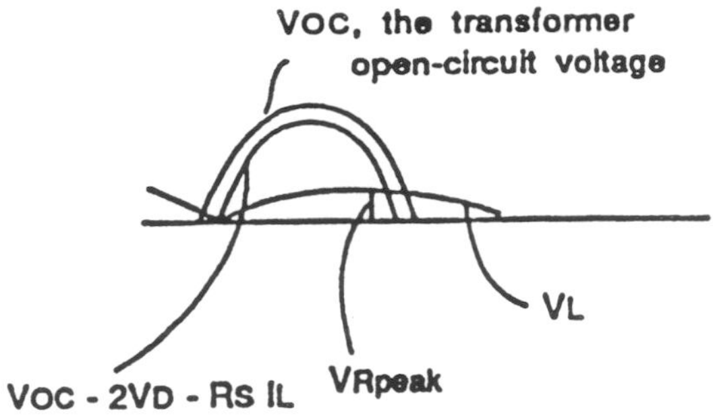

Ideal Case: Peak output = 20 – 0.7 – 0.7 = 18.6V, for which the peak output current = 18.6/200 = 93mA. Ripple voltage is $V = IT/C = \dfrac{93 \times 10^{-3} \times 1/60 \times 1/2}{1000 \times 10^{-6}}$ = **0.775V.**

Diode conduction is from $\sin^{-1}$ ((20 – .775)/20) = 74.0° to 90°, ie 16° or 16/360 = 0.0444 of a cycle, with an average diode current = 93/0.0444 × ½ ⇒ **1.047A**, and an average DC output = 18.6 – 0.775/2 = **18.2V.**

With Source Resistance, R_S, the peak diode current is limited, and the output voltage will drop. As the output voltage drops, the recharging interval will increase, and the ripple will decrease, both because at lower voltages the load current is smaller, and also because during the charging interval(s), the load is supported directly through the diodes.

Now, the average output decreases to 18.2× where $x \le 1.0$, and the ripple to $0.78x$. Thus the peak diode current (near the sine-wave peak and where (say) the ripple is halfway) is + $0.78(x/2)/R_S = (18.6 - 17.8x)/R_S$. Now, assume the current to be triangular in form, flowing for a fraction y of a cycle, twice per cycle. Thus average charge delivered from the supply (through two diodes) in one cycle must be that required by the load in the whole cycle, that is, $2(½)(y)\left(\dfrac{18.6 - 17.8x}{R_S}\right) = \dfrac{18.2x}{200}$ – – – (1)

Now for y: Diodes begin to conduct at $\sin^{-1}\dfrac{18.2x - 0.78x/2}{18.6} = \sin^{-1}(0.958x)$ and cease conduction (following the input peak) at $90° + \left(90° - \sin^{-1}\dfrac{18.2x + 0.78x/2}{18.6}\right) = 180° - \sin^{-1}(0.999x)$,

where $y = \dfrac{180 - \sin^{-1} 0.958y - \sin^{-1} 0.999y}{360}$ – – – (2)

Now, for $R_S = 1.0\Omega$, (1) → $2(½)y\,\dfrac{(18.6 - 17.8x)}{1} = \dfrac{18.2x}{200}$, and $x = 11.0y\,(18.6 - 17.8x)$ – – – (3)

Iterate: Try $x = 0.95$ initially:

(2) → $y = (180 - \sin^{-1} .958(.95) - \sin^{-1} .999\,(.95))/360 = (180 - 65.5 - 71.6)/360 = 0.119$, and

(3) → $x = 11.0\,(.119)\,(18.6 - 17.8\,(.95)) = 2.21$.

Try $x = 1.0$:

(2) → $y = (180 - \sin^{-1} .958 - \sin^{-1} .999)/360 = (180 - 73.3 - 87.4)/360 = 0.0536$,

(3) → $x = 11.0\,(.0536)\,(18.6 - 17.8) = 0.472$.

Try $x = 0.97$:

$y = (180 - \sin^{-1}(.958 \times 0.97)\ \sin^{-1}(.999)\,(.97))/360 = (180 - 68.3 - 75.7)/360 = 0.100$,

$x = 11\,(.100)\,(18.6 - 17.8\,(.97) = 1.47$.

Try $x = 0.99$:

$y = (180 - \sin^{-1}(.958)\,(.99) - \sin^{-1}(.999 \times .99))/360 = (180 - 71.5 - 81.5)/360 = 0.075$,

$x = 11\,(.075)\,(18.6 - 17.8\,(.99)) = 0.807$.

Try $x = 0.98$:

$y = (180 - \sin^{-1}(.958)(.98) - \sin^{-1}(.999 \times .98))/360 = (180 - 69.86 - 78.24)/360 = 0.0886,$

$x = 11\ (.0886)\ (18.6 - 17.8\ (.98) = 1.13.$

Try $x = 0.983$:

$y = (180 - \sin^{-1}(.958 \times .983) - \sin^{-1}(.999)(.983))/360 = (180 - 70.34 - 79.12)/360 = 0.0848,$

$x = 11\ (.0848)\ (18.6 - 17.8\ (.983)) = 1.028.$

Try $x = 0.984$:

$y = \left[180 - \sin^{-1}(.958 \times .984) - \sin^{-1}(.999 \times .984)\right]/360 = (180 - 70.50 - 79.43)/360 = 0.0835,$

$x = 11\ (.0835)\ (18.6 - 17.8\ (.984)) = 0.996.$

Try $x = 0.986$:

$y = \left[180 - \sin^{-1}(.958 \times .986) - \sin^{-1}(.999 \times .984)\right]/360 = (180 - 70.84 - 79.43)/360 = 0.0826,$

$x = 11\ (.0826)\ (18.6 - 17.8\ (.986)) = 0.953.$

Try $x = 0.985$:

$y = \left[180 - \sin^{-1}(.958 \times .985) - \sin^{-1}(.999 \times .985)\right]/360 = (180 - 70.07 - 79.74)/360 = 0.0821,$

$x = 11\ (.0821)\ (18.6 - 17.8\ (.985)) = 0.964.$

Use $x = 0.985$.

One can conclude that:

a) The iterative process is not a very good one, but the result is probably OK.

b) The output voltage decreases to $18.2x = 18.2\ (.985) =$ **17.9V**, a drop of about $18.2 - 17.9 =$ **0.3V**, or $0.3/18.2 \times 100 = 1.6\%$, in the transformer resistance of 1 ohm, implying an "equivalent" current of 300 mA.

SECTION 3.8: LIMITING AND CLAMPING CIRCUITS

3.55 The upper limiting level is 2.3 + 0.7 = **3.0V**. The corresponding input threshold level is 3.0 + (3.0/10k) × 10k = **6.0V**. The corresponding lower values are **−3.0V** and **−6.0V** respectively. The gain K (for linear operation) is 10k/(10k + 10k) = **0.5V/V**. At twice the upper threshold, $V_{in} = 2\ (6.0) = 12\text{V}$, and the current is (12 − 3)/10k = **0.9mA**.

3.56

(a)

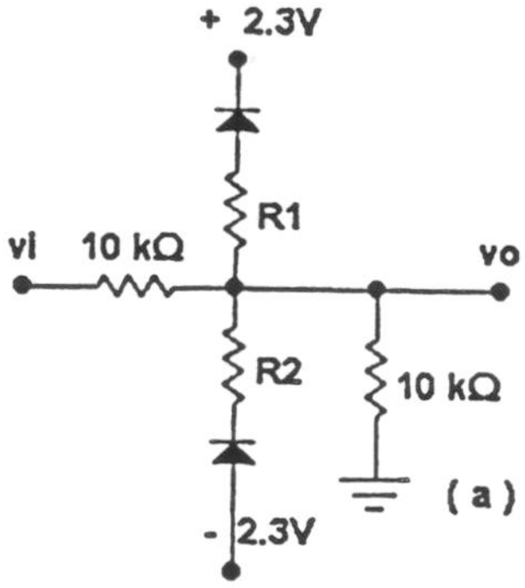

For $v_O \geq 3.0$ V, current begins to flow in R_1.

Thus $K = ¼ = \dfrac{10 \parallel R_1}{10 \parallel R_1 + 10}$, and $4 \dfrac{10\,R_1}{10 + R_1} = 10 + \dfrac{10\,R_1}{10 + R_1}$, $40\,R_1 = 100 + 10\,R_1 + 10\,R_1$, $20\,R_1 = 100$, whence $R_1 = 5\text{k}\Omega$.

For symmetrical operation, $R_2 = 5\text{k}\Omega$ also, note, further that for symmetrical operation, R_1 and R_2 can be combined, as shown.

(b)

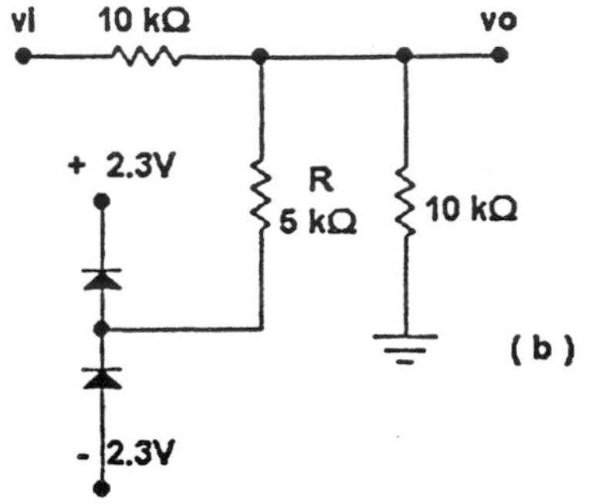

Hard limiting at ±5V can be provided using two additional diodes: D_3, from v_O to +4.3V with anode at v_O, and D_4, from v_O to –4.3V with cathode at v_O.

(c)

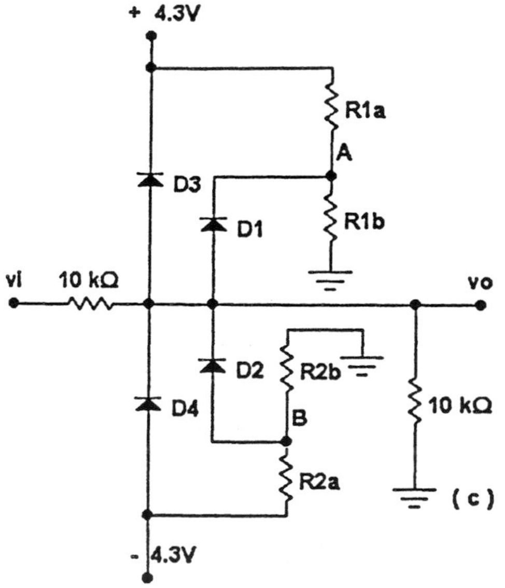

Here, $\dfrac{R_{1a}}{R_{1a} + R_{1b}} \times 4.3 = 2.3$ V and $R_{1a} \parallel R_{1b} = 5\text{k}\Omega$.

Thus $\dfrac{R_{1a}\,R_{1b}}{R_{1a} + R_{1b}} = 5$, or $\dfrac{R_{1a}}{R_{1a} + R_{1b}} = \dfrac{5}{R_{1b}}$.

Substitute the last into the first to get: $\dfrac{5\,(4.3)}{R_{1b}} = 2.3$, or $R_{1b} = \dfrac{5\,(4.3)}{2.3} = 9.35\text{k}\Omega$,

and $R_{1a} = \dfrac{5}{9.35}\,(R_{1a} + 9.35) = 0.535\,R_{1a} + 5 = \dfrac{5}{1 - .535} = 10.75\text{k}\Omega$. Again, a resistor can be saved by replacing R_{1b} and R_{2b} by a single resistor equal to their sum (18.7kΩ) between nodes A and B with no ground connection.

3.57

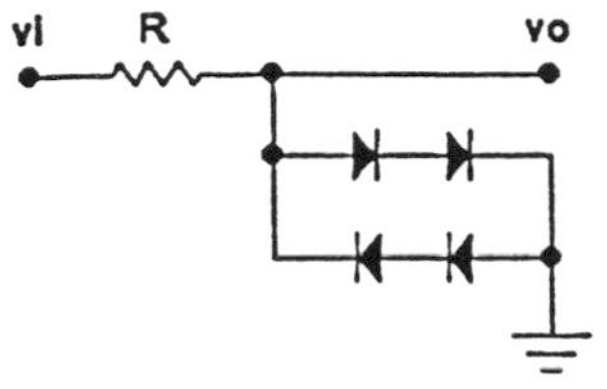

Consider an input $v_i = A \sin \omega t$. Now, $A \sin \Theta$ = 1.4V, where $\Theta/360$ = 5/100. See that $\Theta = 18^\circ$ and $A = \frac{1.4}{\sin 18^\circ}$ = **4.53V peak**, or **9.1Vpp**.

The peak diode current = $\frac{4.53 - 1.4}{R}$ = 10mA. Use a resistor, $R = \frac{4.53 - 1.4}{10}$ = **313Ω.**

3.58

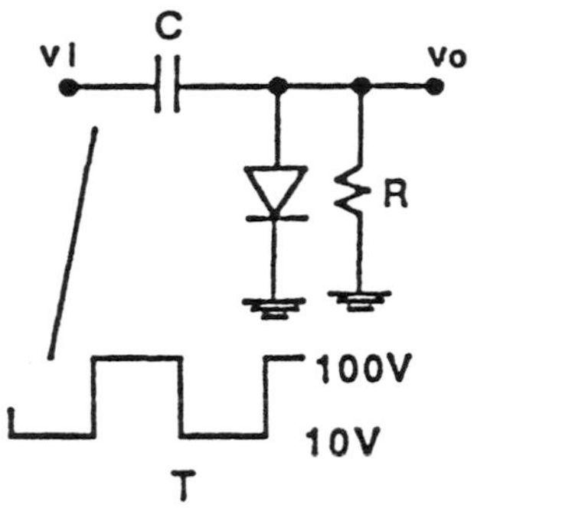

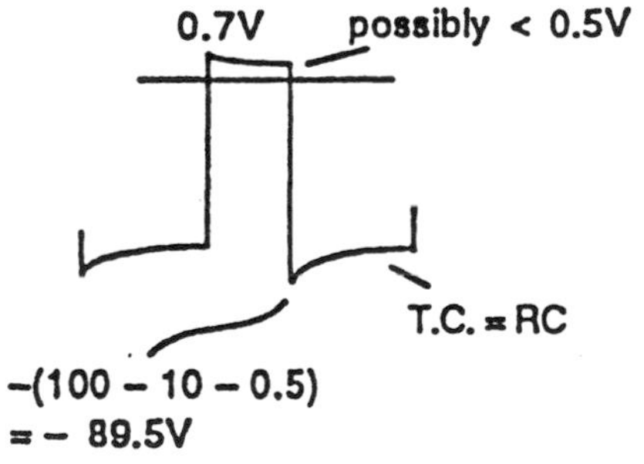

For light load, the output is a square wave of period T going from **+0.5V** to +0.5 –(100 – 10) = **–89.5V.** As the load resistance reduces, the negative side of the waveform is no longer flat at –89.5V, but rather rises toward ground. As well, upon the positive transition, the diode current increases initially, with 0.7V or so positive output at first. For $RC = 2T$, in one half cycle, where $t = T/2$, the output falls to $e^{\frac{-T/2}{2T}} = e^{-1/4} =$ 0.779 of its original value. Thus, assuming the diode to have a constant 0.7V drop when conducting, the waveform initially rises to 0.7V, then drops to 0.7 (.779) = 0.55V, then falls to –90 + .55 = –89.5V, then droops (up) to –89.5 (0.779) = –69.7V, then rises to 0.7 V, and so on. In practice, the diode conducts at voltages lower than 0.7 V. Thus the upper level is not simply an exponential, but will fall more rapidly, to slightly less than 0.55V.

3.59

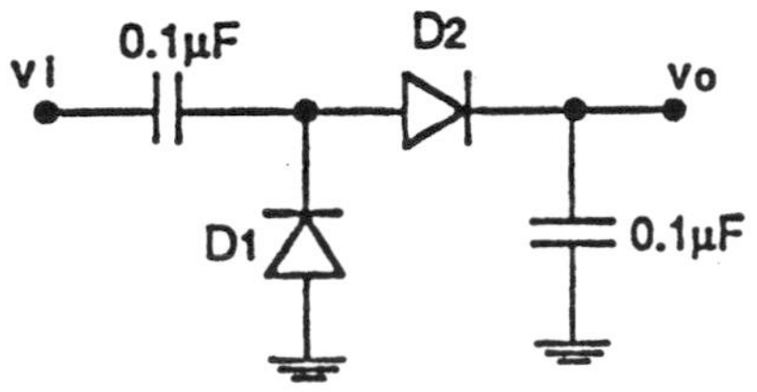

For a 100V-peak sine wave and no load, the output would be 2(100) –0.7 – 0.7 = **198.6V** for a 0.7V diode drop. For a pp ripple of 5% of peak, ripple voltage is (5/100) (198.6) = 9.93V, and the average output voltage = 198.6 – 9.93/2 = **193.6V.** The corresponding load current is $I = \frac{0.1 \times 10^{-6} \times 9.93}{1/(20 \times 10^3)}$ = **19.86mA.**

3.60 Assume ideal diodes having a drop of 0V. Initially, the input capacitor C_1 is charged, with 0V on its internal end and input low, while the output capacitor C_2 is discharged. As the input rises by 100V, charge is dumped through the connecting diode D_2 from C_1 into C_2. Since the capacitors are equal and the input voltage change is 100V, the output rises to **50V**.

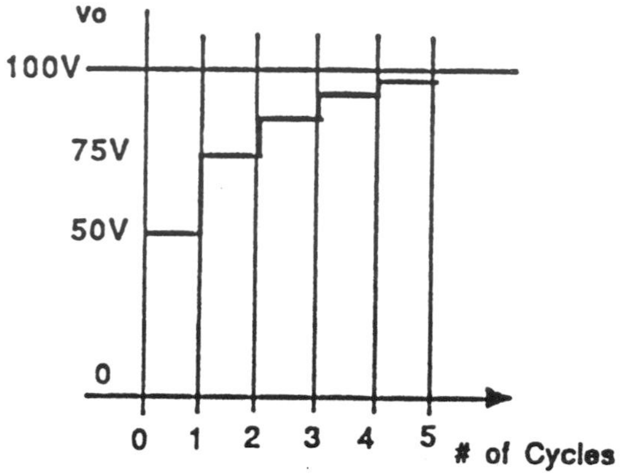

When the input falls, C_1 is recharged through the grounded diode D_1, while the connecting diode D_2 opens, leaving C_2 charged at 50V. Now, when the input rises again, diode D_2 does not conduct until the input rises by 50V. For the remaining 50V change of input, charge is shared equally by C_1 and C_2, while the voltage on C_2 rises by 25V to **75V**. When the input falls, C_1 is recharged. Correspondingly, after 2 cycles, the output becomes $50 + 25 + \frac{25}{2}$ = **87.5V**. After four cycles, the output is $50 + 25 + \frac{25}{2} + \frac{25}{4} + \frac{25}{8}$ = **96.9V**. After eight cycles the output is $50 + 25 + \frac{25}{2} + \frac{25}{4} + \frac{25}{8}\left(1 + \frac{1}{2} + \frac{1}{4} + \frac{1}{8} + \frac{1}{16}\right)$ = **99.8V**.

3.61 Ultimately, an equilibrium is established with the output reaching V_1 at the beginning of a cycle, V_2 half way, and V_3 at the end. Here, $V_2 = V_1\left(1 - \frac{5/100}{2}\right) = 0.975V$, since two capacitors C supply the load. Also $V_3 = V_2(1 - 5/100) = 0.95V_2 = 0.95\,(.975)\,V_1 = 0.92625\,V_1$. But at the start of the next cycle, with $V_1 = (V_0 + V_3)/2$, hence $V_1 = (V_0 + .92625\,V_1)/2 = V_0/2 + .4631V_1$.

$V_1 = \frac{V_0}{2(1-0.4631)} = \frac{V_0}{2(.5369)} = 0.931V_0$, $V_3 = 0.9263V_1 = 0.863V_0$, and the average output = $(V_1 + V_3)/2 = (0.931\,V_0 + 0.863\,V_0)/2 = 0.897V_0$.

Here, including an approximation of the effect of diode drops, the average output would be about 0.897 (100 – 1.4) = **88.4V**, with ripple = $V_1 - V_3$ = (.931 – .863) 100 = **6.70Vpp**, the output ranging from 88.4 + 6.70/2 = 91.8V, to 88.4 – 6.70/2 = 85.0V.

SECTION 3.9: SPECIAL DIODE TYPES

3.62 For a junction diode, $i = I_S e^{v/nV_T}$, and $v = nV_T \ln(i/I_S)$.

Thus, $300 = n(25)\ln 100/I_S$, or $12 = n\ln 100 - n\ln I_S$ – – – (1)

Also, $370 = n(25)\ln 1000/I_S$, or $14.8 = n\ln 1000 - n\ln I_S$ – – – (2)

(2) – (1) = $2.8 = n(\ln 1000 - \ln 100)$,

whence, $n = 2.8/(6.908 - 4.605)$, or n = **1.216**

Now, $I_S = ie^{-v/nV_T} = 1e^{-370/(1.216 \times 25)}$ = **5.18 × 10⁻⁶A**.

Check: $i = 5.18 \times 10^{-6} e^{300/(1.216 \times 25)}$ = 0.1 A, OK.

At 20 A, $v = nV_t \ln(i/I_S) = 1.216(25)\ln[20/(5.18 \times 10^{-6})]$ = 461 mV.

Thus the ohmic drop is $R(20A) = (800 - 461)10^{-3}$, whence $R = 339 \times 10^{-3}/20$ = **16.95 mΩ**.

Check: 20 A and 17 mΩ → $GA\,v_R = 20 \times 17/1000$ = 0.34 V. Thus the series resistance of this power diode is $17 \times 10^{-3}\Omega$ or 17 mΩ.

3.63 For a junction diode, $i = I_S e^{\upsilon/nV_T}$, and $\upsilon = nV_T \ln(i/I_S)$.

For the specified diode having a 10Ω series resistance, $\upsilon_D = 10i + 1.5(25)\ln(i/I_S)$

At 10 mA, $420 = 10(10) + 37.5\ln(10/I_S)$, $\ln(10/I_S) = (420 - 100)/37.5 = 8.533$, $10/I_S = e^{8.533} = 5081$, whence $I_S = 10/5081 = 1.97 \times 10^{-3} mA$.

At 1 mA, $\upsilon_D = 10(1) + 37.5\ln(1/(1.97 \times 10^{-3})) =$ **334 mV.**

At 50 mA, $\upsilon_D = 10(50) + 37.5\ln(50/1.97 \times 10^{-3}) =$ **880 mV.**

3.64 $i = I_S e^{\upsilon/nV_T}$: Taking base e logarithms, $\upsilon = nV_T \ln(i/I_S)$

At 0.1 mA, $\upsilon = 1.1(25)\ln[0.1 \times 10^{-3}/10^{-15}] =$ **696 mV.**

At 10 mA, $\upsilon = 1.1(25)\ln[10 \times 10^{-3}/10^{-15}] =$ **823 mV.**

3.65 For a varactor (or variable-capacitance diode) (from Eq. 3.27), $C_j = C_{j0}(1 + V_R/V_0)^{-m}$.

Here, $33 = C_{j0}(1 + 2/2.2)^{-0.9}$, $C_{j0} = 33(1 + 2/2.2)^{0.9} =$ **59.1 pF.**

Thus, at 0 V, $C =$ **59.1 pF**

At 1 V, $C = 59.1(1 + 1/2.2)^{-0.9} =$ **23.6 pF**

At 10 V, $C = 59.1(1 + 10/2.2)^{-0.9} =$ **7.1 pF**

3.66 In sunlight, $1000 W/m^2 = 1000 \times 10^{-4} W/cm^2 = 100 mW/cm^2$

At 25°C, the diode photocurrent is $I_R = 100 \times 0.7 =$ **70 μA.**

At 125 °C, $I_R = 70 \times 10^{-6}(1.035)^{(125-25)/10} = 70 \times 10^{-6}(1.035)^{10} =$ **98.7 μA**

At 25 °C, the dark current is **1.5 nA**

At 125 °C, the dark current is $1.5 \times 2^{(125-25)/10} = 1536$ nA $=$ **1.54 μA.**

3.67 In direct light, $I_D = 20 \times 0.7 \times 10^{-6} + 1.5 \times 10^{-9} = 14.0 \mu A$

In reflected light, $I_R = 0.5 \times 0.7 \times 10^{-6} + 1.5 \times 10^{-9} = 0.35\ \mu$A

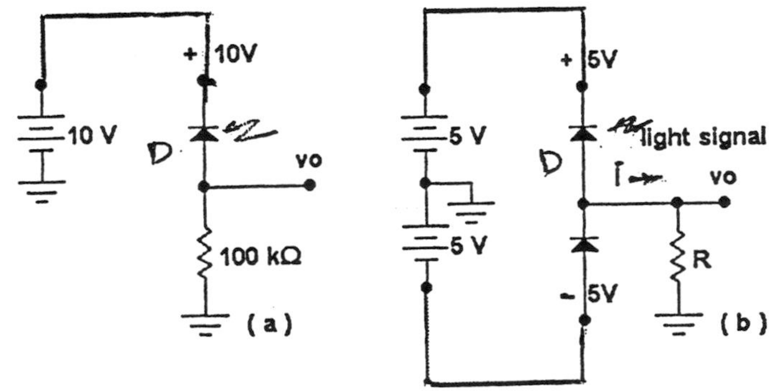

For direct light: $\upsilon_O = 14 \times 10^{-6} \times 10^5 =$ **1.40 V.**

For reflected light: $\upsilon_O = 0.35 \times 10^{-6} \times 10^5 =$ **0.35 V**

In the modified circuit shown, D_2 is biased by light passing through a filter to adjust its intensity to about half the direct light beam applied to D_1.

In operation I_1 is either 14 μA or 0.35 μA, and I_2 is adjusted (by varying the light bias) to balance the two output current magnitudes.

For direct light: $I = I_{1D} - I_2$.

For reflected light: $-I = I_2 - I_{1R}$.

Now, the magnitudes should be equal, that is $|I_{1D} - I_2| = |I_2 - I_{1R}|$.

Thus $I_{1D} - I_2 = \pm (I_2 - I_{1R})$

Take the positive sign: $2I_2 = I_{1D} + I_{1R} = 14 + 0.35$, and $I_2 = $ **7.17** μA.

Check: For $I_1 = 14\mu A$, $I = 14 - 7.17 = 6.83\mu A$. For $I_1 = 0.35\mu A$, $I = 7.17 - .35 = 6.82\mu A$. OK

For a ± 05 V signal, $R = 1/6.83 \times 10^{-6} = $ **146 kΩ**.

Note that in this balanced circuit, if the diodes are matched, dark currents cancel and operation is independent of the operating temperature of the diodes (as long as they are at the *same temperature*).

3.68 See that for a usual op amp, the diode is connected across virtual ground and $\upsilon_O = 0$ V. Thus, if the diode is not exposed to light, the current through it is zero. Note that this is the case with an ordinary diode in a sealed package, where $\upsilon_O = $ **0 V**. When light falls on the junction of D, photo-carriers are produced with the photocurrent polarity (which allows photocurrent to flow from cathode to anode). When an illuminated diode is short-circuited (as is D in Fig. Q3.68); current is extracted from the external circuit. For the connection shown, υ_O rises. For light at 20 mW/cm^2, and the diode rated at 0.7 μA/mW/cm^2, the photo current is $0.7 \times 10^{-6} \times 20 = $ **14** μA. For light applied to the circuit, $\upsilon_O = 5$ kΩ $\times$ 14 μA $= 5 \times 10^3 \times 14 \times 10^{-6} = $ **70 mV**. For no light applied, $\upsilon_O = $ **0 V** (if usual small offsets are ignored).

3.69

As noted in P3.68 above, an illuminated photodiode attempts to conduct photo-current (as seen externally) flowing from cathode to anode. Such a current is shown in Fig. 3.69a) as i. In Fig S3.69b), the situation is redrawn with the external circuit indicated by the dashed lines, where current labels in b) follow from a): Here i_1 is i, i_2 is i_1, i_3 is i_2 is i_1.

We conclude that a photodiode can generate current, which flows out of its anode, just like a photo-operated battery with the anode positive. Certainly, this happens if the external circuit is a short-circuit (This was illustrated in P3.68 above.) But as the resistance in the external circuit rises, the voltage across the diode increases until the internal junction begins to conduct increasingly. Thus an open-circuited silicon photodiode when illuminated has an open-circuit voltage equal to a diode drop (of 0.7 V or so for silicon). Thus, tt behaves like a 0.7 V battery. When an external load is connected, the terminal voltage drops until at (say) 0.5 V, 99% or more of the photo current flows in the external load, and only 1% is internally short-circuited by the (slightly) forward-biased junction.

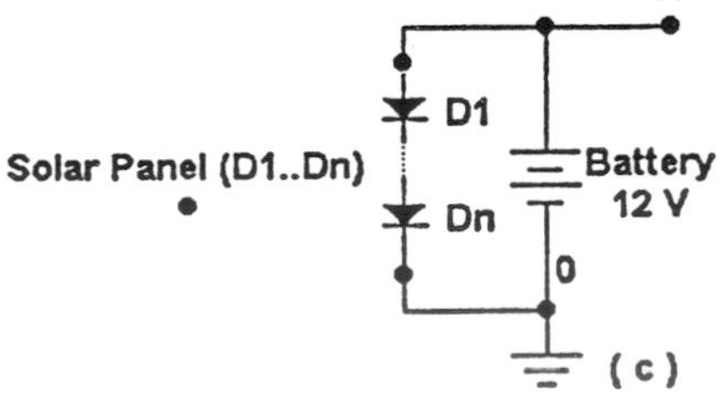

For the specific solar panel:

Output power level is

$P = IV = 100 \times 10^{-3} \times 14.5 = $ **1.45 W**.

Panel power level is $P = IV = 110 \times 14.5 = $ **1.59 W**.

Open-circuit power level is

$P = IV = 110 \times 10^{-3} \times 24 = $ **2.64 W**

For an open circuit, the photo-current of 110 mA or so flows in the series-connected forward-biased junction to create the open-circuit voltage of 24 V. Since each cell has a forward voltage of 0.67 at 110 mA or so, the number of series cells is approximately 24/0.67 = **36 cells.**

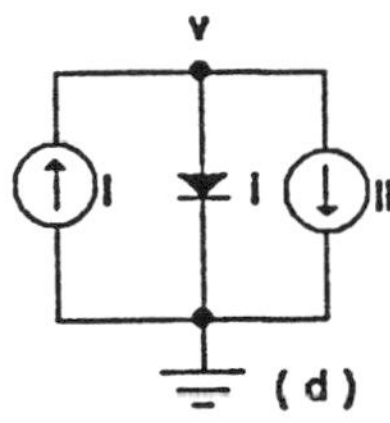

Maximum power calculations: Maximize $P = I_L \upsilon = (I - i)\upsilon$ --- (1)

where $i = I_S e^{\upsilon/V_T}$ --- (2)

Maximize $P = (I - I_S e^{\upsilon/V_T})\upsilon$.

Now $\partial p/\partial \upsilon = I - I_S e^{\upsilon/V_T} - \upsilon I_S/V_T e^{\upsilon/V_T} = I - I_S(1 + (\upsilon/V_T)e^{\upsilon/V_T}$

Maximum power occurs when $\partial p/\partial \upsilon = 0$, or $I = (1 + \upsilon/V_T)I_S e^{\upsilon/V_T} = i(1 + \upsilon/V_T)$, or $I = i + i\upsilon/V_T$, or $\upsilon = (I - i)/iV_T$ --- (3), or $\upsilon = (I/i - 1)V_T$ --- (4).

Now, $P_{max} = (I - i)^2/iV_T$ --- (5)

Combining (2) and (4), P_{max} occurs at current $i = I_S e^{(I/i - 1)} = 0.367 I_S e^{I/i}$ --- (6)

For the specific case, $\upsilon =$ 670 mV, and $i =$ 110 mA, with the initial assumption of $n = 1$:

From Eq.(2), $110 \times 10^{-3} = I_S e^{670/1(25)}$, and $I_S = 110 \times 10^{-3} e^{-670/25} = 2.525 \times 10^{-13} A = 2.525 \times 10^{-10} mA$

Generally, $i = I_S e^{(I/i - 1)}$, and taking ln, $\ln i/I_S = I/i - 1$, or $I/i = \ln i + 1 - \ln I_S$ --- (7), or

$$i = \frac{I}{\ln i + 1 - \ln I_S} \quad \text{--- (8).}$$

For the specific case, $i = \dfrac{I}{\ln i + 1 - \ln 2.525 \times 10^{-10}}$, or $i = \dfrac{I}{\ln i + 23.1}$ --- (9)

Solve Eq.(9) interatively for $I =$ 110 mA:

Try $i =$ 10 mA: $i = 110/(2.30 + 23.1) =$ 4.33 mA

Try $i =$ 4.5 mA: $i = 110/(1.504 + 23.1) =$ 4.47 mA

Try $i =$ 4.47 mA: $i = 110/(1.497 + 23.1) =$ 4.472 mA

Thus, maximum power occurs for an internal diode current $i =$ 4.47 mA or 4.47/110 ≡ 4% of full short-circuit current.

At maximum load power, each diode voltage is (from Eq.(4)):

$\upsilon = (I/i -)V_T = (110/4.47 - 1)25 \times 10^{-3} =$ 590 mV.

For a stack of such diodes (with $n = 1$) with open-circuit voltage of 24 V, the maximum-power voltage would be 590/670(24) = **21.1 V.**

This is clearly much higher than the 17.5 volts specified. Probably $n > 1$, although the detail of the specifications, whether nominal or best/worst case, may also be suspect.

Now if $n = 2$ is assumed,

from Eq.(2), $110 \times 10^{-3} = I_S e^{670/2(25)}$, and $I_S = 110 \times 10^{-3} e^{-670/50} = 1.67 \times 10^{-7} A = 1.67 \times 10^{-4} mA$.

From (8), $i = \dfrac{I}{\ln i + 1 - \ln(1.67 \times 10^{-4})}$, or $i = \dfrac{I}{\ln i + 9.70}$ --- (10).

Solve Eq.(10) iteratively for $I =$ 110 mA:

Try $i =$ 10 mA: $i = \dfrac{110}{\ln 10 + 9.70} =$ 9.16 mA

Try $i =$ 9 mA: $i = \dfrac{110}{\ln 9.2 + 9.70} =$ 9.24 mA

Try $i = 9.2$ mA: $i = \dfrac{110}{\ln 9 + 9.70} = 9.23$ mA

Thus, maximum power occurs for $i = 9.23$ mA, or $9.23/110 \equiv 8.4\%$ of full current, for which (from Eq.(4) extended) the diode voltage is $(I/i = 1)nV_T = (110/9.23 - 1)(2 \times 25 \times 10^{-3}) =$ **546 mV**

For the full stack of diodes, this maps to $546/670 \times 24 =$ **19.5 V**

Clearly, n is more likely to be 2 than 1!

For the operation at 14.5 V, each diode drop is $14.5/24 \times 670 =$ **405 mV**, and $i = I_S e^{\upsilon/2(V_t)} = .67 \times 10^{-4} e^{405/50} = 0.55$ mA.

Thus the current loss at 14.5 is about **0.6 mA**, which is quite small!

3.70 Input power is $P = i\upsilon = 10 \times 10^{-3} \times 1.9 =$ **19 mW**

Now, $i = I_S e^{\upsilon/nV_T}$, and $10 = I_S e^{1900/1.2(25)}$, or $I_S = 10e^{1900/1.2(25)} = 3.12 \times 10^{-27} mA$.

At a power level of $60/2 = 30$ mW, assume the current increases by 50% to 15 mA.

Thus for $i = 15mA$, $\upsilon = nV_t \ln(i/I_S) = 1.2(25)\ln[15/3.12 \times 10^{-27}] = 1.2(25)(63.74) = 1.912$ V,

for which $i\upsilon = 15(1.912) = 28.7$ mW, a bit small.

Now, try $i = 30/1.912 = 15.69$ mA, for which $\upsilon = 1.2(25)\ln[15.69/3.12 \times 10^{-27}] = 1.914$ V, and $i\upsilon = 15.69 \times 1.914 =$ **30.03 mW**.

Thus a current of **15.7 mA** produces an output of **1.91 V** for a dissipation of half rated power, **30 mW**.

For 10 mA operation: $R = (5 - 0.5)(10 \times 10^{-3}) = 450\Omega$

For 15.7 mA operation: $R = (5 - 0.5)(15.7 \times 10^{-3}) = 287\Omega$

From Appendix H, standard 1% resistor values to be used would be **453 Ω** and **287 Ω**.

3.71 The detector is specified to have an open-circuit voltage of 500 mV and a short-cirucit current of 70μA. From the discussion in P3.68 above, this specification relates to operating in the low-voltage or solar-cell mode.

For an emitter current of 5 mA, the output short-circuit currents can range from $5 \times 10^{-3} \times 0.6/100 =$ **30μA** to $5 \times 10^{-3} \times 1.6/100 =$ **80μA**

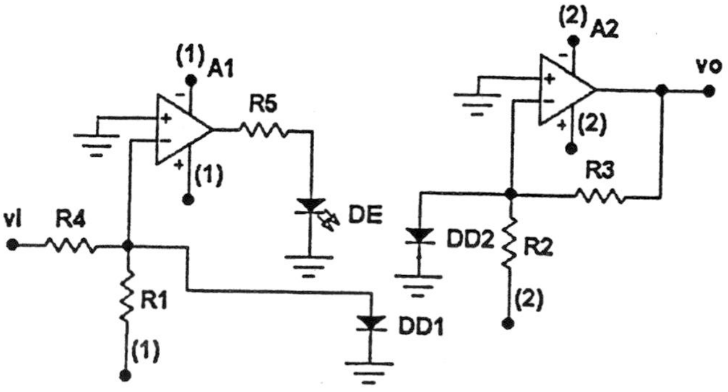

In this circuit, the isolated $\pm 5V$ supplies are labelled (1) and (2) (for primary and secondary). Op amps A_1 and A_2 operate in the inverting mode. Resistor $R_1 (= 100k\Omega)$ where R_1 establishes the operating current for D_{D1} (and D_{D2}) at 50 μA. Initially it causes the output of A_1 to rise driving D_E through R_5, limits the maximum current in D_E to 30 or 40 mA. As the current rises in D_E, so does the current in D_D (and D_D). The current in D_{D1} rises until it equals that in R_1 (50 μA). $R_2 \approx R_1$ ($= 100k\Omega$) ensures that D_{D2} operates at the same current. Diode mismatch can be compensated by minor control of R_2. Resistor R_3 (with R_4) establishes the overall voltage gain to a value R_3/R_4 (here of 100k/10k = 10 V/V). Resistor R_4 isolates the input and controls the range of current variation in D_{D1}, D_{D2} to ± 0.1/V10k = ± 10 μA for ± 0.1 V signals. *Notice* that the current transfer gain is completely compensated by establishing the current in D_{D1} (via R_1) and allowing the current in D_E to seek an appropriate value (limited only by R_5 on those rare occasions on turn-on, or when the loop is broken accidentally.

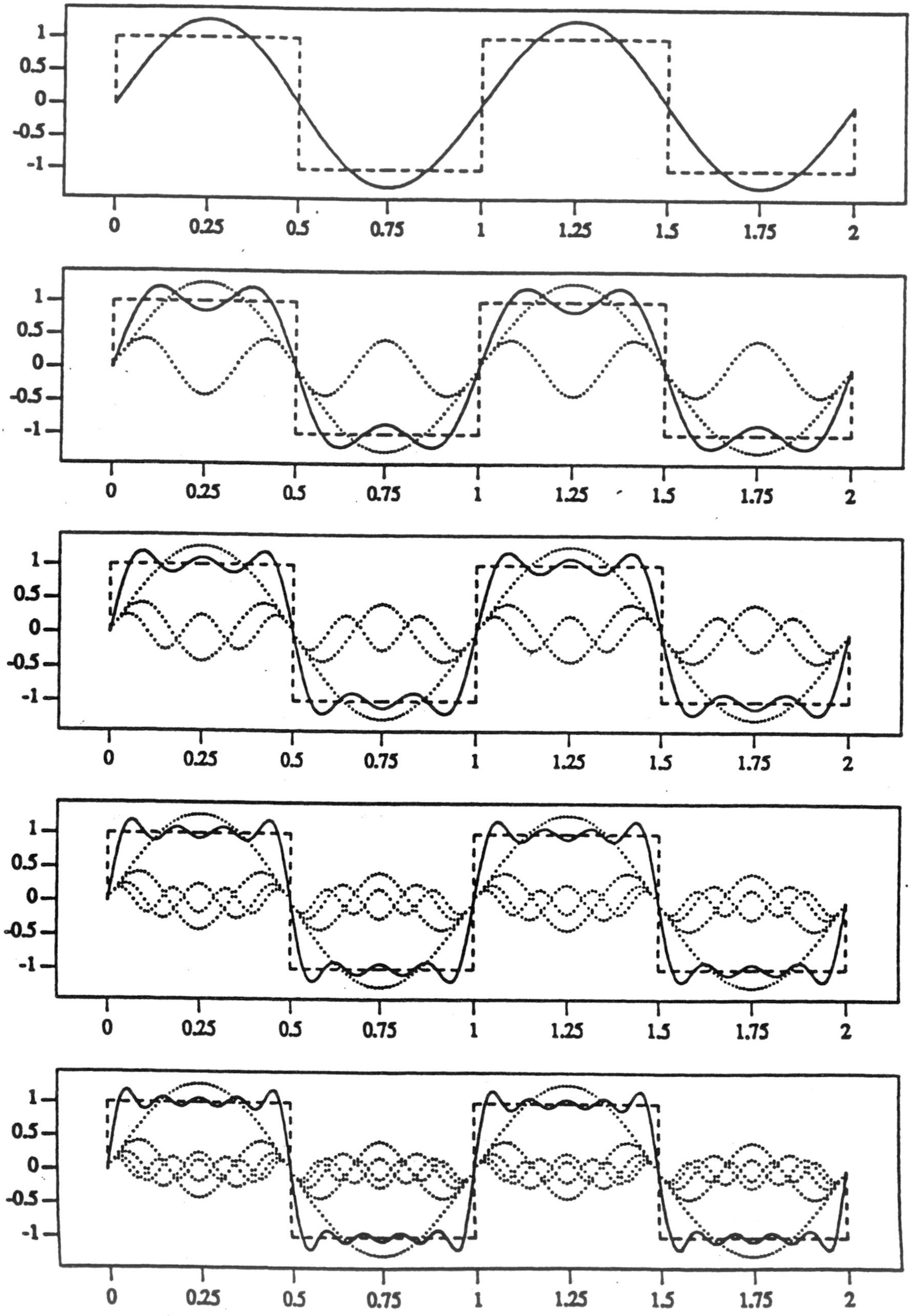
1
0.5
0
-0.5
-1
0
0.25
0.5
0.75
1
1.25
1.5
1.75
2

Chapter 4

Bipolar Junction Transistors (BJTs)

SECTION 4.1: PHYSICAL STRUCTURE AND MODES OF OPERATION

4.1

#	Type	Mode
1	npn	active
2	npn	cutoff
3	pnp	cutoff
4	pnp	saturated
5	pnp	cutoff*
6	npn	saturated

*Cutoff, but nearly active.

4.2 Two junctions, each in two states of conduction, imply $2 \times 2 =$ **4** modes of operation.
Table 4.1 lacks the case: **EBJ Reverse Biased, CBJ Forward** biased.

SECTION 4.2: OPERATION OF THE NPN TRANSISTOR IN THE ACTIVE MODE

4.3 Generally, $i_C = I_S e^{v_{BE}/nV_T}$. Here, $2.0 \times 10^{-3} = I_S e^{700/25}$ and $I_S = 2 \times 10^{-3} e^{-700/25} = 1.38 \times 10^{-15}$**A**. Now, from Eq.4.4 $I_S = A_E q D_n n_i^2/(N_A W)$ and from Eq. 3.12, $D_n/\mu_n = D_p/\mu_p = V_T$. Thus, $D_n = V_T \mu_n = 25 \times 10^{-3} \times 1100 =$ **27.5 cm²/s**.

Now (from Eq.4.4) $A_E = N_A W I_s/(q D_n n_i^2)$, and calculation in cm,

$A_E = 10^{18} \times W \times 1.38 \times 10^{-15}/(1.6 \times 10^{-19} \times 27.5 \times (1.5 \times 10^{10})^2) = 1.39 \mathbf{W} cm^2$

(with W in cm) $=$ **1.39** $\times$ **10⁴Wμm** (with W in μm).

For $W = 2\mu m = 2 \times 10^{-4} cm$,

$A_E = 1.39(2 \times 10^{-4}) = 2.78 \times 10^{-4} cm^2 = 2.78 \times 10^{-4} \times 10^8 = 2.78 \times 10^4 \mu m^2$, being, for example, $167 \times 167 \mu m^2$.

4.4 For a transistor whose EBJ is 100× larger, I_S is 100× larger, namely $I_S = 100 \times 1.38 \times 10^{-15} A$ or $I_S = 1.38 \times 10^{-13}$**A**.

For $v_{BE} =$ 0.70V, $i_C = 100(2) =$ **200 mA**.

For $i_C = 1mA$, $1.0 \times 10^{-3} = 1.38 \times 10^{-13} e^{V_{BE}/V_T}$ and $V_{BE} = 25mV \ln(1 \times 10^{-3})(1.38 \times 10^{13}) =$ **568 mV**.

For $I_S = A_E q D_n n_i^2/(N_A W)$ and $n_i^2 = BT^3 e^{-E_G/KT}$, with $B = 5.4 \times 10^{31}$ and $E_G = 1.12 eV$, $k = 8.62 \times 10^{-5} eV/K$.

Thus $I_S = A_E q D_n B T^3 e^{-E_G/kT}/N_A W = constant \times T^3 e^{-E_G/kT} = K^* \times T^3 e^{-1.12/8.62 \times 10^{-5}/T} = K^* T^3 e^{-1.30 \times 10^4/T}$.

At 300K, $I_S = K^* 300^3 e^{-1.30 \times 10^4/300} = K^* 4.09 \times 10^{-12}$.

At 400K, $I_S = K*400^3e^{-1.30\times10^4/400} = K*4.91\times10^{-7}$.

Thus at 400K, I_S increases by 1.20×10^5 times! to $I_S = 1.2\times10^5\times1.38\times10^{-13}A$ = 1.656×10^{-8}**A**, for which $\upsilon_{BE} = 25\times10^{-3}\ln[1\times10^{-3}/(1.656\times10^{-8})]$ = **275 mV**

4.5 Now, in the base, $n_{p0} = n_i^2/N_A$, and at 25°C, $n_{p0} = (1.5\times10^{10})^2/10^{17} = 2250/cm^3$, and (from Eq.4.1) $n_p(0) = n_{p0}e^{\upsilon_{BE}/V_T} = 2250e^{700/25}$ = **3.25** × 10^{15}/**cm**^{3}. Now (from Eq.4.2), $I_n = A_E qD_n(-n_p(0)/W)$, and using cm,

$I_n = -20\times10^{-4}\times20\times10^{-4}\times1.6\times10^{-19}\times21.3\times3.25\times10^{15}/W = -4.43\times10^{-8}/W$.

For $W = 1\mu m = 1\times10^{-4}cm$, $I_n = -4.43\times10^{-8}/10^{-4} = 4.43\times10^{-4}A$, or – **0.443 mA.**

For $W = 0.1\mu m$, $I_n = -4.43$**mA.**

Now (from Eq.4.4), $I_S = A_E qD_n n_i^2/(N_A W)$, and with all dimensions in cm, $I_S = 20\times10^{-4}\times20\times10^{-4}\times1.6\times10^{-19}\times21.3\times(1.5\times10^{10})^2/10^{17}W = 3.067\times10^{-20}/W$.

For $W = 1\mu m = 1\times10^{-4}cm$, I_S = **3.07** × 10^{-16}**A.**

For $W = 0.1\mu m$, I_S = **3.07** × 10^{-15}**A.**

Check: Now, for υ_{BE} = 700 mV, $i_C = I_S e^{\upsilon_{BE}/V_T}$.

For $W = 1\mu m$, $i_C = 3.07\times10^{-16}e^{700/25}$ = **0.444 mA.** For $W = 0.1\mu m$, i_C = **4.44mA**, both (as expected), the same as I_n (within a factor α).

Now, for the base current (from Eq.4.12), $\beta = 1/\left[\frac{D_p}{D_n}\frac{N_A}{N_D}\frac{W}{L_p} + 1/2\frac{W^2}{D_n\tau_b}\right]$, where $\tau_b = L_n^2/D_n$, by adaptation of Eq.3.30. Here, the base minority-carrier lifetime (using calculations in cm) is $\tau_b = L_n^2/D_n = (19\times10^{-4})^2/21.3$ = 169.5 ns.

Thus,

$$\beta = 1/\left[\frac{1.7}{21.3}\times\frac{10^{17}}{10^{19}}\frac{W}{0.6} + \frac{W^2}{2(21.3\times10^8\times169.5\times10^{-9})}\right] = 1/[1.33\times10^{-3}W + W^2(1.385\times10^{-3})].$$

Now, for $W = 1\mu m$: $\beta = 1/[1.33(1) + 1^2(1.385)]10^{-3}$ = **368**, for which $\alpha = \frac{\beta}{\beta+1}$ = **0.997,**

and for $W = 0.1\mu m$: $\beta = 1/[1.33(0.1) + 0.1^2(1.39)]10^{-3}$ = **6807**, for which α = **0.9999.**

Note that for $W = 0.1\mu m$, β is very very high. Such a transistor would be very difficult to make in practice, and would have low breakdown voltages. (See P4.6, next).

4.6 Generally, from Eq. 4.12, $\beta = 1/\left[\frac{D_p}{D_n}\times\frac{N_A}{N_D}\times\frac{W}{L_p} + \frac{1}{2}\frac{W^2}{D_n\tau_b}\right]$, where $\tau_b = L_n 2/D_n$. From P4.2 in the Text, $N_D = 10^{19}/cm^3$, $N_A = 10^{17}/cm^3$, L_n = 19 μm, $D_n = 21.3cm^2/s$, L_p = 0.6 μm, $D_p = 1.7cm^2/s$.

Thus $\tau_b = (19\times10^{-6})^2/21.3\times(10^{-2})^2 = 16.95\times10^{-8} = 169.5ns$.

Thus $\beta = 1/\left[\frac{1.7}{21.3}\times\frac{10^{17}}{10^{19}}\times\frac{W}{0.6} + \frac{W^2}{2(21.3\times10^8)(169.5\times10^{-9})}\right]$, or

$\beta = 1/[1.33\times10^{-3}W + 1.385\times10^{-3}W^2] = 1/[1.33W + 1.39W^2]10^{-3}$

Now, for β = 1000:

$1.39W^2 + 1.33W = 1 = 0$ or

$$W = \frac{-1.33\pm\sqrt{1.33^2 - 4(1.39)(-1)}}{2(1.39)} = \frac{-1.33\pm2.71}{2(1.39)} = \mathbf{0.469\ \mu m}$$

For β = 2000: $1.39W^2 + 1.33V - 0.5 = 0$ or

$$W = \frac{-1.33\pm\sqrt{1.33^2 - 4(1.39)(-0.5)}}{2.(1.39)} = \frac{1.33\pm2.13}{2(1.39)} = \mathbf{0.288\ \mu m}$$

4.7 $i_C = I_S e^{v_{BE}/V_T}$, whence $I_S = i_c e^{-v_{BE}/V_T} = 10 \times 10^{-6} e^{-650/25} =$ **5.11** $\times 10^{-17}$**A**

For a 0.500 V drop at 10 μA, $I_S = 10 \times 10^{-6} e^{-500/25} = 2.06 \times 10^{-14}$

Junction size increase is $2.06 \times 10^{-14}/(5.11 \times 10^{-17}) =$ **403 times.**

Check: $e - \dfrac{(500 - 650)}{25} = 403$, OK.

At 0.65 V, this large junction has a current of $2.06 \times 10^{-14} e^{650/25} =$ **4.03 mA.**

At 0.70 V, the current is $2.06 \times 10^{-14} e^{700/25} =$ **29.8 mA.**

Check: $4.03 e^{(700 - 650)/25} = 29.8$, OK.

4.8 $i_C = I_S e^{v_{BE}/V_T}$, and $I_S = i_C e^{-v_{BE}/V_T}$.

Here, $I_S = 10 \times 10^{-3} e^{-690/25} =$ **1.03** $\times$ **10**$^{-14}$ **A.**

Also $\beta = i_C/i_B = 10 \times 10^{-3}/75 \times 10^{-6} =$ **133.3** and

$\alpha = \dfrac{\beta}{\beta + 1} = \dfrac{133.3}{134.3} =$ **0.993**, whence

$I_S/\alpha =$ **1.039** $\times 10^{-14}$**A** and $I_S/\beta =$ **7.73** $\times 10^{-17}$**A.**

4.9 Here, $i_E = 0.753 \pm 0.001$ mA and $i_C = 0.749 \pm 0.001$.

Thus $i_B = i_E - i_C = 0.004 \pm 0.002$, a current ranging from 0.002 to 0.006.

Thus β varies from $\dfrac{0.749 + 0.001}{0.002} =$ **375** to $\dfrac{0.749 - 0.001}{0.006} =$ **124.7**, for which $\alpha = \dfrac{\beta}{\beta + 1} =$ **0.9973** to **0.9920**

Directly: α ranges from 0.750/0.752 = 0.9973 to 0.748/0.754 = 0.9920.

Clearly, measurement of the currents which are nearly equal leads to a lot of error in β, although α is relatively insensitive (except that casual measurement of i_E and i_C can easily suggest that α is negative!

4.10 From P4.3, $i_C = 2.0$ mA, $V_{BE} = 0.7$ V, $n = 1$, $N_A = 10^{18}/cm^3$, $\mu_n = 1100 cm^2/V_s$, $I_S = 1.38 \times 10^{-15} A$, $D_n = 27.5$ cm/s. $A_E = 1.39$ W cm (for W in cm) $W = 2\mu m = 2 \times 10^{-4} cm$

The Minority stored charge in the base (with calculations in cm) is (from Eq.4.9)

$$Q_n = \frac{A_E W q n_i^2}{2N_A} e^{v_{BE}/V_T}$$

or $$Q_n = \frac{1.39 \times (2 \times 10^{-4})^2 (1.6 \times 10^{-19})(1.5 \times 10^{10})^2}{2(10^{18})} e^{700/25} 1.446 \times 10^{12} = \mathbf{1.46\ pC}$$

For equation 4.12, assuming recombination to be dominant, $\beta = 1/\left(\dfrac{W^2}{2D_n \tau_b}\right)$

for which, the lifetime $\tau_b = \dfrac{\beta W^2}{2D_n} = \dfrac{120 \times (2 \times 10^{-4})^2}{2(27.5)} =$ **87.3 ns**

4.11 For (ensured) active-mode operation of an npn transistor $v_{CB} \geq 0V$ and $v_{CE} = v_{CB} + v_{BE} \geq$ **700 mV.** In the active mode, $i_C = I_S e^{v_{BE}/V_T}$ (for $n = 1$); $10 \times 10^{-3} = I_S e^{700/25}$; or $I_S = 10 \times 10^{-3} e^{-28} = 6.91 \times 10^{-15}$A. Also, in the active mode, $\beta = \dfrac{i_C}{i_B} = \dfrac{10mA}{100\mu A} = \dfrac{10}{0.1} =$ **100**, and $i_E = i_C + i_B = 10mA + 0.1mA =$ **10.1 mA.**

4.12 For the tabulated devices,

(b) $I_B = \dfrac{I_C}{\beta} = \dfrac{1mA}{50} = 20\mu A$; $I_E = I_C + I_B = 1 + .02 =$ **1.02 mA**; $\alpha = \dfrac{I_C}{I_E} = \dfrac{1.00}{1.02} =$ **0.980.**

(c) $I_C = \alpha I_E = 0.98(2) =$ **1.96mA**; $I_B = I_E - I_C = 2 - 1.96 = .04mA = 40\mu A$; $\beta = \dfrac{I_C}{I_B} = \dfrac{1.96}{0.04} =$ **49.**

(d) $I_C = \beta I_B = \dfrac{\alpha}{1-\alpha} I_B = \dfrac{0.995}{1-0.995} \times 0.01 =$ **1.99 mA**; $I_E = I_C + I_B = 1.99 + .01 =$ **2.00 mA**; $\beta = \dfrac{\alpha}{1-\alpha} = \dfrac{0.995}{1-.995} =$ **199.**

(c) $I_C = \alpha I_E = \dfrac{\beta}{\beta+1} I_E = \dfrac{10}{10+1} \times 110 =$ **100 mA**; $I_B = \dfrac{I_C}{\beta} = \dfrac{100mA}{10} =$ **10 mA**; $\alpha = \dfrac{I_C}{I_E} = \dfrac{100}{110} =$ **0.909.**

(d) $I_C = \beta\, I_B = 1000 \times 0.001 =$ **1mA**; $I_E = I_C + I_B = 1 + 0.001 =$ **1.001mA**; $\alpha = \dfrac{I_C}{I_E} = \dfrac{1}{1.001} =$ **0.999.**

Device #	I_C mA	I_B mA	I_E mA	α	β
a	10	0.1	10.1	0.99	100
b	1	0.02	1.02	0.98	50
c	1.96	0.04	2	0.98	49
d	1.99	0.01	2.00	0.995	199
e	100	10	110	0.909	10
f	1	0.001	1.001	0.999	1000

4.13

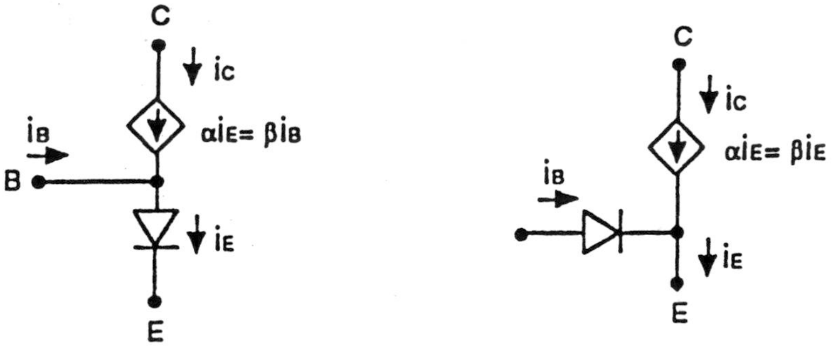

The controlled sources can be labelled α **i**$_E$ or β **i**$_B$ where $\alpha\, i_E = \beta\, i_B$

4.14 Generally, $i_C = I_S\, e^{v_{BE}/nV_T}$.

For $n = 1$, at 1mA, $10^{-3} = I_S\, e^{700/25}$ — — — (1), and at 0.1μA, $10^{-7} = I_S\, e^{v/25}$ — — — (2). Now, dividing, $10^{-3}/10^{-7} = e^{(700-v)/25} = 10^4$. Taking logarithms, $700 - v = 25 \ln 10^4 = 230$. Thus, $v = 700 - 230 = 470$mV, or **0.47V**.

For $n = 2$, see $v = 700 - 2(25) \ln 10^4 = 700 - 2(230) = 240$mV, or **0.24V**.

4.15 Generally, for the base open, $i_C = \beta\, i_B = 100\, I_{CBO}$.
At 25°C, $I_{CBO} = 0.1$nA, and $i_C = 100\,(0.1 \times 10^{-9})\,A$ = **10nA.**
At 95°C, $I_{CBO} = 0.1 \times 10^{-9} \times 2^{\frac{95-25}{10}} = 10^{-10} \times 2^7 A$, whence $i_C = 100 \times 10^{-10} \times 2^7 = 1.28\mu A$.

4.16 α_F/α_R = 100, the relative junction area. Thus $\alpha_R = \frac{\alpha_F}{100} = \frac{150}{150+1}/100 = 0.00993$, and $\beta_R = \frac{\alpha_R}{1-\alpha_R} = \frac{0.00993}{1-0.00993} = 0.0092 \equiv$ **0.01.**

SECTION 4.3: THE PNP TRANSISTOR

4.17 For $\alpha = 0.975$, $\beta = \frac{\alpha}{1-\alpha} = \frac{0.975}{1-.975} = 39$. For $I_B = 10\mu A$, $I_C = 39\,(10)\mu A = 3.90\mu A$ = **0.390mA,** and $I_E = 390 + 39 = 429\mu A$ = **0.429mA.** Generally, $v_{BE2} = v_{BE1} + nV_T \ln i_{C2}/i_{C1}$. Thus at i_C = 0.390mA, $v_{BE} = 700 + 25 \ln \frac{0.390}{1.00}$ = **676mV.**

4.18 For D_B, $I_S/\beta = 10^{-13}$A, and for D_E, $I_S/\alpha = 10^{-11}$A. Thus, $1/\beta/1/\alpha = 10^{-2}$, or $\alpha/\beta = 10^{-2}$. Correspondingly, $\frac{\beta/(\beta+1)}{\beta} = 10^{-2}$, or $\beta + 1 = 100$, and β = **99.**
Now, $i_C = I_S\, e^{643/25(1)} = 99 \times 10^{-13}\, e^{25.7}$ = **1.46A!** Note that this is a large transistor, a 14.5-Ampere device (where $10^{-11}\, e^{700/25} = 14.5$A).

SECTION 4.4: CIRCUIT SYMBOLS AND CONVENTIONS

4.19

(a) Here, $i_E = I$ = **1mA,** independent of v_{BE} or β, V_E ranges from **–0.6V** to **–0.8V**; $i_C = \alpha\, i_E$ ranges from $\frac{10}{10+1}(1)$ = **0.909mA** to $\frac{300}{300+1}(1)$ = **0.997mA.**

(b) $V_C = V_{CC} - R_C I_C$, ranges from 10 – 5 (.909) = **5.45V** to 10 – 5 (.997) = **5.02V.**
No, $\mathbf{V_E}$ **variation has no effect** on I_C or V_C.

(c) For $v_{CB} \geq 0$, at the largest value of i_C, $R_C\, i_C \leq V_{CC} - 0 = 10$V, whence $R_C \leq \frac{10}{.997}$ = **10.03kΩ** (use 10kΩ).

4.20 Here, $\mathbf{i_C = \alpha\,(I + i_c)}$. Now, for high β, $\alpha = 1$ and $i_C = 1 + 0.1$ = **1.1mA.** For $\beta = 10$, $\alpha = \frac{10}{10+1}$ = 0.909, and $i_C = (1.1)\,(.909)$ = **1mA.** For high β, the largest allowed R_C is limited to be $R_C = \frac{10-0}{1.1}$ = **9.90kΩ,** and $v_C = 1(0.1\text{mA})\,(9.09\text{k}\Omega) = 0.909$V peak or **1.82 Vpp.** For $\beta = 10$, $v_C = 0.909\,(0.1\text{mA})\,(9.09\text{k}\Omega) = 0.826$V peak, or **1.65Vpp.**

4.21 For $V_E = 0.7$V, use $R_E = (0 - 0.7 - -10)/1 = 9.3\text{k}\Omega$. Now the largest i_C occurs for small R_E, small v_{BE}, and large β. That is, $i_{C\,max} = \left[\frac{-0.6 - -10}{9.3\,(1-.01)}\right] \times 1$ = **1.021mA.** For active-mode operation, $R_C \leq \frac{10-0}{1.021} = 9.79\text{k}\Omega$. Use **9.7kΩ.** Now for R_C varying by 1%, the lowest possible value of v_C = 10 – (9.7) (1.01) (1.021) = **–2.8mV.** Note that operation is still in the active mode, since $v_E \equiv -0.70$V.

SECTION 4.5: GRAPHICAL REPRESENTATION OF TRANSISTOR CHARACTERISTICS

4.22 Generally, $TC \simeq -2.0\text{mV}/^\circ\text{C}$. Now, for 10mA, at 25°C, $V_{BE} = 700\text{mV}$. Thus, for 10mA, at 0°C, $V_{BE} = 700 + (0 - 25)(-2.0) = 750\text{mV}$. For 10mA, at 50°C, $V_{BE} = 700 + (50 - 25)(-2.0) = 650\text{mV}$. For V_{BE} constant at 620mV, at 25°C, $i = 10e^{(620-700)/25} = \mathbf{0.407mA}$; at 0°C, $i = 10\,e^{(620-750)/25} = \mathbf{0.055mA}$; at 50°C, $i = 10\,e^{(620-650)/25} = \mathbf{3.012mA}$, a nearly 55 to 1 range!!

4.23 Here, $\Delta i_C = 2.19 - 2.10 = 0.09\text{mA}$, for $\Delta V_{CE} = 9 - 2 = 7\text{V}$. Thus $r_o = \frac{\Delta \upsilon}{\Delta i} = \frac{7}{.09} = \mathbf{77.8k\Omega}$, at an average current of (2.19 + 2.10/2) = 2.145mA. Thus $V_A = 77.8 \times 2.145 = \mathbf{167V}$. Correspondingly, at 0.1mA, $r_o = \frac{167}{0.1} = 1.67\text{M}\Omega \equiv \mathbf{1.7M\Omega}$, and at 10mA, $r_o = \frac{167}{10} = 16.7\text{k}\Omega \equiv \mathbf{17\ k\Omega}$.

4.24 At 100μA, $r_o = \frac{200}{0.1mA} = 2\text{M}\Omega$. For an increase in V_{CE} from 5 to 50V, the increase in current is $\frac{50 - 5}{2M} = 22.5\mu\text{A}$. Thus at 50V, the current becomes 100 + 22.5 = **122.5μA**.

SECTION 4.6: ANALYSIS OF TRANSISTOR CIRCUITS AT DC

4.25 In general, assume the active mode initially, with a conducting emitter-base junction, then verify that the collector-base junction is *not* conducting. Since $\beta = 50$, $\alpha = \frac{50}{51} = 0.98$.

(a) $V_B = -4\text{V}$; $V_E = -4 + 0.7 = \mathbf{-3.3V}$; $I_E = \frac{0 - -3.3}{3.3} = \mathbf{1mA}$; $I_C = 0.98(1) = \mathbf{0.98mA}$; $I_B = \frac{0.98}{50} = 19.6\mu\text{A}$; $V_C = -10 + 4.7(0.98) = \mathbf{-5.39V}$. See this is OK; operation is in the active mode.

(b) $V_B = -6\text{V}$; $V_E = -6 + 0.7 = \mathbf{-5.3V}$; $I_E = \frac{0 - -5.3}{3.3} = \mathbf{1.606mA}$; $I_C = 1.606(.98) = 1.574\text{mA}$; $I_B = \frac{1.57}{50} = 31.4\mu\text{A}$; $V_C = -10 + 4.7(1.574) = -2.60\text{V}$. Since V_C is well above V_B, the transistor is saturated.

In practice, in saturation $\upsilon_{EC} \simeq 0.2\text{V}$. Thus, $V_C = V_E - 0.2 = -5.3 - 0.2 = \mathbf{-5.5V}$, and $I_C = \frac{-5.5 - -10}{4.7} = \mathbf{0.957mA}$, with $I_B = I_E - I_C = 1.606 - .957 = \mathbf{0.648mA}$.

(c) $V_B = -2\text{V}$; $V_E = -2 + 0.7 = \mathbf{-1.3V}$; $I_E = \frac{2 - -1.3}{3.3} = \mathbf{1mA}$; $I_C = \mathbf{0.98mA}$; $I_B = \frac{0.98}{50} = 19.6\mu\text{A}$; $V_C = -8 + 4.7(0.98) = \mathbf{-3.39V}$. See OK, active.

(d) $V_B = 0\text{V}$; $V_E = \mathbf{0V}$. Thus the transistor is cutoff.
$V_C = \mathbf{-10V}$, and $I_B = I_C = I_E = \mathbf{0mA}$.

(e) $V_B = -4\text{V}$; $V_E = -4 - 0.7 = \mathbf{-4.7V}$; $I_E = \frac{-4.7 - -10}{4.7k\Omega} = \mathbf{1.128mA}$; $I_C = 1.128(0.98) = \mathbf{1.105mA}$; $I_B = \frac{1.105}{50} = 22.1\mu\text{A}$; $V_C = 0 - 1.105(3.3) = \mathbf{-3.67V}$. Since $V_C > V_B$, operation is in the active mode, as assumed.

(f) $V_B = -6\text{V}$; $V_E = -6 - 0.7 = \mathbf{-6.7V}$; $I_E = \frac{-6.7 - -10}{4.7k\Omega} = \mathbf{0.702mA}$; $I_C = 0.702(0.98) = \mathbf{0.688mA}$; $I_B = \frac{0.688}{50} = 13.8\mu\text{A}$; $V_C = 0 - 0.688(3.3) = \mathbf{-2.27V}$. Since $V_C > V_B$, this is active-mode operation, as assumed.

4.26

(a) $V_B = -4V$; $V_E = -4 + 0.7 = -3.3V$; $I_E = \frac{0 - -3.3V}{R_E} = 0.5mA$; Thus, $R_E = \frac{3.3}{0.5} = $ **6.6kΩ.** Now, $I_C = 0.5 \times 1 = 0.5mA$; $V_C = -10 + R_C\,(0.5) = V_B - V_{BC} = -4 - 0 = -4V$; Thus, $R_C = \frac{1}{0.5}\,(-4 + 10) = $ **12kΩ.**

(b) $V_B = -6V$; $V_E = -6 + 0.7 = -5.3V$; $I_E = \frac{0 - -5.3}{R_E} = 0.5mA$; Thus, $R_E = \frac{5.3}{0.5} = $ **10.6kΩ.** Now, $V_C = -10 + R_C\,(0.5) = V_B - V_{BC} = -6 - 0 = -6V$. Thus $R_C = \frac{1}{0.5}\,(-6 + 10) = $ **8kΩ.**

4.27 Assume active mode, with forward conduction of the base-emitter junction.

(a) $V_E = $ **0V**; $V_B = 0 + 0.7 = $ **0.7V**; $I_B = \frac{10 - 0.7}{100k\Omega} = 93\mu A$; $I_C = 93\mu A \times 10 = $ **0.930mA**; $I_E = 930 + 93 = 1023\mu A = $ **1.023mA**; $V_C = 10 - 2k\,(0.93) = $ **8.14V.**

(b) $V_E = $ **+10V**; $V_B = +10 - 0.7 = $ **9.3V**; $I_B = \frac{9.3 - 0}{100k} = 93\mu A$; $I_C = 93\,(10) = $ **0.930mA**; $I_E = 11\,(93) = $ **1.023mA**; $V_C = 0.93\,(2) = $ **1.86V.**

(c) $V_E = $ **0V**; $V_B = $ **0.7V**; $I_B = \frac{10 - 0.7}{100k\Omega} - \frac{0.7}{10k\Omega} = 93\mu A - 70\mu A = $ **23μA**; $I_C = (23\mu A)\,10 = $ **0.230mA**; $I_E = 11\,(23) = $ **0.253mA**; $V_C = 10 - 2\,(0.23) = $ **9.54V.**

(d) See that $V_E = $ **0V**, and that if the transistor conducts with $V_B = -0.7V$, that the current in the upper 100kΩ will exceed that in the lower, providing no net base current. Thus the transistor is cut off, with $V_B = +10 - (10 - -10)\,\frac{100k\Omega}{100k\Omega + 100k\Omega} = $ **0V**, $V_C = $ **−10V**, and $I_B = I_C = I_E = $ **0mA.**

4.28

(a) Assume active mode, and consider the base-to-emitter circuit: Thus $0 - 10k\Omega\,(I_B) - 0.7 - 10k\Omega\,(I_E) = -10V$. But $I_E = (\beta + 1)\,I_B = 21 I_B$. Thus $10\,I_B + 10\,(21\,I_B) = 9.3V$, whence $I_B = \frac{9.3}{220} = $ **42.27μA.** Thus, $V_B = 0 - 42.27\,(10k) = $ **−0.423V**; $V_E = -0.423 - 0.7 = $ **−1.123V**; $I_E = 21\,(42.27) = $ **0.888mA**; $I_C = 20\,(42.27) = $ **0.845mA**; and $V_C = +10 - 10k\,(0.845) = $ **1.55V.**

(b) As before: $I_B = \frac{10V}{100k + 21(10k)} = 32.26\mu A$; $V_B = 0 - 100k\,(32.26) = $ **−3.23V**; $V_E = -3.23 - 0.7 = $ **−3.93V**; $I_E = 21\,(32.26) = $ **0.677mA**; $I_C = 20\,(32.26) = $ **0.645mA**; $V_C = +10 - 10k\,(0.645) = $ **3.55V.**

(c) Here, $I_E = \frac{10V - 0V}{10k + 1M/21} = \frac{10}{(10 + 47.6)k} = $ **0.174mA**; $V_E = 10 - 10k\Omega\,(0.174mA) = $ **8.26V**; $V_B = 8.26 - 0.70 = $ **7.56V**; $I_C = \frac{20}{21}\,(0.174) = $ **0.166mA**; $I_B = \frac{0.174mA}{21} = $ **8.29 μA**; $V_C = -10 + 0.166\,(10) = $ **−8.34V.**

4.29 For $V_{BE} = 0.7V$, the current in the base-emitter shunting resistor is $0.7V/10k\Omega = 70\mu A$. This flows in the resistor to 10V, creating an equivalent base source of $V_{BB} = 10 - 100k\Omega\,(70\mu A) = 3.0V$, with $R_{BB} = 100k\Omega$. Now, for the base-emitter loop and base current I_B, $3V - (100k\Omega)\,I_B - 0.7V - (\beta + 1)\,(3.3k\Omega)\,I_B = 0$. Thus, $I_B = \frac{2.3}{3.3\,(\beta + 1) + 100}$.

For $\beta = \infty$: $I_B = 0\mu A$; $V_B = $ **3V**; $V_E = 3 - 0.7 = $ **2.3V**; $I_E = \frac{2.3}{3.3k\Omega} = 0.697mA$; $V_C = 10 - 3.3\,(0.697) = 10 - 2.3 = $ **7.7V.**

For $\beta = 100$: $I_B = \frac{2.3}{3.3\,(101) + 100} = 5.35\mu A$; $V_B = 3.0 - 5.35 \times 10^{-6}\,(10^5) = $ **2.465V**; $V_E = 2.465 -$

$0.7 = \mathbf{1.765V}$; $I_C = 100\,(5.35) = 0.535\text{mA}$; $V_C = 10 - 3.3\,(.535) = \mathbf{8.23V}$.

For $\beta = 10$: $I_B = \dfrac{2.3}{3.3\,(11) + 100} = 16.87\mu\text{A}$; $V_B = 3.0 - (16.87)\,(0.1) = \mathbf{1.313V}$; $V_E = 1.313 - 0.7 = \mathbf{0.613V}$; $V_C = 10 - 3.3 \times 10^3\,(16.87 \times 10^{-6})\,(10) = \mathbf{9.44V}$.

4.30 From the solution of P4.29, $V_{BB} = 3\text{V}$, $R_{BB} = 100\text{k}\Omega$.

For $\beta = \infty$: $V_E = 3 - 0.7 = 2.3\text{V}$, with $I_E = \dfrac{2.3V}{3.3k\Omega} = 0.697\text{mA}$.

For β generally, $I_E = (\beta + 1)\,I_B = \dfrac{(\beta + 1)\,(2.3)}{3.3\,(\beta + 1) + 100} = \dfrac{2.3}{3.3 + 100/(\beta + 1)}$.

Thus, $\dfrac{2.3}{3.3 + 100/(\beta + 1)} = 0.8\,(0.697) = .5576$, or $3.3 + \dfrac{100}{\beta + 1} = \dfrac{2.3}{0.5576} = 4.125$, or $100/(\beta + 1) = 0.8248$, $\beta + 1 = 100/.8248 = 120$, whence $\beta = \mathbf{120}$.

4.31 Here, $V_{BB} = \dfrac{10k}{10k + 20k}\,(9\text{V}) = 3\text{V}$; $R_{BB} = 10\text{k} \parallel 20\text{k} = 6.66\text{k}\Omega$.

(a) *For* $\beta = \infty$: $V_B = 3\text{V}$; $V_E = 3 - 0.7 = 2.3\text{V}$; $I_E = \dfrac{2.3V}{1k\Omega} = \mathbf{2.3mA}$; $V_C = 9 - 2\,(2.3) = 4.4\text{V}$. $V_{CE} = V_C - V_E = 4.4 - 2.3 = \mathbf{2.1V}$.

(b) *For* $\beta = 100$: $I_E = \dfrac{3 - 0.7}{1k + 6.6/101} = \mathbf{2.156mA}$; $V_E = 2.156\,(1\text{k}) = 2.156\text{V}$; $V_C = 9 - 2k\,(2.156) \times 100/101 = 4.731\text{V}$; $V_{CE} = 4.731 - 2.156 = \mathbf{+2.575V}$.

(c) *For* $\beta = 10$: $I_E = \dfrac{2.3}{1k + 6.6/11} = \mathbf{1.432mA}$; $V_E = 1.432\text{V}$; $V_C = 9 - 2\,(1.432)\,10/11 = 6.396\text{V}$; $V_{CE} = 6.396 - 1.432 = \mathbf{4.964V}$.

4.32

(a) Let $I_C = i$. From the supply to ground: $10 = 10\text{k}(i + i/\beta + I) + 100\text{k}(I + i/\beta) + 0.7$, or $10\,i + 10\,i/50 + 100\,i/50 = 9.3 - 10I - 100I$. Thus $12.2i = 9.3 - 110I$ — — — (1), or $i = \dfrac{9.3 - 110\,(.02)}{12.2}$, or $i = \dfrac{7.1}{12.2} = \mathbf{0.582mA} = I_C$, for which $V_{CE} = 10 - 10\,(.02 + \frac{51}{50}\,(0.582))$ $= \mathbf{3.86V}$.

(b) For $V_B = 0.7\text{V}$, $I_{560k} = \dfrac{0.7 - -10}{560} = .0191\text{mA}$. From the previous result (1) in (a): $i = \dfrac{9.3 - 110\,(.0191)}{12.2} = \mathbf{0.590mA} = I_C$, and $V_{CE} = 10 - 10\,(.0191 + \frac{51}{50}\,(.590)) = \mathbf{3.79V}$.

(c) For $V_B = 0.7 \rightarrow I_{33k} = \dfrac{0.7V}{33k} = .0212\text{mA}$. Thus, $I_C = \dfrac{9.3 - 110\,(.0212)}{12.2} = \mathbf{0.571mA}$, and $V_{CE} = 10 - 10\,(.0212 + \frac{51}{50}\,(0.571)) = \mathbf{3.96V}$.

(d) Let the base voltage be υ and base current be i. Thus, $\upsilon = 0.7 + (\beta + 1)\,i\,(1\text{k}) = 0.7 + 51\,i$ — — (1); $\upsilon_C = \upsilon + 100\text{k}\,(\upsilon/68\text{k} + i) = \upsilon + 1.47\,\upsilon + 100\,i = 2.47\,\upsilon + 100\,i$ — — — (2); Also $\upsilon_C = 10 - 10k\,((\beta + 1)\,i + \upsilon/68\text{k}) = 10 - 510\,i - 0.147\,\upsilon$ — — — (3). Now (3) with (2) $\rightarrow$ $2.47\,\upsilon + 100\,i = 10 - 510\,i - 0.147\,\upsilon$, or $2.617\upsilon + 610\,i = 10$ — — — (4). Then (4) with (1) $\rightarrow$ $2.617\,(0.7 + 51\,i) + 610\,i = 10$, or $1.832 + 133.5\,i + 610\,i = 10$. Thus, $i = \dfrac{10 - 1.832}{133.5 + 610} = \dfrac{8.18}{743.5} = .011\text{mA}$, whence $\upsilon = 0.7 + 51\,(.011) = 1.261\text{V}$, and $I_C = \beta\,i = 50\,(.011) = \mathbf{0.55mA}$, $V_E = 1k\,(\beta + 1)\,i = 1\,(51)\,(.011) = 0.561\text{V}$, $V_C = 10 - 10\text{k}\,(51(.011) + \frac{1.261}{68k})$ $= 10 - 5.61 - .185 = 4.205\text{V}$. Thus, $V_{CE} = 4.205 - .561 = \mathbf{3.69V}$

4.33

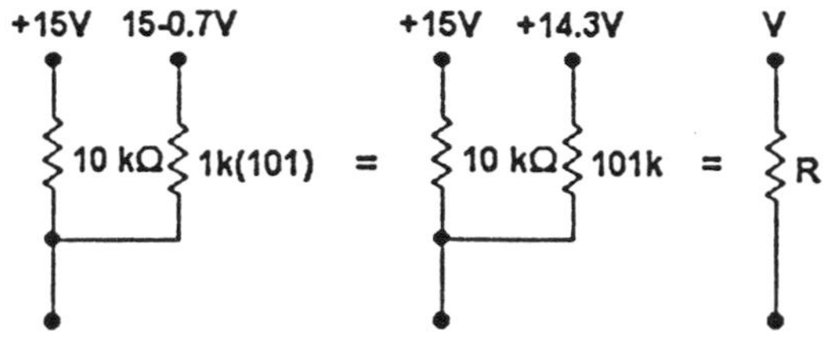

For $\beta = \infty$: $V_A = \frac{100}{100 + 200}(15) = $ **5V**; $V_B = 5 - 0.7 = $ **4.3V**; $V_C = 15 - \frac{10k}{10k}(4.3) = $ **10.7V**; $V_D = 10.7 + 0.7 = $ **11.4V**; $V_E = (\frac{15 - 11.4}{1k})\ 1\text{k} = $ **3.6V**.

For $\beta = 100$: *At node A*: $V_{AA} = 5\text{V}$, and $R_{AA} = 100\text{k}\Omega \parallel 200\text{k}\Omega = 66.7\text{k}\Omega$; $I_E = \frac{5 - 0.7}{10k\Omega + 66.7/101} = .4034\text{mA}$; $V_B = 1\text{k}\ (.4034) = $ **4.03V**; and $V_A = 4.03 + 0.7 = $ **4.73V**.

At node C: $V = 15 - \frac{10}{10 + 101}(0.7) = 14.94$; $R = 10\text{k} \parallel 101\text{k} = 9.10\text{k}\Omega$. Now, from the collector at node C, $I = 0.4034 \times \frac{100}{101} = 0.399\text{mA}$. Thus, $V_C = 14.94 - 0.399\ (9.10) = $ **11.31V**, and $V_D = 11.37 + 0.7 = $ **12.07V**.

Thus, $V_E = 0 + 1\text{k}\ (\frac{15 - 12.07}{1k} \times \frac{100}{101}) = $ **2.90V**.

SECTION 4.7: THE TRANSISTOR AS AN AMPLIFIER

4.34 Generally, $g_m = \frac{I_C}{V_T}$: For 1μA, $g_m = \frac{10^{-6}}{25 \times 10^{-3}} = 40 \times 10^{-6} = $ **40μA/V**; and for 100μA, $g_m = $ **4mA/V**; for 1mA, $g_m = $ **40mA/V**; for 100mA, $g_m = $ **4A/V**.

4.35 At the emitter, $r_e = \frac{V_T}{I_E} = \frac{\alpha V_T}{I_C} = \frac{\alpha}{g_m} \approx \frac{1}{g_m}$. At the base, $r_\pi = \frac{V_T}{I_B} = \frac{V_T}{I_C/\beta} = \frac{\beta}{g_m} = (\beta + 1)r_e$. For $\beta = 100$, $\alpha = 0.99$. Now, for $I_C = 1\mu\text{A}$, $r_e = \frac{0.99}{40 \times 10^{-6}} = 24.75\text{k}\Omega \equiv$ **25kΩ**, $r_\pi = \frac{100}{40 \times 10^{-6}} = $ **2.5MΩ**; for 100μA, $r_e \approx 250\Omega$, $r_\pi = $ **25kΩ**; for 1mA, $r_e \equiv$ **25Ω**, $r_\pi = $ **2.5kΩ**; for 100mA, $r_e = $ **0.25Ω**, $r_\pi = 250\Omega$.

4.36 Gain $= -g_m R_L$, where $g_m = \frac{\alpha I_E}{V_T} \approx \frac{I_E}{V_T}$. The voltage across $R_L = I_C R_L = \alpha I_E R_L$ is a constant, K. Thus $R_L = \frac{K}{\alpha I_E}$, for which the $|$ gain $| = \frac{\alpha I_E}{V_T} \times \frac{K}{\alpha I_E} = \frac{\mathbf{K}}{\mathbf{V_T}}$, a constant! Thus the gain is constant. There is no gain variation possible. The bias current *does not matter, if the load resistor is varied this way!*

4.37 The input resistance at the emitter, $r_e = \frac{V_T}{I_E} = \frac{25 \times 10^{-3}}{100 \times 10^{-6}} = 250\Omega$. The input resistance at the base, $r_\pi = (\beta + 1)r_e = 151\ (250) = $ **37.75kΩ**. The voltage gain, base-to-collector, is $-\frac{\alpha R_L}{r_e} = -\frac{150}{151} \times \frac{10k\Omega}{250} = $ **−39.73V/V**.

4.38 The collector load is the 100kΩ resistor from collector to (grounded) base. Now, I_E = 1mA, and $r_e = \frac{25mV}{1mA} = 25\Omega$. Thus the gain $= -\frac{\alpha R_L}{r_e} \approx -\frac{100k\Omega}{25\Omega}$ = **–4000V/V**. The resistance "seen" by the source υ_S is $r_e = 25\Omega$. Now, for $R_S = 75\Omega$, $\frac{\upsilon_i}{\upsilon_s} = \frac{25}{25+75} = ¼$, and $\frac{\upsilon_o}{\upsilon_s} = \frac{\upsilon_i}{\upsilon_s} \times \frac{\upsilon_o}{\upsilon_i} = -4000 \times$ ¼ = **–1000V/V.**

SECTION 4.8: SMALL-SIGNAL EQUIVALENT-CIRCUIT MODELS

4.39

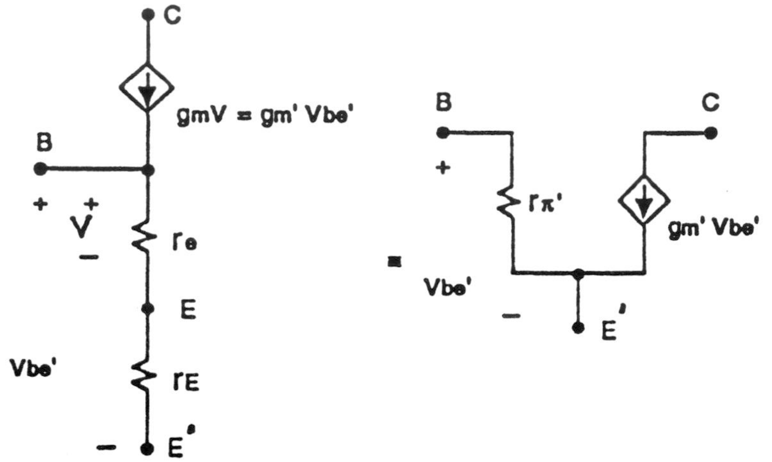

See $\upsilon = \left[\frac{r_e}{r_e + r_E}\right] \upsilon_{be}'$, and $g_m' \upsilon_{be}' = g_m \upsilon = g_m \left[\frac{r_e}{r_e + r_E}\right] \upsilon_{be}'$. Thus, $\mathbf{g_m' = g_m \frac{r_e}{r_e + r_E}}$. Now, $r_\pi' = \frac{\upsilon_{be}'}{i_b} = \upsilon_{be}' / \left[\frac{\upsilon_{be}'}{r_e + r_E} - g_m' \upsilon_{be}'\right] = 1/\left[\frac{1}{r_e + r_E} - g_m'\right] = 1/\left[\frac{1}{r_e + r_E} - g_m \frac{r_e}{r_e + r_E}\right] = \frac{r_e + r_E}{1 - g_m r_e} = \frac{r_e + r_E}{1 - \alpha} = (\beta + 1)(r_e + r_E)$. Now, for I_C = 1mA, $r_e = 25\Omega$, and $r_E = 3r_e = 75\Omega$, $\mathbf{r_\pi' = (\beta + 1)(r_e + r_E)}$ = 101 (25 + 75) = **10.1k Ω**, and $g_m' = \frac{g_m r_e}{r_e + r_E} = \frac{(\alpha/r_e) r_e}{r_e + r_E} = \frac{\alpha}{r_e + r_E} = \frac{100/101}{25 + 75} = 9.9 \times 10^{-3}$A/A = **9.9mA/V.**

4.40

(a)

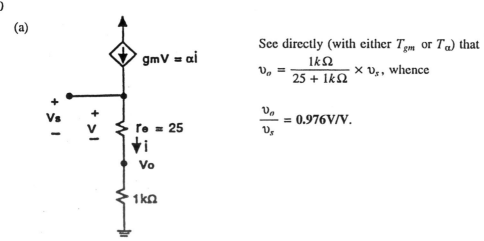

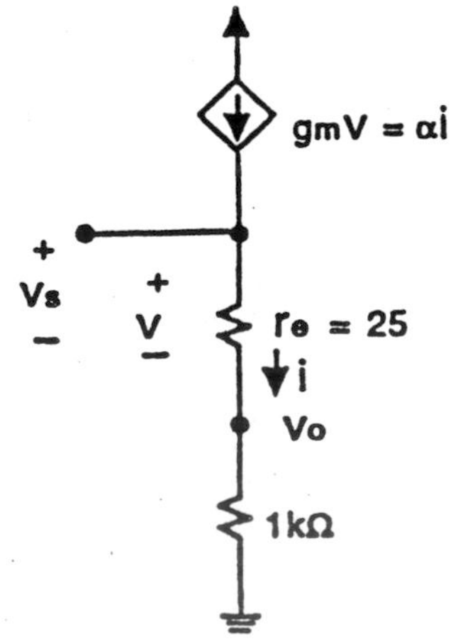

See directly (with either T_{gm} or T_α) that $\upsilon_o = \frac{1k\Omega}{25 + 1k\Omega} \times \upsilon_s$, whence

$\frac{\upsilon_o}{\upsilon_s}$ = **0.976V/V.**

(b)

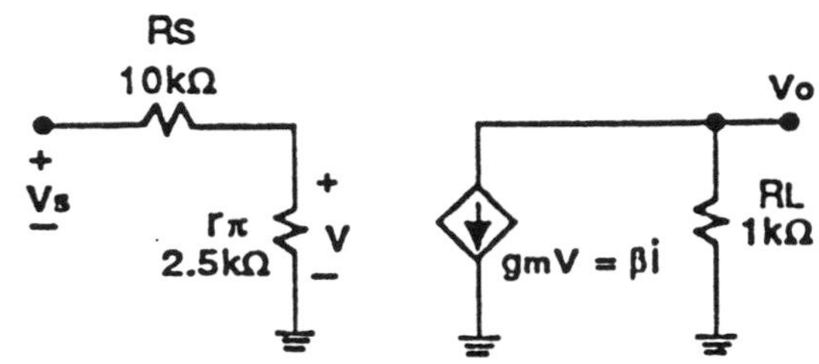

For Π_{gm}, see $v_o = -g_m\, v\,(R_i) = -\dfrac{r_\pi}{r_\pi + R_S}(v_s) \times g_m R_L$, whence $\dfrac{v_o}{v_s} = -\dfrac{2.5}{2.5 + 10} \times 40 \times 1$ = **−8V/V.** *For* Π_β, see $v_o = -\beta(i)\,R_L$ $= -\beta\left[\dfrac{v_S}{R_S + r_\pi}\right] R_L$, whence

$$\frac{v_o}{v_s} = -\frac{\beta R_L}{R_S + r_\pi} = -\frac{100(1)}{(10 + 2.5)} = \mathbf{-8V/V.}$$

(c) As in (b), with $R_S = 0$, $\dfrac{v_o}{v_s} = -\dfrac{g_m r_\pi R_L}{r_\pi + R_S} = -g_m R_L = -40 \times 1 =$ **−40V/V**, or $\dfrac{v_o}{v_s} = -\dfrac{\beta R_L}{R_S + r_\pi} = -\dfrac{100(1)}{0 + 2.5} =$ **−40V/V.**

(d)

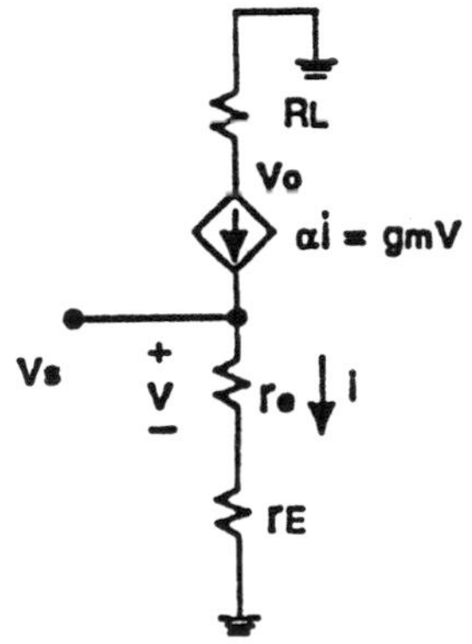

$$\frac{v_o}{v_s} = -\frac{\alpha R_L}{r_e + r_E} = -\frac{0.99(1k)}{25 + 100} = \mathbf{-7.92V/V,}$$

or $\dfrac{v_o}{v_s} = -g_m R_L \times \dfrac{r_e}{r_e + r_E} = -\dfrac{40 \times 25}{25 + 100} =$ **−8V/V**, with the former (using T_α) being more direct

(e)

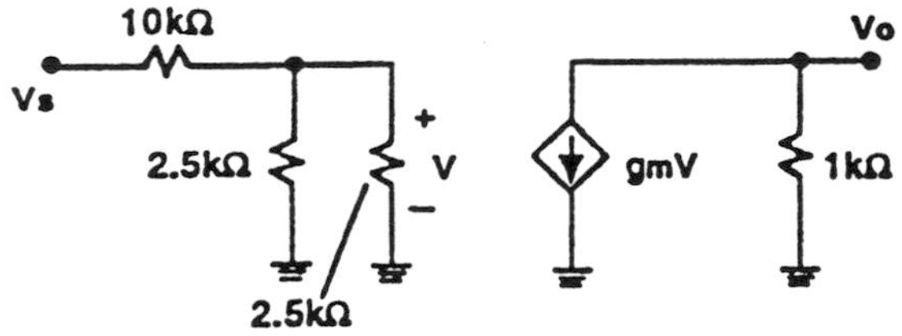

See $\dfrac{v_o}{v_s} = -g_m R_L \times \dfrac{R_B \,||\, r_\pi}{R_B \,||\, r_\pi + R_S} = -40 \times 1 \times \dfrac{2.5/2}{2.5/2 + 10} =$ **−4.44V/V.**

4.41 Since $I_E = 1\text{mA}$, $r_e = \dfrac{25mV}{1mA} = 25\Omega$, and $\dfrac{v_o}{v_s} = -\dfrac{\alpha R_L}{r_e} = -0.99\,\dfrac{(7.5k\Omega)}{25} =$ **−297V/V.** For a signal voltage of 0V, $V_O = 10 - 7.5\,(.99)\,(1) = 2.575\text{V}$. For guaranteed active operation, $v_{CE} \geq 0\text{V}$. Thus the largest allowed sinusoid has a peak value of 2.575 V −0V = **2.575V** at the output, and 2.575/297 = **8.67mV** peak at the input.

4.42 Now, $i = I_S\, e^{v/V_T}$ in general. Thus $\dfrac{i_1}{i_2} = e^{(v_1 - v_2)/V_T}$, and for $i_2 = I\text{mA}$ initially, $i_1 = Ie^{10/25} = 1.49I\text{mA}$, (or $i_1 = Ie^{-10/25} = 0.670I\text{mA}$). That is, the current **increases by 49%**, or **reduces by 33%**, for ±10mV variation around the operating point. For linear operation over a ±10mV input range, a current of 1.49I

must be tolerated. For $\upsilon_{CB} \geq 0$, with $I = 1\text{mA}$, $\dfrac{10-0}{R_C} = 1.49\ I = 1.49\text{mA}$, and $R_C \leq \dfrac{10}{1.49mA} \leq$ **6.7kΩ.**

4.43 For $i_C = 100\mu\text{A}$, $r_e = \dfrac{25mV}{100\mu A} = 250\Omega$, and $r_o = \dfrac{V_A}{i_C} = \dfrac{200V}{100\mu A} = 2 \times 10^6\Omega$. Thus, the gain $\dfrac{\upsilon_o}{\upsilon_i} = -\dfrac{\alpha R_L}{r_e} \approx -\dfrac{2 \times 10^6 \parallel 2 \times 10^6}{250} = \mathbf{-4000V/V}$.

4.44 For υ across the two-terminal device, the voltage across the base-emitter junction is $\upsilon_\pi = \dfrac{R_1 \parallel r_\pi}{R_1 \parallel r_\pi + R_2} \times \upsilon$.

For this situation, the total current, $i = \dfrac{\upsilon}{R_1 \parallel r_\pi + R_2} + g_m\ \upsilon_\pi = \upsilon \left[\dfrac{1}{R_1 \parallel r_\pi + R_2} + \dfrac{g_m\ (R_1 \parallel r_\pi)}{R_1 \parallel r_\pi + R_2} \right]$

$$= \upsilon \left[\frac{1 + \dfrac{g_m\ r_\pi\ R_1}{R_1 + r_\pi}}{R_2 + \dfrac{r_\pi\ R_1}{R_1 + r_\pi}} \right] = \upsilon \left[\frac{R_1 + r_\pi + \beta R_1}{R_1 R_2 + r_\pi R_2 + r_\pi R_1} \right].$$

Thus, resistance $r = \dfrac{\upsilon}{i} = \dfrac{R_1\ R_2 + r_\pi\ (R_1 + R_2)}{r_\pi + R_1\ (\beta + 1)} = \left[\dfrac{R_1\ R_2}{\beta + 1} + r_e\ (R_1 + R_2) \right] / (r_e + R_1)$.

(a) For $R_2 = 0$, $R_1 = \infty$, $r = \dfrac{R_2 + r_\pi\ (1 + R_2/R_1)}{r_\pi/R_1 + (\beta + 1)} = \dfrac{0 + r_\pi}{0 + (\beta + 1)} = \dfrac{r_\pi}{\beta + 1} = \mathbf{r_e}$.

(b) For $R_1 = \infty$, $R_2 = r_\pi$, $r = \dfrac{r_\pi + r_\pi\ (1 + 0)}{0 + \beta + 1} = \mathbf{2r_e}$.

(c) For $R_1 = R_2 = r_\pi$, $r = \dfrac{r_\pi + r_\pi\ (1 + 1)}{\dfrac{r_\pi}{r_\pi} + \beta + 1} = \dfrac{3\ r_\pi}{\beta + 2} \approx \mathbf{3\ r_e}$.

4.45

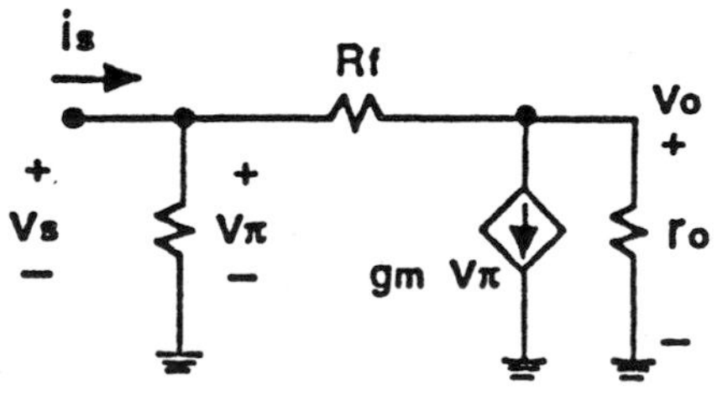

For $\upsilon_b = \upsilon_s$, small compared to υ_o, the gain is $A_\upsilon = \dfrac{\upsilon_o}{\upsilon_s} = \mathbf{-g_m\ (r_o \parallel R_f)}$, and the input current $i_s = \dfrac{\upsilon_s}{r_\pi} + \upsilon_s\ \dfrac{(1 - -g_m\ (r_o \parallel R_f))}{R_f}$,

whence $R_{in} = \dfrac{\upsilon_s}{i_s} = \dfrac{1}{\dfrac{1}{r_\pi} + \dfrac{1}{R_f/(1 + g_m\ (r_o \parallel R_f))}}$

$$= \mathbf{r_\pi \parallel \frac{R_f}{1 + g_m\ (r_o \parallel R_f)}}$$

Now for $R_f = r_o$, the gain is $A_\upsilon = -g_m\ (r_o \parallel r_o) = -\dfrac{\mathbf{g_m\ r_o}}{2}$,

and $R_{in} = r_\pi \parallel \left[\dfrac{r_o}{1 + g_m\ r_o/2} \right] \approx r_\pi \parallel \left[\dfrac{2}{g_m} \right] \approx \mathbf{2r_e}$, that is, very small.

4.46

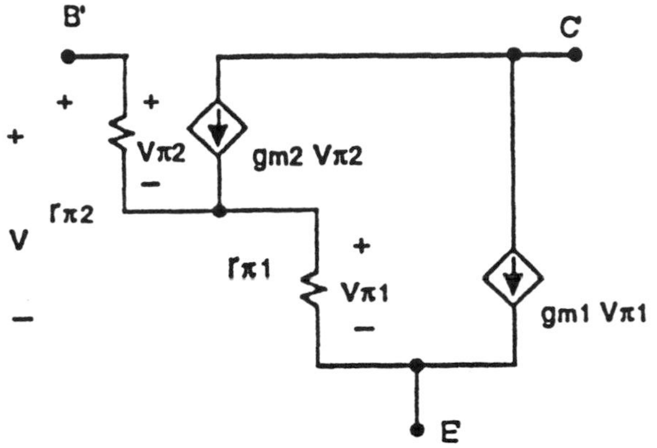

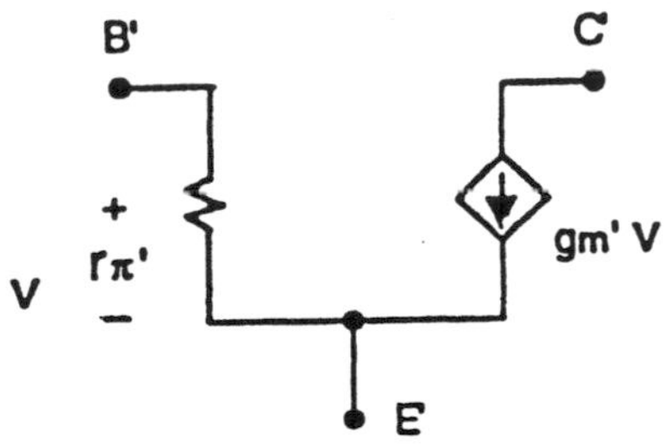

Note that the emitter current of Q_2 is the base current of Q_1, and therefore $r_{e\,2} = r_{\pi\,1}$ and $r_{\pi\,2} = (\beta_1 + 1)\, r_{\pi\,1}$. Correspondingly, $\upsilon_{\pi\,1} \approx \upsilon_{\pi\,2} = \upsilon/2$, and $g_{m2} = g_{m1}/(\beta_1 + 1)$. By considering the input base current, see: $r_\pi' = r_{\pi\,2} + (\beta_1 + 1)\, r_{\pi\,1} = (\beta_1 + 1)\, r_{\pi\,1} + (\beta_1 + 1)\, r_{\pi\,1} = 2\,(\beta_1 + 1)\, r_{\pi\,1}$. By considering the output collector current, see: $g_m'\, \upsilon = g_{m1}\, \upsilon_{\pi\,1} + g_{m2}\, \upsilon_{\pi\,2} = g_{m1}\, \upsilon/2 + (g_{m1}\, \upsilon/2)/(\beta_1 + 1)$
$= (g_{m1}\, \upsilon/2)\, [1 + 1/(\beta_1 + 1)\,] \approx g_{m1}\, \upsilon/2$. Thus $g_m' = g_{m1}/2$.

SECTION 4.9: GRAPHICAL ANALYSIS

4.47

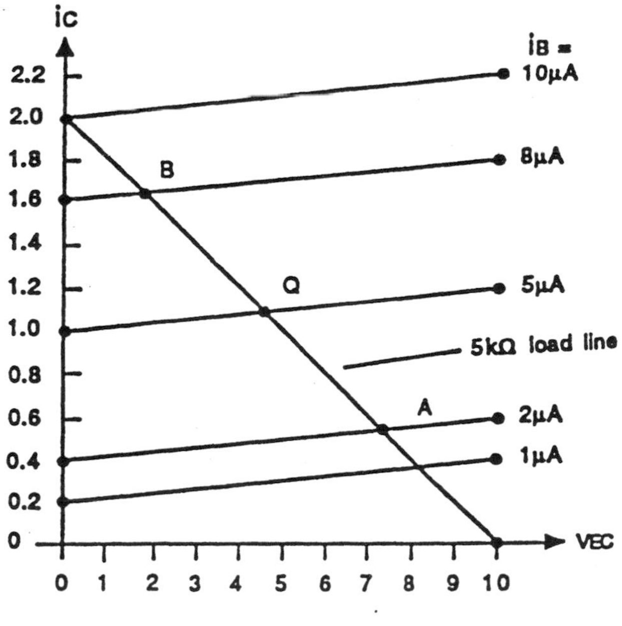

With $\beta = 200$ and $V_A = 100$V, for $i_B = 1, 2, 5, 8$ and 10μA, $i_C = \beta\, i_B = 200, 400, 1000, 1600$ and 2000μA, and $r_o = V_A/i_C = 500$kΩ, 250kΩ, 100kΩ, 62.5kΩ and 50kΩ, with the current at 10V greater than that at 0V, by $10/r_o = 20, 40, 100, 160$ and 200μA (ie by 10% (= (10V/100V) × 100)). For V_{CC} = 10V and R_L = 5kΩ, the intercept is $i_C = 10/5k\Omega = 2$mA.

For the operating point (Q), $i_B = 5\mu$A, $V_{EC} \approx 5$V from the graph, and $i_C \approx (5\,(200))\,(1 + 5/100) \equiv$ **1050µA**, with $\upsilon_{EC} = 10 - 5k\Omega\,(1.05\text{mA}) =$ **4.75V**. *For a* ±3µA peak input wave, operation varies from Q to A to B to Q above. *At A*: $i_B = 2\mu$A, $\upsilon_{EC} \equiv 8$V from the graph, with $i_C \approx 2\,(200)\,(1 + 8/100)\mu$A = 432µA = 0.432mA, for which $\upsilon_{EC} = 10 - 5k\Omega\,(.432) =$ **7.84V**, and $i_C = 2\,(200)\,(1 + 7.84/100) =$ **0.431mA**. *At B*: $i_B = 8\mu$A, $\upsilon_{EC} \equiv 2$V from the graph, with $i_C \approx 8\,(200)\,(1 + 2/100)\mu$A = **1.632mA**, and $\upsilon_{EC} = 10 - 5\,(1.632) =$ **1.84V**.

Thus the output wave has a positive peak of 4.75 − 1.84 = **2.91V**, at an output current of 1.632 − 1.050 = **0.582mA**, {*Check*: R_L' = 2.91/0.582 = 5kΩ}, and a negative peak of 7.84 − 4.75 = **3.09V**, with a current of 1.050 − 0.432 = **0.618mA**, {*Check*: R_L' = 3.09/.618 = 5kΩ}.

Note that the positive and negative peaks are different, indicating that a (small) signal distortion results.

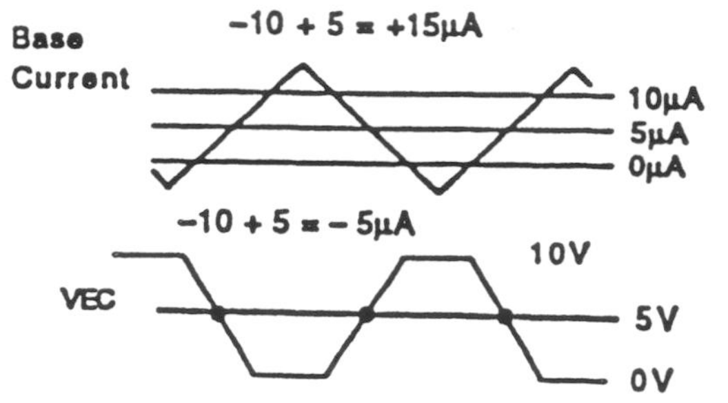

For a ±10μA peak input wave: See from the graph that for base signals of more than +5μA, $i_B \geq 10\mu A$ and the transistor saturates, and that for signals less than −5μA, $i_B \leq 0\mu A$, and the transistor cuts off. For this situation, the output is clipped for **50%** of the cycle.

SECTION 4.10: BIASING THE BJT FOR DISCRETE-CIRCUIT DESIGN

4.48 From Fig. 4.39b), in general, $I_E = (V_{BB} - V_{BE})/\left[R_E + \frac{R_B}{\beta + 1}\right]$. For $\beta = \infty$, $I_E = \frac{V_{BB} - V_{BE}}{R_E} = I$. For $I_E = .99\, I$, $\frac{R_B}{\beta + 1} = \frac{1}{100} R_E$, or $\beta + 1 = 100\,\frac{R_B}{R_E}$. For $R_E = R_B$, $\beta + 1 = 100$ or $\beta = 99$. That is, for $\beta \geq 99$, I_E is within 1% of its maximum value.

Alternatively, one could interpret the situation to mean ±1% of a nominal value, where the largest occurs for $\beta = \infty$, and the nominal for $\beta = 99$, where $\frac{1}{\beta + 1} = \frac{1}{100}$, and the minimum where $\frac{1}{\beta + 1} = \frac{2}{100}$, for which $\beta = \frac{100}{2} - 1 = \mathbf{49}$.

4.49 See that $V_{BB} = \frac{12}{3} = 4V$. Generally, $I_E = \frac{V_{BB} - V_{BE}}{R_E + \frac{R_B}{\beta + 1}}$. Here, $100mA = \frac{4 - 0.7}{R_E + \frac{50}{51} \times \frac{R_E}{10}}$. Thus $R_E = \frac{3.3}{100} / \left[1 + \frac{50}{51} \times \frac{1}{10}\right] = 0.03005k\Omega$. Practically speaking, use $R_E = 30\ \Omega$, with $R_B = \frac{\beta R_E}{10} = 50(30)/10 = 150\Omega$. Now $\frac{R_2}{R_1 + R_2} \times 12 = V_{BB} = \frac{12}{3}$, whence $3R_2 = R_1 + R_2$, or $R_1 = 2R_2$. Now, since $R_1 \parallel R_2 = 150 = \frac{R_1 R_2}{R_1 + R_2}$, $\frac{2R_2 (R_2)}{3R_2} = 150$, or $\frac{2}{3} R_2 = 150$, or $R_2 = \frac{3(150)}{2} = 225\Omega$, and $R_1 = 2R_2 = 450\Omega$. For a conservative design, use smaller values, such as $R_2 = \mathbf{200\Omega}$ and $R_1 = 2(200) = \mathbf{400\Omega}$. Now, for $I_E = 100mA$, $I_C = \frac{50}{51}(100) = 98mA$, and $R_C I_C = \frac{V_{CC}}{3} = 4V$. Thus $R_C = \frac{4V}{98mA} = 0.0408k\Omega$, for which use $R_C = \mathbf{40\ \Omega}$. Now, for the design overall: $R_E = \mathbf{30\ \Omega}$, $R_C = \mathbf{40\ \Omega}$, $R_2 = \mathbf{200\ \Omega}$, $R_1 = \mathbf{400\ \Omega}$, where $R_B = 200 \parallel 400 = \frac{200\,(400)}{600} = 133.3\Omega$, and $V_{BB} = \frac{200}{400 + 200} \times 12 = 4V$. Thus, $I_E = \frac{4 - 0.7}{30 + 133.3/51} = \mathbf{101.2mA}$, $V_{CE} = 12 - 0.1012\left[\frac{50}{51}\right] 40 - 0.1012\,(30) = \mathbf{4.995V}$, and $V_{CB} = 4.995 - 0.7 \equiv 4.30V$.

4.50 Assume I_E varies 5% over the entire range of β, from 20 to ∞. Assume that for ±1V output, the base signal is very small. Further, assume that $V_B \approx 0V$, and that operation is for $v_{CB} \geq 0$. Now, for a −1V output signal and $\beta = \infty$, $\frac{5V - 1V}{1k\Omega} = I_C = I_E = 4mA$. Since $V_E = -0.7V$, $R_E = \frac{-0{,}7 - -5}{4} = 1.075k\Omega$.

In practice one would use $R_E = \mathbf{1.00k\Omega}$, in which case $V_C \geq 5 - 1k(1)\left[\frac{5 - 0.7}{1}\right] = 0.7V$, and the collector goes to −0.3V with a 1V signal peak. For $R_E = 1.075k\Omega$, $I_E = 0.95\,(1mA) = \frac{0 - -5 - 0.7}{1.075 + \frac{R_B}{20 + 1}}$, whence $R_B = 21\left[\frac{4.3}{.95} - 1.075\right] = 72.5k\Omega$. *In practice* use a smaller (standard) value, say **68kΩ**.

4.51 For V_{CB} = 0.5V, and β = 200: $I_E = \frac{5 - 0.7 - 0.5}{3.6k\Omega}$ = 1.056mA, and $I_B = \frac{1.056mA}{201}$ = 5.25μA. Thus, $R_B = \frac{0.5V}{5.25\mu A}$ = 95.2kΩ. {In practice one would use a **100kΩ** resistor.} Now, for β = 50, 5 − 3.6 I_E − 95.2 $\frac{I_E}{50 + 1}$ −0.7 = 0, or $I_E = \frac{5 - 0.7}{3.6 + \frac{95.2}{51}}$ = 0.787mA. That is, with R_B = 95.2kΩ, I_E varies from **1.056mA** to **0.787mA** and, V_{CB} from **0.5V** to $95.2\frac{(0.787)}{51}$ = **1.47V**.

Alternatively, with R_B = 100kΩ: For β = 200: $I_E = \frac{5 - 0.7}{3.6 + \frac{100}{201}}$ = **1.049mA**, and $V_{CB} = 100\,\frac{(1.049)}{201}$ = **0.522V**. For β = 50: $I_E = \frac{5 - 0.7}{3.6 + \frac{100}{51}}$ = **0.773mA**, and $V_{CB} = 100\,\frac{(0.773)}{511}$ = **1.516V**.

4.52 For this situation, the "base" current is $I_B = \frac{I_E}{\beta + 1} + \frac{V_{BE}}{R_\beta} = \frac{I_E}{\beta_{eq} + 1}$. Now for β = ∞ ,and β_{eq} = 200, $R_\beta = \frac{0.7\ (201)}{I_E}$. Now, for V_{CB} = 0.5V, β_{eq} = 200, V_{CC} = 5V, R_C = 3.6kΩ, $I_E = \frac{5 - 0.5 - 0.7}{3..6k\Omega}$ = 1.056mA. Thus, $R_\beta = \frac{0.7(201)}{1.056}$ = 133.2kΩ. In practice, use a (smaller) standard value, 130kΩ, or **120kΩ** as it is more commonly available. *With* R_β = 120kΩ, $I_B = \frac{0.7}{120k}$ = 5.83μA, and for β = ∞ and V_{CB} = 0.5V, $R_B = \frac{0.5}{5.83\mu A}$ = 85.8kΩ. In practice, use a larger standard value, say R_B = **91kΩ**, for which V_{CB} = 91k (5.83μA) = 0.530V, and $I_E = \frac{5 - 0.53 - 0.70}{3.6}$ = 1.047mA. Now for β = 50, base current flows in R_B to produce a voltage drop which combines with a constant voltage drop of V_β = 0.53V in R_B due to R_β. Thus, $I_E = \frac{V_{CC} - V_\beta - V_{BE}}{R_C + R_B/(\beta + 1)} = \frac{5 - 0.53 - 0.7}{3.6 + 91/(51)}$ = 0.700mA, with V_{CB} = 5 −3.6 (0.700) −0.7 = 1.78V. That is, for R_β = 120kΩ, R_B = 91kΩ, and β ≥ 50, I_E varies from **0.700mA** to **1.047mA**, while V_{BC} varies from **1.78V** to **0.530V**.

4.53 For β_{eq} = 100, and using the solution for P4.52, $R_\beta = \frac{0.7(101)}{1.056}$ = 66.95kΩ. Use **68kΩ** as very close (though larger). Now $I_B = \frac{0.70V}{68k}$ = 10.3μA, whence $R_B = \frac{0.50}{10.3}$ = 48.5kΩ. Use **47kΩ** as close (though smaller). For these choices and for β = ∞, $V_{CB} = \frac{0.70}{68} \times 47$ = **0.484V**, and $I_E = \frac{5 - 0.484 - 0.700}{3.6}$ = **1.06mA**. Now for β = 50, $I_E = \frac{5 - 0.484 - 0.7}{3.6 + \frac{47}{51}}$ = **0.844mA**, and V_{CB} = 5 − 3.6 (.844) −0.7 = **1.26V**.

4.54 In Fig. 4.42a) of the Text, I = 1 mA. For β in the range 40 to 200, base current ranges from $1/41 = 24.4\mu A$ to $1/201 = 4.98\mu A$. The lowest the emitter can operate is at −5 V. The lowest the base is allowed to go is − 5 + 0.7 = − 4.3 V. Thus the largest acceptable $R_B = 4.3/24.4 \times 10^{-6}$ = **176 kΩ**. For R_B = 176 kΩ, the base will range from $-176 \times 10^3 \times 24.4 \times 10^{-6}$ = − **4.29 V** to $-176 \times 10^3 \times 4.98 \times 10^{-6}$ = − **0.876 V**.

Now, at low β, $r_\pi = (\beta + 1)r_e = (\beta + 1)V_T/I = 41(25 \times 10^{-3})/1 \times 10^{-3}$ = 1025Ω.

Thus consider R_B = 100(1025) ≈ 100 kΩ, in practice.

The base voltage now ranges from $-100 \times 10^3 \times 24.4 \times 10^{-6} =$ **$-$ 2.44 V** to $-100 \times 10^3 \times 4.98 \times 10^{-6} =$ **$-$ 0.50 V.**

As you will see in subsequent Sections, the $100r_\pi$ design allows 99% or so of the signal currents applied to the base lead to enter the transistor. From the signal point of view, this is a very efficient design.

4.55 For $V_{BE} = 0.7$ V, the drop across R is $(5 - -5 - 0.7) = 9.3$ V. For $I = 1$ mA, use $R = 9.3/1mA =$ **9.3 kΩ.** From a practical point of view (See Appendix H of the Text) a resistor of 9.31 kΩ is available on the 1% scale. If a lower cost resistor were needed, 9.1 kΩ ± 5% unit would be acceptable. The output current I remains essentially constant provided Q_2 does not break down or saturate, from **+ 5 V** or more to **$-$ 4.8 V** or so (if $V_{CE_{sat}} = 0.7 - 0.5 = 0.2$ V)

SECTION 4.11: BASIC SINGLE-STORE BJT AMPLIFIER CONFIGURATIONS

4.56 As is customary, ignore V_A in the bias calculation: Directly, $I_E = 10.0mA$. Thus, $I_C = (100/101)10 =$ **9.90 mA** and $I_C = 10.0/(101) = 0.099mA$. Thus $V_B = 0 - 0.5(0.099) =$ **$-$ 0.049 V**, and $V_E = -0.049 - 0.700 =$ **$-$ 0.749 V**. Now $V_C = V_{CC} - R_C I_C = 10.0 - 0.5(9.90) =$ **5.05 V.**

For this bias situation: $g_m = I_C/V_T = 9.90/25 = 0.396A/V =$ **396 mA/V**, $r_e = V_T/I_E = 25/10 =$ **2.50 Ω**, $r_\pi(\beta + 1)r_e = 101(2.50) =$ **252 Ω**, $r_o = V_A/I_C = 100/(9.90 \times 10^{-3}) =$ **10.1 kΩ.**

4.57 Using the result of P4.56 above: $R_i = r_\pi =$ **252 Ω**, $R_o = R_C \parallel r_o = 0.5k \parallel 10.1k =$ **476Ω**, $A\upsilon = \upsilon_O/\upsilon_S = -\beta(R_C \parallel r_o)/(R_S + r_\pi) = -\beta R_o/(R_s + R_i) = -100(476)/(500 + 252) =$ **$-$ 6.3. V/V**, $A_i = i_o/i_b = -\beta r_o/(r_o + R_C) = -100(10.1k)/(10.1k + 0.5k) =$ **$-$ 95.3 A/A.**

For $A\upsilon$ with a 500 Ω load there are two approaches: a) the direct, and b) the Thevenin:

(a) $A\upsilon = -\beta(R_C \parallel r_o \parallel R_L)/(R_S + r_\pi) = -100(0.5 \parallel 10.1 \parallel 0.5)/(0.5 + 0.252) = -100(244/752) =$ **$-$ 32.4 V/V.**

(b) The amplifier as a Thevenin equivalent voltage gain, $A_T = A\upsilon_O = -63.3V/V$ with a Thevenin equivalent source resistance $R_o = 476Ω$. Thus with load, $A\upsilon = A\upsilon o R_L/(R_L + R_O) = -63.3(500)/(500 + 476) =$ **$-$ 32.4 V/v.**

When comparing with the results of Exercise 4.31 in the Text, we see that with resistor/current scaling that the voltage and current gains are essentially constant. this is reasonable, since we are dealing with a linearized circuit model. Here, even the nonlinearity associated with bias design is eliminated by the use of the constant-emitter-current bias design. In general, for such designs, parameters scale by the same factor, and gains are constant.

4.58 The need for highest-possible gain for a fixed load implies a large bias current: Thus, for $\beta = \infty$, $V_B = 0$ and for ±1V swing, and $\upsilon_C \geq 0$, $V_C = 0 + 1 = 1$V and $I_C = I_E = \frac{9 - 1}{10k} = 0.8$mA, whence $R_E = \frac{0 - 0.7 - -9}{0.8} = 10.38$kΩ.

Now, if we use $R_E =$ **10kΩ** (as a standard value), we see that υ_C falls to 0.7V (for $\beta = \infty$) with $\upsilon_{CB} =$ $-$0.3V, which is often acceptable for linear operation. Otherwise, use $R_E = 11$kΩ.

For $R_E = 10$kΩ: *With* $\beta = \infty$: $V_E =$ **$-$0.7V**, $V_C =$ **+0.7V**, $I_C = \frac{9 - 0.7}{10} =$ **0.83mA**; $r_e = \frac{25mV}{0.83} = 30.1Ω$; $r_\pi = \infty$, $r_o = \frac{100V}{0.83} = 120.5$kΩ, $\frac{\upsilon_o}{\upsilon_s} = -\frac{10k \parallel 120k}{30.1} =$ **$-$307V/V**, and for ±1V output, $\upsilon_b = \upsilon_s = \frac{1}{307}$ **= 3.26mV.**

With $\beta = 90$: $V_E = \frac{-9 + 0.7}{100k + 91(10k)} \times 100k - 0.7 =$ **$-$1.52V**, $I_C = \frac{90}{91}\frac{9 - 1.52}{10} =$ **0.740mA**, $V_C = 9 -$ 10k (.740) = **1.60V**, $r_e = \frac{25mV}{0.740} = 33.8$mA, $r_\pi = 91\ (33.8) = 3.074$kΩ, $r_o = \frac{100V}{0.740} = 135$mA, $\frac{\upsilon_o}{\upsilon_s} =$

$-\dfrac{90\,(10k \parallel 135k)}{3.074k + 10k} = $ **−64.1V/V**. {Alternatively, $\dfrac{\upsilon_o}{\upsilon_s} = -\dfrac{3.07}{10 + 3.07} \times \dfrac{90}{91}\,\dfrac{(10k \parallel 135k)}{33.8} =$ −64.0V/V}. Now for ±1V output, $\upsilon_s = \dfrac{1}{64.1} =$ **15.6mV**, and $\upsilon_b = \dfrac{3.07}{10 + 3.07}(15.6) =$ **3.66mV**.

4.59 Approximately, since the resistor through which base current flows is $R_S \approx R_C$, I_E is essentially fixed at $\dfrac{10 - 0.7}{10k} = 0.930\text{mA}$ for reasonable β. Thus, $r_e = \dfrac{25mV}{0.93mA} = 26.9\Omega$, and $r_o = \dfrac{200V}{0.93} = 215\text{k}\Omega$. Now, for β = 50, $r_\pi = 51\,(26.9) = 1.372\text{k}\Omega$, and $\dfrac{\upsilon_o}{\upsilon_s} = -\dfrac{1.37k \parallel 10k}{1.37k \parallel 10k + 10k} \times \dfrac{50}{51} \times \dfrac{10k \parallel 215k}{26.9} = -0.177 \times 0.98 \times 355 =$ **−37.5V/V**. Now for, β = 150, $r_\pi = 151\,(26.9) = 4.062\text{k}\Omega$, and $\dfrac{\upsilon_o}{\upsilon_s} = -\dfrac{4.06 \parallel 10}{4.06 \parallel 10 + 10} \times \dfrac{150}{151} \times \dfrac{10k \parallel 215k}{26.9} = -0.224 \times 0.993 \times 355 =$ **−79.0V/V**.

4.60 From P4.59 above, $I_E \approx 0.93\text{mA}$, $r_e = 26.9$, and $r_o = 215\text{k}\Omega$. *For* β = 50, $r_{ib} = 51\,(26.9 + 100) = 6.47\text{k}\Omega$, and $\dfrac{\upsilon_o}{\upsilon_s} = -\dfrac{6.47 \parallel 10}{6.47 \parallel 10 + 10} \times \dfrac{50}{51} \times \dfrac{10k \parallel 215k}{100 + 26.9} = -0.282 \times 0.98 \times 75.3 =$ **−20.8V/V**. *For* β = 150; $r_{ib} = 151\,(26.9 + 100) = 19.2\text{k}\Omega$, and $\dfrac{\upsilon_o}{\upsilon_s} = -\dfrac{19.2 \parallel 10}{19.2 \parallel 10 + 10} \times .98 \times 75.3 =$ **−29.3V/V**. We see that the design using unbypassed resistor in the emitter is relatively insensitive to β variation.

4.61 For each transistor, $I_E = 1\text{mA}$, $r_e = 25\Omega$, and $r_\pi = 151\,(25) = 3.78\text{k}\Omega$. Thus $\dfrac{\upsilon_o}{\upsilon_{b2}} = -\dfrac{150}{151}\,\dfrac{10k \parallel 10k}{25} =$ **−199V/V**. Now, $R_{b2} = $ **3.78kΩ**, and $\dfrac{\upsilon_{b2}}{\upsilon_{b1}} = -\dfrac{150}{151}\,\dfrac{10k \parallel 3.78k}{25} =$ **−109V/V**. Now $R_{b1} =$ **3.78kΩ**, and $\dfrac{\upsilon_{b1}}{\upsilon_s} = \dfrac{3.78k}{10k + 3.78k} =$ **0.274V/V**. Thus $\dfrac{\upsilon_o}{\upsilon_s} = .274\,(-109)\,(-199) =$ **5934V/V**.

4.62 $R_i = (\beta + 1)\,(r_e + R_E) = 10\text{k}\Omega$, $r_e = \dfrac{25}{0.2} = 125\Omega$, β = 50. Thus $10\text{k}\Omega = 51\,(125 + R_E)$, whence $R_E = \dfrac{10^4}{51} - 125 =$ **71Ω**. Thus the voltage gain $\upsilon_o/\upsilon_s = -\dfrac{\beta\,(R_C \parallel R_L)}{R_i + R_s} = -\dfrac{50\,(10k \parallel 1k)}{10k = 10k} =$ **−2.27V/V**.

4.63 $I_E = 3\text{mA}$, $r_e = \dfrac{25mV}{3mA} = 8.33\Omega$. Thus $R_i = R_E \parallel \left(r_e + \dfrac{R_B}{\beta + 1}\right) = 3k \parallel \left(8.33 + \dfrac{2k}{151}\right) = 3k \parallel 21.58 =$ **21.4Ω** for β = 150, and **8.3Ω** for β = ∞. Now, the gain from a 100Ω source is: $\upsilon_o/\upsilon_s = \dfrac{\alpha\,(R_L \parallel R_C)}{R_s + R_i}$ in general. For β = 150, $\upsilon_o/\upsilon_s = \dfrac{150}{151}\,\dfrac{(1k \parallel 3k)}{100 + 21.4} =$ **6.14V/V**. For β = ∞, $\upsilon_o/\upsilon_s = \left(\dfrac{1k \parallel 3k}{100 + 8.3}\right) =$ **6.92V/V**. For $R_B = 0$, the results for β = ∞ apply, that is $r_i = 8.3\Omega$ and $\upsilon_o/\upsilon_s =$ **6.92 V/V**.

4.64 For a base current i and $V_{EB} = 0.7\text{V}$, using KVL: $9 - 10k\Omega\,(101\,i + \dfrac{0.7}{10k\Omega}) - 0.7$, whence $-10k\Omega\,(i + \dfrac{0.7}{10k\Omega}) - 10k\Omega\,(101\,i + \dfrac{0.7}{10k\Omega}) = 0$, or $9 - 1010\,i - 0.7 - 0.7 - 10\,i - 0.7 - 1010\,i - 0.7 = 0$, $9 - 4\,(0.7) = 2030\,i$, and $i = 3.054\mu\text{A}$. Thus $I_C = 50\,(3.054) =$ **0.153mA**, $V_E = 9 - 10\text{k}\Omega\,(101\,(3.054 \times 10^{-3}) + \dfrac{0.7}{10k}) = 9 - 3.084 - 0.7 =$ **5.216V**, $V_B = 5.216 - 0.7 =$ **4.516V**, $V_C = 10\text{k}\Omega\,(101\,(3.054 \times 10^{-3}) + \dfrac{0.7}{10k\Omega}) =$ **3.784V**. *Check*: $V_B - V_C = 4.516 - 3.784 = 0.732 \approx 0.7 + 10\text{k}\Omega\,(3.054\mu\text{A})$m as required.

For all designs, all couplings are via capacitors:

(a) Source coupled to B; Load to E; (Ground to C)

(b) Source to B; 10kΩ coupled from E to ground (or (better) 10/3kΩ from E); Load to C.

(c) Source to B; Ground to E, Load to C.

(d) Source to E; Ground to B, Load to C.

4.65 For all designs, I_C = 0.153mA, $r_e = \frac{25}{.153} = 163.4\Omega$;

(a) Since $\upsilon_e \approx \upsilon_b$, the shunt 10kΩ can be ignored, and $A_\upsilon = \frac{10k \parallel 10k}{0.163 + 10k \parallel 10k}$ = **0.968V/V.**

(b)

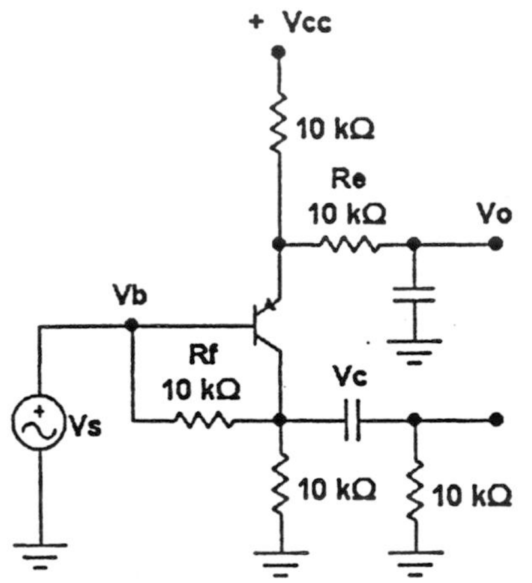

For $\frac{\upsilon_o}{\upsilon_b}$ = –1V/V, and $\upsilon_s = \upsilon$, the voltage across R_f is 2υ. Thus $R_{Leq} = R_L \parallel R_C \parallel R_f/2$ = 10k ∥ 10k ∥ 5k = 2.5kΩ, and the gain = $-\alpha \frac{R_{Leq}}{R_{Eeq}} = -\frac{100}{101} \times \frac{2.5k}{10k \parallel 10k} \approx$ **–0.5V/V.** For $R_e = 10/3k\Omega$, Gain = $-\frac{100}{101} \times \frac{2.5}{10k \parallel \frac{10k}{3}} \equiv$ **–1V/V.**

(c) For K large, the signal across R_f is essentially only due to the output voltage. Thus $R_{Leq} = R_L \| R_C \| R_f = \frac{10k}{3}$ = 3.33kΩ, and the gain = $-\frac{3.33k\Omega}{163.4}$ = **–20.4V/V.**

(d) Base is grounded, and the gain = **+20.4V/V.**

4.66

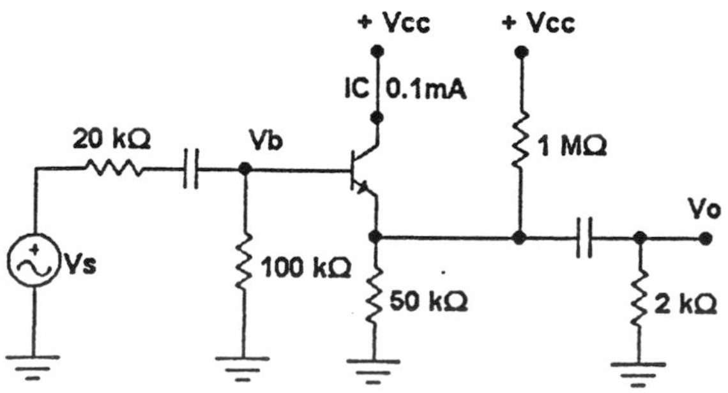

$\beta = 50$; V_A = 100V. Now for I_C = 0.1mA, $r_o = \frac{100V}{0.1}$ = 1MΩ,

$r_e = \frac{25mV}{0.1mA}$ = 250Ω, R_{Leq} = 2k ∥ 50k ∥ 1M = 1.919kΩ,

Gain $\frac{\upsilon_o}{\upsilon_b} = \frac{1.919k}{250 + 1.919}$ = 0.885V/V, R_{inb} = 101 (.25 + 1.919) = 219kΩ, Gain $\frac{\upsilon_b}{\upsilon_s} = \frac{100 \parallel 219}{100 \parallel 219 + 20}$ = 0.775. Thus gain $\frac{\upsilon_o}{\upsilon_s}$ = 0.775 × 0.885 = **0.685V/V.**

4.67 *Iterate*: $V_{B2} = 0.7V$, $V_{B1} = 1.4V$, $I_{B1} \geq \frac{0.7}{68k} \times \frac{1}{101} = 0.102\mu A$. Thus, $V_{C2} \approx 1.4 + 0.1 = 1.5V$, $I_{C2} \approx \frac{5 - 1.5}{1.5} = 2.33mA$, $I_{B2} = \frac{2.33}{101} = 23.1\mu A$, $I_{B1} \approx \left(\frac{0.7}{68k} + 23.1\right)/101 = 0.33\mu A$. Thus, $V_{C2} = 0.7 + 0.7 + 0.33 \times 10^{-6} \times 1 \times 10^{+6} = 1.73V$. Thus, $I_{C2} = \frac{5 - 1.73}{1.5} = 2.18mA$

(a) See with $R = 68k\Omega$ included, $I_{E2} \approx 2.20mA$, $I_{E1} \approx 32\mu A$. Now, $r_{e2} = 25/2.20 = 11.4\Omega$, $r_{\pi 2} \equiv 101$ $(11.4) = 1148\Omega$, $r_{e1} = 25/.032 = 781\Omega$, $r_{\pi 1} = 781\ (101) = 78.9k\Omega$. Thus, $\frac{v_o}{v_b} = -\frac{100}{101}\ \frac{1.5k \parallel 1k}{11.4}$ $= -52.1V/V$, $\frac{v_b}{v_s} = \frac{1.148 \parallel 68}{0.781 + 1.148 \parallel 68} = 0.591V/V$, and $\frac{v_o}{v_s} = -52.1 \times 0.591 =$ **−30.8V/V**, with $R_{in} = (1.148k \parallel 68k + 0.78k)\ 101 =$ **193kΩ.**

(b) Now with $R = 68k\Omega$ removed, the base current in Q_1 reduces slightly, and the collector of Q_2 lowers by 0.1V or so, with I_{C2} increasing by $\frac{0.1}{1.5} \approx 0.07mA$. Thus $I_{E2} \approx 2.3mA$, $I_{E1} \approx 23\mu A$, with $r_{e2} \approx 10.9\Omega$, $r_{\pi 2} \approx 1.098k\Omega$, $r_{e1} \approx 25/.023 = 1087\Omega$, and $r_{\pi 1} \approx 110k\Omega$. Now, $\frac{v_o}{v_b} = -52.1 \times \frac{11.4}{10.9} = -54.5V/V$, $\frac{v_b}{v_s} = \frac{1.098}{1.098 + 1.087} = 0.503$, and $\frac{v_o}{v_s} = -54.5 \times 0.503 =$ **−27.4V/V**, with $R_{in} = (1.098 + 1.087)\ 101 =$ **221kΩ.**

Thus resistance seen by v_s for (a) is $\left(\frac{1M\Omega}{1 - -30.8}\right) \parallel 193k\Omega = 31.4 \parallel 193 =$ **27kΩ.**

SECTION 4.12: THE TRANSISTOR AS A SWITCH – CUTOFF AND SATURATION

4.68 $I_C = \frac{5 - 0.2}{1k\Omega} = 4.8mA$, and $I_B = \frac{5 - 0.7}{R_B} = \frac{4.3}{R_B}$. But $\frac{I_C}{I_B} = 3$. Thus $4.8 = 3\ (4.3)/R_B$, whence

$R_B = \frac{3(4.3)}{4.8} = 2.69k\Omega \approx$ **2.7kΩ.** Now, $\beta_{forced} = \frac{I_C}{I_B} = \frac{\frac{5-0.2}{1k}}{\frac{5-0.7}{R_B}} = R_B\left(\frac{4.8}{4.3}\right) \leq \beta/2$, and

$R_B \leq \frac{4.3}{4.8} \times \frac{\beta}{2} =$ **0.448 βkΩ.**

4.69

(a) $v_I = 0V \rightarrow$ **Q_1 cutoff**, and **Q_2 saturated;**

(b) $v_I = 5V \rightarrow$ **Q_1 saturated**, and **Q_2 cutoff.**

$\beta_{forced} = \frac{I_C}{I_B} = \left(\frac{5 - 0.2}{1k\Omega}\right) / \left(\frac{5 - 0.7}{1k\Omega}\right) = \frac{4.8}{4.3} =$ **1.12.**

4.70 (a) Assume the transistor is saturated. Working on the diagram:

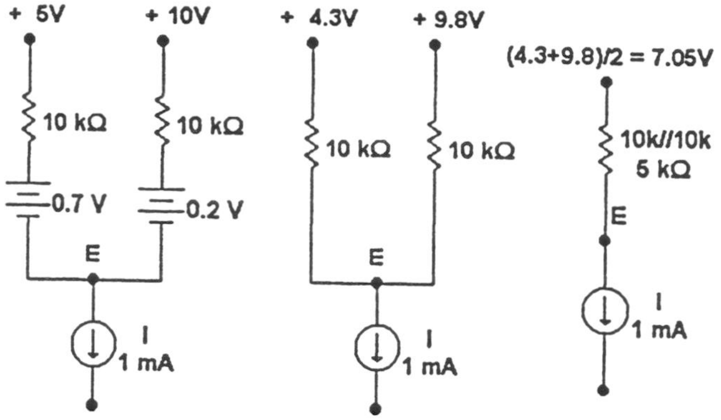

$V_E = 7.05 - 5k\Omega(1mA) = \mathbf{2.05V}$, $V_B = 2.05 + 0.7 = \mathbf{2.75V}$, $V_C = 2.05 + 0.2 = \mathbf{2.25V}$, $I_C = \dfrac{10-2.25}{10k} = .775mA$, $I_B = \dfrac{5-2.75}{10k} = .225mA$, and $\beta_{forced} = \dfrac{0.775}{.225} = \mathbf{3.44}$. For the edge of saturation at $\upsilon = V_E$, $\left[\dfrac{5-0.7-\upsilon}{10k}\right] 100 = \dfrac{10-0.2-\upsilon}{10k}$, for which $430 - 100\,\upsilon = 9.8 - \upsilon$, $99\upsilon = 420.2$, and $\upsilon = \mathbf{4.244V}$, with $I = \dfrac{101}{100}\dfrac{(9.8-4.24)}{10k} = \mathbf{0.561mA}$ (at the edge of saturation).

(b) For saturation: $V_C = \mathbf{5V}$, $V_E = 5 - 0.2 = \mathbf{4.8V}$, $V_B = 4.8 + 0.7 = \mathbf{5.5V}$, $I_C = 1 - 0.1 = 0.9mA$, and $\beta_f = \dfrac{0.9}{0.1} = \mathbf{9}$. For barely linear operation, $I = I_E = (\beta + 1)\, I_B = 101\ (0.1mA) = \mathbf{10.1mA}$.

SECTION 4.13: A GENERAL LARGE-SIGNAL MODEL FOR THE BJT: THE EBERS-MOLL (EM) MODEL

4.71 For $\beta_F = 150$, $\alpha_F = 150/151 = \mathbf{0.9934}$. Now, we are given $I_{SE} = 2 \times 10^{-13}$A

Since the BCJ is 40 times larger than the EBJ, $I_{SC} = 40 I_{SE} = 40(2 \times 10^{-13}) = \mathbf{8 \times 10^{-12}A}$, and $\alpha_R = (1/40)\alpha_F = 0.9934/40 = \mathbf{0.0248}$. Finally, $\beta_R = \dfrac{\alpha_R}{1-\alpha_R} = \dfrac{0.0248}{1-0.0248} = \mathbf{0.0254}$.

4.72 For the forward active mode, $i_E = 100$ mA. From Eq. 4.10, and for diode connection, $i_E = I_{SE}(e^{\upsilon_{BE}/V_T} - 1)$, since $\upsilon_{BC} = 0$.

Thus, $\upsilon_{BE} = V_T \ln[i_E/I_{SE}] = 25 \times 10^{-3} \ln[100 \times 10^{-3}/(2 \times 10^{-13})] = \mathbf{673\ mV}$.

4.73 With this connection, operation is in the **reverse active mode**, with collector current flow in the forward BCJ direction, making i_C in Fig. 4.55 of the Text negative. Now, in Eq. 4.107, $I_C = -100$ mA, $\upsilon_{BE} = 0$, the first term is zero, and the -1 is negligible due to the large value of υ_{BC}: Thus $-100mA = I_{SC} e^{\upsilon_{BC}/V_T}$ where $I_{SC} = 40 I_{SE} = 40(2 \times 10^{-13}A)$. Thus $\upsilon_{EC} = \upsilon_{BC} = 25 \times 10^{-3} \ln(100 \times 10^{-3})/(40 \times 2 \times 10^{-13}) = \mathbf{581\ mV}$.

4.74 $\alpha_F = \dfrac{\beta_F}{\beta_F + 1} = \dfrac{200}{201} = 0.995$, $\alpha_R = \dfrac{\alpha_F}{50} = 0.0199$, and $\beta_R = \dfrac{\alpha_R}{1-\alpha_R} = \dfrac{0.0199}{1-0.0199} = \mathbf{0.0203}$.

For normal saturated operation, with $I_B = 1$ mA, $I_C = 0$, $\beta_{forced} =$,

$V_{CE_{sat}} = V_T \ln[(1 + (\beta_{forced} + 1)/\beta_R)/(1 - \beta_{forced}/\beta_F)] = 25 \times 10^{-3}\ln[(1 + (0 + 1)/0.0203)/(1 - 0)]$
$= 25 \times 10^{-3}\ln 50.25 =$ **97.9 mV**.

4.75 For the collector open, and I_B finite, $\beta_{forced} =$ **0**. From Eq. 4.114, $V_{CE\ sat} = V_T \ln \dfrac{1 + (\beta_{forced} + 1)/\beta_R}{1 - \beta_{forced}/\beta_F}$.
For $\beta_{forced} = 0$, $V_{CE\ sat} = V_T \ln \left[1 + \dfrac{1}{\beta_R}\right]$. Now, for $\beta_F = 70$ to 280, and, correspondingly, $\alpha_F = \dfrac{\beta_F + 1}{\beta_F} = \dfrac{70}{71}$ to $\dfrac{280}{281} = 0.9859$ to 0.9964, with $\alpha_R = \dfrac{\alpha_F}{10} = 0.0986$ to 0.0996, and $\beta_R = \dfrac{\alpha_R}{1-\alpha_R} = \dfrac{.0986}{1-.0986}$ to $\dfrac{.0996}{1-.0996}$, or 0.109 to 0.111, Thus, $V_{CE\ sat} = 25\ln\left[1 + \dfrac{.1}{0.109}\right]$ to $25 \ln \left[1 + \dfrac{1}{0.111}\right]$, or **58.0mV** to **57.7mV**.

4.76 For open-collector operation, $\beta_{forced} = 0$, and $V_{CE\ sat} = V_T \ln \left[1 + \dfrac{1}{\beta_R}\right]$. For $V_{CE\ sat} = 100$ mV, $\ln\left[1 + \dfrac{1}{\beta_R}\right] = \dfrac{100}{25} = 4$, $1 + \dfrac{1}{\beta_R} = 54.6$, and $\beta_R = 0.0187$. Now, for the circuit shown, assume V_{CE} to be 100 mV. Thus $i_C = -\dfrac{100\ mV}{1k + 1k} = -50\mu A$. Thus $\beta_{forced} = \beta_f = \dfrac{-50 \times 10^{-6}}{1 \times 10^{-3}} = -0.05$. Now, $V_{CE\ sat} = V_T \ln \dfrac{1 + (\beta_f+1)/\beta_R}{1 - \beta_f/\beta_F} = 25\ln \dfrac{1 + (1 -.05)/.0187}{1 - (-.05)/100} = 25\ln \dfrac{51.8}{1.0005} = 98.67$ mV. Thus $v_O = \dfrac{98.7}{2} =$ **49.3mV**.

Now for $R_1 = R_2 = 500\Omega$, $i_C \approx -100\mu A$, and $\beta_f = -0.1$, and $V_{CE\ sat} = 25\ln\dfrac{1 + (1 - 0.1)/.0187}{1 - (-0.1)/100} =$
$25\ln\dfrac{49.13}{1.001} = 97.33$ mV, with $v_O = \dfrac{97.3}{2} =$ **48.7mV**.

See $R_{CE\ sat} \approx \dfrac{\Delta V}{\Delta I} = \dfrac{(49.3 - 48.7)\ (mV)}{(100 - 50)\ (\mu A)} = 12\Omega$

4.77 For Table 4.4, $\beta_F = 50$, and $\beta_R = 0.1$. For the required table, $\beta_F = 0.1$, $\beta_R = 50$, $\beta_{forced} = \dfrac{current\ in\ emitter\ lead}{current\ in\ base\ lead} = \beta_f < \beta_F =$ **0.1**. The voltage from emitter lead to collector lead (from Eq. 4.114), is $V_{EC\ sat} = V_T \ln \left\{\dfrac{1 + (\beta_f + 1)/\beta_F}{1 - \beta_f/\beta_R}\right\}$, or $V_{ECsat} = 25 \ln \left\{\dfrac{1 + (\beta_f + 1)/50}{1 - \beta_f/0.1}\right\} =$
$25 \ln \dfrac{1.02 + .02\ \beta_f}{1 - 10\ \beta_f}$. Now, for $\beta_f = 0.1$, $V_{CE\ sat} = \infty$. For $\beta_f = 0.09$, $V_{CE\ sat} = 25 \ln \dfrac{1.02 + .02\ (.09)}{1 - 10(.09)}$
$= 58.1$ mV. For $\beta_f = 0.05$, $V_{CE\ sat} = 25 \ln \dfrac{1.02 + .02(.05)}{1 - .5} = 17.8$ mV. For $\beta_f = 0.02$, $V_{CE\ sat} =$
$25 \ln \dfrac{1.02 + .02(.02)}{1 - .2} = 6.9$ mV. For $\beta_f = 0.01$, $V_{CE\ sat} = 25 \ln \dfrac{1.02 + .02(.01)}{1 - .1} = 3.4$ mV. For $\beta_f =$
0.001. $V_{CE\ sat} = 25 \ln \dfrac{1.02 + .02(.001)}{1 - .01} = 0.75$ mV. For $\beta_f = 0$, $V_{CE\ sat} = 25 \ln \dfrac{1.02 + 0}{1 - 0} = 0.50$ mV.
In summary:

β_{forced}	0.1	0.09	0.05	0.02	0.01	0.001	0.000
$V_{EC\ sat}$ (mV)	∞	58.1	17.8	6.1	3.1	0.75	0.50

4.78 Here, $\alpha_F = 0.995$ implies $\beta_F = \frac{.995}{1-.995} = 199$, and $\alpha_R = \frac{0.995}{5} = 0.199 \simeq 0.2 \rightarrow \beta_R = \frac{.199}{1-.199} = 0.25$. The limiting value of forced β is **199**. For $\beta_{forced} = \beta_f < 199$, $V_{CE\ sat} = 25 \ln \left[\frac{1 + (\beta_f + 1)/0.25}{1 - \beta_f/199}\right] = 25 \ln \left[\frac{5 + 4\beta_f}{1 - .005\beta_f}\right]$.

Now for β_f = 199, 180, 100, 40, 20, 2, 0: For $\beta_f = 199$: $V_{CE\ sat} = \infty$. For $\beta_f = 180$: $V_{CE\ sat} = 25 \ln \frac{5 + 4(180)}{1 - 180/199} = 223$ mV. For $\beta_f = 100$: $V_{CE\ sat} = 25 \ln \frac{5 + 4(100)}{1 - 100/199} = 168$ mV. For $\beta_f = 40$: $V_{CE\ sat} = 25 \ln \frac{5 + 4(40)}{1 - 40/199} = 133$ mV. For $\beta_f = 20$: $V_{CE\ sat} = 25 \ln \frac{5 + 4(20)}{1 - 20/199} = 114$ mV. For $\beta_f = 2$: $V_{CE\ sat} = 25 \ln \frac{5 + 4(2)}{1 - 2/199} = 64$ mV. For $\beta_f = 0$: $V_{CE\ sat} = 25 \ln = 40$ mV.

These results are summarized in the table:

β_{forced}	199	180	100	40	20	2	0
$V_{EC\ sat}$ (mV)	∞	223	168	133	114	64	40

Now, for $I_B = 10$ mA, $I_C = 1$ mA, $\beta_f = \frac{1}{10} = 0.1$. **In normal mode:** $V_{CE\ sat} = 25 \ln \left[\frac{5 + 4(0.1)}{1 - \frac{0.1}{199}}\right] =$ **42.2mV. In inverted mode:** $V_{EC\ sat} = 25 \ln \left[\frac{1 + (0.1+1)/199}{1 - 0.1/0.25}\right] = 25\ln \frac{1.0055}{0.6} =$ **12.9mV.**

4.79

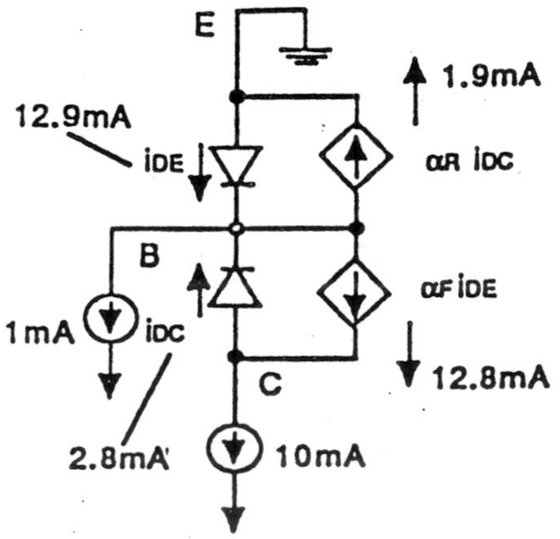

Here, $\alpha_F = \frac{\beta_F}{\beta_F + 1} = \frac{200}{201} = 0.995$, and $\alpha_R = \frac{2}{2+1} = 0.667$.

a) See $i_{DE} - 0.667\ i_{DC} = 10 + 1 = 11$ – – (1), and $0.995\ i_{DE} - i_{DC} = 10.0$ – – (2). From (1), $0.995\ i_{DE} - 0.663\ i_{DC} = 10.945$ – – (3). (2) – (3) $\rightarrow -0.337\ i_{DC} = -0.945$, and $i_{DC} =$ **2.804mA.** >From (1), $i_{DE} = 0.667\ (2.804) + 11 =$ **12.87mA.** *Check* in (2): $0.995\ (12.87) - 2.804 = 10.00$, with $\alpha_R\ i_{DC} =$ **1.87 mA**, and $\alpha_F\ i_{DE} =$ **12.8 mA.**

b) Now, $i = I_S\ e^{v/V_T} \rightarrow v = V_T \ln i/I_S$. Thus $v_{EB} = 25 \ln \frac{12.87 \times 10^{-3}}{10^{-14}} =$ **697mV.**

Now for the collector: $I_{SC} = 10^{-14} \times \frac{.995}{.667} = 1.49 \times 10^{-14}\,A$. Thus, $\upsilon_{CB} = 25 \ln \frac{2.804 \times 10^{-3}}{1.49 \times 10^{-14}} =$ **649mV** and $V_{EC\,sat} = 697 - 649 =$ **48mV.**

c) From Eq. 4.114: $V_{EC\,sat} = V_T \ln \frac{1 + (\beta_f + 1)/\beta_R}{1 - \beta_f/\beta_F} = 25 \ln \frac{1+(10+1)/2}{1+10/200} = 25 \ln \frac{6.5}{1.05} =$ **45.6mV,** nearly the same as in b).

4.80 For $\upsilon_{BE} =$ 700 mA at $i_C =$ 10 mA, with $n = 1$

$I_S = i_c e^{-\upsilon_{BE}/V_T} = 10 \times 10^{-3} e^{-700/25} = 6.91 \times 10^{-15} A$.

From Eq. 4.116: $i_B = \frac{I_S}{\beta_F} e^{\upsilon_{BE}/V_T} + \frac{I_S}{\beta_R} e^{\upsilon_{BC}/V_T}$.

From Eq. 4.117: $i_T = I_S e^{\upsilon_{BE}/V_T} - I_S e^{\upsilon_{BC}/V_T}$.

Here, $\upsilon_{BC} = \upsilon_{BE} - 0.10 = \upsilon - 100$, in mV, and $i_B =$ 100 μA.

From Eq. 4.116:

$100 \times 10^{-6} = 6.91 \times 10^{-15}[(1/50)e^{\upsilon/25} + (1/0.1)e^{(\upsilon - 100)/25} = 6.91 \times 10^{-15}(e^{\upsilon/25})(1/50 + 10e^{-4})$

$= 6.91 \times 10^{-15}(0.185)e^{\upsilon/25}$.

Thus $\upsilon = 25\ln[(100 \times 10^{-6})/(6.91 \times 10^{-15} \times 0.185)]$, or $\upsilon =$ **627.1 mV.**

Check: $691 \times 10^{-15}[1/tpe^{627.1/25} + 1/0.1e^{527.1/15}] = 10.82\mu A + 99.1\mu A \approx 100\mu A$, as expected.

Now, from Eq. 4.117,

$i_T = 6.91 \times 10^{-15}(e^{627.1/25} - e^{527.1/25} = 6.91 \times 10^{-15}(10^{10})(7.83 - 0.14) = 531\mu A$ flowing from the 100 mV supply to ground. Also, from the base, (see *Check* above), 99.1 μA flows into the 100 mV supply.

The net current from the 100 mV supply to ground is 531 – 99 = **432μA**

SECTION 4.14: THE BASIC LOGIC INVERTER

4.81 Model the fanout as a single 0.7 V diode in series with $R = R_B/n$ where n is the fanout.

(a) For a fanout of 10, $R = 450/10 = 45\Omega$:

$V_{OH} = V_{CC} - (R_C/(R_C + R)(V_{CC} - V_{BE}) = 3.0 - (640/(640 + 45))(3.0 - 0.7) =$ **0.85V,**

$V_{OL} = V_{CE_{sat}} =$ **0.3 V**; $V_{IL} = V_{BE} =$ **0.5 V**, roughly.

$V_{IH} = V_{BE} + R_B(V_{DD} - V_{CE_{sat}})/(R_C\beta) = 0.70 + 450(3 - 0.3)/(640 \times 30) = 0.70 + 0.0632 =$ **0.763 V**

$NM_H = V_{OH} - V_{IH} = 0.851 - 0.763 =$ **0.088 V**, and $NM_L = V_{IL} - V_{OL} = 0.5 - 0.3 =$ **0.2 V.**

For $\upsilon_O =$ 1.0 V, $i_C = (3.0 - 1.0)/64 - (0.7 - 0.7)/45 =$ 3.125 mA, for which $r_e \approx 25/3.125 =$ 8 Ω, and the gain is $G = -$ [640 || (450/10)](30/31)(1/8) = **– 5.09 V/V**

(b) For a fanout of 1, $R =$ 450 Ω:

$V_{OH} = 3.0 - (640/(640 + 450)(3.0 - 0.7) =$ **1.65 V,**

$V_{OL} =$ **0.3 V**, $V_{IH} =$ **0.763 V**, $V_{IL} =$ **0.5 V,**

$NM_H = V_{OH} - V_{IH} = 1.65 - 0.76 =$ **0.89 V**, and $NM_L = V_{IL} - V_{OL} = 0.5 - 0.3 =$ **0.2 V**

Gain $G \approx -$ [640 || 450]30/31(1/8) = **– 32 V/V.**

4.82 For saturation, $i_C = (V_{CC} - 0.3)/640 =$ 4.22 mA.

Now, $i_C = I_S e^{\upsilon_{BE}/nV_T}$, and $\upsilon_{BE} = \upsilon_{BEO} + V_T\ln(i_c/i_{C_o}) = 0.70 + 25\ln(4.22/1) =$ 736 mV.

Thus $V_{OH} = V_{CC} - (V_{DD} - V_{BE})R_C/(R_C + R_B) = 3.0 - (3.0 - .736)(640/(640 + 450) =$ **1.67 V.**

Now, for V_{OL}, $\alpha_R = (30/31)/5 = 0.1935$,

and $\beta_R = \alpha_R/(1 - \alpha_R) = 0.1935/1 - 0.1935 = 0.24$.

Collector current with output low (say 0.3 V) is $i_C = (V_{DD} - V_{CE_{sat}})/R_C = (3.0 - 0.3)/640 = 4.22$ mA. Now, the base drive i_B depends on the fanout of the previous gate being largest for a fanout of 1: $i_B = (V_{CC} - V_{BE})/(R_C + R_B)$, or $i_B = (3 - 0.7)/(640 + 450) = 2.11$ mA.

For these conditions, using Eq. 4.114, where $\beta_{forced} = i_C/i_B = (4.22/2.11 = 2.0$, $V_{OL} = V_{CE_{sat}}$

$= V_T \ln[(1 + (\beta_{forced} + 1)/\beta_R)/(1 - \beta_{forced}/\beta_F)] = 25\ln[(1 + (2.0 + 1)/0.24)/(1 - 2/30)] =$ **66.8 mV**

Now, for V_{IL}, operation is where the transistor barely turns on, and the gain is $-$ 1 V/V. For such an arrangement, $G \approx -\beta R_C/(R_B + r_\pi)$.

Thus $\beta R_C/(R_B + r_\pi) = 1$ or $R_B + r_\pi = \beta R_C$ or $r_\pi = \beta R_C - R_B = 30(640) - 450 = 18750$.

Now, $r_\pi = (\beta + 1)re = (\beta + 1)V_T/i_E$, or $i_E = (\beta + 1)V_T/r_\pi = 31(25 \times 10^{-3})/18750 = 41\mu A$, for which $i_B \approx 1.4\ \mu A$, and $\upsilon_I \approx \upsilon_{BE}$ at 41 μA.

Now, $i_C = I_S e^{\upsilon_{BE}/V_T}$, and $\upsilon_{BE_2} = \upsilon_{BE_1} + V_T \ln i_{E2}/i_{E1}$.

Thus at 41 μA, $\upsilon_{BE} = 700 + 25\ln[41/1000] = 620$ mV.

Thus $V_{IL} = \upsilon_{BE} + i_B R_B = 0.620 + 41 \times 10^{-6} \times 450 =$ **0.638 V**.

Now, for V_{IH}, operation is where the transistor is turned on, but past the edge of saturation and the gain is $-$ 1 V/V. This gain results as the base collector junction conducts, shunting R_C. Though it is quite possible to calculate the detail using the Ebers-Moll model, the process is quite tedious. A good worst-case approximation occurs when the incremented resistances of base-emitter and base-collector diodes are about equal, when the base current splits equally between them, while that in the base-emitter is enough to sustain i_C in saturation. For this situation, $i_E \approx (3.0 - 0.3)/640 = 4.2$ mA and for saturation $i_B = 4.2/30 = 0.141$ mA.

Now at $i_C = 4.2$ mA, $\upsilon_{BE} = 700 + 25 ln\, 4.2/1 = 736$ mV and $V_{IH} \approx 0.736 + 450(2)(0.41 \times 10^{-3}) =$ **1.105 V**. A more usual estimate might ignore the base-collector current split (and a factor of 2) and simply use $V_{IH} \approx 0.736 + 450(0.41 \times 10^{-3}) =$ **0.921 V**.

The larger value gives a more conservative view of noise margins:

Now $NM_H = V_{OH} - V_{IH} = 1.67 - 1.105 =$ **0.57 V**, and $NM_L = V_{IL} - V_{OL} = 0.638 - 0.067 =$ **0.57 V**.

Now, the large-signal voltage gain is

$G = (V_{OH} - V_{OL})/(V_{IL} - V_{IH}) = -(1.67 - 0.067)/(1.105 - 0.638)$ or $G =$ **$-$ 3.43 V/V**.

Note that this is low because of the choice of a conservative value for V_{IH}. For the other value $G = (1.67 - 0.067)/0.921 - 0.638) =$ **$-$ 5.63 V/V**.

Now at $\upsilon_O = 0.7$ V, and assuming no effect of fanout load, $i_C(3.0 - 0.7)/640 = 3.59$ mA for which $r_e = 25/3.59 = 6.96\ \Omega$ and $r_\pi = 31(6.96) = 215.9$. Thus, the small-signal gain $= -\beta R_C/(R_B + r_\pi) = -30(640)/(450 + 216) =$ **$-$ 28.8 V/V**.

SECTION 4.15: COMPLETE STATIC CHARACTERISTICS, INTERNAL CAPACITANCES, AND SECOND-ORDER EFFECTS

4.83 For a grounded-base amplifier, the output resistance r_{ob} is approximately $r_\mu \parallel \beta r_o$. Here, $r_{ob} = \dfrac{\Delta \upsilon_C}{\Delta i_C} = \dfrac{10}{50 \times 10^{-9}} = 200 \times 10^6 \Omega$, with $r_o = \dfrac{V_A}{I_C} = \dfrac{200}{0.1 \times 10^{-3}} = 2M\Omega$. Thus $\dfrac{1}{r_{ob}} = \dfrac{1}{r_\mu} + \dfrac{1}{\beta r_o}$, or $\dfrac{1}{200 \times 10^6} = \dfrac{1}{r_\mu} + \dfrac{1}{120 \times 2 \times 10^6}$. That is, for r_μ in MΩ, $\dfrac{1}{r_\mu} = \dfrac{1}{200} - \dfrac{1}{240} = \dfrac{240 - 200}{200\,(240)}$, or $r_\mu = \dfrac{200}{40}(240) = 1200M\Omega =$ **1.2GΩ**

4.84 $V_{Oa} = BV_{CBS} \approx BV_{CBO}$ = **50V**; $V_{Ob} = BV_{CEO}$ = **30V**; $V_{Oc} = BV_{EBO}$ = **7V**.

4.85 $R_{CE\ sat} = \dfrac{\Delta V_{CE}}{\Delta I_C} = \dfrac{0.2 - 0.1}{3 - 1} = \dfrac{0.1V}{2mA}$ = **50Ω**. Generally, $V_{CE\ sat} = V_{CE\ off} + R_{CE\ sat}\ I_C$. Thus at $\dfrac{1+3}{2}$ = 2mA, $V_{CE\ sat} = \dfrac{0.1 + 0.2}{2} = V_{CE\ off} + 0.05(2)$, and $V_{CE\ off} = \dfrac{.3}{2} - .05(2) = .15 - .10 =$ **0.05V**. Otherwise, we could use values at one of the 1mA or 3mA points, to obtain the same result since the same line is involved in all 3 cases.

4.86 At 1.20mA, $h_{FE} = \dfrac{I_C}{I_B} = \dfrac{1.20mA}{11\mu A}$ = **109**, and $h_{fe} = \dfrac{\Delta I_C}{\Delta I_B} = \dfrac{(1.29 - 1.20)mA}{(12 - 11)\mu A} = \dfrac{.09}{.001}$ = **90**. If one assigns the increase in *dc* β to the effect of collector voltage on the base width, then at the particular value of v_{CE} = 10V, $I_C = h_{fe}\ I_B + \left(\dfrac{10}{r_o}\right)$, with $r_o = \dfrac{V_A}{I_C}$. Thus $1.20 = 90\ (11 \times 10^{-3}) + 10\left(\dfrac{1.20}{V_A}\right)$ - - - (1), and also $1.29 = 90\ (12 \times 10^{-3}) + 10\left(\dfrac{1.29}{V_A}\right)$ - - - (2). From (1), $V_A = \dfrac{10\ (1.20)}{(1.20 - .99)}$ = 57.1V, and from (2), $V_A = \dfrac{10\ (1.29)}{1.29 - 90\ (12) \times 10^{-3}}$ = 61.4V. Thus $V_A \approx$ **60V**.

4.87 Assuming saturation, with $I_B = I$ = 1mA, and $V_{CE\ sat} = 0.05 + I_C(0.05)$, we see $v_o = 5 - V_{CE\ sat}$, and $I_C = \dfrac{5 - (0.05 + I_C\ (0.05))}{0.82} - 1.0$. That is, $0.82\ I_C = 5 - .82 - 0.05 - .05\ I_C$, I_C = 4.75mA, and $V_{CE\ sat}$ = .05 + 4.75 (0.05) = 0.287V, with v_O = 5 − .287V = **4.71V**.
Now for I = 4(1) = 4mA, $I_C = \dfrac{5 - (.05 + I_C\ (.05))}{0.82}$ − 4mA, $0.82\ I_C = 5 - .05 - .05\ I_C - 3.28$, $I_C = \dfrac{1.67}{.87}$ = 1.92mA, and $V_{CE\ sat}$ = .05 + 1.92(.05) = 0.146V, with v_O = 5 −.146 = **4.85V**.

4.88 From the graph in Fig. 4.68: *At 1 μA:* β is 200 at 125 °C, 120 at 25 °C, and 70 at − 55 °C.
Thus an average TC is (200 − 70)/(125 − − 55) = *1.38/°C*,
or 1.38/120 × 100 or **1.15%/°C**.
At 1 mA: β is 105 at 125 °C, 190 at 25 °C, 320 at − 55 °C. Thus the average TC is (320 − 105)/(180) = *1.19/°C* or 1.19/190 × 100 = **0.63%/°C**.

4.89 From Eq. 4.120, the stored base charge is $Q_n = W^2 i_C/(2D_n) = \tau_F i_C$
For $W = 1\mu m$, $Q_n = (10^{-6})^2 \times 1 \times 10^{-3}/(21.3 \times (10^{-2})^2)$ = **4.69** × **10⁻¹³C**.
The forward base transit time $\tau_F = W^2/(2D_n) = Q_n/i_c = 4.69 \times 10^{-13}/(1 \times 10^{-3}) = 4.69 \times 10^{10} s$ = **0.47 ns**.
The small-signal emitter diffusion capacitance is (from Eq. 4.121) $C_{de} = \tau_F g_m = \tau_F I_C/V_T = 0.47 \times 10^{-9} \times \dfrac{1 \times 10^{-3}}{25 \times 10^{-3}}$ = **1.88 pF**.
For $W = 5\mu m$: $Q_n = 25(4.69 \times 10^{13} C)$ = **1.17** × 10^{-11}**C**, $\tau_F = 25(0.47)$ = **11.8 ns**, $C_{de} = 25(1.88pF)$ = **47 pF**.
It is apparent that thick (wide) base regions can lead to slower operation!

4.90 From Eq. 3.26, $C_{jo} = A\left[\left(\dfrac{\varepsilon_S 9}{2}\right) \times \left(\dfrac{N_A N_D}{N_A + N_D}\right) \times \left(\dfrac{1}{V_O}\right)\right]^{1/2}$,

where (from Eq. 3.18, $V_O = V_T \ln\left(\dfrac{N_A N_D}{n_{i2}}\right)$.

Here $n_i = 1.5 \times 10^{10}/cm^3 N_A = 10^{17}/cm^3$, $N_D = 10^{19}/cm^3$, $q = 1.60 \times 10^{-19} C$, $\varepsilon_S = 1.04 \times 10^{-12} F/cm$, and $A = 10\mu m \times 10\mu m = 100 \times 10^{-8} cm^2$.

Thus $V_O = 25 \times 10^{-3} \ln[(10^{17} \times 10^{19})/(1.5 \times 10^{10})^2] = 0.909$ V.

Thus $C_{je0} = C_{j0} = 100 \times 10^{-8}[(1.04 \times 10^{-12} \times 1.6 \times 10^{-19}/2) \times 10^{17} \times 10^{19}/(10^{17} + 10^{19})/0.90]^{1/2} =$ $10^{-6}(1.093 \times 10^{-14})^{1/2} = 0.104$ pF, and $C_{je} \approx 2C_{je0} =$ **0.208 pF.**

Now for C_μ ($= C_{jc}$), use Eq. 3.26 with $N_D = 10^{15}/cm^3$ in the collector.

Thus $V_O = V_T \ln(10^{15} \times 10^{17}/(1.5 \times 10^{10})^2) = 0.67$ V, and $C_{jo} = 10 \times 100 \times 10^{-8}[1.04 \times 10^{-12}$ $\times (1.6 \times 10^{-19}/2) \times 10^{17} \times 10^{15}/(10^{17} + 10^{15})/0.67]^{1/2} = 0.157$ pF

and $C_j = C_{jo}/(1 + V_R/V_O)^m = 0.157/(1 + 2.0/0.67)^{0.4} = 0.090$ pF. Thus $C_\mu = 0.090$ pF = **90 fF.**

Now, from P4.89 above for $W = 1\mu m$, $C_{de} = 1.88$ pF, and from directly above, $C_{je} = 0.208$ pF. Thus $C_\pi = 1.88 + 0.208 =$ **2.09 pF**. For operation at 1 mA,

$g_m = I_C/V_T = 1 \times 10^{-3}/25 \times 10^{03} = 40$ mA/V.

Thus, $f_T = (1/2\pi)g_m/(C_\pi + C_\mu) = 40 \times 10^{-3}/((2.09 + 0.090) \times 10^{-12})/(2\pi) =$ **2.92 GHz.**

4.91 Here, f_T is 10 GHz at 10 mA and 7 GHz at 1 mA, where g_m is 400 mA/V, and 40 mA/V respectively.

For each case, $C_\pi + C_\mu = g_m/(2\pi f_T)$

Thus $C_{\pi 10} + C_\mu = 400 \times 10^{-3}/(2\pi \times 10 \times 10^9) =$ **6.37 pF** and $C_{\pi 1} + C_\mu = 40 \times 10^{-3}/(2\pi \times 7 \times 10^9) =$ **0.909 pF.**

Now, $C_\pi = C_{je} + C_{de}$, and $C_{de} \propto i$, such that $C_\pi = C_{je} + iC$.

Thus $C_{je} + 10C + C_\mu = 6.37$ – – – (1), and $C_{je} + 1C + C_\mu = 0.909$ – – – (2)

Subtract (1), (2), $9C = 6.37 \times 0.91$, and $C = 0.607$ pF(/mA).

From (2), $C_{je} + C_\mu = 0.909 - 0.607 =$ **0.302 pF.**

Thus, $C_{de} =$ **6.07 pF** at 10 mA, and **6.1 fF** at 10 μA.

At 10 μA, $f_T = \dfrac{400/1000}{2\pi(0.302 + 0.0061) \times 10^{-12}} =$ **206 MHz**

4.92 We note that the area of all junctions is increased by 4, and that the current density is unchanged. Now, using the results of Ex. 4.440 on page 315 of the Text:

At 4 mA:

From Eq. 4.120, τ_F is unchanged, at **20 ps.**

From Eq. 4.121, C_{de} increases by 4, to 4(0.8) = **3.2 pF**

From Eq. 3.26, C_{je0} increases by 4, to 4(20) = **80 fF**

From Eq. 3.26, $C_{\mu 0}$ increases by 4, to 4(20) = **80 fF.**

From Eq. 3.18, V_{0e} is unchanged at **0.9 V.**

See Eq. 4.124, m_{CBJ} is unchanged at **0.33**

C_{je} increases by 4 to 4(40) = **160 fF**

C_π increases by 4 (since both area and charge are 4 × larger) to 4(0.84) = **3.36 pF**

C_μ increases by 4, to 4(12) = **48 fF**

From Eq. 4.130, f_T at 4 mA is unchanged, at **7.5 GHz.**

For operation at 1 mA, $g_m = 40$ mA/V, $C_{je} = 160$ fF, $C_\mu = 48$ fF, $C_{de} = 3.2/4 = 0.80$ pF.

Thus, $f_T = \dfrac{40 \times 10^{-3}}{2\pi(800 + 160 + 48) \times 10^{-15}} =$ 6.3 GHz.

4.93 For this transistor, $C_\mu = 2$ pF and $C_\pi = 10.7$ pF at 1 mA. If we assume that $C_{je} \approx C_\mu = 2$ pF, Then $C_{de} = 10.7 - 2 = 8.7$ pF at 1 mA, and 8.7/100 = 87 fF at 1 mA/100 = 10 μA, and 8.7 pF at 1 μA.

At 10 μA, $f_T \approx \dfrac{40 \times 10^{-3}/100}{2\pi(2.0 + 2.0 + .087) \times 10^{-12}} =$ **15.6 MHz.**

At 1 μA, $f_T = \dfrac{40 \times 10^{-3}/1000}{2\pi(2.0 + 2.0 + 0.009) \times 10^{-12}} =$ **1.59 MHz.**

{Note that if C_{je} is assumed to be 0, rather than equal to C_μ, f_T values are **30.2 MHz** and **3.17 MHz**}.

The proposed change would reduce the capacitance C_μ (and C_{je}) by a factor of 100 since the junction areas are reduced by that factor. Continue to use $C_{de} = 8.7$ pF at 1 mA for lack of a better choice.

Now at $i_C = 10$ μA,

$$f_T \approx \frac{40 \times 10^{-3}/100}{2\pi(2/100 + 2/100 + 8.7/100) \times 10^{-12}} = \textbf{501 MHz.}$$

Now at 1 $i_C = 1\mu A$,

$$f_T = \frac{40 \times 10^{-3}/1000}{2\pi(2/100 + 2/100 + 8.7/1000) \times 10^{-12}} = \textbf{131 MHz}$$

NOTES

Chapter 5

FIELD-EFFECT TRANSISTORS (FETs)

SECTION 5.1: STRUCTURE AND PHYSICAL OPERATION OF THE ENHANCEMENT-TYPE MOSFET

5.1 *In general,* a channel is induced for $\upsilon_{GS} \geq \upsilon_S + V_t = 0 + 1.5\text{V}$. Hence $\upsilon_{GS} \geq \mathbf{1.5V}$, here.

In general, the drain end of the channel is pinched off for $\upsilon_{GD} \leq V_t = 1.5\text{V}$. Now for $\upsilon_{GS} = 3.0\text{V}$ and $V_S = 0$, $\upsilon_G = 3.0\text{V}$, and the drain is pinched off for $\upsilon_D \geq 3.0 - 1.5 = 1.5\text{V}$. Hence $\upsilon_{DS} \geq \mathbf{1.5V}$, here.

In general, saturation occurs for a given υ_{GS}, when $\upsilon_{DS} \geq \upsilon_{GS} - V_t$, or $\upsilon_{DS} \geq 3.0\text{V} - 1.5\text{V} = 1.5\text{V}$, for which $\upsilon_D \geq \mathbf{1.5V}$, here (which is, of course, when the drain end of the channel is pinched off).

In general, triode operation occurs for $\upsilon_{DS} \leq \upsilon_{GS} - V_t$, for which $\upsilon_D \leq \mathbf{1.5V}$, here.

SECTION 5.2: CURRENT-VOLTAGE CHRACTERISTICS OF THE ENHANCEMENT MOSFET

5.2

(a) $\upsilon_{DS} = \upsilon_D - \upsilon_S = 2.1 - 0 = 2.1\text{V}$; $\upsilon_{GS} = \upsilon_G - \upsilon_S = 3 - 0 = 3\text{V}$; $\upsilon_{GS} - V_t = 3 - 1 = 2.0\text{V} \leq \upsilon_{DS}$ $\rightarrow$ **saturated mode.**

(b) $\upsilon_{DS} = \upsilon_D - \upsilon_S = -0.1 - -2 = 1.9\text{V}$; $\upsilon_{GS} = \upsilon_G - \upsilon_S = 2 - -2 = 4\text{V}$; $\upsilon_{GS} - V_t = 4 - 2 = 2.0 \geq \upsilon_{DS}$ $\rightarrow$ **triode mode.**

(c) $\upsilon_{SD} = \upsilon_S - \upsilon_D = 0 - -3 = 3\text{V}$; $\upsilon_{SG} = \upsilon_S - \upsilon_G = 0 - -1 = 1\text{V} < |V_t| = 2\text{V}$ $\rightarrow$ **cutoff mode.**

(d) $\upsilon_{SD} = \upsilon_S - \upsilon_D = 2 - -1 = 3\text{V}$; $\upsilon_{SG} = \upsilon_S - \upsilon_G = 2 - 0 = 2\text{V}$; $\upsilon_{SG} + V_t = 2 - 1 = 1\text{V} \leq \upsilon_{SD}$ $\rightarrow$ **saturated mode.**

(e) $V_t = 2\text{V}$ $\rightarrow$ **n channel;** $\upsilon_{GS} = 0 - -3 = 3\text{V}$; Since saturated, $\upsilon_{DS} \geq \upsilon_{GS} - V_t = 3 - 2 = 1\text{V}$; $\upsilon_D = \upsilon_{DS} + \upsilon_S \geq 1 - 3 = \mathbf{-2V}$.

(f) $V_t = -2\text{V}$ $\rightarrow$ **p channel;** $\upsilon_{SD} = \upsilon_S - \upsilon_D = 3 - -1 = 4\text{V}$; $\upsilon_{SG} = \upsilon_S - \upsilon_G = 3 - 0 = 3\text{V}$; $\upsilon_{SG} + V_t = 3 - 2 = 1\text{V} \leq \upsilon_{SD}$ $\rightarrow$ **saturated mode.**

(g) $V_t = -2\text{V}$ $\rightarrow$ **p channel;** $\upsilon_S = 3\text{V}$, $\upsilon_D = -3\text{V}$; $\upsilon_S - \upsilon_G \leq -V_t = +2\text{V}$ for cutoff;
$-\upsilon_G \leq 2 - \upsilon_S = 2 - 3 = -1\text{V}$, and $\upsilon_G \geq \mathbf{1V}$.

5.3

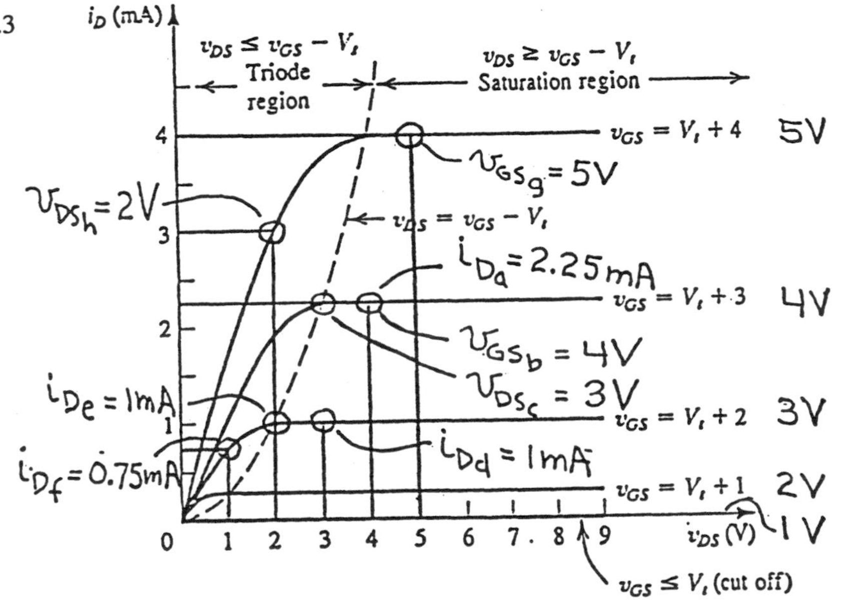

Since $V_t = 1\text{V}$, the υ_{GS} are relabelled as shown. See $\mathbf{i_{Da} = 2.25\ mA}$, $\mathbf{\upsilon_{GSb} = 4\ V}$, $\mathbf{\upsilon_{DSc} => 3\ V}$, $\mathbf{i_{Dd} = 1\ mA}$, $\mathbf{i_{De} = 5\ mA}$, $\mathbf{i_{Df} = 0.75\ mA}$, $\mathbf{\upsilon_{GSg} = 5\ V}$, $\mathbf{\upsilon_{DSh} = 2\ V}$.

5.4

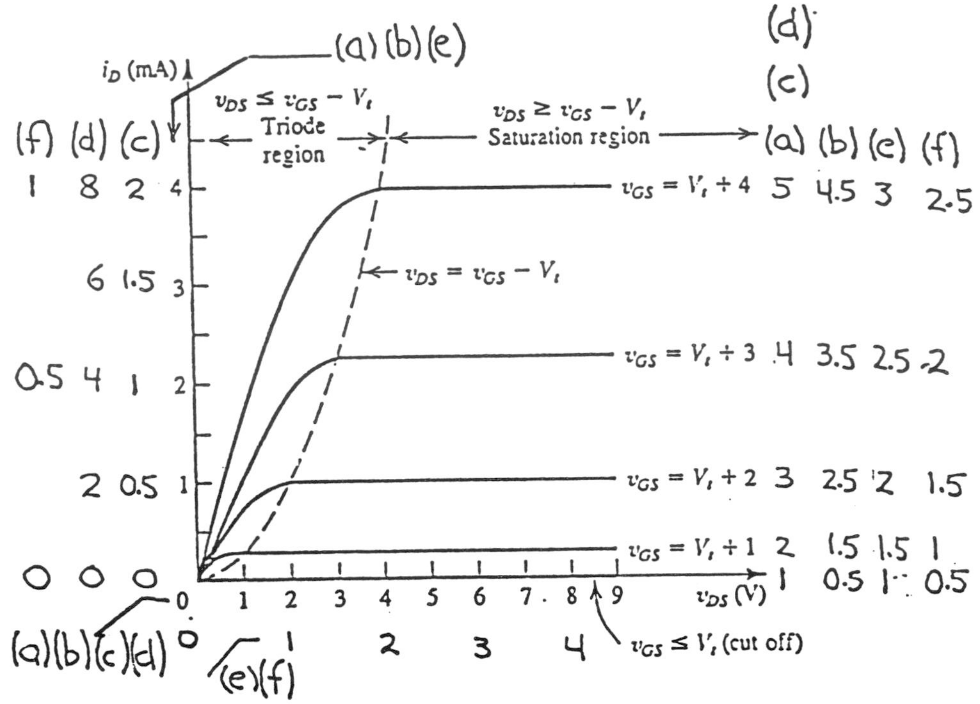

The axes labels are indicated by (a), (b), etc, at the top right and left, and at the right bottom.

5.5 $V_S = 0$V, $V_G = 3$V. Now, for saturation $V_{DS} \ge V_{GS} - V_t$, and for $V_S = 0$, $V_D \ge 3 - 1 = 2$V. Thus the device is in triode operation for $\mathbf{V_D \le 2V}$.

In general, in triode mode, $i_D = k'(W/L)[(v_{GS} - V_t)\, v_{DS} - v_{DS}^2/2d])$.

For $v_{DS} = 2$V, $i_D = 20 \times 10^{-6} \times 10\,[(3-1)\,2 - 2^2/2] = 200\,(\,(2)\,(2) - 2^2/2) = 200\,(\,4-2) = 400\mu$**A**.

Check, in saturation, $i_D = 1/2(20)(10)\,(v_{GS} - V_t)^2 = 100\,(3-1)^2 = 400\mu$A.

For $v_{DS} = 1$V, $i_D = 200\,[(2)\,1 - 1^2/2] = 300\mu$**A**.

For $v_{DS} = 0.5$V, $i_D = 200\,((2)\,0.5 - 0.5^2/2) = 175\mu$**A**.

For v_{DS} very small, $i_D = k'(W/L)\,(v_{GS} - V_t)\,v_{DS}$, whence

$$r_{DS} = \frac{v_{DS}}{i_D} = \frac{1}{k'(W/L)\,(v_{GS} - V_t)} = \frac{1}{200\,(3-1)} = \mathbf{2.5k\Omega}.$$

Now, r_{DS} increases by 1% when $v_{DS}^2/2 = ((v_{GS} - V_t)\,v_{DS})\,\frac{1}{100}$, for which $v_{DS} = \frac{2\,(3-1)}{100} = \mathbf{0.04V}$.

Now r_{DS} increases by 10% when $\frac{1}{((v_{GS} - V_t)\,v_{DS} - v_{DS}^2/2)} = 1.10\,\frac{1}{(v_{GS} - V_t)v_{DS}}$, or $(v_{GS} - V_t) - v_{DS}/2 = 0.909\,((v_{GS} - V_t))$, for which, $v_{DS} = 2\,(1 - 0.909)\,(v_{GS} - V_t) = 2\,(0.091)\,(3-1)$ $= \mathbf{0.36V}$.

5.6 For triode operation at low v_{DS}, $i_D = k'(W/L)\,((v_{GS} - V_t)\,v_{DS})$, whence

$$r_{DS} = \frac{v_{DS}}{i_D} = \frac{1}{k'(W/L)\,((v_{GS} - V_t))} = \frac{10^6}{20(20/2)\,(v_{GS} - 1)}, \quad \text{or} \quad r_{DS} = \frac{5000\Omega}{v_{GS} - 1},$$

whence $v_{GS} = \frac{5k\Omega}{r_{DS}} + 1$.

For $r_{DS} = 1$kΩ: $v_{GS} = \frac{5}{1} + 1 = \mathbf{6V}$, and for $r_{DS} = 1$MΩ: $v_{GS} = \frac{5k\Omega}{10^3 k\Omega} + 1 = \mathbf{1.005V}$.

For υ_{DS} near 0V, and less than a 10% increase in r_{DS}, $(\upsilon_{GS} - 1) - \upsilon_{DS}/2 \geq \frac{1}{1.1}(\upsilon_{GS} - 1)$, or $\upsilon_{DS} \leq 2\,(\upsilon_{GS} - 1)\,(0.0909)$.

For $r_{DS} = 1k\Omega$, $\upsilon_{GS} = 6V$, and $\upsilon_{DS} \leq 2\,(6-1)\,(.0909) =$ **0.909V**, at which $i_D = k'(W/L)\,[(\upsilon_{GS} - V_t)\,\upsilon_{DS} - \upsilon_{DS}^2/2] = 200\,((6-1)\,0.909 - (0.0909)^2/2) = 200\,(4.545) - .413) =$ **826μA.**

For $r_{DS} = 1M\Omega$, $\upsilon_{GS} = 1.005V$, and $\upsilon_{DS} \leq 2\,(1.005 - 1.00)\,(.0909) =$ **0.91 mV**, at which $i_D = 200\,[(1.005 - 1)\,(0.91 \times 10^{-3}) - (.91 \times 10^{-3})^2/2] = 200\,[\,(5 \times 0.91 \times 10^{-6}) - 0.91^2 \times 10^{-6}/2] = 100\,(9.1 \times 10^{-6} - .83 \times 10^{-6}) = 827 \times 10^{-6}\mu A =$ **0.83nA!** or, *easier*, $i_D = \frac{0.91mV}{10^6\Omega} \approx 0.91$ nA, as a quick approximation.

5.7 For saturation, $i_{D_s} = 1/2k'(W/L)\,(\upsilon_{GS} - V_t)^2$, and for triode operation,

For $i_{D_t} = 1.00 i_{DsS}$: $i_{D_t} = k'(W/L)\,[(\upsilon_{GS} - V_t)\,\upsilon_{DS} - \upsilon_{DS}^2/2]$.

$k'(W/L)\,[(\upsilon_{GS} - V_t)\,\upsilon_{DS} - \upsilon_{DS}^2/2] = k'2(W/L)\,(\upsilon_{GS} - V_t)^2$, or $\upsilon_{DS}^2 - 2\,(\upsilon_{DS})\,(\upsilon_{GS} - V_t) + (\upsilon_{GS} - V_t)^2 = 0$,

whence $\upsilon_{DS} = \dfrac{2\,(\upsilon_{GS} - V_t) \pm \sqrt{2^2\,(\upsilon_{GS} - V_t)^2 - 4\,(\upsilon_{GS} - V_t)^2}}{2}$,

or $\upsilon_{DS} = (\upsilon_{GS} - V_t)[1 \pm \sqrt{1-1}] = (uspilon_{GS} - V_t)$

For $i_{Dt} = 0.99\, i_{Ds}$:

$2\,(\upsilon_{GS} - V_t)\,\upsilon_{DS} - \upsilon_{DS}^2 = 0.99\,(\upsilon_{GS} - V_t)^2$, or $\upsilon_{DS}^2 - 2\,(\upsilon_{DS})\,(\upsilon_{GS} - V_t) + 0.99\,(\upsilon_{GS} - V_t)^2 = 0$, whence $\upsilon_{DS} = (\upsilon_{GS} - V_t)\left[1 \pm \sqrt{1 - (.99)^2}\right] = (\upsilon_{GS} - V_t)\,(1 \pm 0.141)$, (where the negative value applies). Thus $\upsilon_{DS} =$ **0.859 ($\upsilon_{GS} - V_t$).**

For $i_{Dt} = 0.90\, i_{Ds}$: $\upsilon_{DS} = (\upsilon_{GS} - V_t)\left[1 \pm \sqrt{1 - .9^2}\right]$. Thus $\upsilon_{DS} =$ **0.564 ($\upsilon_{GS} - V_t$).**

For $i_{Dt} = 0.50\, i_{Ds}$: $\upsilon_{DS} = (\upsilon_{GS} - V_t)\left[1 \pm \sqrt{1 - .5^2}\right]$. Thus $\upsilon_{DS} =$ **0.134 ($\upsilon_{GS} - V_t$).**

Now, for $\upsilon_{GS} = 2\,V_t$ and $V_t = 2V$, $\upsilon_{DS} =$ **2V, 1.72V, 1.13V**, and **0.264V**, for 100%, 99%, 90%, and 50% of saturation value, respectively.

5.8 Here, $r_o = \frac{5V}{(2.2 - 2.1)mA} =$ **50kΩ**; $V_A = r_o\, i_D = 50k\Omega \left(\frac{2.1 + 2.2}{2}\right) =$ **107.5V**; and $\lambda = \frac{1}{V_A} =$ **0.0093V^{-1}.**

5.9 Now $K_p = \frac{1}{2}\mu_p\, C_{ox}\, \frac{W}{L}$. Assuming $\mu_p = \frac{1}{2}\,\mu_n$, $K_p = \frac{1}{2} \times \frac{1}{2} \times 20 \times \frac{100}{3} = 166.7\mu A/V^2$.

For $\upsilon_{GS} = \upsilon_{DS} = -5V$, $i_D = K\,(\upsilon_{GS} - V_t)^2\,(1 + \lambda\,\upsilon_{DS}) = 166.7\,(-5 - -2)^2\,(1 - .01(-5)) = 166.7\,(3^2)\,(1.05) =$ **1.575mA.**

5.10 For the substrate V_B of the PMOS connected to +5V, while the source voltage is varied: $|V_t| = |V_{t0}| + \gamma\,[\sqrt{2\,\Phi_f + V_{SB}} - \sqrt{2\,\Phi_f}] = +1 + 0.6\,[\sqrt{0.6 + V_{SB}} - \sqrt{0.6}]$. For $V_S = +5V$, $V_{SB} = 0V$, and $|V_t| = 1V$, or $V_t =$ **−1 V**. For $V_S = 0V$, $V_{SB} = 5 - 0 = 5V$, and $|V_t| = 1 + 0.6\,[\sqrt{0.6 + 5} - \sqrt{0.6}] = 1.955V$, or $V_t =$ **−1.96 V.**

5.11 In each of the circuits, consider the saturation transconductance factor is $K = 1/2\mu_n\, C_{ox}(W/L) = 1/2 \times 20 \times 10^{-6}(20/2) = 100\mu A/V^2 = 0.1mA/V^2$. Accordingly: (a) $\upsilon_{GS} = \upsilon_{DS} = 5V$; saturated operation, for which $i_D = K\,(\upsilon_{GS} - V_t)^2 = 0.1\,(5-2)^2 = 0.9mA$. Thus $I_a =$ **0.9mA.**

(b) $\upsilon_{SG} = \upsilon_{SD}$; saturated operation, for which $0.4 = 0.1\ (\upsilon_{SG} - -2)^2$, or $\upsilon_{SG} + 2 = \pm\sqrt{4} = \pm\ 2$. Thus $\upsilon_{SG} = 0$, which is not possible, or –4V. Thus V_b = **+4V.**

(c) $V_{GS} = 0$; cutoff, for which I_c = **0mA.**

(d) $\upsilon_{GS} = \upsilon$, $\upsilon_{DG} = 3V$; saturated operation, for which $i_D = 0.9 = 0.1\ (\upsilon - 2)^2$, or $\upsilon - 2 = \sqrt{9} = 3$, or $\upsilon = 5V$. Thus $V_d = \upsilon_G - \upsilon_{GS} = 2 - 5 =$ **–3V.**

Section 5.3: THE DEPLETION – TYPE MOSFET

5.12 For all conditions, the saturatin transconductance factor is $K = 1/2\mu_n C_{ox}(W/L) = 1/2 \times 20 \times 10^{-6} \times 200/2 = 1000 \times 10^{-6} = 1mA/V^2$; $V_t = -4V$ (for an n-channel depletion mode).

(a) $\upsilon_{GS} = -4 - 0 =$ **–4V. cutoff.** i_D = **0mA**, and $\upsilon_{DS} = 5 - 0 = 5V$.

(b) $\upsilon_G = \upsilon_{GS} + \upsilon_S = -2 + 0 = -2V$, and $\upsilon_D = \upsilon_S + \upsilon_{DS} = 0 + 3 = 3V$. This implies saturation for which $i_D = (\upsilon_{GS} - V_t)^2 = 1\ (-2 - -4)^2$ = **4mA.**

(c) $\upsilon_{GS} = \upsilon_G - \upsilon_S = 0 - 0 =$ **0V**, and $\upsilon_{DS} = \upsilon_D - \upsilon_S = 5 - 0 =$ **5V**,

for which $\upsilon_{DG} = 5 - 0 = 5V > |V_t|$. This implies **saturation** operation, for which $i_D = 1\ (0 - -4)^2$ = **16mA.**

(d) $\upsilon_G = \upsilon_{GS} + \upsilon_S = 0 + 0 =$ **0V**, and $\upsilon_D = \upsilon_S + \upsilon_{DS} = 0 + 2 =$ **2V**, for which $\upsilon_{DG} = 2 - 0 < |V_t|$. This implies **triode operation**, for which $i_D = K\ [2\ (\upsilon_{GS} - V_t)\ \upsilon_{DS} - \upsilon_{DS}^2] = 1\ (2\ (0 - -4)\ 2 - 2^2)$ = **12mA.**

(e) $\upsilon_S = \upsilon_G - \upsilon_{GS} = 0 - 1 =$ **–1V**, and $\upsilon_D = \upsilon_S + \upsilon_{DS} = -1V + 5V =$ **4V**, for which $\upsilon_{DG} = \upsilon_D - \upsilon_G = 4V - 0 = |V_t|$. This implies operation at the **edge of saturation**, for which $i_D = 1\ (1 - -4)^2$ = **25mA.**

(f) $\upsilon_{GS} = \upsilon_G - \upsilon_S = 2 - 0 =$ **2V**, and $\upsilon_{DS} = \upsilon_D - \upsilon_S = 5 - 0 =$ **5V**, for which $\upsilon_{DG} = \upsilon_D - \upsilon_G = 5 - 2 = 3V < |V_t|$. This implies **triode operation**, for which, $i_D = K\ [2\ (\upsilon_{GS} - V_t)\ \upsilon_{DS} - \upsilon_{DS}^2] = 1\ [2\ (2 - -4)\ 5 - 5^2] = 60 - 25$ = **35mA.**

(g) $\upsilon_G = \upsilon_{GS} + \upsilon_S = 2 + 0 =$ **2V**, and $\upsilon_{DS} = \upsilon_D - \upsilon_S = 0 - 0 =$ **0V**. See that operation is in the **triode mode**, but with i_D = **0mA.**

(h) $\upsilon_S = \upsilon_{SG} + \upsilon_G = -2 + 0 =$ **–2V**, and $\upsilon_{DS} = \upsilon_D - \upsilon_S = 0 - -2 =$ **2V**. See that $\upsilon_{DG} = \upsilon_D - \upsilon_G = 0 - 0 = 0 < |V_t| \rightarrow$ **triode operation**, for which $i_D = K\left[2\ (\upsilon_{GS} - V_t)\ \upsilon_{DS} - \upsilon_{DS}^2\right] = 1\left[2\ (2 - -4)\ 2 - 2^2\right] = 24 - 4$ = **20mA.**

5.13 *Depletion MOS*: For simplicity, use $K = 1/2\mu_n C_{ox}(W/L) = 1/2(20 \times 10^{-6})(10) = 100 \times 10^{-6} = 0.1mA/V^2$, $|V_t| = 2V$.

(a) Saturation with $\upsilon_{GS} = 0$, for which $I_a = i_D = K\ (\upsilon_{GS} - V_t)^2 = 0.1\ (0 - -2)^2$ = **0.4mA.**

(b) Saturation, for which $0.4 = 0.1\ (\upsilon_{GS} - -2)^2$, and $\upsilon_{GS} + 2 = \pm\sqrt{4} = \pm 2V$, whence $\upsilon_{GS} = -4V$ or 0V. Clearly, 4V is not possible. Thus $V_b = 0 - 0 =$ **0V.**

(c) Saturation, for which $0.9 = 0.1\ (\upsilon_{GS} - -2)^2$, and $\upsilon_{GS} + 2 = \pm\sqrt{9} = \pm 3$, for which $\upsilon_{GS} = -5$ or $+1V \rightarrow$ Clearly –5V is not possible. Thus $\upsilon_{GS} = +1V$, and $V_c = 0 + 1V =$ **+1V.**

(d) Devices are connected symmetrically: $I_d = \frac{1}{2} I = \frac{1.8}{2}$ = **0.9mA.** Also see saturated operation, for which $0.9 = 0.1\ (\upsilon_{GS} - -2)^2$. Thus $\upsilon_{GS} = +1V$ (as in (c)), and V_d = **+1V.**

(e) Here, $\upsilon_{DS} = \upsilon_{GS}$. Thus $\upsilon_{DG} = 0 < |V_t|$ implies triode operation, for which $i_D = 0.4mA = 0.1\left[2\ (\upsilon_{GS} - -2)\ \upsilon_{GS} - \upsilon_{GS}^2\right]$, or $4 = 2\ \upsilon_{GS}^2 + 4\ \upsilon_{GS} - \upsilon_{GS}^2$, $\upsilon_{GS}^2 + 4\ \upsilon_{GS} - 4 = 0$, whence $\upsilon_{GS} = \frac{-4 \pm \sqrt{4^2 - 4(-4)1}}{2} = \frac{-4 \pm 4\sqrt{2}}{2} = -2 \pm 2\sqrt{2}$. But it must be positive. Thus $= -2 + 2\sqrt{2} = 0.828V$, and $V_e = 5 - .828 =$ **4.172V.**

5.14 For convenience, let $K = 1/2k'(W/L)$. Operation is triode mode in both cases: Thus $i_D = K\left[2\,(\upsilon_{GS} - V_t)\,\upsilon_{SD} - \upsilon_{SD}^2\right]$, V_t being negative for a p-channel depletion device, when υ_{SG} is used.

For $\upsilon_D = 4.8$V, $\upsilon_{SD} = 5.0 - 4.8 = +0.2$V, $\upsilon_G = 5.0$V, $\upsilon_{GS} = 5.0 - 5.0 = 0.0$V;

$0.1 = K\left[2\,(0 - V_t)\,(+0.2) - 0.2^2\right]$, or $0.1 = K\left[-0.4V_t - 0.04\right]$ - - - (1).

For $\upsilon_D = 4.95$, $\upsilon_{SD} = 5.0 - 4.95 = +.05$V, $\upsilon_G = 0$V, $\upsilon_{SG} = 5 - 0 = +5$V;

$0.1 = K\left[2\,(+5 - V_t)\,(+.05) - .05^2\right]$, or $0.1 = K\left[+0.5 + 0.1V_t - .0025\right]$,

or $0.1 = K\left[0.4975 + 0.1V_t\right]$ - - - (2).

Now (1)/(2) $\rightarrow 1 = \dfrac{-0.4V_t - 0.04}{+0.4975 + 0.1V_t}$, or $-0.4\,V_t - .04 = +\,0.4975 + 0.1V_t$, or $-0.5V_t = +0.5375$, whence $V_t =$ **−1.075V**. Thus with υ_{GS} as defined, the depletion threshold is 1.075V.

From (1), $0.1 = K\,[\,-0.4V_t - .04]$, $K = \dfrac{0.1}{-0.4(-1.075) - .04} = \dfrac{0.1}{+.43 - .04} = 0.256\text{mA}/V^2$. Now, $i_D = K\,(\upsilon_{GS} - V_t)^2 = I_{DSS}\,(1 - \dfrac{\upsilon_{GS}}{V_t})^2 = \dfrac{I_{DSS}}{V_t^2}\,(\upsilon_{GS} - V_t)^2$. Thus $I_{DSS} = K\,V_t^2 = 0.256 \times (1.075)^2 =$ **0.296mA**.

SECTION 5.4: MOSFET CIRCUITS AT DC

5.15 $K = K_n = \dfrac{1}{2}\,\mu_n\,C_{ox}\,\dfrac{W}{L} = \dfrac{1}{2}(20) \times 40 = 0.4\text{mA}/V^2$. Thus $I_D = \dfrac{V_{DD} - V_D}{R_D} = \dfrac{5-2}{7.5k\Omega} =$ **0.4mA**. In saturation, $i_D = K\,(\upsilon_{GS} - V_t)^2$, or $0.4 = 0.4\,(\upsilon_{GS} - 1)^2$. Thus $\upsilon_{GS} - 1 = 1$, or $\upsilon_{GS} = 2$V. $\upsilon_S = \upsilon_G - \upsilon_{GS} = 0 - 2 =$ **−2V**, and $R_S = \dfrac{\upsilon_S - V_{SS}}{I_D} = \dfrac{-2 - -5}{0.4} = \dfrac{3V}{0.4} =$ **7.5kΩ**.

5.16 Now, $I_D = \dfrac{V_{DD} - V_D}{R_D} = \dfrac{5-2}{7.5} = 0.4$mA, and $V_S = -5 + 7.5(0.4) = -2$V. Thus $V_{GS} = 2$V, whence $0.4 = K\,(2 - 1)^2$, and $K =$ **0.4mA/V²**.

Now for $K = \dfrac{0.4}{2} = 0.2$mA/V, in the source circuit: $\dfrac{V_S - -5}{7.5k} = i_D = 0.2\,(-V_S - 1)^2$, or $V_S + 5 = 1.5\,(V_S + 1)^2 = 1.5\,V_S^2 + 3V_S + 1$, or $1.5V_S^2 + 2V_S - 4 = 0$, whence

$V_S = \dfrac{-2 \pm \sqrt{2^2 - 4(-4)\,(1.5)}}{2\,(1.5)} = \dfrac{-2 \pm \sqrt{4 + 24}}{3} = \dfrac{-2 \pm 5.29}{3} = -2.43$ or $+1.10$ (too small). Thus $V_S = -2.43$V, and $V_D =$ **+2.43V**, with a corresponding change of $\dfrac{2.43 - 2.00}{2.00} =$ **21.5%**.

5.17 For Fig. 5.25, $V_{DD} = 10$V, $V_{DS} = V_{GS}$, and operation is in saturation. Thus $V_D = V_{DD} - I_D\,R$, or $V_D = 10 - 0.4R$. Also $I_D = K\,(V_{GS} - V_t)^2$, where $K = 1/2\mu_n C_{ox} \times W/L = ½ \times 20 \times 40 = 0.4\text{mA}/V^2$, and $V_{GS} = V_D$. Thus $0.4 = 0.4\,(V_D - 1)^2$, or $V_D - 1 = 1$, or $V_D = 2$V. Thus $R = \dfrac{10-2}{0.4} =$ **20kΩ**.

5.18 For $V_D = 2$V, $I_D = \dfrac{5-2}{150k} = 20\mu$A.

Also $I_D = 1/2k'(W/L)\,(\upsilon_{GS} - V_t)^2 = 0.5 \times 10^{-3}\,(\upsilon_{GS} - 1)^2 = 20 \times 10^{-6}$, or $\upsilon_{GS} - 1 = (40 \times 10^{-3})^{½} = 0.2$, whence $\upsilon_{GS} = 1.2$V. Now, since $V_G = 5$V, $V_S = 5 - 1.2 = 3.8$V, see $R_S = \dfrac{3.8}{20\mu A} = 190$kΩ. Thus, to one significant digit, $R_S =$ **200kΩ**.

Now, $I_D = 500\ (\upsilon_{GS} - 1)^2$ and $I_D = \frac{5 - \upsilon_{GS} - 0}{0.2M\Omega}$, or $\frac{5 - \upsilon_{GS}}{0.2} = 500\ (\upsilon_{GS} - 1)^2$, or $5 - \upsilon_{GS} = 100\ \upsilon_{GS}^2 - 200\ \upsilon_{GS} + 100$, or $100\ \upsilon_{GS}^2 - 199\ \upsilon_{GS} + 95 = 0$, or $\upsilon_{GS}^2 - 1.99\ \upsilon_{GS} + 0.95 = 0$, whence $\upsilon_{GS} = \frac{1.99 \pm \sqrt{1.99^2 - 4(.95)}}{2} = \frac{1.99 \pm \sqrt{.1601}}{2} = \frac{1.99 \pm .4}{2} = 1.195\text{V}$, or < 1. Thus, $V_S = 5 - 1.195\text{V} = 3.805\text{V}$, $I_D = \frac{3.805}{0.2} = 19.025\mu\text{A}$, and $V_D = 5 - .15 \times 19.025 =$ **2.15V**.

5.19 $V_G = \frac{10M}{10M + 10M} \times 5\text{V} = 2.5\text{V}$. Now for $\upsilon_{GS} = \upsilon$, $i_D = \frac{5 - \upsilon - 2.5}{1k} = i$, and $i = K\ (\upsilon_{GS} - V_t) = 0.5\ (\upsilon - 1)^2$, $0.5\ (\upsilon - 1)^2 = \frac{2.5 - \upsilon}{1k}$, $(\upsilon - 1)^2 = 5 - 2\ \upsilon$, $\upsilon^2 - 2\upsilon + 1 = 5 - 2\upsilon$, $\upsilon^2 = 4$, $\upsilon = 2\text{V}$. Thus $V_S = V_G + \upsilon_{GS} = 2.5 + 2.0 = 4.5\text{V}$, and $i_D = \frac{5 - 4.5}{1k} = 0.5\text{mA}$, $V_D = 0 + 4\text{k}\ (0.5\text{mA}) =$ **2V**.

We actually find $\frac{90}{100}$ (2V) = 1.8V, in which case i_D is reduced to 90% or 0.45mA.

For $K = 1/2k'(W/L) = 0.5\text{mA}/V^2$ and V_t varying, $5 - 1\text{k}\Omega\ (0.45\text{mA}) - \upsilon_{GS} = 2.5\text{V}$, whence $\upsilon_{GS} = 5 - 0.45 - 2.5 = 2.05\text{V}$. Now, $0.45 = 0.5\ (2.05 - V_t)^2$, or $V_t - 2.05 = \pm\ (0.90)^{1/2} = +.949$, whence $V_t = 2.05 \pm .949 = 1.101\text{V}$. That is, V_t could have raised by **10.1%**.

For $V_t = 1\text{V}$ and K varying, again $\upsilon_{GS} = 2.05\text{V}$, but now, $0.45\text{mA} = K(2.05 - 1)^2$, or $K = \frac{0.45}{1.05^2} = 0.408\text{mA}/V^2$. Thus K could have dropped by $\frac{0.5 - .408}{0.5} =$ **18.4%**.

Note that the effect of V_t is essentially direct, a 10% change in current resulting from a 10% change in V_t. However, the change in current is only about 10/18.4 or about 54% of that in K, due to negative feedback included in the circuit. (See Chapter 8.)

5.20 For the Depletion Device, $K = 1/2\mu_p C_{ox}(W/L) = 1/2(8 \times 10^{-6})(500/2) = 0.1mA/V^2$ and $\upsilon_{GS} = 0$. Thus $i_D = I_{DSS} = K(\upsilon_{GS} - V_t)^2 = 1\text{mA}/V^2(0 - 2)^2 =$ **4mA**. Thus $V_S = 15 - 1\text{k}\Omega\ (4\text{mA}) =$ **11V** $= V_G$, and $V_D = 0 + 2\text{k}\Omega\ (4\text{mA}) =$ **8V**, whence $\upsilon_{SD} = 11 - 8 = 3\text{V} > V_t$. Thus the device operates in saturated mode.

Triode operation begins for $\upsilon_{SD} = V_t = 2\text{V}$, in which case $V_{SS} = 8 + 2 + 4 = 14\text{V}$, with operation being saturated for $\mathbf{V_{SS} \geq 14V}$.

5.21 Here $K = 1/2\mu_p C_{ox}(W/L) = 1/2(8)(250) = 1mA/V^2$. See $V_G = \frac{4}{4 + 1} \times 5 = 4\text{V}$. Now $\upsilon_{SG} = 5 - 1\text{k}\Omega\ (i_D) - 4 = 1 - i_D$, or $i_D = 1 - \upsilon_{SG} = 1 + \upsilon_{GS}$. Assuming saturation, $i_D = K\ (\upsilon_{GS} - V_t)^2 = 1\ (\upsilon_{GS} - 2)^2$, or $1 + \upsilon_{GS} = (\upsilon_{GS} - 2)^2 = \upsilon_{GS}^2 - 4\upsilon_{GS} + 4$, or $\upsilon_{GS}^2 - 5\upsilon_{GS} + 3 = 0$, whence $\upsilon_{GS} = \frac{5 \pm \sqrt{5^2 - 4(3)}}{2} = \frac{5 \pm 3.61}{2}$, or $\upsilon_{GS} = 0.697\text{V}$ (or 4.305V (too large)). Now $V_S = V_G - \upsilon_{GS} = 4 - 0.697 = 3.30\text{V}$, and $V_D = 0 + \frac{5 - 3.30}{1} \times 1 = 1.70\text{V}$. Thus $V_{DS} = 1.70 - 3.30 =$ **−1.60V**, and $V_{GD} = 4 - 1.70 = 2.30\text{V} > V_t$. Thus, operation is in saturation.

5.22 For operation at $i_D = 150\mu A$ in saturation, $i_D = (1/2)\mu_n C_{ox}(W/L)(\upsilon_{GS} - V_t)^2$ or $150 \times 10^{-6} = 0.5(20 \times 10^{-6})(30/2)(\upsilon_{GS} - 1)^2$, or $\upsilon_{GS} - 1 = 1^{1/2}$, whence $\upsilon_{GS} = 2$ V.

Now $R_1 = \frac{5 - 2}{150 \times 10^{-6}} = 20\text{k}\Omega$. If $R_2 = R_1 = 20k\Omega$, then $\upsilon_{D2} =$ **+ 2 V** (also).

If M_3 is joined to M_2 with corresponding elements connected, then the current in R_2 will tend to double. If R_2 is reduced to $10\ \text{k}\Omega$, $\upsilon_{DS2} = \upsilon_{DS1} = 2$ V

SECTION 5.5: THE MOSFET AS AN AMPLIFIER

5.23 Here $i_D = 1/2k'(W/L)(\upsilon_{GS} - V_t)^2 = 0.5(2 \times 10^{-3})(5 - 2)^2 =$ **9 mA**, and $\upsilon_D = V_{DD} - i_D R_D = 12 - 0.5 \times 10^3(9 \times 10^{-3}) =$ **7.5 V**.

For $\upsilon_{GS} = 5 + 0.5 = 5.5$ V, $i_D = 1(5.5 - 2)^2 =$ **12.25 mA**, and for $\upsilon_{GS} = 5 - 0.5 = 4.5$ V, $i_D = 1(4.5 -)^2 =$ **6.25 mA**.

That is, current reduces by $9 - 6.25 = 2.75$ mA, or increases by $12.25 - 9 = 3.25$ mA.

Total variation in drain current is $2.75 + 3.25 =$ **6.0 mA**.

Thus $\upsilon_d = i_d R_D = 6.0mA \times 0.5k\Omega =$ **3.0 V**. [Note that the gain is $-3/(2 \times 0.5) = -3.0$ V/V.]

5.24 Here, from P5.23 above, $K = 1/2k'(W/L) = 1/2(2) = 1mA/V^2$.

For $\upsilon_{gs} = \pm 0.5$V, and $V_{GS} = 5$ V, the largest value of υ_{GS} is $5 + 0.5 = 5.5$ V for which $i_D = 12.25$ mA. For saturation, the smallest value of υ_{DS} is $\upsilon_{GS} - V_t = 5.5 - 2 = 3.5$ V. Thus the largest value of R_L that can be used is $R_L = (12 - 35)/12.25 = 694\Omega$.

Now, for a 1 kΩ load resistor, and $\upsilon_{GS} = 5.5$ V, operation is in the triode mode where $i_D = K(2(\upsilon_{GS} - V_t)\upsilon_{DS} - \upsilon_{DS}^2)$, and $i_D = (V_{DD} - \upsilon_{DS})/R_L$. Now with $\upsilon_{DS} = \upsilon$, for simplicity, $(12 - \upsilon)/1 = 1(2(5.5 - 2)\upsilon - \upsilon^2)$, or $12 - \upsilon = 7\upsilon - \upsilon^2$, or $\upsilon^2 - 8\upsilon + 12 = 0$.

Thus $\upsilon = (- \; - 8 \pm \sqrt{8^2 - 4(12)})/2 = (8 \pm \sqrt{64 - 48}/2 = (8 \pm 4)/2 = 2$ V. Thus $\upsilon_{DS} =$ **2 V**.

Now for zero signal, $\upsilon_{GS} = 5$ V, $i_D = 9$ mA, and $\upsilon_{DS} = 12 - 9/1 = 3$ V, with operation just at the edge of saturation. Correspondingly, the negative output swing for a + 0.5 V input is $3 - 2 = 1$ V.

For a − 0.5 V signal, $\upsilon_{GS} = 4.5$ V, $i_D = 6.25$ mA and $\upsilon_{DS} = 12 - 6.25(1) =$ **5.75 V**, with output swing for − 0.5 V input being $5.75 - 3 =$ **2.75 V**. Thus the ratio of peak voltage outputs is 2.75 to 1 or **2.75 V/V**.

5.25 $K = 1/2k'(W/L) = 1$mA/V^2, $V_t = 2$V, $V_{DD} = 12$V, $R_L = 0.5$kΩ, $\upsilon_{GS} = 5$V ± 0.5V. Thus $i_D = K(\upsilon_{GS} - V_t)^2 = 1(5 - 2)^2 =$ **9mA**, and $\upsilon_D = V_{DD} - R_D i_D = 12 - 0.5(9) =$ **7.5V**. See, from Eq. 5.44, that $g_m = 2\sqrt{K}\sqrt{I_D} = 2\sqrt{1}\sqrt{9} =$ **6mA/V**. From Eq 5.43, $g_m = 2K(V_{GS} - V_t) = 2(1)(5 - 2) =$ **6mA/V**, in correspondence. From Eq 5.40, $\upsilon_d/\upsilon_{gs} = -g_m R_D = -6$mA/V (0.5kΩ) = **−3.0V/V**. For a ±0.5V input, expect a ±0.5(−3.0) = **±1.5V** output signal. From Eq 5.35, $i_D = K[(V_{GS} - V_t)^2 + 2K(V_{GS} - V_t)\upsilon_{gs} + K\upsilon_{gs}^2]$. Thus for $\upsilon_{gs} = \pm 0.5$V, $i_D = 1(5 - 2)^2 + 2(1)(5 - 2)0.5 + (1)0.5^2 = 9 + 3 + 0.25 = 12.25$mA, for which $\upsilon_D = 12 - 0.5(12.25) =$ **+5.875V**, and $i_D = 1(5 - 2)^2 + 2(1)(5 - 2)(-0.5) + (1)(-0.5)^2 = 9 - 3 + 0.25 = 6.25$mA, for which $\upsilon_D = 12 - 0.5(6.25) =$ **8.875V**. This is to be contrasted with 7.5 − 1.5 = **6.0V**, and 7.5 + 1.5 = **9.0V**, as calculated from a linearized model.

5.26 $K_p = \frac{1}{2}\mu_p C_{ox}\frac{W}{L} = \frac{1}{2}(10)(300/3) = 500\mu A/V^2$. $g_m = 2\sqrt{K}\sqrt{I_D}$ from Eq 5.4 or $g_m = 2\sqrt{0.500}\sqrt{4} = 2\sqrt{2} =$ **2.83mA/V**. Generally, gain $= -g_m R_L = -10$V/V implies that $R_L = \frac{10}{g_m} = \frac{10}{2.83} =$ **3.53kΩ**. Operation is reasonably linear for $\upsilon_{gs} << 2(V_{GS} - V_t)$, but $I_D = K(V_{GS} - V_t)^2 \rightarrow (V_{GS} - V_t) = \sqrt{I_D/K}$. Thus, linear for $\upsilon_{gs} << 2(V_{GS} - V_t) = 2\sqrt{I_D/K}$, that is $\upsilon_{gs} << 2\sqrt{4/0.5} = 2\sqrt{8} =$ **5.66V**. For 1% nonlinearity, $\upsilon_{gs} \approx$ **0.06V** peak. For 10% nonlinearity $\upsilon_{gs} \approx$ **0.6V** peak.

5.27 Gain $= -g_m(R_L \parallel r_o)$, where $g_m = 2\sqrt{K I_D}$, and $r_o = \frac{V_A}{I_D}$. Now, at 1mA, $9.091 = 2\sqrt{K(1)}\left[10k \parallel \frac{V_A}{1}\right]$ --- (1), and at 0.25mA, $4.808 = 2\sqrt{K/4}\left[10k \parallel \frac{V_A}{1/4}\right]$ --- (2). Thus (1)/(2)

$$\equiv \frac{9.091}{4.808} = \frac{2\,\dfrac{10\,V_A}{10 + V_A}}{\dfrac{10\,V_A\,(4)}{10 + 4\,V_A}} = \frac{1}{2}\,\frac{4\,V_A + 10}{V_A + 10}.$$

whence $9.091V_A + 90.91 = 9.616V_A + 24.040$, and

$V_A = \dfrac{90.91 - 24.04}{9.616 - 9.091}$ = **127V**. From (1): $9.091 = 2\sqrt{K}\,(10 \| 127)$, whence $K = \left[\dfrac{9.091}{2}\right]^2 \dfrac{1}{(10 \| 127)^2}$ = **0.240mA/V²**. **Check:** At 0.25mA, gain $= 2\sqrt{.24/4}\left[10 \| \dfrac{127}{1/4}\right] \approx 2\sqrt{.06} \times \dfrac{10 \times 508}{10 + 508} = 4.804$ V/V → OK.

For an output distortion of 10%, as stated, from Eq 5.35, $K\, v_{gs}^2 = \dfrac{1}{10}\, 2K\, (V_{GS} - V_t)\, v_{gs}$, or $(V_{GS} - V_t) = \dfrac{10\, v_{gs}}{2} = 5 v_{gs} = 5\,(0.5) = \mathbf{2.5V}$.

5.28 *Generally:* $K = 1/2 k'(W/L) = 2/2 = 1 mA/V^2$. $I_D = I$ **mA**, $I = K\,(v_{GS} - V_t)^2 = 1\,(V_D - 1)^2$. $V_D = (\sqrt{I} + 1)$**V**, $g_m = 2K\,(v_{GS} - V_t) = 2\,(V_D - 1) = 2\sqrt{I}$ **mA/V**, $r_o = V_A/I = 50/I$ **kΩ**, whence $\dfrac{v_o}{v_i} = -g_m\, R_L \| R_G \| r_o \approx -g_m R_L = -2R\sqrt{I}$ **V/V**, and $R_i = \dfrac{R_G}{1 - gain} = \dfrac{10^4}{1 + 2R\sqrt{I}}$ **kΩ**. *For* I = 1mA and $R_L = R_G$: $v_o/v_i = -2\sqrt{I}\,(R_G \| R_G \| r_o) = -2 \times (10^4 \| 10^4 \| 50) \approx$ **−100V/V**, $R_i \approx \dfrac{10^4}{1 + 100}$ = **99kΩ**. For $R_L = r_o$: $v_o/v_i = -2\,(10^4 \| 50 \| 50) \approx$ **−50V/V**, $R_i = \dfrac{10^4}{1 + 50}$ = **196kΩ**. For $R_L = R_i$: $v_o/v_i = -2\,(10^4 \| 50 \| R_i) \approx -2\,(50 \| R_i)$, $R_i \approx \dfrac{10^4}{1 + 2\,(50 \| R_i)}$, or

$R_i\left[1 + 2\dfrac{(50)R_i}{50 + R_i}\right] = 10^4$, $R_i + \dfrac{100\, R_i^2}{R_i + 50} = 10^4$, $R_i^2 + 50\, R_i + 100\, R_i^2 = 10^4\, R_i + 50 \times 10^4$, $101 R_i^2 - 9950\, R_i - 50 \times 10^4 = 0$, $R_i^2 - 98.5\, R_i - 4950 = 0$, whence $R_i = \dfrac{98.5 \pm \sqrt{98.5^2 - 4\,(-4950)}}{2} = \dfrac{98.5 \pm 171.8}{2}$ = **135kΩ**, and $v_o/v_i \approx -2\,(50 \| 135)$ = **−73.0V/V**.

5.29 Note that while the lower end of r_o is not actually grounded, the signal there is small. Assume it to be zero. For R_S = 1kΩ, gain $\dfrac{v_o}{v_i} = -\dfrac{R_C \| R_L \| r_o}{1/g_m + R_S} = -\dfrac{10 \| 10 \| 100}{1/1 + 1} = \dfrac{-4.76}{2}$ = **−2.38V/V**. For R_S = 0Ω, gain $\dfrac{v_o}{v_i} = \dfrac{-4.76}{1}$ = **−4.76V/V**. For R_S = 3.76kΩ, gain $\dfrac{v_o}{v_i} = \dfrac{-4.76}{1 + 3.76}$ = **−1V/V**.

5.30 For the T-model, the equivalent resistor in the source is $r_s = 1/g_m = 1/0.725 mA/V$ = 1.38 kΩ. The output resistance of the follower (with body effect ignored) is $R_{out} = r_o \| r_s = 47 \| 1.38$ = 1.34kΩ.

Thus, the no-load gain is $G_0 = \dfrac{47}{1.38 + 47}$ = **0.971 V/V**.

For load R_L, the gain is $G = (R_L/(R_L + R_{out})) \times G_0$ for which $GR_L + GR_{out} = G_0 R_L$, and $R_L = GR_{out}/(G_o - G)$.

For G = 0.95 V/V, $R_L = 0.99(1.34)/(0.971 = 0.95)$ = 63kΩ.

For G = 0.90 V/V, $R_L = 0.90(1.34)/(0.971 - 0.90)$ = 17.0kΩ.

5.31 For the situation in which I_S is fixed, the g_m of the transistor is independent of V_t: This follows from the fact that: $i_D = 1/2 k'(W/L)(v_{GS} - V_t)^2$, $g_m = \partial i_D/\partial v_{GS}$, or

$g_m = 1/2(2)k'(W/L)(v_{GS} = V_t) = k'(W/L)(2i_D/k'(W/L))^{1/2}$. Thus $g_m = (2k'(W/L)i_D)^{1/2}$, depending only on $i_D = i_S$.

Now, for χ = 0.2, g_{mb} = 0.2 (0.725) = 0.145 mA/V. Thus the additional load on the source is $1/g_{mb} = 1/0.145$ = 6.90 kΩ.

Now the follower output resistance is $1.34k\Omega \parallel 6.90k\Omega =$ 1.12 kΩ amd the no-load gain is $G_o = \frac{6.90 \parallel 47}{1.38 + 6.90 \parallel 47} =$ **0.814 V/V.**

For a gain of 50%, $0.5 = 0.814 \times R_L/(R_L + 1.12)$ or $0.50R_L + 0.56 = 0.814R_L$, or $0.314R_L = 0.56$ or $R_L =$ **1.78 kΩ**

SECTION 5.6: BIASING IN MOS AMPLIFIER CIRCUITS

5.32

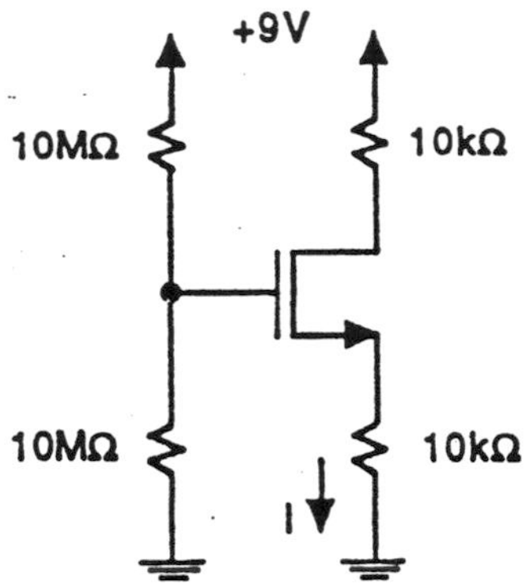

For this device, $V_t = 2V$, $K = 1/2k'(W/L) = 0.5mA/V^2$. Assume saturation: *For* $V_t = 2V$, $V_G = 1/2\ (9) = 4.5V$, $I = K(v_{GS} - V_t)^2 = 0.5\ (v_{GS} - 2)^2$ --- (1), and $v_{GS} = 4.5 - 10\ I$ --- (2). Substitute (2) in (1) → $2\ I = (4.5 - 10\ I - 2)^2 = (2.5 - 10\ I)^2 = 6.25 - 50\ I + 100\ I^2$, or $100\ I^2 - 52\ I + 6.25 = 0$,

whence $I = \frac{52 \pm \sqrt{52^2 - 4\ (100)\ (6.25)}}{2\ (100)} =$ **0.189mA**, (or too large a value). $v_S = 10k\ (0.189) = 1.89V$, $v_{GS} = 4.5 - 1.89 = 2.61V$, $v_D = 9 - (10K\Omega)\ (0.189mA) = 9 - 1.89 = 7.11V$, and $v_{DS} = 7.11 - 1.89 =$ **5.22V** → OK, saturation. Operation remains in saturation until $v_{GD} \geq V_t = 2V$, ie, for $v_{GS} = 2.61V$, and $v_{DS} \geq 2.61 - 2 = 0.61V$. Thus the peak negative-going output signal allowed is $7.11 - (1.89 + 0.61) =$ **4.61V.** But note that the largest positive-going output signal (for cutoff) is 1.89V.

Now, *For* $V_t = 1V$, $I = 0.5\ (v_{GS} - 1)^2$, and $v_{GS} = 4.5 - 10I$. $2I = (4.5 - 10\ I - 1)^2 = (3.5 - 10I)^2 = 12.25 - 70I + 100I^2$, and $100I^2 - 72I + 12.25 = 0$.

Thus $I = \frac{72 \pm \sqrt{72^2 - 4\ (100)\ (12.25)}}{200} =$ **0.276mA.** $v_S = 10\ (.276) = 2.76V$, $v_{GS} = 4.5 - 2.76 = 1.74V$, $v_D = 9 - 2.76 = 6.24V$, $v_{DS} = 6.24 - 2.76 =$ **3.48V**. Now, saturation prevails while $v_{GS} \leq 1.0V$. Thus the maximum negative swing is $6.24 - 4.5 + 1 =$ **2.74V**. The largest positive-going output signal (for cutoff) is $9 - 6.24 = 2.76V$.

5.33

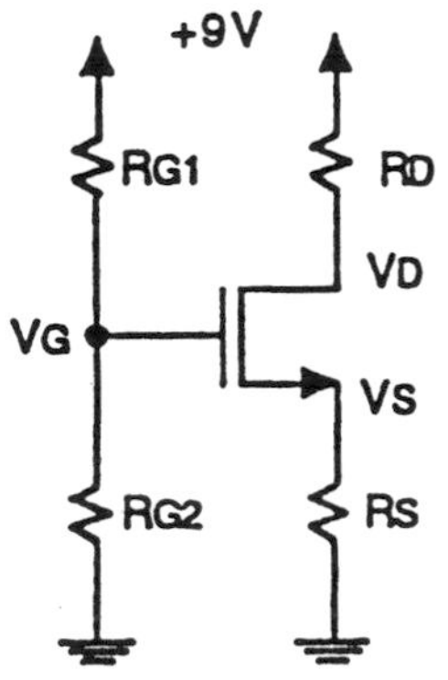

The design is required to endure the following variations: V_t from 1 to 2V, K from 0.3 to 0.5mA/V^2, and I_D from 0.5 to 1mA: Largest current occurs when V_t smallest (1V) and K largest (0.5mA/V^2). Thus $1 = 0.5\ (v_{GS} - 1)^2$, or $v_{GS} = \pm\sqrt{2} + 1 = 2.414V$ --- (1). Smallest current when V_t largest (2V) and K smallest (0.3mA/V^2). Thus $0.5 = 0.3\ (v_{GS} - 2)^2$, and $v_{GS} = \sqrt{1.67} + 2 = 1.29 + 2 = 3.29V$ --- (2). From (1), $\frac{V_{GG} - 2.414}{R} = 1mA$, where $R = R_S$. From (2), $\frac{V_{GG} - 3.29}{R} = 0.5mA$. See $V_{GG} - 2.414 = R$, and $2\ V_{GG} - 6.58 = R$. Subtracting, $-V_{GG} + 4.166 = 0 \rightarrow V_{GG} = 4.166V$. For $V_{GG} = 4.166V$, $R_{G1} =$ **10M Ω**, $R_{G2} = 4.166/(\frac{9 - 4.166}{10}) =$ **8.6MΩ** (Use 8.2 MΩ), and $R_S = R = 4.166 - 2.414 =$ **1.75kΩ** (Use 1.8 kΩ).

Check: $\frac{4.166 - 3.29}{1.75} = 0.507mA$. Now for $I_D = 1mA$, $V_t = 1V$, and a 0.5V signal, $v_D > V_{GG} - V_t$. That is, $v_D > 4.166 - 1.0 = 3.166$, and $R_D = \frac{9 - (3.166 + 0.5)}{1mA} = 5.33k\Omega$. (Use **5.1kΩ**).

5.34

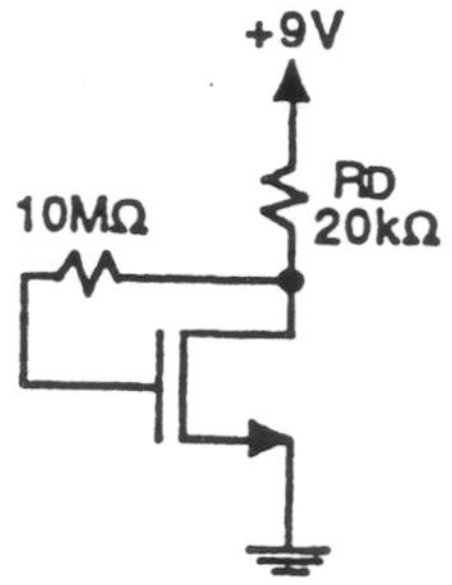

For V_t = 2V, and $k'(W/L) = 1.0mA/V^2$, $i_D = 1/2k'(W/L)(v_{GS} - V_t)^2$, $= 1/2(1)(v_{GS} = 2)^2 = 0.5\,(v_{GS} - 2)^2$, and $i_D = \frac{9 - v_{GS}}{20k\Omega}$. Thus $9 - v_{GS} = 20\,(0.5)\,(v_{GS} - 2)^2$, $9 - v_{GS} = 10\,v_{GS}^2 - 40\,v_{GS} + 40$, $10\,v_{GS}^2 - 39\,v_{GS} + 31 = 0$, whence $v_{GS} = \frac{39 \pm \sqrt{39^2 - 4\,(10)\,(31)}}{2\,(10)}$ = 2.79V. v_{DS} = **2.79V**, and $I_D = \frac{9 - 2.79}{20}$ = **0.311mA.** For negative peak outputs of up to **2Vp**, operation remains in saturation. *For* V_t = 1V, $i_D = 0.5\,(v_{GS} - 1)^2$, and $i_D = \frac{9 - v_{GS}}{20}$.

$9 - v_{GS} = 10\,(v_{GS} - 1)^2 = 10\,v_{GS}^2 - 20\,v_{GS} + 10$, and $10\,v_{GS}^2 - 19\,v_{GS} + 1 = 0$, whence $v_{GS} = \frac{19 \pm \sqrt{19^2 - 4\,(10)\,(1)}}{2\,(10)} = \frac{19 \pm 17.9}{20}$ = 1.845V. Thus v_{DS} = **1.85V**, $I_D = \frac{9 - 1.85}{20}$ = **0.358mA,** with a **1V peak** signal allowed.

5.35 For R_{G2} = 10MΩ from gate to source, $v_{DS} = 2\,v_{GS}$. *For* V_t = 2V, $i_D = 1/2(1.0)(v_{GS} - 2)^2$ and $i_D = \frac{9 - 2\,v_{GS}}{20}$. Thus $9 - 2\,v_{GS} = 10\,(v_{GS} - 2)^2 = 10\,v_{GS}^2 - 40\,v_{GS} + 40$, and $10\,v_{GS}^2 - 38\,v_{GS} + 31$ = 0, whence $v_{GS} = \frac{38 \pm \sqrt{38^2 - 4\,(10)\,(31)}}{2\,(10)} = \frac{38 \pm 14.28}{20}$ = 2.61V.

Thus v_{DS} = **5.22V,** and $I_D = \frac{9 - 5.22}{20}$ = **0.189mA.** For negative peak outputs, 5.22 –2.61 + 2 = **4.61V** is allowed for operation in saturation.

For V_t = 1V, $i_D = 0.5\,(v_{GS} - 1)^2$ and $i_D = \frac{9 - 2\,v_{GS}}{20}$. Thus $9 - 2\,v_{GS} = 10\,(v_{GS} - 1)^2 = 10\,v_{GS}^2 - 20\,v_{GS} + 10$, and $10\,v_{GS}^2 - 18\,v_{GS} + 1 = 0$, whence $v_{GS} = \frac{18 \pm \sqrt{18^2 - 40}}{20} = \frac{18 \pm 16.85}{20}$ = 1.74V. Thus v_{DS} = **3.48V** and $I_D = \frac{9 - 3.48}{20}$ = **0.276mA,** with 3.48 –1.74 + 2 = **3.74V** negative output peaks allowed, while saturated operation prevails.

5.36

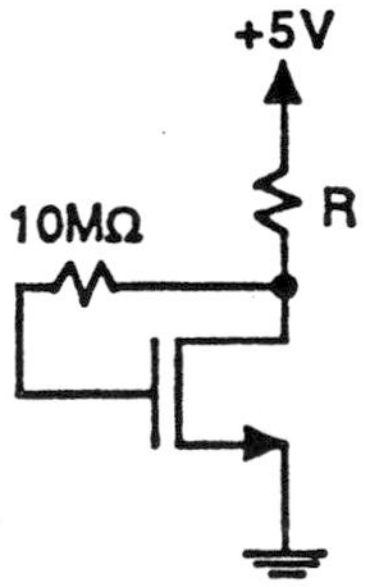

Here, V_t varies from 1 to 2V, $K = 1/2k'(W/L)$ varies from 0.3 to 0.5mA/V^2, I_D varies from 0.5 to 1mA. The largest current (1mA) occurs for the smallest V_t (1V), and largest K (0.5mA/V^2). Thus $1 = 0.5\,(v_{GS} - 1)^2$. v_{GS} = 2.414V, and $R \geq \frac{5 - 2.414}{1mA} \geq$ 2.59kΩ. The smallest current (0.5mA) occurs for the largest V_t (2V) and smallest K (0.3mA/V^2). Thus $0.5 = 0.3\,(v_{GS} - 2)^2$. $v_{GS} = \sqrt{1.67} + 2$ = 3.29V, and $R \leq \frac{5 - 3.29}{0.5} \leq$ 3.42kΩ. Use $R = \frac{2.59 + 3.42}{2}$ = **3.0kΩ.** Because of feedback, the effect of variation is reduced.

The circuit automatically allows a negative signal = V_t > 1V but the gain is smaller than in P 5.33, since R here (3.0kΩ) is less than R_D there (5.2kΩ). Raising R to (say) **3.3kΩ** would be allowed here, and would improve the gain by 10%.

5.37 In P5.36 above, the minimum negative-going signal is 1V. Here it should be 1.5V. That is, we want $v_{DS} = v_{GS}$ + 0.5V for the case in which V_t = 1V.

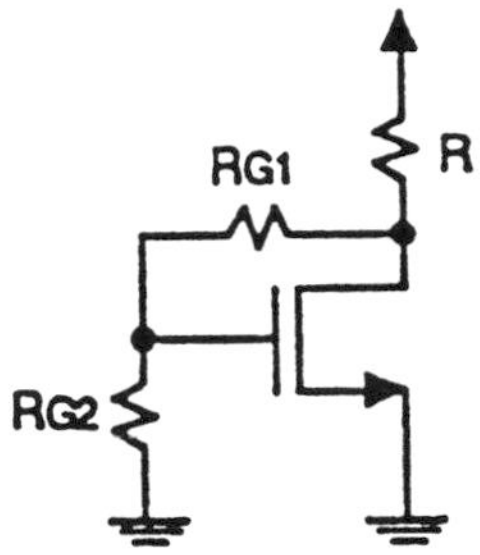

For smallest V_t (and also the largest current), from the results in P5.36, υ_{GS} = 2.414V, and υ_{DS} = 2.414 + 0.5 = 2.914, with $R > \frac{5 - 2.914}{1mA}$ = 2.09kΩ. If we use **R_{G2} = 10MΩ**, $R_{G1} = \frac{10}{2.414}$ (0.5) = 2.07MΩ. Use **R_{G1} = 2.0MΩ**. Now for V_t = 2V (and also the smallest current), from the solution for P5.36 we see, υ_{GS} = 3.29V, and $\upsilon_{DS} = \frac{3.29}{10} \times 2.0 + 3.29 = 3.95$V. Thus $R < \frac{5 - 3.95}{0.5} \leq 2.1$kΩ.

Notice that a solution *barely exists*, using R = **2.1kΩ**, essentially as a consequence of a demand for large signal swings with a limited supply voltage.

5.38

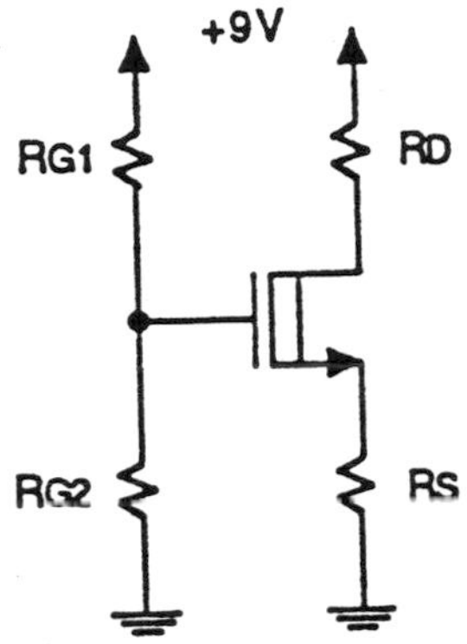

Here, V_p = –4V, I_{DSS} = 32mA. For I_D = 8mA, $8 = 32\,(1 - \frac{\upsilon_{GS}}{V_p})^2$. Thus $(1 - \frac{\upsilon_{GS}}{V_p})^2 = 1/4$, $-\frac{\upsilon_{GS}}{V_p} = \pm\, 1/2 - 1 = -1.5$ or –0.5, and υ_{GS} = $0.5V_p$ = –2V. Now for a negative swing of 2V to the edge of saturation, $\upsilon_D \geq \upsilon_G + |\,V_t\,|$. Now for the largest possible value of R_D, υ_D will be lowest and υ_G lowest. The lowest possible υ_G is 0V, with $R_{G1} = \infty$ and R_{G2} = **10MΩ**, in which case R_S = 2V/8mA = **0.25kΩ**, and $V_D \geq 0 + 4 = 4$V for lowest swing or-4 + 2 = 6V for no signal, with $R_D \leq$ 9–6/8 = **0.375kΩ**. Note that for a 2V positive output swing, υ_D rises to 6 + 2V = 8V, and the transistor is not yet cut off. OK. Note that this design with no biassing supply is relatively sensitive to device variability, all as a result of wanting a large signal swing with a small supply.

5.39 Here, $i_D = 1/2\mu_n C_{ox}(W/L)(\upsilon_{GS} - V_t)^2 = 1/2 \times 20 \times 10^{-6}(4/2)(2.5 - 1)^2 = 45\mu A$.

Thus, assuming the Early effect to be negligible, Q_2 operates at **45 μA**. For a 5 V supply $R = (5 - 2.5)/(45 \times 10^{-6})$ = **55.6 kΩ.**

Transistor Q_2 remains in saturation for $\upsilon_O > (\upsilon_{GS} - V_t) = 2.5 - 1 =$ 1.5 V or $\upsilon_O \geq$ **1.5V.**

In triode mode, $i_D = \mu_n C_{ox}(W/L)[(\upsilon_{GS} - V_t)\upsilon_{DS} - \upsilon_{DS}^2/2]$. Now for i_D reduced to half, for $\upsilon_{DS} = \upsilon$, $45/2 = 20(4/2)[(2.5 - 1)\upsilon - \upsilon^2/2]$, or $0.5625 = 1.5\upsilon - \upsilon^2/2$, or $\upsilon^2 - 3\upsilon + 1.125 = 0$, whence

$\upsilon = (-\ -3 \pm \sqrt{3^2 - 4(1.125)})/2 = (3 \pm 2.121)/2 =$ 0.439 V.

Thus the current reduces to 1/2 normal for $\upsilon_{DS} = \upsilon_O$ = **0.439 V.**

Check: $i_D = 20 \times 2[(2.5 - 1)(.439) - .439^2/2]$ = **22.5 μA.**

5.40 For a 1 MΩ output resistance at 100 μA output, $V_A = 10^6 \times 100 \times 10^{-6} =$ 100 V. This requires that the output transistor have a channel length $L = 100V/10V/\mu m$ = **10μm**. Use this for both transistors. For the edge of saturation at υ_O = 0.5 V, $\upsilon_{GS} = 0.5 + V_t =$ 1.5 V. Now for Q_1, $25 = 20(W/10)(1.5 - 1)^2$, or $W_1 = 25 \times 10/(10(0.5)^2)$ = **100μm.**

Correspondingly, $W_2 = (100/25)W_1 = 4(100)$ = **400μm.** The output current will be exactly 100 μA when Q_1 and Q_2 operate identically, with $\upsilon_{DS} = \upsilon_{GS}$ = 1.5 V and υ_O is **1.5 V above** the negative supply. For operation at υ_O = 5 V above the negative supply,

$i_O = 100 \times 10^{-6} + (5 - 1.5)/10^6$ = **103.5μA.**

5.41

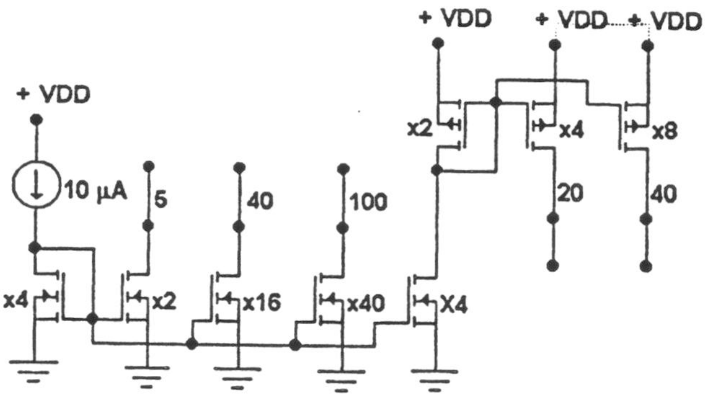

S5-41(a)

Topology A: This design is relatively straightforward, except that it p-channel devices operating at twice the current density of the n-channel. Widths are as indicated: **8** transistors are needed. The total width of the NMOS is $\Sigma W_n = 4 + 2 + 16 + 40 + 4 =$ **66μm**. The PMOS width is $\Sigma W_p + 2 + 4 + 8 =$ **14μm**. If the same current density is used in the PMOS, the total PMOS width would be double, namely 2(14) = **28μm**. Notice, incidentally, that neither of these designs, nor the others to follow, compensate for the μ_n/μ_p ratio r. For such compnesation, multiply all the PMOS widths by r.

Topology B: If both ends of $I_{REF}(= 10\mu A)$ are available:

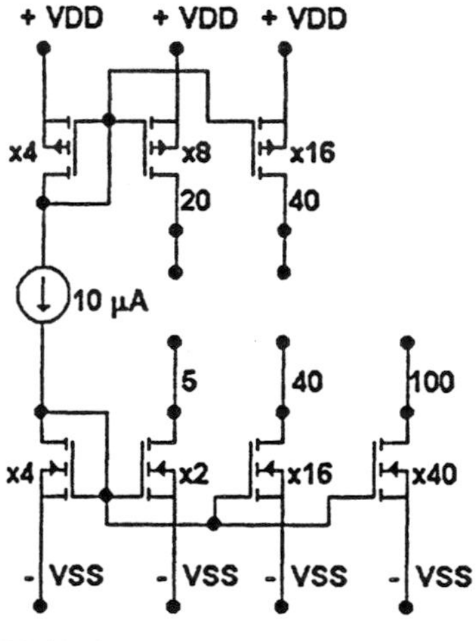

S5-41(b)

This is a design for equal current densities in the PMOS and NMOS transistors: $\#T =$ **7**, $\Sigma W_n =$ **62 μm**, $\Sigma W_p =$ **28 μm**.

Topology C: The attempt here is to create the 5 μA output separately without making all the other transistors twice as large.

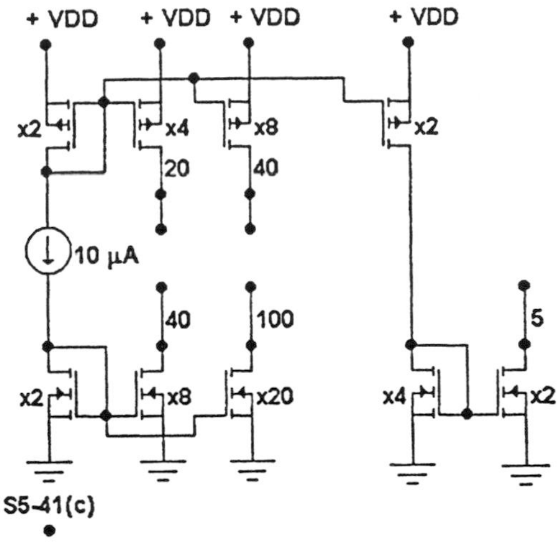

This uses **9** transistors but it is smaller: ΣW_n = **36 μm** and ΣW_p = **16 μm.**

SECTION 5.7: BASIC CONFIGURATIONS OF SINGLE-STAGE IC MOS AMPLIFIERS

5.42 *For Fig. 5.44a):*

Input: Operates at υ_{GS} = **2 V.**

Output: Source saturates at 5 – 0.5 = **4.5 V**; amplifier saturates at $\upsilon_{GS} - V_t = 2 - 1 =$ **1 V.**

Overall: Input: **2 V.** Output: **1 V to 4.5 V.**

For Fig. 5.44b):

Input: Operates at **– 2 V.**

Output: Operates from **+ 4.5 V** to **– 1 V** where the drain falls V_t below the gate.

Overall: Input **– 2 V** . Output: **– 1 V to + 4.5 V.**

For Fig. 5.44c):

Input: Saturation at $\upsilon_I = 5 + 1 =$ **6 V.**

Output: Sink saturates at – 5 + 0.5 = **– 4.5 V.** Driver saturates at $5 + V_t - V_{GS}$ = **4 V.**

Input: For output at – 4.5 V, input is limited to – 4.5 + 2.0 = **– 2.5 V.**

Overall: Input: **– 2.5 V to + 6 V**; Output: **– 4.5 V to 4 V.**

5.43 Eq 5.101 indicates $A_\upsilon = -\dfrac{\sqrt{K}\,|V_A|}{\sqrt{I_{REF}}}$. Let $I = I_{REF}$. For I = 25μA: $A_\upsilon = -\dfrac{\sqrt{10}\,|100|}{\sqrt{25}}$ = **–63.2V/V.**
For I = 2.5μA: $A_\upsilon = -\dfrac{\sqrt{10}\,(100)}{\sqrt{2.5}}$ = **– 200V/V.** For I = 0.25μA: $A_\upsilon = -\dfrac{\sqrt{10}\,(100)}{\sqrt{0.25}}$ = **–632V/V.**

5.44 For the diode-connected NMOS (call it Q_4) of half the width of Q_1, its width is also half that of Q_3. Thus, Q_4, Q_3 have the same k, and the same V_t. Thus $\upsilon_{SG3} = \upsilon_{SG2} = 5/2 = 2.5$V.

$I_{REF} = i = (1/2) \times 20 \times \dfrac{100/2}{10}(2.5 - 1)^2 = 112.5\mu A$, with a total supply current = 2 (112.5) = **225μA.**
For Q_1, i = 112.5 = $(1/2) \times 40 \times 100/10\ (\upsilon_{GS} - 1)^2 = 200(\upsilon_{GS} - 1)^2$. Thus $\upsilon_O = \upsilon_{GS} = 1 +$

$(112.5/200)^{1/2}$ = **1.75V**, and $g_{m1} = k(v_{GS} - V_t) = 2\ (20)\ (100/10)\ (1.75 - 1) = 300\mu A/V$. Thus the gain $v_o/v_i = -g_m\ (r_{01} \parallel r_{02})$, where $r_{01} = r_{02} = r_o = 50/112.5 = 0.444M\Omega$. Thus, $v_o/v_i = -300 \times 0.444/2 =$ **−66.7V/V**. Max positive output = (5 − 2.5 + 1) = **3.5V**. Max negative output = 1.75 − 1 = **0.75V**.

5.45 Current in $Q = I_{REF} = 50\ \mu A$.

Since $i_D = 1/2\mu_n C_{ox}(W/L)(v_{GS} - V_t)^2$. $50 = 1/2(20)(100/10)(v_{GS} - 1)^2$, $v_{GS} - 1 = 0.5^{1/2}$, whence v_{GS} = 1.707 V. Now for average input of 0 V, V_{BIAS} must be **1.707 V**.

Since the average source voltage is zero, no change of threshold results, but a g_{mb} effect applies.

For Q_1, $g_{m1} = 2(1/2\mu_n C_{ox}(W/L)(v_{GS} - V_t) = 20(100/10)(1.707 - 1) = 141\ \mu/V$, and $g_{mb} = \chi g_m = 0.2(141) = 28.3\mu V$.

For Q_1, $r_{o1} = 50V/50\mu A = 1\ M\Omega$. For Q_2, $r_{o2} = 1\ M\Omega$ also. Thus the Voltage gain, v_o/v_i, (from Eq. 5.63) $v_o/v_i = +(141 + 28.3 + 1/1)(1 \parallel 1) =$ **+ 85.2 V/V**.

Input resistance at the source is $R_i = \dfrac{1}{g_{m1} + g_{mb1}}(1 + r_{o2}/r_{o1}) = \dfrac{1}{141 + 28.3}(1 + 1/1) =$ **11.8 kΩ**.

5.46 For input high, the saturation edge is reached when $v_{DS1} = v_{GS1} = V_t$, at $i_D = 50\ \mu A$.

Here, $50 \times 10^{-6} = 1/2(20 \times 10^{-6})(100/10)(v_{GS} - V_t)^2$, or $(v_{GS} - V_t)^2 = 50 \times 2/(20 \times 10) = 0.5$, or $(v_{GS} - V_t) = 0.707$.

Thus linear (saturation-mode) operation is available for v_O up to 5 − 0.707 = **4.29 V**, with v_I up to $4.29 + v_{GS} = 4.29 + 0.707 + V_t = V_{DD} + V_t = 5 + V_t$.

Now, because of the back-bias effect, (from Eq. 5.30),

$V_t = V_{t0} + \gamma[\sqrt{2\Phi_f + V_{SB}} - \sqrt{2\Phi_f}]$ where (from Eq. 5.51), $\gamma = 2\chi\sqrt{2\Phi_f + V_{SB}}$.

Thus $V_t = V_{t0} + 2\chi(2\Phi_f + V_{SB}) - 2\chi(2\Phi_f + V_{SB})^{1/2}(2\Phi_f)^{1/2}$.

Note in practice that the substrate of Q_1 must be connected to the − 5 V supply, to ensure that the channel-to-substrate junction is always reverse-biased. Thus, here, $V_{SB} = 4.68 + 5.0 = 9.68$ V.

Here, $V_t = 1 + 0.2(2)(0.6 + 9.68) - 0.2(2)(0.6 + 9.68)^{1/2}(0.6)^{1/2} = 1 + 4.112 - 0.993 =$ **4.12 V**!.

Thus highest v_I is 4.29 + 4.12 = **8.41 V** with the correspondingly highest v_O = 4.29 V.

For input low: Q_2 with $V_{SB} = 0$, saturates when $v_{DS2} = v_{GS2} - V_t$.

Here, $50 = 1/2(20)(100/10)(v_{GS2} - 1)^2$, or $(v_{GS2} - 1)^2 = 50 \times 2/(20 \times 10) = 0.5$, whence $v_{GS2} = 1.707$ V. Saturation occurs at $v_O = -5 + 1.707 - 1.0 =$ **− 4.29 V** at which time V_t, is nearly 1, since V_{SB} is 0.707 V. More exactly $V_t = 1 + 0.2(2)(0.6 + 0.707) - 0.2(2)(0.6 + 0.707)^{1/2}(0.6)^{1/2} = 1 + 0.523 - 0.354 =$ 1.17 V.

Thus for $v_O = -4.29$, $v_I = -4.29 + 1.17 =$ **− 3.12 V**.

For signal gain: For $i_D = K(v_{GS} - V_t)^2$,

$g_m = 2K(v_{GS} - V_t) = 2K(i_D/K)^{1/2} = 2(Ki_D)^{1/2} = (2k'(W/L)i_D)^{1/2} = (2(20)(100/10) \times 50)^{1/2} =$ **141.4μA/V**.

(Note that this is independent of the average signal level.)

Now, $g_{mb} = \chi g_m = 0.2(141.4) = 28.3\ \mu A/V$, and $r_{o1} = r_{o2} = V_A/i_D = 50V/50\mu A = 1\ M\Omega$.

Thus the "internal load" on the follower is $R = (1/g_{mb}) \parallel r_{o1} \parallel r_{o2} = (1/28.3) \parallel 0.5M\Omega = 0.0353 \parallel 0.5 =$ 35.1 kΩ. Thus, the "no-load" gain $v_o/v_i = R/(1/g_m + R) = 35.1 \times 10^3/(1/141.4 \times 10^{-6} + 35.1 \times 10^3)$, or $v_o/v_i =$ **0.832 V/V**, with output resistance $R_o = 35.1k\Omega \parallel (1/141.4 \times 10^{-6}) = 35.1 \parallel 7.07 =$ **5.88 kΩ**.

For a load R_L, the gain is reduced by a factor of 2 when $R_L = R_o =$ **5.88 kΩ**.

5.47 $R_{out} = 1/g_m \parallel r_o$, $r_o \propto 1/I$, $g_m \propto \sqrt{I}$. Originally $r_o \parallel 1/g_m = 0.952$. With four times the current, $(r_o/4) \parallel 1/(\sqrt{4}g_m = (r_o/4) \parallel (1/2g_m)) = 0.455$. Now, $\dfrac{r_o/g_m}{r_o + 1/g_m} = 0.952$, $\dfrac{r_o}{1 + g_m r_o} = 0.952$, or $r_o = 0.952 + 0.952 g_m r_o$ − − − (1). Also $\dfrac{(r_o/4)(1/2g_m)}{r_o/4 + 2/(2g_m)} = 0.455 = \dfrac{r_o}{4 + 2g_m r_o}$, or $r_o = 1.82 + .91 g_m r_o$ − − − (2). From (1), $20.62 = .952 + .952(20.62)g_m$, whence g_m = **1.002mA/V**.

If the original current is 1mA, V_A = 1mA × 20.6kΩ = **20.6V**. At 1mA, R_{out} = 952Ω, Gain = $R_L/(R_L + .952) \geq 0.900$ V/V, $R_L \geq 0.9\, R_L + 0.857$, $R_L \geq 0.857/0.1 \geq$ **8.57kΩ**.

5.48

Type	S	G	D	Joined Terminals	Positive End	Saturated Operation Possible
Enhancement	+	−	−	GD	S	Yes
	−	−	+	SG	D	Yes
Depletion	+	+	−	SG	S	Yes
	+	−	−	GD	S	No
	−	+	+	GD	D	Yes
	−	−	+	SG	D	No

There are **2 Enhancement** Configurations, and **4 Depletion** Configurations, with **6 Configurations in total**, for which current flows. Of these, **4 allow saturated operation.**

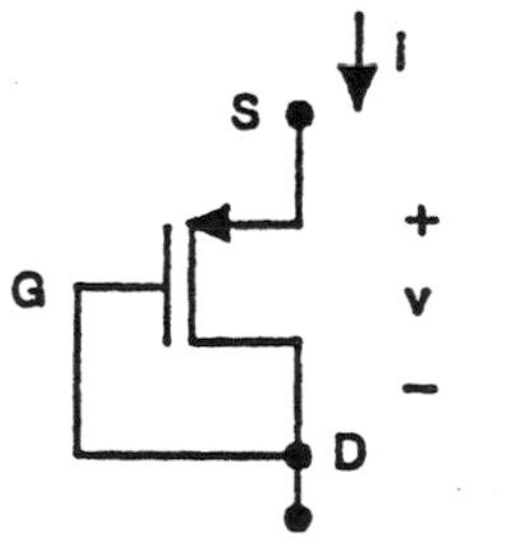

$\upsilon \geq |V_t|$, saturated
$\upsilon < |V_t|$, cutoff

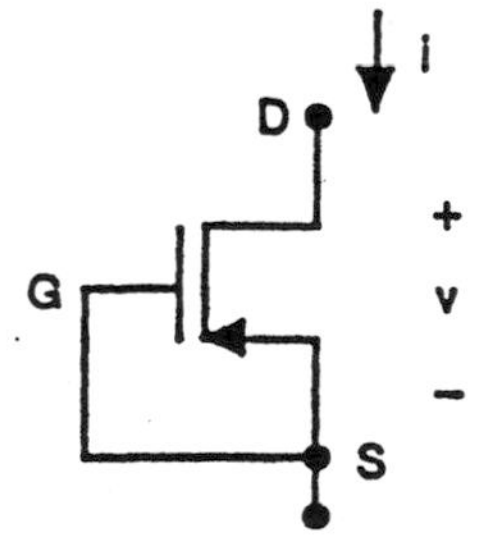

$\upsilon \geq |V_t|$, saturated
$\upsilon < |V_t|$, cutoff

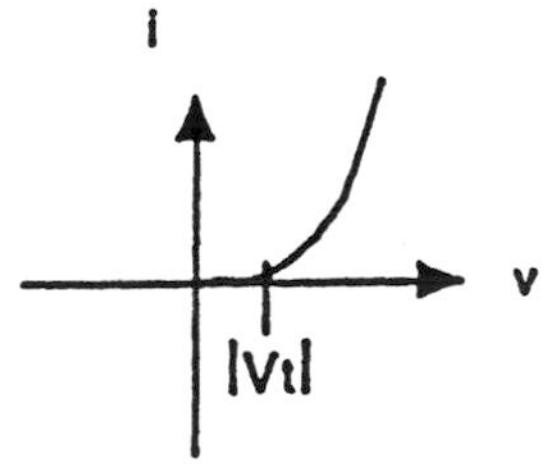

5.48 (continued)

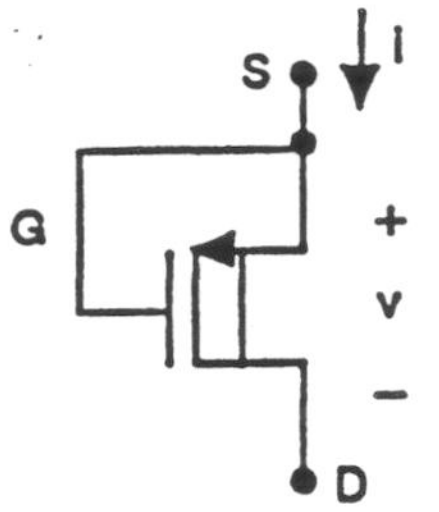

$\upsilon \geq |V_t|$ saturated
$\upsilon < |V_t|$ triode
$\upsilon < 0$ triode

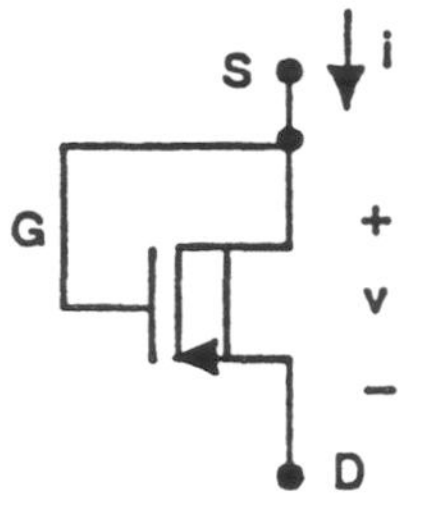

$\upsilon \geq |V_t|$ saturated
$\upsilon < |V_t|$ triode
$\upsilon < 0$ triode

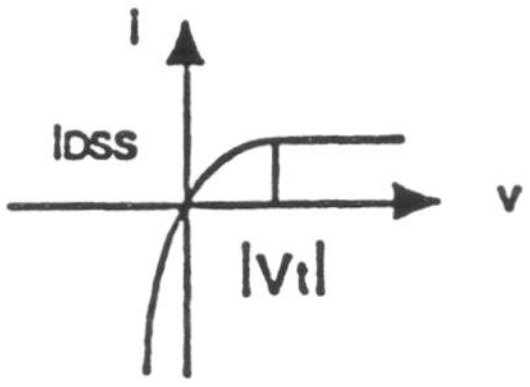

5.49 Assuming no back-bias effect:

(a) Devices identical, $V_1 = 5/2 =$ **2.5V**, $i = K(\upsilon_{GS} - V_t)^2 = 1\,(2.5 - 2)^2 = 0.25$mA. Thus $I_1 =$ **0.25mA.**

(b) Lower transistors operate as one with $K = 2$mA/V^2. Here for $V_2 = \upsilon$, $i = 1\,(5 - \upsilon - 2)^2 = 2\,(\upsilon - 2)^2$, $(3 - \upsilon)^2 = 2\,(\upsilon - 2)^2$, $3 - \upsilon = \pm\sqrt{2}\,(\upsilon - 2) = \pm\,1.414\,\upsilon \pm 2.828$. Thus, $3 - \upsilon = 1.414\,\upsilon - 2.828$, or $3 - \upsilon = -1.414\,\upsilon + 2.828$. Correspondingly $-2.414\,\upsilon = -5.828$, $\upsilon = 2.414$V, for which $i = 2\,(2.414 - 2)^2 = 0.343$mA. Thus $V_2' =$ **2.414V**, $I_2 =$ **0.343mA.**

(c) I_3 operates at I_{DSS} if the upper transistor is in saturation, in which case, $i = K(\upsilon_{GS} - V_t)^2 = 1\,(-\,-2)^2 = 4\text{mA} = I_3$. Also $i = K(\upsilon_{GS} - V_t)^2$, or $4 = 1\,(\upsilon_{GS} - 2)^2$, $\upsilon_{GS} - 2 = \pm 2$, whence $\upsilon_{GS} = 4$V (which is too high to allow the upper transistor to saturate).

Thus, the upper is in triode mode, the lower in saturation. For $\upsilon = V_3$, $i = K\,(2\,(0 - -2)\,(5 - \upsilon) - (5 - \upsilon)^2) = K\,(\upsilon - 2)^2$. $20 - 4\,\upsilon - 25 + 10\,\upsilon - \upsilon^2 = \upsilon^2 - 4\upsilon + 4$, $2\upsilon^2 - 10\upsilon + 9 = 0$, $\upsilon = \dfrac{+10 \pm \sqrt{10^2 - 4(2)\,(9)}}{2\,(2)} = \dfrac{10 \pm 5.29}{4} = 3.82$V (or an impossibly low value), for which $i = 1\,(3.82 - 2)^2 = 3.31$mA. Thus $I_3 =$ **3.31mA** and $V_3 =$ **3.82V.**

(d) From Symmetry, $V_4 =$ **2.5V**, and $I_4 = 1\,(2.5 - 2)^2 =$ **0.25mA.**

(e) From Symmetry, $V_5 = 5/2 =$ **2.5V**, and $I_5 = 1\,(2\,(2.5 - -2)\,(2.5) - 2.5^2) = 5\,(4.5) - 2.5^2 =$ **16.25mA.**

5.50

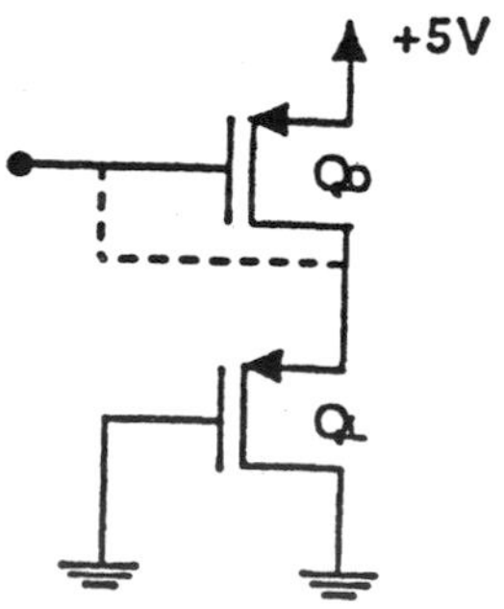

Here $V_t = -1$V, $K_D = 90\mu$A/V^2, and $K_L = 10\mu$A/V^2. Ignoring the body effect, and assuming both in saturation, $i = K(\upsilon_{GS} - V_t)^2$. For the driver, $g_{mD} = 2K_D\,(\upsilon_{GS} - V_t) = 2K_D\sqrt{i/K_D} = 2\sqrt{K_D\,i}$. For the load, $g_{mL} = 2\sqrt{K_L\,i}$. Ignoring r_o, (and back-bias effects), gain $\upsilon_o/\upsilon_i = -g_{mD}\dfrac{1}{g_{mL}} = -\dfrac{2\sqrt{K_D\,i}}{2\sqrt{K_L\,i}} = -\sqrt{K_D/K_L} - \sqrt{90/10} =$ **−3V/V.**
For $\upsilon_O = V_{DD}/2$, $i = K_L\,(-2.5 - -1)^2 = 10\,(1.5^2) = 22.5\mu$A. Now, for Q_D, $22.5 = 90\,(-(5 - \upsilon_I) - -1)^2 = 90\,(\upsilon_I - 4)^2$, or $\upsilon_I - 4 = \pm\sqrt{22.5/90} = \pm\,0.5$, $\upsilon_I = 3.5$ or 4.5 (not possible). Thus υ_I = **3.5V** for $\upsilon_O = V_{DD}/2$. Equation 5.78 applies while devices are both in saturation: ie for υ_O down to $|V_t|$, or $\upsilon_O =$ **1V**, where $\upsilon_I =$ **+4V** (at cutoff), and υ_O up to $\upsilon_O + 1 = \upsilon_I$.

For Q_D: $i = 90\,(5 - \upsilon_I - 1)^2$, For Q_L: $i = 10\,(\upsilon_O - 1)^2$. Thus $90\,(4 - \upsilon_I)^2 = 10\,(\upsilon_O - 1)^2$, or $3\,(4 - \upsilon_I) = \upsilon_O - 1$, or $\upsilon_O = 13 - 3\upsilon_I$. *Check*: For $\upsilon_I = +4$, $\upsilon_O = 13 - 3(4) = 1$V as found before. Now for $\upsilon_I = \upsilon_O + 1$, $\upsilon_O = 13 - 3(\upsilon_O + 1) = 10 - 3\upsilon_O$. $4\upsilon_O = 10$V, $\upsilon_O =$ **2.5V**, for which $\upsilon_I = 2.5 + 1 =$ **3.5V**. *Check*: For υ_I varying from 4 V to 3.5 V, υ_O, varies from 1 V to 2.5 V, respectively. That is a 1.5 V change results from a − 0.5 V change, consistent with a gain of − 3 V/V.

5.51

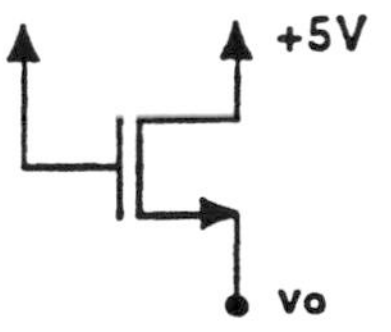

Eq 5.25: $V_t = V_{t0} + \gamma\,[\sqrt{2\Phi_f + V_{SB}} - \sqrt{2\Phi_f}\,]$. Eq 5.84: $\chi = \dfrac{\gamma}{2\sqrt{2\Phi_f + V_{SB}}}$ Eq 5.83: $g_{mb} = \chi\, g_m$. *The upper output voltage* limit is $V_{DD} - V_t = 5 - V_t$ at which $V_{SB} = 5 - V_t$. Thus $V_t = 0.9 + 0.5\,(\sqrt{0.6 + 5 - V_t} - \sqrt{0.6})$, or $V_t = 0.513 + 0.5\sqrt{5.6 - V_t}$.

Iterate: Try $V_t = 2$V, $V_t = 0.513 + 0.5\sqrt{5.6 - 2} = 0.513 + .949 = 1.46$, $V_t = 0.513 + 0.5\sqrt{5.6 - 1.46} = 1.53$V, $V_t = 0.513 + 0.5\sqrt{5.6 - 1.53} = 1.52$V.

Thus the upper output voltage is $5 - 1.52 =$ **3.48V**, at which $V_t =$ **1.52V**, and $\chi = \gamma/(2\sqrt{2\Phi_f + V_{SB}}) = 0.5/(2\sqrt{0.6 + 1.52}) =$ **0.172**, and $g_m =$ **0**, since i_D is zero at the upper limit. At $\upsilon_O = 0$V, $V_{SB} = 0$ and $V_t =$ **0.9V**. Thus $i = K\,(\upsilon_{GS} - V_t)^2 = 10\,(5 - 0 - 0.9)^2 = 6.168$mA, for which $g_m = 2K\,(\upsilon_{GS} - V_t) = 2\,(10)\,(5 - 0 - 0.9) = 82\mu$A/V, and $\chi = \gamma/(2\sqrt{2\,\Phi_f + V_{SB}}) = 0.5/(2\sqrt{0.6 + 0}) =$ **0.323**.

5.52 From P5.51: At $V_O = 2.5$V, $V_{SB} = 2.5$V, $V_t = V_{to} + \gamma\,(\sqrt{2\Phi_f + V_{SB}} - \sqrt{2\Phi_f}) = 0.9 + 0.5\,(\sqrt{0.6 + 2.5} - \sqrt{0.6}) = 1.39$V. Now, $g_{mL} = 2K\,(\upsilon_{GS} - V_t) = 2\,(10)\,(5 - 2.5 - 1.39) = 22.2\mu$A/V; $i = K\,(\upsilon_{GS} - V_t)^2 = 10\,(5.0 - 2.5 - 1.39)^2 = 12.3\mu$A, $\chi = \dfrac{\gamma}{2\sqrt{2\Phi_f + V_{SB}}} = \dfrac{0.5}{2\sqrt{0.6 + 2.5}} = 0.142$. Thus, $g_{mb} = \chi\, g_{mL} = .142 \times 22.2\mu$A/V $= 3.15\mu$A/V. Also $g_{mD} = 2\,K\sqrt{i/K} = 2\sqrt{Ki} = 2\sqrt{90 \times 12.3} = 66.54\mu$A/V Now the overall voltage gain $= -g_{mD}\,(1/g_{mL} \parallel 1/g_{mb}) = -\,66.54\,((1/22.2) \parallel (1/3.15)) = -\,66.54\mu$A/V $(45.0$k $\parallel 317$k$) =$ **− 2.62V/V**, rather than $-\sqrt{90/10} = -3$V/V that the basic calculation would indicate.

5.53 Assuming V_t includes the back-bias effect, $i_L = K\,(\upsilon_{GS} - V_t)^2 = 22.5\,(0 - 2.0)^2 = 90\mu$A, $g_{mL} = 2K\,(\upsilon_{GS} - V_t) = 2\,(22.5)(2) = 90\mu$A/V, $g_{mB} = \chi\, g_{mL} = 0.2\,(90) = 18\mu$A/V, $r_o = V_A/i_L = 50/90\mu$A $= 556$kΩ, $g_{mD} = 2\sqrt{K\, i_L} = 2\sqrt{90 \times 90} = 180\mu$A/V. Thus, gain (around $V_D = 2.5$V), is $-g_m\,(r_o \parallel r_o \parallel 1/g_{mB}) = -\,180\mu$A/V $(556$k$/2 \parallel 1/18\mu$A/V$) = -0.180\,(278$k $\parallel 55.6$k$) =$ **−8.34V/V**.

This again applies reasonably well until the load enters triode operation at $\upsilon_O = V_{DD} - |V_t| = 5 - 2 = 3.0$V, or until the driver enters the triode region at $\upsilon_O = \upsilon$. Now for the lower level, $i_L = 90\mu$A, that is $90 = 90\,(\upsilon_{GS} - 1)^2 \rightarrow \upsilon_{GS} = 2$V. Now for $\upsilon_{GS} = 2$V, triode operation begins at $\upsilon_{DS} = \upsilon_{GS} - V_t = 2 - 1 = 1$V. In actual fact, as υ_O falls from the middle, that is from 2.5V to 1V, input must rise by $(2.5 - 1)/8.34 \simeq 0.2$V. Thus, the output range is from **3.0V** to about **1.2V**.

SECTION 5.8: THE CMOS DIGITAL INVERTER

5.54 For the NMOS, $K_n = 1/2\mu_n C_{ox}(W/L) = 1/2(100)(2/1) = 100\mu A/V^2$. Since the inverter is matched, $K_p = 100\mu A/V^2$ also. With a 3.3 V supply, $V_{OH} =$ **3.3 V** and $V_{OL} =$ **0 V**.

Now $V_{IH} = (1/8)(5V_{DD} - 2V_t) = [5(3.3) - 2(0.8)]/8 =$ **1.86 V**, and $V_{IL} = 3.3 - 1.86 =$ **1.44 V**.

Thus $NM_H = V_{OH} - V_{IH} = 3.3 - 1.86 =$ **1.44 V**, and $NM_L = V_{IL} - V_{OL} = 1.44 - =$ **1.44 V**.

Now, by symmetry, $V_{th} = V_M = V_{DD}/2 = 3.3/2 =$ **1.65 V**.

Peak current from the supply is $i_D = 100 \times 10^{-6}(3.3/2 - 0.8)^2 =$ **72.25 μA**.

Current is half the peak at $\upsilon_{GS} = \upsilon$ when one of the transistors is in triode operation, and one is in saturation, where $72.25/2 = 100(\upsilon - 0.8)^2$ or $\upsilon = 0.8 + 0.362^{1/2} =$ 1.40 V.

Thus by symmetry, half-current operation occurs at $\upsilon_I =$ **1.40 V** and $3.3 - 1.4 =$ **1.90 V**.

Current is one-tenth the peak value where $\upsilon = 0.8 + ((72.25/10)/100)^{1/2} = 1.07$ V.

Thus 1/10 peak current occurs at $\upsilon_I =$ **1.07 V** and 3.3 – 1.07 = **2.23 V**.

For output resistances at the output limits:

For triode mode operation, $i_D = k'(W/L)[(\upsilon_{GS} - V_t)\upsilon_{DS} - \upsilon_{DS}^2/2]$.

For small υ_{DS}, $i_D = k'(W/L)(\upsilon_{GS} - V_t)\upsilon_{DS}$, and

$r_{DS} = 1/[k'(W/L)(\upsilon_{GS} - V_t) = 1/(100(2)(3.3 - 0.8)) = 2 \times 10^3\Omega$.

Thus the output resistances in each extreme state are **2 kΩ**.

For output currents of peak value = 72.25µA and triode operation with full input signals: $72.25 = 200[(3.3 - 0.8)\upsilon_{DS} - \upsilon_{DS}^2/2]$, or $\upsilon_{DS}^2/2 - 2.5\upsilon_{DS} + 0.36125 = 0$, or $\upsilon_{DS}^2 - 5\upsilon_{DS} + 0.7225 = 0$

and $\upsilon_{DS} = (- - 5 \pm \sqrt{5^2 - 4(0.7225)})/2 = (5 \pm 4.70)/2 = 0.15$ V.

Thus, the peak current of 72.25 µA flows at $\upsilon_O =$ **0.15 V** and **3.15 V**

5.55 From Eq. 5.102, $t_{PHL} = t_{PLH} = \dfrac{1.6C}{k_n'(W/L)_n V_{DD}} = \dfrac{1.6 \times 50 \times 10^{-15}}{100 \times 10^{-6}(2/1)3.3} = 12.12 \times 10^{-11} = 0.121$ ns

and the average propagation delay is $t_P = (t_{PLH} + t_{PHL})/2 =$ **0.121 ns.**

Alternatively: For saturation mode operation initially, $i_{D1} = (1/2)100(2/1)(3.3 - 0.8)^2 = 0.6125$ mA, and at the mid point: $i_{D2} = 100(2/1)[(3.3 - 0.8)(3.3/2) - (3.3/2)^2/2] = 200(4.125 - 1.36) = 553$ µA.

Thus the average current is (553 + 612.5)/2 = 583 µA

and $t_p \approx \dfrac{50 \times 10^{-15} \times 3.3/2}{583 \times 10^{-6}} = 0.142 \times 10^{-9}s =$ **0.142 ns.**

For operation a) with an ideal (0-ns) input, dissipation is entirely dynamic, $P_d = fCV_{DD}^2$, or $P_d = \dfrac{CV_{DD}^2}{4t_p} = \dfrac{50 \times 10^{-15} \times 3.3^2}{4(0.121 \times 10^{-9})} =$ **1.125 mW** at a frequency of $1/(4(121{\times}10^{-12})) = 2.07$ GHz.

b) with transition times of $2t_p$, an additional power is lost due to device-to-device current whose peak value is 72.25 µA (from P5.54 above). Average power loss per transition is the product of the half the peak current and the supply voltage for the duration of the active part of two transitions, while both devices conduct as the input goes from 0.8 V to 3.3 – 0.8 = 2.5 V. Thus the average power loss in two transitions/cycle is $(72.25/2) \times 3.3 \times 2((2.5 - 0.8)2 \times t_p/3.3)/4t_p = 61.4$ µW. Thus the total power loss is 1.125 + 0.061 = 1.186 = **1.19 mW.**

See that for this logic, at this supply voltage, at this frequency, that the dominant loss is due to load capacitance charging, (ie fCV_{DD}^2). The delay-power product is, $DP = 0.121 \times 10^{-9} \times 1.19 \times 10^{-3} =$ **0.14 pJ.** For the approximation on page 435 of the Text, $DP \simeq CV_{DD}^2 = 50 \times 10^{-15}(3.3)^2 = 0.54$ pJ. Note that this is essentially 4x the earlier value, simply because in the Text, f is assumed to be $1/t_P$. $f = (2t_p)^{-1}$ would have been a better choice.

SECTION 5.9: THE MOSFET AS AN ANALOG SWITCH

5.56 Now, for 10mV to ground with a 3.3V, 2.1kΩ source, $i_D = (3.30 - 0.01)/\ 2.1\text{k}\Omega = 3.29\text{V}/21\text{k}\Omega =$ 0.157mA. Now in the triode region, $i_D = K\left[2(\upsilon_{GS} - V_t)\,\upsilon_{DS} - \upsilon_{DS}^2\right] \simeq K\,(2\,(\upsilon_{GS} - V_t)\,\upsilon_{DS})$, or $0.157 \times 10^{-3} = 1/2 \times 20 \times 10^{-6} \times W/10 \times 2\,(5 - 1)\,(10 \times 10^{-3})$, or $0.157 = 10 \times 10^{-6} \times 2\,(4)W$. Thus $W = 0.157/80 \times 10^{-6} = 1963\mu\text{m}$, which is quite a large device! Now, if 0.10V were acceptable: $W = \dfrac{(.30 - .10)/21}{800} \times 10^6 = 190\mu\text{m}$.

5.57 $K_n = 1/2\,(20 \times 10^{-6})\,(50L/L) = 500\mu\text{A/V}^2$. Assuming $\mu_p = 1/2\mu_n$ with $W_p = 2W_n$,

then $K_p = K_n = 500\mu\text{A/V}$. Now, for operation in the triode mode,

$i_D = K\left[2(\upsilon_{GS} - V_t)\,\upsilon_{DS} - \upsilon_{DS}^2\right] \approx 2K\,(\upsilon_{GS} - V_t)\,\upsilon_{DS}$, and $r_{DS} = \upsilon_{DS}/i_D = 1/(2K\,(\upsilon_{GS} - V_t))$. Now, for $V_I = -5V$ with $V_{Gn} = +5V$ and $V_{Gp} = -5V$, only the n-channel device conducts with $r_{DS} = 1/(2(500 \times 10^{-6})\,(5 - -5 - 2)) = 125\Omega$. Now, with 5kΩ load, ac loss in the switch is 125/(125 + 5000) = .0244 or **2.4%**. Now, for $V_I = +5V$, with $K_p = K_n$, the result is the same and the loss is **2.4%** in the switch. Now, for $V_I = 0V$, with $V_{Gn} = +5V$, $V_{Gp} = -5V$, both switches conduct (equally), with $r_{DS} = 1/(2\,(500 \times 10^{-6})\,(5 - 0 - 2)) = 333.3\Omega$ each. Thus the total switch resistance is 333.3/2 = 167Ω, and the ac switch loss is 167/(167 + 5000) = 0.0323, or **3.2%**.

SECTION 5.10: THE MOSFET INTERNAL CAPACITANCES AND HIGH-FREQUENCY MODEL

5.58 Gate oxide capacitance ranges from 1.75 fF/μm^2 for 20 nm oxide to 0.35 fF/μm^2 for 100 nm oxide. For a 1 pF capacitance of area W^2 in the thin-oxide technology, $1 \times 10^{-12} = W^2 \times 1.75 \times 10^{-15}$ and $W =$ **23.9μm**.

For the thick oxide, $W = [(1 \times 10^{-12})/(0.35 \times 10^{-15})]^{1/2} =$ **53.5μm.**

For the minimum-size MOS, the area is $W \times L = 2.4 \times 1.2 = 2.88\ \mu m^2$.

Over the range of oxide thicknesses, from $23.9^2/2.88 =$ **198 transistors**, to $53.5^2/2.88 =$ **994 transistors** would be required.

5.59 Here, $L_{ov} = 0.15\ \mu m$, $C_{sb0} = C_{db0} = 40$ fF for a 10 μm wide device, $V_0 = 0.8$ V. From Table 5.1 on page 364 of the Text, for $t_{ox} = 20$ nm, $C_{ox} = 1.75$ fF/μm^2.

a) *For (W/L) = – (100* μm/2.4 μm), and $L_{ov} = 0.15\ \mu m$

$C_{ov} = WL_{ov}C_{ox} = 100 \times 10^{-6} \times 0.15 \times 10^{-6} \times 1.75 \times 10^{-15}/10^{-12} =$ **26.25 fF**

Basic $C_{gs}' = 2/3(WL)C_{ox} = 2/3 \times 100 \times 10^{-6} \times 2.4 \times 10^{-6} \times 1.75 \times 10^{-15}/10^{-12} = 280$ fF, in saturation.

Including overalap, $C_{gs} = C_{gs}' + C_{ov} =$ **306 fF**. Basic $C_{gd}' \approx$ **0.0fF**, in saturation.

Including overlap, $C_{gd} = C_{gd}' + C_{ov} =$ **26 fF.**

Now assuming the source and drain areas are proportional to the device width, then $C_{sb0} = C_{db0} = (40/10)100 = 400$ fF and since $|V_{DB}| = |V_{SB}| = 2$ V, from Eq. 5.11,

$C_{sb} = C_{db} = 400/(1 + 2.0/0.8)^{1/2} =$ **214 fF**. {This can be seen directly from Ex. 5.41, since $|V_{SB}| = |V_{DB}|$ is also 2 V there.}

b) *For* (W/L) = (10/24):

$C_{ov} = 10 \times 10^{-6} \times 0.15 \times 10^{-6} \times 1.75 \times 10^{-15}/10^{-12} =$ **2.62 fF.**

$C_{gs}' = 2/3 \times 10 \times 10^{-6} \times 24 \times 10^{-6} \times 1.75 \times 10^{-15}/10^{-12} = 280$ fF

$C_{gs} = C_{ov} + C_{gs}' =$ **282.6 fF**, and $C_{gd} = C_{ov} + C_{gd}' =$ **2.6 fF**, and $C_{db} = 214 \times 10/100 =$ **21.4 fF**, and $C_{sb} =$ **21.4 fF**.

5.60 *For the wide transistor at 100* μA:

From Eq. 5.44, $g_m = (2k_n'(W/L)I_D)^{1/2} = (2 \times 100 \times 10^{-6}(100/2.4)100 \times 10^{-6})^{1/2} = 912\mu A/V$.

and $C_{gs} = 306$ fF and $C_{gd} = 26$ fF.

Thus $f_T = g_m/(2\pi(C_{gs} + C_{gd})) = 912 \times 10^{-6}/(2\pi \times (306 + 26) \times 10^{-15}) =$ **437 MHz.**

For the longer transistor at 100 μA:

$g_m = (2 \times 100 \times 10^{-6}(10/24)100 \times 10^{-6})^{1/2} = 91.3\mu A/V$.

and $C_{gs} = 283$ fF, and $C_{gd} = 2.6$ fF.

Thus $f_T = 91.3 \times 10^{-6}/(2\pi(283 + 2.6) \times 10^{-15} =$ **50.9 MHz.**

Now at 10 μA, a reduction by a factor of 10, g_m will reduce by $\sqrt{10}$ = 3.16, and so will f_T to **138 MHz** and **29.9 MHz** for the two transistors.

5.61 From Exercises 5.41, 5.42, C_{gs} = 30.6 fF, C_{gd} = 2.6 fF and f_T = 1.38 GHz. For drain and source grounded, the input capacitance is $C = 30.6 + 2.6 =$ 33.2 fF, whose impedance at f_T is $Z = 1/(2\pi f_T C) = 1/(2\pi \times 1.38 \times 10^9 \times 33.2 \times 10^{-15}) =$ **3.47 kΩ.**

Now for a voltage gain of – 2 V/V, the input capacitance becomes C_{in} = 30.6 + 2.6(1 – – 2) = 38.4 pF. At $f_T/10$, $Z_{in} = 1/(2\pi \times (1.38/10) \times 38.4) =$ **30.03 kΩ.**

SECTION 5.11: THE JUNCTION FIELD-EFFECT TRANSISTOR (JFET)

5.62 For $V^+ = 4$V, operation is in saturation, and $i_D = I_{DSS}\,(1 - \frac{v_{GS}}{V_p})^2 = 10\,(1 - \frac{0}{-2}) =$ **10mA.**

For $V^+ = 2$V, operation is at the edge of saturation and i_D = **10mA**, also.

For $V^+ = 1$V, operation is in triode mode, and $i_D = \frac{I_{DSS}}{V_p^2}\left[2\,(v_{GS} - V_p)\,v_{DS} - (v_{DS})^2\right]$, or $i_D = \left[2\,(0 - -2)\,1 - 1^2\right] = \frac{10}{4}\,(4 - 1) =$ **7.5mA.**

For $i_D = \frac{I_{DSS}}{2} = 5\text{mA} = \frac{10}{4}\left[2\,(2)\,v_{DS} - v_{DS}^2\right]$, $2 = 4\,v_{DS} - v_{DS}^2$, $v_{DS}^2 - 4\,v_{DS} + 2 = 0$, and $v_{DS} = \frac{- -4 \pm \sqrt{4^2 - 4(2)}}{2} = \frac{4 \pm 2\sqrt{2}}{2} = 2 \pm \sqrt{2} = 3.414$ or 0.586V. Clearly $V^+ =$ **0.586V.**

Check: $i_D = \frac{10}{4}\left[2\,(2)(.586) - .586^2\right] = 5$mA.

5.63 For triode-mode operation: $i_D = \frac{I_{DSS}}{V_p^2}\left[2\,(v_{GS} - V_p)\,v_{DS} - v_{DS}^2\right]$.

Now, for i_D = 5mA, $5 = \frac{10}{4}\left[2\,(v_{GS} + 2)\,1 - 1^2\right]$, whence $2 = 2\,(v_{GS} + 2) - 1 = 2\,v_{GS} + 4 - 1$.

Thus $v_{GS} = \frac{2 - 3}{2} =$ **–0.5V.**

Check: $i_D = \frac{10}{4}\left[2\,(-0.5 - -2)\,1 - 1\right] = 2.5\,[2\,(1.5) - 1] = 5$mA.

Now, for i_D = 1mA, $1 = \frac{10}{4}\left[2\,(v_{GS} + 2)\,1 - 1^2\right]$, or $0.4 = 2\,v_{GS} + 4 - 1$, whence $v_{GS} = \frac{0.4 - 3}{2} = \frac{-2.6}{2} =$ **–1.3V.**

5.64 For triode operation: $i_D = \frac{I_{DSS}}{V_p^2}\left[2\,(v_{GS} - V_p)\,v_{DS} - v_{DS}^2\right]$.

For small v_{DS}, $i_D = \frac{2I_{DSS}}{V_p^2}\,(v_{GS} - V_p)\,v_{DS}$, and $r_{DS} = \frac{v_{DS}}{i_D} = \frac{1}{\frac{2I_{DSS}}{V_p^2}\,(v_{GS} - V_p)} = \frac{V_p^2}{2I_{DSS}\,(v_{GS} - V_p)}$.

Now for I_{DSS} = 10mA, V_p = –2V: For v_{GS} = 0V, $r_{DS} = \frac{2^2}{2 \times 10\,(0 - -2)} = 100\Omega$. For v_{GS} = –1V, $r_{DS} = \frac{2^2}{2 \times 10\,(-1 - -2)} = 200\Omega$. For v_{GS} = –2V, $r_{DS} = \frac{2^2}{2 \times 10\,(-2 - -2)} = \infty$. (Of course, since at v_{GS} = –2V, the switch is cut off!)

5.65 For $\upsilon_{DS} = 2V$, and $V_p = -2V$, the JFET is just at the edge of saturation, for which $i_D = K\,(\upsilon_{GS} - V_t)^2 = \frac{I_{DSS}}{V_p^2}(\upsilon_{GS} - V_p)^2 = I_{DSS}\left[1 - \frac{\upsilon_{GS}}{V_p}\right]^2$, or $5 = 10\left[1 - \frac{\upsilon_{GS}}{-2}\right]^2$, and $1 + \frac{\upsilon_{GS}}{2} = \sqrt{½} = .707$. Thus $\upsilon_{GS} = 2\,(0.707 - 1) = \mathbf{-0.586V}$.

Check: $i_D = 10\left[1 - \frac{-.586}{-2}\right]^2 = 5.00mA$.

For $\upsilon_{GS} = -0.586V$, $\upsilon_{DS} = 7V$, $5.10 = 5.00\,(1 + \lambda\,(7 - 2))$, or $0.10 = 25\lambda$, whence $\lambda = \frac{0.1}{25} = \mathbf{0.004V^{-1}}$. Thus $V_A = \frac{1}{\lambda} = \frac{1}{0.004} = \mathbf{250V}$, and r_o (at 5mA) $= \frac{250V}{5mA} = \mathbf{50k\Omega}$.

More painstakingly: In saturation, $i_D = I_{DSS}\left[1 - \frac{\upsilon_{GS}}{V_p}\right]^2(1 + \lambda\upsilon_{DS})$. Thus, at $\upsilon_{DS} = 2V$, $5 = 10\left[1 - \frac{\upsilon_{GS}}{-2}\right]^2(1 + \lambda\,2)$, and at $\upsilon_{DS} = 7V$, $5.1 = 10\left[1 - \frac{\upsilon_{GS}}{-2}\right]^2(1 + \lambda\,7)$.

Divide: Thus $\frac{5.1}{5} = \frac{1 + 7\lambda}{1 + 2\lambda} = 1.02$, or $1 + 7\lambda = 1.02 + 2.04\lambda$, whence $\lambda = \frac{.02}{7 - 2.04} = \mathbf{0.00403V^{-1}}$.

Thus $5 = 10\left[1 - \frac{\upsilon_{GS}}{-2}\right]^2(1 + 2\,(.00403))$, or $\left[1 - \frac{\upsilon_{GS}}{-2}\right] = \sqrt{\frac{1}{2(1.00806)}} = 0.7043$, whence $\upsilon_{GS} = -2\,(.2957) = \mathbf{0.591V}$, for which $V_A = \frac{1}{.00403} = \mathbf{248V}$, and $r_o = \left[I_{DSS}\,(1 - \frac{\upsilon_{GS}}{V_p})^2\,\lambda\right]^{-1} = (10\,(.7043)^2\,(.00403))^{-1} = \mathbf{50.02k\Omega}$.

5.66 Now, $i_D = \frac{I_{DSS}}{V_p^2}\left[2\,(\upsilon_{GS} - V_p)\,(\upsilon_{DS}) - \upsilon_{DS}^2\right]$ in triode mode, or $i_D = I_{DSS}\left[1 - \frac{\upsilon_{GS}}{V_p}\right]^2$ in saturation.

(a) For p-channel; $\upsilon_{GS} = 0$; $\upsilon_{GD} = 5V \rightarrow$ saturation. Thus $I_a = I_{DSS} = \mathbf{4mA}$.

(b) For n-channel; $\upsilon_{DG} = 5V \rightarrow$ saturation. Thus $i_D = I_{DSS}\left[1 - \frac{\upsilon_{GS}}{V_p}\right]^2$, or $1 = 4\left[1 - \frac{\upsilon_{GS}}{-2}\right]^2$, or $1 - \frac{\upsilon_{GS}}{-2} = \pm ½$, whence $\upsilon_{GS} = 2\,(\pm ½ - 1) = -1$ or -3 (cutoff). Thus $V_b = \upsilon_G - \upsilon_{GS} = 0 - -1 =$ **1V**.

(c) For n-channel; $i_D < I_{DSS} \rightarrow$ triode. Thus $1 = \frac{4}{2^2}(2\,(0 - -2)\,\upsilon_{DS} - \upsilon_{DS}^2)$, or $\upsilon_{DS}^2 - 4\,\upsilon_{DS} + 1 = 0$, whence $\upsilon_{DS} = \frac{4 \pm \sqrt{4^2 = 4(1)}}{2} = 0.268V$, or very large. Thus $V_c = \mathbf{+0.268V}$.

(d) The p-channel device is operating with the gate somewhat forward-biassed in the triode mode. Thus $i_D = \frac{I_{DSS}}{V_p^2}(2\,(\upsilon_{GS} - V_p)\,\upsilon_{DS} - \upsilon_{DS}^2)$, or $1 = \frac{4}{2^2}(2\,(V_d - 2)\,(V_d) - (V_d)^2) = 2V_d^2 - 4V_d - V_d^2$, or $V_d^2 - 4V_d - 1 = 0$, whence $V_d = \frac{4 \pm \sqrt{16 - 4(-1)}}{2} = \frac{4 \pm 4.472}{2} = \mathbf{-0.236V}$.

5.67 For the lower device, assumed to be in saturation, $i_D = I_{DSS}\left[1 - \frac{\upsilon_{GS}}{V_p}\right]^2$, and $\upsilon_{GS} = -i_D\,(1k\Omega) = -i_D$. Thus, $i_D = 4\left[1 - \frac{-i_D}{-2}\right]^2 = (2 - i_D)^2$, $4 - 4\,i_D + i_D^2 = i_D$, $i_D^2 - 5i_D + 4 = 0$, whence $i_D = \frac{+5 \pm \sqrt{5^2 - 4(4)}}{2} = \frac{+5 \pm 3}{2} = 1$ or 4mA (not acceptable). $I_D = \mathbf{1mA}$ and $\upsilon_{GS} = -1V$. Now, the

upper circuit is the same. Thus, the since the gate is at 0V, source is at +1V and V_O = **0V.**
Now, if *both* resistors are raised to 2kΩ, I_D reduces, but it is the same in both cases, and V_o = 0V is retained. Here $v_{GS} = -2i_D$, and $i_D = 4\left[1 - \dfrac{-2i_D}{-2}\right]^2 = (2 - 2i_D)^2 = 4 - 8i_D + 4i_D^2$, for which $4i_D^2 - 9i_D + 4 = 0$, and $i_D = \dfrac{9 \pm \sqrt{81 - 4(4)(4)}}{2(4)} = \dfrac{9 \pm 4.123}{8}$ = **0.61mA**. Now for I_D = 0.61mA, v_{GS} = −1.22V, but V_O remains at **0V**.

5.68 Current in the 1MΩ network can be ignored. Thus $I_D = I$ = **10mA**. Now, $I_D = I_{DSS}\left[1 - \dfrac{v_{GS}}{V_p}\right]^2$, or $10 = 10\left[1 - \dfrac{v_{GS}}{V_p}\right]^2 \rightarrow v_{GS}$ = 0V (as could be seen directly). Since v_{GS} = 0, v_G = 0, $v_D = 0 + \dfrac{0 - -5}{1M} \times 1M$ = **5V**, $r_o = \dfrac{V_A}{I_D} = \dfrac{50V}{10mA}$ = **5kΩ**, $g_m = \dfrac{2\,I_{DSS}}{-V_p}\left[1 - \dfrac{v_{GS}}{V_p}\right] = \dfrac{2\,(10)}{2}(1 - 0)$ = **10mA/V**. For $R_L = \infty$, $v_o/v_i = -g_m\, r_o = -10 \times 10^{-3}\,(5 \times 10^3)$ = **−50V/V**. For $R_L = r_o$, $v_o/v_i = -g_m\,(r_o \parallel R_L) = -10 \times \dfrac{5}{2}$ = **−25V/V**. Now R_i = 1MΩ ∥ (1MΩ/(1 − gain)) in general, or R_i = 1 ∥ 1/(1 − −50) = **19.2kΩ**, or 1 or ∥ 1/(1 − −25) = **37kΩ**, in the two cases.

SECTION 5.12: GALLIUM-ARSENIDE (GaAs) DEVICES – THE MESFET

5.69 Here, from Eq 5.108 and 5.109, $g_m = 2\,\beta\,(V_{GS} - V_t)\,(1 + \lambda\,V_{DS})$, $r_o \approx 1/(\lambda\beta\,(V_{GS} - V_t)^2)$, and the highest available gain is $\mu = g_m\, r_o$.

For v_{GS} = +0.2V, g_m = 2 (10^{-4}) × 100 (0.2 − −1.0) (1 + 0.2(3)) = 200 × 10^{-4} × 1.2 (1.6) = **38.4mA/V**, $r_o \approx$ 1/(.2 × 100 × 10^{-4} (.2 − − 1.0)^2) = 347Ω, and μ = 38.4 × 10^{-3} × 347 = **13.3V/V**.

For v_{GS} = −0.2V, g_m = 2 (10^{-4}) × 100 (−0.2 − −1.0) (1 + 0.2(3)) = 20 × 10^{-3} × 0.8 × 1.6 = **25.6mA/V**, r_o = 1/(20 × 10^{-4} (.8)^2) = **781Ω**, and μ = 25.6 × 10^{-3} × 781 = **20.0V/V**.

For v_{GS} = 0V, g_m = 2 (10^{-4}) × 100 (0 − −1) (1.6) = 20 × 10^{-3} (1) (1.6) = **32.0mA/V**, r_o = 1/(20 × 10^{-4} (1.0)^2) = 500Ω, and μ = 32.00 × 500 = **16.0V/V**.

5.70

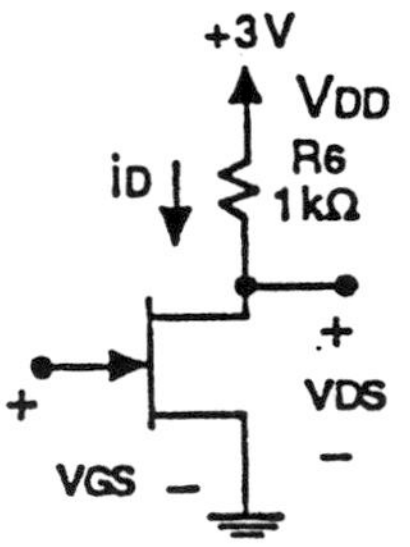

β = 100 × 10^{-4}A/V^2 = 10mA/V^2 for a 100μm device. From Eq 5.107, $i_D = \beta\,(v_{GS} - V_t)^2\,(1 + \lambda v_{DS})$ assuming operation is in saturation, and $v_{DS} = V_{DD} - i_D\,R_L$, $i_D = (V_{DD} - v_{DS})/R_L$. For v_{GS} = **+0.2V**, (3 − v_{DS})/0.1 = 10 (0.2 − −1)^2 (1 + 0.2 v_{DS}), or 3 − v_{DS} = 1 (1.2)^2 (1 + 0.2 v_{DS}) = 1.44 + 0.288 v_{DS}, and 1.288 v_{DS} = 3 − 1.44 = 1.56, v_{DS} = 1.56/1.288 = **1.211V**. Now this exceeds (0.2 − − 1.0) = 1.2V, OK. For v_{GS} = **−0.2V**, (3 − v_{DS})/0.1 = 10 (−0.2 − −1)^2 (1 + 0.2 v_{DS}), or 3 − v_{DS} = 0.64 (1 + 0.2 v_{DS}) = 0.64 + 0.128v_{DS}, 1.128v_{DS} = 2.36, v_{DS} = 2.36/1.128 = **2.092V**.

For v_{GS} = 0V, 3 − v_{DS} = 1 (1)^2 (1 + 0.2v_{DS}) = 1 + 0.2v_{DS}, 1.2v_{DS} = 2, and v_{DS} = **1.67V**.

Voltage gains: For v_{GS} = 0.2V to −0.2V, "gain" = (1.211 − 2.092)/(0.2 − −0.2) = **−2.2V/V**. For v_{GS} = 0.0V to −0.2V, gain = (1.67 − 2.092)/(0 − − 0.2) = **−2.13V/V**.

5.71 Now, $\beta_1 = \beta_2$ = 10mA/V^2. Assume that the dc output is stabilized at half the supply voltage. That is, $v_{DS1} = v_{DS2}$ = 5V. Now, $I_{D2} = \beta_2\,(v_{GS2} - V_t)^2\,(1 + \lambda\,v_{DS2})$ = 10 (0 − 1)^2 (1 + 0.1 × 5) = 15mA, and $I_{D1} = I_{D2}$ = **15mA**, with v_{GS1} = **0V** as well. Thus g_{m1} = 2 (10) (0 − − 1) (1 + 0.1 × 5) = **30mA/V**, r_{01} = 1/(0.1 (10) (0 − − 1)^2) = 1kΩ, r_{02} = 1/(0.1 (10) (0 − − 1)^2) = 1kΩ, A_v = 30 (1kΩ ∥ 1kΩ) = **−15V/V**.

5.72 For V_O = +3V, υ_{DS2} = 10 − 3 = 7V, I_{D2} = 10 (0 − 1)^2 (1 + 0.1 (7)) = 17mA = I_{D1}. Now for Q_1: 17 = 10 (υ_{GS1} − − 1)^2 (1 + 0.1(3)), or (υ_{GS1} + 1)^2 = 17/(10 (1 + .3)) = 1.308, υ_{GS1} = ± 1.144 − 1 = −2.144 (cutoff) or **+.144V**. Now g_{m1} = 2 × (10) (.144 − − 1) (1 + 0.1(3)) = **29.7mA/V**, r_{01} = 1/(0.1 (10) (1.144)^2) = 0.764kΩ, r_{02} = 1/(0.1 (10) (0 − − 1)^2) = 1kΩ. Thus, the gain: = − g_m r_{01} ∥ r_{02} = −29.7 (1kΩ ∥ 0.764kΩ) = **−12.9V/V**.

Chapter 6

DIFFERENTIAL AND MULTISTAGE AMPLIFIERS

SECTION 6.1: THE BJT DIFFERENTIAL PAIR

6.1 Eq.6.7, 6.8: $i_{E1} = \dfrac{I}{1 + e^{(\upsilon_{B2} - \upsilon_{B1})/V_T}} = \dfrac{I}{1 + e^{-\upsilon_d/V_T}}$, $i_{E2} = \dfrac{I}{1 + e^{(\upsilon_{B1} - \upsilon_{B2})/V_T}} = \dfrac{I}{1 + e^{\upsilon_d/V_T}}$,

(a) $i_{C2} = \alpha\, i_{E2} \approx i_{E2} = 0.99I$, when $\dfrac{1}{1 + e^{\upsilon_d/V_T}} = 0.99$, or $e^{\upsilon_d/V_T} = 1/0.99 - 1 = 0.0101$, or $\upsilon_d = V_T \ln .0101 = -4.595V_T = -$ **115mV**. That is, υ_{B1} must be **lower** than υ_{B2} by 115mV.

(b) $i_{C1} = \alpha\, i_{E1} \approx i_{E1} = 0.95I$, when $\dfrac{1}{1 + e^{-\upsilon_d/V_T}} = 0.95$, or $e^{-\upsilon_d/V_T} = 1/0.95 - 1 = 0.05263$, or $-\upsilon_d/V_T = -2.94$, or $\upsilon_d = 2.94\ V_T =$ **73.6mV**. That is, υ_{B1} must be **higher** than υ_{B2} by 73.6mV.

(c) For $i_{C1} = 9.0\ i_{C2}$ with $i_{C1} + i_{C2} = I$, $I - i_{C2} = 9i_{C2}$, or $i_{C2} = I/10 = 0.1I$, for which $i_{C1} = 0.9I$. Therefore, $\dfrac{I}{1 + e^{-\upsilon_d/V_T}} = 0.9I$, $e^{-\upsilon_d/V_T} = 1/0.9 - 1 = 0.1111$, $-\upsilon_d/V_T = -2.197$, or $\upsilon_d =$ **54.9mV**. That is, υ_{B1} must be **higher** than υ_{B2} by 54.9 mV.

6.2

Case	υ_{B1} V	υ_{B2} V	$\upsilon_{E1,2}$ V	υ_{C1} V	υ_{C2} V
a	0	**0**	– 0.7	**6**	6
b	2	2	**1.3**	6	**6**
c	**2.0**	1	1.3	**2**	**10**
d	–2	**1.0**	0.3	10	**2**
e	1	**3.5**	2.8	**10**	3
f	–4	–4	**–4.7**	**4**	8
g	**4.0**	0	+3.3	**3.5**	**10**
h	1	3.5	**2.8**	**10**	**3**

(a) $\upsilon_{B1} = 0V$, $\upsilon_{E12} = -0.7V \rightarrow \upsilon_{B2} = -0.7 + 0.7 =$ **0V** (or lower). $\upsilon_{C2} = 6V \rightarrow \upsilon_{RC2} = 10 - 6 = 4V$, and $i_{C2} = 4V/4k\Omega = 1.0mA$. Therefore $i_{C1} = 2.0 - 1.0 = 1.0mA$, and $\upsilon_{C1} = 10 - 4(1) =$ **6V**. For equal current split, $\upsilon_{B1} = \upsilon_{B2} =$ **0V**.

(b) $\upsilon_{B1} = \upsilon_{B2} = 2.0\ \upsilon \rightarrow \upsilon_{E12} = 2.0 - 0.7 =$ **1.3V**. Now $\upsilon_{C1} = 6V \rightarrow i_{C1} = (10-6)/4 = 1mA$, and $i_{C2} = 2.0 - 1.0 = 1.0mA$, and $\upsilon_{C2} = 10 - 4(1) =$ **6V**.

(c) $\upsilon_{E12} = 1.3V$. Thus one of υ_{B1}, $\upsilon_{B2} = 1.3 + 0.7 = 2.0V$. Therefore, $\upsilon_{B1} =$ **2.0V**. Now, since $\upsilon_{B2} = 1.0V$, $i_{C1} = 2.0mA$, and $\upsilon_{C1} = 10 - 4(2) =$ **2V**. Also $i_{C2} = 0mA$, so $\upsilon_{C2} =$ **10V**.

(d) $\upsilon_{E12} = 0.3V$. Thus one of υ_{B1}, $\upsilon_{B2} = 0.3 + 0.7 = 1.0V$. Therefore $\upsilon_{B2} =$ **1.0V**, Q_2 conducts 2mA and $\upsilon_{C2} = 10 - 2(4) =$ **2V**, with $\upsilon_{C1} = 10V$.

(e) $\upsilon_{E12} = 2.8V$. Thus one of υ_{B1}, υ_{B2} must be $2.8 + 0.7 = 3.5V$. Therefore $\upsilon_{B2} =$ **3.5V** and $i_{C2} = 2mA$. Thus $\upsilon_{C2} = 10 - 2(4) = 2V$, possibly. Therefore Q_2 is saturated with $\upsilon_{C2} = 2.8 + 0.2 = 3.0V$, with extra current flowing in the base of Q_2. But Q_1 is cut off and $\upsilon_{C1} =$ **+10V**.

(f) $\upsilon_{B1} = \upsilon_{B2} = -4.0\ V \rightarrow \upsilon_{E1,2} = -4.0 - 0.7 =$ **–4.7V**, $\upsilon_{C2} = 8V \rightarrow i_{C2} = (10-8)/4k\Omega = 0.5mA$. Thus $i_{C1} = 2.0 - 0.5 = 1.5mA$, and $\upsilon_{C1} = 10 - 1.5(4) =$ **4V**.

(g) v_{E12} = 3.3V. Thus one of v_{B1}, v_{B2} is at 3.3 + 0.7 = **4.0V**. Thus v_{B1} = 4.0V, with Q_1 conducting, Q_2 cut off, v_{C2} = **10V**. For i_{C1} = 2mA, v_{C1} = 10 − 2(4) = 2V. But v_{E1} = 3.3V. Thus Q_1 is saturated with v_{C1} = 3.3 + 0.2 = **3.5V**.

(h) $v_{E1,2}$ = 3.5 − 0.7 = **2.8V** with Q_1 cut off and Q_2 conducting. Thus v_{C1} = **+10V**, and v_{C2} possibly as low as 10 − 2(4) = 2V. But v_{E2} = 2.8V. Thus v_{C2} = 2.8 + 0.2 = **3.0V**.

6.3

Case	I mA	v_{B1} V	v_{B2} V	v_E V	v_{C1} V	v_{C2} V
a	0.2	0.00	**0.00**	−0.700	**9.60**	9.60
b	0.2	0.01	0.00	**−0.695**	**9.52**	**9.68**
c	0.2	0.00	0.05	**−0.664**	**9.90**	**9.30**
d	0.2	**0.037**	0.00	−9.675	**9.35**	**9.85**
e	0.2	−1.00	**−1.05**	**−1.714**	**9.30**	9.90
f	2.0	−1.00	−1.00	**−1.758**	**6.00**	**6.00**
g	2.0	0.01	0.00	**−0.752**	**5.21**	**6.79**
h	2.0	0.00	0.05	**−0.722**	**9.05**	**2.95**
i	2.0	1.00	**0.951**	**0.228**	3.00	**9.00**

(a) v_{C2} = 9.60V → i_{C2} = (10.0 − 9.6)/4k = 0.1mA, that is $i_{C2} = I$ − 0.1 = 0.2 − 0.1 = 0.1mA → v_{C1} = **9.60V**. Since $i_{C1} = i_{C2}$, $v_{B1} = v_{B2}$ = **0.00V**. **Note** that for i_C = 0.1mA, v_{BE} = 0.700V.

(b) $v_d = v_{B1} - v_{B2}$ = .01 − .00 = 10mV. Thus $i_{E2} = \dfrac{0.2}{1 + e^{10/25}} = \dfrac{0.200}{1 + 1.492}$ = .0803mA, and i_{E1} = 0.200 − .0803 = 0.1197mA, v_{BE1} = 0.700 + 25 ln (0.1197/0.100) = 0.7045V. Thus v_E = 0.010 −.7045 = **−0.695V**, v_{C1} = 10 − 4(.1197) = **9.52V**, and v_{C2} = 10 − 4(.0803) = **9.68V**.

(c) $v_d = v_{B1} - v_{B2}$ = −0.05V, $i_{E2} = \dfrac{0.2}{1 + e^{-50/25}}$ = 0.176mA, i_{E1} = 0.200 − 0.176 = 0.024mA, v_{C1} = 10 − 4(.024) = **9.90V**, v_{C2} = 10 − 4(.176) = **9.30V**, v_{BE2} = 0.700 + 25 ln (0.176/0.1) = .714V. Thus $v_E = v_{B2} - v_{BE2}$ = .050 − .714 = **−0.664V**.

(d) Assuming v_{B1} > 0 and i_{E2} < 0.1mA, $i_{E2} = 0.1\ e^{(675-700)/25}$ = 0.0368mA, i_{E1} = 0.200 − 0.0368 = 0.1632mA, v_{BE1} = 700 + 25 ln (0.1632/0.100) = 712.2mV. Thus v_{B1} = 712.2 − 675 = + **0.037V**, and v_{C1} = 10 − 4(.1632) = **9.35V**, v_{C2} = 10 − 4(.0368) = **9.85V**.

(e) i_{C2} = (10 − 9.90)/4 = 0.025mA. Thus i_{C1} = .200 −.025 = 0.175mA, and v_{C1} = 10 − 4 (.175) = **9.30V** See from (c) that $v_{BE\,2}$ = 0.664V, and $v_{BE\,1}$ = 0.664V + .050V, and since v_{B1} = −1.00V, v_{B2} = −1.00 − 0.050 = **−1.05V**, and v_E = −1.05 − 0.664 = **−1.714V**.

(f) Inputs equal: current splits equally and $i_{C1} = i_{C2}$ = 2.0/2 = 1.0mA, v_{C1} = 10 − 4(1) = **6.00V**, v_{C2} = 10 − 4(1) = **6.00V**. Thus v_{BE} = 700 + 25 ln (1.00/0.1) = 757.6mV, and v_E = −1.00 −.758 = **−1.758V**.

(g) $v_d = v_{B1} - v_{B2}$ = 10mV. Thus $i_{E2} = \dfrac{I}{1 + e^{v_D/v_T}} = \dfrac{2.0}{1 + e^{10/25}} = \dfrac{2.0}{2.492}$ = 0.803mA, and i_{E1} = 2.00 − .803 = 1.197mA, $v_{BE\,2}$ = 700 + 25 ln (0.803/0.100) = 752mV, v_E = 0 − .752 = **−0.752V**, v_{C1} = 10 − 4 (1.197) = **5.21V**, v_{C2} = 10 − 4 (0.803) = **6.79V**.

(h) $v_d = v_{B1} - v_{B2}$ = 0 − .05 = −0.05V = −50mV. Now $i_{E2} = \dfrac{2.0}{1 + e^{-50/25}}$ = 1.762mA, and $v_{BE\,2}$ = 700 + 25 ln (1.762/0.1) = 772mV, that is $v_E = v_{B2} - v_{BE\,2}$ = 50 − 772 = − 722mV = **−0.722V**. Now i_{E1} = 2.0 − 1.762 = 0.238mA, and v_{C1} = 10 − 4 (.238) = **9.05V**, and v_{C2} = 10 − 4 (1.762) = **2.95V**.

(i) $v_{C1} = 3.00V \rightarrow i_{C1} = (10-3)/4 = 1.75$mA. Thus $i_{C2} = 2.00 - 1.75 = 0.25$mA, $v_{C2} = 10 - 4\,(.25)$ = **9.00V**, $v_{BE\,1} = 700 + 25 \ln (1.75/0.1) = 771.6$mV, $v_{BE\,2} = 700 + 25 \ln (0.25/0.1) = 722.9$mV. Thus $v_E = 1.00 - .772 =$ **0.228V**, and $v_{B2} = .228 + .723 =$ **0.951V**.

SECTION 6.2: SMALL-SIGNAL OPERATION OF THE BJT DIFFERENTIAL AMPLIFIER

6.4 Using $e^x \simeq 1 + x + x^2/2$ in Eq. 6.11: $i_{C1} = \dfrac{\alpha I\, e^{v_d/2V_T}}{e^{v_d/2V_T} + e^{-v_d/2V_T}} = \alpha I \left[1 + \dfrac{v_d}{2V_T} + \dfrac{v_d^2}{8V_T^2}\right]$

$$/ \left[1 + \frac{v_d}{2V_T} + \frac{v_d^2}{8V_T} + 1 - \frac{v_d}{2V_T} + \frac{v_d^2}{8V_T}\right] = \frac{\alpha I}{2}\left[1 + \frac{v_d}{2V_T} + \frac{v_d^2}{8V_T^2}\right] / \left[1 + \frac{v_d^2}{8V_T^2}\right]$$

$$\simeq \frac{\alpha I}{2}\left[1 + \frac{v_d}{2V_T} + \frac{v_d^2}{8V_T^2}\right] \times \left[1 - \frac{v_d^2}{8V_T^2}\right]$$

$$= \frac{\alpha I}{2}\left[1 + \frac{v_d}{2V_T} + \frac{v_d^2}{8V_T^2} - \frac{v_d^2}{8V_T^2} - \frac{v_d^3}{16V_T^3} - \frac{v_d^4}{64V_T^4}\right] = \frac{\alpha I}{2}\left[1 + \frac{v_d}{2V_T} - \frac{v_d^3}{16V_T^3} - \frac{v_d^4}{64V_T^4}\right]$$

$$= \frac{\alpha I}{2} + \frac{\alpha I}{2V_T}\left[\frac{v_d}{2}\right]\left\{1 - \frac{1}{2}\left[\frac{v_d}{2V_T}\right]^2 - \frac{1}{4}\left[\frac{v_d}{2V_T}\right]^3\right\} --- (1).$$

Now for $v_d/2 = 10$mV, $\left\{1 - \dfrac{1}{2}\left[\dfrac{v_d}{2V_T}\right]^2 - \dfrac{1}{4}\left[\dfrac{v_d}{2V_T}\right]^3\right\} = 1 - \dfrac{1}{2}\left[\dfrac{10}{25}\right]^2 - \dfrac{1}{4}\left[\dfrac{10}{25}\right]^3 =$

$1 - 0.08 - 0.016 = 0.904$. {Rather than 1.000 for a linear system.}

That is, we see that the higher-order approximation implies a reduction in output current by about **10%** from that derived from the linear one.

Alternatively (and directly) at $v_d/2 = 10$mV: $i_{C1} = \dfrac{\alpha I}{2}\,\dfrac{2\,e^{10/25}}{e^{10/25} + e^{-10/25}} = \dfrac{\alpha I}{2}\left[\dfrac{2\,(1.492)}{1.492 + 0.670}\right] =$ $1.380\,\dfrac{\alpha I}{2}$, of which the signal part is $0.380\,\dfrac{\alpha I}{2}$, whereas from Equation 6.12: $i_{C1} = \dfrac{\alpha I}{2}\left[1 + \dfrac{v_d/2}{V_T}\right] = \dfrac{\alpha I}{2}\left[1 + \dfrac{10}{25}\right] = 1.400\,\dfrac{\alpha I}{2}$ of which the signal part is $0.400\,\dfrac{\alpha I}{2}$. Thus the linear approximation produces a result which is high by $(0.400 - 0.380)/0.380 = 0.053$, or about **5%**.

For specified errors, using the result (1) above, but only the term in v_d^2, we see that the error is about $\dfrac{1}{2}\left[\dfrac{v_d}{2}/V_T\right]^2 = \varepsilon$. Now for $\varepsilon = 10\%$, $\dfrac{v_d}{2}/V_T = (2\,(0.1))^{1/2} = 0.447$, and $v_d/2 = 0.447\,(25) =$ **11mV**. For 5%, $v_d/2 = (2\,(.05))^{1/2}\,(25) =$ **7.9mV**. For 1%, $v_d/2 = (2\,(.01))^{1/2}\,(25) =$ **3.5mV**.

Check with the original (Eq. 6.11): $i_{C1} = \dfrac{\alpha I}{2}\left[\dfrac{2\,e^{v_d/2V_T}}{e^{v_d/2V_T} + e^{-v_d/2V_T}}\right]$. For $v_d/2 = 3.5$mV,

$i_{C1} = \dfrac{\alpha I}{2}\left[\dfrac{2e^{3.5/25}}{e^{3.5/25} + e^{-3.5/25}}\right] = \dfrac{\alpha I}{2}\,\dfrac{2\,(1.150)}{1.150 + .869} = 1.139\,\dfrac{\alpha I}{2}$, whereas, from Eq. 6.12: $i_{C1} = \dfrac{\alpha I}{2}\,(1 + v_d/2\,V_T) = \dfrac{\alpha I}{2}\,91 + 3.5/25) = \dfrac{\alpha I}{2}\,(1.140)$. Thus the error is $(0.140 - 0.139)/0.139 \equiv 0.7\%$.

6.5 For a differential input of 0V, $i_{C1} = i_{C2} = I/2 = 100\mu$A, and $v_{C1} = 3 - 0.1\,(10) = 2V = v_{C2}$. For $I_E = 100\mu$A, $r_e = \dfrac{V_T}{I_E} = \dfrac{25mV}{0.1mA} = 250\Omega$. For differential output, $\dfrac{v_{od}}{v_d} = \dfrac{\alpha\,(10k + 10k)}{250 + 250} \simeq$ **40V/V**. For outputs individually, the gain magnitude is **20V/V**. For $v_{CB} = -0.4$V and very small signals, $v_{C1} = v_{C2} =$ 2V, and $v_{B1} = v_{B2} = 2.0 - -0.4 = 2.4$V. That is, the upper limit of the input range is **+2.4V**.

6.6 For I = 200μA, r_e = 25mV/100μA = 250Ω. For differential output, gain is $\alpha \frac{R_L}{r_e} \approx \frac{100k\Omega}{0.25k\Omega}$ = **400V/V**. Differential input resistance is $(\beta + 1)(r_e + r_e)$ = 151 (0.25 + 0.25) = **75.5kΩ**. To double the input resistance, add 0.25kΩ resistors in series with each emitter, at which point the gain (for differential output) is (100k+100k)/(250+250+250+250) = **200V/V**.

6.7

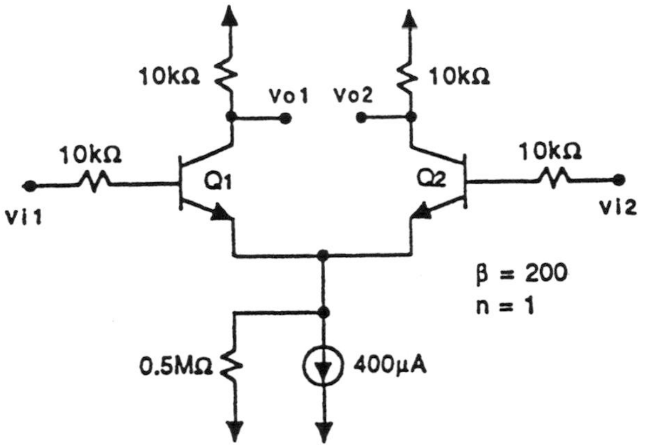

For each transistor, I_E = 200μA and r_e = 25mV/0.2mA = 125Ω. Thus the differential input resistance is 2 (201) (0.125kΩ) = **50.25kΩ**.

For differential output: Differential gain from bases = 200/201 (10k/0.125k) = 79.6V/V. Differential gain from input sources = 50.25/(10+10+50.25) × 79.6 = **56.94V/V**. Common-mode gain = **0V/V**. CMRR = 56.9/0 = ∞, as ratio and in dB. Common-mode input resistance = $(\beta + 1)(R)$ = 201 (0.5MΩ) ≈ **100MΩ**.

For single-ended output:

$|A_{ds}| = \frac{200}{201}\ \frac{10k\Omega}{0.125 + 0.125} \times \frac{50.25}{70.25} = \frac{56.94}{2}$ = **28.5V/V**, $A_{cm} = -\frac{200}{201}\ \frac{10k\Omega}{2\ (0.5M\Omega)}$ = **–.00995V/V**, CMRR = 28.5/0.00995 = – 9.95 mV/V. **2864V/V ≡ 69.1dB.**

6.8 From P6.7 above, for outputs taken differentially, A_d = 56.9V/V. For A_{cm}: **For matched loads,** it is 10kΩ/1MΩ – 10kΩ/1MΩ = 0V/V. **For ± 1% loads,** it is 10k(1.01)/1M – 10k(0.99)/1M = 0.02(10k)/1M = 2×10^{-5}**V/V ≡ –94dB**, for which CMRR = 56.9/(2×10^{-5}) = 2.85×10^{6}**V/V ≡ 129dB. For ± 10% loads**, correspondingly, A_{cm} = 2×10^{-4}**V/V ≡ –74dB**, and CMRR = **2.85×10^{5}V/V ≡ 109dB**

6.9

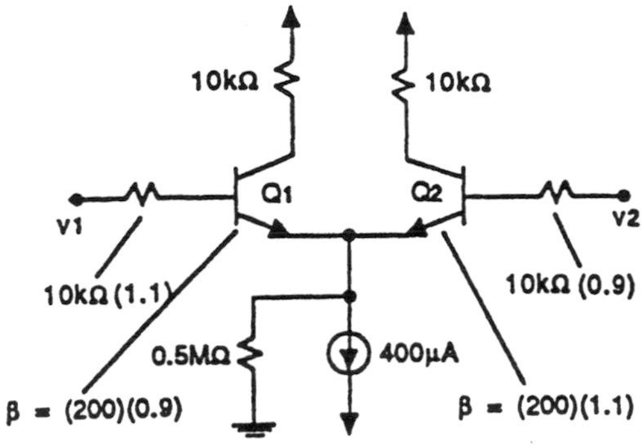

Note that the collector resistors are ideal (that is both are exactly 10kΩ).

For DC Bias: Assume that junction voltages are adequately modelled by $r_{e1}, r_{e2} \approx$ 25mV/200μA = 0.125kΩ nominally. Now for i in Q_1 and $0.400 - i$ in Q_2, $10\,(1.1)\,\frac{i}{200\,(.9)} + 0.125\,i = 0.125\,(0.4 - i) + 10\,(.9)\,\frac{(.4 - i)}{200\,(1.1)}$. Now multiplying by 100: $6.11\,i + 12.5i = 5 - 12.5i + 1.636 - 4.09i$, $35.2i = 6.636$, $i = 0.1885$. That is i_{E1} = 188.5μA, and i_{E2} = 211.5μA.

For signals: Now r_{e1} = 25mV/188.5μA = 132.6Ω, r_{e2} = 25/211.5 = 118.2Ω, $r_{\pi 1}$ = ((200) (.9) +1) 132.6 = 24.0kΩ, $r_{\pi 2}$ = (200 (1.1) +1) 118.2 = 26.1kΩ. Now for common-mode input, base currents split, with 26.1/(24.0 + 26.1) = 0.52 of the input change in the base of Q_1 (and 0.48 in the base of Q_2). Correspondingly, for a total base current i, $i_{C1} = \beta_1 i_{b1}$ = (200) (0.9) (0.52)i = 93.6i, and $i_{C2} = \beta_2 i_{b2}$ = (200) (1.1) (.48)i = 105.6i. Thus i_{C2}/i_{C1} = 105.6/93.6 = 1.128. **Note** that the dc current ratio is 211.5/188.5 = 1.122, essentially the same. From either point of view, there will be about a 12% total mismatch in the two output voltages. $A_{cm} \approx \dfrac{10k\Omega}{2\,(500k\Omega)} \times \dfrac{12}{100}$ = 12 × 10⁻⁴V/V. Since $A_d \approx \dfrac{10k\Omega + 10k\Omega}{.133 + .118}$ = **79.7V/V**, CMRR = $\dfrac{79.7}{12 \times 10^{-4}}$ = 66.4 × 10³V/V ≡ **96.4dB.**

6.10 For I_{bias} = 400μA, $I_{E1} = I_{E2}$ = 200μA nominally, and r_e = 25mV/0.2mA = 125Ω, R_E = 9 (125) = **1.125kΩ**, and $R_E + r_e$ = 1.25kΩ. Using the dc analysis from P6.9 above with $i_{E1} = i$, see that 10 (1.1i)/(200(.9)) + 1.25i = 1.25(0.4 –i) + 10(0.9) (0.4 – i)/(200(1.1)). Multiplying by 100, 6.11i + 125i = 50 – 125i + 1.64 – 4.09i, or 260.2i = 51.64, and i = 0.1985. That is, i_{E1} = 198.5μA and i_{E2} = 201.5μA, with $\dfrac{i_{E2}}{i_{E1}} = \dfrac{201.5}{198.5}$ = 1.015. Thus there is about a 1.5% mismatch, such that $A_{cm} \approx \dfrac{10k\Omega}{1M\Omega} \times$ 1.5/100 = **1.5 × 10⁻⁴V/V** with A_d = **79.9V/V**, the same as before, and CMRR = $\dfrac{79.7}{1.5 \times 10^{-4}}$ = 53.3 × 10⁴ ≡ **114.5dB**. Note that there is a nearly 20dB improvement due to the balancing effect of the emitter resistors.

SECTION 6.3: OTHER NON-IDEAL CHARACTERISTICS OF THE DIFFIERENTIAL AMPLIFIER

6.11 For I_{bias} = 200μA, $r_{e1} = r_{e2}$ = 25mV/0.1mA = 250Ω. **For the Basic Amplifier,** differential gain = $\dfrac{R_C + R_C}{0.25k\Omega + 0.25k\Omega}$ = 4 R_CV/V. Now ±5% variation in R_C produces an output offset of 0.1mA (1.05R_C – 0.95R_C) = 0.1 (0.1)R_C = 0.01R_C. Corresponding input offset (to reduce output to zero) is $V_{OS} = V_O/\text{gain} = \dfrac{.01\,R_C}{4\,R_C}$ = **2.5mV**, or from equation 6.49: $|V_{OS}| = V_T \left(\dfrac{\Delta R_C}{R_C}\right)$ = 25mV (2 (5/100)) = **2.5mV**. **For emitter resistors** $R_E = 9r_e$: Here, the differential gain = $\dfrac{R_C}{0.25\,(1+9)}$ = 0.4R_C. Now to compensate an output offset of 0.01R_C, we need $V_{OS} = \dfrac{.01R_C}{0.4R_C}$ = **25mV.**

6.12 From P6.11 above, to compensate for ±5% R_C variation, one needs a 2.5mV input offset with no emitter resistors. This involves an increase in one of the collector currents to 105μA and decrease in the other to 95μA. Now with $R_E = 9\,r_e$ = 9 (250) = 2250Ω nominally, but actually ranging from 0.95(2250) = 2.1375kΩ, to 1.05(2250) = 2.3625kΩ, equivalent offset can reach .105(2.3625) –.095(2.1375) = .2481 – .2031 = 45mV. Total maximum offset is approximately 2.5 + 45 = **47.5mV.**

For uncorrelated variation: For nominal R_E (from P6.11 above), acquire 25mV due to R_C variation. For nominal R_C and varying R_E, to achieve 100μA in each transistor, we need an offset of 0.1(2.3625) – 0.1(2.1375) = 22.5mV. (See the worst case is again 22.5 + 25 = 47.5mV, as an alternative approach). **For no correlation,** $V_{OS} = (22.5^2 + 25^2)^{1/2}$ = **33.6mV. For collector resistors trimmed:** Collector currents will be both 100μA and, as above, V_{OS} = **22.5mV.**

6.13 **For equal 2mV offsets:** $V_{OS} = (2^2 + 2^2 + 2^2 + 2^2)^{1/2} = (16)^{1/2}$ = **4mV.**

For unequal offsets: $V_{OS} = (0.5^2 + 1^2 + 2^2 + 4^2)^{1/2} = (.25 + 1 + 4 + 16)^{1/2} = (21.25)^{1/2}$ = **4.61mV.**

6.14 For the offset totally compensated, the collector currents in both transistors will be 100/2 = 50μA. Assume β_1 is 5% high and β_2 is 5% low, while R_{S1} is 5% low and R_{S2} is 5% high. Thus the total offset

is $I_{B1} R_{S1} - I_{B2} R_{S2}$ = 50/(105 + 1) × 100 (.95) − 50 /(95 + 1) × 100 (1.05) or
$\frac{50 \times 10^{-6} \times 100 \times 10^3}{100} \left(\frac{.95}{1.06} - \frac{1.05}{.96} \right) = 50 \times 10^{-3} (.896 - 1.094) = -50 \times 10^{-3} (.198) = -9.9\text{mV}.$
Thus the offset can be as large as **9.9mV**.

6.15 For each transistor, r_e = 25mV/150μA = 0.1667kΩ. For υ_{be} = 10mV, υ_c = 60kΩ/.1667kΩ × 10mV = 3.6V. Thus, the lowest collector voltage is 15V − 150μA x 60kΩ −3.6V, or 15 − 9 − 3.6 = 2.4V. For bare saturation, the base voltage can exceed this by 0.7 − 0.4 = 0.3V. Thus, the highest usable common-mode input is 2.4 + 0.3 = **2.7V**.

SECTION 6.4: BIASING IN BJT INTEGRATED CIRCUITS

6.16 As seen from the emitter, I_E = 100μA and r_e = 25mV/100μA = 250 Ω. Thus the resistance between the terminals is 250Ω.

For two in parallel, the current divides (say equally) with each r_e = 25mV/50μA = 500Ω. The parallel resistance is then 500Ω‖500Ω = 250Ω, as before. One can see this directly since the junctions are bigger, but the current is the same.

For two in series, the current in each is the same; the resistance of each is the same, and the total resistance is 250 + 250 = 500Ω.

6.17 From Eq.6.63, $\frac{I_O}{I_{REF}} = \frac{1}{1 + 2/\beta}$. For 1% error, $\frac{1}{1 + 2/\beta}$ = 0.99, 1 = 0.99 + 1.98/β, β = 1.98/0.01 = **198.**
For 0.1% error, $\frac{1}{1 + 2/\beta}$ = 0.999, $\beta = \frac{1.998}{1 - 0.999}$ = **1998.**

6.18 At 1mA, V_{BE} = 700 + 25 ln (1mA/10mA) = 642.4mV.

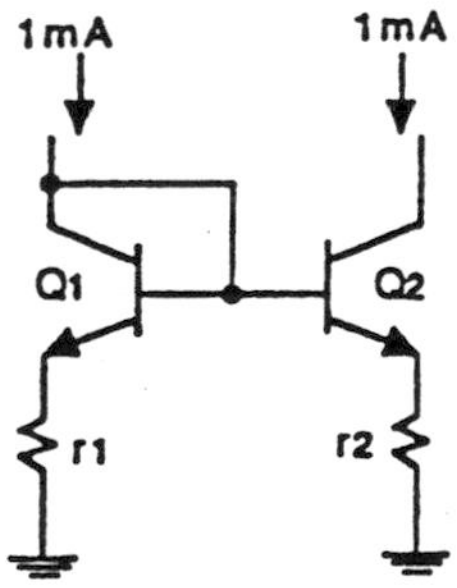

Required r = (642.4mV)(0.1)/1mA = 64.2Ω. Use r = **60 Ω. For β = 90, I$_R$ = 1mA** and I_O = 1mA, I_{B2} = 1/90, I_{E1} = (1 − 1/90) = 0.9889, and I_{C1} = 0.9778 mA, I_{E2} = 91/90(1) = 1.0111, and I_{C2} = 1.0000mA. V_{BE1} = 700 + 25 ln (0.9778/10) = 641.9mV, V_{BE2} = 700 + 25 ln (1/10) = 642.4mV, V_{r1} = 60 (.9889) = 59.3mV, V_{r2} = 60 (1.0111) = 60.67mV, $V_{BE1} + V_{r1}$ = 641.9 + 59.3 = 701.2mV, $V_{BE2} + V_{r2}$ = 642.4 + 60.67 = 703.1 mV. Thus r_1 must be increased to $\frac{703.1 - 641.9}{0.9889}$ = **61.89Ω. For β = 90, I = 0.5mA:** $I_{B2} \approx$ 0.5/90, and $I_{E1} \approx$ 0.5 − 0.5/90 = 0.5 (0.9889) = 0.4944mA, and I_{C1} = 90/91 (.494) = 0.489mA, and $V_{BE1} + V_{r1}$ = 700 + 25 ln (0.489/10) + 0.4944 (61.89) = 624.55 + 60.67 = 655.1 mV.

Assume $I_{C2} \approx$ 0.5mA → V_{BE2} = 700 + 25 ln (0.5/10) = 625.1mV. Thus $I_{E2} \approx \frac{655.1 - 625.1}{60}$ = 0.500 mA, and I_{C2} = 90/91 (0.500) = 0.4945mA. Gain = .4945/.5000 = **0.990A/A.**

For β = 90, I$_R$ = 2.0mA: $I_{B2} \approx$ 2/90, I_{E1} = 2 − 2/90 = 2 (0.9889) = 1.9778mA, and I_{C1} = 90/91 (1.9778) = 1.9561mA, and $V_{BE1} + V_{r1}$ = 700 + 25 ln (1.9561/10) + 1.9778 (61.89) = 781.6 mV. Assume $I_{C2} \approx$ 2mA → V_{BE2} = 700 + 25 ln (2/10) = 659.8mV, $I_{E2} = \frac{781.6 - 659.8}{60}$ = 2.030mA, I_{C2} = 90/91 (2.03) = 2.008mA. Gain = 2.009/2.00 = **1.004A/A.**

For β = 70, 0.5mA: $I_{B2} \approx$ 0.5/70, $I_{E1} \approx$ 0.5 − 0.5/70 = 0.5 (0.9857) = 0.4929mA, I_{C1} = 70/71 (0.4929) = 0.4859mA, $V_{BE1} + V_{r1}$ = 700 + 25 ln (0.4859/10) + 0.4929 (61.89) = 654.9 mV. Assume $I_{C2} \approx$ 0.5mA → V_{BE2} = 700 + 25 ln (0.5/10) = 625.1mV, $I_{E2} = \frac{654.9 - 625.1}{60}$ = 0.4967 mA, I_{C2} = 70/71 (0.4967) = 0.4897 mA, gain = 0.4897/0.5000 = **0.979A/A.**

And at 1.0mA; $I_{B2} \approx 1.0/70$, $I_{E1} \approx 1.0 - 1.0/70 = 1.0\ (0.9857) = 0.9857\text{mA}$, $I_{C1} = 70/71\ (0.9857) = 0.9718\text{mA}$, $V_{BE1} + V_{r1} = 700 + 25 \ln (0.9718/10) + 0.9857\ (61.89) = 702.7$ mV. Assume $I_{C2} \approx 1.0 \rightarrow V_{BE2} = 700 + 25 \ln (1/10) = 642.43\text{mV}$. Thus $I_{E2} = \dfrac{702.7 - 642.4}{60} = 1.005\text{mA}$, $I_{C2} = 70/71\ (1.005) = 0.991$, gain = .991/1.00 = **0.991A/A.**

And at 2.0mA: $I_{B2} \approx 2.0/70$, $I_{E1} = 2.0\ (0.9857) = 1.9714\text{mA}$, $I_{C1} = 1.9714\ (70/71) = 1.944$, $V_{BE1} + V_{r1} = 700 + 25 \ln (1.944/10) + 1.9714\ (61.89) = 781.1$. Assume $I_{C2} \approx 2.0\text{mA} \rightarrow V_{BE2} = 700 + 25 \ln (2/10) = 659.8\text{mV}$. Thus $I_{E2} = \dfrac{781.1 - 659.8}{60} = 2.021\text{mA}$, $I_{C2} = 70/71\ (2.022) = 1.993\text{mA}$, gain = 1.993/2.0 = **0.996A/A.**

6.19 $\dfrac{I_O}{I_R} = \dfrac{1}{1 + 2/\beta}$, $I_O = 100\mu A \dfrac{1}{1 + 2/150} = 98.684\mu\text{A}$. Thus I_O is low by 100 – 98.684 = 1.316μA, for which $r_O = 150\text{V}/100\mu\text{A} = 1.5\text{M}\Omega$. To compensate, $\dfrac{V_{out}}{1.5 \times 10^6} = 1.316 \times 10^{-6}$, or $V_{out} = 1.5\ (1.316) =$ **1.974V.** For a net error of <1%, output current must range from 99μA to 101μA.

At 99μA, r_o contributes 99 – 98.684 = 0.316μA, for which $V_{out} = .316 \times 1.5 =$ **0.474V.**

At 101μA, r_o contributes 101 – 98.684 = 2.316μa, for which $V_{out} = 2.316 \times 1.5 =$ **3.474V.**

6.20 For a change from 25°C to 75°C, V_{BE} drops by (75 – 25)2 = 100mV. For 100mV to be a 5% change, the drop in R must be 100mV/.05 = 2V. That is, V_{CC} = **2.7V**, and R = 2/100μA = **20kΩ.**

6.21

(a)

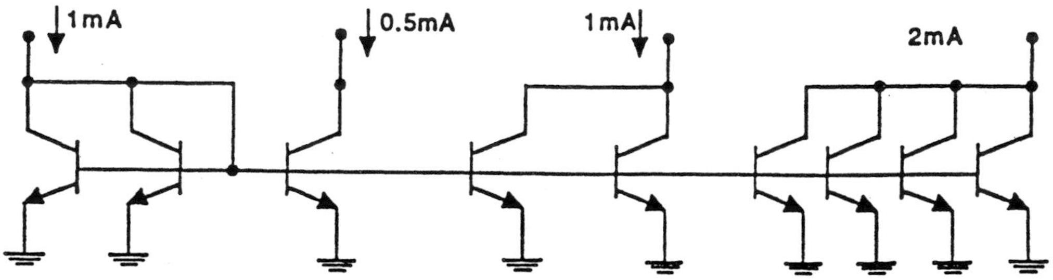

Need **9 BJTs.**

(b)

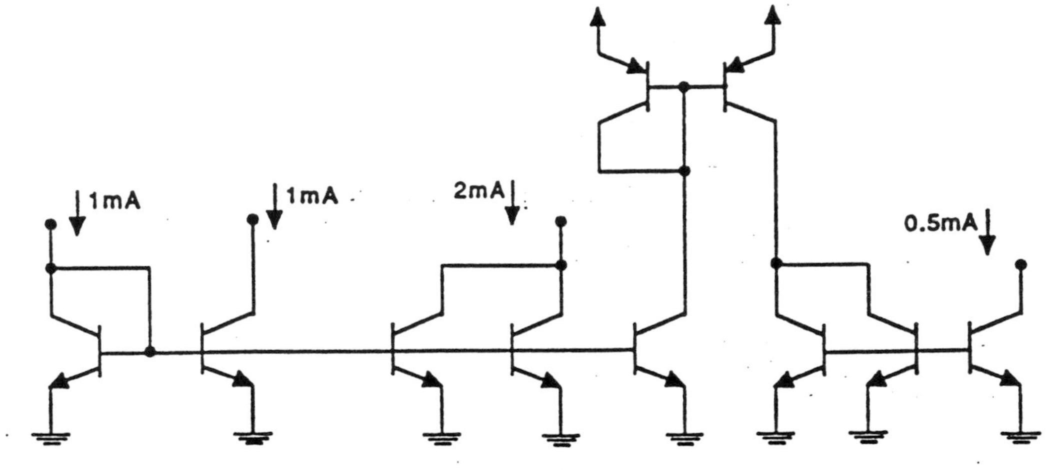

Need **10 BJTs.**

(c) For both ends of I_R available, T_A is not needed. Require **9 BJTs.**

6.22

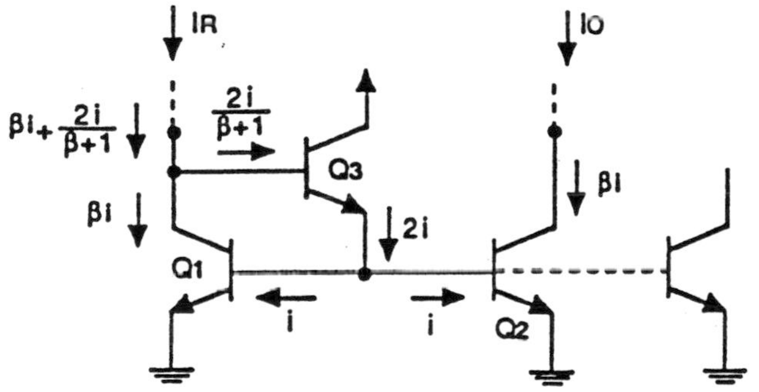

$$\frac{I_O}{I_R} = \frac{\beta i}{\beta i + \dfrac{2\,i}{\beta+1}} = \frac{\beta}{\beta + \dfrac{2}{\beta+1}},$$

or

$$\frac{I_O}{I_R} = \frac{1}{1 + \dfrac{2}{\beta^2+\beta}},$$ as noted before.

For two outputs, I_O:

$$I_R = \beta i + \frac{3i}{\beta+1}, \text{ and}$$

$$\frac{I_O}{I_R} = \frac{1}{1 + \dfrac{3}{\beta^2+\beta}}$$

6.23

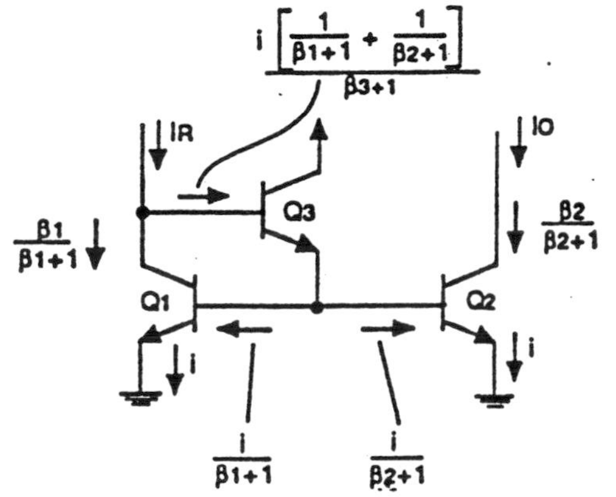

$$a = \frac{i\left[\dfrac{1}{\beta_1+1} + \dfrac{1}{\beta_2+1}\right]}{\beta_3+1}$$

$$\frac{I_O}{I_R} = \frac{\dfrac{\beta_2}{\beta_2+1}\,i}{\dfrac{\beta_1}{\beta_1+1}\,i + i\left[\dfrac{1/(\beta_1+1) + 1/(\beta_2+1)}{\beta_3+1}\right]}$$

$$= \frac{1}{\dfrac{\beta_1(\beta_2+1)}{\beta_2(\beta_1+1)} + \dfrac{\dfrac{\beta_2+1}{\beta_1+1} + 1}{\beta_2\,(\beta_3+1)}} \quad ---\ (1)$$

For the optimal location: See a) Q_3 provides a second-order effect, and any β would be OK. b) For Q_1 with lower β, i_{C1} is fixed and i_{E1} will increase, causing V_{BE1} to increase, and I_O to increase. c) For Q_2 with higher β, I_O is larger for a fixed V_{BE}. Conclude that one should make $\beta_1 = (1 - \mathbf{k})\,\beta$, $\beta_2 = (1 + \mathbf{k})\,\beta$, $\beta_3 = \beta$.

Substitute in (1)

$$\frac{I_O}{I_R} = \frac{1}{\dfrac{(1-k)\,\beta\,(\beta + k\beta + 1)}{(1+k)\,\beta\,(\beta - k\beta + 1)} + \dfrac{\dfrac{\beta + k\beta + 1}{\beta - k\beta + 1} + 1}{(\beta + k\beta)\,(\beta + 1)}} = \frac{1}{\dfrac{(1-k)\,(\beta + k\beta + 1)}{(1+k)\,(\beta - k\beta + 1)} + \dfrac{\beta + k\beta + 1 + \beta - k\beta + 1}{\beta\,(k+1)\,(\beta+1)\,(\beta - k\beta + 1)}}$$

$$= \frac{1}{\dfrac{(1-k)\,(\beta + k\beta + 1)}{(1+k)\,(\beta - k\beta + 1)} + \dfrac{2\,(\beta + 1)}{\beta\,(k+1)\,(\beta+1)\,(\beta - k\beta + 1)}}$$

$$= \frac{1}{\dfrac{(1-k)\,(\beta + k\beta + 1)}{(1+k)\,(\beta - k\beta + 1)} + \dfrac{2}{\beta\,(1+k)\,(\beta - k\beta + 1)}} = \frac{(\beta + k\beta)(\beta - k\beta + 1)}{(\beta - k\beta)(\beta + k\beta + 1) + 2}$$

$$= \frac{\beta^2 - k\beta^2 + \beta + k\beta^2 - k^2\,\beta^2 + k\,\beta}{\beta^2 + k\beta^2 + \beta - k\beta^2 - k^2\,\beta^2 - k\,\beta + 2} = \frac{\beta^2 + \beta + k\beta - k^2\beta^2}{\beta^2 + \beta - k\beta + 2 - k^2\beta^2}.$$

This becomes one, if $k\beta = 2 - k\beta$, or $k\beta = 1$, or $\mathbf{k} = 1/\beta$!

Note that for $k = 1/\beta$, $\beta_1 = (1 - 1/\beta)\beta = \beta - 1$, and $\beta_2 = (1 + 1/\beta = \beta + 1$, with $\beta_3 = \beta$.

6.24

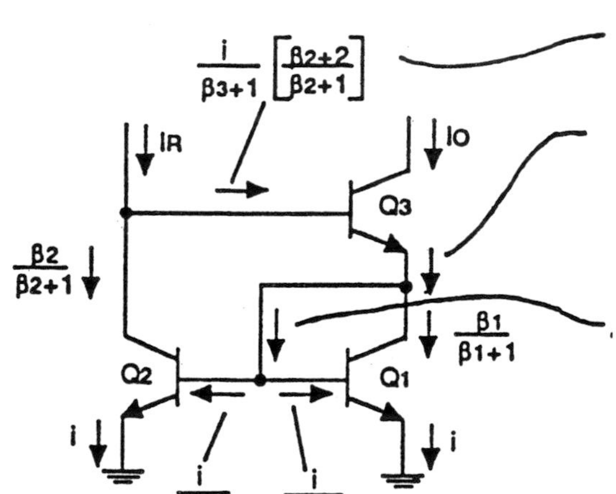

$$a = \frac{i}{\beta_3 + 1}\left(\frac{\beta_2 + 2}{\beta_2 + 1}\right)$$

$$b = \left(\frac{\beta_1}{\beta_1 + 1} + \frac{1}{\beta_1 + 1} + \frac{1}{\beta_2 + 1}\right) i = \left(\frac{1}{\beta_2 + 1} + 1\right) i = \left(\frac{\beta_2 + 2}{\beta_2 + 1}\right) i$$

$$c = \left(\frac{1}{\beta_2 + 1} + \frac{1}{\beta_1 + 1}\right) i$$

$$\frac{\mathbf{I_O}}{\mathbf{I_R}} = \frac{\dfrac{\beta_3}{\beta_3+1}\left(\dfrac{\beta_2+2}{\beta_2+1}\right)}{\dfrac{1}{\beta_3+1}\left(\dfrac{\beta_2+2}{\beta_2+1}\right) + \dfrac{\beta_2}{\beta_2+1}} = \frac{\beta_3\,(\beta_2+2)}{\beta_2+2 + \beta_2\,(\beta_3+1)} = \frac{\beta_2\,\beta_3 + 2\beta_3}{\beta_2\,\beta_3 + 2\beta_2 + 2} = \frac{1}{1 + \dfrac{2\,(\beta_2 - \beta_3 + 1)}{\beta_2\,\beta_3 + 2\beta_3}}$$

For optimal placement of transistors: See (a) that Q_1, being diode-connected, β_1 does not matter, (b) that for Q_2 with low beta, i_{C2} being fixed, i_{E2} increases, V_{BE} increases, and i_{E1} increases, (c) that for β_3 high, I_O increases. Thus, use $\beta_1 = \beta$, $\beta_2 = (1 - \mathbf{k})\,\beta$, $\beta_3 = (1 + \mathbf{k})\,\beta$. Correspondingly, $\dfrac{I_O}{I_R}$ =

$\frac{\beta_2\beta_3 + 2\beta_3}{\beta_2\beta_3 + 2\beta_2 + 2} = \frac{(1+k)(1-k)\beta^2 + 2(1+k)\beta}{(1+k)(1-k)\beta^2 + 2(1-k)\beta + 2} = \frac{\beta^2 - k^2\beta^2 + 2k\beta + 2\beta}{\beta^2 - k^2\beta^2 - 2k\beta + 2\beta + 2}$. This is unity if $2k\beta = -2k\beta + 2$, or $k\beta = ½$, or $\mathbf{k} = \frac{1}{2\beta}$.

6.25 Assume very large β: For the 100μA reference: V_{BE} = 700 + 25 ln (0.1/1) = 642.4mV. For the 1μA output: V_{BE} = 700 + 25 ln (10^{-3}/1) =527.3mV. For the 10μA output: V_{BE} = 700 + 25 ln (10^{-2}/1) = 584.9mV. Thus, for 1μA output, $R_E = \frac{(642.4 - 527.3)\times 10^{-3}}{10^{-6}}$ = **115kΩ**. For 10μA output, $R_E = \frac{(642.4 - 584.9)\ 10^{-3}}{10^{-5}}$ = **5.75kΩ**

SECTION 6.5: THE BJT DIFFERENTIAL AMPLIFIER WITH ACTIVE LOAD

6.26 Each transistor conducts 50μA, for which r_o = 150/50μA = 3MΩ, and r_e = 25mV/50μA = 500Ω, G_m = 2/(500+500) = **2mA/V**, $A_\upsilon = -\frac{2}{500+500}(3M\Omega \| 3M\Omega) = -2\times 10^{-3} \times \frac{3}{2} \times 10^6$ = **−3000V/V**. $R_o = 3M\Omega \| 3M\Omega$ = **1.5MΩ**, R_{in} = (75 + 1) (500 + 500) = **76kΩ**. For a 76kΩ load, $A_\upsilon = -2\times 10^{-3}(1.5M\Omega \| 76k\Omega)$ = **−144.7V/V**.

6.27 For each of the transistors, with an emitter resistor $R_E = r_e$, the output resistance increases to $R_o = r_o(1 + g_m R_E') \simeq 3\times 10^6 (1 + 500/(500))$ = 6MΩ. Thus, the output resistance is approximately 6MΩ || 6MΩ = **3MΩ**. The overall transconductance is 2/(500+500+500+500) = **1mA/V**. The open-circuit voltage gain is

− 1mA/V × 3MΩ = **−3000V/V**.

6.28 I = 100μA → Collector current for all transistors is 50μA, for which r_o = 75V/50μA = 1.5MΩ, and r_e = 25mV/50μA = 500Ω, and r_π = (75 + 1) (0.50) = 38kΩ. For the cascode transistors, $R_o \simeq r_o(1 + g_m r_\pi) = r_o(\beta + 1) = 1.5\times 10^6 (76)$ = 114MΩ. Also $r_\mu \simeq 10\,(75)\,(1.5\times 10^6)$ = 1125MΩ. Thus the output resistance (in MΩ) is 114 || 114 || 1125 || 1125 = 114/2 || 1125/2 = **51.8MΩ**. Overall, $g_m = G_m \simeq$ 2/(500+500) = **2mA/V**, A_υ = −2mA/V × 51.8MΩ = **103.5 × 10³V/V**.

SECTION 6.6: MOS DIFFERENTIAL AMPLIFIERS

6.29 Here, $g_m = \frac{I}{V_{GS} - V_t}$ for each transistor − − − (1), with a maximum at

$i_{D1} = i_{D2} = I/2 = K(V_{GS} - V_t)^2$ − − − (2), where $K = 1/2k = 1/2k_n'(W/L)$, and $V_{GS} - V_t = \left[\frac{KI}{2}\right]^{½}$,

and $g_{m\ max} = \frac{I}{(KI/2)^{½}} = \sqrt{2I/(K)}$, or $g_{m\ max}$ = 2K $(V_{GS} - V_t)$, (from (1) with (2)). In general, $i_D = K(\upsilon_{GS} - V_t)^2$, and $g_m = 2K(\upsilon_{GS} - V_t) = 2K(V_{GS} \pm \frac{\upsilon_{id}}{2} - V_t)$. For a 10% g_m drop, $2K(V_{GS} \pm \frac{\upsilon_{id}}{2} - V_t)$ = 0.9 (2K) $(V_{GS} - V_t)$, when $V_{GS} \pm \frac{\upsilon_{id}}{2} - V_t = 0.9 V_{GS} - 0.9V_t$, or $0.1(V_{GS} - V_t) = \pm\frac{\upsilon_{id}}{2}$, or $\frac{\upsilon_{id}}{V_{GS} - V_t}$ = ± **0.2**. For 5%: $\frac{\upsilon_{id}}{V_{GS} - V_t}$ = ± 2(.05) = ± **0.1**. For 1%: $\frac{\upsilon_{id}}{V_{GS} - V_T}$ = ± 2(.01) = **±0.02**.

6.30 $i_D = 1/2k_p'(W/L)(\upsilon_{GS} - V_t)^2 = K(\upsilon_{GS} - V_t)^2$, where $K = ½\,\mu_p C_{ox}(W/L) = ½ \times 10 \times 120/6 = 100\mu A/V^2$. For equal current division, 25/2 = 100 $(\upsilon_{GS} - 1)^2$, $\upsilon_{GS} - 1 = \pm(1/8)^{½}$, and υ_{GS} = 1.354V. Thus V_{GS} = **1.35V**. Now, $g_m = 2K(\upsilon_{GS} - V_t)$ = 2 (100) (1.354 − 1) = **70.8μA/V**, and $r_o = \frac{V_A}{I_D} = \frac{50}{25/2}$ = 4MΩ.

Maximum gain will occur for outputs taken differentially: (a) For ideal loads: $A_v = \frac{2\,(4 \times 10^6)}{2\,(1/70.8 \times 10^{-6})} =$ **283.2V/V**, (b) For loads with V_A = 50V: $A_v = \frac{2\,(4 \times 10^6 \,||\, 4 \times 10^6)}{2\,(1/70.8 \times 10^{-6})} = 283.2/2 =$ **141.6V/V**.

6.31 **With balanced loads,** $i_D = \frac{25}{2} = 100(v_{GS} - V_t)^2$, with v_{GS} = 1.354V (from P6.30 above). Now for i_D = 0.9 (25/2) = 100 $(v_{GS} - 1)^2$, $v_{GS} = (\frac{0.9}{8})^{1/2} + 1 = 1.3354$V. For i_D = 1.1 (25/2) = 100 $(v_{GS} - 1)^2$, $v_{GS} = (\frac{1.1}{8})^{1/2} + 1 = 1.371$V. Thus the input offset = 1.371 – 1.335 = **36mV**.

6.32 From Exercise 6.15, $k_n' = \mu_n C_{ox} = 20\mu A/V^2$, $(W/L) = (120/6) = 20$, and $1/2 k_n'(W/L) = 200\mu A/V^2$, with V_{GS} = 1.25V, g_m = 100μA/V, and V_t = 1V. For an R_D mismatch of ± 1%, $V_{OS} = \left[\frac{V_{GS} - V_t}{2}\right] \frac{\Delta R_D}{R_D} = \frac{1.25 - 1}{2} \times \frac{1}{100}$ = **±1.25mV**. For a (W/L) mismatch of ±1%, $V_{OS} = \left[\frac{V_{GS} - V_t}{2}\right] \frac{\Delta(W/L)}{(W/L)} = \frac{1.25 - 1}{2} \times \frac{1}{100}$ = **±1.25mV**. For V_t tolerance of ±0.6mV, V_{OS} = **±0.6mV**. Thus the worst-case offset is 1.25 + 1.25 + 0.6 = **3.1mV**. The likely offset = $\sqrt{1.25^2 + 1.25^2 + 0.6^2}$ = **1.87mV**.

6.33 Here, $I_O \approx I_{REF}$ = 100μA, and $K = 1/2 k'(W/L) = 1/2(200) = 100\mu A/V^2$.
Roughly: $i_D = K\,(v_{GS} - V_t)^2$, $100 = 100\,(v_{GS} - 1)^2 \rightarrow v_{GS}$ = 2V. Now, for Q_1, Q_4, $v_{DS} \approx$ 2V and $(1 + \frac{v_{DS}}{V_A}) = 1 + \frac{2}{20} = 1.1$. This could be ignored, but let us include it in a more basic calculation:

More precisely: For Q_1, Q_4, $v_{DS} = v_{GS} = v$, and $100 = 100\,(v - 1)^2\,(1 + v/20)$, or $1 = (v^2 - 2v + 1)\,(1 + .05v)$. Thus $v^2 - 2v + 1 + .05v^3 - 0.1v^2 + .05v = 1$ or $.05\,v^3 + 0.9\,v^2 - 1.95\,v = 0$, or $.05\,v^2 + 0.9\,v - 1.95 = 0$, $v = (-0.9 \pm \sqrt{0.9^2 + 4(1.95)\,(.05)})/(2(.05)) = (-.9 \pm 1.095)/0.1$ = 1.954V. Thus V_{GS1} = **1.954V**. Now for $V_O = V_{D3} = V_{D4}$, see I_O = **100μA**, from symmetry arguments. For each device, $r_o \approx \frac{V_A}{I_D} = \frac{20}{100\mu A}$ = 200kΩ, and $g_m = 2K\,(v_{GS} - V_t) \approx 2\,(100 \times 10^{-6})\,(2 - 1)$ = 200μA/V. For the whole mirror, (from Eq. 6.116) $R_{out} \approx g_{m3}\,r_{o3}\,r_{o2} = 200 \times 10^{-6} \times 200 \times 10^3 \times 200 \times 10^3 = 8 \times 10^6\Omega$ = **8MΩ**. Thus, for V_O = 12V, with the standard output being at 2 + 2 = 4V, the extra current = (12 – 4)/8MΩ = 8/8 = 1μA. Thus I_O = **101μA** for V_O = +12V.

6.34 Assume $I_O \approx I_{REF}$ = 100μA. From results of P6.33 above, $V_{GS} \approx$ 2.0V. Thus $V_{D1} \approx$ 2.0 + 2.0 = 4V. Correspondingly, for Q_1, with $v_{GS} = v \approx$ 2.0V and $v_{DS} \approx$ 2V, $100\mu A = 100\mu A/V^2\,(v - 1)^2\,(1 + \frac{2v}{20})$, $1 = (v^2 - 2v + 1)\,(1 + 0.1v) = v^2 - 2v + 1 + 0.1v^3 - 0.2v^2 + 0.1v$, or $+ 0.1v^3 + 0.8v^2 - 1.9v = 0$, $0.1v^2 + .8v - 1.9 = 0$, $v = \frac{-0.8 \pm \sqrt{.8^2 + 4(1.9)\,(.1)}}{2(.1)}$ = 1.916V. Thus, $v_{GS1} \approx$ 1.916V = **1.92V**. Now for Q_2, $i_D = 100\,(1.916 - 1)^2\,(1 + \frac{1.916}{20})$ = **91.9μA**. Now, for Q_3 with $V_O \approx$ 2 (1.92) = 3.84V, I_O = 91.9μA. From P6.91 of the Text, $R_o \approx (g_m\,r_o)r_o \approx$ **8MΩ** using the results of P6.33 above. Thus for V_O = 12V, I_O = 91.9 + (12 – 3.84)/8 = **92.9μA**.

6.35

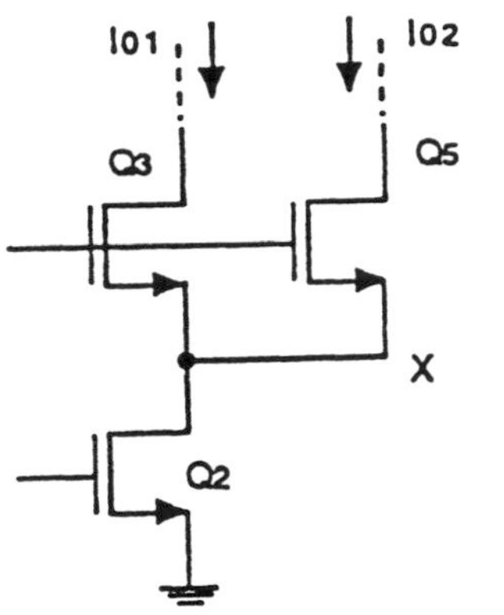

Generally speaking, see that currents in Q_3 and Q_5 will be the same, provided the output voltages are the same. Further, the current in each will slightly exceed 100/2 μA since the voltage at node X will rise slightly due to doubling of the equivalent K of Q_3, Q_5. Since the current in each of Q_3, Q_5 is only slightly more than half what it was, each output resistance will be twice as large. In particular, with outputs joined, the output resistance will be only slightly smaller than before: **Now to check these ideas:** $i_{D2} \approx 100\mu A \approx (100 + 100)(v_{GS} - 1)^2$. Thus $v_{GS} = 1 \pm \sqrt{½} = 1.707V$, rather than 2.0V previously. Thus node X will rise about 0.3V, and i_{D3} will increase by 0.3/200kΩ = 1.5μA. The current in Q_3, Q_5 will be about **50.8μA**, for which gm = 2K $(v_{GS} - V_t) = 2\,(100 \times 10^{-6})\,(1.707 - 1) = 141\mu A/V$. Now for outputs joined, $r_{035} = 20/(100 + 1.5) = 197k\Omega$.

Consider r_{02} in two parts of 400kΩ each, with $R_{03} = R_{05}$ = 197k × 400k × 141μA/V = 11.1MΩ. Together, R_{035} = 11.1/2 = **5.6MΩ**. When outputs operate independently, the output resistance decreases a lot since node X is grounded via $1/g_m$ of the other output. From Eq. 6.116 (full version) $R_{03} = r_{03} + 1/g_{m5} + g_{m3}\, r_{03}\, 1/g_{m5} \approx 2\, r_{03}$ = **800kΩ**. An improved circuit would split both Q_3 and Q_2 into two parts, with each pair of half-size transistors driven from the gate of Q_4, Q_1 respectively. The total device width needed would be same as in the original design, that is **4W** for each original transistor of width W. The version with only Q_3 duplicated uses a total width of **5W** and has poorer performance!

6.36 For I_B = 200μA, I_D for each transistor = 100μA, $r_o = V_A/I_D$ = 20/100μA = 200kΩ, $i_D = 1/2k'(W/L)(v_{GS} - V_t)^2$, $100 \times 10^{-6} = 1/\times 200 \times 10^{-6}(v_{GS} - 1)^2 \rightarrow v_{GS}$ = 2V, $g_m = k'(W/L)(v_{GS} - V_t) = (200)(2 - 1) = 200\mu A/V$. Gain $A_v = -\dfrac{2\,(200k \parallel 200k)}{1/0.2 + 1/0.2}$ = **−20V/V**. Gain reduces by a factor of two for a load of **100kΩ**.

SECTION 6.7: BiCMOS AMPLIFIERS

6.37

(a) Here, i_C = 10μA, $g_m = \dfrac{10 \times 10^{-6}}{25 \times 10^{-3}}$ = **400μA/V**, r_o = 100V/10μA = **10MΩ**, $R_i = \beta/g_m$ = 100/400 × 10^{-6} = **250kΩ**, $A_v = -400 \times 10^{-6} \times 10 \times 10^6$ = **−4000V/V**.

b) For i_D = 10μA = ½ × 20 × 20/2 $(v_{GS} - V_t)^2 = 100\,(v_{GS} - V_t)^2$. Thus $v_{GS} - V_t = (10/100)^{½}$ = 0.316, $g_m = k'(W/L)(v_{GS} - V_t) = (200 \times 10^{-6})\,0.316$ = **63.2μA/V**, r_o = 20/10μA = **2MΩ**, $R_i = \infty$, $A_v = -63.2 \times 10^{-6} \times 2 \times 10^6$ = **−126.4V/V**, that is much, much less!

6.38 For I = 10μA, g_{m2} = 400μA/V, $r_{\pi 2} = \dfrac{\beta_2}{g_{m2}} = \dfrac{100}{400 \times 10^{-6}}$ = 250kΩ, r_{02} = 10MΩ, g_{m1} = 63.2μA/V, r_{01} = 2MΩ. From Eq. 6.116, $R_{out} \approx g_{m2}\, r_{02}\,(r_{01} \parallel r_{\pi 2}) = 400 \times 10^{-6} \times 10 \times 10^6 \times (2 \times 10^6 \parallel 0.25 \times 10^6)$ = $888 \times 10^6\Omega$. Thus $\dfrac{v_o}{v_i} = -63.2 \times 10^{-6} \times \dfrac{100}{101} \times .888 \times 10^9$ = **−55.6 × 10³V/V**.

6.39 For I = 100μA, see by scaling from P6.38 above, g_{m2} = 4mA/V, $r_{\pi 2}$ = 25kΩ, r_{02} = 1MΩ, r_{01} = 200kΩ, and $v_{GS1} - V_t = (100/100)^{½} = 1$. Thus $g_{m1} = 2\,(100 \times 10^{-6})\,1$ = 200μA/V, $R_{out} \approx 4 \times 10^{-3} \times 10^6 \times (200 \times 10^3 \parallel 25 \times 10^3) = 88.8 \times 10^6$, $v_o/v_i = -200 \times 10^{-6} \times 100/101 \times 88.8 \times 10^6$ = **−17.6 × 10³V/V**.

6.40 From P6.37 at 10μA: For the BJTs, g_m = 400μA/V, r_π = 250kΩ, r_o = 10MΩ, r_μ = 10 (100)10^7 = 10 × 10^9Ω. For the MOS, g_m = 63.2μA/V, r_o = 2MΩ.

For the circuit as shown, $R_{out} \simeq 2 \times 10^6 \times 63.2 \times 10^{-6} \times (10 \times 10^6 \times 400 \times 10^{-6} \times 250 \times 10^3) \parallel (10 \times 10^9) = 126.4 \times (4000 \times 250 \times 10^3) \parallel 10^{10} = 126.4 \times .909 \times 10^9 =$ **115 × 10^9Ω**. *With Q_3, Q_6 not used,* $R_{out} \simeq 10^{10} \parallel (10^7 \times 400 \times 10^{-6} \times (.25 \times 10^6) \parallel (10 \times 10^6) = (10 \parallel .976)\ 10^9 =$ **0.889 × 10^9Ω**. *With Q_5, Q_2 eliminated,* $R_{out} \simeq 2 \times 10^6 \times 63.2 \times 10^{-6} \times (10 \times 10^6 \parallel 10 \times 10^9) = 1.264 \times 10^9$Ω.

SECTION 6.8: GaAs AMPLIFIERS

6.41 For a 1μm long, 1μm wide *GaAs* device, V_t = –1.0V, β_1 = 100μA/V^2, λ = 0.1V^{-1}, V_A = 10V, $I_S = 10^{-15}$A, n = 1.1.

(a) Symmetry would indicate that V_a = 5/2 = **2.5V**, $\beta = \beta_{10}$ = 10 (100) = 1mA/V^2, $i_D = \beta\,(v_{GS} - V_t)^2\,(1 + \lambda\, v_{DS}) = 1 \times 10^{-3}\,(0 - -1)^2\,(1 + 0.1\,(2.5))$. Thus I_a = **1.25mA**.

(b) Assume operation is in saturation. Lower transistor (Q_1) operates with v_{GS} = 0. Upper one (Q_2) is 1.5 × larger: Thus $1.5\,(v_{GS2} - -1)^2 = (0 - -1)^2$, or $v_{GS2} \simeq (1/1.5)^{1/2} - 1 = -0.18$V. Thus, $i_{D1} = 20 \times 100 \times 10^{-6}\,(0 - 1)^2\,(1 + 0.1\,(5 - .18)) \simeq 2.964$mA. See that the drain of Q_2 is at about 5 – 1kΩ (2.96) = 2.04V, while the gate is at 0V. See that operation is indeed in saturation. *Check*: $2.964 = 30 \times 100 \times 10^{-6}\,(v_{GS} - -1)^2\,(1 + 0.1\,(2.96 - -0.18))$ $(v_{GS} + 1)^2 = 0.752$, $v_{GS} = 0.867 - 1 = -0.132$V OK. $V_b \simeq$ **–0.13V**, and $I_b \simeq$ **2.95mA.**

(c) See Q_1 operates at v_{GS} = 0 with $V_{c1} \simeq 0$V. $I_{c1} \simeq 20 \times 100 \times 10^{-6}\,(0 - -1)^2\,(1 + 0.1(5)) =$ **3.0mA**. Then $I_{S3} = I_{S2} \simeq 3.0/2 =$ **1.5mA** $= I_{c2}$, and $V_{C2} \simeq 5 - 2(1.5) =$ **2V**. Thus, $1.5 = 10\,(100 \times 10^{-6})\,(v_{GS} - -1)^2\,(1 + 0.1(2))$, or $(v_{GS} + 1)^2 = 1.5/(1(1.2)) = 1.25$, $v_{GS} = 1.12 - 1 = 0.12$V. Thus $V_{c1} \simeq$ **–0.12V**. *Check*: $I_{c2} = 1\,(0.12 + 1)^2\,(1 + 0.1(2)) = 1.505$mA.

(d) See that Q_1 is near cutoff, though it is larger. Assume V_d is near +5V, say at 5V, in which case $i_{D1} = 2 \times 100 \times 10^{-6}\,(-0.8 - -1)^2)\,(1 + .1\,(5 - 0)) = 0.12$mA. Now $i_{D2} = \beta\left[2\,(v_{GS} - V_t)\,v_{DS} - v_{DS}^2\right](1 + \lambda\,v_{DS})$. Let $v_{DS} = v$, which is small, such that the λ term can be ignored. Thus $0.12 \times 10^{-3} = 10 \times 100 \times 10^{-6}\,[2\,(0 - -1)\,v - v^2]$, or $0.12 = (2v - v^2)$, $v^2 - 2v + 0.12 = 0$, and $v = \dfrac{- -2 \pm \sqrt{4 - 4\,(.12)}}{2} = 0.062$V. Thus $V_d = 5 - .062 =$ **4.94V** and I_d = **0.12mA.**

(e) See that Q_1 is turned on with V_e near 0V. For Q_2, $i_D = 10 \times 100 \times 10^{-6}\,(0 - -1)^2\,(1 + 0.1\,(5)) = 1.5$mA. For Q_2, $i_D = 20 \times 100 \times 10^{-6}\,(0.2 - -1)^2\,(1 + 0.1\,(0)) = 2.88$mA. Thus, Q_2 is in triode mode with $v_{DS} = v$, assumed small and ignored. Correspondingly, 1.5mA = 2.0mA $[2\,(0.2 - -1)\,v - v^2]$, and $v^2 - 2.4v + .75 = 0$, $v = \dfrac{2.4 \pm \sqrt{2.4^2 - 4(.75)}}{2} = 0.37$V. Thus V_e = **0.37V**, and I_e = **1.5mA.**

6.42 Here, $\beta_1 = 100 \times 10^{-6} \times 5 = 0.5$mA/$V^2$:

(a) $i_{D1} = 0.5\,(0 - -1)^2\,(1 + 0.1\,(1)) = 0.55$mA, for $v_{DS1} = |V_t|$. For V_O = –3V, V_{SS} = –5V, V_{S2} = –4V, $i_{D2} = 0.55 \times 10^{-3} = 20\,(100) \times 10^{-6}\,(v_{GS} - -1)^2\,(1 - 0.1\,(-3 - -4))$. Thus $(v_{GS} + 1)^2 = 0.25$, $v_{GS} = \pm .5 - 1 = -0.5$V.
$V_{bias} = -4 - 0.5 =$ **–4.5V.**

(b) Lowest $V_O = -4V + |V_t| =$ **–3V.**

(c) For V_O = –3V, r_o of Q_2 does not matter, the current being established by Q_1 at a value I_O = **0.55mA.**

(d) From Eq. 5.122: $r_o = \dfrac{1}{\lambda\,\beta\,(V_{GS} - V_t)^2}$. Thus $r_{o1} = \dfrac{1}{0.1 \times 0.5\,(0 - -1)^2}$ = 20kΩ, and $r_{o2} = \dfrac{1}{0.1 \times 2.0\,(-0.5 - -1)^2}$ = 20kΩ. From Eq 5.121: $g_m = 2\beta\,(v_{GS} - V_t)\,(1 + \lambda\,v_{DS})$. Thus $g_{m2} = 2\,(2)\,(-0.5 - -1)\,(1 + 0.1\,(1)) = 2.2$mA/V. Thus $R_{out} = g_{m2}\,r_{o2}\,r_{o1} = 2.2 \times 10^{-3} \times 20 \times$

$10^3 \times 20 \times 10^3 =$ **880kΩ.**

(e) Now for the output raised from −3V to +1V, ie by 4V, $\Delta I = \frac{4V}{880 \times 10^3} =$ **4.5μA.**

6.43

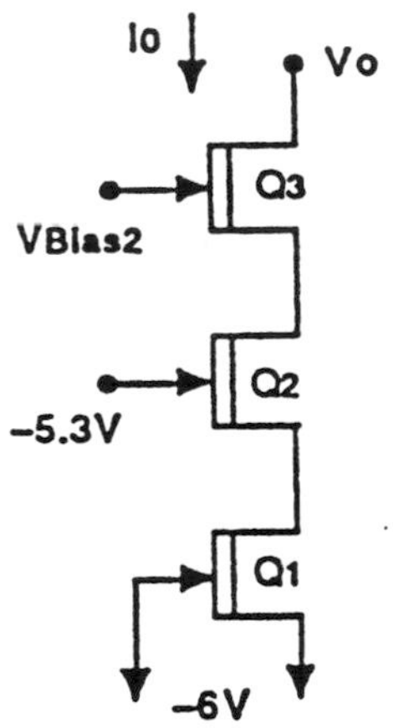

Add Q_3 with $W_3 = 20\mu m$, $\beta_3 = 2mA/V^2$, $\beta_2 = 2mA/V^2$, $\beta_1 = 1mA/V^2$, $V_t = -1V$, $\lambda = 0.1V^{-1}$,

(a) Use the results of Ex. 6.24. Since Q_2, Q_3 are the same size, $V_{bias\ 1} = -5.3V$ permits $\upsilon_{DS1} = 1V$. Thus $V_{bias\ 2} = V_{bias\ 1} + 1.0 =$ **−4.3V**. *Check*: Now, for $\upsilon_{DS1} = 1V$, $i_{D1} = 1mA/V^2\ (0 - -1)^2\ (1 + 0.1(1)) = 1.1mA$. Now for Q_2, $1.1mA = 2mA/V^2\ (\upsilon_{GS} - -1)^2\ (1 + 0.1(1))$, or $1 = 2\ (\upsilon_{GS} + 1)^2$, and $\upsilon_{GS} = \sqrt{½} - 1 = .707 - 1 \approx -0.3V$. OK.

(b) Now, Q_3 remains in saturation for $\upsilon_{DS} \geq \upsilon_{GS} - V_t$ or $\upsilon_{DS} > -0.3 - -1 = 0.7V$. Thus, the output can be as low as −4.3 + 1.0 = **−3.3V.**

(c) Now as Q_3 is operating just as Q_2, V_O can go as low as −4.3 + 0.3 + 1 = −3.0V, at which point $I_O =$ **1.1mA.**

(d) $r_{01} = \frac{1}{0.1 \times 1 \times (0 - -1)^2} = 10k\Omega$, $r_{02} = \frac{1}{0.1 \times 2 \times (-0.3 - -1)^2} = 10.2k\Omega = r_{03}$, $g_{m2} = 2\ (2)\ (-0.3 + 1) = 2.8mA/V = g_{m3}$. Thus, $R_{02} = g_{m2}\, r_{02}\, r_{01} = 2.8 \times 10.2 \times 10.0 = 285.6k\Omega$, and $R_{03} = g_{m3}\, r_{03}\, R_{02} = 2.8 \times 10.2 \times 285.6 =$ **8.16MΩ**

(e) For a 4V change in output voltage, $\Delta I = \frac{4}{8.16 \times 10^6} =$ **0.49μA.**

6.44 See Q_1 and Q_3 have the same width. Thus for $V_{DD} = 5V$ and $V_A = 2V$, $\upsilon_{DS3} = \upsilon_{DS1} = (5-2)/2 = 1.5V$, and $V_E = V_B = 5 - 1.5 = 3.5V$. See $5 \times 10^{-3} = W \times 100 \times 10^{-6}\ (0 - -1)^2\ (1 + 0.1\ (1.5))$. Thus $W_1 = W_3 =$ **43.5μm**, and $W_2 = 43.5/2 =$ **21.7μm**. Note that the diodes are sized to give 0.75V drop each for $I/2 = 2.5mA$. Now, $r_o = \frac{1}{\lambda\, \beta\, (\upsilon_{GS} - V_t)^2}$, $g_m = 2\beta\ (\upsilon_{GS} - V_t)\ (1 + \lambda\, \upsilon_{DS})$. Thus $r_{01} = \frac{1}{0.1\ (4.35)\ (0 + 1)^2} = 2.3k\Omega = r_{03}$, and $r_{02} = \frac{1}{0.1\ (2.17)\ (0 + 1)^2} = 4.61k\Omega$. Also $g_{m1} = 2\ (4.35)\ (0 + 1)\ (1 + 0.1\ (1.5)) = 10mA/V = g_{m3}$, and $g_{m2} = 2\ (2.17)\ (0 + 1)\ (1 + 0.1\ (5 - 2)) = 5.64mA/V$.

Now, from Eq. 6.132: $\alpha = \frac{\upsilon_b}{\upsilon_a} = \frac{10 \times 2.30 \times \frac{5.64 \times 4.61}{5.64 \times 4.61 + 1} + \frac{2.30}{2.30}}{10 \times 2.30 + \frac{2.30}{2.30} + 1} = \frac{23.15}{25} =$ **0.926V/V**. From Eq. 6.133: $R_o = \frac{r_{01}}{1 - \alpha} = \frac{2.3}{1 - .926} =$ **31.1kΩ**. From Eq. 6.134: $R_o = r_{01}\ (g_{m3})\ \frac{r_{03}}{2} = 2.3 \times 10 \times \frac{2.3}{2}$ = **26.5kΩ.**

6.45 $I_{DSSeq} = 0.5\text{mA}$, $\upsilon_{DS} = 3\text{V}$, $\upsilon_{DS1} = 0.7\text{V}$. Now operate with $\upsilon_{GS1} = 0\text{V}$.

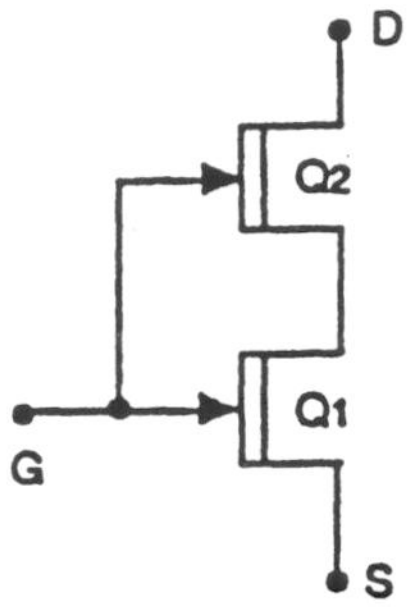

Thus $\upsilon_{GS2} = -0.7\text{V}$, with $\upsilon_{DS2} = 3 - 0.7 = 2.3\text{V}$, $0.5 = W_1\ (0.1)\ (0 - 1)^2 \times (1 + 0.1\ (0.7))$, or $W_1 = \frac{0.5}{0.1}\ \frac{1}{1^2}\ \frac{1}{1.07} = \mathbf{4.67\mu m}$. Also $0.5 = W_2\ (0.1)\ (-0.7 - -1)^2\ (1 + 0.1\ (2.3))$, or $W_2 = \frac{0.5}{0.1} \times \frac{1}{(0.3)^2} \times \frac{1}{1.23} = \mathbf{45.2\mu m}$. For $\upsilon_{DS} = 6\text{V}$, the current I_{DSS} will increase, with $\upsilon_{DS1} = \upsilon$ increasing as well. Thus $i = 0.467\ (0 - -1)^2\ (1 + 0.1\ \upsilon)$, and $i = 4.52\ (-\upsilon + 1)^2\ (1 + 0.1\ (6 - \upsilon))$, $(1 + 0.1\upsilon) = 9.68\ (\upsilon^2 - 2\upsilon + 1)\ (1.6 - 0.1\upsilon)$, $1 + 0.1\upsilon = 15.5\ \upsilon^2 - 31\upsilon + 15.5 - .968\ \upsilon^3 + 1.94\ \upsilon^2 - .968\ \upsilon$, $0.968\upsilon^3 - 17.34\upsilon^2 + 32.07\upsilon - 14.5 = 0$, $\upsilon^3 - 18\upsilon^2 + 33.13\ \upsilon - 14.98 = 0$.

Solve this cubic iteratively: $\upsilon = \frac{18\upsilon^2 + 14.98 - \upsilon^3}{33.13} = 0.543\upsilon^2 + 0.452 - 0.030\upsilon^3$. Now, for $\upsilon = 0.7$: $\upsilon = .543\ (.7)^2 + .452 - .03\ (.7)^3 = 0.266 + 0.452 - .01 = 0.708$; for $\upsilon = 0.71$: $\upsilon = 0.543\ (0.71)^2 + 0.452 - 0.030\ (0.71)^3 = 0.274 + 0.452 - 0.011 = 0.715$. See that the process does not converge: *Thus reformulate it*: $\upsilon^2 = \frac{\upsilon^3 + 33.13\upsilon - 14.98}{18}$, $\upsilon = (0.056\upsilon^3 + 1.84\upsilon - 0.832)^{½}$. Now, for $\upsilon = 0.7$: $\upsilon = (0.056\ (.7^3) + 1.84\ (.7) - 0.832)^{½} = (0.0192 + 1.288 - 0.832)^{½} = (0.4752)^{½} = 0.689$. See that now it will diverge for lower values: Also for $\upsilon = 0.8$: $\upsilon = (0.056\ (.8^3) + 1.84\ (.8) - 0.832)^{½} = (0.027 + 1.472 - 0.832)^{½} = 0.818$. Again we see that it also diverges, but that a solution lies between 0.7 and 0.8, but nearer to 0.7. Now, for $\upsilon = 0.75$: $\upsilon = (0.056\ (0.75)^3 + 1.84\ (0.75) - 0.832)^{½} = (0.024 + 1.38 - 0.832)^{½} = 0.756$; for $\upsilon = 0.74$: $\upsilon = (0.056\ (0.74)^3 + 1.84\ (0.74) - 0.832)^{½} = 0.743$; and for $\upsilon = 0.73$: $\upsilon = (0.022 + 1.343 - 0.832)^{½} = 0.7302$.

Now conclude $\upsilon = \upsilon_{DS1} = 0.73\text{V}$, for which Q_1, $i_D = 0.467\ (1^2)\ (1 + .1\ (0.73)) = 0.501\text{mA}$, and for Q_2, $i_D = 4.52\ (-.73 - -1)^2\ (1 + 0.1\ (6 - 0.73)) = 4.52\ (0.0729)\ (1.527) = 0.503\text{mA}$. This calculation is very sensitive due to squared term. Use $i_D = \mathbf{0.502mA}$.

Small-Signal analysis: $r_o = \frac{1}{\lambda\ \beta\ (\upsilon_{GS} - V_t)^2}$, $g_m = 2\ \beta\ (\upsilon_{GS} - V_t)\ (1 + \lambda\ \upsilon_{DS})$. Here $R_{out} = r_{02}\ g_{m2}\ r_{01}$, for which $r_{02} = \frac{1}{0.1\ (45.2 \times 0.1 \times 10^{-3})\ (-0.7 - -1)^2} = 24.6\text{k}\Omega$, $r_{01} = \frac{1}{0.1\ (.467 \times 10^{-3})\ (0 + 1)^2} = 21.4\text{k}\Omega$, $g_{m2} = 2\ (4.52 \times 10^{-3})\ (-0.7 - -1)\ (1 + 0.1\ (3 - 0.7)) = 3.34\text{mA/V}$. Thus $R_{out} = 24.6 \times 10^3 \times 21.4 \times 10^3 \times 3.34 \times 10^{-3} = \mathbf{1.76M\Omega}$. Now, the expected current increase as υ_{DS} rises from 3V to 6V is $(6-3)/(1.76 \times 10^6) = 1.7\mu\text{A}$. The earlier estimate of change from 0.500mA to 0.502mA by 2μA is quite consistent. The small-signal scheme is certainly more straightforward!

6.46 From the results of P6.45 above, for $V_{DD} = 6\text{V}$ and $V_O \approx 6/2 = 3\text{V}$, the upper composite would conduct 0.5mA, while the lower one would be biased at $V_I = \mathbf{0V}$ to correspond. For each composite, the output resistance is 1.76MΩ. For the amplifier, the output resistance is 1.76/2 = **.88MΩ**, and the gain is $-g_m\ R_{out}$. Here, $g_m = g_{m1} = 2\beta_1\ (\upsilon_{GS1} - V_t) = 2(0.467 \times 10^{-3})(0 - -1) = 0.934\text{mA/V}$, and the gain is $-0.934 \times 10^{-3} \times -0.88 \times 10^6 = \mathbf{-822V/V}$.

6.47 Though it is not required explicitly by the specification, make all devices the same size. For $V_{cm} \approx 0\text{V}$, $V_{DD} = 5\text{V}$, and Q_1 and Q_3 matched, the voltages across each will be the same, and $\upsilon_{DS1} = \upsilon_{DS3} = \frac{5 - 0 - 0.5}{2} \approx 2.25\text{V}$. Thus $0.5 \times 10^{-3} = W_1\ (0.1 \times 10^{-3})\ (-0.5 - -1)^2\ (1 + 0.1\ (2.25))$, or $W_1 = \frac{0.5}{0.1} \times \frac{1}{(0.5)^2} \times \frac{1}{1.225} = \mathbf{16.3\mu m} = W_2 = W_3$. Use $I = \mathbf{1.0mA}$: Now for 0.5mA in Q_1, the voltage at the sources of Q_1 and Q_2 is 0.5V, and that at the source of Q_3 is 0.5 + 2.25 = 2.75V, with 2.75 − 0.5 = 2.25V at its gate. Thus $\upsilon_{DS2} \simeq 2.25 - 0.5 = 1.75$V. Now, $0.5 \times 10^{-3} \simeq 16.3\ (0.1 \times 10^{-3})\ (\upsilon_{GS} - -1)^2\ (1 + 0.1\ (1.75))$. Thus $\upsilon_{GS2} = \pm \left[\frac{0.5}{1.63} \times \frac{1}{1.175} \right]^{1/2} - 1 = 0.5109 - 1 = -0.489$V (rather than 0.500V). Thus, there would be an **11mV offset** to maintain V_O at 0.5 + 1.75 = **2.25V**. Note that since the drain voltages of Q_1 and Q_2 are different, as required by Q_3 operating at $\upsilon_{GS3} \neq 0$, while Q_1 and Q_2 are matched and share a source connection, there must be an input offset voltage. Otherwise, for gate inputs connected, and the drain supply exactly half that of the source bias supply, the output will rise until Q_3 enters the triode mode and $\upsilon_{GS3} = 0$. For the offset acceptable, and nominal operation with $i_D \simeq 0.5$mA and $\upsilon_{DS} \simeq 2.25$V, and $\upsilon_{GS} = -0.5$V, $g_m = 2\beta\ (\upsilon_{GS} - V_t)\ (1 + \lambda\ \upsilon_{DS}) = 2\ (0.1 \times 16.3 \times 10^{-3})\ (-0.5 - -1)\ (1 + 0.1\ (2.25)) = 2.00$mA/V. $r_o = \frac{1}{\lambda\ \beta\ (\upsilon_{GS} - V_t)^2} = \frac{1}{0.1\ (1.63 \times 10^{-3})\ (-0.5 + 1)^2} = 24.5\text{k}\Omega$. Now from Eq. 6.139 of the Text: $\frac{\upsilon_o}{\upsilon_i} = \frac{-g_{m1}\ r_{01}}{\left[\frac{g_{m1}\ r_{01} + 1}{g_{m2}\ r_{02} + 1} - \frac{g_{m3}\ r_{03}}{g_{m3}\ r_{03} + 1} \right]} \simeq \frac{-2.00 \times 24.5}{\frac{2 \times 24.5 + 1}{2 \times 24.5 + 1} - \frac{2 \times 24.5}{2 \times 24.5 + 1}} \simeq -\frac{49}{\frac{50}{50} - \frac{49}{50}} = -49\ (50) =$ **2450V/V.**

Alternatively, for $W_1 = W_2$, but W_3 smaller, so $\upsilon_{DS1} = \upsilon_{DS2} = 2.0$V, and $\upsilon_{GS3} = 0$V with $\upsilon_{GS1} = \upsilon_{GS2} = -0.5$V, all operating at 0.5mA nominally. Find: $0.5 \times 10^{-3} = W_1\ (0.1 \times 10^{-3})\ (-0.5 + 1)^2\ (1 + 0.1\ (2.0))$, $W_1 = W_2 = \frac{0.5}{0.1} \times \frac{1}{(0.5)^2} \times \frac{1}{1.2} =$ **16.7μm.** Now for W_3: $0.5 \times 10^{-3} = W_3\ (0.1 \times 10^{-3}\ (0 + 1)^2\ (1 + 0.1\ (5 - 2))$, or $W_3 = \frac{0.5}{0.1} \times \frac{1}{1^2} \times \frac{1}{1.3} =$ **3.85μm.** Now $g_{m3} = 2\ (0.1)\ (3.85)\ (0 - -1)\ (1 + 0.1\ (3)) = 1$mA/V, $r_{03} = \frac{1}{0.1\ (0.1)\ (3.85)\ (1^2)} = 25.97\text{k}\Omega$, and $g_{m1} = g_{m2} = 2\ (1.67)\ (-0.5 - -1)\ (1 + 0.1)\ (2)) = 2.004$mA/V, with $r_{01} = r_{02} = \frac{1}{0.1\ (1.67)\ (-0.5 + 1)^2} = 23.95\text{k}\Omega$. Thus the gain is $\frac{\upsilon_o}{\upsilon_i} = -\frac{2.00\ (24.0)}{\frac{2.00\ (24.0) + 1}{2.00\ (24.0) + 1} - \frac{1\ (26)}{1\ (26 + 1)}} = -\frac{48}{\frac{49}{49} - \frac{26}{27}} =$ **−1296V/V.**

SECTION 6.10: MULTISTAGE AMPLIFIERS

6.48 **DC bias for V_{BE} = 0.7V**, $\beta = \infty$, and ±15V supplies, $I_{E9} = \frac{15 - 0.7}{28.6} = 0.5$mA. Thus $I_{C3} = 0.5$mA, $I_{C1} = I_{C2} = 0.25$mA $\rightarrow r_e = 100\Omega$, $I_{C6} = 2.0$mA, $I_{C4} = I_{C5} = 1.0$mA $\rightarrow r_e = 25\Omega$. $V_{B7} = 15 - 3(1) = 12$V, $V_{E7} = 12 + 0.7 = 12.7$V, $I_{E7} = \frac{15 - 12.7}{2.3} = 1$mA $\rightarrow r_e = 25\Omega$. $V_{B8} = -15 + 1\ (15.7) = 0.7$V, $V_O = 0$V, $I_{E8} = \frac{0 - 15}{3k} = 5$mA $\rightarrow r_e = 5\Omega$.

AC for $\beta = \infty$: $R_{in} = \infty$, $R_{out} = 3k\Omega \parallel 5\Omega = 5\Omega$, Gain $= \frac{20k + 20k}{100 + 100} \times \frac{3k}{25 + 25} \times \frac{15.7k}{2.3k + .025k} \times \frac{3k}{3k + .005k} = 200 \times 60 \times 6.75 \times .998 =$ **80.9 × 10³V/V.**

AC for β = 50: $R_{in} = 51\ (100 + 100) =$ **10.2kΩ**, $R_{out} = 3k\Omega \parallel 5 + \frac{15.7k}{51} = 3k\Omega \parallel .313\text{k}\Omega =$ **283Ω**

For gain: Second-stage $R_{i2} = 51\ (25 + 25) = 2.55\text{k}\Omega$, third-stage $R_{i3} = 51\ (2.3 + .015) = 117.6\text{k}\Omega$, fourth-stage $R_{i4} = 3\text{k}\ (51) = 153\text{k}\Omega$. Thus the gain $= \frac{40k \parallel 2.55k}{0.2k} \times \frac{50}{51} \times \frac{3k \parallel 117.6k}{(.025k) \times 2} \times \frac{50}{51} \times \frac{15.7k \parallel 153k}{2.325k} \times \frac{50}{51} \times \frac{3k}{3.005k} = \frac{2.397}{0.2} \times \frac{50}{51} \times \frac{2.925}{.05} \times \frac{50}{51} \times \frac{14.24}{2.325} \times \frac{50}{51} \times \frac{3}{3.005} =$ **4040V/V.**

6.49 To raise i_{C7} from 1mA to 2mA, and i_{C8} from 5mA to 20mA, reduce R_4 to 2.3kΩ/2 = **1.15kΩ**, R_5 to 15.7/2 = **7.85kΩ**, R_6 to 3/4 = **0.75kΩ**. Now R_{i4} = 101 (25/20 + 750) = 75.85kΩ,

$$A_3 \approx -\frac{R_5 \parallel R_{i4}}{r_{e7} + R_4} = -\frac{7.85k \parallel 75.85k}{12.5 + 1150} = \mathbf{-6.12V/V},\ A_4 \approx \frac{R_6}{r_{e8} + R_6} = \frac{750}{\frac{25}{50} + 750} = \mathbf{0.998V/V}.$$ Now,

$A = A_1\,A_2\,A_3\,A_4$ = 22.4 × (−59.2) × (−6.12) × 0.998 = **8099V/V**, and $R_o = R_6 \parallel (r_{e8} + R_5/(\beta + 1))$ = 750 ∥ (1.25 + 7850/101) = 750 ∥ 78.97 = **71.4Ω**. For load R_L, loss is $\frac{R_L}{R_L + 71.4}$ = 0.8. Thus R_L = 0.8R_L + 57.12, 0.2R_L = 57.12, R_L = **286Ω**. For a 286Ω load, the upper swing is limited by Q_7 saturating. Assume 0V between the emitter and collector of Q_7, and look down from the collector of Q_7. See an equivalent load resistance at the emitter of Q_8 of 286Ω||750Ω = 207Ω connected to a supply of (.286/(.286+.75)) (−15) = −4.14V. At its base, see a resistor of 101 (207) = 20.9kΩ to a supply of −4.14 + 0.7 = −3.44V. The equivalent circuit is as shown:

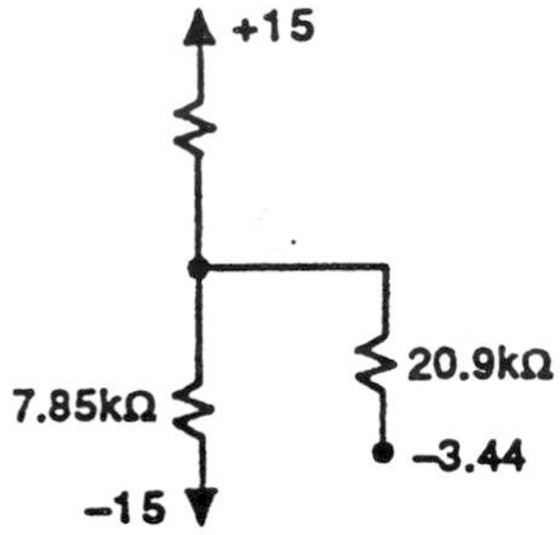

Looking down from the collector of Q_7, see 20.9 ∥ 7.85 = 5.71kΩ to −3.44 + (20.9/(20.9+7.85)) (−15+3.44) = −11.84V.

Thus, $V_A = +15 - \frac{1.16k}{1.16 + 5.71}$ (15 − − 11.84) = 10.47V (for which Q_5 does not saturate). The maximum positive output is 10.47 − 0.7 = **+9.77V**. For negative output, with Q_8 cutoff, the output is (286/(286+750)) (−15) = **−4.14V**. Thus with a 286Ω load, the output can swing from **9.8V** to **−4.1V**.

6.50 *For the resistor values:*

The reference resistor, R_0 = (0 − − 10 − 0.7)/0.5 = **18.6** kΩ; R_6 = (0 − − 10)/5 = **2** kΩ; R_5 = (+ 0.7 − − 10)/1 = **10.7** kΩ; $R_1 = R_2$ = 3/(0.5/2) = **12** kΩ; R_3 = 2/(2/2) = **2** kΩ; R_4 = (2 − 0.7)/1 = **1.3 kΩ**.

For the emitter resistances:

For Q_1, Q_1, $r_{e1} = r_{e2}$ = 25 mV/0.25 mA = 100 Ω. For Q_4, Q_5, $r_{e4} = r_{e5}$ = 25 mV/1 mA = 25 Ω. For Q_7, r_{e7} = 25 mV/1 mA = 25 Ω. For Q_8, r_{e8} = 25 mV/5 mA = 5 Ω.

For the input resistances:

$r_\pi = (\beta + 1)r_e = 51r_e$, in general; $r_{\pi 1} = r_{\pi 2} = 5.1k\Omega$; $r_{\pi 4} = r_{\pi 5} = r_{\pi 7}$ = 51(25) = 1.275 kΩ; $r_{\pi 8}$ = 51(5) = 0.255 kΩ.

Now $R_{i1} = R_{id} = r_{\pi 1} + r_{\pi 2}$ = 2(5.1) **10.2 kΩ**; $R_{i2} = r_{\pi 4} + r_{\pi 5}$ = 2(1.275) = **2.55 kΩ**;

$R_{i3} = r_{\pi 7} + (\beta + 1)(R_4)$ = 1.275 + 51(1.3) = **67.575** kΩ; $R_{i4} = r_{\pi 8} + (\beta + 1)R_6$ = 0.255 + 51(2) = **102.26 kΩ**.

As noted on page 557 of the Text; $v_o/v_{id} = (R_6/R_{i1})(i_{e8}/i_i) = (2/10.2)(i_{e8}/i_i) = 0.196 i_{e8}/i_i$,

where $i_{e8}/i_i = i_{e8}/i_{b8} \times i_{b8}/i_{c7} \times i_{c7}/i_{b7} \times i_{b7}/i_{c5} \times i_{c5}/i_{b5} \times i_{b5}/i_{c2} \times i_{c2}/i_i$.

Here, $i_{e8}/i_{b8} = \beta + 1$ = 51; $i_{b8}/i_{c7} = R_5/(R_5 + R_{i4})$ = 10.7/(10.7 + 102.26) = 0.0947; $i_{c7}/i_{b7} = \beta$ = 50; $i_{b7}/i_{c5} = R_3/(R_3 + R_{i3})$ = 2/(2 + 67.6) = 0.0287; $i_{c5}/i_{b5} = \beta$ = 50;

$$i_{b5}/i_{c2} = \frac{R_1 + R_2}{(R_1 + R_2) + R_{i2}} = \frac{12 + 12}{12 + 12 + 2.55} = 0.904;\ i_{c2}/i_i = \beta = 50.$$

Thus overall, v_o/v_{id} = 0.196 × 51 × 0.0947 × 50 × 0.0287 × 50 × 0.904 × 50 = **3070 V/V**.

Clearly this method fails if β = ∞, although it is generally OK for large (but finite) β, in which case the current-divider factors become smaller and smaller as β rises.

NOTES

Chapter 7

FREQUENCY RESPONSE

SECTION 7.1: S-DOMAIN ANALYSIS: POLES, ZEROS, AND BODE PLOTS

7.1 Using the voltage-divider rule: $T(s) = \dfrac{V_o(s)}{V_i(s)} = \dfrac{Z_{shunt}}{Z_{shunt} + Z_{series}}$. Here $Z_{series} = \dfrac{1}{C_1 s} || R_1 = \dfrac{R_1/C_1 s}{R_1 + 1/C_1 s} = \dfrac{R_1}{1 + R_1 C_1 s}$. $T(s) = \dfrac{V_o(s)}{V_1(s)} = \dfrac{1/C_2 s}{1/C_2 s + R_1/(1 + R_1 C_1 s)} = \dfrac{1}{1 + R_1 C_2 s/(1 + R_1 C_1 s)}$, or $T(s) = \dfrac{1 + R_1 C_1 s}{1 + R_1 (C_1 + C_2) s}$. That is, we see a single-time-constant response with a time constant $\tau = R_1 (C_1 + C_2)$, where the resistor R_1 "sees" $C_1 + C_2$, the parallel capacitance of C_1 and C_2 when the source is shorted.

Type of STC:

a) See that the circuit passes dc directly via R_1, with the signal reduced at high frequencies. Thus it is a low-pass (LP) STC circuit.

b) See for $s = 0$, $V_o(s)/V_i(s) = 1$, and for $s = \infty$, $V_o(s)/V_i(s) = C_1/(C_1 + C_2) < 1$.

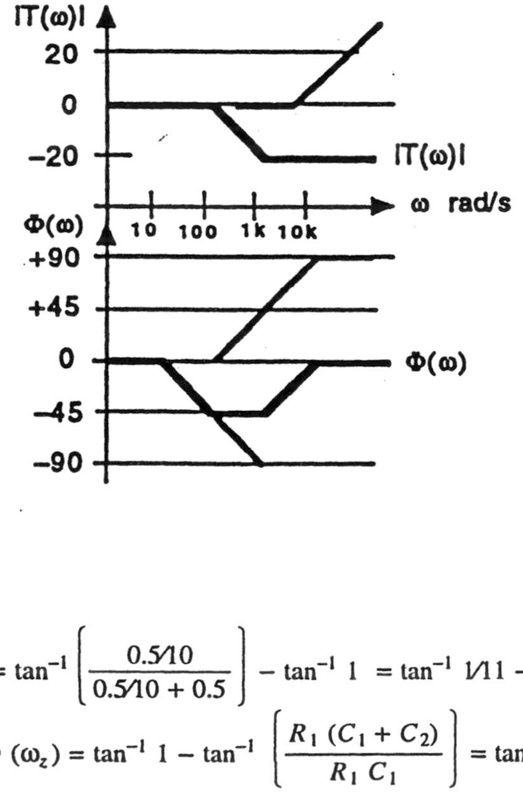

Now, for $R_1 = 10^4\Omega$, $C_1 = (0.5/10)\mu F$, and $C_2 = 0.5\mu F$. There is a pole at $\omega_p = 1/(R_1 (C_1 + C_2)) = 1/(10^4(0.05+0.5)10^{-6})$ = **182 rad/s**, and a zero at $\omega_z = 1/(R_1 C_1) = 1/(10^4(0.05)10^{-6})$ = **2000 rad/s.** From the Figure, or directly, see $T(\omega) = \dfrac{1 + j\,\omega R_1 C_1}{1 + j\omega R_1 (C_1 + C_2)}$. Now, $|T(\omega)| = \left[\dfrac{1 + (\omega R_1 C_1)^2}{1 + (\omega R_1 (C_1 + C_2))^2}\right]^{1/2}$. Thus, $|T(0)| = 1 \equiv 0dB$, $|T(\infty)|\ 1 = \dfrac{C_1}{C_1 + C_2} = \dfrac{0.5/10}{0.5/10 + 0.5} = \dfrac{1/10}{1/10 + 1} = \dfrac{0.1}{0.1 + 1} = \dfrac{1}{11} \equiv -20.8\text{dB}$.

$\Phi(\omega) = \tan^{-1} \omega R_1 C_1 - \tan^{-1} \omega R_1 (C_1 + C_2)$. $\Phi(0) = 0 - 0 = 0°$, $\Phi(\infty) = 90 - 90 = 0°$.

$\Phi(\omega_p) = \tan^{-1} \dfrac{R_1 C_1}{R_1 (C_1 + C_2)} - \tan^{-1} 1$

$= \tan^{-1} \left[\dfrac{0.5/10}{0.5/10 + 0.5}\right] - \tan^{-1} 1 = \tan^{-1} 1/11 - \tan^{-1} 1 = 5.19° - 45° = -39.8°$.

$\Phi(\omega_z) = \tan^{-1} 1 - \tan^{-1} \left[\dfrac{R_1 (C_1 + C_2)}{R_1 C_1}\right] = \tan^{-1} 1 - \tan^{-1} 11 = 45° - 84.8° = -39.8°$.

For $\omega_m = (\omega_p\, \omega_z)^{1/2} = (182 \times 2000)^{1/2} = 603$ rad/s, $\Phi(\omega_m) = \tan^{-1} \dfrac{603}{2000} - \tan^{-1} \dfrac{603}{182} = 16.8° - 73.2° = -56.4°$.

7.2

(a) $T(s) = \dfrac{10^{14}\,(s)\,(s + 10)}{(s + 1)\,(s + 100)\,(s + 10^5)\,(s + 10^6)}$

$= \dfrac{10^{14}\,(10)\,(s)\,(1 + s/10)}{(100)\,(10^5)\,(10^6)\,(1 + s)\,(1 + s/100)\,(1 + s/10^5)\,(1 + s/10^6)}$

or $\mathbf{T(s)} = \dfrac{10^2\,s\,(1 + s/10)}{(1 + s)\,(1 + s/100)\,(1 + s/10^5)\,(1 + s/10^6)}$.

(b) As $s \to 0$, $|T(0)| = \dfrac{10^2\,(0)\,(1)}{(1)\,(1)\,(1)\,(1)}$ = **0V/V**, and $\Phi(0) = \tan^{-1} 1/0 + \tan^{-1} 0/1 - 4\tan^{-1} 0/1 =$ **90°**.

As $s \to \infty$, $|T(\infty)| = \dfrac{(\infty)^2}{(\infty)^4} = \dfrac{1}{\infty^2}$ = **0V/V**, $\Phi(\infty) = 2\tan^{-1}(\infty) - 4\tan^{-1}(\infty) = -2\tan^{-1}(\infty)$ = **−180°**.

(c) Poles at $s = -1, -100, -10^5, -10^6$ **rad/s**. Zeros at s = **0, −10,** ∞, ∞ rad/s.

(d) As the frequency rises, the zeros increase the gain while the poles reduce it. We see that the gain increases from 0 to the first pole at 1 rad/s, then begins to rise again at 10 rad/s until 100 rad/s where it flattens, beginning to fall again at 10^5 rad/s, and more at 10^6 rad/s. Thus, we see that the gain is greatest from 100 rad/s to 10^5 rad/s. Thus at $\omega = 10^3$ rad/s, $|T(10^3)|$

$= \dfrac{10^2\,(10^3)\,(10^2)}{10^3\,(10)\,(1)\,(1)} = 10^3 \Phi\,(10^3) = 90 + \tan^{-1}10^2 - \tan^{-1}10^3 - \tan^{-1} 10 - \tan^{-1} 10^{-2} - \tan^{-1} 10^{-3}$

$\simeq 90 + 90 - 90 - 84 - 0 - 0 =$ **6°**. Also at $\omega = 10^4$ rad/s, $|T(10^4)| = \dfrac{10^2\,(10^4)\,(10^3)}{10^4\,(10^2)\,(1.1)\,(1)}$

$= 0.9 \times 10^3$, and

$\Phi\,(10^4) \simeq 90 + \tan^{-1}10^3 - \tan^{-1}10^4 - \tan^{-1} 10^2 - \tan^{-1} 10^{-1} - \tan^{-1} 10^{-2}$

$\simeq 90 + 90 - 90 - 90 - 0 - 0 =$ **0°**. Thus the greatest gain is 10^3**V/V**, and the corresponding phase is **0°**, occuring from about 10^3 to 10^4 rad/s.

(e) Gain at 10^3 rad/s,

$T\,(\omega = 10^3) = \dfrac{(10^2)(10^3)(1 + (10^3/10)^2)^{1/2}}{(1 + (10^3/1)^2)^{1/2}\,(1 + (10^3/100)^2)^{1/2}\,(1 + (10^3/10^5)^2)^{1/2}\,(1 + (10^3/10^6)^2)^{1/2}}$

$\simeq \dfrac{10^2\,(10^3)\,(10^2)}{10^3\,(10)\,(1)\,(1)} = 10^{2+3+2-3-1} = 10^3\text{V/V} \equiv$ **60dB.**

At 10^5 rad/s, $T\,(\omega = 10^5) = \dfrac{10^2\,(10^5)\,(10^4)}{10^5\,(10^3)\,(1 + (10^5/10^5)^2)^{1/2}\,(1 + (10^5/10^6)^2)^{1/2}}$

$= \dfrac{10^2\,(10^5)\,(10^4)}{10^5\,(10^3)\,(1.414)\,(1)} = 0.707 \times 10^3 \equiv$ **57dB.**

It certainly would have been better to prepare the Bode plots earlier, certainly by the end of part (a). It is actually possible to sketch the pole and zero locations immediately, and to sketch the *shape* of the magnitude plot. However the absolute magnitude requires a calculation like the conversion done in part (a) (see P7.4 following). Probably easiest after (c), most useful before (d), and usefully possible before (b).

(See the Bode Plot on the following page.)

7.2 (Continued)

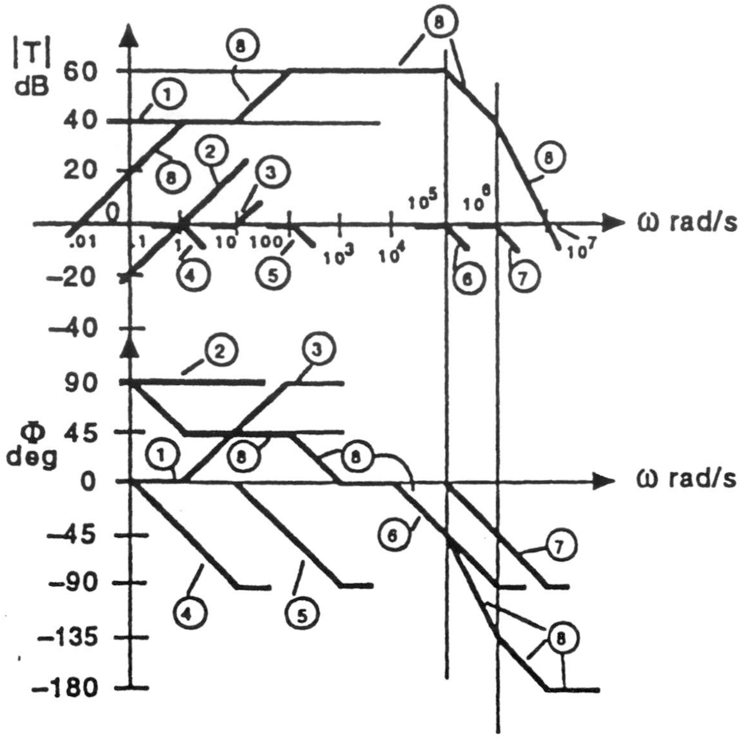

7.3 From P7.2 above, $T(s) = \dfrac{10^2\, s\ (1 + s/10)}{(1+s)\,(1+s/100)\,(1+s/10^5)\,(1+s/10^6)}$.

For $\omega = 100$ rad/s, $|T(\omega)| = \dfrac{10^2\ 10^2\ (1 + (100/10)^2)^{1/2}}{(1+100^2)^{1/2}\ (1+(100/100)^2)^{1/2}\ (1+(100/10^5)^2)^{1/2}\ (1+(100/10^6)^2)^{1/2}}$

$= \dfrac{10^2 \times 10^2 \times 10^1}{10^2\ (2^{1/2})\ (1)\ (1)} = 0.707 \times 10^{2+2+1-2} = 0.707 \times 10^3 \equiv$ **57dB**. $\Phi(\omega) = 90 + \tan^{-1} 10 - \tan^{-1} 100 - \tan^{-1} 1 - \tan^{-1} 10^{-3} - \tan^{-1} 10^{-4} = 90 + 84.3 - 89.4 - 45 - 0.1 - 0 =$ **39.8°**.

For $\omega = 2 \times 10^5$ rad/s,

$$|\ T(\omega)\ | = \frac{10^2 \times 2 \times 10^5\ (1 + (\frac{2\times 10^5}{10})^2)^{1/2}}{(1 + (2\times 10^5)^2)^{1/2}\ (1 + (\frac{2\times 10^5}{100})^2)^{1/2}\ (1 + (\frac{2\times 10^5}{10^5})^2)^{1/2}\ (1 + (\frac{2\times 10^5}{10^6})^2)^{1/2}}$$

$$= \frac{10^2 \times 2 \times 10^5\ (2\times 10^4)}{2\times 10^5\ (2\times 10^3)\ (1+4)^{1/2}\ (1+0.2^2)^{1/2}} = \frac{4\times 10^{11}}{4\ (\sqrt{5})\ (\sqrt{1.04}) \times 10^8}$$

$= 0.4385 \times 10^3 \equiv$ **52.8dB**.

$\Phi(\omega) = 90 + \tan^{-1}(2\times 10^4) - \tan^{-1}(2\times 10^5) - \tan^{-1}(2\times 10^3) - \tan^{-1}(2) - \tan^{-1}(0.2) = 90 + 90 - 90 - 90 - 63.4 - 11.3 =$ **–74.7°**.

SECTION 7.2: THE AMPLIFIER TRANSFER FUNCTION

7.4 From P7.2, $$T(s) = \frac{10^8\, s\,(s+10)}{(s+1)(s+100)(s+10^5)(s+10^6)} = \frac{s\,(s+10)}{(s+1)(s+100)} \times \frac{1}{(1+\frac{s}{10^5})(1+\frac{s}{10^6})} \times \frac{10^8}{10^5 \times 10^6} = \frac{s\,(s+10)}{(s+1)(s+100)} \times 10^3 \times \frac{1}{(1+\frac{s}{10^5})(1+\frac{s}{10^6})}.$$

Thus, $\mathbf{A_M} = 10^3$, $\mathbf{F_L(s)} = \dfrac{s\,(s+10)}{(s+1)(s+100)}$, $\mathbf{F_H(s)} = \dfrac{1}{(1+\frac{s}{10^5})(1+\frac{s}{10^6})}$,

$\mathbf{A_L(s)} = \dfrac{10^3\, s\,(s+10)}{(s+1)(s+100)}$, $\mathbf{A_H(s)} = \dfrac{10^3}{(1+\frac{s}{10^5})(1+\frac{s}{10^6})}$.

7.5 From P7.4, $\mathbf{F_L(s)} = \dfrac{s\,(s+10)}{(s+1)(s+100)} \approx \dfrac{s}{s+100}$, $\mathbf{F_H(s)} = \dfrac{1}{(1+\frac{s}{10^5})(1+\frac{s}{10^6})} \approx \dfrac{1}{1+\frac{s}{10^5}}$,

with the dominant-pole responses indicated by $\approx$. Thus $\mathbf{A(s)} \approx \dfrac{10^3\, s}{(s+100)(1+\frac{s}{10^5})}$.

7.6 For the low-frequency response of the transfer function given in P7.2, there are 2 low-frequency poles, at 1 rad/s and 100 rad/s, and zeros at 0 and 10 rad/s.

For the 3dB frequency:

(a) *Dominant Pole*: $\omega_L \approx$ **100 rad/s.**

(b) *Root Squares*: $\omega_L = (100^2 + 1^2 - 2\,(10^2)\,)^{1/2} = 10^2\,(1 + 10^{-4} - 2 \times 10^{-2})^{1/2} =$ **99 rad/s.**

(c) Exactly: $$\frac{(\omega^2 + 10^2)(\omega^2)}{(\omega^2 + 1^2)(\omega^2 + 100^2)} = \frac{1}{2} \quad ---\ (1),$$

$2\omega^4 + 2(10)^2\,\omega^2 = \omega^4 + 100^2\,\omega^2 + \omega^2 + 100^2$,

$\omega^4 - \omega^2\,(9801) - 10000 = 0$, $\omega^2 = [\,- -9801 \pm \sqrt{9801^2 - 4(-10000)})]/2 = [9801 \pm 9803]/2$, that is $\omega^2 = 9802$, or $\omega_L =$ **99.00 rad/s.** Note above in (1) that the zero at zero frequency *must* be included in the calculation. Why?

7.7 For the upper 3dB frequency of the transfer function in P7.2, two poles, at 10^5 and 10^6 rad/s.

For the 3dB cutoff:

(a) *Dominant pole*: $\omega_H = 10^5$ **rad/s.**

(b) *Root squares*: $$\omega_H = 1/\sqrt{\frac{1}{(10^5)^2} + \frac{1}{(10^6)^2}} = 1/\left[\frac{1}{10^5}\,(1 + \frac{1}{10^2})^{1/2}\right] = \frac{10^5}{(1 + 1/100)^{1/2}},$$ or

$\omega_H = 0.995 \times 10^5$ **rad/s.**

(c) *Exactly*: $F_H(s) = \dfrac{1}{(1 + s/10^5)(1 + s/10^6)}$, $T^2(\omega) = \dfrac{1}{(1 + (\omega/10^5)^2)(1 + (\omega/10^6)^2)} = 1/2$,

$1 + \omega^2\,(1/10^{10} + 1/10^{12}) + \omega^4/10^{22} = 2$, $\omega^4 + \omega^2\,(10^{12} + 10^{10}) - 10^{22} = 0$,

$\omega^4 + 101 \times 10^{10}\,\omega^2 - 10^{22} = 0$,

$$\omega^2 = \frac{-101 \times 10^{10} \pm \sqrt{101^2 \times 10^{20} + 4 \times 10^{22}}}{2},$$

$$\omega^2 = \frac{-101 \times 10^{10} \pm 102.96 \times 10^{10}}{2} = 0.9805 \times 10^{10} \text{ and } \omega_H = 0.990 \times 10^5 \textbf{ rad/s.}$$

7.8 Old $T(s) = \frac{s(s+10)}{(s+1)(s+100)} \times 10^3 \times \frac{1}{(1+s/10^5)(1+s/10^6)}$.

Modified $T(s) = \frac{s(s+10)}{(s+1)(s+100)} \times 10^3 \times \frac{1}{(1+s/10^5)(1+s/10^6)} \times \frac{(1+s/10^5)}{(1+s/(2\times 10^6))}$.

Note the *pole-zero cancellation* with the result that T(s) = $\frac{10^3\, s(s+10)}{(s+1)(s+100)(1+s/10^6)(1+s/(2\times 10^6))}$, and $\omega_H \approx 1/\sqrt{(1/(1\times 10^6))^2 + (1/(2\times 10^6))^2} = 10^6 (1 + 1/4)^{-1/2} = 0.894 \times 10^6$ **rad/s.**

Exactly, $\frac{1}{(1+(\omega/10^6)^2)(1+(\omega/(2\times 10^6))^2} = 1/2$, $1 + \frac{\omega^2}{10^{12}}(1+\frac{1}{4}) + \frac{\omega^4}{2(10^{24})} = 2$,

$\omega^4 + 2 \times 10^{12}\ \omega^2 - 2(10^{24}) = 0$, $\omega^2 = \frac{-2\times 10^{12} \pm \sqrt{4\times 10^{24} - 4(-2\times 10^{24})}}{2} = \frac{10^{12}(-2 \pm 3.46)}{2} = 0.73 \times 10^{12}$. Thus $\omega_H = 0.856 \times 10^6$ **rad/s.**

7.9

(1) *Using open-circuit time constants:* For C_1, $R_{eq} = R_1$, and $\tau_1 = R_1 C_1$. For C_2, $R_{eq} = R_1 + R_2$, and $\tau_2 = C_2 (R_1 + R_2)$.

(a) For $R_1 = R_2 = 10k\Omega$, $C_1 = C_2 = 100\text{pF}$, $\tau_1 = 10^4 \times 100 \times 10^{-12} = 10^{-6}\text{s}$, $\tau_2 = 100 \times 10^{-12}(10^4 + 10^4) = 2 \times 10^{-6}\text{s}$, $\omega_H = \frac{1}{\tau_1 + \tau_2} = \frac{1}{1\times 10^{-6} + 2\times 10^{-6}} = 1/3 \times 10^{-6} = 333 \times 10^3$ rad/s, or = 0.333×10^6**rad/s.** Alternatively, the sum of squares approach yields $\omega_H = 1/(1^2 + 2^2)^{1/2} \times 10^6 =$ **0.447** $\times 10^6$**rad/s.**

(b) For $R_1 = 10\text{k}\Omega$, $R_2 = 100\text{k}\Omega$, $C_1 = 100\text{pF}$, $C_2 = 10\text{pF}$, $\tau_1 = 10^4 \times 100 \times 10^{-12} = 10^{-6}\text{s}$, $\tau_2 = 10 \times 10^{-12}(10 \times 10^4 + 1 \times 10^4) = 10 \times 10^{-8}(11) = 1.1 \times 10^{-6}\text{s}$,

$\omega_H = \frac{1}{(1+1.1)10^{-6}} = 0.476 \times 10^6$ **rad/s.** Alternatively, the sum of squares approach yields $\omega_H = 1/(1^2 + 1.1^2)^{1/2} \times 10^6 = 0.673 \times 10^6$ **rad/s.**

(c) $R_1 = 10\text{k}\Omega$, $R_2 = 100\text{k}\Omega$, $C_1 = C_2 = 10\text{pF}$, $\tau_1 = 10^4 \times 10 \times 10^{-12} = 0.1 \times 10^{-6}\text{s}$, $\tau_2 = 10 \times 10^{-12} \times (10^5 + 10^4) = 10 \times 10^{-12} \times 10^4(11) = 1.1 \times 10^{-6}\text{s}$, $\omega_H = 1/((0.1 + 1.1)\ 10^{-6}) =$ **0.833** $\times 10^6$ rad/s. Alternatively, using the sum of squares, $\omega_H = 1/(0.1^2 + 1.1^2)^{1/2} \times 10^6 = 0.905 \times 10^6$ **rad/s.**

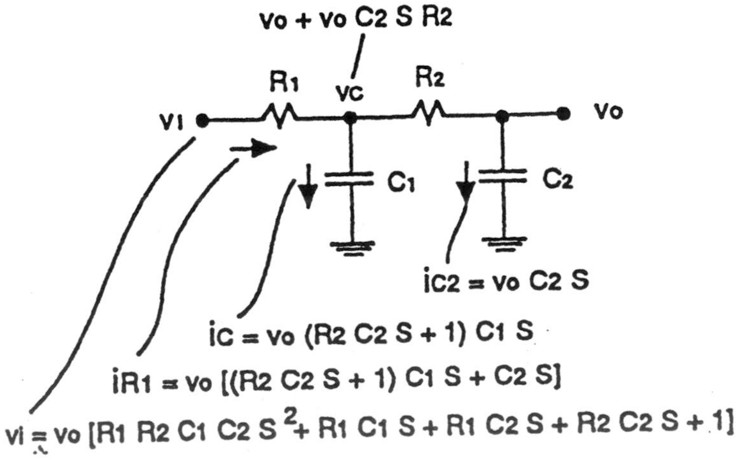

(2) *Exactly*: $T(s) = \dfrac{\upsilon_o}{\upsilon_i} = \dfrac{1}{s^2 R_1 R_2 C_1 C_2 + s\ (R_1 C_1 + R_1 C_2 + R_2 C_2) + 1}$,

$T(j\omega) = \dfrac{1}{j\omega\,(R_1 C_1 + R_1 C_2 + R_2 C_2) + (1 - \omega^2 R_1 R_2 C_1 C_2)}$. Now, response is 3dB down when $\omega^2 (R_1 C_1 + R_1 C_2 + R_2 C_2)^2 + (1 - \omega^2 R_1 R_2 C_1 C_2)^2 = 2$. Now, for particular cases:

(a) $R_1 = R_2 = 10\text{k}\Omega$, $C_1 = C_2 = 100\text{pF}$. Now, $\omega^2 (3 \times 10^4 \times 10^{-10})^2 + (1 - \omega^2 \times 10^4 \times 10^4 \times 10^{-10} \times 10^{-10})^2 - 2 = 0$, $9 \times 10^{-12}\ \omega^2 + (1 - \omega^2\ 10^{-12})^2 - 2 = 0$, $9\ \omega^2\ 10^{-12} + 1 - 2\ \omega^2\ 10^{-12} + \omega^4\ 10^{-24} - 2 = 0$, $\omega^4 + \omega^2\ (7 \times 10^{12}) - 10^{24} = 0$, $\omega^2 = \dfrac{-7 \times 10^{12} \pm 10^{12}\sqrt{7^2 - -4}}{2} = 10^{12}\dfrac{(-7 \pm 7.28)}{2} = 0.14 \times 10^{12}$, and $\omega_H = 0.374 \times 10^6$ **rad/s.**

(b) $R_1 = 10\text{k}\Omega$, $R_2 = 100\text{k}\Omega$, $C_1 = 100\text{pF}$, $C_2 = 10\text{pF}$. Now, $\omega^2\ (10^4 \times 10^{-10} + 10^4 \times 10^{-11} + 10^5 \times 10^{-11})^2 + (1 - \omega^2 \times 10^4 \times 10^5 \times 10^{-10} \times 10^{-11})^2 - 2 = 0$, or $\omega^2\ (10^{-6} + 10^{-7} + 10^{-6})^2 + (1 - \omega^2\ 10^{-12})^2 - 2 = 0$, or $\omega^2\ (2.1 \times 10^{-6})^2 + (1 - 10^{-12}\ \omega^2)^2 - 2 = 0$, $4.41 \times 10^{-12}\ \omega^2 + 1 - 2 \times 10^{-12}\ \omega^2 + \omega^4\ 10^{-24} - 2 = 0$,

$\omega^4\ 10^{-24} + 2.41 \times 10^{-12}\ \omega^2 - 1 = 0$, $\omega^4 + 2.41 \times 10^{12}\ \omega^2 - 10^{24} = 0$,
$\omega^2 = 10^{12}\ (-2.41 \pm \sqrt{2.41^2 - -4})/2 = 10^{12}\ (-2.41 \pm 3.132)/2 = 0.309 \times 10^{12}$.

Thus $\omega_H = 0.601 \times 10^6$ **rad/s.**

(c) $R_1 = 10\text{k}\Omega$, $R_2 = 100\text{k}\Omega$, $C_1 = 10\text{pF}$, $C_2 = 10\text{pF}$. Now, $\omega^2\ (10^4 \times 10^{-11} + 10^4 \times 10^{-11} + 10^5 \times 10^{-11})^2 + (1 - \omega^2 \times 10^4 \times 10^5 \times 10^{-11} \times 10^{-11})^2 - 2 = 0$. Thus $\omega^2\ (1.2 \times 10^{-6})^2 + (1 - \omega^2\ (0.1)\ 10^{-12})^2 - 2 = 0$, $1.44 \times 10^{-12}\ \omega^2 + 1 - 2\ \omega^2\ (0.1)\ 10^{-12} + \omega^4\ (0.01)\ 10^{-24} - 2 = 0$, $\omega^4 + 1.24 \times 10^{+14} - 10^{+26} = 0$,
$\omega^2 = (-1.24 \times 10^{14} \pm \sqrt{1.24^2 \times 10^{28} - 4\ (-10^{26})})/2 = 10^{14}\ (-1.24 \pm \sqrt{1.24^2 + .04})/2 = 0.00801 \times 10^{14} = 0.801 \times 10^{12}$. Thus ω_H = **0.895 × 10^6** rad/s.

7.10 Using short-circuit time constants: For C_1, $R_{eq} = R_1 \parallel R_2$, and $\tau_1 = C_1 R_1 R_2 / (R_1 + R_2)$. For C_2, $R_{eq} = R_2$, and $\tau_2 = C_2 R_2$.

(a) For $R_1 = R_2 = 10\text{k}\Omega$, $C_1 = C_2 = 1\mu\text{F}$, $\tau_1 = 1 \times 10^{-6} \times 10^4/2 = 0.5 \times 10^{-2}\text{s}$, and $\omega_1 = 1/0.5 \times 10^{-2} = 200$ rad/s; $\tau_2 = 10^{-6} \times 10^4 = 10^{-2}\text{s}$, and $\omega_2 = 100$ rad/s. Thus $\omega_L \approx 100 + 200 =$ **300 rad/s.** Using the sum of squares idea, $\omega_L = (100^2 + 200^2)^{1/2} =$ **224 rad/s.**

(b) For $R_1 = 10\text{k}\Omega$, $R_2 = 100\text{k}\Omega$, $C_1 = 1\mu\text{F}$, $C_2 = 0.1\mu\text{F}$, $\tau_1 = 10^{-6} \times 10k\Omega \parallel 100k\Omega = 0.909 \times 10^{-2}\text{s}$, $\tau_2 = 10^5 \times 10^{-7} = 10^{-2}\text{s}$. Thus $\omega_L = (1/0.909 + 1/1) \times 1/10^{-2} =$ **210 rad/s.** Alternatively, $\omega_L = ((1/.909)^2 + (1/1)^2)^{1/2} \times 100 =$ **147 rad/s.**

(c) For $R_1 = 10\text{k}\Omega$, $R_2 = 100\text{k}\Omega$, $C_1 = C_2 = 0.1\mu\text{F}$, $\tau_1 = 10^{-7} \times .909 \times 10^{+4} = 0.909 \times 10^{-3}\text{s}$, $\tau_2 = 10^{-7} \times 10^5 = 10^{-2}\text{s}$. Thus $\omega_L = 1/(0.909 \times 10^{-3}) + 1/10^{-2} = 1100 + 100 =$ **1200 rad/s.** Alternatively, $\omega_L = (1/.909^2 + 1/10^2)^{1/2} \times 1000 =$ **1054 rad/s.**

SECTION 7.3: LOW-FREQUENCY RESPONSE OF THE COMMON-SOURCE AND COMMON-EMITTER AMPLIFIERS

7.11 *For Midband Gain*: all capacitors are ac short circuits. Thus, $A_\upsilon = \dfrac{-R_{G1} \parallel R_{G2}}{R + R_{G1} \parallel R_{G2}} \times g_m R_L \parallel R_D = -\dfrac{10M \parallel 22M}{100k + 10M \parallel 22M} \times 2 \times 10^{-3}\ (20k \parallel 10k) = -\dfrac{6.875}{6.975} \times 10^{-3} \times 6.667 \times 10^3 =$ **− 13.14 V/V.**

For C_{C1}: $R_{C1} = 100\text{k}\Omega + 6.875\text{M}\Omega = 6.975 \times 10^6\Omega$, and $f_{p1} = 1/(2\pi \times 6.975 \times 10^6 \times 0.01 \times 10^{-6}) =$ **2.28Hz.** *For* C_{C2}: $R_{C2} = 10\text{k}\Omega + 20\text{k}\Omega = 30\text{k}\Omega$, and $f_{p2} = 1/(2\pi \times 30 \times 10^3 \times 0.1 \times 10^{-6}) =$ **53.1Hz.**

For C_S: $R_S' = R_S \parallel 1/g_m = 10k\Omega \parallel (1/(2 \times 10^{-3})) = 0.5k \parallel 10k = 476\Omega$, and $f_{pS} = 1/(2\pi \times .476 \times 10^3 \times 1 \times 10^{-6}) =$ **334.3Hz**, with $f_{zS} = 1/(2\pi \times 10 \times 10^3 \times 1 \times 10^{-6}) =$ **15.9Hz.**

7.12 Use the largest capacitor at the source, where the resistance level is least. Now, $C_S = 1/(2\pi \times .476 \times 10^3 \times 10) = 33.4\mu F$. Use $C_S = 30\mu F$. Now, $C_{C1} = 1/(2\pi \times 6.975 \times 10^6 \times 1) = 0.0236\mu F$. Use $C_{C1} = 0.02\mu F$. Now, $C_{C2} = 1/(2\pi \times 30 \times 10^3 \times 1) = 5.3\mu F$. Use $C_{C2} = 5\mu F$.

Actual critical frequencies are: $f_{pS} = 1/(2\pi \times .476 \times 10^3 \times 30 \times 10^{-6}) =$ **11.1Hz**, $f_{p1} = 1/(2\pi \times 6.975 \times 10^6 \times 0.02 \times 10^{-6}) =$ **1.14Hz**, $f_{p2} = 1/(2\pi \times 30 \times 10^3 \times 5 \times 10^{-6}) =$ **1.06Hz**, $f_{zS} = 1/(2\pi \times 10 \times 10^3 \times 30 \times 10^{-6}) =$ **0.53Hz.**

7.13 Modify Equations 7.35 through 7.36 for the addition of r_S in series with C_S: $I_d(s) = I_S(s) = \dfrac{V_g(s)}{1/g_m + Z_S} = g_m V_g(s) \dfrac{Y_S}{g_m + Y_S}$. Here, $Y_S = \dfrac{1}{Z_S} = \dfrac{1}{R_S} + \dfrac{1}{1/(s C_S + r_S)}$

$$= \frac{1}{R_S} + \frac{s C_S}{1 + s r_S C_S} = \frac{s C_S (R_S + r_S) + 1}{R_S (s C_S r_S + 1)}.$$

Thus $i_d(s) =$

$$g_m V_g(s) \frac{\dfrac{s C_S (R_S + r_S) + 1}{R_S (s C_S r_S + 1)}}{g_m + \dfrac{s C_S (R_S + r_S) + 1}{R_S (s C_S r_S + 1)}} = g_m V_g(s) \frac{s C_S (R_S + r_S) + 1}{s C_S (g_m R_S r_S + R_S + r_S) + R_S g_m + 1}$$

$$= V_g(s) \frac{(R_S + r_S)}{g_m R_S r_S + R_S + r_S} \times \frac{s + \dfrac{1}{C_S (R_S + r_S)}}{s + \dfrac{1 + R_S g_m}{C_S (g_m R_S r_S + R_S + r_S)}}.$$

See that there is a zero at $\omega_z = 1/(C_S (R_S + r_S))$, and a pole at $\omega_p = \dfrac{1}{C_S \left[\dfrac{R_S + r_S (1 + g_m R_S)}{1 + g_m R_S}\right]}$, where the resistor associated with C_S at the pole frequency is $r_S + 1/g_m \parallel R_S = r_S + \dfrac{R_S/g_m}{1/g_m + R_S} = r_S + \dfrac{R_S}{1 + g_m R_S} = \dfrac{r_S (1 + g_m R_S) + R_S}{1 + g_m R_S}$, as noted. Now the equivalent **transconductance** is $\dfrac{1}{1/g_m + R_S \parallel r_S} = \dfrac{g_m}{1 + \dfrac{g_m R_S r_S}{R_S + r_S}} =$

$\mathbf{\dfrac{g_m (R_S + r_S)}{g_m R_S r_S + R_S + r_S}}$, as noted.

Now for the situation in P7.11, with $g_m = 2mA/V$, $R_S = 10k\Omega$, the gain is reduced by a factor of two, when $\dfrac{R_S + r_S}{g_m R_S r_S + R_S + r_S} = 1/2$, or $1 + \dfrac{g_m R_S r_S}{R_S + r_S} = 2$, or $g_m R_S r_s = R_S + r_S$, or $r_S = \dfrac{R_S}{g_m R_S - 1} = \dfrac{10 \times 10^3}{2 \times 10^{-3} \times 10 \times 10^3 - 1} = \dfrac{10 \times 10^3}{19} =$ **526Ω.** *For the new pole*: $R_p = r_S + (1/g_m) \parallel R_S = 526 + (1/(2 \times 10^{-3})) \parallel (10 \times 10^3) = 526 + 476 = 1000\Omega$. This should have been obvious since the gain was to have been reduced by a factor of 2, and $1/g_m = 500\Omega$. Thus $f_{pS} = 1/(2\pi \times 1000 \times 1 \times 10^{-6}) =$ **159Hz**, and

$f_{zS} = 1/(2\pi \times 1 \times 10^{-6} (10k\Omega + 0.526k\Omega)) =$ **15.1Hz.**

SECTION 7.4: HIGH-FREQUENCY RESPONSE OF THE COMMON-SOURCE AND COMMON-EMITTER AMPLIFERS

7.14 $r_e = \dfrac{25 \times 10^{-3}}{0.15 \times 10^{-3}} = 166.7\Omega$, $r_\pi = 166.7 \times 151 = 25.17\text{k}\Omega$, $A_M = -\dfrac{25.17 \parallel 40}{25.17 \parallel 40 + 10} \times \dfrac{9.1 \parallel 10 \parallel 500}{166.7} = -\dfrac{15.45}{25.45} \times \dfrac{4.93 \times 10^3}{166.7} = -0.607 \times 25.6 = -17.95 \approx$ **−18V/V.**

For C_{C1}: $R_{C1} = 10k\Omega + 40k\Omega \parallel 25.2k\Omega = 10k\Omega + 15.5k\Omega = 25.5k\Omega$,

$$f_{p1} = \frac{1}{2\pi \times 25.5 \times 10^3 \times 1 \times 10^{-6}} = \mathbf{6.24Hz.}$$

For C_{C2}: $R_{C2} = (9.1 \parallel 500k\Omega + 10k\Omega) = 18.94\text{k}\Omega$,

$$f_{p2} = \frac{1}{2\pi \times 18.94 \times 10^3 \times 1 \times 10^{-6}} = \mathbf{8.4Hz.}$$

For C_E: $R_{CE} = \left[167 + \dfrac{50 + 40k \parallel 10k}{151} \right] \parallel 8.2k = 220.3 \parallel 8.2k = 214\Omega$,

$$f_{pE} = \frac{1}{2\pi \times 214 \times 10 \times 10^{-6}} = \mathbf{74.4Hz,} \qquad f_{zE} = \frac{1}{2\pi \times 8.2 \times 10^3 \times 10 \times 10^{-6}} = \mathbf{1.94Hz.}$$

Overall: $f_L = \left[6.24^2 + 8.4^2 + 74.4^2 - 2\,(1.94)^2 \right]^{1/2} = (39+71+5535-7.5)^{1/2} =$ **75.1Hz.**

7.15 Use data from P7.14 above, where for C_{C1}, $R_{C1} = 25.5\text{k}\Omega$; for C_{C2}, $R_{C2} = 18.9\text{k}\Omega$; for C_E, $R_{CE} = 0.214\text{k}\Omega$. Now, for $f_{pE} = 20\text{Hz}$, $C_E = \dfrac{1}{2\pi\,(.214 \times 10^3) \times 20} =$ **37.2μF,** for which $f_{ZE} = \dfrac{1}{2\pi \times 8.2 \times 10^3 \times 37.2 \times 10^{-6}} =$ **0.521Hz.** Now, for $f_{p1} = 0.521\text{Hz}$, $C_{C1} = \dfrac{1}{2\pi \times 25.5 \times 10^3 \times 0.521} =$ **12.0μF.** Now, for $f_{p2} = 2\text{Hz}$, $C_{C2} = \dfrac{1}{2\pi \times 18.9 \times 10^3 \times 2} =$ **4.2μF.**

Alternatively, for $f_{p2} = 0.521\text{Hz}$, $C_{C2} = \dfrac{1}{2\pi \times 18.9 \times 10^3 \times .521} =$ **16.2μF,** and for $f_{p1} = 2\text{Hz}$, $C_{C1} = \dfrac{1}{2\pi \times 25.5 \times 10^3 \times 2} =$ **3.12μF.** This arrangement is **better,** since it makes the pole - zero cancellation independent of β, although it takes a larger total capacitance.

7.16 Here, as in P7.14 above, $r_e = (25 \times 10^{-3})/(0.15 \times 10^{-3}) = 166.7\ \Omega$.

Now, $r_E = 350\ \Omega$, and the resistance looking into the base, (call it r'_π) is $r'_\pi = 151(166.7 + 350) = 78.0k\Omega$.

Now $A_M = \dfrac{-78.0 \parallel 40}{78.0 \parallel 40 + 10} \times \dfrac{150}{151} \times \dfrac{(9.1 \parallel 10 \parallel 500) \times 10^3}{166.7 + 350} = \dfrac{-26.4}{36.4} \times \dfrac{150}{151} \times \dfrac{4.72 \times 10^3}{516.7} =$ **− 6.58 V/V.**

For C_{C1}: $R_{C1} = 10k\Omega + 40k\Omega \parallel 78k\Omega = 10k\Omega + 26.4k\Omega = 36.4k\Omega$,

and $f_{p1} = 1/(2\pi \times 36.4 \times 10^3 \times 1 \times 10^{-6}) =$ **4.37 Hz**

For C_{C2}: $R_{C2} = 9.1 \parallel 500k\Omega + 10k\Omega = 8.94k\Omega + 10k\Omega = 18.9k\Omega$

and $f_{p2} = 1/(2\pi \times 18.9 \times 10^3 \times 1 \times 10^{-6}) =$ **8.42 Hz**

For C_E: $R_{CE} = 350 + [167 + (40k\Omega \parallel 10k\Omega)/151] \parallel 8.2k\Omega$

$= 350 + [167 + 53] \parallel 8.2k\Omega = 350 + 214 = 564\Omega$

and $f_{pE} = 1/(2\pi \times 564 \times 10 \times 10^{-6}) =$ **28.2 Hz**

with $f_{zE} = 1/(2\pi(8.2 \times 10^3 + 350)10 \times 10^{-6}) =$ **1.86 Hz**

For f_L: $f_L = [4.37^2 + 8.42^2 + 28.2^2 - 2(1.86)^2]^{1/2} = [19.1 + 70.9 + 795.2 - 6.9]^{1/2} =$ **29.6 Hz**

7.17 From Eq. 5.115, $f_T = g_m / [\,2\pi\,(C_{gs} + C_{gd})\,]$.

(a) $i_D = I_{DSS}(1 - \frac{\upsilon_{GS}}{V_p})^2$ - - - (1).

$1 = 4\,(1 - \frac{\upsilon_{GS}}{-2})^2 \rightarrow 1 + \frac{\upsilon_{GS}}{2} = (\frac{1}{4})^{1/2} = 1/2$. Thus $\frac{\upsilon_{GS}}{2} = -1/2$, $\upsilon_{GS} = -1$. Now $g_m = \frac{2I_{DSS}}{-V_p}(1 - \frac{\upsilon_{GS}}{V_p}) = \frac{2(4)}{--2}(1/2) = 2\text{mA/V}$. Thus $f_T = \frac{2\times 10^{-3}}{2\pi(2 + 0.2)\times 10^{-12}} =$ **144.7MHz.**

(b) $i_D = 1/2k'(W/L)(\upsilon_{GS} - V_t)^2$, $200 = 100\ (\upsilon_{GS} - 1)^2 \rightarrow (\upsilon_{GS} - 1) = \sqrt{2}$, $g_m = k'(W/L)(\upsilon_{GS} - V_t) = (200 \times 10^{-6})\sqrt{2} = 283\mu\text{A/V}$, $C_{gs} = 0.15 \times 10^{-12} + 20 \times 10^{-15} + 0.1 \times 10^{-12} = 0.27\text{pF}$, and $C_{gd} = 20 \times 10^{-15}\text{F} = 0.02\text{pF}$. Thus $f_T = \frac{283 \times 10^{-6}}{2\pi\,(0.27 + 0.02) \times 10^{-12}} =$ **155.3MHz.**

(c) $f_T = \frac{10 \times 10^{-3}}{2\pi\,(0.15 + 0.015) \times 10^{-12}} =$ **9.65GHz.**

7.18 For convenience, $K = 1/2\mu\,C_{ox}\,(W/L) = 1/2\,(.05 \times 10^{12}) \times 1 \times 10^{-15} \times 27/3 = .225 \times 10^{-3}\text{A/V}^2 = 225\mu\text{A/V}^2$; $g_m = 2K\,(\upsilon_{GS} - V_t) = 2\,(225 \times 10^{-6})\,(2.5 - 0.5) = 900\mu\text{A/V}$; $C_{gd} = L_d\,W\,C_{ox} = .3 \times 27 \times 1 \times 10^{-15} =$ **8.1fF**. $C_{gs} = 2/3\,WL\,C_{ox} + L_d\,W\,C_{ox} = 2/3 \times 27 \times 3 \times 1 \times 10^{-15} + 8.1 \times 10^{-15} = 54 + 8.1 =$ **62.1fF**. Thus $f_T = \frac{g_m}{2\pi\,(C_{gs} + C_{gd})} = \frac{900 \times 10^{-6}}{2\pi\,(62.1 + 8.1) \times 10^{-15}} =$ **2.04GHz.**

7.19 $C_g = C_{gs} + C_{gd}\,(1 - \text{gain}) = 200 + 20\,(1 - -1) =$ **240fF**, for gain of −1V/V, or = 200 + 20 (1 − − 100) = **2220fF**, for gain of −100V/V. $C_d = C_{db} + C_{gd}\,(1 - 1/\text{gain}) = 100 + 20\,(1 - - 1/1) =$ **140fF**, for gain of −1V/V, or = 100 + 20 (1 − − 1/100) = **120.2fF**, for gain of −100V/V.

7.20 Gain, gate to source, is $-g_m\,(r_o \parallel R_D \parallel R_L) = -1 \times 10^{-3}\,(50 \parallel 10 \parallel 30) \times 10^3 = -6.52\text{V/V}$. $C_g = 1\text{pF} + 0.5\text{pF}\,(1 - -6.52) =$ **4.76pF**, $C_d = 0.5\,(1 - - 1/6.52) =$ **0.577pF**.

Input pole: $f_{pg} = 1/(2\pi\,(4.76 \times 10^{-12}) \times (100k\Omega \parallel 1M\Omega)) =$ **0.368MHz.**

Output pole: $f_{pd} = 1/(2\pi \times 0.577 \times 10^{-12}\,(50 \parallel 10 \parallel 30) \times 10^3) =$ **42.3MHz.**

Upper 3dB frequency $f_{H1} = f_{pg} =$ **0.37MHz.** For R_s reduced to zero, the output pole dominates, and $f_{H2} = f_{pd} =$ **42.3MHz.** Now, for $f_{H3} = 0.9\,(42.3) = 38\text{MHz}$, with R_s non-zero and for $f_{pg}' = f_g$. Now $\left[\frac{1}{f_g^2} + \frac{1}{f_{pd}^2}\right]^{1/2} = \frac{1}{38.0}$, $\frac{1}{f_g^2} + \frac{1}{42.3^2} = \frac{1}{38.0^2}$, $\frac{1}{f_g} = (0.000693 - 0.000559)^{1/2}$. Thus $f_g = 86.4\text{MHz}$.

Now $f_g = 1/(2\pi\,(4.76 \times 10^{-12})R_s)$, whence $R_s = 1/(2\pi\,(4.76 \times 10^{-12})\,(86.35 \times 10^6)) =$ **387Ω**

7.21 Using the results of P7.20 and Equations 7.63 and 7.64,

$$\omega_{P1} = 1/\left[\,C_{gs} + C_{gd}\,(1 + g_m\,R_L') + C_{gd}\,(R_L'/R_s)\,\right] R_s$$

$$= \frac{1}{[\,1.0 + 0.5\,(1 + 6.52) + 0.5\,(6.52/90.9)]\ 90.9 \times 10^3 \times 10^{-12}} = \frac{1}{[4.76 + .036] \times 90.9 \times 10^{-9}}$$

= 2M rad/s, or

$f_{p1} =$ **0.365MHz.** $\omega_{p2} = \frac{C_{gs} + C_{gd}\,(1 + g_m\,R_L') + C_{gd}\,(R_L'/R')}{C_{gs}\,C_{gd}\,R_L'} = \frac{4.796 \times 10^{-12}}{1 \times 10^{-12} \times 0.5 \times 10^{-12} \times 6.52 \times 10^3} = 1.47\text{Grad/s}$, or $f_{p2} =$ **234MHz.** $f_z = \frac{g_m}{2\pi\,C_{gd}} =$

$\dfrac{1 \times 10^{-3}}{(2\pi\,(0.5 \times 10^{-12}))}$ = **318MHz**, for which $f_H \approx f_{p1}$ = **0.365MHz.** Now, for R_S = 1kΩ, f_z = **318MHz,** $f_{p1} = 1/(2\pi\,(4.76 + 0.5 \times (6.52/1))\,10^{-12} \times 10^{3})$ = **19.8MHz,** f_{p2} = $\dfrac{(4.76 + 3.26) \times 10^{-12}}{2\pi\,(1 \times 0.5 \times 10^{-24} \times 6.52 \times 10^{3})}$ = **392MHz,** for which $f_H \approx$ **19.8MHz.**

7.22 Generally, $i_D = K\,(\upsilon_{GS} - V_t)^2 \rightarrow 1 = K\,(2-1)^2 = K$; $g_m = 2K\,(\upsilon_{GS} - V_t) = 2\,(1)\,(2-1)$ = 2mA/V. Thus $f_T = \dfrac{g_m}{2\pi\,(C_{gs} + C_{gd})} \rightarrow 10^9 = \dfrac{2 \times 10^{-3}}{2\pi\,(C_{gs} + C_{gd})}$. Thus $C_{gs} + C_{gd} = \dfrac{2 \times 10^{-3}}{2\pi\,(10^9)}$ = 0.318pF. Now, if $C_{gd} = 0.2\,C_{gs} = C_{db}$, then $1.2\,C_{gs}$ = 0.318pF, C_{gs} = 0.265pF, $C_{gd} = C_{db}$ = 0.053pF. Thus C_{in} = 0.265 + 0.053 (1 – –3) = **0.477pF**. For input source, resistance, $3/g_m$ = $3/(2 \times 10^{-3})$ = 1.5kΩ, and capacitance = 4 (0.053pF) = 0.212pF, $f_H \approx \dfrac{1}{2\pi\,(0.212 \times 10^{-12})\,(1.5 \times 10^{-3})}$ = **500MHz.** Problem P7.50 of the Text provides the topology for which this high performance is possible.

7.23 $R_{in} = \dfrac{R_f}{1 - -gain} = \dfrac{R_f}{1+3} = \dfrac{R_f}{4}$, $\dfrac{R_{in}}{R_{in} + 10k\Omega}$ = 0.95 = $\dfrac{R_f/4}{R_f/4 + 10}$. Thus $R_f/4 = 0.95\,R_f/4 + 9.5$, $(R_f/4)(1 - 0.95) = 9.5$, whence R_f = 4(9.5)/0.05 = **760kΩ**! See that R_f must be very high, even in such a low-impedance circuit!

7.24 $f_\beta = 1/(2\pi\,(C_\pi + C_\mu)\,r_\pi)$. For I_C = 2mA, g_m = 2mA/25mV = 80mA/V, $r_\pi = \beta/g_m$ = 200/80 = 2.5kΩ. Now, $12.7 \times 10^6 = 1/(2\pi\,(C_\pi + 0.5) \times 10^{-12} \times 2.5 \times 10^3)$.

C_π = 5.01 –0.5 = **4.51pF**, and $f_T = \beta_o\,\omega_\beta$ = 200 (12.7 × 10^6) = **2.54GHz.** For I_C = 10mA, g_m = 10/2 × 80 = 400mA/V, r_π = 2/10 × 2.5 = 0.5kΩ, C_π = 10/2 × 4.51 = **22.55pF**, f_β = 1/(2π (22.55 + 0.5) × 10^{-12} × 0.5 × 10^3) = **13.8MHz.** For $C_\pi = C_\mu$ = 0.5pF, $\dfrac{4.51pF}{2mA} = \dfrac{0.5pF}{I_C}$. Thus $I_C = \dfrac{0.5 \times 2}{4.51}$ = **0.222mA.** That is, f_T is maintained at 2.5GHz for currents > 0.22mA or so.

7.25 $r_e = \dfrac{25 \times 10^{-3}}{0.15 \times 10^{-3}}$ = 166.7Ω, r_π = 166.7 × 151 = 25.17kΩ, $A_M = -\dfrac{25.17 \parallel 40}{25.17 \parallel 40 + 10} \times \dfrac{9.1 \parallel 10 \parallel 500}{166.7} = -\dfrac{15.45}{25.45} \times \dfrac{4.93 \times 10^3}{166.7}$ = –0.607 × 25.6 = –17.95 ≈ **–18V/V.** Now, $C_\pi + C_\mu$ = $\dfrac{g_m}{2\pi f_T} = \dfrac{\frac{150}{151} \times \frac{1}{166.7}}{2\pi \times 10^9}$ = 0.948pF, C_π = 0.948 –0.30 = 0.648pF.

Input Pole: C_T = 0.648 + 0.3 (1 – –25.6) = 8.63pF, $R_T = 25.17k\Omega \parallel 40k\Omega \parallel 10k\Omega + 50$ = 6.12kΩ,

$$f_{p1} = \frac{1}{2\pi \times 8.63 \times 10^{-12} \times 6.12 \times 10^3} = 3.00\text{MHz}.$$

Output Pole: C_T = 0.3pF, $R_T = 9.1k\Omega \parallel 10k\Omega \parallel 500k\Omega$ = 4.93kΩ,

$$f_{p2} = \frac{1}{2\pi \times 0.3 \times 10^{-12} \times 4.93 \times 10^3} = 108\text{MHz}.$$

The upper 3dB frequency is $f_H \approx$ **3.0MHz.**

7.26 For R = 350 in series with C_E, using data from P7.25 above, the total equivalent emitter resistance becomes 166.7 + 350 ∥ 8.2k = (0.167+0.336) kΩ = 0.503kΩ, and at the base, $R_{in} = 40k\Omega \parallel (151\;(0.503k\Omega))$ = 40 ∥ 75.95 = 26.2kΩ. Thus $A_M = -\dfrac{26.2}{26.2 + 10} \times \dfrac{(9.1 \parallel 10 \parallel 500)k\Omega}{0.503k\Omega} = -\dfrac{26.2}{32.4} \times \dfrac{4.72}{0.503}$ = –0.724 × 9.38 = **–6.79V/V.**

Now, from P7.25: $C_\pi = 0.65\text{pF}$, $C_\mu = 0.3\text{pF}$.
Input Pole: $C_T = 0.65\ (1 - (0.336/0.503)) + 0.3\ (1 - -9.38) = 0.216 + 3.114 = 3.33\text{pF}$,
$R_T = 26.2k\Omega \parallel 10k\Omega = 7.24\text{k}\Omega$, $f_{p1} = \dfrac{1}{2\pi \times 3.33 \times 10^{-12} \times 7.24 \times 10^3} = \mathbf{6.60MHz.}$
Output Pole: As before, $f_{p2} = \mathbf{108MHz}$. Thus f_{p1} dominates, and $f_H \approx \mathbf{6.6MHz.}$

7.27 The output pulse is *positive* with amplitude = 50 (50 × 10^{-3}) = **2.5V** and duration of 50μs. Its transition times are $\dfrac{2.2}{2\pi f_H} = \dfrac{2.2}{2\pi \times 50 \times 10^6} = \mathbf{7ns}$. Its sag (or droop) = $\dfrac{\Delta V}{V} = \dfrac{t_p}{\tau_L} = 2\pi f_L t_p$ $= 2\pi \times 50 \times 50 \times 10^{-6} = 0.0157$, or **1.6%**.

SECTION 7.5: THE COMMON-BASE, COMMON-GATE AND CASCODE CONFIGURATIONS

7.28 $r_e = 25/0.2 = 125\Omega$, $f_T = \dfrac{\alpha/r_e}{2\pi\ (C_\pi + C_\mu)} \rightarrow C_\pi = \dfrac{\alpha/r_e}{2\pi f_T} - C_\mu = \dfrac{\frac{150}{151} \times \frac{1}{125}}{2\pi \times 1 \times 10^9} - 0.3 = 1.26 - .3$ $= 0.96\text{pF}$. $A_M = \alpha \times \dfrac{r_e \parallel R_E}{r_e \parallel R_E + R_S} \times \dfrac{R_C \parallel R_L \parallel r_o}{r_e} = (0.993)\ \dfrac{.125 \parallel 8.2}{.125 \parallel 8.2 + .1} \times \dfrac{9.1 \parallel 10 \parallel 400}{.125}$ $= (.993)\ \dfrac{.1231}{.2231} \times \dfrac{4.71}{.125} = 20.65\text{V/V} \approx \mathbf{20.7V/V}$. $f_{p1} = \dfrac{1}{2\pi C_\pi (r_e \parallel R_E \parallel R_S)} =$ $\dfrac{1}{2\pi\ (.96 \times 10^{-12})\ (.125 \parallel 8.2 \parallel .100) \times 10^3} = 3\text{GHz}$. $f_{p2} = \dfrac{1}{2\pi C_\mu (R_C \parallel R_L \parallel r_o)} =$ $\dfrac{1}{2\pi\ (0.3 \times 10^{-12})\ (4.71 \times 10^3)} = 112.6\text{MHz}$. Thus $f_H \approx \mathbf{113MHz}$, with $A_M = \mathbf{20.7V/V}$.

7.29 $I_E = 0.15\text{mA}$, $r_e = 25/0.15 = 167\Omega$. Thus $C_\pi = \dfrac{\alpha/r_e}{2\pi f_T} - C_\mu = \dfrac{\frac{150}{151} \times \frac{1}{167}}{2\pi \times 10^9} - 0.3 = 0.946 - 0.3 = 0.65\text{pF}$.

At the input: $C_T = C_\pi + C_\mu\ (1 - -1) = 0.65 + 0.3\ (2) = 1.25\text{pF}$,
$R_T = 15k\Omega \parallel 10k\Omega \parallel (151(167)) = 15 \parallel 10 \parallel 25.2 = 4.85\text{k}\Omega$,

$$f_{p1} = \frac{1}{2\pi\ (4.85 \times 10^3)\ (1.25 \times 10^{-12})} = 26.3\text{MHz}.$$

At the emitter of Q_2: $C_T \approx C_\pi + C_\mu\ (1 - -1/1) = 0.63 + 2\ (0.3) = 1.25\text{pF}$, $R_T = 167\Omega$,

$$f_{p2} = \frac{1}{2\pi\ (167)\ (1.25 \times 10^{-12})} = 953\text{MHz}.$$

At the collector: $C_T \approx C_\mu = 0.3\text{pF}$, $R_T = 9.1k\Omega \parallel 10\text{k}\Omega = 4.76\text{k}\Omega$.

$f_{p3} = \dfrac{1}{2\pi \times 4.76 \times 10^3 \times 0.3 \times 10^{-12}} = 111\text{MHz}$. Thus $f_H \approx \left[\dfrac{1}{26.3^2} + \dfrac{1}{111^2} + \dfrac{1}{953^2}\right]^{-1/2} =$ **25.6MHz.**

Midband gain $A_M = -\dfrac{15k \parallel (151 \times 167)}{15k \parallel (151 \times 167) + 10k} \times \left[\dfrac{150}{151}\right]^2 \times \dfrac{9.1k\Omega \parallel 10k\Omega}{0.167k\Omega}$

$$= -\frac{15 \parallel 25.2}{15 \parallel 25.2 + 10} \times \left[\frac{150}{151}\right]^2 \times \frac{9.1 \parallel 10}{0.167} = -\frac{9.40}{19.40} \times \left[\frac{150}{151}\right]^2 \times \frac{4.76}{0.167} = \mathbf{-13.6V/V}.$$

SECTION 7.6: FREQUENCY RESPONSE OF THE EMITTER AND SOURCE FOLLOWERS

7.30 $r_e = \frac{V_T}{I_E} = \frac{25mV}{0.15mA} = 167\Omega$, $g_m = \frac{150}{151} \times \frac{1}{167} = 5.95\text{mA/V}$, $r_\pi = 151\ (167) = 25.2\text{k}\Omega$, $C_\pi = \frac{\frac{150}{151} \times \frac{1}{167}}{2\pi \times 10^9} - 0.3 = 0.947 - .3 = 0.65\text{pF}$.

See $A_M = \frac{8.2 \parallel 10}{8.2 \parallel 10 + 0.167} \times \frac{151\ (0.167 + 8.2 \parallel 10)}{151\ (0.167 + 8.2 \parallel 10) + 10} = \frac{4.5}{4.667} \times \frac{151\ (4.667)}{151\ (4.667) + 10} =$ **0.963V/V.**

For f_H: $C_T = C_\mu + C_\pi\ (1 - (4.5/4.667)) = 0.3 + 0.65\ (1 - 0.964) = 0.323\text{pF}$,

$R_T = 10k\Omega \parallel (151\ (0.167 + 8.2 \parallel 10)k\Omega) = 10\ k\Omega \parallel 705k\Omega = 9.86\text{k}\Omega$,

$f_H \approx f_{p1} = \frac{1}{2\pi \times 9.86 \times 10^3 \times 0.323 \times 10^{-12}} =$ **50.0MHz.**

For f_L: $C_{C1} = 1\mu\text{F}$, $R_{C1} = 10k\Omega + 8.2k\Omega \parallel (.167 + \frac{10k\Omega}{151}) = 10k\Omega + 8.2k\Omega \parallel .233k\Omega = 10.23\text{k}\Omega$, $f_L = \frac{1}{2\pi \times 1 \times 10^{-6} \times 10.23 \times 10^3} =$ **15.6Hz.**

7.31 $r_s = 1/g_m = 1\text{k}\Omega$, $A_m = \frac{10k\Omega}{(10 + 1)\ k\Omega} =$ **0.909V/V.**

For f_H: $C_T = 1 + \frac{1}{1 + 1 \times 10} = 1.09\text{pF}$, $R_T = 100\text{k}\Omega$, $f_H = \frac{1}{2\pi \times 10^5 \times 1.09 \times 10^{-12}} =$ **1.46MHz.**

SECTION 7.7: THE COMMON-COLLECTOR-COMMON-EMITTER CASCADE

7.32 Since $V_{BE} \approx 0.70\text{V}$, $I_R \approx 0.70/70\text{k}\Omega = 10\mu\text{A}$. Thus $I_{E2} \approx 160 - 10 = 150\mu\text{A}$, $r_{e2} = \frac{V_T}{150\mu A} = 167\Omega$, $g_{m2} = \frac{150}{151} \times \frac{1}{167} = 5.95\text{mA/V}$, $r_{\pi 2} = 151\ (167) = 25.2\text{k}\Omega$, $I_{B2} = \frac{150}{151} \approx 1\mu\text{A}$. Now $I_{E1} = 1 + 10 = 11\mu\text{A}$, $r_{e1} = \frac{25mV}{11\mu A} = 2.27\text{k}\Omega$, $r_{\pi 1} = 151\ (2.27) = 343\text{k}\Omega$.

For Q_2: $C_\pi = \frac{5.95 \times 10^{-3}}{2\pi \times 10^9} - 0.3 = 0.65\text{pF}$. For Q_1: $C_\pi = 0.3\text{pF}$. For A_M: $R_{in2} = 70k\Omega \parallel 25.2k\Omega = 18.5\text{k}\Omega$. $A_M = \frac{9.1 \parallel 10}{.167} \times \frac{150}{151} \times \frac{18.5}{18.5 + 2.37} \times \frac{(18.5 + 2.27)\ 151}{(18.5 + 2.27)\ 151 + 100}$

$= -\frac{4.76}{0.167} \times .993 \times \frac{18.5}{20.77} \times \frac{(20.77)\ 151}{3136 + 100} =$ **−24.4V/V.**

For f_H: *At the base of* Q_2: $C_T = 0.65 + 0.3\ (1 + 4.76 \times 5.95) = 9.45\text{pF}$, $R_T = 70k\Omega \parallel 25.2k\Omega \parallel (2.27k\Omega + \frac{100k\Omega}{151}) = 70 \parallel 25.2 \parallel 2.93 = 18.5 \parallel 2.93 = 2.53\text{k}\Omega$, $f_{p1} = \frac{1}{2\pi \times 2.53 \times 10^3 \times 9.45 \times 10^{-12}} = 6.66\text{MHz}$.

At the collector of Q_2: $f_{p2} = \frac{1}{2\pi \times 4.76 \times 10^3 \times 0.3 \times 10^{-12}} = 111\text{MHz}$.

At the base of Q_1: $C_T \approx 0.3 + 0.3\ (1 - (18.5/18.5 + 2.27)) = 0.333\text{pF}$, $R_T = 100k\Omega \parallel (151\ (2.27 + 18.5)) = 100\text{k}\Omega \parallel 3.1\text{M}\Omega = 96.9\text{k}\Omega$, $f_{p3} = \frac{1}{2\pi \times 96.9 \times 10^3 \times .333 \times 10^{-12}} = 4.93\text{MHz}$.

$$f_H \approx \left[\frac{1}{4.93^2} + \frac{1}{6.66^2} + \frac{1}{111^2} \right]^{-1/2} = \mathbf{3.96MHz.}$$

For f_L: *For* C_{C1} $R_{C1} = 10k + 9.1k = 19.1k\Omega$, $f_{p1} = \dfrac{1}{2\pi \times 1 \times 10^{-6} \times 19.1 \times 10^3} = 8.33Hz$.

For C_E: $R_{CE} = 167 \parallel 70k + \dfrac{70k \parallel (2.27k + 100k/151)}{151} = 0.167 \parallel 70 + 0.019 = 186\Omega$,

$$f_{p2} = \frac{1}{2\pi \times 186 \times 10 \times 10^{-6}} = 85.6Hz, \text{ and } f_z = 0Hz.$$

$f_L = (8.33^2 + 85.6^2)^{1/2} =$ **86Hz.**

7.33 a) **For R = 14kΩ:** $I_R = \dfrac{0.700}{14} = 50\mu A$, $I_{E2} = 160 - 50 = 110\mu A$, $r_{e2} = \dfrac{25}{110} = 227\Omega$, $g_{m2} = \dfrac{150}{151} \times \dfrac{1}{227} = 4.38mA/V$, $r_{\pi 2} = 34.3k\Omega$, $I_{B2} = \dfrac{110}{151} = 0.73$, $I_{E1} = 0.73 + 50 = 50.7\mu A$, $r_{e1} = \dfrac{25}{50.7} = 493\Omega$, $g_{m1} = \dfrac{150}{151} \times \dfrac{1}{493} = 2.01mA/V$, $r_{\pi 2} = 151\ (0.493) = 74.5k\Omega$.

For Q_2: $C_\pi = \dfrac{4.38 \times 10^{-3}}{2\pi \times 10^9} - 0.3 = 0.40pF$.

For Q_1: $C_\pi = \dfrac{2.01}{2\pi \times 10^9} - 0.3 \rightarrow$ Use 0.30pF.

For A_M: $R_{in2} = 14k\Omega \parallel 34.3k\Omega = 9.94k\Omega$, $A_M = -\dfrac{4.76}{0.227} \times 0.993 \times \dfrac{9.94}{9.94 + 0.493} \times \dfrac{(9.94 + 0.493)\ 151}{(10.4)\ 151 + 100} = -\dfrac{4.76}{0.227} \times 0.993 \times \dfrac{9.94}{10.4} \times \dfrac{1570.4}{1670.4} =$ **−18.7V/V.**

For f_H: *At the base of* **Q_2**: $C_T = 0.40 + 0.3\ (1 + 4.76 \times 4.38) = 6.95pF$, $R_T = 14 \parallel 34.3 \parallel (0.493 + \dfrac{100}{151}) = 14 \parallel 34.3 \parallel 1.16 = 9.94 \parallel 1.16 = 0.963k\Omega$,

$$f_{p1} = \frac{1}{2\pi \times 0.963 \times 10^3 \times 6.95 \times 10^{-12}} = 23.8MHz. \text{ Also } f_{p2} = 111MHz.$$

At the base of Q_1:

$C_T \approx 0.3 + 0.3\ (1 - \dfrac{9.94}{9.94 + 0.493}) = 0.314pF$, $R_T = 100 \parallel (151\ (0.493 + 9.94)) = 100k\Omega \parallel 1.58M\Omega = 94.0k\Omega$, $f_{p3} = \dfrac{1}{2\pi \times 94.0 \times 10^3 \times 0.314 \times 10^{-12}} = 5.39MHz$.

Thus $f_H = \left[\dfrac{1}{5.39^2} + \dfrac{1}{23.8^2} \right]^{-1/2} =$ **5.26×10^6Hz.**

For f_L: $f_{p1} = 8.33Hz$. Now, *For* C_E: $R_{CE} = .227k \parallel 14 + \dfrac{(.493k + \dfrac{100}{151}) \parallel 14k}{151} = 0.234k\Omega$,

$f_{p2} = \dfrac{1}{2\pi \times 234 \times 10 \times 10^{-6}} = 68.0Hz$. Thus $f_L = (68^2 + 8.33^2)^{1/2} =$ **68.5Hz.**

b) **For R = ∞:** $I_R = 0$, $I_{E2} = 160\mu A$, $r_{e2} = \dfrac{25}{160} = 0.156k\Omega$, $g_{m2} = \dfrac{150}{151} \times \dfrac{1}{156} = 6.37mA/V$, $r_{\pi 2} = 151\ (0.156) = 23.6k\Omega$, $I_{B2} = \dfrac{160}{150} = 1.07\mu A$, $I_{E1} = 1.07\mu A$, $r_{e1} = \dfrac{25 \times 10^{-3}}{1.07 \times 10^{-6}} = 23.4k\Omega$, $r_{\pi 1} = 151\ (23.4) = 3.5M\Omega$. *For* Q_2: $C_{\pi 2} = \dfrac{6.37 \times 10^{-3}}{2\pi \times 10^{-9}} - 0.3\times10^{-12} = 0.714pF$. *For* Q_1: $C_{\pi 1} = 0.3pF$.

For A_M: $R_{in2} = r_{\pi 2} = 23.6k\Omega$, $A_M = \dfrac{-4.76}{0.156} \times 0.993 \times \dfrac{23.6}{23.6 + 23.4} \times \dfrac{(23.6 + 23.4)\ 151}{(23.6 + 23.4)\ 151 + 100} = \dfrac{-4.76}{0.156} \times 0.993 \times \dfrac{23.1}{47.0} \times \dfrac{7097}{7197} =$ **−15.0V/V.**

For f_H: *At the base of Q_2*: $C_T = 0.714 + 0.3\,(1 + 6.37 \times 4.76) = 10.1\text{pF}$, $R_T = 23.6 \parallel (23.4 + 100/151) = 11.92\text{k}\Omega$, $f_{p1} = 1/(2\pi \times 11.92 \times 10^3 \times 10.1 \times 10^{-12}) = 1.32\text{MHz}$, $f_{p2} = 111\text{MHz}$.

At the base of Q_1: $C_T = 0.3 + 0.3\,(1 - \frac{23.6}{23.6 + 23.4}) = 0.449\text{pF}$, $R_T = 100\text{k}\Omega \parallel (151\,(23.6 + 23.4)) = 99\text{k}\Omega$, $f_{p3} = 1/(2\pi \times 99 \times 10^3 \times 0.449 \times 10^{-12}) = 3.58\text{MHz}$. Thus $f_H = \left[\frac{1}{1.32^2} + \frac{1}{3.58^2}\right]^{-1/2} = \mathbf{1.24MHz}$.

For f_L: $f_{p1} = 8.33\text{Hz}$. Now *For C_E*: $R_{CE} = 0.156 + \frac{(23.4 + (100/151))}{151} = 0.315\text{k}\Omega$, $f_{p2} = 1/(2\pi \times 315 \times 10 \times 10^{-6}) = 50.5\text{Hz}$. Thus $f_L = (50.5^2 + 8.33^2)^{1/2} = \mathbf{51.2Hz}$.

Summary (also using P7.32)

R(kΩ)	A_M(V/V)	f_L(Hz)	f_H(MHz)
14	−18.7	68.5	5.26
70	−24.4	86.0	3.96
∞	−15.0	51.2	1.24

See that R is important in improving gain **and** bandwidth. The worst idea is to use $R = \infty$. Perhaps a good design would be at $R = \sqrt{14 \times 70} = 31\text{k}\Omega$. *(You can check this for interest).*

SECTION 7.8: FREQUENCY RESPONSE OF THE DIFFERENTIAL AMPLIFIER

7.34 For each, $I_E = 300/2 = 150\mu\text{A}$, $r_e = 166.7\Omega$, $g_m = 150/151 \times 1/166.7 = 5.96\text{mA/V}$, $r_\pi = 166.7 \times 151 = 25.2\text{k}\Omega$, $C_\pi = \frac{5.96 \times 10^{-3}}{2\pi \times 10^9} - 0.3 \times 10^{-12} = 0.65\text{pF}$.

For Gain: $\frac{v_o}{v_s} = \frac{\frac{150}{151}((4 + 4) \parallel 10)}{2\,(0.1667)} \times \frac{2\,(25.2)}{10 + 2\,(25.2)} = \mathbf{11.05V/V}$. *For f_H*: (From Eq.7.96),

$$f_H = \frac{1}{2\pi\,(\frac{10}{2} \parallel 25.2) \times 10^3(0.65 + 0.3\,(1 + 5.96 \times 4 \parallel (\frac{10}{2}) \times 10^{-12}} = \frac{1}{2\pi\,(4.172)\,4.92 \times 10^{-9}} =$$

7.75MHz.

7.35 As in P7.34: $r_e = 166.7\Omega$, $g_m = 5.96\text{mA/V}$, $r_\pi = 25.2\text{k}\Omega$, $C_\pi = 0.65\text{pF}$, $C_\mu = 0.3\text{pF}$.

For R_L connected to the collector of the input transistor: Gain $= -\frac{150}{151}\frac{(4 \parallel 10)}{2\,(0.1667)} \times \frac{2\,(0.1667)\,151}{2\,(0.1667)\,151 + 10} = -\frac{150}{151}\frac{(2.86)}{0.333} \times \frac{50.3}{60.3} = \mathbf{-7.12V/V}$.

Gain from base to collector of the input transistor $= -\frac{150}{151}\frac{2.86}{0.333} = -8.53\text{V/V}$, (from which base there is a Miller-multiplied C_μ, and a voltage divider with two C_π in series). Thus, $C_T = 0.65/2 + 0.3\,(1 - -8.53) = 3.18\text{pF}$. $R_T = 10k\Omega \parallel (2\,(0.1667)\,151 = 10k\Omega \parallel 50.3k\Omega = 8.34\text{k}\Omega$, $f_H \approx 1/(2\pi \times 8.34 \times 10^3 \times 3.18 \times 10^{-12}) = \mathbf{6MHz}$.

For R_L connected to the collector of the grounded-base amplifier: Gain = **+7.12V/V**. Gain from base to collector of the input transistor is − (150/151) (4)/0.333 = −11.92V/V. At the input base $C_T = 0.65/2 + 0.3\,(1 - -11.92) = 4.20\text{pF}$, $R_T = 8.34\text{k}\Omega$, $f_H \approx 1/(2\pi \times 8.34 \times 10^3 \times 4.2 \times 10^{-12}) = \mathbf{4.54MHz}$.

7.36 Parameters as in P7.35: Gain = **+7.12V/V**, $C_T = 0.65/2 + 0.3 = 0.625\text{pF}$, $R_T = 8.34\text{k}\Omega$, $f_{p1} = 1/(2\pi \times 8.34 \times 10^3 \times 0.625 \times 10^{-12}) = 30.5\text{MHz}$. Now check the output pole, where $C_T = 0.3\text{pF}$,

$R_T = 4k \parallel 10k = 2.857k\Omega$, $f_{p2} = 1/(2\pi \times 2.857 \times 10^3 \times 0.3 \times 10^{-12}) = 186MHz$. Thus, $f_H \approx 1/(1/30.5 + 1/186) \approx$ **26.2**

7.37 From P7.36, $r_e = 166.7\Omega$, $g_m = 5.96mA/V$, $r_\pi = 25.2k\Omega$, $C_\pi = 0.65pF$, $C_\mu = 0.3pF$. With r_e added in each emitter lead, and load taken differentially, $A_M = \dfrac{\frac{150}{151}[(4+4) \parallel r\ 10]}{4\ (0.1667)} \times \dfrac{4\ (25.2)}{10 + 4\ (25.2)} = \dfrac{4.44 \times 0.993}{4\ (0.1667)} \times \dfrac{100.8}{110.8} = \mathbf{6.02\ V/V}$.

At each input, gain from base to collector of $Q_1 = \dfrac{1}{2} \times \dfrac{4.44 \times 0.993}{2\ (0.1667)} = -6.68V/V$, that is $C_T \approx 0.65/2 + 0.3\ (1 - -6.68) = 2.63pF$, $R_T = (10/2) \parallel (2(25.2)) = 4.55k\Omega$, and $f_H = 1/(2\pi \times 4.55 \times 10^3 \times 2.63 \times 10^{-12}) = \mathbf{13.3MHz}$.

7.38 For the current source, $r_o = 200V/300\mu A = 667k\Omega$, $C_T = 0.3 + 0.5 = 0.8pF$. Thus $V_{out} = \dfrac{4 \times 10^3 \times \frac{150}{151}}{2 \times 667 \times 10^3} \times 5V = \mathbf{14.9\ mV}$. For 1V peak on the load ends, $\dfrac{4k\Omega}{2\ Z_C} \times 5V = 1V \rightarrow Z_C = 10k\Omega$, and $f = 1/(2\pi(0.8) \times 10^{-12} \times 10 \times 10^3) = \mathbf{20MHz}$.

For saturation: Quiescent collector voltage $= V_{CC} - I/2\ (4k) = 10 - 0.15 \times 4 = 9.4V$. For 5V peak on the bases, saturation begins for a collector voltage of 5.0V, for which the peak load signal is $9.4 - 5V = 4.4V$. This occurs at a frequency of $(4.4/1) \times 20 =$ **88MHz.**

7.39 Here, $I_E = 150\mu A$, $r_e = 25 \times 10^{-3}/(150 \times 10^{-6}) = 166.7\Omega$,

$$C_\pi = \frac{(150/151) \times (1/166.7)}{2\pi \times 10^9} - 0.3 = 0.65pF,$$

$$A_M = \frac{(150/151)2.7 \times 10^3}{2(166.7)} \times \frac{2 \times 166.7 \times 151}{2 \times 166.7 \times 151 + 10 \times 10^3} = \frac{2.68 \times 10^3}{333.3} \times \frac{50.3}{60.3} = \mathbf{6.71\ \ V/V}.$$

For f_H: $C_T = 0.65/2 + 0.3 = 0.625pF$,

$R_T = [(2 \times 166.7)151] \parallel (10 \times 10^3) = (50.3 \parallel 10)10^3 = 8.34k\Omega$.

Now, $f_H = 1/(2\pi \times 8.34 \times 10^3 \times 0.625 \times 10^{-12}) =$ **30.5 MHz.**

Note that the gain-bandwidth product $GB = 6.71 \times 30.5 \times 10^6 = 204.6MHz$

7.40 $I_E = 150\mu A$, $r_e = (25 \times 10^{-3})/(150 \times 10^{-6}) = 166.7\Omega$, Add $r_E = r_e =$ **166.7Ω** to double the input resistance. Now, $C_\pi = \dfrac{150/151 \times 1/166.7}{2\pi \times 10^9} - 0.3pF = 0.65pF$, $A_M = \dfrac{2.7 \times 10^3 \times \frac{150}{151}}{4\ (166.7)} \times \dfrac{4 \times 166.7 \times 151}{4 \times 166.7 \times 151 + 10 \times 10^3} = \dfrac{2.68 \times 10^3}{667} \times \dfrac{100.7k\Omega}{110.7k\Omega} = \mathbf{3.66V/V}$.

For f_H: $C_T = 0.65/4 + 0.3 = 0.4625pF$, $R_T = (4\ (166.7)\ (151)) \parallel 10k = 100.7k \parallel 10k = 9.10k\Omega$, $f_H = 1/(2\pi \times 9.10 \times 10^3 \times 0.4625 \times 10^{-12}) = \mathbf{37.8MHz}$.

Note here that $GB = 3.66 \times 37.8 = 138.3$ MHz. Notice that this is considerably smaller than GB in P7.39 above. Why? [Hint: Although one might expect feedback through r_E would allow gain and bandwidth to be exchanged (see Section 2.7 of the Text), we note here that C_μ, which dominates, is outside of the feedback loop.].

7.41 From Eq. 5.115, $f_T = g_m/(2\pi(C_{gs} + C_{gd})) = 1 \times 10^{-3}/(2\pi(200 + 20)10^{-15}) =$ **723 MHz.**

The midband gain $A_M = +R_D/(1/g_m + 1/g_m) = +5 \times 10^3/(2(1 \times 10^{-3})^{-1}) =$ **2.5 V/V.**

At the input: $C_T = C_{gs}/2 + C_{gd} = 200/2 + 20 = 120fF$, and $R_T =$ 10 kΩ, for which,

$f_{Pin} = 1/(2\pi \times 10^4 \times 120 \times 10^{-15}) = 132.6MHz$

At the output: $C_T = C_{db} + C_{dg} = 100 + 20 = 120fF$, and $R_T =$ 5 kΩ, for which $f_{Pout} = 1/(2\pi \times 5 \times 10^3 \times 120 \times 10^{-15}) =$ 265 MHz.

Overall, $f_{3dB} = (132.6^{-2} + 265^{-2})^{-1/2} =$ **118.6 MHz.**

7.42 For all transistors, $r_e = V_T/I_E = 25 \times 10^{-3}/(150 \times 10^{-6}) = 166.\dot{6}\Omega$, or $0.1\dot{6}k\Omega$, and $g_m = \beta/r_\pi = (\beta/\beta + 1)/r_e = \alpha/r_e = (150/151)/166.7 = 5.96mA/V$. Since $f_T = g_m/(2\pi(C_\pi + C_\mu))$, $C_\pi + C_\mu = 5.96 \times 10^{-3}/(2\pi \times 10^9) = 0.949pF$, and $C_\pi = 0.949 - 0.3 = 0.649pF$.

For the gain:

Now, for differential signals, consider half-circuits: Input resistance at the base is $(\beta + 1)(r_e + r_e) = 151(2)(0.16)10^3 = 50.3k\Omega$

Thus, the gain $\dfrac{\upsilon_o/2}{\upsilon_s/2} = \upsilon_o/\upsilon_s = \dfrac{50.\dot{3}}{10 + 50.\dot{3}} \times \dfrac{(150/151)2.7 \times 10^3}{2(0.1\dot{6}6 \times 10^3)} = 0.834 \times 8.05 =$ **6.71 V/V.**

For the cutoff frequency:

At the input, $C_T = C_\pi/2 + C_\mu = 0.649/2 + 0.3 = 0.625pF$, and $R_T = 10k\Omega \parallel 50.3k\Omega = 8.34k\Omega$.

Thus, $f_{Pin} = 1/(2\pi \times 8.34 \times 10^3 \times 0.625 \times 10^{-12}) = 30.5MHz$.

At the output: $C_T = C_\mu = 0.3pF$, and $R_T = 2.7k\Omega$.

Thus, $f_{Pout} = 1/(2\pi \times 2.7 \times 10^3 \times 0.3 \times 10^{-12}) = 196.5MHz$.

Overall, $f_{3dB} = (30.5^{-2} + 196.5^{-2})^{-1/2} =$ **30.61 MHz**

7.43 Comparing this with the solution of P7.42 above, the emitter resistances remain at $0.1\dot{6}$ kΩ, but $g_m = \alpha/r_e$ reduces g_{mp} to $(50/51)/0.16 = 5.88mA/V$ (while g_{mn} stays at *5.96 mA/V*.

For this reason and the change in f_T of the pnp,

$(C_\pi + C_\mu)_p = 5.88 \times 10^{-3}/(2\pi \times 0.3 \times 10^9) = 3.12pF$.

Thus $C_{\pi p} = 3.12 - 1 = 2.12pF$.

For the gain: $\upsilon_o/\upsilon_s = \dfrac{50.\dot{3}}{10 + 50.\dot{3}} \times \dfrac{(50/51)2.7 \times 10^3}{2(0.\dot{1}6 \times 10^3)} = 0.834 \times 0.794 =$ **6.63 V/V.**

For the cutoff frequency:

At the input: $C_T = (0.649 \cdot\cdot 2.12) + 0.3 = 1/(1/0.649 + 1/2.12) + 0.3 = 0.797pF$ {where " ·· " indicates a series connection}, and $R_T = 10k\Omega \parallel 50.3k\Omega = 8.34k\Omega$,

whence $f_{Pin} = 1/(2\pi(3.34)10^3 \times 0.797 \times 10^{-12}) = 23.9MHz$.

At the output: $C_T = C_\mu = 1pF$, and $R_T = 2.7k\Omega$.

Thus, $f_{Pout} = 1/(2\pi \times 2.7 \times 10^3 \times 1 \times 10^{-12}) = 58.9MHz$

Overall, $f_{3dB} = (23.9^{-2} + 58.9^{-2})^{-1/2} =$ **22.1 MHz.**

Notice, in comparsion with the results of P7.42, that the details of the specifications of the pnp are quite important!

Chapter 8

FEEDBACK

SECTION 8.1: THE GENERAL FEEDBACK STRUCTURE

8.1 From Fig. 8.1, see $x_i = x_s - x_f = 1.00 - 0.99 =$ **0.01V**. Thus $A = x_o/x_i = 3.00/.01 =$ **300V/V**. Thus $\beta = x_f/x_o = 0.99/3.0 =$ **0.33V/V**. The open-loop gain is $A =$ **300V/V**. The amount of feedback is $(1 + A\beta) = 1 + 300\ (.33) =$ **100**. The closed-loop gain is $A_f = x_o/x_s = A/(1 + A\ \beta)$, where, directly, $A_f = 3.0/1.00 =$ **3.0V/V**, and, indirectly, is $300/(1 + 300\ (.33)) = 300/100 =$ **3.0V/V**. For the β network disconnected, $\upsilon_i = \upsilon_s$, and υ_o tends toward $A\ \upsilon_s = 300\ (1.0) =$ **300V**. This value would *not* be measured, since much before it is reached, the **output would** typically **limit** (or saturate).

8.2 See from $A_f = A/(1 + A\ \beta)$, that $8 = 10^2/(1 + 10^2\ \beta)$, or $1 + 10^2\ \beta = 10^2/8$, whence $\beta = (10^2/8 - 1)/10^2 =$ **0.115**. Now $\beta = R_1/(R_1 + R_2) = 0.115$, or $(R_1 + R_2)/R_1 = 1/0.115 = 8.696 = 1 + R_2/R_1$. Thus $R_2/R_1 = 8.696 - 1 =$ **7.696**. Amount of feedback is $1 + A\ \beta = A/A_f = 10^2/8 =$ **12.5V/V**. $\equiv 20\log_{10} 12.5 =$ **22dB**. For $V_s = 0.125$V, $V_o = A_f\ V_s = 8\ (.125) =$ **1V**, $V_f = \beta\ V_o = 1 \times .115 =$ **0.115V**, $V_i = V_o/A = 1/10^2 =$ **0.01V**, or $V_i = V_s - V_f = 0.125 - 0.115 =$ **0.01V**, as expected. For A increasing by 100%, A becomes $2(10^2) = 200$, and $A_f = A/(1 + A\ \beta)$ becomes $200/(1 + 200\ (0.115)) = 8.33$, rather than the former value of 8. That is, A_f increases by $(8.33 - 8)/8 \times 100 =$ **4.12%**.

SECTION 8.2: SOME PROPERTIES OF NEGATIVE FEEDBACK

8.3 For the original design, $A_f = A/(1 + A\ \beta) = 10^3/(1 + 10^3 \times 10^{-2}) = 90.9$. For the fabricated design, $A_f = 0.5 \times 10^3/(1 + 0.5 \times 10^3 \times 10^{-2}) =$ **83.3** results. Now, the desensitivity factor, $1 + A\ \beta$, as designed, was $1 + 10^3 \times 10^{-2} =$ **11**, and, as fabricated, is $1 + 0.5\ (10^3)\ (10^{-2}) =$ **6**. Thus for (small) changes around the original design, one would expect the original 50% reduction in A to result in a 50/11 = 4.5% reduction in A_f. Using the changed desensitivity factor, a 50/6 = 8.33% change would be expected. Now, conceptually (for small changes), the sensitivity, $\dfrac{\partial A_f/A_f}{\partial A/A} = \dfrac{1}{1 + A\ \beta} = \dfrac{1}{11} = .091 \equiv$ **−20.8dB**.

Actually, $\dfrac{\partial A_f/A_f}{\partial A/A} \approx \dfrac{\Delta A_f/A_f}{\Delta A/A} = \dfrac{(90.9 - 83.3)/90.9}{(10^3 - 0.5\ (10^3))/10^3} = \dfrac{7.6/90.9}{500/1000} = 0.167 \equiv$ **16.7%** The resulting manufactured closed-loop gain is (as calculated above) **83.3**.

8.4 We know that $A\beta = 89$, and $A_f = A/(1 + A\beta)$. Thus, ideally, $99 = A/(1 + 89)$, for which $A = 99\ (90) = 8910$, and $\beta = 89/8910$. After a time, $A_f = 98 = A/(1 + A\ (89/8910))$, or $A = 98 + 98\ (A)\ (89)/8910 = 98 + .9789A$. $A = 98/(1 - 0.9789) = 4645$, lower by more than a factor of 2. *Check*: $A_f = \dfrac{4645}{1 + 4645\ (89/8910)} = 98.000$. Thus gain A has reduced by $(8910 - 4645)/8910 \times 100 =$ **47.9%**.

8.5 For the closed loop, $1 + A\beta = A/A_f = 10^4/10^2 = 100$, and $\beta = (100 - 1)/10^4 = 99 \times 10^{-4}$. Thus the closed-loop 3dB frequency is $(10^4)\ 100 =$ **10 ^{6}Hz**. For the basic amplifier, $GB = 10^4 \times 10^4 =$ **10^8Hz**. For the feedback arrangement, $GB = 100 \times 10^6 =$ **10^8Hz**, the same! (as expected!)

8.6 For f_{3dB} of A reduced to 2×10^3Hz, the 3dB closed-loop frequency reduces to $2 \times 10^3 \times 10^2 =$ **2 × 10^5Hz**. On the surface, it appears that the desensitivity idea is not working, since the percentage change of the open-loop and closed-loop 3dB frequencies are the same. For the original amplifier at 10^4Hz, the open-loop gain is down by 3dB, to $10^4/1.414 =$ **0.707 × 10^4**, a percentage reduction of about 30%. Correspondingly, the percentage reduction in closed-loop gain $= 30/1 + A\beta = 30/100 = 0.3\%$, from 100 to $100\ (1 - .3/100) \approx 99.7$. For the manufactured amplifier, for which $f_{3dB} = 2 \times 10^3$Hz, the gain at

10^4Hz is $A_M = \dfrac{A_m}{(1^2 + (f/f_H)^2)^{½}} = \dfrac{10^4}{(1 + (10^4/(2 \times 10^3))^2)^{½}} = \dfrac{10^4}{(1 + 25)^{½}} = 0.196 \times 10^4$. Now $A_f = \dfrac{A}{1 + A\,(\beta)} = \dfrac{0.196 \times 10^4}{1 + .196 \times 10^4\,(99 \times 10^{-4})} = 96.06$. This corresponds to a drop of about 4% in gain for a change of f_{3dB} (from 10^4Hz to 2×10^3Hz) of 80% in frequency, an improvement of about 20 times. Note that at 10^4Hz, $(1 + A\,\beta)$ is 26, correspondingly. Thus the desensitivity factor is still at work, maintaining the gain for frequencies above the cutoff.

8.7 With a low-noise preamplifier of gain A_2, $S/N = (V_s/V_n)\,A_2$. For an improvement of 40dB, $A_2 = 10^{40/20}$ = **100V/V**. New S/N = –3dB + 40dB = **37dB**.

8.8

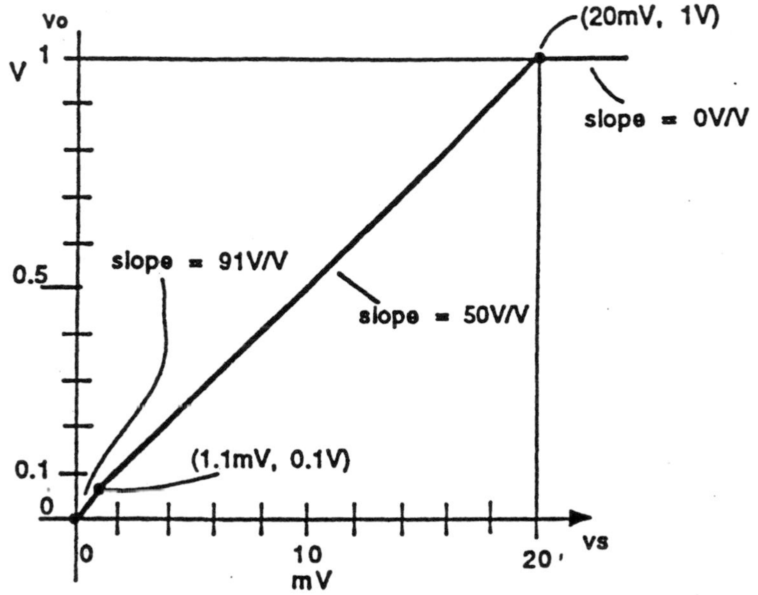

For $\upsilon_o \le 0.1$V, and $A = 10^3$V/V, $A_f = \dfrac{10^3}{1 + 10^3\,(.01)} = 90.9$V/V. Now for $\upsilon_o = 0.1$V, $\upsilon_s = 0.1/90.9 \equiv$ **1.1mV**, and $\upsilon_i = 0.1/10^3 = 0.1$mV. For $0.1 < \upsilon_o < 1.0$, and $A = 10^2$V/V, $A_f = \dfrac{10^2}{1 + 10^2\,(.01)} = 50$V/V. Now for $\upsilon_o = 1.0$V, $\upsilon_s = 1.0/50 \equiv$ **20mV**, and $\upsilon_i = 1.0/10^3 = 1$mV. For $\upsilon_i > 1$mV, υ_o limits at 1V, and $A = 0$, for which $A_f = 0$, as well.

SECTION 8.3: THE FOUR BASIC FEEDBACK TOPOLOGIES

8.9 For the circuits shown, the feedback type is:

(a) **Shunt-Shunt:** $\beta = \dfrac{I_f}{V_o} = -\dfrac{V_o/R_2}{V_o} = -\dfrac{1}{\mathbf{R_2}}$.

(b) **Shunt-SeriesK:** Now, assuming $R_2 >> r$, $\beta = \dfrac{I_f}{I_o} \approx -\dfrac{I_o\,r/R_2}{I_o} = -\dfrac{\mathbf{r}}{\mathbf{R_2}}$.

(c) **Series-Shunt:** $\beta = \dfrac{V_f}{V_o} = \dfrac{R_1}{R_1 + R_2} \times \dfrac{V_o}{V_o} = \dfrac{\mathbf{R_1}}{\mathbf{R_1 + R_2}}$.

(d) **Series-Series:** Now, assuming $(R_1 + R_2) >> r$, $\beta = \dfrac{V_f}{I_o} \approx \dfrac{R_1}{R_1 + R_2} \times \dfrac{I_o\,r}{I_o} = \dfrac{\mathbf{r\,R_1}}{\mathbf{R_1 + R_2}}$.

For (b), assuming $R_2 \approx r$, $I_f = \dfrac{-r}{r + R_2} \times I_o$, and $\beta = \dfrac{I_f}{I_o} = \dfrac{-\mathbf{r}}{\mathbf{r + R_2}}$.

For (d), assuming $(R_1 + R_2) \simeq r$, $V_r = (r \,||\, (R_1 + R_2))\, I_o = \frac{r\,(R_1 + R_2)}{r + R_1 + R_2} \times I_o$, and

$$\beta = \frac{V_f}{I_o} = \frac{\frac{R_1}{R_1 + R_2} \times V_r}{I_o} = \frac{R_1}{R_1 + R_2} \times \frac{r(R_1 + R_2)}{r + R_1 + R_2} \times \frac{I_o}{I_o} = \frac{\mathbf{r\,R_1}}{\mathbf{r + R_1 + R_2}}.$$

8.10 Assuming g_m is very high, the basic gain A is high, and $r_s = 1/g_m \simeq 0$.

(a) **Series-Series:** $v_f = r\, i_o$, and $\beta = v_f/i_o = \mathbf{r}$.

(b) **Shunt-Shunt:** $i_f = -v_o/R_F$, and $\beta = i_f/v_o = \mathbf{-1/R_F}$.

(c) **Series-Shunt:** $v_f = v_o$, and $\beta = v_f/v_o = \mathbf{1}$.

(d) **Series-Series:** $v_f = i_o\, r$, and $\beta = v_f/i_o = \mathbf{r}$.

8.11 Here, $V_i = V_s - V_f = 1\text{V} - 0.1\text{V} = 0.9\text{V}$. Thus $A = I_o/V_i = 2A/0.9\text{V} = 2.22\text{A/V} = \mathbf{2.22S}$, and $\beta = V_f/I_o = 0.1\text{V}/2\text{A} = .05\text{V/A} = .05\Omega = \mathbf{50m\Omega}$.

SECTION 8.4: THE SERIES-SHUNT FEEDBACK AMPLIFIER

8.12 For the β network, $\beta = 2/(2 + 18) = 0.1$, as suggested. Also, $R_{11} = 2\text{k}\Omega \,||\, 18k\Omega = 1.8\text{k}\Omega$, and $R_{22} = 2\text{k}\Omega + 18\text{k}\Omega = 20\text{k}\Omega$. For the A circuit, there are losses at the input and output. *For a zero impedance source and no load:* Thus

$$A = \frac{R_{ia}}{R_{ia} + R_S + R_{11}} \times A_a \times \frac{R_L \parallel R_{22}}{R_L \parallel R_{22} + R_{oa}} = \frac{10}{10 + 0 + 1.8} \times 100 \times \frac{\infty \parallel 20}{\infty \parallel 20 + .01} = 84.7\text{V/V}$$

Now for $A = 84.7\text{V/V}$, $\beta = 0.1\text{V/V}$, $A\beta = 84.7 \times 0.1 = 8.47$, and $1 + A\beta = 9.47$. Thus $A_f = A/(1 + A\beta) = 100/9.47 = \mathbf{10.6V/V}$.

Now, $R_i = R_S + R_{ia} + R_{11} = 0 + 10\text{k}\Omega + 1.8\text{k}\Omega = 11.8\text{k}\Omega$, and $R_o = R_L \parallel R_{oa} \,||\, R_{22} = \infty \parallel 10\Omega \parallel 20\text{k}\Omega = 10\Omega$. Thus with feedback, $R_{if} = R_i\,(1 + A\beta) = 11.8\,(9.47) = \mathbf{11.2k\Omega}$, and, $R_{of} = R_o/(1 + A\beta) = 10/9.47 = \mathbf{1.06\Omega}$. *For a 0.1V rms, 10kΩ* source and 100Ω load, $v_{out} = 0.1 \times \frac{112}{10 + 112} \times 10.6 \times 100/(100 + 1/.06) = \mathbf{0.963V}$ rms.

8.13 *For the feedback network*: $\beta = 10/(1290 + 10) = \mathbf{0.05V/V}$, $R_{11} = 10\text{k}\Omega \parallel 190\text{k}\Omega = \mathbf{9.5k\Omega}$, and $R_{22} = 10\text{k}\Omega + 190\text{k}\Omega = \mathbf{200k\Omega}$.

For the A circuit: $A = \frac{R_{ia}}{R_{ia} + R_S + R_{11}} \times A_a \times \frac{R_L \parallel R_{22}}{R_L \parallel R_{22} + R_{oa}} = \frac{20}{20 + 10 + 9.5} \times 900 \times \frac{1 \parallel 200}{1 \parallel 200 + 1} = \mathbf{227V/V}$. $R_i = R_{ia} + R_S + R_{11} = 20 + 10 + 9.5 = 39.5\text{k}\Omega$

$R_o = R_{oa} \parallel R_L \parallel R_{22} = 1 \parallel 1 \parallel 200 = 0.499\text{k}\Omega$.

Overall, $A_f = \frac{A}{1 + \beta A} = \frac{227}{1 + 227 \times .05} = \frac{227}{12.35} = \mathbf{18.4V/V}$, $R_{if} = R_i\,(1 + A\beta) = 39.5 \times 12.35 = 488\text{k}\Omega$, $R_{of} = R_o/(1 + A\beta) = 0.499/12.35 = 40.4\Omega$. Resistance seen by the source is $R_{if} - R_S = 488 - 10 = \mathbf{478k\Omega}$. Resistance seen by the load is $R_{of} \parallel (-R_L) = 40.4\Omega \parallel (-1k\Omega) = \frac{0.0404 \times 1}{1 - 0.0404} = 0.0421\text{k}\Omega = \mathbf{42.1\Omega}$. Overall gain, $V_o/V_s = A_f = \mathbf{18.4V/V}$.

8.14 *For the β circuit* (between nodes F and B): $\beta = 100\text{k}\Omega/(1\text{M}\Omega + 100\text{k}\Omega) = 0.0909$, $R_{11} = 100\text{k}\Omega \parallel 1\text{M}\Omega = 90.9\text{k}\Omega$, $R_{22} = 100\text{k}\Omega + 1\text{M}\Omega = 1.1\text{M}\Omega$.

For the A circuit (between nodes A, B to F): For Q_1, Q_2, $I_E = 100\mu\text{A}$. Thus $r_e = 25/0.1 = 250\Omega$, $r_\pi = 250\,(120 + 1) = 30.25\text{k}\Omega$. Thus $R_{id} = 2\,(30.25) = 60.5\text{k}\Omega$. For Q_5, assuming $V_o \simeq 0\text{V}$, $I_E = 2\text{mA}$, and $r_e = 25/2 = 12.5\Omega$, with $r_\pi = 12.5\,(121) = 1.51\text{k}\Omega$. Input resistance to the right at node C is $R_{ic} = (120)$

[12.5 + 1kΩ + 10kΩ || 1.1MΩ] = 1.32MΩ. Now ignoring device r_o, the gain from S to F, is $A = \frac{60.5}{100 + 90.9 + 60.5} \times \frac{120}{121} \times 2 \times \frac{1.32 \times 10^6}{250 + 250} \times \frac{10k\Omega \parallel 1.1M\Omega}{10k\Omega \parallel 1.1M\Omega + 1k\Omega + 12.5\Omega}$ or A = 1143V/V, with R_i = (100 + 90.9 + 60.5)kΩ = 251.4kΩ, and R_o = 10kΩ || 1.1MΩ || (R_{Q5} + 1kΩ). Now, since r_o of the devices has been ignored, the resistance driving node C is infinite, and thus the output resistance at the emitter of Q_5 is also infinite (however strange that may be!). {Note that if V_A = 200V, $r_{02} \simeq r_{01} = \frac{200}{100(10^{-6})}$ = 2MΩ, and r_{oc} = 1MΩ, and $R_{Q5} \simeq 10^6/121$ = 8.3kΩ, which with 1kΩ, will reduce R_o by about a factor of 2, not a large effect}. Ignoring the latter effect, R_o = 10kΩ || 1.1MΩ = 9.91kΩ. Now, A = 1143V/V, β = 0.0909, $A\beta$ = 103.9 and 1 + $A\beta$ = 104.9. Thus, $\frac{v_o}{v_s} = \frac{A}{1 + A\beta} = \frac{1143}{104.9}$ = **10.9V/V**. Now, $R_{if} = R_i\,(1 + A\beta)$ = (251.4kΩ) (103.9) = 26.1MΩ, and $R_{in} = R_{if} - R_S$ = 26.1 − .1 = **26MΩ**. Now, $R_{of} = R_o/(1 + A\beta)$ = 9.91kΩ/103.9 = 95.4Ω, and $R_{out} = R_{of}$ || ($-R_L$) = 95.4 || (−10kΩ) = **96.3Ω**. For source resistance and load resistance considered separately in the reduction of v_o/v_s to ½, the required R_S = **26MΩ** and the required R_L = **96.3Ω**. Clearly in practice, reduction of gain through loading will predominate.

8.15 Assume $v_o \simeq 0$ and that current splits equally between Q_1, Q_4. Now for Q_1, Q_2, Q_3, Q_4, i_D = (200μA)/2 = 100μA = $K\,(v_{GS} - V_t)^2$. Thus, v_{GS} − 1 = $(100/100)^{½}$ = 1, v_{GS} = 2V, $g_{m1} = 2K\,(v_{GS} - V_t)$ = 2 (100 × 10^{-6}) (2 − 1) = 200μA/V, $r_{01} = V_A/I_D$ = 20/100 × 10^{-6} = 200kΩ.

For Q_5, i_D = 200μA, (v_{GS} − 1) = $(200/100)^{½}$ = 1.414, v_{GS} = 2.414V, g_{m5} = 2 (100 × 10^{-6}) (1.414) = 282μA/V, r_{o5} = 20V/200μA = 100kΩ. For the β circuit, β = 1.0, R_{11} = 0Ω, $R_{22} = \infty$. For the A circuit, $R_i = R_S + R_{iu} + R_{11}$ = 1MΩ + ∞ + 0 = ∞, $R_o = r_{o5} \parallel \frac{1}{g_{m5}} = 10^5 \parallel \frac{1}{282 \times 10^{-6}}$ = 100kΩ || 3.55kΩ = 3.36kΩ, $A \simeq \frac{2\,(r_{03} \parallel r_{04})}{1/g_{m1} + 1/g_{m4}} \times \frac{r_{05}}{r_{05} + 1/g_{m5}} = \frac{2\,(200k\Omega/2)}{2\,(1/200 \times 10^{-6})} = \frac{100k\Omega}{100k\Omega + 3.55k\Omega}$ = 19.3V/V. Thus $A\beta$ = 19.3 × 1 = 19.3, and 1 + $A\beta$ = 20.3. Thus $A_f = \frac{A}{1 + A\beta}$ = 19.3/20.3 = **0.951V/V** = v_s/v_o, $R_{of} = R_o/(1 + A\beta)$ = 3.36/20.3 = 166Ω, that is R_{out} = 166Ω. Now $R_{if} = \infty$ (20.3) = ∞, that is R_{in} = ∞Ω. Now for 1kΩ load, the overall gain, v_o/v_s = 1kΩ/(166 + 1kΩ) × .951 = **0.816V/V**.

Concerning offset: For V_o = 0V, V_{G5} = 2.41V, V_{G2} = 5 − 2 = 3V, $V_{S1} = V_{S2}$ = 0 − 2 = −2V, that is current in $r_{01} = \frac{3 - -2}{200k\Omega}$ = 25μA, in $r_{02} = \frac{5 - 3}{200}$ = 10μA, in $r_{03} = \frac{5 - 2.41}{200k\Omega}$ = 12.95μA, in $r_{04} = \frac{2.41 - -2}{200k\Omega}$ = 22μA. Thus, assuming an ideal mirror, the net current offset at the gate of Q_5 is 25 − 10 + 12.95 − 22 = 5.95μA. Input offset to compensate is $V_{os} = I_{os}/g_m$. Here $G_m = \frac{2}{r_{s1} + r_{s2}} = \frac{2}{1/g_{m1} + 1/g_{m2}} = \frac{1}{1/g_{m1}} = g_{m1}$ = 200μA/V. Thus V_{os} = 5.95μA/200μA/V = **29.8mV**.

8.16 See that 100μA in the drain of Q_1 forces i_{D1} = 100μA. Since I = 200μA, thus i_{D2} = 200 − 100 = 100μA. Since $i_{D1} = i_{D2}$, for v_s = 0, then v_o = 0, and i_{R_L} = 0. Thus i_{D3} = 100μA also. Now 100μA = 100μA $(v_{GS} - 1)^2$, whence v_{GS} = 2V. Also g_m = 2K ($v_{GS} - V_t$) = 200μA/V, $r_s = 1/g_m$ = 5kΩ, and r_o = 20/100μA = 200kΩ.

For Q_2 seen as part of A, the feedback is a wire for which β = 1, R_{11} = 0Ω, and R_{22} = ∞Ω. Now, the A circuit consists, at the input, of v_s connected via 1MΩ to the gate of Q_1 and the gate of Q_2 grounded (through R_{11} = 0Ω). At the output, R_L only is connected to v_o. Thus $A = \frac{200k\Omega}{5k\Omega + 5k\Omega} \times \frac{10k\Omega \parallel 200k\Omega}{5k\Omega}$ = 38.1V/V, $R_i = \infty$, and $R_o = r_{03} \parallel R_L$ = 200kΩ || 10kΩ = 9.52kΩ. Thus $A\beta$ = 38.1 × 1 = 38.1, and 1 + $A\beta$ = 39.1, and $A_f = \frac{A}{1 + A\beta} = \frac{38.1}{39.1}$ = **0.974V/V**, $R_{if} = R_i\,(1 + A\beta) = \infty$, $R_{of} = R_o/(1 + A\beta)$ = 9.52/39.1 = 243Ω, that is R_{out} = 243 || (−10kΩ) = 249Ω, and v_o/v_s = **0.974V/V**.

Now for Q_2 seen as part of β, the feedback is a resistor r_{s2} = 5kΩ, for which β = 1, and R_{11} = 5kΩ, and R_{12} = ∞Ω. Now the A circuit, at the input, includes R_{11} = 5kΩ to ground from the source of Q_1, and otherwise is as before. Thus A = 38.1V/V, and β = 1, as before, with the **same results.**

Now for Q_2 and Q_3, both 10 × *wider with I* = 1.1mA: i_{D1} = 100μA as before, while $i_{D2} = i_{D3}$ = 1mA. But K_2, K_3 are each 1mA/V, and $\upsilon_{GS1} = \upsilon_{GS2} = \upsilon_{GS3}$ as before, but $g_{m2} = g_{m3}$ = 10 (200μA/V) = 2mA/V, $r_s = 1/g_m$ = 0.5kΩ and r_{03} = 20/1mA = 20kΩ. Now using the first idea (ie Q_2 part of A), $A = \dfrac{200k\Omega}{5k\Omega + 0.5k\Omega} \times \dfrac{10k\Omega \parallel 20k\Omega}{0.5k\Omega}$ = 484.8V/V, $A\beta$ = 484.8, $A\beta + 1$ = 485.8, R_o = 10kΩ ∥ 20kΩ = 6.66kΩ. Thus $\dfrac{\upsilon_o}{\upsilon_s} = A_f = \dfrac{A}{1 + A\beta} = \dfrac{484.8}{485.8}$ = **0.998V/V**, R_{of} = 6.66kΩ/484.8 = 13.75Ω, R_{out} = 13.75 ∥ (−10kΩ) = **13.8Ω**. See a great improvement in performance as a unity-gain buffer.

8.17 Following the idea in P8.16 above, see $\upsilon_o \approx$ 0V and $i_{D1} = i_{D4} = i_{D2} = i_{D3}$ = 100μA, for which $g_{m1} = g_{m2} = g_{m3} = g_{m4}$ = 200μA, r_s = 5kΩ, and r_o = 200kΩ for each. Here, consider the A circuit to consist of Q_1 with source grounded through R_{11}, and Q_3 loaded by R_{22} and R_L. Correspondingly, the β circuit consists of Q_2 and Q_4 with $r_{s2} = r_{s4}$ = 5kΩ, where β = 5/(5 + 5) = ½, R_{11} = 5kΩ ∥ 5kΩ = **2.5kΩ**, R_{22} = 5kΩ + 5kΩ = **10kΩ**. Thus, $A = \dfrac{200k\Omega}{5k\Omega + 2.5k\Omega} \times \dfrac{10k\Omega \parallel 10k\Omega \parallel 200k\Omega}{5k\Omega}$ = **26.0V/V**. Now, R_o = 10kΩ ∥ 10kΩ ∥ 200kΩ = 4.88kΩ, $A\beta$ = 26.0 × ½ = 13.0, $A\beta + 1$ = 14.0, $A_f = A/(A\beta + 1)$ = 26/14 = **1.857V/V**, R_{of} = 4.88kΩ/14 = 0.349kΩ, R_{out} = 0.349 ∥ (−10kΩ) = 362Ω.

SECTION 8.5: THE SERIES-SERIES FEEDBACK AMPLIFIER

8.18 For the A circuit: At the input, R_S, R_{ia}, and R_{11} are in series such that R_i = 10kΩ + 20kΩ + 10kΩ = 40kΩ. At the output, R_{oa}, R_L, and R_{22} are in series such that R_o = 1kΩ + 1kΩ + 0.2kΩ = 2.2kΩ.

Now $A = \dfrac{i_o}{\upsilon_s} = \dfrac{R_{ia}}{R_i} \times \dfrac{A_\upsilon}{R_o} = \dfrac{20}{40} \times \dfrac{900}{2.2k\Omega}$ = **204.5mA/V**. Now β = 50V/A = 0.05V/mA, and $A\beta$ = 204.5 × .05 = 10.23, and $A\beta + 1$ = 11.2, A_f = 204.5/11.2 = **18.3mA/V**. R_{if} = 40kΩ × 11.2 = 448kΩ, $R_{in} = R_{if} - R_S$ = 448 − 10 = **438kΩ**, R_{of} = 2.2kΩ × 11.2 = 24.64kΩ, $R_{out} = R_{of} - R_L$ = 24.64 − 1 = **23.6kΩ**.

8.19 Using the results of Example 8.2 as much as possible: As before $\beta = V_f'/I_o'$ = 11.9Ω , with I_o' now in the emitter of Q_3. $A = I_o'/V_i'$ = 20.51/.99 = **20.7A/V**, with I_o' corrected for being in the emitter of Q_3. R_i = 13.65kΩ, $R_o = R_{22} + R_L + R_{03}$. Here $R_{22} = R_{E2} \parallel (R_F + R_{E1})$ = 100 ∥ (640 + 100) = 88.1Ω. R_L = 600Ω, and $R_{03} = r_{e3} + R_{C2}/(\beta + 1)$ = 6.25 + 5kΩ/101 = 55.8Ω. Thus R_o = 88.1 + 600 + 55.8 = 744Ω. Now A = 20.7A/V, β = **11.9Ω**, $A\beta$ = 20.7 × 11.9 = 246.3, $A\beta + 1$ = 247.3, $A_f = A/(1{+}A\beta)$ = 20.7/247.3 = **83.7mA/V**. Thus R_{if} = 13.65 × 247.5 = **3.38MΩ** = R_{in} (since R_S = 0), R_{of} = (744) (247.3) = **184kΩ**, and R_{out} = 184kΩ − 0.6kΩ = **183kΩ**, as seen by R_L.

8.20 *For Q_1, Q_2, Q_3, Q_4,* $i_E \approx$ (200μA)/2 = 100μA for balanced operation. Thus r_e = 25mV/0.1mA = 250Ω, r_π = 101 (250) = 25.25kΩ, and r_o = 200V/0.1mA = 2MΩ.

For Q_5, i_E = 1mA. Thus r_e = 25Ω, r_π = 101 (25) = 2.525kΩ and r_o = 200V/1mA = 200kΩ. Now, consider the 10kΩ at the base of Q_2 (included to compensate for the dc drop in R_S) to be part of the A circuit. Thus the β circuit consists only of the 10Ω resistor, and $\beta = \upsilon_f/i_o$ = 10Ω, for which R_{11} = 10Ω and R_{22} = 10Ω.

For the A circuit: For the input series connection, R_i = 10kΩ + 2(25.25kΩ) + 10kΩ + 10Ω = 70.5kΩ, and for the output series connection, $R_o = R_{22} + R_L + (r_e + (1/(\beta + 1))\,(r_{04} \parallel r_{02})) \parallel r_{05}$ = 10 + 10^3 + (25 + $\dfrac{2 \times 10^6}{2(101)}$) ∥ (200 × $10^{3)}$ = 10.5kΩ. Now, $A = \dfrac{I_{E2}}{V_i} \approx \dfrac{2\,(25.25k\Omega)}{2\,(25.25k\Omega) + 10k\Omega + 10k\Omega + 10}$

$$\times \frac{2\,\frac{100}{101}\,(2M\Omega \parallel 2M\Omega \parallel (101\,(1k\Omega + 10\Omega + 25\Omega)\,)\,)}{2\,(250)} \times \frac{1}{(1k\Omega + 10\Omega + 25\Omega)}$$

$= 0.716 \times 376.2 \times .966 \times 10^{-3} = 0.260\text{A/V} = \mathbf{260mA/V}$.

$A\beta = .260\text{A/V} \times 10\text{V/A} = 2.60$, $A\beta + 1 = 3.60$, $A_f = \dfrac{I_o}{V_s} = \dfrac{A}{1+A\beta} = \dfrac{0.260}{3.60} = .0722\text{A/V}$ or $\mathbf{72.2mA/V}$, $R_{of} = R_o\,(1 + A\beta) = 10.5\text{k}\Omega \times 3.6 = 37.8\text{k}\Omega$, whence $R_{out} = R_{of} - R_L = 37.8 - 1 = \mathbf{36.8k\Omega}$. Now $R_{if} = R_i\,(1 + A\beta) = 70.5\,(3.6) = 254\text{k}\Omega$, whence $R_{in} = R_{if} - R_S = 254 - 10 = \mathbf{244k\Omega}$.

SECTION 8.6: THE SHUNT-SHUNT AND SHUNT-SERIES FEEDBACK AMPLIFIER

8.21 *For the basic amplifier*: $R_m = \dfrac{v_o}{i_i} = \dfrac{A_v v_i}{v_i/R_i} = A_v R_i = 900\text{V/V} \times 20\text{k}\Omega = \mathbf{18M\Omega}$. For the purposes of shunt-shunt analysis, convert the input into a current source $I_S = V_S/10\text{k}\Omega$ with a shunt $R_S = 10\text{k}\Omega$.

For the β circuit: $\beta = \dfrac{I_f'}{V_o'} = -\dfrac{1}{R_f} = -\dfrac{1}{100 \times 10^3} = -10^{-5}\text{A/V}$, $R_{11} = \mathbf{100k\Omega}$, and $R_{22} = \mathbf{100k\Omega}$.

For the A circuit: The input network consists of $10\text{k}\Omega \parallel 100\text{k}\Omega \parallel 20\text{k}\Omega = 6.25\text{k}\Omega$. The gain is $R_m = 18\text{M}\Omega$ (for input current flowing into the $20\text{k}\Omega$ input resistor). The output network consists of a series resistor of $R_{oa} = 1\text{k}\Omega$ and a shunt load of $R_{22} \parallel R_L = 100\text{k}\Omega \parallel 1\text{k}\Omega$. Now $A = \dfrac{V_o'}{I_i'} = -\dfrac{10k \parallel 100k}{10k \parallel 100k + 20k} \times 18 \times 10^6 \times \dfrac{1k \parallel 100k}{1k + 1k \parallel 100k}\ \text{V/A} = -\dfrac{9.09}{9.09 + 20} \times 18 \times 10^6 \times \dfrac{0.99}{1 + 0.99} = -2.80 \times 10^6\text{V/A} = -2800\text{V/mA} = \mathbf{-2.8V/\mu A}$, $A\beta = -2.8 \times 10^6 \times -10^{-5} = 28$, $1 + A\beta = 29$, $A_f = \dfrac{V_o}{I_s} = \dfrac{-A}{1 + \beta A} = \dfrac{-2.8 \times 10^6}{29} = \mathbf{-96.6V/mA}$. For A, $R_i = 6.25\text{k}\Omega$, $R_{if} = 6.25\text{k}\Omega/29 = 216\Omega$, $R_{in} = 216 \parallel (-10\text{k}\Omega) = \mathbf{221\Omega}$, $R_o = 1\text{k}\Omega \parallel 100\text{k}\Omega \parallel 1\text{k}\Omega = 498\Omega$, $R_{of} = 498/29 = 17.16\Omega$, $R_{out} = 17.16 \parallel (-1\text{k}\Omega) = \mathbf{17.5\Omega}$.

Now, for a load of $1k/2 = 500\Omega$, the gain reduces to $500/(500 + 17.5) \times (-96.6\text{V/mA}) = \mathbf{-93.3V/mA}$. To compensate at the input, seen as a fixed current source, we require that the source resistance increase from $10\text{k}\Omega$ in order that more of the available input current enters the amplifier. For a source I_S, R_S and $R_{in} = 221\Omega$, we want the same output voltage for the original and new loads. That is, $\dfrac{R_S}{221 + R_S}\,I_S \times 93.3 = \dfrac{10k\Omega}{221 + 10k\Omega} \times 96.6$, or $R_S = (221 + R_S)\,(1.013)$, or $R_S = 221/(-.013) = -17\text{k}\Omega$. Thus it is not possible for normal input circuits to compensate if I_S is fixed. Alternatively, if the input voltage is fixed at $V_S = I_S\,(10\text{k}\Omega)$, then we may lower R_S (from $10\text{k}\Omega$) so that the output is the same for the original and new loads. That is, $\dfrac{221}{221 + R_S} \times I_S\,(10k) \times \dfrac{93.3}{221} = \dfrac{221}{221 + 10k} \times I_S\,(10k) \times \dfrac{96.6}{221}$, or $\dfrac{93.3}{221 + R_S} = \dfrac{96.6}{221 + 10k}$, $221 + 10\text{k}\Omega = (221 + R_S)\,1.035$, whence $R_S = 10.221/1.035 - 221 = \mathbf{9.65k\Omega}$.

8.22 At low frequencies, with the R_1, R_2, R_3 loop viewed as defining the voltage at node A, operation is as a voltage regulator, with the reference voltage being V_{BE1}. Here, R_1 acts as a resistive-wire connection to the input with a voltage comparison being made across the base-emitter of Q_1. Thus the feedback is of the **series-shunt variety**. As a result of it, $V_A \approx 0.7\text{V}$, $I_{R2} = 0.7/700 = 1\text{mA} = I_{E2}$, $V_B = 0.7 + 1(1) = 1.7\text{V}$, $V_C = 1.7 + 0.7 = 2.4\text{V}$, $I_{C1} \approx (5-2.4)/2.7\text{k}\Omega \approx 1\text{mA}$, with $r_{e1} = r_{e2} = 25\text{mV}/1\text{mA} = 25\Omega$, and $r_{\pi 1} = r_{\pi 2} \approx 2.5\text{k}\Omega$.

At high frequencies, feedback is of the **shunt-series variety**, with I_o as output and $I_s = V_s/R_S$ as input. For this, the β network consists of R_3 and R_5 with $\beta = \dfrac{I_f'}{I_o'} = \dfrac{-R_3}{R_3 + R_5} = \dfrac{-1}{1 + 5} = \mathbf{-0.166A/A}$, $R_{11} = R_5 + R_3 = \mathbf{6k\Omega}$, $R_{22} = R_5 \parallel R_3 = (5 \times 1)/(5 + 1) = \mathbf{0.833k\Omega}$.

The A circuit consists at the input, of $R_S \parallel R_{11} \parallel r_{\pi 1} = 10\text{k}\Omega \parallel 6\text{k}\Omega \parallel 2.5\text{k}\Omega$ at the base of Q_1 fed by I_i and I_o' emerging from the emitter of Q_2 and connected to R_{22} to ground. The output resistance associated with I_o' is $R_{02} = r_{e2} + R_4/(\beta + 1) = 25 + 2.7\text{k}\Omega/101 = 51.7\Omega$. Now, using a current-divider approach with device β, $A = \frac{I_o'}{I_i'} = -\frac{10 \parallel 6}{10 \parallel 6 + 2.5} \times 100 \times \frac{2.7}{2.7 + 101\,(.025 + .833)} \times 101 =$ $-3.75/6.25 \times 100 \times 2.7/89.4 \times 101 =$ **−183A/A**. That is, $A\beta = -183 \times (-0.166) = 30.5$, $1 + A\beta = 31.5$, $A_f = \frac{I_o}{I_s} = \frac{-183}{31.5} = 5.81$A/A. Now $A_f' = \frac{I_o}{V_S} = \frac{I_o}{I_S R_S} = 5.81/10^4 =$ **0.581mA/V**. Now $R_i = R_3 \parallel R_{11} \parallel r_{\pi 1} = 10\text{k}\Omega \parallel 6\text{k}\Omega \parallel 2.5\text{k}\Omega = 1.5\text{k}\Omega$. Thus, $R_{if} = 1.5\text{k}\Omega/31.5 = 47.6\Omega$, $R_{in} = 47.6 \parallel (-10\text{k}\Omega) \approx 50\Omega$. Also $R_o = R_{22} + R_{02} = 833.3 + 51.7 = 885\Omega$, and $R_{of} = 885\,(31.5) = 27.9\text{k}\Omega$. This is the resistance seen (for example) by a low-resistance load inserted between node B and the emitter of Q_2. Since V_A is assumed large and $r_{02} = \infty$, then R_{out} is also **infinite** (independent of the feedback detail). But, even if r_{02} were very low, R_{out} would still be extremely high (see Eq. 6.78 on page 519 of the Text), since R_{of} is quite high.

8.23 For the dc loop, a series-shunt configuration, $\beta =$ **−1.0** (via R_1) with $R_{11} =$ **10kΩ**, and $R_{22} = \infty\Omega$. For A, $R_i = r_{\pi 1} \approx 2.5\text{k}\Omega$, $R_o = R_3 + r_{e2} + \frac{R_4}{\beta + 1}$, or $R_o = (700\Omega) \parallel (1k\Omega + 25\Omega + \frac{2.7k\Omega}{101}) = 700 \parallel 1051.7 =$ 420Ω, and the output of the A circuit consists of a 700Ω load. Thus $A = -\frac{100}{101} \times \frac{2.7k\Omega \parallel (101\,(25\Omega + 1.7k\Omega))}{25\Omega} \times \frac{700}{1725} = -43.4$V/V. Thus, $A\beta = 43.3$, $A\beta + 1 = 44.4$, $A_f = \frac{A}{1 + A\beta} = \frac{-43.4}{44.4} =$ **−0.977V/V**. At node A, $R_o = 420\Omega$, $R_{of} = \frac{420}{44.4} = 9.46\Omega$. Thus the resistance seen by C_3 is 9.46Ω! For a 100μF capacitor, $f_H = \frac{1}{2\pi \times 9.46 \times 100 \times 10^{-6}} =$ **168Hz**. Note that before your exposure to the effects of feedback, you may have considered the resistance seen by R_3 to be (roughly) $700 \parallel 1k\Omega \parallel 10k\Omega \approx 0.4\text{k}\Omega$, with $f_H = \frac{1}{2\pi \times 0.4 \times 1^{-3} \times 100 \times 10^{-6}} =$ **3.98Hz!**

8.24 When C_5 is removed, we end up with a complex feedback network, consisting of R_3, R_2, R_1, R_5 in a shunt-series feedback loop. Refer to P8.22 above for the basic calculations. For β, I_o' sees two paths to I_f', one through R_5 and one through R_3. The resistance in the R_3 path is $1\text{k}\Omega + 0.7\text{k}\Omega \parallel 10\text{k}\Omega =$ $1.654\text{k}\Omega$. $I_f' = -I_o' \left[\frac{1.654}{5 + 1.654} + \frac{5}{5 + 1.654} \times \frac{0.7}{10 + 0.7}\right]$, or $I_f/I_o' = \beta = -(.2486 + .0492) =$ **−0.298A/A** and $R_{11} = ((R_5 + R_3) \parallel R_1) + R_2 = (5 + 1) \parallel 10 + .7 =$ **4.45kΩ**, $R_{22} = 1.654\text{k}\Omega \parallel 5\text{k}\Omega$ $=$ **1.24kΩ**. Using current ratios directly, $A = \frac{I_o'}{I_i'} = -\frac{10 \parallel 4.45}{10 \parallel 4.45 + 2.5} \times 100 \times$ $\frac{2.7}{2.7 + 101\,(.025 + 1.24)} \times 101 = -\frac{3.08}{5.58} \times 100 \times \frac{2.7}{130.4} \times 101 =$ **−115.4A/A**. Now, $A\beta = -115.4 \times$ $(-.298) = 34.4$, $A\beta + 1 = 35.4$, $A_f = \frac{I_o}{I_s} = -\frac{115.4}{34.4} = 3.35$, $A_f' = \frac{I_o}{V_S} = \frac{3.35}{10k\Omega} =$ **0.335mA/V**.

8.25 Using results from P8.23: *For* C_1, $R_{inD} \approx (r_{\pi 1} \parallel R_1)(A\beta + 1) = (2.5k\Omega \parallel 10k\Omega) \times 43.4 = 86.8\text{k}\Omega$, $R_{source} \approx R_S \parallel R_5 = 10\text{k} \parallel 5\text{k} = 3.33\text{k}\Omega$. Now, for 1Hz cutoff, $C_1 \approx \frac{1}{2\pi \times 1 \times (3.3 + 86.8) \times 10^3} =$ **1.77μF**. *For* C_2, $R_T \approx R_5 + R_S \parallel R_{inD} = 5\text{k} + 10\text{k} \parallel 86.8\text{k} = 14\text{k}\Omega$. For 10Hz cutoff, $C_2 = \frac{1}{2\pi \times 10\,(14 \times 10^3)} =$ **1.14μF**.

8.26 For Fig. Q8.10 d) in P8.10 above, with g_m = 2mA/V, r_o = 10kΩ, R_S = 100kΩ, r = 1kΩ, for $R_L = R \geq 0$, find I_o/V_S and R_{out} (facing R_L). This is a series-series feedback circuit, with the output current monitored by r and a corresponding voltage fed to the source of the transistor. Thus $\beta = V_f'/I_o' = (I_o' r)/I_o'$ $= r$ = 1kΩ, for which $R_{11} = r$ = 1kΩ, and $R_{22} = r$ = 1kΩ. The A circuit is as shown, where:

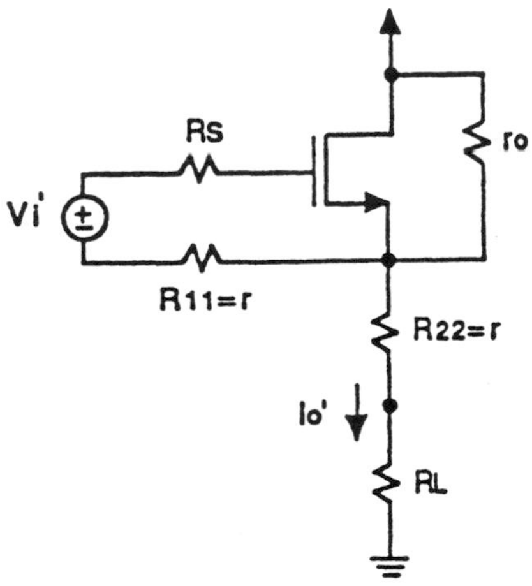

$$A = \frac{I_o'}{V_i'} = \frac{g_m V_i'}{V_i'} \times \frac{r_o}{r_o + R_L + r} = 2\text{mA/V} \times (10/11 + R_L)$$

where $R = R_L$ in kΩ and $R_i = \infty$.

For R_L = 0, $A = 2\frac{10}{11}$ = 1.82mA/V. For R_L = 10kΩ, $A = 2\frac{10}{11 + 10}$ = 0.95mA/V. For R_L = 100kΩ, $A = 2\frac{10}{11 + 100}$ = 0.09mA/V. Generally, $\beta = \frac{V_f'}{I_o}$ $= r$ = 1kΩ, Thus $A\beta$ = 1.82 × = 1.82 to 0.95 to 0.09 for R_L from 0 to 10kΩ to 100kΩ respectively, and $1 + A\beta$ = 2.82, to 1.95, to 1.09 correspondingly, with $A_f = \frac{I_o}{V_S} = \frac{1.82}{2.82}$ = .645mA/V for R_L = 0, or $\frac{0.95}{1.95}$ = .487mA/V for R_L = 10kΩ, or $\frac{0.09}{1.09}$ = .082mA/V for R_L = 100kΩ.

See the gain is 3dB down from the 0-ohm value, when $\dfrac{2\frac{10}{11 + R_L}}{1 + \frac{2\,(10)}{11 + R_L}} = \frac{1}{\sqrt{2}}$ (.645) = .456. This is of the form $2x$ = .456 + .456 $(2x)$. Thus $\frac{10}{11 + R_L} = \frac{.456}{2\,(1 - .456)}$ = 0.419, for which $R_L = \frac{10}{.419} - 11$ = **12.85kΩ**. It is a further 3dB down (ie 6dB altogether) at R_L = 31kΩ. For the output resistance of A, see that since the transistor gate and source are joined through the (floating) signal source, the MOS output resistance is just r_o. Thus $R_o = R_L + r + r_o = 11 + R$. Now $1 + A\beta = 1 + 1\,(\frac{20}{11 + R}) = \frac{R + 31}{11 + R}$, and $R_{of} = R_o\,(1 + A\beta) = (11 + R)\,(\frac{R + 31}{11 + R}) = (R + 31)$kΩ. Now the resistance seen by R_L is $R_{out} = R_{of} - R = R + 31 - R$ = **31kΩ** (!). The limit on the value of R_L used depends on the degree to which transconductance constancy is required, and on the size of output voltage that can be tolerated while maintaining linear operation. Certainly for loads from 0kΩ to 12.9kΩ, the transconductance varies only by 3dB, and by another 3dB for R_L up to 31kΩ.

8.27 Note the solution of P8.23. See that R_1 = 10kΩ = $R_{1a} + R_{1b}$ with a tap at x. Let $R_{1a} = R$. Now at the tap, without feedback, the resistance is $R_{ox} \approx (R_{1a} + r_{\pi 1}) \parallel (R_{1b} + R_2 \parallel (R_3 + r_{e2} + \frac{R_4}{\beta + 1}))$, or R_{ox} = $(R + 2.5) \parallel (10 - R + 0.7 \parallel (1 + 0.05)) = (R + 2.5) \parallel (10.42 - R) = \frac{(2.5 + R)\,(10.42 - R)}{12.92}$. Now, $\frac{\partial R_{ox}}{\partial R}$ = 0, when $(2.5 + R)\,(-1) + (10.42 - R)\,1 = 0$, $1042 + 2.5 - 2R = 0$, or R = 12.92/2 = 6.5kΩ, for which $\frac{R_{1a}}{R}$ = **0.65**, $R_{ox} = \frac{(2.5 + 6.5)\,(10.4 - 6.5)}{12.9}$ = 2.72kΩ, and $R_{of} = \frac{R_{ox}}{1 + A\beta} = \frac{2.72}{44.4}$ = **61.3Ω**, with the low-frequency loop closed. For this situation, for 100μF, $f_H = \frac{1}{2\pi \times 61.3 \times 100 \times 10^{-6}}$ = **26Hz.** For f_H = 168Hz, as in P8.23, C_3 = 100μF × 26/168 = **15.5μF**, or **15.5%** of the (100μF) capacitor needed with R_1 untapped.

SECTION 8.7: DETERMINING THE LOOP GAIN

8.28 At 1kHz, $A\beta = \frac{1.27}{20 \times 10^{-3}}$ = **63.5V/V**. At 10Hz, $A\beta = \frac{3.1}{2 \times 10^{-3}}$ = **1550V/V**. Assume the amplifier to be direct-coupled, and that the feedback network employs a single capacitor to ground which limits the loop gain at higher frequencies by maintaining β lower there. Further, assume that the loop-gain frequency response around 1kHz is relatively flat, and (separately) that at 10Hz, it falls as frequency rises. Since a single-pole response is postulated, it must fall (proportional to $1/f$) at 20dB/decade, from 1550V/V at 10Hz to reach 63.5V/V at 10 × 1550/63.5 = 244Hz. Thus the corner (3dB) frequency for the feedback network seems to be at **244Hz**. Now, as frequency is lowered below 244Hz, β increases and $A\beta$ increases, reaching 1550 at 10Hz and only 2 × 1550 = 3100V/V at 1Hz. Thus there must be an associated pole of β at about 10Hz/2 = 5Hz with the loop gain at dc being about 3100V/V. Since the loop gain $A\beta$ is 3100 at dc, and 63.5 at 1kHz, while β is frequency-dependent, it is likely that at 1kHz, β is at most 63.5/3100 = 0.0205, and it may even be less if feedback is not unity at dc. Now assuming the worst case, that is that A is at least **3100**: Then at 1kHz, $A_f = \frac{A}{1 + A\beta} = \frac{3100}{1 + 63.5}$ = **48V/V**. For the closed loop, the lower 3dB frequency is a 3dB frequency of the β network, occuring in particular where β is 3dB high, at which the closed-loop gain is essentially 3dB down (for $A\beta >> 1$). For a closed-loop pole at 244Hz with $C = 1\mu F$, $R_{eq} = \frac{1}{2\pi \times 244 \times 1 \times 10^{-6}} = 650\Omega$.

8.29 At high frequencies, $\beta = \frac{R_1}{R_1 + R_2} = \frac{2}{2 + 47}$ = 0.0408V/V, and $A\beta$ = 0.0408 (1550) = **63.3V/V**. At very low frequencies, β = 1V/V, and $A\beta$ = 1 (1550) = **1550V/V**. Now β has a zero when the magnitude of the reactance of C equals the resistance of R_1, for $C = \frac{1}{2\pi \times 2 \times 10^3 \times 2.45}$ = **32.5μF**. The associated pole is at $f_p = \frac{1}{2\pi\,(R_1 + R_2)\,C} = \frac{1}{2\pi\,(49k\Omega) \times 32.5 \times 10^{-6}}$ = **0.100Hz**. See at high frequencies, that the gain will be $\frac{1550}{1 + 1550\,(.0408)}$ = 24.1V/V, and at low frequencies is essentially 1.00V/V. See that f_L is the frequency at which β begins to rise, ie at the frequency at which β has a zero, ie at 2.45Hz, ie f_L = **2.45Hz**. Generally speaking, $A_f = \frac{A}{1 + A\beta} \approx \frac{1}{\beta}$. Thus a pole of the closed-loop response will occur at a zero of β. For the capacitor reduced from 32.5μF to 10μF, resistors must be raised by 32.5/10 = 3.25 times, to R_1 = 2 (3.25) = 6.5kΩ (use **6.8kΩ**), and R_2 = 47 (3.25) = 152.8kΩ (use **150kΩ**).

8.30 Loop gain = 1.2V/10mV = **120V/V**. From the β network, $\beta = \frac{100\Omega}{100\Omega + 10k\Omega}$ = 0.0099. Thus A = 120/.0099 = **1.21 × 10⁴V/V**, is the basic op-amp gain.

8.31 $|A\beta\,(f)| = 10$, $|A_f\,(f)| = 10$. Now $A_f = \frac{A}{1 + A\beta}$. Assume that at a particular frequency there is no phase shift in A, but that $\beta = a + jb$, and $A_f = x + jy$. Thus $x + j\,y = \frac{A}{1 + A\,(a + j\,b)} = \frac{1}{a + 1/A + j\,b}$ $= \frac{a + 1/A - j\,b}{(a + 1/A)^2 + b^2}$. Thus $x = \frac{a + 1/A}{(a + 1/A)^2 + b^2}$, $y = \frac{-b}{(a + 1/A)^2 + b^2}$. Now $x^2 + y^2 = 10^2$, and $A^2\,(a^2 + b^2) = 10^2$. Combining: $\frac{(a + 1/A)^2 + b^2}{((a + 1/A)^2 + b^2)^2} = 100 = A^2\,(a^2 + b^2)$, whence $a^2 + b^2 \approx \frac{1}{1 + A^2\,(a^2 + b^2)}$, (say).

$A^2\,(a^2 + b^2)^2 + (a^2 + b^2) - 1 = 0$, $a^2 + b^2 = \frac{-1 \pm \sqrt{1 - 4(-1)\,A^2}}{2A^2}$, or

$A^2\,(a^2 + b^2) = \frac{-1 \pm \sqrt{1 + 4A^2}}{2}$. Thus $\frac{-1 \pm \sqrt{1 + 4A^2}}{2} = 100$, $-1 \pm \sqrt{1 + 4A^2} = 200$, $1 + 4A^2 = (201)^2$, $4A^2 = 40400$, $A^2 = 10100$, $A = 100.5$. Now $A^2\,(a^2 + b^2) = 10^2$, $a^2 + b^2 = \frac{0.10}{(100.5)^2}$. Thus

$|\beta| = (a^2 + b^2)^{1/2} = 10/100.5 = .0995$. Correspondingly, at this frequency, A = **100.5V/V** and $|\beta|$ = **0.0995.**

An *alternative approach* is simply to ignore the possibility of phase shifts in either L or A_f: Directly, $A\beta =$ 10 and $A_f =$ 10. Thus, $A_f = A/(1 + A\beta) = A/(1 + 10) =$ 10, whence $A =$ **110,** and $\beta = A\beta/A = 10/110 =$ **0.091.**

8.32 In the circuit of Fig. Q8.15 in P8.15 above, open the loop at the gate of Q_4 and inject a signal υ. Now, for Q_1, Q_2, Q_3, Q_4, i_D = 100μA, that is $i_D = K\,(\upsilon_{GS} - V_t)^2$, or $100 = 100\,(\upsilon_{GS} - 1)^2$, whence υ_{GS} = 2V, for which $g_m = 2K\,(\upsilon_{GS} - V_t) = 2\,(100)\,(1) = 200\mu A/V$, $r_s = 1/g_m$ = 5kΩ and r_o = 20/100μA = 200kΩ. Also for Q_5, $200 = 100\,(\upsilon_{GS} - V_t)^2$, and $(\upsilon_{GS} - V_t) = \sqrt{2} = 1.414V$, whence $g_m = 2\,(100)\,1.44$ = 282μA/V, $r_s = 1/g_m$ = 3.546kΩ, and r_o = 20/200μA = 100kΩ. Now, *for no load*, $A\beta = -\dfrac{\upsilon_o}{\upsilon} = \dfrac{2\,(200k \parallel 200k)}{2\,(5k)} \times \dfrac{100k}{100k + 3.55k} = 20 \times 0.966 =$ **19.3V/V**. Since $\beta = 1$, $A_f = \dfrac{A}{1 + A\beta} = \dfrac{19.3/1}{1 + 19.3} =$ **0.951V/V**, the same as originally found. Now, *for* 1kΩ *load*, $A\beta = \dfrac{2\,(200k \parallel 200k)}{2\,(5k)} \times \dfrac{100k \parallel 1k}{100k \parallel 1k + 3.55k} = 20 \times \dfrac{0.99}{0.99 + 3.55} =$ **4.36V/V**, and with $\beta = 1$, $A_f = \dfrac{4.36/1}{1 + 4.36} =$ **0.813V/V**, where the earlier calculation yields $A_f = 0.816$V/V.

SECTION 8.8: THE STABILITY PROBLEM

8.33 At frequency ω, $\Phi = \tan^{-1}\omega/10^3 + 2\tan^{-1}\omega/10^5 = 180°$. Now at $\omega = 10^5$rad/s: $\Phi = \tan^{-1}100 + 2\tan^{-1}1 = 89.4 + 2\,(45) = 179.4°$. Thus the phase shift is 180° at (slightly above) **10^5 rad/s.**

$$|A\,(10^5)| = \frac{10^3}{\left[\left[1 + \left(10^5/10^3\right)^2\right]\left[1 + \left(10^5/10^5\right)^2\right]\left[1 + \left(10^5/10^5\right)^2\right]\right]^{1/2}} = \frac{10^3}{10^2\,(\sqrt{2})\,(\sqrt{2})}$$

$$= \frac{10}{2} = \mathbf{5V/V.}$$

Now for $\beta < 1/5$, $A\beta < 1/5 \times 5 = 1$, when $\Phi = 180°$, with no margins. For a 20dB gain margin, $\beta = 1/10 \times 1/50 =$ **0.02 or less.** For $\beta = 0.02$, $A\beta = 1$, when $A = 50$. Now $20\log_{10} 50 = 34$dB. Since the midband gain is 60dB, the gain drop is 60 − 34 = 26dB, implying a 3dB frequency of 26/20 or 1.3 decades above 10^3rad/s, that is at $10^3 \times 10^{1.3} = 2 \times 10^4$rad/s.

Check: At 2×10^4rad/s, $\dfrac{10^3}{\left[1 + \left[\dfrac{2 \times 10^4}{10^3}\right]^2\right]^{1/2}\left[1 + \left[\dfrac{2 \times 10^4}{10^5}\right]^2\right]^{1/2 + 1/2}} = \dfrac{10^3}{(20.02)\,(1.04)} = 48$, for which $\Phi = \tan^{-1}20 + 2\tan^{-1}0.2 = 87.14 + 2\,(11.3) = 110°$, for a phase margin of 70°.

8.34

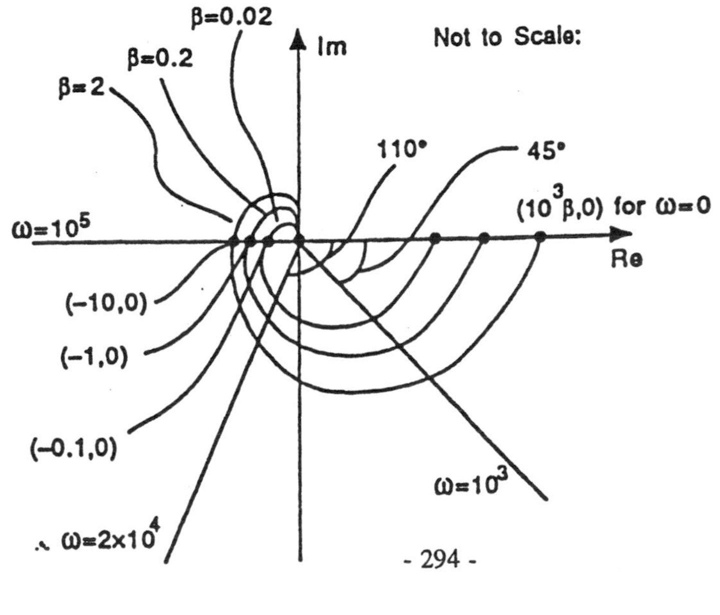

From P8.33: $A = 10^3$, $\omega_{p1} = 10^3$rad/s, $\omega_{p2} = \omega_{p3} = 10^5$rad/s, $\beta_{critical} = 0.2$. Now $A\beta(\omega)$ $= \dfrac{10^3\,\beta}{(1 + j\omega/10^3)(1 + j\omega/10^5)(1 + j\omega/10^5)}$. Now from the solution of P8.33: Know at 10^5rad/s, $|A\beta| = 5\beta$,

$\Phi(A\beta) = 180°$, and at 10^3rad/s, $|A\beta| = 0.7 \times 10^3\,\beta$, $\Phi(A\beta) = 45°$, and at 2×10^4rad/s, $|A\beta| = 48\beta$, $\Phi(A\beta) = 110°$. The plot is for $\beta = 0.2$, 2, and 0.02. Note that for 3 poles, the maximum shift is 270°.

8.35 For $\beta = 1.0$, $|A\beta| = 1$ at $\omega = 10^3 \times 10^6 = 10^9$rad/s, at which the total phase shift is $\tan^{-1}(10^9/10^3) + \tan^{-1}(10^9/10^8) = 90° + 84.3° =$ **174.3°**. Thus, **oscillation does not occur**, although there is only 5.7° phase margin. Now for $\beta = 0.5$, $A = 2$, when $A\beta = 1$, whence $A = \dfrac{10^6}{\left[1 + \left(\omega/10^3\right)^2\right]^{1/2}\left[1 + \left(\omega/10^8\right)^2\right]^{1/2}} = 2$, for ω just above 10^8rad/s. At 10^8rad/s, $A = \dfrac{10^6}{10^5(1+1)^{1/2}} = 7.07$. At 3×10^8rad/s, $A = \dfrac{10^6}{3 \times 10^5(1 + 3^2)^{1/2}} \approx 1$. At 2×10^8rad/s, $A = \dfrac{10^6}{2 \times 10^5(1 + 2^2)^{1/2}} = 2.24$. At 2.1×10^8rad/s, $A = \dfrac{10^6}{2.1 \times 10^5(1 + 2.1^2)^{1/2}} = 2.05$. Now at 2.1×10^8rad/s, $\Phi = 90° + \tan^{-1}\dfrac{2.1 \times 10^8}{10^8} = 90° + 64.5° = 154.5°$, for which the phase margin is $180° - 154.5° = 25.5°$. Now with an input capacitance, the β network provides a 3rd pole at ω_p. Oscillation can begin if $\tan^{-1}\left(\dfrac{2.1 \times 10^8}{\omega_p}\right) = 25.5°$, or $\dfrac{2.1 \times 10^8}{\omega_p} = 0.477$, or $\omega_p = \dfrac{2.1 \times 10^8}{0.477} = 4.4 \times 10^8$rad/s. Now for $C =$ 5pf, $\omega = \dfrac{1}{RC}$, $R = \dfrac{1}{\omega C} = \dfrac{1}{4.4 \times 10^8 \times 5 \times 10^{-12}} =$ **454Ω**. Thus for the β network consisting of two equal resistors R, $R/2 < 454$, or $R < 910\Omega$. To ensure stability, even smaller resistors would be needed, say 100Ω.

SECTION 8.9: EFFECT OF FEEDBACK ON AMPLIFIER POLES

8.36 The low-frequency amplifier gain, $A = \dfrac{f_t}{f_p} = \dfrac{20 \times 10^6}{5 \times 10^3} = 4 \times 10^3$V/V. With feedback, $A_f = \dfrac{A}{1 + A\beta} = \dfrac{4 \times 10^3}{1 + 4 \times 10^3 \times 0.125} = \dfrac{4000}{501} =$ **7.98V/V**. $f_{3dB} = f_p(1 + A\beta) = 5 \times 10^3 \times 501 =$ **2.5MHz**. The unity-gain frequency = 2.5 × 7.98 = 19.95MHz, or **20MHz**. Thus the pole is shifted by the amount of the feedback factor, that is by about 500 times.

8.37 From P8.36 above, for A = 4000V/V and f_p = 5kHz. To achieve $f_{pf} = 10^6$Hz, the amount of feedback, $1 + A\beta = \dfrac{10^6}{5 \times 10^3} =$ **200**. Correspondingly, the loop gain is $A\beta =$ **199**, and the feedback factor, $\beta = 199/4000 = 0.04975$. Use $\beta =$ **0.05**, for which the low-frequency gain is $A_f = \dfrac{4000}{1 + 4000 \times 0.5} =$ **19.9V/V**.

8.38 $A(s) = \dfrac{10^4 K}{(1 + s/10^5)(1 + s/(10^6/K))}$. For loop closure with a feedback factor β, the poles are: $s = -1/2(\omega_{p1} + \omega_{p2}) \pm 1/2\,[(\omega_{p1} + \omega_{p2})^2 - 4(1 + A_o\beta)\,\omega_{p1}\,\omega_{p2}]^{1/2}$ See that the poles are coincident at 5×10^5 when $5 \times 10^5 = 1/2(10^5 + 10^6/K)$, $10 = 1 + 10/K$, or $10/K = 9$, $K = 10/9 =$ **1.11**. The poles are coincident when $(\omega_{p1} + \omega_{p2})^2 - 4(1 + A\beta)\,\omega_{p1}\,\omega_{p2} = 0$, or $(10^5 + 10^6/K)^2 = 4(1 + A\beta)\,10^5 \times 10^6/K$, or $1 + A\beta = \dfrac{(1 + 10/K)^2}{4(10/K)} = \dfrac{(1+9)^2}{4(9)} = \dfrac{10^2}{4(9)} = 2.778$. Thus $A\beta = 1.778$, and $\beta =$ **1.778 × 10⁻⁴**. Thus the dc open-loop gain of the amplifier is $10^4 K =$ **1.11 × 10⁴V/V**, with poles of **10⁵Hz** and **9 × 10⁵Hz**. The Low-frequency closed-loop gain is $A_f = \dfrac{A}{1 + \beta A} = \dfrac{1.11 \times 10^4}{2.778} =$ **4 × 10³V/V**.

8.39 *Note that for convenience of notation we will use ω as the variable which denotes frequency in Hz!*

For a maximally-flat design, $Q = \mathbf{0.707}$, where from Eq. 8.38 in the Text, $Q = \frac{((1 + A\beta)\,\omega_{p1}\,\omega_{p2})^{1/2}}{(\omega_{p1} + \omega_{p2})}$ --- (1) and $\omega_{p1} + \omega_{p2} = \omega_o/Q$, with poles at $-1/2\,(\omega_{p1} + \omega_{p2}) \pm 1/2\,((\omega_{p1} + \omega_{p2})^2 - 4\,\omega_{p1}\,\omega_{p2}\,(1 + A_o\beta))^{1/2}$. That is, poles are at $\frac{\omega_{p1} + \omega_{p2}}{2} \pm 1/2\,((\omega_{p1} + \omega_{p2})^2\,(1 - 4\,Q^2))^{1/2}$, or $\frac{\omega_{p1} + \omega_{p2}}{2}\,(1 \pm j\,(4\,Q^2 - 1)^{1/2})$, or $\frac{10^6 + 20 \times 10^6}{2} \times (1 \pm j\,(4\,(.707)^2 - 1)^{1/2}) = \mathbf{10.5 \times 10^6\,(1 \pm j)}$ **Hz** (as seen directly from Fig. 8.32 in the Text, and the text following), and $\omega_o = Q\,(\omega_{p1} + \omega_{p2}) = .707\,(21 \times 10^6) = 14.85 \times 10^6$Hz $=$ **14.85MHz** (that is $\sqrt{2} \times 10.5$MHz). Now from the characteristic equation (Eq. 8.37 of the Text), $s^2 + s\,(\omega_{p1} + \omega_{p2}) + (1 + A_o\beta)\,\omega_{p1}\,\omega_{p2} = 0$, and (1) above, $s^2 + s\,(\omega_{p1} + \omega_{p2}) + Q^2\,(\omega_{p1} + \omega_{p2})^2$ is the denominator of the transfer characteristic, which for $s = j\omega$, is $(-\omega^2 + j\omega\,(21 \times 10^6) + (21 \times 10^6)^2/2)$. Now at the 3 dB frequency (ω_c): $[(\omega_c(21 \times 10^6)) + ((21 \times 10^6)^2/2 - \omega_c^2)]^2 = 1/2$. Normalizing to 21×10^6Hz, $\omega_c^2 + (1/2 - \omega_c^2)^2 = 1/2$, or $\omega_c^2 + 1/4 + (-\omega_c^2) + \omega_c^4 = 1/2$. Thus $\omega_c^4 = 1/4$, $\omega_c^2 = 1/2$, $\omega_c = 0.707$ (normalized), and, in general, $\omega_c = 0.707 \times 21 \times 10^6 =$ **14.85MHz** (that is, ω_o). Now, from (1), $(1 + A_o\beta) = \frac{(Q\,(\omega_{p1} + \omega_{p2}))^2}{\omega_{p1}\,\omega_{p2}} = \frac{(0.707\,(21 \times 10^6))^2}{(1 \times 10^6)\,(20 \times 10^6)} = \frac{2\,(21)\,(21)}{1 \times 20} = 44.1$. Thus $A_o\beta = 43.1$, $\beta = 43.1/10^3 = \mathbf{0.0431}$, and $A_f = \frac{A}{1 + A\beta} = 10^3/44.1 =$ **22.7V/V**.

8.40 $A_o = 10^3$, $\omega_{p1} = 10^3$rad/s, $\omega_{p2} = \omega_{p3} = 10^5$rad/s, and $A(s) = \frac{10^3}{(1 + s/10^3)\,(1 + s/10^5)^2}$.

Thus $A_f\,(s) = \dfrac{\dfrac{10^3}{(1 + s/10^3)\,(1 + s/10^5)^2}}{1 + \dfrac{\beta\,10^3}{(1 + s/10^3)\,(1 + s/10^5)^2}} = \dfrac{10^3}{(1 + s/10^3)\,(1 + s/10^5)^2 + \beta\,10^3}$.

Denominator is $(1 + s/10^3)\,(1 + 2s/10^5 + s^2/10^{10}) + \beta\,10^3 = 1 + 2s/10^5 + s^2/10^{10} + s/10^3 + 2s^2/10^8 + s^3/10^{13} + \beta\,10^3 = (\beta\,10^3 + 1) + s\,(2/10^5 + 1/10^3) + s^2\,(1/10^{10} + 2/10^8) + s^3/10^{13}$. Normalize to $s/10^5$, that is $10^3\,\beta + 1 + s\,(102) + s^2\,(201) + 100\,s^3$. Now divide by 100 and set equal to 0: $s^3 + 2.01\,s^2 + 1.02s + 10\,\beta + .01 = 0$ --- (1), which is in the form $(s + a)\,(s + b - jc)\,(s + b + j\,c) = 0 = (s + a)\,(s^2 + sb + jsc + bs + b^2 + jbc - jcs - jbc + c^2) = (s + a)\,(s^2 + 2\,sb + b^2 + c^2) = (s^3 + 2s^2\,b + sb^2 + sc^2 + as^2 + 2abs + ab^2 + ac^2) = s^3 + s^2\,(a + 2b) + s\,(b^2 + c^2 + 2ab) + a\,(b^2 + c^2)$ --- (2). Now, see $a + 2b = 2.01$, or $a = 2.01 - 2b$ --- (3), $b^2 + c^2 + 2ab = 1.02$ --- (4), $ab^2 + ac^2 = 10\beta + .01 = x$ --- (5). Now, (4) + (3) $\rightarrow b^2 + c^2 + 2b\,(2.01 - 2b) = 1.02$, $b^2 + c^2 + 4.02b - 4b^2 = 1.02$, $c^2 + 4.02b - 3b^2 = 1.02$ --- (6).

Special Cases: (i) Two poles are coincident when $c = 0$

(3) $\rightarrow a = 2.01 - 2b$ --- (7),

(4) $\rightarrow b^2 + 2ab = 1.02$ --- (8),

(5) $\rightarrow ab^2 = x$ --- (9).

(9) $\rightarrow$ (8) $b^2 + 2x/b = 1.02 \rightarrow b^3 + 2x = 1.02b$ --- (10)

(9) $\rightarrow$ (8) $x/a + 2ab = 1.02$, $x + 2a^2b = 1.02a$

(9) $\rightarrow$ (7) $b^2\,(2.01 - 2b) = x$, $2.01b^2 - 2b^3 = x \rightarrow b^3 = 2.01/2\,b^2 - x/2$ --- (11)

(10) + (11) $\rightarrow 2.01/2\,b^2 - x/2 + 2x = 1.02b$, $2.01b^2 - 2.04b + 3x = 0$

$b = 2.04 \pm \frac{\sqrt{2.04^2 - 4\,(+3x)2.01}}{2\,(2.10)} = 0.507 \pm \sqrt{.258 - 1.49}$. The two are identical when $0.258 = 1.49x = 1.49\,(10\beta + 0.01)$, or $\beta = \frac{0.173 - 0.01}{10} = \mathbf{0.0163}$, at which $b = 0.507$ and $a = 2.01 - 2\,(.507) = 0.996$.

Denormalizing, the poles are at about **-1×10^5, -0.5×10^5**, and **-0.5×10^5 rad/s.**

(ii) Now the $j\omega$ axis is reached when $b = 0$, for which:

(3) $\rightarrow a = 2.01$, and

(4) $\rightarrow c^2 = 1.02$, $c = 1.01$,

where from (5), $x = 10\beta + .01 = ac^2 = 2.01 \times 1.02$, whence $\beta = \dfrac{2.05 - .01}{10} = $ **0.204**. In this case, the poles are approximately at **-2×10^5 rad/s** and **$\pm$ j 10^5 rad/s.**

(iii) Now, $Q = 0.707$ for the complex pole pair implies that $b = c$, (4) $\rightarrow 2b^2 + 2ab = 1.02$, (5) $\rightarrow$ $2ab^2 = x$, (6) $\rightarrow b^2 + 4.02b - 3b^2 = 1.02$, $2b^2 - 4.02b + 1.02 = 0$,

$b = \dfrac{4.02 \pm \sqrt{4.02^2 - 4(2)\,(1.02)}}{2\,(2)} = \dfrac{4.02 \pm 2.83}{4} = 1.71$ or 0.298. Now, from (3), for $a = 2.01 - 2b$ positive, $b = 0.298$ for which $a = 2.01 - 2\,(0.298) = 1.41$, and $x = ab^2 + ac^2 = 2ab^2 = 2\,(1.41)\,(0.298)^2$ $= 0.250$. Thus $x = 10\beta + 0.01 = 0.250$, and $\beta = \dfrac{0.240}{10} \approx$ **0.024**. Accounting for the initial normalization, the pole locations for $Q = 0.707$ are at **$-$ 1.41 $\times$ 10^5** rad/s, and at **($-$0.298** $\pm j$ 0.298) $\times$ 10^5 rad/s.

Now for Q = 0.707, for which $\beta = 0.024$, using the **normalized frequency, w = ω/10^{5}rad/s**, see

$$T(w) = \frac{10^3}{(1 + j100w)\,(1 + jw)^2 + 24} = \frac{10^3}{(1 + j100w)\,(1 - w^2 + 2jw) + 24}$$

$$= \frac{10^3}{1 - w^2 + 2jw + j100w - j100w^3 - 200w^2 + 24} = \frac{10^3}{25 - 201w^2 + j\,(102w - 100w^3)} \text{ - - - (12).}$$

Thus $|T| = \dfrac{10^3}{((25 - 201w^2)^2 + (102w - 100w^3)^2)^{1/2}}$. Now at the closed-loop unity-gain frequency, $|T| =$ 1.

Thus (squaring), $10^6 = (25 - 201w^2)^2 + (102w - 100w^3)^2 = 625 - 10050w^2 + 40401w^4 + 10404w^2 - 20400w^4 + 10000w^6$, whence $w^6 + 2w^4 + 0.035w^2 - 100.06 = 0$. Solve $w^6 + 2w^4 - 100 = 0$ by trial. For $w = 2$, $2^6 + 2 \times 2^4 - 100 = 64 + 32 - 100 = -4$. For $w = 2.02$, (2.02)6 + 2 $\times$ (2.02)4 $-$ 100 = 67.9 + 33.3 $-$100 = 1.2. Use $w = 2$ as an approximate solution. Now for $w = 2$, from (12), $\Phi(w) = -\tan^{-1}$ $\dfrac{102(2) - 100(2^3)}{25 - 201(2^2)} = -\tan^{-1}\dfrac{204 - 800}{25 - 804} = -\tan^{-1}\dfrac{596}{779} = -37.4°$. Thus the phase margin for $\beta = 0.024$, with $Q = 0.707$, occurring at about 2 $\times$ 10^5 rad/s, is about 180 $-$ 37.4 or **143°**.

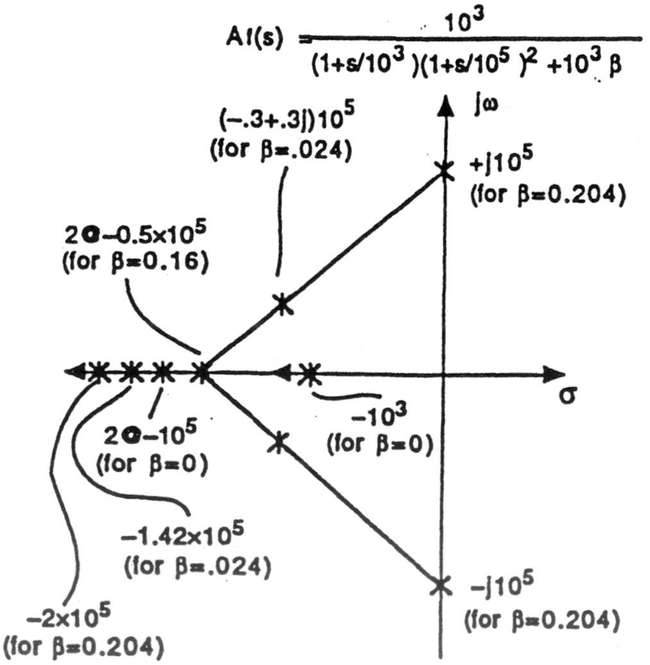

SECTION 8.10: STABILITY USING BODE PLOTS

8.41 From Eq.8.48: At the unity-loop-gain frequency, $A_f\ (j\omega) = \dfrac{1/\beta\ e^{-j\Theta}}{1 + e^{-j\Theta}}$, and $|A_f\ (j\omega)| = \left|\dfrac{1/\beta}{|1 + e^{-j\Theta}}\right| = \dfrac{1/\beta}{|1 + \cos\Theta - j\sin\Theta|} = \dfrac{1/\beta}{((1 + \cos\Theta)^2 + \sin^2\Theta)^{1/2}} = \dfrac{1/\beta}{(1 + \cos^2\Theta + 2\cos\Theta + \sin^2\Theta)^{1/2}} = \dfrac{1/\beta}{(2\ (1 + \cos\Theta))^{1/2}}$. Now as noted on page 726 of the Text, for a margin of 45°, Θ = 180–45 = 135°, and

$$|A_f| = \frac{1/\beta}{(2\ (1 + \cos 135°))^{1/2}} = 1.307/\beta.$$

There is no peak when $(2\ (1 + \cos x))^{1/2} = 1$, or $1 + \cos x = 1/2$, $\cos x = -1/2$, x = 120°, and phase margin = 180 − 120 = **60°**, (and, of course, greater). *For a peaking factor of 2*: $2 = 1/(2 + 2\cos\Theta)^{1/2}$, or $2 + 2\cos\Theta = 0.25$, $\cos\Theta = -1.75/2 = -0.875$, or Θ = 151°, for which the margin is **29°**. *For a peaking factor of 10*: $2 + 2\cos\Theta = 1/100$; $\cos\Theta = -1.99/2 = -0.995$, Θ = 174.3°, for which the margin is **5.7°**.

8.42

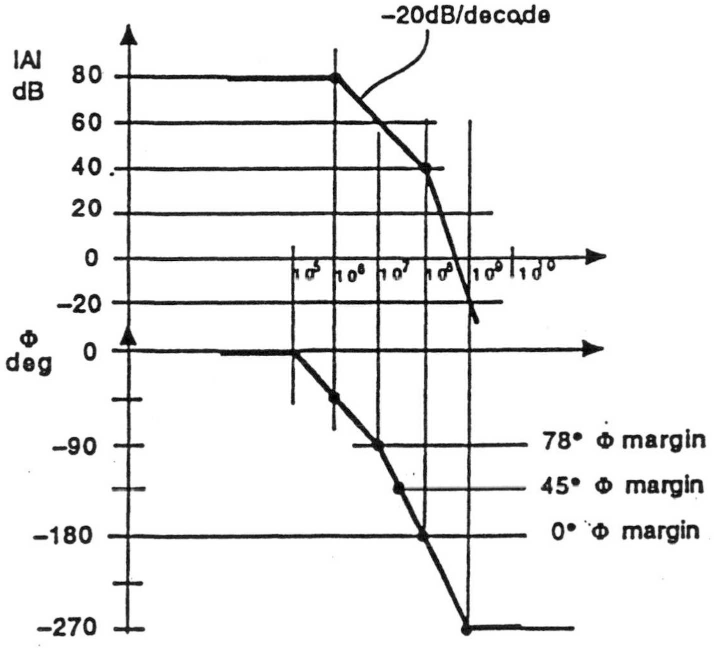

See that margins at 10^8Hz, are likely to be zero for $1/\beta \equiv$ 40dB, or $1/\beta = 10^{40/20} = 100$, or β = **0.01**. Moreover, the phase margin is about 78° at 10^7Hz, where P_1 contributes 90° and P_2 , P_3 each 6° to the total shift. The corresponding $1/\beta \equiv$ 60dB, or β = **0.001**. The phase margin is 45° at 3×10^7Hz, for which $1/\beta \equiv$ 50dB, and $\beta = 1/10^{50/20}$ = **0.0032**. For β = .001, $A_f = \dfrac{A}{1 + A\beta} = \dfrac{10^4}{1 + 10^4 \times 10^{-3}}$ = **909V/V**. For β = .0032, $A_f = 10^4/(1 + 10^4(.0032))$ = **306.5V/V**. For $\beta = 3 \times 10^{-2}$, $1/\beta$ = 100/3 = 33.3, where 20 log 33.3 = 30.5dB ≃ 30dB. See from the figure that $f = 10^8$ (10 × 10/60) = 1.67×10^8Hz, and the phase margin = − 1/6 (90) = **−15°**.

8.43 For the situation in P8.42, $A = \dfrac{10^4}{(1 + j\ f/10^6)\ (1 + j\ f/10^8)^2}$, for which Φ = − **tan⁻¹ f/10⁶ − 2 tan⁻¹ f/10⁸**. For Φ = −180°, check $f = 10^8$. See $\Phi = -\tan^{-1} 10^8/10^6 - 2\tan^{-1} 10^8/10^8$ = −89.43 −2(45) = 179.4°.

For $f = 1.1 \times 10^8$ $\Phi = -\tan^{-1} 110 - 2\tan^{-1} 1.1$ = −89.5 −2 (47.7) = 185°.

For $f = 1.01 \times 10^8$, $\Phi = -\tan^{-1} 101 - 2 \tan^{-1} 1.01 = -89.4 - 2\,(45.29) = 180°$.

Thus margins are zero at $f = \mathbf{1.01 \times 10^8}$**Hz.**

Now, at $f = 3 \times 10^7$, $\Phi = -\tan^{-1} \dfrac{3 \times 10^7}{10^6} - 2 \tan^{-1} \dfrac{3 \times 10^7}{10^8} = -88.1 - 2\,(16.7) = 121.5$ for a margin of $180 - 121.5 = 58.5°$ (rather than 45°).

For $f = 4 \times 10^7$Hz, $\Phi = -\tan^{-1} 40 - 2 \tan^{-1} .4 = -88.6 - 2(21.8) = -132.2°$, for a margin of $180° - 132.2° = 47.8°$.

For $f = 4.2 \times 10^7$Hz, $\Phi = -\tan^{-1} 42 - 2 \tan^{-1} .42 = 88.6 - 2(22.8) = 134.2°$, for a margin of $180° - 134.2° = 45.8°$.

Thus margins are 45° at $f \approx \mathbf{4.3 \times 10^7}$**Hz.**

Now at $f = 10^7$, the phase margin is $180 - \tan^{-1} \dfrac{10^7}{10^6} - 2 \tan^{-1} \dfrac{10^7}{10^8} = 180 - 84.3 - 11.4 = 84.3°$ (not 78° as suggested). For $f = 1.2 \times 10^7$, the phase margin is $180 - 85.2 - 13.7 = 81.1°$. For $f = 1.4 \times 10^7$, the phase margin is $180 - 85.9 - 15.9 = 78.2°$. Thus the margin is 78° at $f \approx \mathbf{1.4 \times 10^7}$ **Hz.**

Now, at $f = 1.4 \times 10^7$Hz, $|A| = \dfrac{10^4}{(1 + 14^2)^{1/2}\,(1 + .14^2)} = \dfrac{10^4}{14.04\,(1.02)} = 698$V/V. Thus β can be 1/698 $= \mathbf{0.00143}$ (where the Φ margin = 78°).

Now at $f = 4.3 \times 10^7$Hz, $|A| = \dfrac{10^4}{(1 + 43^2)^{1/2}\,(1 + .43^2)} = \dfrac{10^4}{(43.01)\,(1.185)} = 196.2$V/V. Thus $\beta = 1/196.2$ $= \mathbf{0.0051}$ (where the phase margin is 45°).

8.44 The available amplifier has a gain of 10^4K with poles at 10^5Hz and 10^6/K Hz.

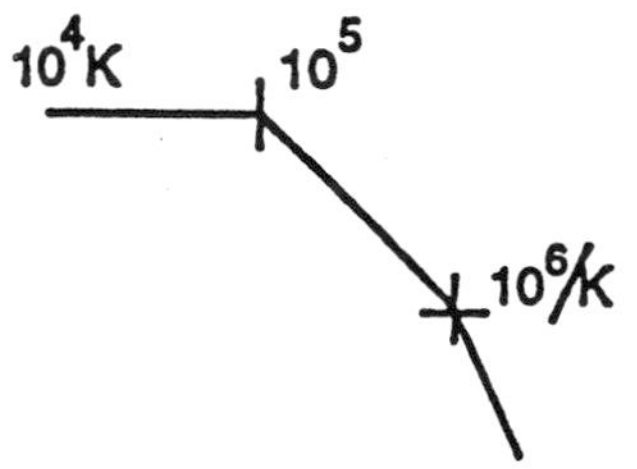

(A) For 20dB of feedback, $10^{20/20} = 1 + A\beta$. Thus $1 + 10^4 K\,(\beta) = 10$, $\beta = 9 \times 10^{-4}/K$. From the rate-of-closure rule, the $1/\beta$ and A lines should intersect at 10^6/K Hz, where $A \approx 10^4 K \times \dfrac{10^5}{10^6/K} = 10^3 K^2$, and where $A\beta = 1$, or $\beta = \dfrac{1 \times 10^{-3}}{K^2}$. Now $\beta = \dfrac{1 \times 10^{-3}}{K^2} = \dfrac{9 \times 10^{-4}}{K}$, for $K = \dfrac{1 \times 10^{-3}}{9 \times 10^{-4}} = \mathbf{1.11}$, for which the bandwidth is $10^6/K = \mathbf{0.9}$**MHz,** and the low-frequency gain is $A_f = \dfrac{A}{1 + A\beta} = \dfrac{10^4 K}{1 + 10^4 K\,(9 \times 10^{-4}/K)} = \dfrac{10^4\,(0.9)}{10} = \mathbf{900}$**V/V.**

The second pole is at $10^6/K = 10^6/1.11 = \mathbf{0.9}$**MHz.** Now from Eq. 8.36, the closed-loop poles are at $-1/2\,(\omega_{p1} + \omega_{p2}) \pm \sqrt{(\omega_{p1} + \omega_{p2})^2 - 4(1 + A_o\,\beta)\,\omega_{p1}\,\omega_{p2}}$.

Here $A_o = 0.9 \times 10^4$V/V, $\beta = 1 \times 10^{-3}$V/V, $\omega_{p1} = 10^5$Hz, $\omega_{p2} = 9 \times 10^5$Hz. Thus the closed-loop poles are at $-\dfrac{(1 + 9) \times 10^5}{2} \pm \sqrt{(\dfrac{(1 + 9)}{2} \times 10^5)^2 - (1 + 0.9 \times 10^4 \times 1 \times 10^{-3})\,(10^5)\,(9 \times 10^5)}$, or $-5 \times 10^5 \pm [(5 \times 10^5)^2 - 10 \times 9 \times 10^{10}]^{1/2}$, or $-5 \times 10^5 \pm 10^5\,(25 - 90)^{1/2}$, or $-10^5\,(-5 \pm 8.06j)$.

(B) Now, for $A_f = \dfrac{A}{1 + \beta A} = \dfrac{10^4 K}{1 + \beta\,10^4 K} = 10$, $10^3 K = 1 + \beta\,10^4 K$, $\beta = \dfrac{10^3 K - 1}{10^4 K} = 0.1 - 10^{-4}/K$. Now, using the rate-of-closure rule, the intersection of the $1/\beta$ and A lines will be at $10^6/K$ Hz, where $A = 10^4 K \times \dfrac{10^5}{10^6/K} = 10^3 K^2$. Now, at $10^6/K$ Hz, $A\beta = 1$. Thus $10^3 K^2\,(0.1 - 10^{-4}/K) = 1$, $100 K^2 - 0.1 K - 1 = 0$, for which $K \approx 0.10$. That is, $K = \mathbf{0.10}$, dc gain = **10V/V**, $\beta = 0.1 - 10^{-4}/0.1 = .099$, $A = 10^4 K = 10^3$, and bandwidth $= 10^6/0.1 =$ **10MHz.** The second pole is at $10^6/K = 10^6/0.1 =$ **10MHz.**

SECTION 8.11: FREQUENCY COMPENSATION

8.45 $A = 10^4$, $f_{p1} = 10^6$Hz, $f_{p2} = f_{p3} = 10^8$Hz. For an added dominant pole at f_{p0}, the second pole would be at 10^6Hz. Now for $A_f = 10$ (and A very high), $\beta \simeq 1/A_f = 0.1$. Thus $1/\beta \equiv 20$dB, while $A \equiv 80$dB. Thus the dominant pole must drop the response by 80 – 20 = 60dB, using three decades.
$f_{p0} = f_{p1}/10^3 = 10^6/10^3 =$ **10^3Hz**. Similarly for $A_f = 1$, $f_{p0} =$ **10^2Hz**. In both cases the closed-loop 3dB frequency would be **10^6 Hz**. s.p

8.46 For the existing pole lowered, from $f_{p1} = 10^6$Hz to f_{p1}', consider the effective second pole at f where $\tan^{-1} f/10^8 = 22.5°$, or $f = 0.414 \times 10^8$Hz. Now for a gain of 10, must lower f_{p1} by a factor of 1000 below 0.414×10^8Hz to $f_{p1}' =$ **4.14×10^4Hz**. Now for a gain of one, similarly need $f_{p1}' =$ **4.14×10^3Hz**. In each case, $f_{3dB} \simeq$ **40MHz.**

8.47 Now $f_{p1} = \dfrac{1}{2\pi\, C_1 R_1} = \dfrac{1}{2\pi \times C \times 10^6}$, and $f_{p2} = \dfrac{1}{2\pi\, C_2 R_2} = \dfrac{1}{2\pi \times C \times 10^4}$, a frequency 100 times larger. That is, $\dfrac{1}{2\pi\, C \times 10^6} = 10^5$, or $C = \dfrac{1}{2\pi \times 10^5 \times 10^6} =$ **1.59pF.**

Now, for Miller compensation, $f_{p1}' \simeq \dfrac{1}{2\pi\,(C_1 R_1 + C_2 R_2 + C_f\,(g_m R_1 R_2 + R_1 + R_2))}$, or

$$10^4 \simeq \frac{1}{10 \times 10^{-12} \times 10^6 + 10 \times 10^{-12} \times 10^4 + 6.28\, C_f\,(100 \times 10^6 + 10^6 + 10^4)},$$ or

$10^4 \simeq \dfrac{1}{10^{-5} + 10^{-7} + 6.8\, C_f\,(1.01 \times 10^8)}$, or $10^{-1} + 10^{-3} + 6.87\, C_f\,(10^{12}) = 1$. Thus $C_f = \dfrac{1}{6.87 \times 10^{12}}$ $(1 - 0.1 - 0.001)$, or $C_f =$ **0.146pF.** Check (From Eq.8.58): $C_f \simeq \dfrac{1}{2\pi\, g_m R_2 R_1 f_{p1}} = \dfrac{1}{2\pi \times 10^2 \times 10^6 \times 10^4} =$ **0.159pF.**

Now $f_{p2}' = \dfrac{g_m C_f}{2\pi\,(C_1 C_2 + C_f\,(C_1 + C_2))} = \dfrac{100/10^4 \times .146 \times 10^{-12}}{2\pi\,((1.59 \times 10^{-12})^2 + 0.146 \times 10^{-12}\,(2)\,(1.59 \times 10^{-12}))}$

$= \dfrac{0.146}{1.59 \times 10^{-9} + 0.292 \times 10^{-9}} =$ **77.6MHz**, and f_{p3} remains at **10MHz**, with f_{p1} reduced to 10^4Hz, or 10kHz. Thus the closed-loop cutoff frequency raises to **10MHz.** Note that the pole split lowered the dominant pole by a factor of 10 and raised the upper pole by a factor of 8 or so. Had the upper pole remained double at 10MHz, and assuming a double pole behaves as a single pole at a frequency for which each contributes 45°/2, that is at $10^7 \tan^{-1}(22.5°) = .414 \times 10^7$Hz, the cutoff would have been at 4.1MHz. Thus pole splitting allows the same phase shift at 10MHz as formerly at 4.14MHz. Thus it seems that the dominant pole could be raised by the factor 10/4.14 = 2.4, to 2.4×10^4Hz for roughly the same margins. For this situation,

$f_{p1}' \simeq \dfrac{1}{2\pi\, g_m R_2 C_f R_1}$, and $C_f = \dfrac{1}{2_\pi \times 100 \times 2.4 \times 10^4 \times 10^6} =$ **0.066pF**, for which

$f_{p2}' = \dfrac{g_m C_f}{C_1 C_2 + C_f\,(C_1 + C_2)} = \dfrac{100/10^4 \times 0.066 \times 10^{-12}}{2\pi\,(1.59 \times 10^{-12})\,(1.59 \times 10^{-12}) + 1.59 \times 2 \times 10^{-12}\,(0.066 \times 10^{-12})} =$

$\dfrac{0.066}{1.59 \times 10^{-9} + 0.066 \times 10^{-9}} =$ **39.8MHz.** Thus the poles are at 24kHz, 10MHz and 39.8MHz. Again, the closed-loop cutoff frequency will be at about **10MHz**, whereas the original frequency for which the phase margin is 45° would have been about **4.1MHz.**

Chapter 9

OUTPUT STAGES AND POWER AMPLIFIERS

SECTION 9.1: CLASSIFICATION OF OUTPUT STAGES

9.1 Peak voltages applied are: 1.414V, 14.14V, and 141.4V. Peak load currents for a 1kΩ load are: 1.4mA, 14mA, 141mA. With a 50mA bias current, corresponding operating modes are **A, A, AB**, respectively. For a load of 0.25kΩ, the peak load currents are 5.7mA, 57mA, and 566mA, with operation in modes **A, AB, AB** respectively. For an nearly-normal large output at zero bias current, **class B** operation is apparently possible.

SECTION 9.2: CLASS A OUTPUT STAGE

9.2 $I_{max} = \dfrac{0 - -3 - 0.7}{1.5k\Omega}$ = 1.53mA. For a 1kΩ load, this will support a negative output peak of −1.53V, and for 10kΩ, a peak of −15.3V. In the latter case, saturation will occur earlier at −3 + 0.3 = −2.7V. For positive inputs, the positive peak is 3.0 − 0.3 = 2.7V, independent of load. Thus for a 1kΩ load, the largest sine wave is **1.53V peak**, and for a 10kΩ load, it is **2.7V peak.** For a negative output at −2.7V with I = 1.53mA, R ≥ 2.7V/1.53mA = **1.76kΩ**. For a second device connected in parallel with Q_2, I doubles to 3.06mA, and load resistances down to 1.76/2 = **0.88kΩ** can be accommodated with a −2.7V peak signal.

9.3 The output signal is voltage-limited by the saturation of Q_1 to $v_o = V_{CC} - 0.2 - 0.7 = V_{CC} - 0.9 = 5 - 0.9 = 4.1$V peak, or current-limited to $v_O = 10\ IR_L$. Thus for large enough I, the largest possible zero-average unclipped output is **4.1V peak.** For $I_{E2} \geq I_{E1}$, there is $10I - I = 9I$ available to the load. For a 4.1V peak, $9\,I \geq 4.1/100\Omega$, $I \geq 4.56$mA. Thus the minimum I required is **4.56mA.**

9.4 (a) The largest-possible sine-wave output is 9 − 0.3 = 8.7V peak. The smallest-possible load resistance is 8.7V/10mA = 870Ω.
Load power = $\dfrac{(8.7/\sqrt{2})^2}{0.870}$ = **43.5mW**. Supply power = 2 (9) × 10 = **180mW**. Conversion efficiency = 43.5/180 × 100 = **24.2%.**

(b) For a signal of 8.7/2 = 4.35V across a load of 870/2 = 435Ω, the load power = $\dfrac{(4.35/\sqrt{2})^2}{0.435}$ = **21.75mW**. Supply power = 2 (9) (10) = **180mW**. Conversion Efficiency = 21.75/180 × 100 = **12.1%.**

(c) The loss in Q_3 and R is 9 V × 10mA = 90mW, the supply power = 2 (9) 10 + 9 (10) = **270mW**. For (a), efficiency = 43.5/270 × 100 = **16.1%.**

(d) For (b), efficiency = 21.75/270 × 100 = **8.06%.**

9.5 For matched FETs, no load, and $v_O = 0$, $I_{D2} = I_{DSS} = 10$mA $= I_{D1}$ and $v_{GS1} = 0$V. Thus v_I = **0V**. For operation in saturation, $v_{DG} \geq |V_p| = 2$V. Thus the negative limit of v_O is −9 + 2 = **−7V**, for which I_L = 7V/1kΩ = 7mA, and I_{D1} = 10 − 7 = 3mA. Generally, $i_D = I_{DSS}\,(1 - \dfrac{v_{GS}}{V_p})^2$.

$3 = 10\,(1 - \dfrac{v_{GS}}{-2})^2$, $1 + \dfrac{v_{GS}}{2} = (\dfrac{3}{10})^{1/2} = .548$, $v_{GS} = 2\,(.548 - 1) = -0.905$V. Thus the corresponding input is −7 −.90 = **−7.9V.** For positive outputs, the input limit for saturation is v_I = 9 − 2 = **7V**, for which $v_O = v$.

$10 + \dfrac{v}{1k\Omega} = 10\,(1 - \dfrac{7 - v}{-2})^2 = 10\,(1 + 3.5 - v/2)^2$, or $10 + v = 10\,(4.5 - v/2)^2 = 202.5 - 45v + 2.5v^2$. Thus, $2.5v^2 - 46v + 192.5 = 0$, $v = \dfrac{46 \pm \sqrt{46^2 - 4(2.5)(192.5)}}{2\,(2.5)} = \dfrac{46 \pm 13.8}{5}$ = **6.43V** $= v_O$.

Check: $i_{D1} = 10\,(1 - \dfrac{7 - 6.43}{-2})^2 = 16.5$mA. Compare with 6.43/1kΩ + 10 = 16.4. OK.

Now for a fixed (dc) output signal of υ_O = 6.43V: Load power = $\dfrac{(6.43)^2}{1k\Omega}$ = **41.3mW**. Supply power = 9 (10 + 6.43) + 9 (10) = 147.9 + 90 = **238mW**. Efficiency = 41.3/238 × 100 = **17.4%**.

Now for a dc output of υ_O = –7.0V. Load power = 7^2/1kΩ = **49mW**. Supply power = 9 (10 – 7) + 9 (10) = 27 + 90 = **117mW**. Efficiency = 49/117 × 100 = **41.9%**

Largest-possible relatively-undistorted sine wave output is **6.43V peak**, for which Load power = $\dfrac{(6.43/\sqrt{2}^2)}{1k\Omega}$ = **20.7mW**. Supply power = 9(10 + 6.43/π + 10) = 198.5mW, where 6.43/ π is the average value of the 6.43V half-sine current pulse. Efficiency = 20.7/198.5 × 100 = **10.4%**.

SECTION 9.3: CLASS B OUTPUT STAGE

9.6 *For* $R_L = \infty$: $i_D = 0$ and $\upsilon_{GS} = V_t$ as υ_I varies. For $|\upsilon_I| \le$ 1V, υ_O = 0. For $1 \le |\upsilon_I| \le$ 11V, $|\upsilon_O| = |\upsilon_I| - 1$.

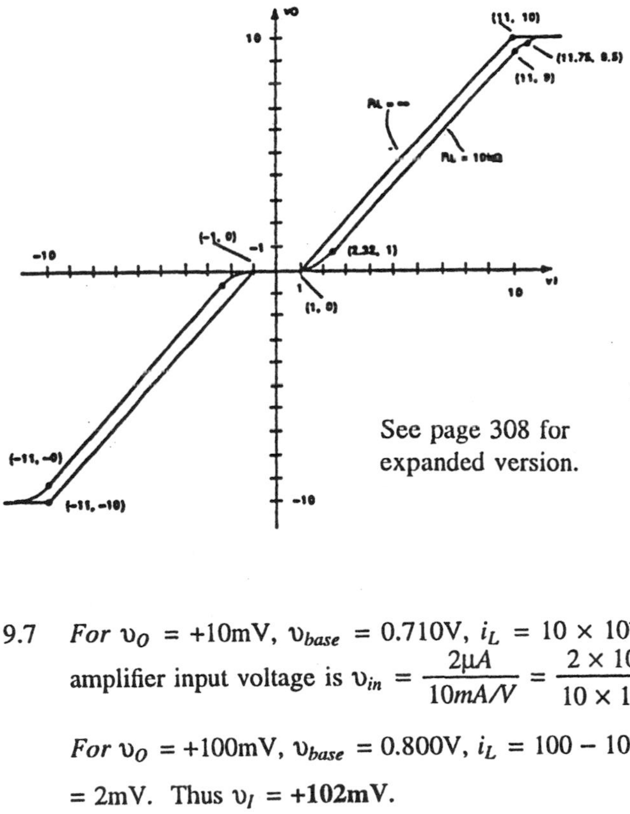

See page 308 for expanded version.

For $|\upsilon_I| \ge$ 11V, $|\upsilon_O|$ = 10V. *For* R_L = 10kΩ: For υ_O = 1V, i_D = 1/10⁴ = 100μA = 1 $(\upsilon_{GS} - 1)^2$, υ_{GS} = ± (1/10)^½ + 1 = 1.32V, υ_I = 2.32V. For υ_O = 9V, $i_D = 9/10^4 = 1 \times 10^{-3}\,(\upsilon_{GS} - 1)^2$, $\upsilon_{GS} = \pm (.9)^{1/2} + 1$ = 1.95V, υ_I = 10.95V ≃ 11V. For υ_O = 10V, i_D = 10/10⁴ = 1mA = 1 [2 $(\upsilon_{GS} - 1)\,\upsilon_{DS} - \upsilon_{DS}^2$]. Say υ_{DS} = 0.1. Thus 1 ≃ 2 $(\upsilon_{GS}$ – 1) (0.1), υ_{GS} = 1/(2(0.1)) + 1 = 6V, υ_I = 16V. For υ_O = 9.5V, i_D ≃ 1mA = 1mA (2 $(\upsilon_{GS}$ – 1) 0.5 – 0.5²), υ_{GS} – 1 = 1 + .25, υ_{GS} = 2.25V, υ_I = 2.25 + 9.5 = 11.75. For Q_1, Q_2 in saturation, the largest possible sinewave output is **9V peak** or **18Vpp**. The corresponding input voltage is **20Vpp** for no load, and **22Vpp** for 10kΩ load. Equivalent gain is 18/20 = **0.9V/V**, or 18/22 = **0.82V/V** respectively. Supply power is **0mW** for no load, and 10 (9/√2)/10k = **6.36mW**, for 10kΩ load. Load power is **0mW**, or $\dfrac{(9/\sqrt{2})^2}{10k\Omega}$ = **4.05mV** respectively. Efficiency is ∞ (for no load), or 4.05/6.36 × 100 = **63.7%** for a 10kΩ load.

9.7 *For* υ_O = +10mV, υ_{base} = 0.710V, $i_L = 10 \times 10^{-3}/100$ = 0.1mA, i_{base} = 0.1/50 = 2μA, for which the amplifier input voltage is $\upsilon_{in} = \dfrac{2\mu A}{10mA/V} = \dfrac{2 \times 10^{-6}}{10 \times 10^{-3}}$ = 0.2mV. Thus υ_I = 10 + 0.2 = **+10.2mV**.

For υ_O = +100mV, υ_{base} = 0.800V, $i_L = 100 - 10^{-3}/100$ = 1mA, $i_{base} = \dfrac{10^{-3}}{50}$ = 20μA, $\upsilon_{in} = \dfrac{20 \times 10^{-6}}{10 \times 10^{-3}}$ = 2mV. Thus υ_I = **+102mV**.

For υ_O = +1V, υ_{base} = 1.7V, i_L = 1/100 = 10mA, i_{base} = 10/50 = 200μA, $\upsilon_{in} = \dfrac{200 \times 10^{-6}}{10 \times 10^{-3}}$ = 20mV. Thus υ_I = **+1.02V**.

9.8 Assuming $\upsilon_{CE\,sat}$ = 0V, the largest possible undistorted output is **6V peak** or 6/√2 = **4.24V rms**. Corresponding output power = $(6/\sqrt{2})^2/16$ = **1.125W**. Current from the supply is a half sinewave of 6/16 = 0.375A peak, whose *average value* is 0.375/π = 0.119A. That is supply power = 12 (0.119) = **1.43W**. Efficiency = 1.125/1.43 × 100 = **78.7%**. Power loss in both transistors is 1.43 –1.125 = **0.305W**. Power loss in each transistor is the same = 0.305/2 = **0.153W**.

For 4V peak output: Output power = $(4/\sqrt{2})^2/16$ = **0.5W**. Supply power = $(12 \times 4/16 \times 1/\pi)$ = **0.95W**. Total device dissipation = 0.95 −0.5 = **0.45W**. Efficiency = 0.5/.95 × 100 = **52.6%**.

For a +14.5V Supply, and 6V peak output: Output power = **1.125W**. Supply power = 14.5 (.119) = **1.73W**. Device dissipation = 1.73 −1.125 = **0.60W**. Efficiency = 1.125/1.73 × 100 = **65%**.

SECTION 9.4: CLASS AB OUTPUT STAGE

9.9 For $r_{out} \le 5\Omega$, $r_{eN} || r_{eP} = 5\Omega = r_e/2$. Thus $r_e = 10\Omega$, and I_E = 25mV/10Ω = **2.5mA**, the quiescent current, for which $V_{BB}/2$ = 690 + 25 ln (2.5/10) = 655mV, and V_{BB} = 2 (655) = **1.31V**. For 5V peak output and 50Ω load, I_L = 5/50 = 100mA, υ_{BE} = 690 + 25 ln 100/10 = 748mV. Thus υ_I = 5.00 + .748 − .655 = **5.09V**. Large-signal gain = 5.00/5.09 = **0.982V/V**. For small changes around 0V and a 50Ω load, gain = 50/(5 + 50) = **0.91V/V**. For small changes around +5V, and a 50Ω load, r_e = 25mV/100mA = 0.25Ω, gain = 50/(50 + .25) = **0.995V/V**.

9.10 For each device, biased at current I, $I = 200\,(\upsilon_{GS} - 1)^2$, and $g_m = 2\,(200)\,(\upsilon_{GS} - 1)$ mA/V with $r_s = 1/g_m$. For a 100Ω load, gain = $\frac{100}{100 + r_s/2}$ = 0.99, 100 = 99 + $.495 r_s$, r_s = 1/.495Ω, or $1/r_s$ = .495 A/V = 495mA/V. Thus 2 (200) (υ_{GS} −1) = 495, υ_{GS} − 1 = 495/400 = 1.237, υ_{GS} = 2.237V, V_{BB} = 2 (2.237) = **4.48V**.

SECTION 9.5: BIASING THE CLASS AB CIRCUIT

9.11 For each junction, V_j = 0.675V, to maintain an output quiescent current of 2.5mA for which I_B = 2.5/31 = 81μA. Correspondingly, the quiescent current of the biassing junctions is 2.5/4 = 625μA. Thus I = 625 + 81 = 706μA. Now for a short-circuit output, the maximum available current = 706μA (30 + 1) = **21.9mA**. For junction bias reduced to 0.1 (625μA) , available base current = 706 − 0.1 (625) = 643.5μA, for which i_E = 31 (643.5) = 19.9mA, and υ_O = 50 × 19.9 = **1.00V** across 50Ω. Thus the peak output for 50Ω is **1.00V**.

9.12 I_Q = 2.5mA for which υ_{BE} = 675mV and i_B = 81μA. For a 1V positive output across 50Ω, i_L = 1V/50 = 20mA, for which i_B = 20/31 = 645μA. Now for a normal bias network current I, $(I - 645)/I \times 100$ = 20, $I - 645 = 0.2I$, $0.8I = 645$, I = 806μA. Thus the resistor-network current level = 1/2 (20%) 806 = 80.6μA. Here V_{BB} = 2 (0.675) = 1.35V, and $R_1 + R_2$ = 1.35/80.6μA = 16.75kΩ. Now the normal current in the bias transistor is 806 − 80.6 − 81 = 644μA, for which V_{BE1} = 690 + 25 ln (0.644/10) = 621mV, and I_{B1} = 644/30 = 21.5μA. That is R_1 = 621/80.6 − 21.5 = 10.5kΩ. Use **10kΩ**, for which I_{R2} = 621/10 + 21.5 = 83.6μA, R_2 = 1.350 − 0.621/83.6 = **8.7kΩ**. In practice, use 8.2kΩ with a variable (or fixed) series resistor. Now for υ_O = 1V, i_{EN} = 20mA, i_{BN} = 645μA, and i_{E1} = 806 − 645 − 80.6 = 80.4μA. That is, r_{e1} = 25mV/80.4μA = 311Ω, $r_{\pi 1}$ = 31 (311) = 9.64kΩ, and for the multiplier with input υ, current i is $\frac{\upsilon}{8.7k + 10k \parallel 9.64k} \times \left[1 + 31\left[\frac{10}{10 + 9.64}\right]\right] = \frac{\upsilon\,(16.78)}{13.6}$, or $r_{eq} = \upsilon/i$ = 16.78/13.6 = 1.23kΩ. Now, at the peak output of 1V, r_{eN} = 25mV/20mA = 1.25Ω, and $r_{\pi N}$ = 31 (1.25Ω) = 38.75Ω. Gain at the peak = $\frac{50}{50 + 1.25} \times \frac{31\,(1.25 + 50)}{31\,(1.25 + 50) + 1.23k\Omega}$ = 0.976 × 0.564 = **0.55V/V**.

For signals around 0V, with a 50Ω load, R_O = 5Ω as before, but i_{E1} rises to 644μA for which r_{e1} = 25/.644 = 38.8Ω, and r_{eq} reduces to about 38.8/311 or 0.12 of the previous, or 0.12 (1.23kΩ) or about 150Ω. Thus the gain ≈ $\frac{50}{50 + 5} \times \frac{(50 + 5)31}{(50 + 5)31 + 150}$ = 0.91 × 0.92 = **0.84V/V**.

9.13 From the sequel to Eq. 9.33 of the Text, neglecting β for the bias situation, $k = 1 + R_2/R_1 = (R_1 + R_2)/R_1$. Now, for a junction voltage υ_{BE}, the emitter current of the multiplying transistor = $I - \frac{k\,V_{BE}}{R_1 + R_2}$, where $I_E = I - V_{BE}/R_1$. Thus $r_e = \frac{V_T}{I_E} = \frac{V_T}{I - V_{BE}/R_1}$. Now for a small incremental voltage

υ applied across the multiplier, a current i_B flows in the R_1, R_2 network, where $i_B = \frac{\upsilon}{R_2 + R_1 \parallel r_\pi} = \frac{(R_1 + r_\pi)\upsilon}{R_1 r_\pi + R_2 (R_1 + r_\pi)}$, and i_C in the transistor collector is $i_C = \frac{R_1 \parallel r_\pi}{R_2 + R_1 \parallel r_\pi} \times \upsilon \times \frac{\beta}{\beta + 1} \times \frac{1}{r_e} = \frac{R_1 r_\pi}{R_1 r_\pi + R_2 (R_1 + r_\pi)} \times \frac{\beta\upsilon}{r_\pi} = \frac{\beta R_1 \upsilon}{R_1 r_\pi + R_2 (R_1 + r_\pi)}$.

Thus $i = i_B + i_C = \left\{ \frac{\beta R_1 + R_1 + r_\pi}{R_1 r_\pi + R_2 (R_1 + r_\pi)} \right\} \upsilon$, $r_{eq} = \frac{\upsilon}{i} = \frac{R_1 r_\pi + R_2 (R_1 + r_\pi)}{(\beta + 1)R_1 + r_\pi} = \frac{(\beta + 1)r_e R_1 + R_1 R_2 + (\beta + 1)r_e R_2}{(\beta + 1)R_1 + (\beta + 1)r_e} = \frac{\mathbf{r_e (R_1 + R_2) + R_1 R_2/(\beta + 1)}}{\mathbf{R_1 + r_e}}$, with $r_e = V_T/(I - V_{BE}/R_1)$, and $k = (R_1 + R_2)/R_1$. Now, $k = 1 + R_2/R_1$, or $R_2/R_1 = k - 1$, or $R_2 = (k - 1) R_1$, and $R_1 + R_2 = k R_1$.

Thus, $\mathbf{r_{eq}} = \frac{\mathbf{r_e (k R_1) + R_1^2 (k - 1)/(\beta + 1)}}{\mathbf{R_1 + r_e}}$, with $r_e = V_T/(I - V_{BE}/R_1)$.

Now for $k = 2$, $I = 1\text{mA}$, $\beta \geq 50$, $R_1 = r_\pi = (\beta + 1)r_e$, $r_e = \frac{25}{1 - 700/R_1}\ \Omega$, with R_1 in ohms, $r_{eq} = \frac{2 r_e R_1 + R_1^2/51}{R_1 + r_e} = \frac{2 r_e (51) r_e + (51 r_e)^2/51}{51 r_e + r_e} = r_e \left\{ \frac{102 + 51}{52} \right\} = 2.94\ r_e$, where $r_e = \frac{25}{1 - 700/(51 r_e)}$, $25 = r_e - 700/51 \rightarrow r_e = 25 + 700/51 = 38.7\Omega$,

$r_{eq} = 38.7\ (2.94) = \mathbf{114\Omega}$.

SECTION 9.6: POWER BJTs

9.14 At 30°C, the junction drop at current I is 630mV. At 10 times that current the drop would be 630 + 25 ln (10I/I) = 687.6mV. Now at T°C, the junction drop is 500mV. The new temperature, T = 30°C + $\frac{(687.6 - 500)mV}{2mV/C^\circ}$ = **123.8°C**. For a total dissipation of 45W, the thermal resistance, junction-to-ambient is (123.8 − 30)/45 = **2.08°C/W**. For a junction temperature of 180C°, total dissipation could be (180 − 30)/2.08 = **72.1W**, and the new current would be 72.1/45 × 10I = **16I**, that is 16 times the original test current. At 30°C, at this current, V_{BE} = 630 + 25 ln 16 = 699.3mV, or 700mV. At 180°C, V_{BE} = 699.3 − 2 (180 − 30) = **399.3mV**, or **400mV**.

9.15 $P_{max} \leq$ (150 − 55)/1.1°C/W = **86.4W**. For 86.4/2 = 43.2W, the junction-to-case rise is 1.1 (43.2) = 47.5°C, and for T_j = 150, T_C = 150 − 47.5 = **102.5°C**. For T_A = 30°, the thermal resistance of the heat sink required is (55 − 30)/86.4 = 0.289°C/W in the first case, and (102.5 − 30)/43.2 = 1.68°C/W in the second. For a heat-sink length L, the rating is 3/L °C/W. In the first case 3/L = 0.289 and L = 3/.289 = **10.4cm**. In the second case 3/L = 1.68, and L = 3/1.68 = **1.79cm**. Now for a potential error of 20% in **all** thermal measurements, but with 86.4W applied, T_C should be 150 − 86.4W (1.1°C/W) (1.2) = 35.95°, for which the thermal resistance of the sink must be (35.95 − 30)/86.4 = 0.0689°C/W, for which L = 3 (1+0.2)/0.0689 = **52.3cm**. Note the dramatic impact of measurement error on the adequacy of a design!

9.16 For a device dissipating W watts: 40° + 1/10 × W + 0.5W + 2W = 150°, W (0.1 + 0.5 + 2) = 110°, W = 110/2.6 = **42.3Watts**. For a heat sink twice as long, $W = \frac{110}{2.5 + 1/20}$ = **43.1W**. We now conclude that the heat sink is already quite large, the major problem lying in the transistor itself, with its dominating thermal resistance. For an infinite heat sink, the maximum rating would only be 44W!

9.17 For I_E = 5A, I_B = 0.2A, β = (5 − 0.2)/0.2 = 24, and r_e = 25mV/5A = 5mΩ. Thus r_π = (24 + 1) 5 × 10^{-3} = 125 × 10^{-3} = 0.125 Ω. For R_{ib} = .72Ω, r_x = .72 − .125 = 0.595Ω. At I_E = 3A, r_e = 25/3 = 8.3 mΩ, r_π = 25 × 8.3 = 208.3 mΩ = 0.208Ω. Thus, R_{ib} = .595 + .208 = 0.80Ω.

SECTION 9.7: VARIATIONS ON THE CLASS AB CONFIGURATION

9.18 For $I_{L\,max}$ = 100 mA and a standing current i, the maximum base current occurring in the pnp transistor, is $(100 + i)/81$. Thus $I_{E2} \geq (100 + i)/81$, or $\geq$ 1.5mA. Thus, use $I = 2\,(100 + i)/81 = 2.47 + .025i$ — — — (1), or $I = (100 + i)/81 + 1.5 = 2.73 + .0123i$ — — — (2). Now for the quiescent state, $i = I_{E3} = I_{E4} = I_{E1} = I_{E2} = I$. Now, for (1) above $I = 2.47 + .025I$, or $I = 2.47/(1 - .025) = 2.53$mA, for which $I_{E2} = (100 + 2.53)/81 < 1.5$, and thus use (2), for which $I = 2.73 + .0123\,(I)$, for which $I = 2.73/(1 - .0123) =$ **2.76mA**. Now, since the output transistors are 5 times larger than the bias transistors, $V_{R3} = V_{R4} = 25 \ln 5 = 40.2$mV at a current $I = 2.76$mA. That is, $R_3 = R_4 = 40.2\text{mV}/2.76\text{mA} = 14.6\Omega$; Use 15Ω. Now for outputs near zero volts, $I_{E3} = I_{E4} = 2.8$mA, and $r_e = 25\text{mV}/2.8\text{mA} \approx 9\Omega$. Thus, $R_{out} = (9 + 15)/2 = 12\Omega$. For a gain of 0.90 (dominated by the output coupling) $\dfrac{R_L}{R_L + 12} = 0.90$, where $R_L = 12/(1 - .90) = 120\Omega$. Near ±10V, where the situations are essentially the same for a particular R_L, one transistor is likely to be cut off and $I_L = 10/R_L$. Thus $r_e = \dfrac{25mV}{10/R_L} = 2.5R_L\ \Omega$, with R_L in kΩ, and $R_{out} = (2.5R_L + 15)\ \Omega$. Thus, $\dfrac{R_L}{R_L + (2.5R_L + 15) \times 10^{-3}} = 0.90$, or $R_L = 0.90R_L + 0.00225\,R_L + 0.0135$. Thus $R_L = 0.0135/(1 - 0.90225) = 138\Omega$. Thus for loads in excess of 140Ω, the gain of the output stage can exceed 0.90, for outputs of ±10V.

9.19 For a standing current of 10.0mA in the output:

(a) $I_{E2} \approx I_{E4} = 10$mA, and $I_{E1} = 10/100 = 0.1$mA, and $I_{E3} = 10/100 = 0.1$mA. That is $V_{BB} = 700 + 25 \ln(10/100) + 700 + 25 \ln(0.1/1) + 700 + 25 \ln(0.1/1) = 2100 - 57.6 - 57.6 - 57.6 = (2100 - 172.8)\text{mV} \equiv$ **1.93V**. Now, for all β increased by 10 and $V_{BB} = 1.93$V, all currents will increase to maintain the voltage. The current, say 10k mA, will be such that the currents in Q_2, Q_1, Q_3 will change to 10k, 10k/(1000) = .01k, and 10k/(100 × 10) = 0.01k respectively. Now − 172.8/25 = (ln 10k/100 + ln 0.01k/1 + ln 0.01k) = −6.91. Try k = 10: ln 1 + ln .1 + ln .1 = −2.3 − 2.3 = −4.60. Try k = 3: ln 0.3 + ln .03 + ln .03 = −1.20 − 3.51 − 3.51 = −8.22. Try k = 5: ln 0.5 + ln 0.05 + ln 0.05 = −0.693 − 2.995 − 2.995 = −6.68. Thus, the standing current increases by more than 5 times!.

(b) $I_{E4} \approx I_{E2} \approx 10 - 1 = 9$mA, $I_{E1} \approx I_{E3} = 1$mA. Thus, $V_{BB} = 700 + 25 \ln(9/100) + 2\,(700) =$ **2.04V**. Now for all β increased by 10, the output current increases slightly, but the base-shunt current, established by resistors and a V_{BE} which changes only slightly, stay essentially the same. Thus for a factor-of-10 change in I_{B2}, from 0.1mA to .01mA, I_{E1} changes from 1.00 to .91 mA for the same output current. Thus it is likely that the standing current changes by a few tens of %. A great improvement!

9.20 For $I_O = 25\text{mA} = I_{E1}$, $I_{B1} = 25/100 = .25$mA. Thus $I_{C5} = 1 - .25 = .75$mA, $\upsilon_{BE\,5} \approx 700 + 25 \ln(0.75/1) = 693$mV. Thus $R_{E1} \approx 693\text{mV}/25\text{mA} = 27.7\Omega$. Use 27Ω. Without Q_5, the peak load current could be 1mA × 100 = **100mA**.

9.21 For **both** devices having β = 50, $I_{E2} = 10$mA, $I_{B2} = I_{E1} = 10/50 = .2$mA, $r_{e2} = \dfrac{25}{(51/50)(10)} = 2.45\Omega$, $r_{\pi2} = 51\,(2.45\Omega) = 125\ \Omega$, $r_{e1} = 25/.2 = 125\Omega$, $r_{\pi1} = 51\,(125) = 6.375$kΩ. For υ_2 at the base of Q_2, $i_{b2} = \upsilon_2/127.5 = i_{e1}$ and $i_{c2} = \beta\,\upsilon_2/r_{\pi2} = 50\,\upsilon_2/127.5$. Now voltage υ_1 at the base of $Q_1 = \upsilon_2 + i_{e1}\,r_{e1}$, or $\upsilon_1 = \upsilon_2 + \upsilon_2/125 \times 125 = 2\,\upsilon_2$, and $i_{c1} = 50/51\ i_{e1} = 50/51 \times \upsilon_2/125$. Thus, $g_{m\,eq} = \dfrac{i_{c1} + i_{c2}}{\upsilon_1} =$

$$\frac{\frac{50}{51}\frac{\upsilon_2}{125} + \frac{50\,\upsilon_2}{125}}{2\,\upsilon_2} = \frac{50}{2\,(125)}(1/51 + 1) = \mathbf{203.9mA/V}.$$

For Q_2, $r_{o2} = \dfrac{100}{(.98)\,(10)} = 10.2\text{k}\Omega$, $r_{\mu2} = 10\,(50)\,10.2 = 5.1\text{M}\Omega$, $r_{o1} \approx 100/.2 = 500\text{k}\Omega$, $r_{\mu1} \approx 250\text{M}\Omega$. Now, for a rise in output of υ, with the input *short-circuited*, the total current is approximately $i = (\upsilon/1\text{M}\Omega) + (\upsilon/250\text{M}\Omega) + (\upsilon/(0.5\text{M}\Omega \parallel 5.1\text{M}\Omega)\ 50$

+ (υ/10.2kΩ). Thus $1/R_{out}$ = 1 + .004 + 109.8 + 98.0 = 208.8μA/V, and R_{out} = **4.79kΩ.** Thus the gain $\upsilon_o/\upsilon_i = -g_{m\ eq}\ R_{out}$ = − 202 × 10^{-3} × 4.79 × 10^3 = **−977V/V**, and R_{in} ≈ (1MΩ/(1 + 977)) ∥ (51 (2 (125))) = 1.022kΩ ∥ 12.75kΩ = **0.946kΩ.**

For **both** devices having β = 150, $I_{B2} = I_{E1}$ = 10/150 = 0.066mA. Thus $r_{\pi 2} = r_{e1}$ = 25/.066 = 375Ω, $r_{\pi 1}$ = 151 (375) = 56.6kΩ. Thus $g_{m\ eq} = \dfrac{\frac{150}{151}\frac{\upsilon_2}{375} + \frac{150\ \upsilon_2}{375}}{\upsilon_2 + \upsilon_2} = \dfrac{150}{2(375)}\left(1 + \dfrac{1}{151}\right)$ = **206.6mA/V**, that is, almost the same as with β = 50.

For Q_2, $r_{o2} = \dfrac{100}{\frac{150}{151}(10)}$ = 10.07kΩ, $r_{\mu 2}$ = 10 (150) (10.07) = 15.1MΩ, r_{01} ≈ 100/.066 = 1.501MΩ.

R_{out} = 1MΩ ∥ ∞ ∥ $\dfrac{(1.5M\Omega \parallel 15M\Omega)}{150}$ ∥ 10.07kΩ = 1MΩ ∥ 10kΩ ∥ 100kΩ ∥ 10.07kΩ = **4.75kΩ.**

Now the gain = −206.6 × 10^{-3} × 4.75kΩ = **−981V/V**, and R_{in} ≈ 1MΩ/[(1 + 981) ∥ (151 (2 (375)))] = 1.018kΩ ∥ 113.3kΩ = **1.01kΩ.** Overall, for β = 50 to 150, g_m ranges from 204mA/V to 207mA/V, the gain ranges from −977V/V to 981V/V, R_{in} ranges from 946Ω to 1010Ω, R_{out} ranges from 4.75kΩ to 4.79kΩ. That is, there is *very little effect.*

9.22 At 125°C, and 100μA, V_{BE} = 700 − 2 (125 − 25) = 500mV . Thus $R_1 = \dfrac{10V - 0.5}{100\mu A}$ = **95kΩ**, and i_B = 100μA/100 = 1μA. Thus i_{R2} = 100 − 1 = 99μA, and R_2 = .500/99μA = **5.05kΩ.** Now, at 25°C (with i_{C2} low) and i_{B2} ≈ 0, $\upsilon_B = \dfrac{5.05}{5.05 + 95.0}$ × 10 = 504.7mV. Now i_2 = 100 × 10^{-6} $e^{\frac{-700 + 504.7}{25}}$ = **0.0405μA.** For doubling, that is i_2 = 0.0810μA, υ_B = 504.7 + 25 ln 2 = 522mV. Thus the supply voltage = (0.522/5.05) (5.05 + 95.0) = **10.34V.** At i_{C2} = 50μA, V_{BE} = 700 + 25 ln $\dfrac{50 \times 10^{-6}}{100 \times 10^{-6}}$ = 682.7mV. Now at 100°, V_{BE} = 682.7 − (100 − 25)2 = 532.7mV, for which the supply voltage ≈ (0.533/5.05)(100.05) = **10.56V.**

SECTION 9.8: IC POWER AMPLIFIERS

9.23 For the circuit shown in Fig. 9.30, $I_{R1} \approx \dfrac{25 - 0.7 - 2(0.7)}{25k\Omega + 25k\Omega}$ = 0.458mA. Thus $I_{bias} = \dfrac{0.458}{20 \times 20}$ = **1.15μA.** To reduce this to 0.5μA, raise R_1 to (1.15/0.5) × 50 = **115kΩ** with **57.5kΩ** in each half. Now for the same gain and to maintain the same assumptions for the gain calculation, raise R_2 and R_3 by the same factor [(= 1.15/0.5 = 2.3)] to **2.3kΩ** and **57.5kΩ** respectively. Because of the change, the current in Q_{10}, Q_{11} and Q_{12} all reduce by a factor of 2.3, that in Q_9 reduces, but not by as large a factor due to R_6, R_7.

9.24 For the calculation of A, include Q_{12}, Q_{11}, Q_7, Q_8, Q_9, as driven by the output resistance of Q_6 and Q_4. With a 27V supply, bias current = (27 − 3(0.7))/50 = 0.5mA. Thus $I_{C4} = I_{C6} = I_{C10} = I_{C11} = I_{C12}$ = 0.5mA, I_{C9} = 10 I_{C11} = 5mA. Now r_{e12} = 25mV/0.5mA = 50Ω, r_{o12} = 100V/0.5mA = 200kΩ = r_{o11}. Now Q_8 operates as a follower with β = 100 × 20 = 2000, while Q_7 operates as a follower with β = 100, where at the output the only load is $r_{o7} \parallel r_{o9} \approx r_{o7}/2$ where r_{o7} = 100V/5mA = 20kΩ. Thus the net load on the collector of Q_{12} is $r_{o12} \parallel r_{o11}$ ∥ ((101) (20kΩ/2)) = 200kΩ ∥ 200kΩ ∥ 1MΩ = 90.9kΩ. The gain from the base of Q_{12} to the collector of Q_{12} is about −90.9kΩ/50Ω = −1818V/V. Follower gain for no load is nearly 1V/V. Thus the overall gain A is − 1818V/V. Equivalent input capacitance is C_T = 10 × 10^{-12} (1 + 1818) = 1.82 × $10^{-8}F$. Corresponding input resistance is $R_T = r_{o6} \parallel r_{o4}$, where at 0.5mA, r_o = 100V/0.5 = 200kΩ. Thus, the cutoff frequency = $\dfrac{1}{2\pi\ (200 \times 10^3/2)\ (1.82 \times 10^{-8})}$ = **87.5Hz.**

9.25 For equal sharing, each conducts 50/2 = 25mA. For V_{EB5} = 0.70V, R_3 = 700mV/25mA = **28Ω**. Note that a specification of 1.0V at 1A is given for Q_3. However a lot of this V_{EB} is likely due to resistive effects in the base. Thus we use 0.7V as above. (Note, that we get a higher result for R_3 if we use the 0.1V/decade idea, in which case V_{EB5} = 1.00 − 0.1 log 25/1000 = 0.84V). Now at I_{out} = 1A, V_{BE} = 1.00V, and I_{R3} = 1.00/28 = 35.7mA. As well, I_{B5} = 1A/30 = 33.3mA. Thus I_{C3} = 35.7 + 33.3 = 69mA. For a load change from 50mA to 1A, a factor of 20, the current in Q_3 varies from 25mA to 69mA, a factor of **2.76**. For Q_1, Q_2 operating at 1mA, $|V_{BE}|$ = 0.700V. For Q_3, Q_4 operating at 2mA, $|V_{BE}|$ = 700 + 0.1 log (2/10) = 0.630V. Thus $R_5 = R_6$ = (0.700 − 0.630)/2mA = **35Ω**.

9.26 For ±12V supplies and 2V saturation, outputs of ±10V are available. Thus a 20V peak signal is possible. Input provided is 0.1V peak. Required input resistance = 10kΩ. Thus, R_3 = 10kΩ, and R_4 = 10V/0.1V × 10kΩ = 1MΩ. For the highest possible input resistance, use R_4 = **10MΩ** and R_3 = **100kΩ** for a 100kΩ input resistance. For the positive side, 10/0.1 = 1 + R_2/R_1, or $R_2 = 99R_1$. Use $R_2 = R_4$ = 1MΩ and R_1 = 1MΩ/99 = **101kΩ**, a 100kΩ and 1kΩ in series, as a quick solution.

9.27 There are several choices:

(a) One is to drive A_1 as shown, but with R_3 connected to the output of A_1, with $R_4 = R_3$ and 1 + R_2/R_1 = 20/2 = 10, $R_2 = 9R_1$, or $R_1 = R_2/9$. Use R_1 = 10kΩ, R_2 = 90kΩ, $R_4 = R_3$ = 100kΩ.

(b) Modify (a) above to merge R_3 and R_1 into R_{13} = 10kΩ, with R_2 = 90kΩ, and R_4 = 100kΩ using only 3 resistors in all.

SECTION 9.9: MOS POWER TRANSISTORS

9.28 $K = 1/2\ \mu_n\ C_{ox}\ W/L = 1/2 \times 30 \times 10^{-6} \times 10^5/5 = 0.3\text{A/V}^2$. At low υ_{GS}, $i_D = K(\upsilon_{GS} - V_t)^2 = 0.3(\upsilon_{GS} - V_t)^2$. At high υ_{GS}, $i_D = 1/2\ C_{ox}\ W\ U_{sat}\ (\upsilon_{GS} - V_t) = 1/2 \times \frac{30 \times 10^{-6}}{5 \times 10^{-2}} \times 10^5 \times 10^{-6} \times 5 \times 10^4\ (\upsilon_{GS} - V_t) = 1.5 \times (\upsilon_{GS} - V_t)$. These currents are equal when $0.3(\upsilon_{GS} - V_t)^2 = 1.5 \times (\upsilon_{GS} - V_t)$, or $\upsilon_{GS} - V_t = 5$, or $\upsilon_{GS} = 5 + 2 =$ **7V**. For υ_{GS} = 7V, $i_D = K(\upsilon_{GS} - V_t)^2 = 0.3(7-2)^2$ = **7.5A**, for which g_m = 2 (0.3) (7 − 2) = **3A/V**. For υ_{GS} = 3.5V, $i_D = 0.3(3.5-2)^2$ = **675mA** or **.675A**, for which g_m = 2 (0.3) (3.5 − 2) = **0.9A/V**. For υ_{GS} = 14V, i_D = 1.5 (14 − 2) = **18A**, for which $g_m = 1.5(\upsilon_{GS} - V_t)$ = 1.5(14 − 2) = 18 A/V. Note that from the velocity-saturated relationship at 7V, the g_m would be 1.5A/V (rather than 3A/V). In practice, the transition between modes begins to occur at lower values of υ_{GS} and g_m.

9.29 For 5mA bias: $i_D = 5 = K(\upsilon_{GS} - V_t)^2 = 200(\upsilon_{GS} - 2)^2$, $\upsilon_{GS} = (5/200)^{½} + 2$ = 2.158V. Thus V_{14} = 4 (0.7) + 2 (2.158) = **7.12V**, R = 2(2.158)/5mA = 863Ω. Total FET TC = 2 (−3) = −6mV/°C. Total BJT TC = 4 (−2) = −8mV°C. Total TC = −14mV/°C. Thus 6/14 = **0.429** of V_{14} must appear across Q_6.

For outputs around zero: $g_m = 2K(\upsilon_{GS} - V_t)$ = 2 (200 (2.158 − 2)) = 63.2mA/V. Thus, R_{out} of follower = $(1/g_m) \parallel (/g_m)$ = (1/63.2)/2 = 7.9Ω. The gain for 100Ω load ≈ 100/(7.9 + 100) = **0.927V/V**.

For outputs around +20V: I_L ≈ 20/.100 = 200mA, whence 200 = 200 $(\upsilon_{GS} - 2)^2$, $\upsilon_{GS} = (200/200)^{½} + 2$ = 3V, and g_m = 2 (200) (3 − 2) = 400mA/V. Thus R_{out} = (1/400)/2 = 1.25. The incremented gain = 100/(1.25 + 100) = **0.988V/V**. Alternatively, note that the input must be 20 + 3 − 2.158 = 20.84V at the peak. The corresponding overall gain = 20/20.84 = **0.960V/V**.

9.30 See that Q_3, Q_4 are 100 times larger than Q_1, Q_2. Thus for 10mA in Q_3, Q_4 require I = 10/100 = **0.1mA** in Q_1, Q_2. For I_Q = 5mA, 5 = 100 $(\upsilon_{GS} - 1)^2$. Thus $\upsilon_{GS} = (5/100)^{½} + 1$ = 1.224V, g_m = 2 (100) (1.224 − 1) = 44.8mA/V, $R_{out} = 1/g_m \parallel 1/g_m = (\frac{1}{44.8})/2$ = 11.2Ω. *For* V_O = 0V, the gain with 100Ω load is 100/(100 + 11.2) = **0.899V/V**, with V_D = **0V**, V_B = **1.22V**, V_C = **−1.22V**, V_A = **0V**. See that $\upsilon_{GD1} = \upsilon_{GD2}$ = 0 and all transistors operate in saturation mode. *For* V_O = +10V, with 100Ω load, i_{D3} = 10/0.1 = 100mA, that is 100 = 100 $(\upsilon_{GS} - 1)^2$, υ_{GS} = 1 + 1 = 2V. Thus V_D = **+10V**, and V_B =

+12V. Now Q_1 continues to conduct 0.1mA with V_{GS} = 1.224V, that is V_A = 12 – 1.224 = **10.78V**. For Q_2, V_D = 10V, V_G = 10.78V, and operation is in triode mode with υ_{GD} = 0.78V, where, 0.1 = 1 (2 $(\upsilon_{GS} - 1)(\upsilon_{DS}) - \upsilon_{DS}^2$). Now, since $\upsilon_{GD} = \upsilon_{GS} - \upsilon_{DS}$ = 0.78, and $\upsilon_{GS} = 0.78 + \upsilon_{DS}$, $1 = 20\,((.78 + \upsilon_{DS} - 1)\,\upsilon_{DS} - 10\,\upsilon_{DS}^2) = -4.4\,\upsilon_{DS} + 20\,\upsilon_{DS}^2 - 10\,\upsilon_{DS}^2$, or 10 $\upsilon_{DS}^2 - 4.4\,\upsilon_{DS} - 1 = 0$. Thus $\upsilon_{DS} = \dfrac{4.4 \pm \sqrt{4.4^2 - 4(-1)(10)}}{2\,(10)}$ = 0.605V, and υ_{GS} = .78 + .605 = 1.38V. Correspondingly, V_C = +10 – 0.605 ≃ **9.40V**. Thus Q_1, Q_3 operate in saturation, Q_4 cuts off, and Q_1 is in triode mode. Overall, the gain is V_D/V_A = 10/10.78 = **0.928V/V**. Incrementally, for Q_3, g_m = 2 (100) (2 – 1) = 200mA/V, and $R_{out} = 1/g_m$ = 1/200 = 5Ω. Thus the gain υ_d/υ_a = 100/(100 + 5) = **0.952V/V**.

9.6 (contined)

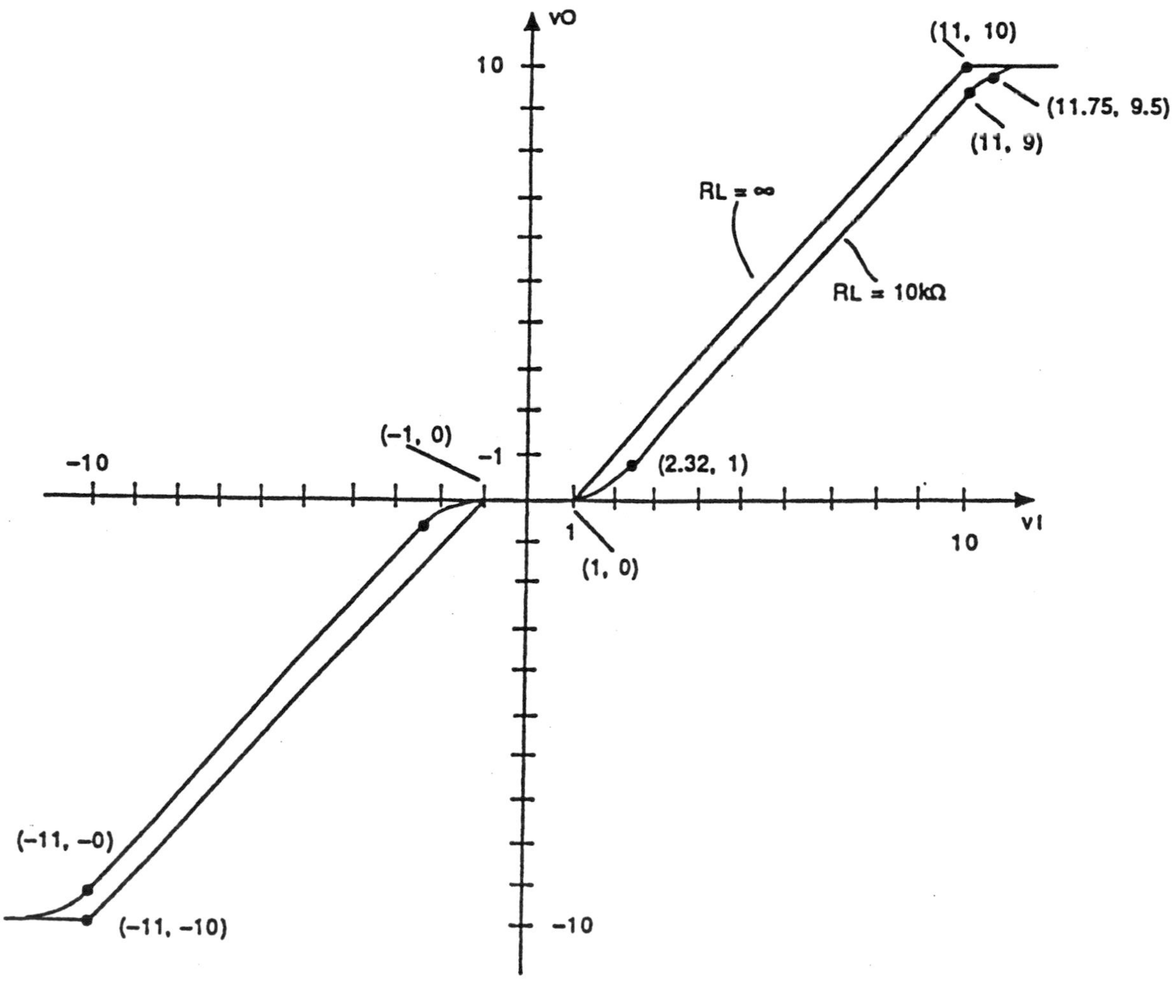

Chapter 10

ANALOG INTEGRATED CIRCUITS

SECTION 10.1: THE 741 OP-AMP CIRCUIT

10.1 For ±15V supplies, I_{REF} = [+15 − −15 − 0.7 − 0.7]/39kΩ = **0.733mA**, or **733μA**. For ±5V supplies, I_{REF} = [+5 − −5 − 1.4]/39kΩ = **0.2205mA**. For ±5V supplies and I_{REF} = .73mA, R_5 = (10 − 1.4)/0.733 = **11.73kΩ**. One could use **12kΩ** as a standard value.

10.2 For ±15V supplies, I_{REF} = 733μA. Thus $\upsilon_{BE11} = n\ V_T \ln (I/I_S) = 25 \ln (.733 \times 10^{-3}/10^{-4})$ = 625.5mV. Now $\upsilon_{BE10} = \upsilon$ = 625.5 + 25 ln (i/.733), with i in mA, where iR_4 = −25 ln i/.73 = 5000i, or i = −.005 ln i/0.73.

Iterate: For i = 0.1 (mA), i = −.005 ln 0.1/.73 = .01mA. For i = .01 (mA), i = −.005 ln .01/.73 = .0215mA. For i = .015 (mA), i = .005 ln .015/.73 = .0195mA. For i = .018 (mA), i = .005 ln .018/.73 = .0185mA. Thus $i \approx$ **18.3μA.**

For ±5V supplies, I_{REF} = .2205mA, $\upsilon_{BE11} = 25 \ln ((.2205/10^{-14}) \times 10^{-3})$ = 595.4mV, and υ_{BE10} = 595.4 × 25 ln i/.2205, where iR_1 = −25 ln i/.2205 = 5000i, or i = −.005 ln i/.2205.

Iterate: For i = .01mA, i = .005 ln .01/.2205 = .0155mA. For i = .015mA, i = .0133mA. For i = .014mA, i = .0138mA. Thus $i \approx$ **13.9μA** (reduced from 18.3μA).

For i = 18.3μA as before, υ_{BE10} = 625.5 + 25 ln (.0183/.733) = 533.2mV, (or, υ_{BE10} = 595.4 + 25 ln (.0183/.2205) = 533.2mV) and, for ±5V supplies, υ_{BE11} = 595.4mV. Thus $R_4 = \dfrac{(595.4 - 533.2)mV}{18.3\mu A}$ = **3.40kΩ.**

10.3 Replace R_a by a transistor Q_{25} whose collector is connected to the emitter of Q_{16}, emitter to a resistor R_{12} connected to $-V_{EE}$, and base to either the base of Q_{11} or the emitter of Q_7. From page 820 of the Text, note that the voltage $V_{B17} \approx$ 618mV + 550μA (100Ω) = 618 + 55 = 673mV, and I_{R9} = 673mV/50kΩ = 13.5μA. Now at 13.5μA, $V_{BE25} = 25 \ln (13.5\mu A/10^{-14} A)$ = 525.6mV. *For the connection to the base of* Q_{11}, $V_{BE11} = 25 \ln \dfrac{730 \times 10^{-6}}{10^{-14}}$ = 625.3mV. Thus $R_{12} = \dfrac{(625.3 - 525.6)mV}{13.5\mu A}$ = **7.4kΩ.** *For the connection to the emitter of* Q_7, $V_{E7} = V_{BE6} + R_2 (I_{E6})$ and, (from page 706), V_{E7} = 517 + 9.5 × 10^{-6} × 1 × 10^3 = 517 + 9.5 = 526.5mV, for which $R_{12}' = \dfrac{(525.6 - 526.5)mV}{13.5\mu A} \approx 0$. To avoid an unusual load on Q_7, it would be best to include a resistor R_{12} = **1kΩ** say (= R_1, R_2), in which case the current extracted from Q_7 is about 9.5μA. The latter has the advantage of using a smaller resistor and, as well provides a (small) signal component of a reinforcing polarity from the collector of Q_5 via Q_7.

10.4 *For inputs limited to the supply range*, the worst case is for one of I_n+ and I_n− connected to +15V (say I_n+) and the other (say I_n−) to −15V, in which situation the collector junction of Q_2 has 15 − 0.7 − −15 = 29.3V reversed across it. OK. Also the EBJ of Q_2 and Q_4 are reversed in series, with a combined rating of 7 + 50 = 57 > 30. OK.

Now for inputs outside the supply range: For I_n+ or I_n− positive, the base-collector diode of Q_2 (say) conducts, reversing the EBJ of Q_8, Q_9. The greatest allowed voltage is 7 + 2 (0.7) = **8.4V** above the positive supply.

For I_n+ or I_n− negative, the most-negative input stresses the CB junction of Q_2 (say) to 50V when V_{in} = 30 − 0.7 − 50 = **−20.7V** below the negative supply.

SECTION 10.2: DC ANALYSIS OF THE 741

10.5 From the preamble to Eq. 10.1 on page 816 of the Text, $V_{BE11} - V_{BE10} = I_{C10} R_4$, or $V_T \ln \frac{R_{REF}}{I_S} - V_T \ln \frac{I_{C10}}{k\, I_S} = I_{C10} R_4$, or $V_T \ln \frac{k\, I_{REF}}{I_{C10}} = I_{C10} R_4$ – – – (1). For I_{REF} = 730μA, R_4 = 5000Ω, and k = 0.5, $25 \times 10^{-3} \ln \frac{0.5(730 \times 10^{-6})}{I_{C10}} = I_{C10} \times 5000$. Now, for $I_{C10} = i$ in μA, ln 365/i = .2i, or i = 5 ln 365/i. Use a process of *trial and success*: i = 10μA → i = 5 ln 365/10 = 17.98, i = 15μA → i = 5 ln 365/15 = 15.95, i = 15.5μA → i = 5 ln 365/15.5 = 15.8, i = 15.7μA → i = 5 ln 365/15.7 = 15.7. Thus, I_{C10} = **15.7μA**.

Normally for k = 1, 25 ln 730/i = 5i, or i = 5 ln 730/i, i = 19 → i = 5 ln 730/19 = 18.24, i = 18.5 → i = 5 ln 730/18.5 = 18.4. Thus for I_{REF} = 730.00μA, I_{C10} = 18.45μA (or so). Thus for k = 0.5, from (1). $R_4 = \frac{25 \times 10^{-3}}{18.45 \times 10^{-6}} \ln \frac{0.5(730)}{18.45}$ = 4.04kΩ. Use **4kΩ**.

10.6 For Q_1 through Q_4, $I_S = 10^{-14}$A, n = 1, and I_C = 19/2 = 9.5μA. Thus $V_{BE} = V_T \ln I/I_S = 25 \ln \frac{9.5 \times 10^{-6}}{10^{-14}}$ = 516.8V. Accordingly, the voltage on the bases of Q_3, Q_4 is –2 (516.8) = **–1.034V**.

10.7 Use the fact that V_{BE6} = 517mV and $I_{E6} \simeq I_{E5}$ = 9.5μA. Correspondingly, $V_{E7} = 0.517 + 9.5 \times 10^{-6} \times 1 \times 10^3$ = **0.5265V**, $I_{E7} = (I_{C6} + I_{C5})/\beta$ + .5265/50kΩ = $I_{C6}/2\beta$ + 11.3μA, and $I_{B7} = I_{C6}/(2\beta^2) + 11.3/\beta$, $I_{C3} = I_{C5} + I_{B7} = I_{C6} + I_{B7} = I_{C6} + I_{C6}/(2\beta^2) + 11.3/\beta$, whence $I_{C3}/I_{C6} = 1 + 1/(2\beta^2) + \frac{11.3}{\beta\, I_{C6}} \simeq 1 + 1/(2\beta^2) + \frac{1}{\beta}(11.3/9.5) = 1 + 1/2\beta^2 + 1.19/\beta$. Thus $\mathbf{I_{C6}/I_{C3} = 1/(1 + 1/(2\beta^2) + 1.19/\beta)}$. For β = 200,

$$I_{C6}/I_{C3} = \left(1 + \frac{1}{2\,(200)^2} + \frac{1.19}{200^{-1}}\right) = 1/(1 + (.0025 + 1.19)/200) = \mathbf{0.994}.$$

10.8 For high β, ignore the base current of Q_7. For $I_{C5} = I$, $I_{C6} = 19 - I$, for I in μA. Now, $V_{BE5} = 25 \ln I/I_S = 25 \ln 10^{14} I$, and $V_{BE6} = 25 \ln (19 - I)\, 10^{14}$.

For R_1 shorted: $25 \times 10^{-3} \ln (10^{14} I) = 25 \times 10^{-3} \ln ((19 - I)10^{14}) + 1 \times 10^3 (19 - I)$, whence I = 19 –25 ln (I/(19 – I)) with I in μA. Try I = 10 → I = 19 – 25 ln (10/(19 – 10)) = 16.37. Try I = 12 → I = 19 – 25 ln 12/7 = 5.52, try I = 11 → I = 19 – 25 ln 11/8 = 11.04. Thus I_{C5} = 11.0μA, for which I_{C6} = 19 – 11.0 = 8.0μA, and I_{C6}/I_{C5} = 8/11 = **0.727**.

For R_2 shorted: $V_{BE5} + R_1 I = V_{BE6}$, $25 \ln (I/I_S) + R_1 I = 25 \ln ((19 - I)/I_S)$, $R_1 I = 25 \ln ((19 - I)/I)$, $I = 25 \ln ((19 - I)/I)$, for I in μA. Try I = 8μA → I = 25 ln (19 – 8)/8 = 7.96. Thus I_{C5} = 8.0μA, for which I_{C6} = 19 – 8 = 11.0μA, and I_{C6}/I_{C5} = 11/8 = **1.375**.

10.9 For 8μA in Q_1, and 11μA in Q_2, $V_{BE1} = 25 \ln \frac{8 \times 10^{-6}}{1 \times 10^{-14}}$ = 512.5mV, $V_{BE2} = 25 \ln \frac{11 \times 10^{-6}}{1 \times 10^{-14}}$ = 520.5mV. Now the offset (due to the npn + pnp devices) = 2 (520.5 – 512.5) = **16mV**, the negative input being higher with R_2 shorted. The offset is **–16mV** for R_1 shorted.

10.10 At present, V_{BE17} = 618mV, and $I_{E16} \simeq$ 16.2μA. Thus, the revised $R_9' = \frac{618 \times 10^{-3}}{16.2 \times 10^{-6}}$ = **38.1kΩ**. Now, for I_{C17} = 550μA, I_{B17} = 550/200 = 2.75μA, for I_{R9} = 4 (2.75) = 11μA, $R_9'' = \frac{618 \times 10^{-3}}{11 \times 10^{-6}}$ = **56kΩ**.

10.11 For the current in Q_{14}, Q_{20}, increased to 1.5 (154) = 231μA, $V_{BE} = 25 \ln \frac{231 \times 10^{-6}}{3 \times 10^{-14}}$ = 569.1mV, and

$V_{R6} = 27 \times 231 \times 10^{-6} = 6.23\text{mV}$. For $R_6 = R_7 = 0\Omega$, $V_{BB} = 2\ (569.1) = 1138.2\text{mV}$. For $R_6 = R_7 = 27$, $V_{BB} = 2\ (6.23 + 569.1) = 1150.7\text{mV}$. Now $V_{BB} = V_{BE18} + V_{BE19}$, and with high β, $V_{BB} = 25 \ln \dfrac{(180 \times 10^{-6} - V_{BE18}/R_{10})}{10^{-14}} + 25 \ln \dfrac{(V_{BE18}/R_{10})}{10^{-14}}$, or for V_{BE18}/R_{10} in μA, $V_{BB} = 25 \ln \dfrac{(180 - V_{BE18}/R_{10})}{10^{-8}}$ $+ 25 \ln \dfrac{(V_{BE18}/R_{10})}{10^{-8}}$ - - - (1).

Now *for* $R_6 = R_7 = 0\Omega$, $V_{BB} = 1138\text{mV}$, $V_{BE18} = \upsilon$, $R_{10} = R$. From (1), $10^{-16} \ln^{-1} 1138/25 = (180 - \upsilon/R) \times (\upsilon/R) = 5876$, or $(\upsilon/R)^2 - 180\ (\upsilon/R) + 5876 = 0$, $\dfrac{\upsilon}{R} = \dfrac{+180 \pm \sqrt{180^2 - 4(5876)}}{2} = \dfrac{180 - 94.3}{2}$ $= 42.8\mu\text{A}$. Thus $I_{E18} = 180 - 42.8 = 138.2\mu\text{A}$, $V_{BE18} = 25 \ln \dfrac{138.2 \times 10^6}{10^{-14}} = 583.7\text{mV}$. Thus $R_{10} = 583.7/42.8 =$ **13.6kΩ**. *Check*: $V_{BE19} = 25 \ln 42.8 \times 10^8 = 554.4\text{mV}$, and $V_{BB} = 583.7 + 554.4 = 1138.1\text{mV}$, as required.

Now for $R_6 = R_7 = 27\Omega$, $V_{BB} = 1151\text{mV}$, $V_{BE18} = \upsilon$, $R_{10} = R$. From (1) $10^{-16} \ln^{-1} 1151/25 = (180 - \upsilon/R) \times (\upsilon/R) = 9766$.

Thus $(\upsilon/R)^2 - 180\ (\upsilon/R) + 9766 = 0$, and $\dfrac{\upsilon}{R} = \dfrac{180 \pm \sqrt{180^2 - 4(9786)}}{2} = \dfrac{180 \pm j81}{2}$.

See the result is **imaginary**, that is it is not possible to provide the desired operation by varying R_{10}. To check this fact, note that the largest real value of $\upsilon/R = 180/2 = 90\mu\text{A}$, at which $V_{BE} = 25 \ln (90 \times 10^{+8}) = 573\text{mV}$, for which $V_{BB\ max} = 2\ (573) = 1146\text{mV} < 1151\text{mV}$ required.

SECTION 10.3: SMALL-SIGNAL ANALYSIS OF THE 741 INPUT STAGE

10.12 From Eq. 10.4 of the Text, $R_{id} = 4\ (\beta_N + 1)\ r_e$, whence $r_e = \dfrac{3.6 \times 10^6}{4\ (180 + 1)} = 4.97\text{k}\Omega$, for which $I_E = \dfrac{25 \times 10^{-3}}{4.97 \times 10^3} =$ **5.03μA**, reduced from the present value of 9.5μA. For this change, $G_{m1} = \dfrac{\alpha}{2r_e} = \dfrac{180/181}{2(4.97)} =$ **0.1mA/V**.

10.13 Generally, $R_0 = r_o\ [1 + g_m\ (R_E \parallel r_\pi)]$. Here, $r_e = 2.63\text{k}\Omega$, and $r_\pi = (\beta + 1)r_e = 201\ (2.63) = 528.6\text{k}\Omega$, with $r_o = 5.26\text{M}\Omega$, where $g_m = \dfrac{\alpha}{r_e} = \dfrac{200}{201\ (2.63) \times 10^3} = 0.378\text{mA/V}$. Thus, $R_{06} = 10.5 \times 10^6 = 5.26 \times 10^6\ [1 + .378 \times 10^{-3}\ (R_2 \parallel 528.6k\Omega)]$.

$R_2 \parallel 529k\Omega = (\dfrac{10.5}{5.26} - 1)\ \dfrac{1}{.378 \times 10^{-3}} = 0.996/(0.378 \times 10^{-3}) = 2.635\text{k}\Omega$, and

$R_2 = 2.635 \parallel (-529)k\Omega =$ **2.65kΩ**. Thus $R_{01} = R_{06} \parallel R_{04} = 10.5/2 =$ **5.25MΩ**. For $G_{m1} = 1/5.26\text{mA/V}$, and for the new situation, $A'_{\upsilon 1} = \dfrac{1}{5.26} \times 10^{-3} \times 5.25 \times 10^6 \simeq$ **1000V/V**. For the old, $R_{06} = 5.26 \times 10^6\ [1 + .378 \times 10^{-3}\ (1\text{k}\Omega \parallel 528k\Omega)] = 7.25\text{M}\Omega$, and $R_{01} = R_{06} \parallel R_{04} = 10.5 \parallel 7.25 = 4.29\text{M}\Omega$, for which $A_{\upsilon 1} = \dfrac{1}{5.26} \times 10^3 \times 4.29 \times 10^6 =$ **815V/V**.

10.14 Use $R_1 = R_2 = 2.65\text{k}\Omega = R$. From Example 10.1, $V_{OS} = \dfrac{\Delta R\ (1/G_{m1})}{R + \Delta R + r_e}$, with $\Delta R = .02\ (2.65) = 53\Omega$, and $r_e = \dfrac{25 \times 10^{-3}}{9.5 \times 10^{-6}} = 2.63\text{k}\Omega$, where $G_{m1} = 1/5.26\text{mA/V}$. Thus $V_{OS} = \dfrac{53 \times 5.26 \times 10^3 \times 9.5 \times 10^{-6}}{(2.65 + .053 + 2.63) \times 10^{13}}$ $=$ **0.497mV**. This is $.497/.3 \simeq 1.67$ of the original, or **67% larger**, although the resistors are only (1 – 2.65/2), or 33% larger.

10.15 From Ex. 10.9 of the Text, $G_{mcm} = \dfrac{\beta_p}{2R_0} \times \dfrac{\Delta R}{R + r_{e5}}$, where $\beta_p = 50$, $r_{e5} = 25\text{mV}/9.5\mu\text{A} = 2.63\text{k}\Omega$, and

$R_o = R_{o9} \parallel R_{o10} = 2.43\text{M}\Omega$ (From Ex. 10.10), and $R_1 = R = 1\text{k}\Omega$. For CMRR = 80dB, $G_{m1}/G_{mcm} = 10^4$, or $G_{mcm} = 10^{-4}\, G_{m1}$, where $G_{m1} = (1/5.26)\text{mA/V}$, or $G_{mcm} = \frac{10^{-4}}{5.26} = \frac{1}{5.26 \times 10^4}$ mA/V,

$$\frac{1}{5.26 \times 10^4} = \frac{50}{2\,(2.43 \times 10^6)} \times \frac{\Delta R}{(1 + 2.63) \times 10^3},$$

$\Delta R = \frac{3.63 \times 10^3 \times 2\,(2.43 \times 10^6)}{5.26 \times 10^6 \times 50} = 67.1\Omega$. The corresponding tolerance is $67/1000 \times 100 =$ **6.7%** or **±3.4%**.

10.16 Add resistors of value R_E in series with the emitters of both Q_8 and Q_9. Now, for the p-channel devices, $V_A = 50\text{V}$, $\beta = 50$, (rather than 125Vand 200 for Q_{10}). Now, since $I_{C10} = I_{C9} = I_{C8} = 19\mu\text{A}$, $r_e = \frac{V_T}{I_E} \approx \frac{25mV}{19\mu A\ (51/50)} = 1.29\text{k}\Omega$, $r_\pi = (\beta + 1)\, r_e = 51\,(1.29) = 65.8\text{k}\Omega$, $g_m = \alpha/r_e = (50/51)/1.29 = 0.760\text{mA/V}$, $r_o = V_A/I_C \approx (50/(19) \times 10^{-6}) = 2.63\text{M}\Omega$. From Equation 10.7, want $R_0 = r_o\,(1 + g_m\,(R_E \parallel r_\pi))$ to be $31.1\text{M}\Omega$. Thus $31.1 = 2.63\,(1 + .760 \times 10^{-3}\,(R_E \parallel 65.8 \times 10^3))$, $\frac{R_E\,(65.8)\,(.760)}{R_E + 65.8 \times 10^3} = 31.1/2.63 - 1 = 10.83$. Thus $50\,R_E = 10.83\,R_E + 712.6$, and $R_E = 712.6/(50 - 10.83) =$ **18.2kΩ**. Thus use resistors $R_E =$ **18.2kΩ**. In this case, $R_{09} = R_{010} = 31.1\text{M}\Omega$, and $R_0 = 31.1/2 =$ **15.6MΩ**. Now $G_{mcm} = \frac{\beta_p}{2R_0}\,\frac{\Delta R}{R + r_{e5}}$, where $\beta_p = 50$, $\frac{\Delta R}{R} = .02$, whence $\Delta R = .02\,(1\text{k}\Omega) = 0.02\text{k}\Omega$. Now $I_{C5} = 9.5\mu\text{A}$, and $r_e \approx 25\text{mV}/9.5\mu\text{A} = 2.63\text{k}\Omega$. Thus $G_{mcm} = \frac{50}{2\,(15.6 \times 10^6)} \times \frac{.02 \times 10^{-3}}{(1 + 2.63) \times 10^{-3}} =$ **0.0088μA/V**. Now $G_{m1} = 1/5.26\text{mA/V}$ as noted below Eq. 10.6 in the Text. Thus CMRR $= \frac{G_{m1}}{G_{mcm}} = \frac{1/(5.26 \times 10^{-3})}{0.0088 \times 10^{-6}} = 21.6 \times 10^3 \equiv$ **86.7dB.**

SECTION 10.4: SMALL-SIGNAL ANALYSIS OF THE 741 SECOND STAGE

10.17 The situation is one in which the base of Q_{25} is joined to the emitter of Q_7, with 1kΩ connecting the emitter of Q_{25} to $-V_{EE}$, and the collector of Q_{25} connected to the emitter of Q_{16} joined to the base of Q_{17}. Here, the collector current of Q_{25} is the same as that in Q_6, namely 9.5μA, for which $r_{025} = V_A/I_C = 125/(9.5 \times 10^{-6}) = 13.2\text{M}\Omega$. Now $I_{C11} = 730\mu\text{A}$, $I_{B11} = 730/200 = 3.65\mu\text{A}$. Thus the emitter current of $Q_{16} = 9.5 + 3.65 = 13.15\mu\text{A}$, for which $r_{e16} = 25\text{mV}/13.15\mu\text{A} = 1.90\text{k}\Omega$. Also $r_{e7} = 25\text{mV}/730\mu\text{A} = 34.2\Omega$. From Eq. 10.12, $R_{i2} = (\beta + 1)\,[r_{e16} + r_{o25} \parallel [(\beta + 1)\,(r_{e17} + R_8)]] = 201\,[(1.90 \times 10^3 + (13.2 \times 10^6) \parallel (201\,(34.2 + 100))]$, or $R_{i2} = 201\,(1.9 \times 10^3 + 27.0 \times 10^3) =$ **5.81MΩ**, (rather than the 4.0MΩ found previously).

10.18 For $R_{i2} = (\beta + 1)\,[r_{e16} + r_{o25} \parallel ((\beta + 1)\,(r_{e17} + R_8))]$, or $R_{i2} = 4 \times 10^6 = 201\,[1.90 \times 10^3 + 13.2 \times 10^6 \parallel ((201\ (34.2) + R_8))]$, $19.9 \times 10^3 = 1.90 \times 10^3 + 6.87 \times 10^3 + 201\,R_8$, $R_8 = \frac{(19.9 - 1.9 - 6.87) \times 10^3}{201} =$ **55.4Ω**. Now $G_{m2} = \frac{i_{C17}}{v_{i2}} = \frac{\alpha\,\frac{R_9 \parallel R_{i17}}{R_9 \parallel R_{17} + r_{e16}}}{r_{e17} + R_8}$, where $R_{i17} = (\beta_{17} + 1)\,(r_{e17} + R_8)$. Here, $r_{e16} = 1.90\text{k}\Omega$ (from the solution to P.10.17 above). Thus, $R_{i17} = 201\,(34.2 + 55.4) = 18.01\text{k}\Omega$, and $G_{m2} = \frac{\frac{200}{201} \times \frac{18.01}{18.01 + 1.90}}{34.2 + 55.4} = 0.01\text{A/V} =$ **10.0mA/V**. This is to be compared to 6.5mA/V found previously. Now $R_{o2} = R_{o13B} \parallel R_{o17}$, where $R_{o13B} = r_{o13B} = 90.9\text{k}\Omega$, and $R_{o17} = r_{o17}\,(1 + g_{m17}\,(R_8 \parallel r_{\pi17}))$, where $r_{o17} = \frac{125V}{550 \times 10^{-6}} = 0.227\text{M}\Omega$, such that $R_{o17} = .227 \times 10^6\,(1 + 200/201 \times (1/34.2)\,(55.4\ \parallel\ (201\,(34.2))) =$ **00.59MΩ**. Thus $R_{o2} = 90.9\text{k}\Omega \parallel 590\text{k}\Omega =$ **78.8kΩ**, and the open-circuit voltage gain is $-G_{m2}\,R_{o2} = -10.0 \times 10^{-3} \times 78.8 \times 10^3 =$ **−788V/V**, compared to −526.5V/V found previously). Thus the change of bias network produces a gain increase of $(788/526.5 - 1) = 0.50$, or **50%**!

SECTION 10.5: ANALYSIS OF THE 741 OUTPUT STAGE

10.19 For the basic design, $R_{o2} = 81\text{k}\Omega$. Now, for $R_L = 2\text{k}\Omega$, $R_{i3} = \beta_{23}\,(74\text{k}\Omega) = 4(81\text{k}\Omega)$ for which, $\beta_{23} = (81/74)\,(4) = $ **4.38**. Now the second-stage gain is $A_2 = -G_{m2}\,R_{o2}\,\dfrac{R_{i3}}{R_{i3} + R_{o2}}$, where (from the bottom of page 716 of the Text) $G_{m2} = 6.5\text{mA/V}$, and $R_{o2} = 81\text{k}\Omega$.

Thus $A_2 = -6.5 \times 10^{-3} \times 81 \times 10^3 \times \dfrac{4\,(81 \times 10^3)}{4\,(81 \times 10^3) + 81 \times 10^3} = -6.5 \times 81 \times 4/5 = \mathbf{-421V/V}$

(as contrasted with -515V/V available with high β_{23}). Now $R_0 = \dfrac{\dfrac{R_{o2}}{\beta_{23} + 1} + r_{e23}}{\beta_{20} + 1} + r_{e20} + R_7$, where (using the data on page 833 of the Text), $R_{o2} = 81\text{k}\Omega$, $\beta_{23} = 50$, $r_{e23} = 139\Omega$, $\beta_{20} = 50$, $r_{e20} = 5\Omega$, $R_7 = 27\Omega$.

$$R_o = \frac{\dfrac{81 \times 10^3}{50 + 1} + 139}{51} + 5 + 27 = \mathbf{65.9\Omega}.$$

10.20 Assume the base current of Q_4 to be i. Thus the collector current in Q_{15} is $180 - i$ and the base current of Q_{15} is $(180 - i)/\beta$. Thus the load current is $(\beta + 1)\,i + 180 - i = \beta i + 180$, while the current in R_6 is $(\beta + 1)\,i - \left(\dfrac{180 - i}{\beta}\right) \approx (\beta + 1)\,i$, for which $V_{BE\,15} = R_6\,(\beta + 1)\,i$, with $I_{C15} = 180 - i$. Thus $27\,(\beta + 1)\,i \times 10^{-6} = 25 \times 10^{-3} \ln\left[\dfrac{(180 - i)\,(10^{-6})}{10^{-14}}\right]$, with i in μA.

For $\beta = 400$, i in μA: $10^{-6} \times 27\,(401)\,i = 25 \times 10^{-3}\,(\ln(180 - i) + \ln 10^8)$, or $10.83 \times 10^{-3}\,i = 25 \times 10^{-3}\,(\ln(180 - i) + 18.42)$, $i = 2.31 \ln(180 - i) + 42.6$.

Try $i = 100$, $i = 2.31 \ln(100) + 42.6 = 52.2\mu\text{A}$;

Try $i = 52$, $i = 2.31 \ln(127.8) + 42.6 = 53.8\mu\text{A}$;

Try $i = 53.8$, $i = 2.31 \ln(126.2) + 42.6 = 53.8\mu\text{A}$, for which $I_L = 400\,(53.8) + 180 = $ **21.7mA.**

Note that for low β, below about 100, the output will be β-limited:

For $\beta = 200$: $10^{-6} \times 27\,(201)\,i = 25 \times 10^{-3}\,(\ln(180 - i) + \ln 10^8)$, $i = 4.60 \ln(180 - i) + 87.74$.

Try $i = 50$, $i = 4.6 \ln(130) + 87.7 = 110$;

Try $i = 110$, $i = 4.6 \ln(70) + 87.7 = 107.2$;

Try $i = 107.2$, $i = 4.6 \ln(72.8) + 87.7 = 107.4$, for which $I_L = 200\,(107.4) + 180 = $ **21.6mA.**

For $\beta = 100$: $10^{-6}\,(27)\,(101)\,i = 25 \times 10^{-3}\,(\ln(180 - i) + \ln 10^8)$, $i = 9.17 \ln(180 - i) + 168.9$

Try $i = 150$, $i = 9.17 \ln(30) + 168.9 = 200$;

Try $i = 175$, $i = 9.17 \ln(5) + 168.9 = 183.6$;

Try $i = 179$, $i = 9.17 \ln(1) + 168.9 = 168.9$;

Try $i = 178$, $i = 9.17 \ln(2) + 168.9 = 175.2$;

Try $i = 177$, $i = 9.17 \ln(3) + 168.9 = 178.9$;

Try $i = 177.5$, $i = 9.17 \ln(2.5) + 168.9 = 177.3$, for which $I_L = 100\,(177.3) + 180 = $ **17.9mA.**

SECTION 10.6: GAIN AND FREQUENCY RESPONSE OF THE 741

10.21

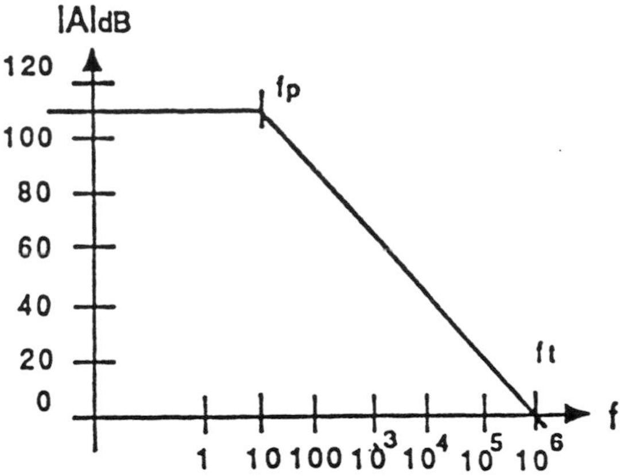

Using the results of the solution of P10.18 above, the overall gain is $A_v = -G_{m1} (R_{o1} \| R_{i2}) \times (-G_{m2} R_{o2}) \mu \dfrac{R_L}{R_L + R_o}$, assuming that R_{i3} is very large, where $G_{m1} = 1/5.26\text{mA/V}$, $R_{o1} = 6.7\text{M}\Omega$, $R_{i2} = 4.0\text{M}\Omega$, $G_{m2} = 10.0\text{mA/V}$, $R_{o2} = 78.8\text{k}\Omega$, $\mu = 1.0$, $R_L = 2\text{k}\Omega$, $R_o = 39 + 27 = 66\Omega$ (from page 833 of the Text). Thus $A_v = + \dfrac{1}{5.26} \times 10^{-3} (6.7 \times 10^6 \| 4 \times 10^6) (10.0 \times 10^{-3} \times 78.8 \times 10^3) \times 1 \times \dfrac{2 \times 10^3}{2 \times 10^3 \times 66} = 362.6 \times 10^3 \equiv$ **111.2dB.** For the pole associated with C_C, at the base of Q_{16}, $C_i = C_C (1 - \text{gain}) = 30 \times 10^{-12} (1 - -G_{m2} R_{o2}) = 30 \times 10^{-12} (1 + 10 \times 10^{-3} \times 78.8 \times 10^3) = 23 \times 10^{-9}\text{F}$, $R_i = R_{o1} \| R_{o2} = 6.7\text{M}\Omega \| 4.0\text{M}\Omega = 2.5 \times 10^6\Omega$. Thus $f_p = \dfrac{1}{2\pi \times 23 \times 10^{-9} \times 2.5 \times 10^6}$ = **2.77Hz**, for which $f_t = 363 \times 10^3 \times 2.77 =$ **1.00MHz.** Note that f_t is the same, since the increased gain is in the Miller compensation stage!

10.22 Using the gain result on page 832 of the Text, see that for customized compensation with capacitor C_C, $C_i = C_C (1 - - \text{gain}) = C_C (1 + 515) = 516 C_C$, where $f_p = \dfrac{1}{2\pi\, C_i\, R_i}$, or $516\, C_C = \dfrac{1}{2\pi\, f_p\, R_i} = \dfrac{1}{2\pi\, f_p\, (2.5 \times 10^6)}$, whence $C_C = 1.23 \times 10^{-10}/f_p$.

(a) 45° *phase margin* occurs when the first pole contributes 90° and the second 45°, where $f = f_2 \tan 45° = f_2$, at the frequency of the second pole, *say at* f_t = 1MHz. *Now for* $A_f = 10^3$, for which $\beta \simeq 10^{-3}$, the first pole should be at $f_p = \dfrac{10^6}{10^6/10^3}$ = 1kHz, for which $C_C = 1.23 \times 10^{-10}/10^3$ = **0.123pF**. The corresponding 3dB frequency is at $10^3 (1 + A\beta) = 10^3 (1 + 243 \times 10^3 \times 10^{-3})$ = **244kHz.** *Now for* $A_f = 10^4$, for which $\beta \simeq 10^{-4}$, the first pole should be at $f_p = \dfrac{10^6}{10^6/10^4} = 10^4$Hz, for which C_C = **0.0123pF**. The corresponding 3dB frequency is at $10^4 (1 + 243 \times 10^3 \times 10^{-4})$ = **253kHz.**

(b) 60° *phase margin* occurs when the first pole contributes 90° and the second 30°, where $f = f_2 \tan 30° = 0.58 f_2$, or $f = 0.58 f_t$ = 580kHz. *Now for* $A_f = 10^3$, for which $\beta \simeq 10^{-3}$, the first pole should be at $f_p = \dfrac{.58 \times 10^6}{10^6/10^3}$ = 0.58kHz, for which $C_C = 1.23 \times 10^{-10}/.58 \times 10^3$ = **0.212pF,** and $f_{3dB} = 0.58 \times 10^3 (244)$ = **142kHz.** *Now for* $A_f = 10^4$, for which $\beta \simeq 10^{-4}$, the first pole should be at $f_p = \dfrac{.58 \times 10^6}{10^6/10^4}$ = 5.8kHz, for which C_C = **0.0212pF,** and f_{3dB} = 5.8kHz (25.3) = **146kHz.**

10.23 $SR = 2I/C_C$, where I = 9.5μA or C_C = 0.123pF and .0123pF, for which $SR = \dfrac{2\,(9.5) \times 10^{-6}}{.123 \times 10^{-12}}$ = **154V/μsec.** and **1540V/μsec.** respectively. Now, from Eq. 2.33 on page 100 of the Text, the full-power bandwidth is $f_M = \dfrac{SR}{2\pi\, V_{O\,max}} = \dfrac{154 \times 10^6}{2\pi \times 10}$ = **2.45MHz,** and **24.5MHz** respectively.

10.24 The output stage is a **Class AB** type. Its standing current is defined by the current in Q_5, being I_{C5} = 500μA, and the fact that Q_6 is four times larger than Q_5, being I_{C6} = 5 × 500μA = **2.5mA.**

For the gain: $I_1 = I_2 = I_3 = I_4$ = 50μA/2 = 25μA, $r_{e1} = r_{e2}$ = 25mV/25μA = 1kΩ, $r_{02} = r_{04}$ = 200V/25 × 10^{-6} = 8MΩ. I_5 = 500μA, r_{e5} = 25 × 10^{-3}/500 × 10^{-6} = 50Ω, $r_{\pi 5}$ = 121 (50) = 6.05kΩ, r_{05} = 200V/500μA = 400kΩ. I_6 = 1.0mA, r_{e6} = 25mV/1.0mA = 25Ω, $r_{\pi 6}$ = 121 (25) = 3.025kΩ, r_{06} = 200V/1.0mA = 200kΩ. For output-stage operation, assume the gain to be controlled primarily through the Q_5, Q_7 connection. Now with I_7 = 1.0mA, r_{e7} = 25Ω, r_{06} = 200kΩ, and with a 10kΩ load, R_L' = 10kΩ ∥ 200kΩ ∥ 200kΩ = 9.90kΩ. At the base of Q_7, R_{in} = 121 (9.09kΩ) = 1.1MΩ, and the load on $Q_5 = r_{05} \parallel R_{in}$ = 400kΩ ∥ 1.1MΩ = 283kΩ. Thus, the gain from the base of Q_5 Q_6 to the output $\approx \dfrac{293 \times 10^3}{50} \times \dfrac{9.09 \times 10^3}{10 + 9.09 \times 10^3}$ = −5860V/V. Now at the base of Q_5, the resistance is R_T = $r_{04} \parallel r_{02} \parallel r_{\pi 5} \parallel r_{\pi 6}$ = 8 × 10^6 ∥ 8 × 10^6 ∥ 6.05kΩ ∥ 3.025kΩ = 2.017kΩ. Thus the gain of stage 1 = $\dfrac{2(2.017 \times 10^3)}{1k\Omega + 1k\Omega}$ = 2.02V/V. Correspondingly, the overall gain = 2.02 × 5860 = **11.8 × 10³V/V.** At the base of Q_5, R_T = 2.017kΩ, $C_T = C(1 - -5860) = 5861C$. Since $f_p = \dfrac{1}{2\pi R_T C_T}$, $C_T = 5861C = \dfrac{1}{2\pi \times 2.017 \times 10^3 \times 1 \times 10^3}$, whence C = **13.5pF.**

SECTION 10.7: CMOS OP AMPS

10.25 For Q_8, Q_5, Q_7, I_D = **25μA,** K = 1/2 (10 × 10^{-6}) (200/10), = 100μA/V², $r_0 = V_A/I_D$ = 25V/25μA = **1MΩ.** 25 = 100 $(v_{GS} - 1)^2$, or $v_{GS} = (25/100)^{1/2} + 1$ = **1.504V,** $g_m = 2K(v_{GS} - V_t)$ = 2 (100) (0.50) = **100μA/V.**

For Q_1, Q_2, I_D = 25/2μA = **12.5μA.** K = 100μA/V², r_o = 25/12.5μA = **2MΩ.** 12.5 = 100 $(v_{GS} - 1)^2$, or $v_{GS} = (12.5/100)^{1/2} + 1$ = **1.354V/V,** g_m = 2 (100) (.354) = **70.8μA/V.**

For Q_3, Q_4, I_D = **12.5μA,** K = 1/2 (20 × 100/10) = 100μA/V², r_o = 25/12.5μA = **2MΩ,** 12.5 = 100 $(v_{GS} - 1)^2$, or v_{GS} = **1.354V,** and g_m = **70.8μA/V.**

For Q_6, I_D = **25μA,** K = 1/2 (20 × 10^{-6}) × 200/10 = **200mA/V²,** r_0 = 25V/25μA = **1MΩ,** 25 = 200 $(v_{GS} - 1)^2$, or $v_{GS} = (25/200)^{1/2} + 1$ = **1.354V,** g_m = 2 (200) (.354) = **141.6μA/V.**

	Q_1	Q_2	Q_3	Q_4	Q_5	Q_6	Q_7	Q_8
I_D (μA)	12.5	12.5	12.5	12.5	25	25	25	25
$\lvert V_{GS}\rvert$(V)	1.35	1.35	1.35	1.35	1.50	1.35	1.50	1.50
g_m (μA/V)	70.8	70.8	70.8	70.8	100	141.6	100	100
r_o (MΩ)	2	2	2	2	1	1	1	1

For Gains: $A_1 = \dfrac{r_{o4} \parallel r_{o2}}{1/g_{m1} + 1/g_{m2}} = \dfrac{2 \times 10^6/2}{2\,(1/70.8 \times 10^{-6})} = \dfrac{10^6\,(10^{-6})\,70.8}{2}$ = **35.4V/V.** *For no load*, $A_2 = g_{m6}\,(r_{o6} \parallel r_{o7})$ = 141.6 × 10^{-6} × 1 × 10^6/2 = **70.8V/V.** Overall, the open-loop gain = $A_1 A_2$ = 35.4 (70.8) = **2506V/V.**

For the Input Common-Mode Range: Input High: V_{G5} = 5 − 1.5 + 1 = 4.5V, V_{GS1} = 1.35. Thus, $V_I \le$ 4.5 − 1.35 = **+3.15V.** Input Low: V_{D3} = −5 + 1.35 = −3.65V, V_{GD1} = −1V, $V_I \ge$ −3.65 − 1 = **−4.65V**

For the Output Common-Mode Range: For triode operation of the output devices, the output range is ±5V. For saturated-mode operation, $V_o \le 5 - 1.5 + 1 =$ **4.5V**, $V_o \ge -5 + 1.35 - 1 =$ **–4.65V**.

10.26 For $I_{REF} = 12\mu A$: Reduce all currents in the Table on page 842 of the Text, by the factor 12/25 = 0.48.

	Q_1	Q_2	Q_3	Q_4	Q_5	Q_6	Q_7	Q_8
I_D (μA)	6	6	6	6	12	12	12	12
$\lvert V_{GS}\rvert$ (V)	1.28	1.28	1.35	1.35	1.4	1.35	1.4	1.4
g_m (μA/V)	42.5	42.5	34.6	34.6	60	69.2	60	60
r_o (MΩ)	4.2	4.2	4.2	4.2	2.1	2.1	2.1	2.1

For Q_8, Q_5, Q_7, $K = 1/2\,(10 \times 10^{-6})\,(150/10) = 75\mu A/V^2$, $I_D = 12\mu A = 75\mu A/V^2\,(\upsilon_{GS} - 1)^2$, $\upsilon_{GS} = (12/75)^{1/2} + 1 =$ **1.4V**, $g_m = 2\,(75)\,(1.4 - 1) =$ **60μA/V**, $r_o = 25/12 =$ **2.08MΩ**.

For Q_1, Q_2, $K = 1/2\,(10 \times 10^{-6})\,(120/8) = 75\mu A/V^2$, $I_D = 12/2 = 6\mu A = 75\,(\upsilon_{GS} - 1)^2$, $\upsilon_{GS} = (6/75)^{1/2} + 1 =$ **1.283V**, $g_m = 2\,(75)\,(.283) =$ **42.5μA/V**, $r_o = 25/6 =$ **4.17MΩ**.

For Q_3, Q_4, $K = 1/2\,(20 \times 10^{-6})\,(50/10) = 50\mu A/V^2$, $I_D = 6\mu A = 50\,(\upsilon_{GS} - 1)^2$, $\upsilon_{GS} = (6/50)^{1/2} + 1 =$ **1.346V**, $g_m = 2\,(50)\,(\upsilon_{GS} - 1) = 2\,(50)\,(.346) =$ **34.6μA/V**. $r_o = 25/6 =$ **4.17MΩ**.

For Q_6, $K = 1/2\,(20 \times 10^{-6})\,(100/10) = 100\mu A/V^2$, $I_D = 12\mu A = 100\,(\upsilon_{GS} - 1)^2$, $\upsilon_{GS} = (12/100)^{1/2} + 1 =$ **1.346V**, $g_m = 2\,(100)\,(.346) =$ **69.2μA/V**, $r_o = 25/12 =$ **2.08MΩ**.

Now $A_1 = -g_{m1}\,(r_{02} \parallel r_{04}) = -42.5 \times 10^{-6}\,(4.2 \parallel 4.2) \times 10^6 =$ **–89.25V/V**, and $A_2 = -g_{m6}\,(r_{06} \parallel r_{07} = -69.2 \times 10^{-6}\,(2.1/2) \times 10^6 =$ **–72.7V/V**. Thus $A_0 = A_1 A_2 = (89.25)\,(72.7) =$ **6488V/V**. Also $\upsilon_{I\,CM\,max} = V_{DD} - \lvert V_{GS5}\rvert + \lvert V_t\rvert - \lvert V_{GS1}\rvert = 5 - 1.4 + 1 - 1.28 =$ **3.32V**, $\upsilon_{I\,CM\,min} = -V_{SS} + \lvert V_{GS3}\rvert - \lvert V_{t1}\rvert = -5 + 1.35 - 1 =$ **–4.65V**, $\upsilon_{O\,max} = V_{DD} - \lvert \upsilon_{GS7}\rvert + \lvert V_t\rvert = 5 - 1.4 + 1 =$ **4.60V**, $\upsilon_{O\,min} = -V_{SS} + \lvert \upsilon_{GS6}\rvert - \lvert V_t\rvert = -5 + 1.35 - 1 =$ **–4.65V**.

10.27 For Q_6, $(W/L)_6 = 50/10$, $K = 1/2\,(20)\,50/10 = 50\mu A/V^2$, $I_D = 25 = 50\,(\upsilon_{GS} - 1)^2$, $\upsilon_{GS} = (1/2)^2 + 1 =$ **1.707V**, $g_m = 2\,(50)\,(.707) = 70.7\mu A/V$, $r_o = 25/25 = 1M\Omega$. $A_2 = -70.7\,(1/2) =$ **–35.4V/V**, $A_1 = -62.5V/V$ (from Example 10.2), $A_0 = A_1 A_2 = -62.5\,(-35.4) =$ **2212V/V**. Now for $I_6 = I_7 = 25\mu A$, $\upsilon_{GS6} = 1.707V$. But $\upsilon_{GS4} = 1.50V$. Thus, the input offset $= \dfrac{\upsilon_{GS6} - \upsilon_{GS4}}{A_1} = \dfrac{1.707 - 1.5}{62.5} =$ **3.3mV**.

10.28 From the solution of P10.26 above, and the development following Eq. 10.50 in the Text, $R_1 = r_{o2} \parallel r_{o4} = 4.2/2 = 2.1M\Omega$, $R_2 = r_{o7} \parallel r_{o6} = 2.1/2 = 1.05M\Omega$, $G_{m1} = g_{m1} = 42.5\mu A/V$, $G_{m2} = g_{m2} = 69.2\mu A/V$, $f_t = \dfrac{1}{2\pi} \times \dfrac{G_{m1}}{C_C}$, $C_C = \dfrac{42.5 \times 10^{-6}}{2\pi \times 10^6} =$ **6.76pF**. For a zero at ∞, $R = 1/G_{m2} = 1/g_{m2} = 1/69.2 \times 10^{-6} =$ **14.5kΩ**. For C_2, the 10pF output capacitance, and $C_2 >> C_1$, the second pole is at $f_2 \approx \dfrac{G_{m2}}{2\pi\,C_2} = \dfrac{69.2 \times 10^{-6}}{2\pi \times 10 \times 10^{-12}} =$ **1.10MHz**. Excess phase at 1MHz is $\tan^{-1}(1/1.1) =$ **42.3°**. For 6° excess phase at f_t', $\tan^{-1} \dfrac{f_t'}{1.1} = 6°$, or $f_t' = 1.1 \times .1051 =$ **0.116MHz**, for which $C_C = \dfrac{G_{m1}}{2\pi\,f_t'} = \dfrac{42.5 \times 10^{-6}}{2\pi \times .116 \times 10^6} =$ **58.3pF** is required. *Slew rate*, $SR = 2I/C_C$ where $2I = 12\mu A$. For the case, of 42.3° excess phase, $SR = \dfrac{12 \times 10^{-6}}{6.76 \times 10^{-12}} =$ **1.78V/μs**. For 6° excess phase, $SR = \dfrac{12 \times 10^{-6}}{58.3 \times 10^{-12}} =$ **0.206V/μs**.

10.29 For $I_B = 5\ \mu A$, and using the rearranged form of Eq. 10.58 on page 849 of the Text:
$R_B = [2/(2 \times 20 \times 10^{-6} \times 2m \times 5 \times 10^{-6})^{1/2}] \times [(2m/2)^{1/2} - 1] = (1/m^{1/2})(1 \times 10^5)(m^{1/2} - 1) = 1 \times 10^5 \times (1 - 1/m^{1/2})$.
Now $g_{m12} = (2\mu_n C_{ox}(W/L)_{12} I_B)^{1/2}$, $= (2 \times 20 \times 10^{-6} \times 2m \times 5 \times 10^{-6})^{1/2} = 20 \times 10^{-6} m^{1/2}$,
and with the currents in Q_{12} and Q_9 equal, with $\mu_n = 2.5\ \mu_p$,
$g_{m9} = g_{m12}[(\mu_p/\mu_n)(W/L)_{12}]^{1/2} = 20 \times 10^{-6} m^{1/2}(2.5m)^{1/2} = 12.65 \times 10^{-6} A/V$.

(a) *For* $m = 2$, $R_B = 1 \times 10^5(1 - (1/2)^{1/2}) =$ **29.3 kΩ**,
and $g_{m12} = 20 \times 10^{-6}\sqrt{2} =$ **28.3μA/V**, and $g_{m9} = 12.7 \times 10^{-6} =$ **12.7μA/V**.

(b) *For* $m = 5$, $R_B = 1 \times 10^5(1 - (1/5)^{1/2}) =$ **55.3 kΩ**, and $g_{m12} = 20 \times 10^{-6}\sqrt{5} =$ **44.7μA/V**, and $g_{m9} =$ **12.7μA/V**.

For the loop gain: {Note that the cascode transistors Q_{10} and Q_{11} provide unity current gain.} Assume a fixed bias current I_B and inject voltage υ at the gate of Q_9 and measure the return as υ_{gs8}.
Thus $i_9 = g_{m9}\upsilon$, and $\upsilon_{gs13} = i_9(1/g_{m13}) = i_9/g_{m13}$, $i_{12} = \upsilon_{gs13}/(1/g_{m13} + R_B)$, and
$\upsilon_{gs8} = i_{12}/g_{m8} = (1/g_{m8})(1/(1/g_{m12} + R_B))(1/g_{m13})(g_{m9}\upsilon)$.
Thus the loop gain is $L = \upsilon_{GS8}/\upsilon = (g_{m9}/g_{m8})/(g_{m13}/g_{m12} + g_{m13}R_B)$.
Now, $g_{m8} = g_{m9}$, $g_{m13} = g_{m12}(1/m)^{1/2}$, and $R_B = (2/g_{m12})(\sqrt{m} - 1)$.
Thus $g_{m13}R_B = 2(1 - 1/m^{1/2})$, and $L = 1/(1/(1/m^{1/2} + 2 - 2/m^{1/2}) = 2 - 1/m^{1/2}$.
Overall, $\mathbf{L = 2 - 1/m^{1/2}}$.
Now for $m = 2$, $L =$ **1.29 V/V**, and for $m = 5$, $L =$ **1.55 V/V**.
Note that the loop gain is less than 1 for small m where $2 - 1/m^{1/2} = 1$, $1/m^{1/2} = 1$, or $m = 1$.
See that as m gets larger, R_B gets larger, a marginal disadvantage, but that as m approaches 1, R_B approaches 0, implying lack of control and sensitivity to minor variations in device parameters. Clearly m must be large enough (for example) to exceed the uncertainty in mirror gain due to the effect, for example, of V_A.

SECTION 10.8: ALTERNATIVE CONFIGURATIONS FOR CMOS AND BICMOS OP AMPS

10.30 For the Wilson mirror (Fig. 10.26) in the Text, $V_{BIAS2} - |V_t|_{1C} + V_{GS3C} + V_{GS4} = -V_{SS}$, or $V_{BIAS2} = -V_{SS} + V_t - 2V_{GS}$. Thus the minimum voltage between V_{BIAS2} and V_{SS} is $\mathbf{2V_{GS} - V_t}$. Now from Eq. 10.62, $R_o = g_{m4C}\, r_{o4C}\, r_{o3} = \mathbf{g_m r_o^2} = k'(W/L)(V_{GS} - V_t)\, r_o^2$. Now for the cascode mirror (Fig. 6.32b)), the bias situation is essentially the same, with $-V_t + V_{GS3C} + V_{GS3}$ between V_{BIAS2} and $-V_{SS}$, ie $\mathbf{2V_{GS} - V_t}$. As well, R_o is the same: $R_o = g_{m4C}\, r_{o4C}\, r_{o4} = \mathbf{g_m r_o^2}$, as before. Thus as measured by the output resistance and output-voltage overhead, the cascode and (modified) Wilson are the same. However, if Q_{3C} were eliminated, V_{BIAS2} could be reduced to $2V_{GS} - 2V_t$ above V_{SS}, but the input offset voltage would be effected, as well.

10.31 For $2I = 10\mu A$, $I_D = 5\mu A$, $K_n = 1/2\mu_n C_{ox}(W/L)_n = 1/2 \times 20 \times 60/8 = 75 = K$, $K_p = 1/2 \times 10 \times 120/8 = 75 = K$, $r_o = 25/5\mu A = 5.0M\Omega$, $5 \times 10^{-6} = 75 \times 10^{-6}(\upsilon_{GS} - 1)^2$, $\upsilon_{GS} = 1 + (5/75)^{1/2} = 1.258V$, $g_m = 2K(\upsilon_{GS} - V_t) = 2(75)(.258) = 38.7\mu A/V$. Now $R_o = 1/2\,(g_m\, r_o^2) = 1/2\,(38.7 \times 10^{-6} \times 5 \times 10^6 \times 5 \times 10^6) =$ **484MΩ**, and $A_1 = g_m R_o = 38.7 \times 10^{-6} \times 484 \times 10^6 =$ $\mathbf{18.7 \times 10^3}$**V/V**.

10.32 For the double cascode, $R_{o4CC} \approx (g_{m4CC}\, r_{o4CC})\, R_{o4C} \approx \mathbf{g_{m4CC}\, r_{o4CC}\, g_{m4C}\, r_{o4C}\, r_{o3}}$, $R_{o2C} \approx (\mathbf{g_{m2C}\, r_{o2C}})\, \mathbf{r_{o2}}$. Now $R_o = R_{o2C} \parallel R_{o4CC}$. For conditions as in Ex. 10.28, $2I = 25\mu A$, $I = 12.5\mu A$, $K_n = K_p = 1/2\,(20)(60/8) = 75\mu A/V^2$, $12.5 = 75(\upsilon_{GS} - 1)^2$, $\upsilon_{GS} = (12.5/75)^{1/2} + 1 = 1.408V$, $g_m = 2(75 \times 10^{-6}).408 = 61.2\mu A/V$, $r_o = 25/12.5 \times 10^{-6} = 2M\Omega$, $R_{o4CC} = g_m^2\, r_o^3 = (61.2 \times 10^{-6})^2 \times (2 \times 10^6)^3 = 29.96G\Omega$, $R_{o2C} = g_m\, r_o^2 = 61.2 \times 10^6 \times (2 \times 10^6)^2 =$ **245MΩ**, $R_o = (0.245 \parallel 30)G\Omega =$ **243MΩ**, $G_{m1} = g_m = 61.2 \times 10^{-6} A/V$. Thus $A_1 = 61.2 \times 10^{-6} \times 243 \times 10^6 =$ $\mathbf{14.9 \times 10^3}$**V/V**. Total voltage from input to supply is $-1 + 1.41 + 1.41 + 1.41 =$ **3.23V**.

10.33 Here, $2I = I_B = 10\mu A$, $|V_t| = 1V$, and $K = 1/2k'(W/L)$ in general.

For Q_6, Q_7: $I_6 = 10\mu A$, $K_6 = 1/2 \times 20 \times 8/8 = 10\mu A/V^2$, $10 = 10\ (\upsilon_{GS} - 1)^2$, or $\upsilon_{GS} = 2V$. Thus $V_{BIAS3} = 2 + -5 =$ **−3V**.

For Q_{1C}, Q_{2C}: $I_{1C} = 10 - 10/2 = 5\mu A$, $K_{1C} = 1/2 \times 20 \times 60/8 = 75\mu A/V$, $5 = 75\ (\upsilon_{GS} - 1)^2$, $\upsilon_{GS} = (5/75)^{1/2} + 1 = 1.258V$, also $g_m = 2\ (75)\ (.258) =$ **38.7μA/V**. Thus, $V_{BIAS2} = -3 - 1 + 1.258V =$ **−2.74V**. Use **−2.75V**.

For Q_5: $I_5 = 10\mu A$, $K_5 = 1/2 \times 20 \times 150/10 = 150\mu A/V^2$, $10 = 150\ (\upsilon_{GS} - 1)^2$, $\upsilon_{GS} = (1/15)^{1/2} + 1 = 1.258V$, $V_{BIAS1} = +5 - 1.26 =$ **3.74V**. Use **3.75V**.

For Q_1, Q_2: $I_1 = 5\mu A$, $K_1 = 1/2 \times 10 \times 120/8 = 75\mu A/V^2$, $5 = 75\ (\upsilon_{GS} - 1)^2$, $\upsilon_{GS} = (5/75)^{1/2} + 1 = 1.258V$, $g_{m1} = 2\ (75)\ (\upsilon_{GS} - 1) = 2\ (75)\ (.258) =$ **38.7μA/V**.

For Q_{4C}: $I_{4C} = 5\mu A$, $K_{4C} = 1/2 \times 10 \times 120/8 = 75\mu A/V^2$, $5 = 75\ (\upsilon_{GS} - 1)^2$, $\upsilon_{GS} = 1.258$, $g_m = 2\ (75)\ .258 =$ **38.7μA/V**.

Output resistance: $r_{o1} = r_{o2} = 25/5\mu A = 5M\Omega$, $r_{o6} = r_{o7} = 25/10\mu A = 2.5M\Omega$, $r_{o1C} = r_{o2C} = r_{o4C} = r_{o3} = 25/5 = 5M\Omega$. Thus, $R_{o4C} = g_{m4C}\ r_{o4C}\ r_{o3} = 38.7 \times 10^{-6} \times 5 \times 10^6 \times 5 \times 10^6 =$ **967MΩ**, and $R_{o2C} = g_{m2C}\ r_{o2C}\ r_{o7} \,\|\, r_{o2} = 38.7 \times 10^{-6} \times 5 \times 10^6 \times (2.5 \times 10^6 \,\|\, 5 \times 10^6) =$ **322MΩ**. Correspondingly, $R_o = 967 \,\|\, 322 =$ **242MΩ**. Gain $A_0 = g_{m1}\ R_o = 38.7 \times 10^{-6} \times 242 \times 10^6 =$ **9.36 × 10³V/V**.

10.34 For all (Q_1, Q_2, Q_{1C}, Q_{2C}, Q_{3C}, Q_{4C}), I_D, K, V_A, and $r_0 = 25/10 = 2.5M\Omega$ are the same. From Eq. 10.69, $f_t = \dfrac{g_{m1}}{2\pi\ C_L}$, or $g_{m1} = 2\pi\ C_L\ f_t = 2\pi \times 10 \times 10^{-12} \times 10^6 =$ **62.8μA/V**. Now $I_{D7} = 10 + 10 = 20\mu A$, and $r_{o7} = 25/20 = 1.25M\Omega$. Thus $R_{o4C} = g_m\ r_o^2 = 62.8 \times 10^{-6} \times (2.5 \times 10^6)^2 = 393M\Omega$, $R_{o3C} = 62.8 \times 10^{-6} \times 2.5 \times 10^6 \times (1.25 \times 10^6 \,\|\, 2.5 \times 10^6) = 131M\Omega$. Correspondingly, $R_0 = 131M\Omega \,\|\, 393M\Omega = 393/4 =$ **98.25MΩ**. Also, $A_0 = g_{m1}\ R_o = 62.8 \times 10^{-6} \times 98.25 \times 10^6 =$ **6170V/V**. The dominant-pole frequency, $f_D = \dfrac{1}{2\pi\ C_L\ R_o} = \dfrac{1}{2\pi \times 10 \times 10^{-12} \times 98.25 \times 10^6} =$ **162Hz** Now $SR = \dfrac{2I}{C_L} = \dfrac{2 \times 10 \times 10^{-6}}{10 \times 10^{-12}} =$ **2V/μs**.

10.35 Here, $I_B = 2I = 800\mu A$, $I_{D1} = I_{D2} = 400\mu A$. Assume $\mu_n\ C_{ox} = 2\mu_p\ C_{ox} = 20\mu A/V^2$, $V_t| = 1V$, $V_A = 25V$ and $K = 1/2\mu m C_{ox}(W/L)$. Thus $K_1 = K_2 = 1/2 \times 10 \times 600/10 = 300\mu A/V^2$, $400 = 300\ (\upsilon_{GS} - 1)^2$, $\upsilon_{GS} = (4/3)^{1/2} + 1 = 2.155V$, $g_{m1} = 2\ (300)\ (2.155 - 1) = 693\mu A/V$, and the output pole is at $f_t = \dfrac{g_{m1}}{2\pi\ C_L} = \dfrac{693 \times 10^{-6}}{2\pi \times 2 \times 10^{-12}} =$ **55.15MHz**. For the parasitic pole located at the folding node, $I_{E1C} \simeq 400\mu A$ and $g_{m1C} = \dfrac{400 \times 10^{-6}}{25 \times 10^{-3}} = 16mA/V$. Now at the emitter of Q_{1C}, the total capacitance is $C_{\pi 1C} + C_{\mu 6} = C_\pi + C_\mu$, with the corresponding pole at $f_{t1C} = \dfrac{g_{m1C}}{2\pi\ (C_\pi + C_\mu)}$, and $C_\pi + C_\mu = \dfrac{16 \times 10^3}{2\pi\ f_{t1C}}$. For this parasitic pole to be $10 \times$ higher than the output pole $f_{t1C} = 10 \times 55.2 =$ **550MHz**, a BJT unity-gain frequency which is relatively easy to achieve.

SECTION 10.9: DATA CONVERTERS – AN INTRODUCTION

10.36 For a 100kHz sampling frequency, the highest frequency signal component that can be sampled "adequately", as noted by Shannon, is at $f = 100kHz/2 =$ **50kHz**. This means that for a square wave at 50kHz, the fundamental would be adequately represented, but the waveshape-specific harmonics would not. Note that for sampling at frequency $f_s = f$, that for an input signal at f, output is at f, for input at $0.5f$, output is at $f/2$, at $2f/2$, output is dc, and at $1.1f/2$, output is at $1.1f/2$ for 9 or so cycles with a break and corresponding phase reversal occurring at a rate of $(1.1 - 1.0)/2 = 0.1f/2$. The figure illustrates input and output waves at various frequencies f_i with sampling at $f_s = 100kHz$. For sampling in a 10ns interval, with source resistance R and capacitor C, $1 - e^{-t/RC} = 0.99$, or $e^{-t/RC} = 0.01$, $-t/RC = -4.6$, $RC = \dfrac{10 \times 10^{-9}}{4.6}$, $R = \dfrac{10 \times 10^{-9}}{4.6 \times 100 \times 10^{-12}} =$ **21.7Ω**. Thus the switch resistance should not exceed 21.7Ω.

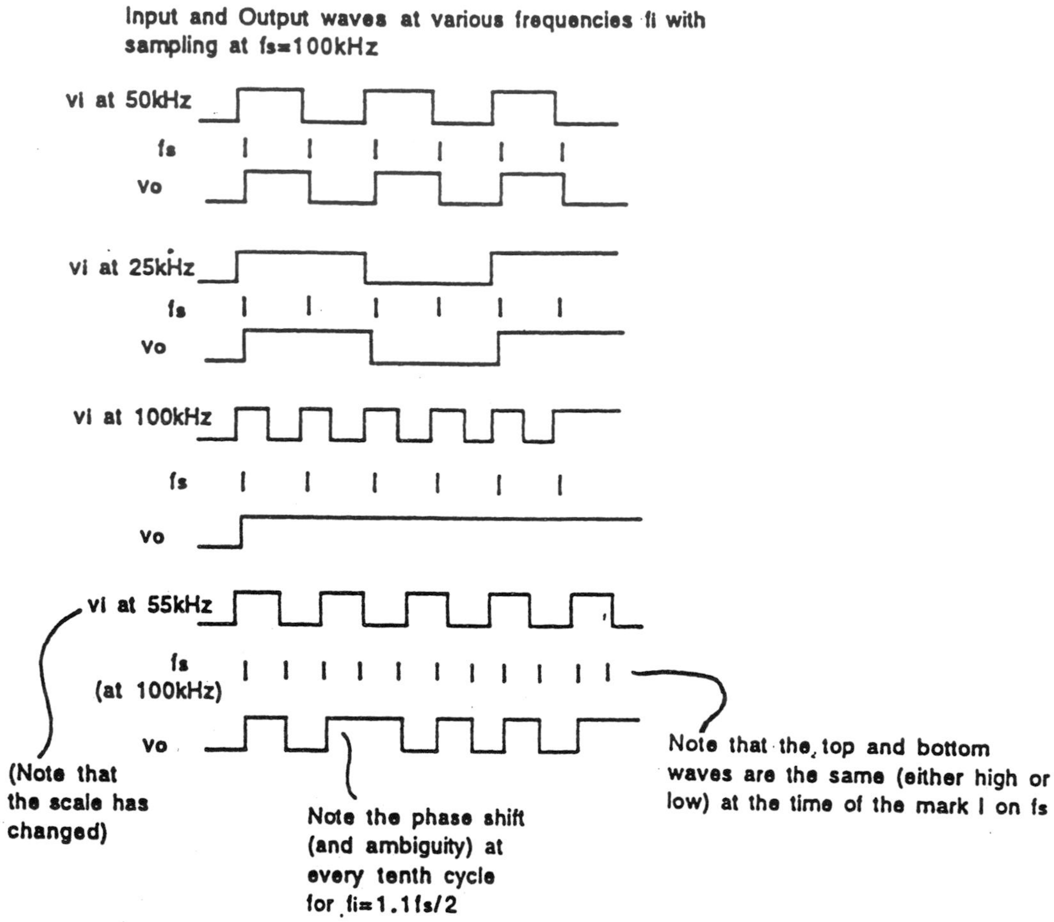

10.37 See that the required resolution is 0.1V in 2(5) = 10V or 1 in 100. Thus $2^n > 100$. Now, $2^7 = 128 > 100$. Thus need **7 bits**. For a 10-bit converter, $2^{10} = 1024$, and the resolution would be 10V/1024 = **9.77mV.**

SECTION 10.10: D/A CONVERTER CIRCUITS

10.38 Notice that there are two interpretations for this question depending on whether $R_f = R/2$ is included in the specification. If not, for R = 1kΩ and n = 8, $2^{n-1}R = 2^{8-1}R = 2^7 R$ = **128kΩ.** *But* if $R/2$ is 1kΩ, the largest resistor required is 2 (128) = **256kΩ.** The LSB current is $V/(2^7 R)$ = 7.81VμA. Correspondingly, the allowed error is 1/2 (7.81V) = 3.91V μA (or 1.95V μA for the second view of R). Here, the MSB current is V/R nominally, or $\dfrac{V}{R + \Delta R}$, for a switch resistance ΔR. Now $\dfrac{V}{R} - \dfrac{V}{R + \Delta R} < \dfrac{V}{2\,(2^7)\,R}$, or $\dfrac{1}{R} - \dfrac{1}{R + \Delta R} \le \dfrac{1}{2^8 R}$, $1 - \dfrac{R}{R + \Delta R} = \dfrac{1}{2^8}$, $\dfrac{R}{R + \Delta R} = 1 - 2^{-8}$, $R = R + \Delta R - 2^{-8} R - 2^{-8}\,\Delta R$ $\Delta R = 2^{-8}R + 2^{-8}\Delta R$. Thus, $\dfrac{\Delta R}{R} = \dfrac{2^{-8}}{1 - 2^{-8}} \simeq 2^{-8}$. Now, for an MSB resistor of 1kΩ, $\Delta R < 2^{-8}\, 10^3$ = **3.91Ω**, or for an MSB resistor of 2kΩ, $\Delta R <$ **7.81Ω.** Now for both resistor error and switch resistance each contributing a half, switch resistances less than 3.91/2 = **1.95 Ω**, (or 7.81/2 =

3.91 Ω) are acceptable. Also, for a perfect R_f (say it is trimmed to the correct value), resistor tolerance allowed is $\left(\frac{100}{2^8}\right)/2 = 0.39\%$. For R_f also variable, allowed resistor-tolerance = 0.39/2 ≈ **0.2%**.

10.39 The resistance of an $R - 2R$ ladder as seen from the supply is $2R \parallel 2R = R$. For a 10V reference, and 1mA, R = 10V/1mA = **10kΩ**. See that the current in the LSB switch is $\frac{1}{2^{n-1}} = \frac{1}{128}$ that in the MSB switch. Thus $\frac{1}{2R} - \frac{1}{2R + \Delta R} = \frac{1}{2R}\left(\frac{1}{128}\right)/2$, $1 - \frac{1}{1 + \Delta R/2R} = \frac{1}{256}$, $\frac{1}{1 + \Delta R/2R} = 1 - \frac{1}{256} = \frac{255}{256}$, or $1 + \Delta R/2R = \frac{256}{255} = 1 + \frac{1}{255}$, that is $\Delta R/2R = \frac{1}{255}$, $\Delta R = \frac{2R}{255} = \frac{2\,(10k\Omega)}{255}$ = **78.4Ω**. Now if $2R$ is reduced by 78.4Ω to compensate, doubling of the nominal switch resistance of **78Ω** to 156Ω would again produce an 1/2 LSB error.

10.40 For device junction area 1% in error, in both Q_{ref} and Q_1, the output current error may be as much as 2% I_{ref}. For n bits, the LSB current = $\frac{I_{ref}}{2^{n-1}}$. Now, $\frac{2}{100} = \frac{1}{2^{n-1}} \times \frac{1}{2}$, when $2^n = \frac{100}{2} = 50$, for which **n < 5 bits**. If the absolute value of the output current is not critical, 6 bits is available.

SECTION 10.11: A/D CONVERTER CIRCUITS

10.41 The requirement is for ±1V signals ≡ 2V range, with 2 bits. Now, the range is divided into 2^2 = 4 parts, each of 1/2 volt. Thus, use **3 comparators**, with references at **–1/2V, 0V, +1/2V**, as shown, first in a parallel connection, or alternatively, as 2 in cascade. Output codes for the two circuits are shown in the Table.

V_I	B_1	B_0	H_2	H_1	H_0
+0.75	1	1	1	1	1
+0.25	1	0	0	1	1
–0.25	0	1	0	0	1
–0.75	0	0	0	0	0

a)

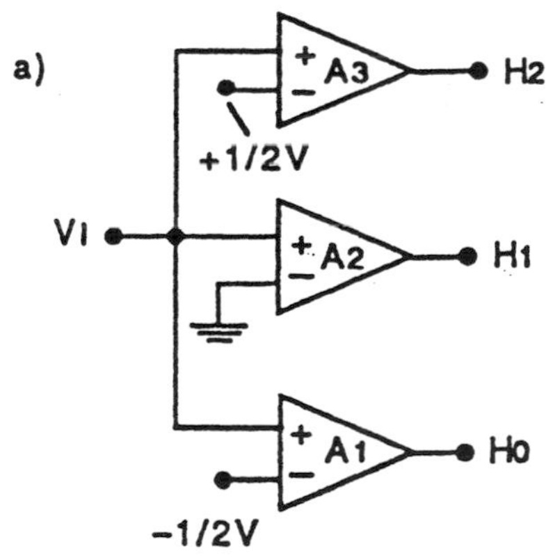

b)

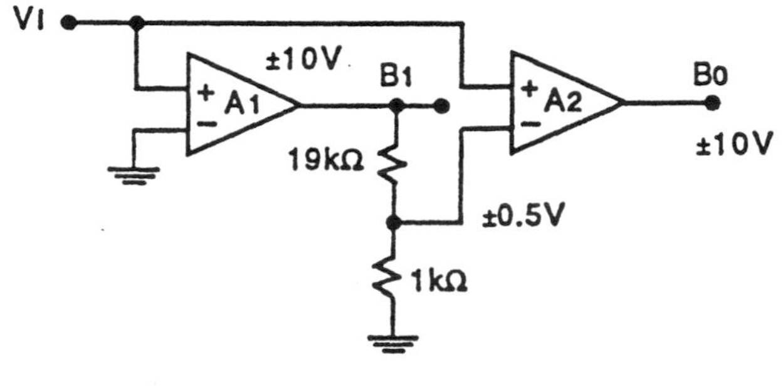

10.42 (a) During Φ_A, $V_X = V_A$, $V_O = V_Y = 0$V.

(b) Following Φ_A, before Φ_B, $V_X = V_A$, $V_Y = 0$, $V_O = 0$V, held by V_Y (and by stray feedback capacitance across the switch connecting output to input).

(c) During Φ_B, $V_X = V_{REF}$, $V_Y = 0 + \dfrac{V_A - V_{REF}}{2}$, V_O saturates at ±10V with the sign reversed from that at node Y.

(d) Following Φ_B, $V_X = V_{REF}$, $V_Y = \dfrac{V_A - V_{REF}}{2}$, V_O stays saturated.

Specifically, for

i) $\mathbf{V_A > V_{REF}}$, (a) $\mathbf{V_O = 0}$, (b) $\mathbf{V_O = 0}$, (c) $\mathbf{V_O = -10V}$, (d) $\mathbf{V_O = -10V}$.

ii) $\mathbf{V_A < V_{REF}}$, (a) $\mathbf{V_O = 0}$, (b) $\mathbf{V_O = 0}$, (c) $\mathbf{V_O = +10V}$, (d) $\mathbf{V_O = +10V}$.

Thus the circuit operates as a comparator of V_A against V_{REF}.

NOTES

Chapter 11

FILTERS AND TUNED AMPLIFIERS

SECTION 11.1: FILTER TRANSMISSION, TYPES, AND SPECIFICATION

11.1 $T(s) = \frac{s}{s + \omega_o}$, $T(j\omega) = \frac{j\omega}{j\omega + \omega_o}$, $|T| = |T(j\omega)| = \frac{\omega}{(\omega^2 + \omega_o^2)^{1/2}}$, $\Phi(\omega) = 90 - \tan^{-1}\omega/\omega_o$, $G = 20 \log |T|$ dB, $A = -20 \log |T|$ dB.

For $\omega = \infty$, $|T|$ **= 1**, $\Phi = 90 - 90 =$ **0°**, $G = 20\log(1) =$ **0dB**, $A =$ **0dB.**

For $\omega = 2\omega_o$, $|T| = \frac{2\omega_o}{((2\omega_o)^2 + \omega_o^2)^{1/2}} = \frac{2}{(5)^{1/2}} =$ **0.894**, $\Phi = 90 - \tan^{-1}\frac{2\omega_o}{\omega_o} =$ **26.6°**, $G = 20 \log(.894) =$ **−0.969dB**, $A = -G =$ **0.969dB.**

For $\omega = \omega_o$, $|T| = \frac{\omega_o}{(\omega_o^2 + \omega_o^2)^{1/2}} = \frac{1}{\sqrt{2}} =$ **0.707**, $\Phi = 90 - \tan^{-1}\frac{\omega_o}{\omega_o} =$ **45°**, $G = -3$dB, $A =$ 3dB.

ω rad/s	$\|T\|$ V/V	Φ °	G dB	A dB
∞	1	0	0	0
$2\omega_o$	0.894	26.6	−0.969	+0.969
ω_o	0.707	45	−3	+3
$\omega_o/2$	0.447	63.4	−6.99	6.99
$\omega_o/5$	0.196	78.7	−14.2	14.2
$\omega_o/10$	0.0995	84.3	−20	20
$\omega_o/100$	0.010	89.4	−40	40
$\omega_o/1000$	0.001	89.94	−60	60

11.2

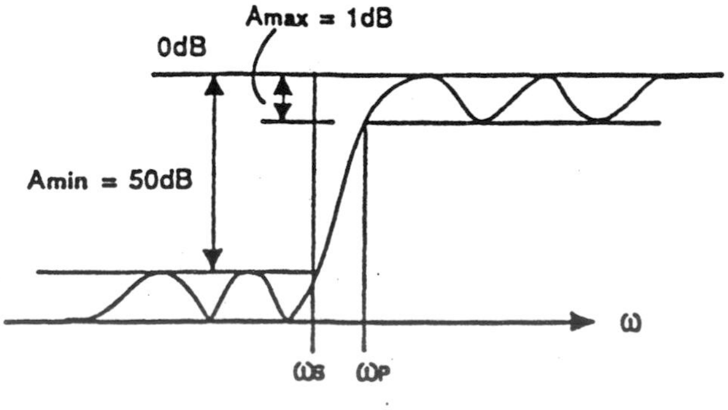

For $\omega = \infty$, $|T|$ = **0 dB,**

$\omega = \omega_p$, $|T|$ = **−1 dB,**

$\omega = \omega_s$, $|T|$ = **−50 dB.**

11.3 Here, ±5% transmission variation ≡ 0.95 ± 0.05. Thus A_{max} = |20log(0.9)| = **0.915 dB.** Now, A_{min} = |20log0.05| = **26 dB**. The selectivity factor (high pass) = $\frac{\omega_p}{\omega_s} = \frac{f_p}{f_s}$ = 4/3.2 = **1.25.**

11.4 See $T(s) = \frac{s}{s + 1/\tau} = \frac{s}{s + 10^3}$, for a high-pass filter with ω_o = 10^3rad/s. Now, $|T(\omega)| = \frac{\omega}{(\omega^2 + (10^3)^2)^{1/2}}$. For A_{max} = 0.5 dB, $|T|$ = $10^{-0.5/20}$ = 0.944. Thus, $\frac{\omega}{(\omega^2 + (10^3)^2)^{1/2}}$ = 0.944, ω^2 = 0.891 (ω^2 + 10^6), ω^2 = 8.195 × 10^6, or ω_p = 2.86 × 10^3 **rad/s.** For A_{min} = 20 dB, $|T|$ = 0.1. Thus $\frac{\omega}{(\omega^2 + (10^3)^2)^{1/2}}$ = 0.1, 100 ω^2 = ω^2 + 10^6, 99ω^2 = 10^6, ω^2 = 10.1 × 10^3, ω_s = **100.5 rad/s.** The selectivity factor (high-pass) = $\frac{\omega_p}{\omega_s} = \frac{2.86 \times 10^3}{100.5}$ = **28.5.**

Now, for a modified filter for which f_p = 10^3 Hz, ω_p = 2 π f_p = 6.283 × 10^3 rad/s. Formerly, for the same shape and ω_o = 10^3 rad/s, ω_p = 2.86 × 10^3 rad/s. Thus the revised ω_o is $\omega_o = \frac{6.283 \times 10^3}{2.86 \times 10^3} \times 10^3$ = 2.197 × 10^3 rad/s, for which τ = 1/(2.197 × 10^3) = **0.455 × 10^{-3}s**, and $f_{3dB} = f_o = \frac{\omega_o}{2\pi} = \frac{2.197 \times 10^3}{2\pi}$ = **0.35 kHz.** At 100Hz, $|T(f)| = \frac{f}{(f_o^2 + f^2)^{1/2}} = \frac{100}{(350^2 + 100^2)^{1/2}}$ = .275 ≡ 20 $_{10}$ .275 = −11.2 dB. Thus, A_{100} = **11.2 dB.**

SECTION 11.2: THE FILTER TRANSFER FUNCTION

11.5 $$T(s) = \frac{s^2 (s - -0.1)}{(s - -1)(s - (-0.5 - j0.8))(s - (-0.5 + j0.8))} = \frac{s^2 (s + 0.1)}{(s + 1)(s + 0.5 + j0.8)(s + 0.5 - j0.8)} =$$
$$\frac{s^2 (s + 0.1)}{(s + 1)(s^2 + 0.5s - j0.8s + 0.5s + 0.25 - j0.4 + j0.8s + 0.4j + 0.64)} = \frac{s^2 (s + 0.1)}{(s + 1)(s^2 + s + 0.89)},$$
or $T(s) = \frac{s^3 + 0.1s^2}{s^3 + 2s^2 + 1.89s + 0.89}$. Note that $|T|$ = 1 as $s \to \infty$.

11.6 Following the preamble to Equation 11.9:

$$T(s) = \frac{a_9\, s\, (s^2 + 1 \times 10^6)(s^2 + 4 \times 10^6)(s^2 + 36 \times 10^6)(s^2 + 144 \times 10^6)}{s^{10} + b_9 s^9 + b_8 s^8 + - - - + b_0},$$

where the filter order is N = 10. From Fig. 11.4:

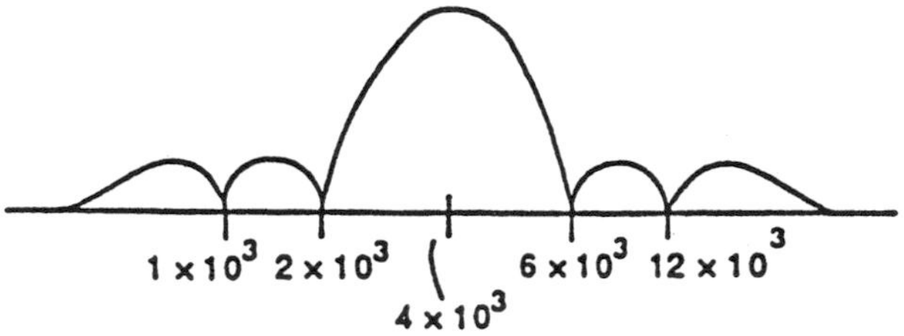

SECTION 11.3: BUTTERWORTH AND CHEBYSHEV FILTERS

11.7 For a Butterworth filter of order N, $\varepsilon = (10A_{max}/10 - 1)^{1/2} = (10^{0.5/10} - 1)^{1/2} = \mathbf{0.349}$. Now, at the edge of the stopband $A(\omega_s) = -20 \log (1 + \varepsilon^2 (\omega_s/\omega_p)^{2N})^{-1/2} = 10 \log (1 + \varepsilon^2 (\omega_s/\omega_p)^{2N})$, or $40 = 10 \log (1 + .349^2 (1.6)^{2N})$, $1 + .1218 (1.6)^{2N} = 10^4$, $(1.6)^{2N} = 8.21 \times 10^4$. Try $N = 10$, or $2N = 20$, whence $(1.6)^{20} = 1.209 \times 10^4$ rather than 8.21×10^4. Try $N = 12$, $2N = 24$, $(1.6)^{24} = 7.92 \times 10^4$. Try $N = 13$, $2N = 26$, $(1.6)^{26} = 20.28 \times 10^4$. That is, 13th order will clearly do the job. For **N = 13**, $A(\omega_s) = -20 \log (1 + \varepsilon^2 (\omega_s/\omega_p)^{2N})^{-1/2} = -20 \log (1 + .349^2 (1.6)^{26})^{-1/2} = -20 \log (1 + .1218 \times 20.28 \times 10^4)^{-1/2} = \mathbf{43.93\ dB}$.

Now, for $A_{min} = 40$ dB exactly, with $N = 13$, $A_{min} = 10 \log (1 + \varepsilon^2 (\omega_s/\omega_p)^{2N})$, or $40 = 10 \log (1 + \varepsilon^2 (1.6)^{26})$, $1 + \varepsilon^2 (1.6)^{26} = 10^4$, $\varepsilon = (10^4/(1.6)^{26})^{1/2} = 10^2/(1.6)^{13} = \mathbf{0.222}$. Now $.222 = (10^{A_{max}/10} - 1)^{1/2}$, $10^{A_{max}/10} = .0493 + 1 = 1.0493$, $A_{max} = 10 \log 1.0493 = \mathbf{0.209\ dB}$. *Alternatively*, if A_{max} is raised from 0.5 dB to 0.6 dB, $\varepsilon = (10^{0.6/10} - 1)^{1/2} = .385$, and $A_{min} = 10 \log (1 + \varepsilon^2 (\omega_s/\omega_p)^{2N}) = 10 \log (1 + .385^2) (1.6)^{26}) = \mathbf{44.8\ dB}$. Now, we can check whether the filter order could be reduced for $A_{min} = 40$ dB. See that $40 = 10 \log (1 + .385^2 (1.6)^{2N})$, or $1.6^{2N} = 6.75 \times 10^4$. Taking logs, $2N = \log (6.75 \times 10^4)/\log 1.6 = 4.829/.204 = 23.6$. Thus, *12th order would suffice*!

11.8 Now, the (low-pass) selectivity ratio is $f_s/f_p = 30/20 = 1.5$, $A_{max} = 1$ dB, $\varepsilon = (10^{1/10} - 1)^{1/2} = 0.5088$, and $A(f_s) = 10 \log (1 + \varepsilon^2 (f_s/f_p)^{2N})$ or $20 = 10 \log (1 + (.509)^2 (1.5)^{2N})$, $1 + .259 (1.5)^{2N} = 100$, $(1.5)^{2N} = 99/.259 = 382.3$. Try $N = 6$, $2N = 12$, $(1.5)^{12} = 129.7$. Try $N = 7$, $2N = 14$, $(1.5)^{14} = 292$. Try $N = 8$, $2N = 16$, $(1.5)^{16} = 657$. Thus use 8th order, **N = 8**. The poles all have the same frequency $\omega_o = 2\pi f_p (1/\varepsilon)^{1/n}$, or $\omega_o = (2\pi)\ 20 \times 10^3 \left(\frac{1}{.5088}\right)^{1/8} = \mathbf{136.7\ krad/s}$. The first pole (or natural mode) p_1 is given by $p_1 = \omega_o (-\cos (90 - 11.25) + j \sin (90 - 11.25)) = \omega_o (-.1951 + j0.9808)$. Combining p_1 with its conjugate p_8 yields the factor $(s^2 + s\,0.3902\omega_o + \omega_o^2)$. Likewise $p_2 = \omega_o (-\cos (90 - \frac{3\pi}{2(8)}) + j \sin (90 - 33.75)) = \omega_o (-0.5556 + j\ 0.8315)$ with factor $(s^2 + s\,1.1111\omega_o + \omega_o^2)$ and $p_3 = \omega_o (-\cos (33.75) + j \sin (33.75)) = \omega_o (-.8315 + j\ 0.5556)$ with factor $(s^2 + s\,1.663\ \omega_o + \omega_o^2)$, and $p_4 = \omega_o (-\cos (11.25) + j \sin (11.25)) = \omega_o (-0.9808 + j\ 0.1951)$ with factor $(s^2 + s\,1.9616\ \omega_o + \omega_o^2)$. Thus

$$T(s) = \frac{\omega_o^8}{(s^2 + 0.3902\ \omega_o s + \omega_o^2)(s^2 + 1.111\ \omega_o s + \omega_o^2)(s^2 + 1.663\ \omega_o s + \omega_o^2)(s^2 + 1.962\ \omega_o s + \omega_o^2)}$$

Generally, $|T| = (1 + \varepsilon^2 (\omega/\omega_p)^{2N})^{-1/2}$. Now with $f_p = 20$kHz, $N = 8$, $\varepsilon = .5088$. At 25kHz, $|T| = (1 + (.5088)^2 (25/20)^{16})^{-1/2} = (1 + .2589 \times 35.53)^{-1/2} = 0.313 \equiv 20 \log .313 = -10.1$ dB, or $A = \mathbf{10.1dB}$ at 25kHz. At 40kHz, $|T| = (1 + .2589 \times (40/20)^{16})^{-1/2} = .00768 \equiv 20 \log .00268 = -42.3$ dB. Thus $A = \mathbf{42.3dB}$ at 40kHz.

11.9

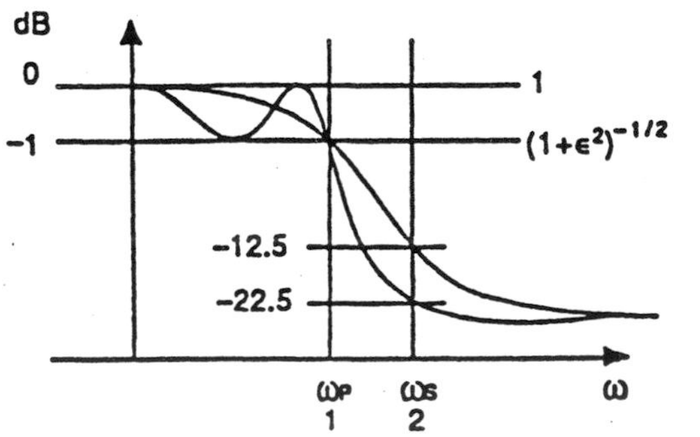

Here, at $\omega_s = 2\omega_p$, $\omega_s/\omega_p = 2$, $N = 3$. For Butterworth and Chebyshev, $\varepsilon = (10^{A_{max}/10} - 1)^{1/2} = (10^{1/10} - 1)^{1/2} = 0.509$. For Butterworth, $A(\omega_s) = 10 \log [1 + \varepsilon^2 (\omega_s/\omega_p)^{2N}] = 10 \log [1 + .509^2 (2)^6] = \mathbf{12.45\ dB}$. For Chebyshev, $A(\omega_s) = 10 \log [1 + \varepsilon^2 \cosh^2 (N \cosh^{-1} \omega_s/\omega_p)] = 10 \log [1 + .509^2 \cosh^2 (3 \cosh^{-1} 2)] = 10 \log [1 + .259 \cosh^2 (3 (1.317))] = 10 \log [1 + .259 \cosh^2 3.951] = 10 \log [1 + .259 (26.0)^2] = \mathbf{22.5\ dB}$.

11.10 Required that $A_{max} = 0.5$dB, $A_{min} \geq 40$ dB, $\omega_s/\omega_p = 1.6$ for a Chebyshev filter.
From Equation 11.21: $\varepsilon = \sqrt{10^{A_{max}/10} - 1} = \sqrt{10^{0.5/10} - 1} = \sqrt{1.122 - 1} = \sqrt{.122} = 0.3493$. From Equation 11.22, at the stopband edge, where $\omega = \omega_s$, $A(\omega_s) = 10\log[1 + \varepsilon^2 \cosh^2(N \cosh^{-1} \omega_s/\omega_p)] = 10\log[1 + .3493^2 \cosh^2(N \cosh^{-1} 1.6)] = 10\log[1 + 0.122 \cosh^2(1.04697N)$. Now for $N = 10$, $A = 10\log[1 + 0.122 \cosh^2(1.047 \times 10)] = 75.8$dB, much greater than required. For $N = 6$, $A = 10\log[1 + .122 \cosh^2(1.047 \times 6)] = 39.4\text{dB} < A_{min}$. For $N = 7$, $A = 10\log[1 + .122 \cosh^2(1.047 \times 7)] = 48.5\text{dB} > A_{min}$. Thus, use **N = 7**, for which A_{min} = **48.5dB**. Now for $N = 7$ and $A_{min} = 40$dB, $40 = 10\log[1 + \varepsilon^2 \cosh^2(7 \cosh^{-1} 1.6)]$. Thus $1 + \varepsilon^2 \cosh^2(7.329) = 10^4$, or $\varepsilon^2 = (10^4 - 1)/0.5802 \times 10^6 = 0.01723$, or $\varepsilon = 0.1312$, for which $A_{max} = 10\log(1 + \varepsilon^2) =$ **0.074dB** is possible. *Check*: $A_{min} = 10\log[1 + \varepsilon^2 \cosh^2(N \cosh^{-1} \omega_s/\omega_p)] = 10\log[1 + .01723 \cosh^2(7 \cosh^{-1} 1.6)] = 40$dB; OK. Now for A_{max} raised to $0.074 + 0.1 = 0.174$dB, $\varepsilon = (10^{A_{max}/10} - 1)^{1/2} = 0.202$, and $A_{min} = 10\log[1 + 0.202^2 \cosh^2(7.329)] =$ **43.8dB**, an increase of nearly 4dB of stopband attenuation in return for a 0.1dB increase in passband ripple!!

11.11 Consider the question to refer to the initial specification in P11.7 and P11.10 above, for which $A_{max} = 0.5$dB and $A_{min} \geq 40$dB, with $\omega_s/\omega_p = 1.6$. Here $\omega_p = 10^3$rad/s, for which $\omega_s = 1.6 \times 10^3$rad/s, and, the dc gain is 1.

Now for the Butterworth filter, $N = 13$, with $\varepsilon = 0.349$, the poles are on a circle with radius $\omega_o = \omega_p (1/\varepsilon)^{1/N} = 10^3 (1/0.349)^{1/13} = 1.084 \times 10^3$rad/s, at an angular separation of $\pi/N = 13.85°$, at angles (from the negative real axis) of 0°, ±13.85°, ±27.69°, ±41.54°, ±55.38°, ±69.23°, ±83.08°. Now $p_1 = \omega_o(-\cos 83.08° + j \sin 83.08°) = 1.084 \times 10^3 (-0.1205 + j\,0.9927) = 10^3(-0.131 + j\,1.076)$. Correspondingly, $p_1, p_{13} =$ $\mathbf{10^3}$ $(-0.131 \pm j\,1.076)$ rad/s; $p_2, p_{12} = \omega_o(-\cos 69.23° \pm j \sin 69.23°) = 1.084 \times 10^3 (-.3546 \pm j\,0.9350) =$ $\mathbf{10^3}$ $(-0.384 \pm j\,1.013)$ rad/s; $p_3, p_{11} = \omega_o(-\cos 55.38° \pm j \sin 55.38°) =$ $\mathbf{10^3}$ $(-0.616 \pm j\,0.892)$ rad/s; $p_4, p_{10} = \omega_o(-\cos 41.54° \pm j \sin 41.54°) =$ $\mathbf{10^3}$ $(-0.811 \pm j\,0.719)$ rad/s; $p_5, p_9 = \omega_o(-\cos 27.69° \pm j \sin 27.69°) =$ $\mathbf{10^3}$ $(-0.960 \pm j\,0.504)$ rad/s; $p_6, p_8 = \omega_o(-\cos 13.85° \pm j \sin 13.85°) =$ $\mathbf{10^3}$ $(-1.052 \pm j\,0.259)$ rad/s; $p_7 =$ $\mathbf{1.08 \times 10^3}$ rad/s.

Now for the Chebyshev filter, $N = 7$, with $\varepsilon = 0.349$, and $\omega_p = 10^3$ rad/s, the poles are (for $k = 1$ to 7):

$$p_k = \omega_p \left\{ -\sin\left[\frac{2k-1}{N}\,\frac{\pi}{2}\right] \sinh\left[\frac{1}{N} \sinh^{-1} 1/\varepsilon\right] + j \cos\left[\frac{2k-1}{N}\,\frac{\pi}{2}\right] \cosh\left[\frac{1}{N} \sinh^{-1} 1/\varepsilon\right] \right\}$$

$$= 10^3 \left\{ -\sin\left[\frac{2k-1}{7}(90°)\right] \sinh\left[\frac{1}{7} \sinh^{-1}\left[1/0.349\right]\right] + j \cos\left[\frac{2k-1}{7}(90°)\right] \cosh\left[\frac{1}{7} \sinh^{-1}\left[1/0.349\right]\right] \right\}$$

$$= 10^3 \left[-0.2563 \sin((2k-1)(12.86°)) + 1.032j \cos((2k-1)(12.86°))\right].$$

Now, $p_1, p_7 =$ $\mathbf{10^3}$ $[-0.057 \pm j\,1.006]$ rad/s; $p_2, p_6 =$ $\mathbf{10^3}$ $[-0.160 \pm j\,0.807]$ rad/s; $p_3, p_5 =$ $\mathbf{10^3}$ $[-0.231 \pm j\,0.448]$ rad/s; $p_4 =$ $\mathbf{10^3}$ $[-0.256]$ rad/s.

SECTION 11.4: FIRST-ORDER AND SECOND-ORDER FILTER FUNCTIONS

11.12

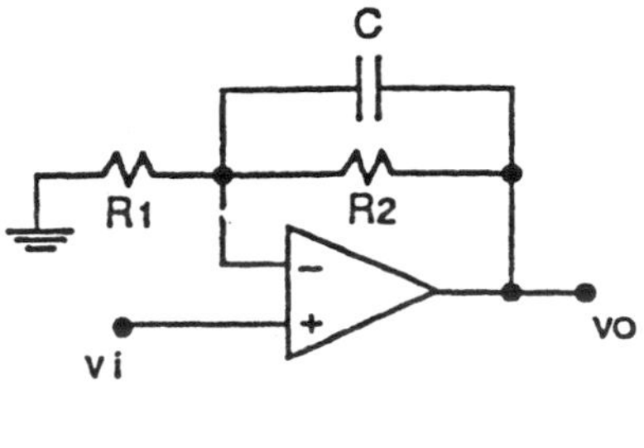

For infinite input resistance, the circuit must be driven as shown. Use $R_1 =$ **10kΩ**. The dc gain is $1 + R_2/R_1 = 11$ V/V. Thus $R_2/R_1 = 10$, and $R_2 = 10\,R_1 =$ **100kΩ**. For a 3dB frequency of 10kHz, $CR_2 = 1/\omega_o$, or $C = \dfrac{1}{2\pi \times 10^4 \times 10^5} =$ **159.2pF**.
Thus, the zero frequency is approximately at

$$f_z = \frac{1}{2\pi C R_1} = \frac{1}{2\pi \times 159.2 \times 10^{-12} \times 10^4} \approx$$

100kHz.

At very high frequencies, C is a short circuit, and v_o follows v_i, such that v_o/v_i = **+1V/V.**

From First Principles: $V_o/V_i = 1 + \dfrac{R_2 \parallel \dfrac{1}{Cs}}{R_1} = \dfrac{1}{R_1}\left\{R_1 + \dfrac{\dfrac{R_2}{Cs}}{R_2 + 1/Cs}\right\} = \dfrac{1}{R_1}$
$\left\{R_1 + \dfrac{R_2}{R_2\,Cs + 1}\right\} = \dfrac{R_2 + R_1 + R_1 R_2 Cs}{R_1 + R_1 R_2 Cs} = \dfrac{s + \left\{\dfrac{R_1 + R_2}{R_1 R_2}\right\}\dfrac{1}{C}}{s + 1/R_2 C}$. Thus the zero is more precisely at $f_z = \dfrac{1}{2\pi C\,(R_1 \parallel R_2)} = \dfrac{1}{2\pi \times 159.2 \times 10^{-12}\,(10^4 \parallel 10^5)}$ = **110kHz.** *Note* that $A_L f_p = A_H f_z$, ie 11 (10kHz) = 1 (110kHz). Now for A_H = 1V/V, f_z/f_p = 100, $A_L = A_H f_z/f_p$ = **100V/V**, for which R_1 = **10kΩ**, and $R_2 = R_1$ (100–1) = **990kΩ**.

11.13 The Bandpass requirement suggests the combination of Fig. 11.13a and 11.13b as shown, with $C_1 R_1 = \dfrac{1}{2\pi\,(100)}$, and $C_2 R_2 = \dfrac{1}{2\pi\,(1000)}$. The midband gain is $-R_2/R_1$ = 1V/V, with $R_1 \approx R_{in} \approx$ 10kΩ. Now, if use $R_1 = R_2$ = 10kΩ, $C_1 = \dfrac{1}{2\pi\,(100)\,(10^4)}$ = 0.159μF. Chose C_1 = **0.1μF**, and C_2= **0.01μF.** Now $R_1 = \dfrac{1}{2\pi \times 0.1 \times 10^{-6} \times 100}$ = **15.9kΩ.** For gain = –1V/V, R_2 = **15.9kΩ**, as well, with an input resistance at midband of about **15.9kΩ.**

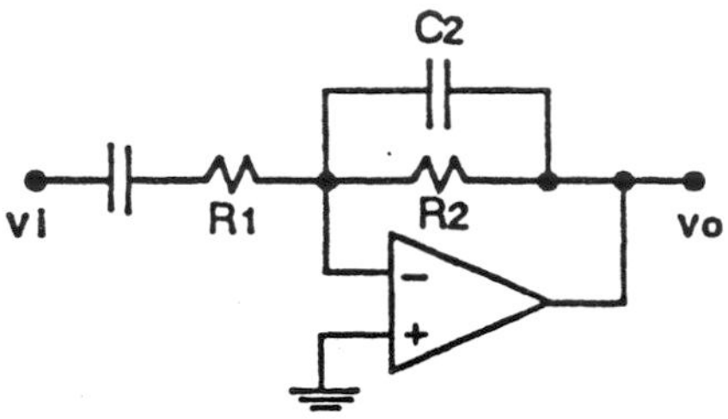

11.14 $T(s) = -\dfrac{R_{2a} + R_{2b} \parallel \dfrac{1}{C_2 s}}{R_{1a} + R_{1b} \parallel \dfrac{1}{C_1 s}} = -\dfrac{R_{2a} + \dfrac{R_{2b}}{C_2 s\,(R_{2b} + 1/C_2 s)}}{R_{1a} + \dfrac{R_{1b}}{C_1 s\,(R_{1b} + 1/C_1 s)}} = -\dfrac{R_{2a} + \dfrac{R_{2b}}{1 + R_{2b} C_2 s}}{R_{1a} + \dfrac{R_{1b}}{1 + R_{1b} C_1 s}}$, or

$$\mathrm{T(s)} = \frac{1 + R_{1b} C_1 s}{1 + R_{2b} C_2 s} \times \frac{R_{2a} + R_{2b} + R_{2a} R_{2b} C_2 s}{R_{1a} + R_{1b} + R_{1a} R_{1b} C_1 s} = \frac{R_{1b} C_1}{R_{2b} C_2} \times \frac{R_{2a} R_{2b} C_2}{R_{1a} R_{1b} C_1} \times \left\{\frac{s + \dfrac{1}{R_{1b} C_1}}{s + \dfrac{1}{R_{2b} C_2}}\right\}$$

$$\times \left\{\frac{s + \dfrac{1}{(R_{2a} \| R_{2b})\,C_2}}{s + \dfrac{1}{(R_{1a} \| R_{1b})\,C_1}}\right\} = \frac{\mathbf{R_{2a}}}{\mathbf{R_{1a}}} \times \left\{\frac{s + \dfrac{1}{\mathbf{R_{1b} C_1}}}{s + \dfrac{1}{\mathbf{R_{2b} C_2}}}\right\} \times \left\{\frac{s + \dfrac{1}{(\mathbf{R_{2a} \parallel R_{2b}})\,\mathbf{C_2}}}{s + \dfrac{1}{(\mathbf{R_{1a} \parallel R_{1b}})\,\mathbf{C_1}}}\right\}$$, with zeros at

$\dfrac{1}{R_{1b} C_1}$ and $\dfrac{1}{(R_{2a} \parallel R_{2b})\,C_2}$, with poles at $\dfrac{1}{R_{2b} C_2}$ and $\dfrac{1}{(R_{1a} \parallel R_{1b})\,C_1}$, and with a high-frequency gain of R_{2a}/R_{1a}, a low-frequency gain of $(R_{2a} + R_{2b})/(R_{1a} + R_{1b})$, a mid-band gain of $(R_{2a} + R_{2b})/R_{1a}$ or $R_{2a}/(R_{1a} + R_{1b})$, depending on the relative locations of the poles and zeros. For a midband gain of –10V/V, gains at low and high frequencies of –1V/V, and 3dB points at 100Hz and 1000 Hz, the corresponding Bode plot is as shown:

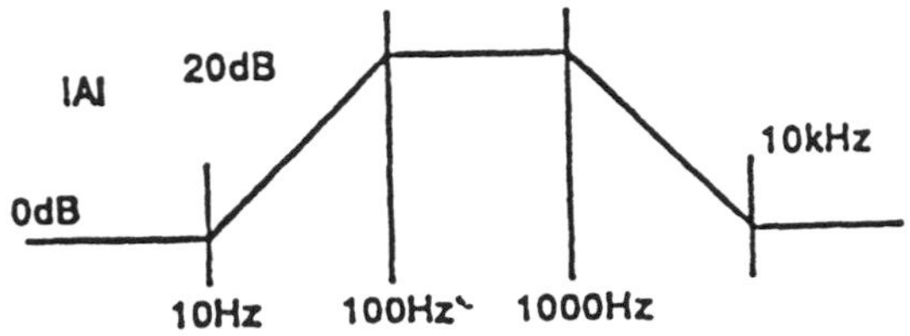

Now, at low frequencies, $\frac{R_{2a} + R_{2b}}{R_{1a} + R_{1b}} = 1$; at high frequencies, $\frac{R_{2a}}{R_{1a}} = 1$; at midband, $\frac{(R_{2a} + R_{2b})}{R_{1a}} = 10$. Thus $R_{2a} + R_{2b} = R_{1a} + R_{1b} = 10\ R_{1a} = 10\ R_{2a}$, and $R_{1b} = 9R_{1a}$, $R_{2b} = 9R_{2a}$, $R_{1a} = R_{2a}$, and $R_{1b} = R_{2b}$. For $R_{in} \approx 10\text{k}\Omega$ at midband, $R_{1a} \approx 10\text{k}\Omega$ and $R_{1b} \approx 90\text{k}\Omega$.

For the zero at 10Hz, $\frac{1}{2\pi R_{1b} C_1} = 10$, $C_1 = \frac{1}{2\pi R_{1b} \times 10} = \frac{1}{2\pi \times 10 \times 90 \times 10^3} = 0.177\mu\text{F}$.

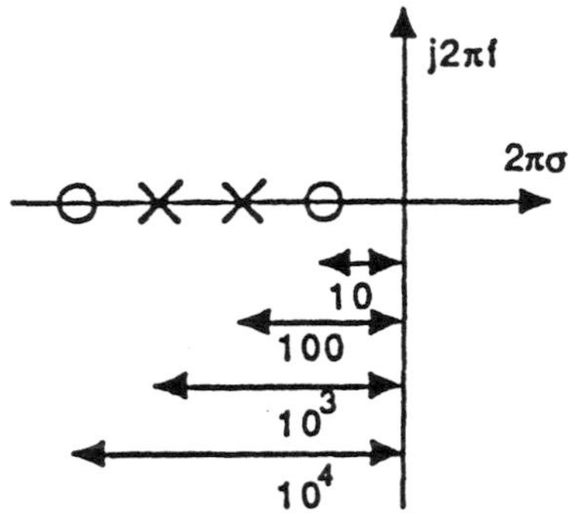

Use $C_1 = \mathbf{0.2\mu F}$, for which $R_{1b} = \frac{1}{2\pi C_1 \times 10} = \frac{1}{2\pi \times 0.2 \times 10^{-6} \times 10} = \mathbf{79.58k\Omega} = R_{2b}$, and $R_{1a} = R_{2a} = 79.58/9 = \mathbf{8.84k\Omega}$. Now for pole the at 1kHz: $\frac{1}{2\pi C_2 R_{2b}} = 10^3$, $C_2 = \frac{1}{2\pi \times 10^3 \times 79.58 \times 10^3} = 0.002\mu\text{F}$. Now, check the second pole, to be at $f_{p2} = \frac{1}{2\pi (R_{1a} \parallel R_{1b})\, C_1}$. Here $R_{1a} \parallel R_{1b} = (8.84 \parallel 79.58)\text{k}\Omega = 7.956\text{k}\Omega$, and $f_{p2} = \frac{1}{2\pi (7.956 \times 10^3) \times 0.2 \times 10^{-6}} = \mathbf{100Hz}$, with f_{p1} at **1kHz.** Correspondingly, f_{z1} = **10Hz** and f_{z2} = **10kHz.** The pole-zero plot is as shown:

11.15 The required response is as shown in the Bode plot below:

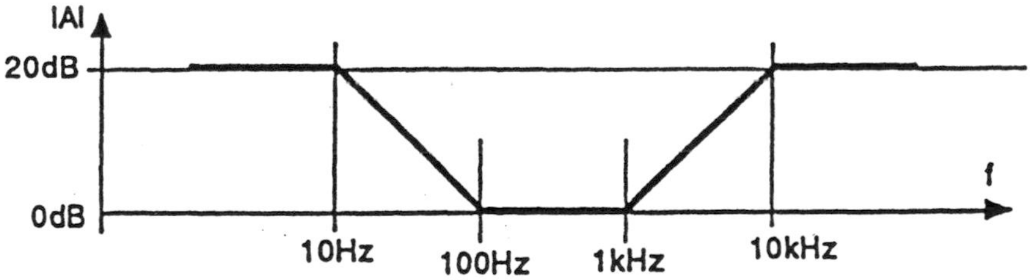

For the circuit in P11.14 to have a gain of −10V/V at dc, $\frac{R_{2a} + R_{2b}}{R_{1a} + R_{1b}} = 10$ − − − (1). For a gain of −10V/V at high frequencies, $\frac{R_{2a}}{R_{1a}} = 10$ − − − (2). Now for a *lower* gain at midband, C_2 must provide a zero at 100Hz, and C_1 a zero at 1kHz, with R_{2b} shorted there (while C_1 is still (relatively) open). Thus in the midband, the gain is $\frac{R_{2a}}{R_{1a} + R_{1b}} = 1$ − − − (3).

The pole associated with C_2 must be at 100Hz/10 = 10Hz. The pole associated with C_1 must be at (1kHz) × 10 = 10kHz. Now, $R_{in} \geq 10\text{k}\Omega$, $R_{1a} \geq 10\text{k}\Omega$. For $R_{1a} = 10\text{k}\Omega$, $R_{2a} = 10\ (10) = 100\text{k}\Omega$. From (3), $R_{2a} = R_{1a} + R_{1b}$, or $R_{1b} = R_{2a} - R_{1a} = 100\text{k}\Omega - 10\text{k}\Omega = 90\text{k}\Omega$. From (1), $R_{2a} + R_{2b} = 10\ R_{1a} + 10\ R_{1b}$, or $R_{2b} = 10\ (10) + 10\ (90) - 100 = 100 + 900 - 100 = 900\text{k}\Omega$. That is (tentatively), $R_{1a} = 10\text{k}\Omega$, $R_{1b} = 90\text{k}\Omega$, $R_{2a} = 100\text{k}\Omega$, $R_{2b} = 900\text{k}\Omega$. For a zero at 1kHz, $\frac{1}{2\pi R_{1b} C_1} = 1\text{kHz}$, or $C_1 = \frac{1}{2\pi \times 10^3 \times 90 \times 10^3} = .00177\mu\text{F}$. Now, for a one-significant-digit capacitor, use $C_1 = \mathbf{.001\mu F}$, or 1nF, to maintain $R_{in} > 10\text{k}\Omega$. *Conclude*: $R_{1b} = \frac{1}{2\pi \times 10^3 \times .001 \times 10^{-6}} = \mathbf{159k\Omega}$, $R_{1a} = (10/90)$

$159 = \mathbf{17.7k\Omega}$, $R_{2a} = 10\ (17.7) = \mathbf{177k\Omega}$, $R_{2b} = 10\ (159) = \mathbf{1.591M\Omega}$.

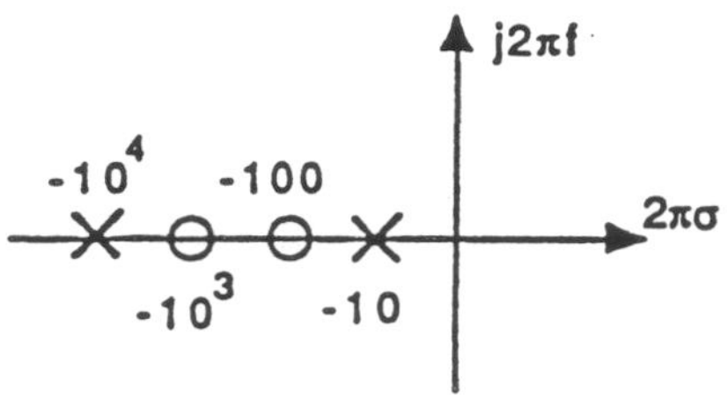

Now for a zero at 100Hz associated with C_2, use $C_2 = \dfrac{1}{2\pi \times 10^2 (R_{2a} \parallel R_{2b})} = \dfrac{1}{2\pi \times 10^2\ (0.177 \parallel 1.591)\ 10^6} = .999 \times 10^{-8} = \mathbf{10nF}$. *Check*: The pole from C_2 is $\dfrac{1}{2\pi C_2 R_{2b}} = \dfrac{1}{2\pi \times 10 \times 10^{-9} \times 1.591 \times 10^6} = \mathbf{10Hz}$, and the pole from C_1 is $1/[2\pi C_1 (R_{1a} \parallel R_{1b})] = 1/[2\pi \times .001 \times 10^{-6} \times (17.7 \parallel 159) \times 10^3] = 0.0099 \times 10^6 = \mathbf{10kHz}$. Overall, there are zeroes at **100Hz, 1kHz**, and poles at **10Hz, 10kHz**, for which the pole-zero plot is shown.

11.16 For Fig. 11.14 modified:

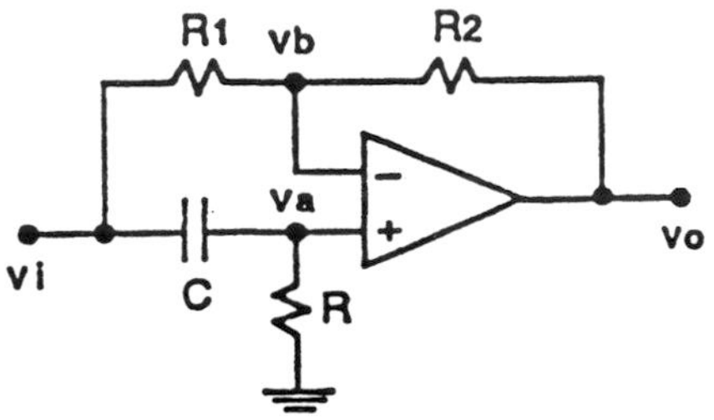

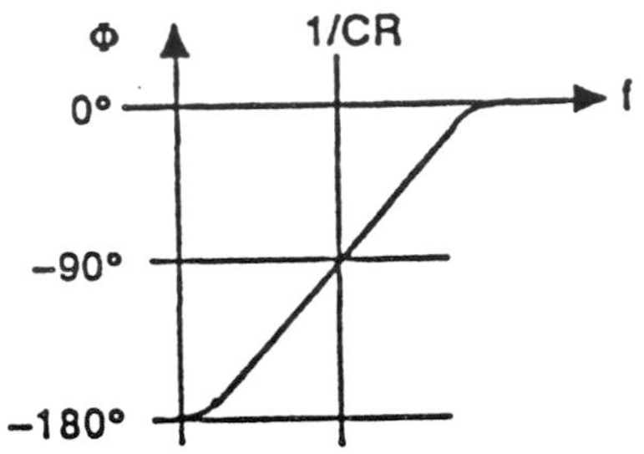

$v_a = \dfrac{R}{R + 1/Cs} v_i = v_b$, $v_o = v_b + v_b - v_i = 2v_b - v_i$, and $T(s) = \dfrac{v_o}{v_i} = \dfrac{2R}{R + 1/Cs} - 1 = \dfrac{2R - R - 1/Cs}{R + 1/Cs} = \dfrac{RCs - 1}{RCs + 1}$, or $\mathrm{T(s)} = \dfrac{s - 1/RC}{s + 1/RC}$. See a pole at $-1/RC$ and a zero at $1/RC$. Now at $\omega_o = 1/RC$, phase shift is $+\tan^{-1}(-1) - \tan^{-1}(1) = \mathbf{-90°}$.

Now, for $f_o = 10^4$, $C = 1.59 \times 10^{-9}$F, and $|\Phi| = 90°$, $R = \dfrac{1}{2\pi \times 1.59 \times 10^{-9} \times 10^4} = 10^4\Omega = \mathbf{10k\Omega}$.

For Φ in general, $\Phi = \tan^{-1}\left[-\dfrac{\omega_o}{1/RC}\right] - \tan^{-1}\left[\dfrac{\omega_o}{1/RC}\right] = -2\tan^{-1}(\omega_o RC)$, or $R = \dfrac{1}{2\pi f_o C}\tan(-\Phi/2)$. Now, at $f_o = 10^4$Hz, for $\Phi = -90°$, $R = \dfrac{1}{2\pi \times 10^4 \times 1.59 \times 10^{-9}}\tan(- - 90/2) = 10^4(1) = 10^4\Omega$.

For $|\Phi| = 6°$, $R = 10^4\tan(6/2) = 524\Omega$; for $12°$, $R = 10^4\tan(12/2) = 1051\Omega$; for $30°$, $R = 10^4\tan(30/2) = \mathbf{2.68k\Omega}$; for $60°$, $R = 10^4\tan(60/2) = \mathbf{5.77k\Omega}$; for $90°$, $R = 10^4\tan(90/2) = \mathbf{10k\Omega}$; for $120°$, $R = 10^4\tan(120/2) = \mathbf{17.3k\Omega}$; for $150°$, $R = 10^4\tan(150/2) = \mathbf{37.3k\Omega}$; for $168°$, $R = 10^4\tan(168/2) = \mathbf{95.1k\Omega}$; for $174°$, $R = 10^4\tan(174/2) = \mathbf{190.8k\Omega}$.

11.17 For $\omega_o = 10^3$rad/s, and 3dB bandwidth is of $\omega_o/Q = 200$ rad/s, see $Q = 10^3/200 = 5$. Now peak gain is $\dfrac{a_1 Q}{\omega_o} = 1$. Thus, $a_1 = \omega_o/Q = 10^3/5 = 200$, and $\mathrm{T(s)} = \dfrac{a_1 s}{s^2 + \dfrac{s\,\omega_o}{Q} + \omega_o^2} = \dfrac{200s}{s^2 + 200s + 10^6}$. At ω, $T(j\omega) = \dfrac{200\,j\omega}{200\,j\omega + 10^6 - \omega^2}$. At $A_{min} = -20$dB, $\dfrac{200\,\omega}{(200\,\omega^2 + (10^6 - \omega^2)^2)^{1/2}} = \dfrac{1}{10}$, $200^2\,\omega^2 = 200\,\omega^2 + 10^{12} - 2 \times 10^6\,\omega^2 + \omega^4$, or $\omega^4 + \omega^2(-4 \times 10^6 - 2 \times 10^6 + 200) + 10^{12} = 0$, or $\omega^4 - 6 \times 10^6\,\omega^2 + 10^{12} = 0$, or $\omega^2 = \dfrac{+6\times10^6 \pm \sqrt{(6\times10^6)^2 - 4\times10^{12}}}{2} = \dfrac{6\times10^6 \pm 5.657\times10^6}{2} = .1716\times10^6$, and 5.829×10^6, where $\omega =$ **414 rad/s** and **2.414 k rad/s**. *Check*: $0.414\times2.414 = 0.999$ krad/s $\approx 10^3$ rad/s, as expected.

11.18

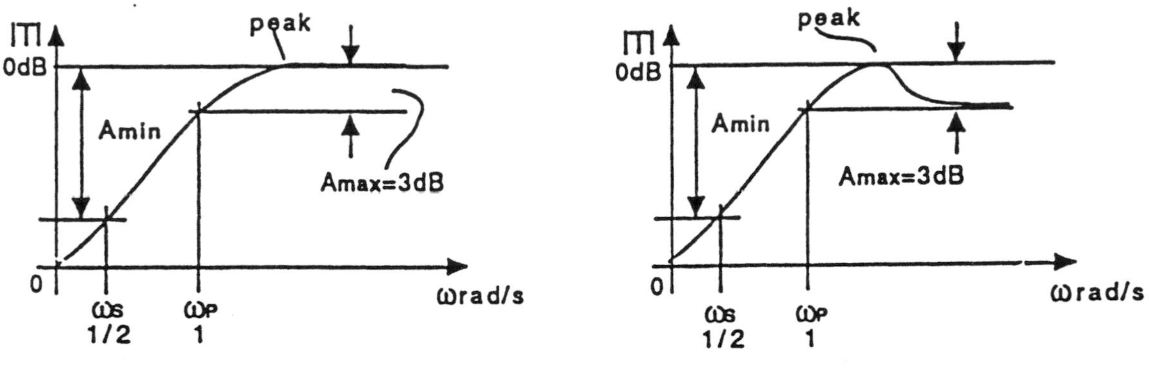

The response characteristics described above are as follows: For both designs, we require $A_{max} = 3$dB at $\omega_p = 1$ rad/s, with maximum gain = 1V/V, and A_{min} to be measured at $\omega_s = 0.5$ rad/s.

a) From Fig. 11.16b) for the response shown first, arrange that the peak amplitude is such that $a_2 = 1$, that is that $\dfrac{a_2 Q}{(1 - 1/4\,Q^2)^{1/2}} = a_2$, $Q^2 = 1 - \dfrac{1}{4\,Q^2}$, $4\,Q^4 = 4\,Q^2 - 1$, $4\,Q^4 - 4\,Q^2 + 1 = 0$,

$Q^2 = \dfrac{+4 \pm \sqrt{16 - 4\,(4)}}{2\,(4)} = 1/2$, and $Q = \dfrac{1}{\sqrt{2}} = 0.707$. Now, $T(s) = \dfrac{a_2 s^2}{s^2 + s\,\omega_o/Q + \omega_o^2}$ in general,

For $a_2 = 1$, $Q = 0.707$, $\omega = \omega_p = 1$ for $|T| = -3\text{dB} \equiv 0.707$. Thus $T(j1) = \dfrac{1\,(j1)^2}{(j1)^2 + j1\,\omega_o/.707 + \omega_o^2}$

and $\dfrac{1}{((\omega_o^2 - 1)^2 + 2\omega_o^2)^{1/2}} = .707$, or $\omega_o^4 + 2\omega_o^2 + 1 - 2\,\omega_o^2 = 2$, or $\omega_o^4 = 1$, and $\omega_o = 1$. Thus **T(s)** = $\dfrac{\mathbf{s^2}}{\mathbf{s^2 + 1.414\,s + 1}}$, with $Q =$ **0.707**, and $\omega_o = \omega_{3dB} =$ **1 rad/s**. Now, for $A_{min} = -\,|T(j\,1/2)|$, $T\,(j/2) =$

$\dfrac{(j/2)^2}{(j/2)^2 + 1.414\,(j/2) + 1} = \dfrac{-1}{4\,[\,(-1/4) + 1 + .707j\,]} = \dfrac{-1}{2.828j + 3}$. $A_{min} = -\,20\,\log \left[\dfrac{1}{((2.828)^2 + 3^2)^{1/2}}\right] = -20\log 0.2425 =$ **12.3dB.**

(b) From the second response, see $a_2 = 0.707$, and at the peak, $\dfrac{a_2 Q}{\sqrt{1 - 1/(4\,Q^2)}} = 1$, or $(0.707\,Q)^2 = 1 - \dfrac{1}{4\,Q^2}$, $\dfrac{Q^2}{2} = 1 - \dfrac{1}{4\,Q^2}$, $Q^2 = 2 - \dfrac{1}{2\,Q^2}$, $2\,Q^4 = 4\,Q^2 - 1$, $Q^4 - 2\,Q^2 + 0.5 = 0$, $Q = \dfrac{--2 \pm \sqrt{2^2 - 4\,(0.5)}}{2} = \dfrac{2 \pm \sqrt{2}}{2} =$ **1.707** for $Q > 1$. Now, in general, $T(s) = \dfrac{a_2 s^2}{s^2 + s\,\omega_o/Q + \omega_o^2}$,

and for $Q = 1.707$, $a_2 = 0.707$, $\omega = \omega_p = 1$ for $|T| = 0.707$. $T(j1) = \dfrac{.707\,(j1)^2}{(j1)^2 + j1\,\omega_o/1.707 + \omega_o^2} =$

$\dfrac{.707}{(\omega_o^2 - 1) + .5858j\,\omega_o}$. Now, $0.707 = \dfrac{0.707}{[((\omega_o^2 - 1)^2 + .5858^2\,\omega_o^2)]^{1/2}}$, or $\omega_o^4 - 2\,\omega_o^2 + 1 + .343\,\omega_o^2 = 1$,

$\omega_o^4 = 1.657\,\omega_o^2$, $\omega_o^2 = 1.657$, $\omega_o =$ **1.287 rad/s**. Thus $T(s) = \dfrac{.707\,s^2}{s^2 + s\,1.287/1.707 + (1.287)^2}$, or **T(s)** = $\dfrac{\mathbf{.707\,s^2}}{\mathbf{s^2 + 0.754s + 1.657}}$, with $\omega_o = 1.287$ rad/s, and $\omega_{3dB} =$ **1 rad/s.** Now $T(j/2) =$

$\dfrac{-.707\,(1/4)}{-1/4 + .754\,j/2 + 1.657} = \dfrac{-.707}{5.628 + 1.508j} = \dfrac{-1}{7.960 + 2.133j}$, and $A_{min} = -20\log \dfrac{1}{(7.96^2 + 2.133^2)^{1/2}}$

$= -20\log(.1213) =$ **18.3dB**, an improvement of 6dB over the arrangement in (a).

11.19 Use the description in Fig. 11.16d for the notch. Here f_o = 60Hz, and $\omega_o = 2\pi(60) =$ **377.0 rad/s.** Now $T(s) = \dfrac{s^2 + 377^2}{s^2 + s\dfrac{377}{Q} + 377^2}$. Now, assuming the required 1Hz band to be centered, arrange that the attenuation is 20dB at 60 + 1/2 = 60.5Hz, or 380.1 rad/s. Thus, $\left|\dfrac{(j380.1)^2 + 377^2}{(j380.1)^2 + j380.1\,(377/Q) + 377^2}\right| = \dfrac{1}{10}$, or $\left|\dfrac{-2347}{-2347 + 143298j/Q}\right| = \left|\dfrac{1}{1 - \dfrac{61.06j}{Q}}\right| = \dfrac{1}{10}$, or $\left[1 + (61.06/Q)^2\right] = 10^2$, $(\dfrac{61.06}{Q})^2 = 99$, $\dfrac{61.06}{Q} = 9.95$, $Q = \dfrac{61.06}{9.95} =$ **6.136**. Now the 3dB bandwidth is $\dfrac{f_o}{Q} = \dfrac{60}{6.14} =$ **9.78Hz.**

For the 3dB frequencies: $\left|\dfrac{(j\omega)^2 + 377^2}{(j\omega)^2 + j\omega\dfrac{377}{6.14} + 377^2}\right| = 0.707$, $\left|\dfrac{377^2 - \omega^2}{377^2 - \omega^2 + j\omega\,61.4}\right| = 0.707$,

$\left|\dfrac{1}{1 + \dfrac{j\omega\,61.4}{377^2 - \omega^2}}\right| = 0.707$, $1^2 + \left[\dfrac{61.4\omega}{377^2 - \omega^2}\right]^2 = \left[\dfrac{1}{.707}\right]^2 = 2$, $\dfrac{61.1\,\omega}{377^2 - \omega^2} = (2 - 1)^{½} = \pm 1$ – – – (1). Now, $377^2 - \omega^2 = \pm 61.1\omega$, $\omega^2 \pm 61.1\omega - 377^2 = 0$, $\omega = \dfrac{\pm 61.1 \pm \sqrt{61.1^2 - -4\,(377)^2}}{2} = \dfrac{\pm 61.1 \pm 756.5}{2}$, of which the relevant solutions are $\dfrac{756.5 - 61.1}{2} = 347.7$ rad/s, and $\dfrac{756.5 + 61.1}{2} =$ 408.8 rad/s, or **55.3Hz** and **65.06Hz**, for a 3dB bandwidth of 65.1 – 55.3 = **9.8Hz.** (As noted.)

For the 1dB frequencies: 20 log $x = -1 \rightarrow x = .891$V/V, and $\left|\dfrac{(j\omega)^2 + 377^2}{(j\omega)^2 + j\omega\dfrac{377}{6.14} + 377^2}\right| = 10^{-1/20}$ = 0.891,

$\left|\dfrac{377^2 - \omega^2}{377^2 - \omega^2 + 61.4\,j\omega}\right| = .891$, $\left|\dfrac{1}{1 + \dfrac{61.4\omega}{377^2 - \omega^2}j}\right| = .891$, $1^2 + \left[\dfrac{61.4\omega}{377^2 - \omega^2}\right]^2 = \left[\dfrac{1}{.891}\right]^2$ = 1.2596,

$61.4\omega = \pm(377^2 - \omega^2)(.2596)^{½} = \pm(377^2 - \omega^2)\,.5096$, $\omega^2 \pm 120.5\omega - 377^2 = 0$,

$\omega = \dfrac{\pm 120.5 \pm \sqrt{120.5^2 + 4\,(377)^2}}{2} = \dfrac{\pm 120.5 \pm 763.6}{2}$, for which relevant solutions are $\dfrac{120.5 + 763.6}{2} =$ 442rad/s or **70.35Hz,** and $\dfrac{736.6 - 120.5}{2} =$ 308.1 rad/s or **49.0Hz.**

For the 1% frequencies (from (1) above) $\dfrac{61.1\omega}{377^2 - \omega^2} = \left[\left[\dfrac{1}{.99}\right)^2\right] - 1\right]^{½} = \pm .142$, or $\omega^2 \pm 428.8\omega - 377^2 = 0$, $\omega = \dfrac{\pm 428.8 \pm \sqrt{428.8^2 + 4\,(377^2)}}{2} = \dfrac{\pm 428.8 \pm 754.3}{2}$, of which appropriate solutions are 591.6 rad/s or **94.1Hz,** and 162.8 rad/s or **25.9Hz.**

SECTION 11.5: THE SECOND-ORDER LCR RESONATOR

11.20 Equation 11.29 indicates that: $\omega_1, \omega_2 = \omega_o \left\{ \left[1 + \frac{1}{4Q^2} \right]^{1/2} \pm \frac{1}{2Q} \right\}$.

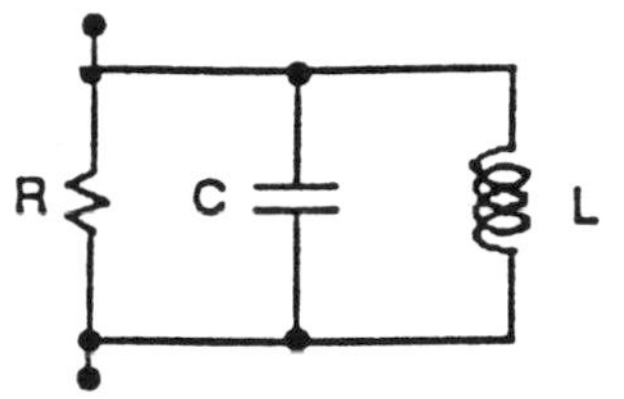

Now $BW = \frac{\omega_o}{Q} = 2\pi(20 \times 10^3)$, whence $Q = \frac{10^6}{20 \times 10^3} = 50$.

Now $Q = \omega_o CR$, and $C = \frac{Q}{\omega_o R} = \frac{50}{2\pi \times 10^6 \times 10^4} = \mathbf{796pF}$.

Accordingly, $L = \frac{1}{\omega_o^2 C} = \frac{1}{(2\pi \times 10^6)^2 \times 796 \times 10^{-12}} = 264\mu H$.

For a 1mA rms input at the 1MHz resonant frequency, $\upsilon_o = 10^4 \times 10^{-3} =$ **10V rms.**

11.21 Here, $\omega_o^2 = \frac{1}{LC} = (2\pi \times 99.9 \times 10^6)^2 = (6.2769 \times 10^8)^2 = 3.940 \times 10^{17}$, $L = \frac{1}{C \times 3.94 \times 10^{17}}$. Now the response is 3dB down at 99.9 −2 = 97.9MHz. Thus $\frac{f_o}{Q} = 2(2 \times 10^6)$, $Q = \frac{99.9 \times 10^6}{4 \times 10^6} = 24.98$.

Now $Q = \omega_o CR$, $C = \frac{Q}{\omega_o R} = \frac{24.98}{2\pi \times 99.9 \times 10^6 \times 75} =$ **530.6pF**, and $L = \frac{1}{530.6 \times 10^{-12} \times 3.94 \times 10^{17}} = 0.0048\mu H$.

For off-tuning by 100kHz, $\omega_1 = 2\pi(99.9 \times 10^6 + 0.1 \times 10^6) = 200\pi \times 10^6 = 6.2822 \times 10^8$rad/s. From Eq. 11.40 for the notch (with $a_2 = 1$), $T(s) = \frac{s^2 + \omega_o^2}{s^2 + s(\omega_o/Q) + \omega_o^2} = \frac{1}{1 + \frac{s\,\omega_o/Q}{s^2 + \omega_o^2}}$, and $T(\omega_1) =$

$$\frac{1}{1 + \frac{j6.2832 \times 10^8 (6.2769 \times 10^8)/24.98}{-(6.2832 \times 10^8)^2 + (6.2769 \times 10^8)^2}} = \frac{1}{1 - \frac{j1.579 \times 10^{16}}{.07913 \times 10^{16}}} = \frac{1}{1 - j19.95}.$$

Now, $|T(\omega_1)| = \frac{1}{(1^2 + 19.95^2)^{1/2}} = .050$. Thus the attenuation expected would be $-20\log(.05) =$ **26dB.**

11.22 For a maximally flat response, $Q = 1/\sqrt{2} = 0.707$. Now, for the high-pass filter, $T(s) = \frac{a_2 s^2}{s^2 + s(\omega_o/Q) + \omega_o^2}$ where $a_2 = 1$, and $T(j\omega) = \frac{-\omega^2}{-\omega^2 + 1.414 j\omega\,\omega_o + \omega_o^2}$. For a 3dB frequency of 100kHz:

$$\left| \frac{-(10^5)^2}{-(10^5)^2 + 1.414j\,10^5 f_o + f_o^2} \right| = \frac{1}{\sqrt{2}},\ \sqrt{2} \times 10^{10} = \left((f_o^2 - 10^{10})^2 + (1.414 \times 10^5)^2 f_o^2 \right)^{1/2},$$

or $2 \times 10^{20} = f_o^4 - 2 \times 10^{10} f_o^2 + 10^{20} + 2 \times 10^{10} f_o^2$, or $f_o^4 = (2-1)10^{20} = 10^{20}$, or $f_o = 10^5$Hz (as could be seen directly). Now, $LC = 1/(2\pi \times 10^5)^2 = 2.533 \times 10^{-12}$. For an ideal coil, $Q = \omega_o CR$, and $C = \frac{0.707}{10^4 \times 2\pi \times 10^5} =$ **112.5pF**, with $L = \frac{1}{\omega_o^2 \times C} = \frac{1}{(2\pi \times 10^5)^2 \times 112.5 \times 10^{-12}} =$ **22.5mH.** ow for the coil available, $Q = \frac{\omega_o L}{r}$, or $r = \frac{2\pi \times 10^5 \times 22.5 \times 10^{-3}}{50} =$ **283Ω** (in series with L). The equivalent parallel resistance is $R_p = \frac{Q}{\omega_o C} = \frac{50}{2\pi \times 10^5 \times 112.5 \times 10^{-12}} =$ **707kΩ.** Since $R_p >> R$, one can ignore it.

SECTION 11.6: SECOND-ORDER ACTIVE FILTERS BASED ON INDUCTOR REPLACEMENT

11.23 For the Inductance-Simulator of Fig. 11.20, $L = C_4 R_1 R_3 R_5/R_2$. Use $R_1 = R_3 = R_2 = 10k\Omega$, and $R_5 \approx 10k\Omega$ to accommodate the lack of capacitor choice. Now for $L = 10H$ and $R_5 = 10k\Omega$, $C_4 = \frac{L \times 10^4}{10^4 \times 10^4 \times R_5} = \frac{L}{10^4 R_5} = \frac{L}{10^4 \times 10^4} = 10 \times 10^{-8} = 0.1\mu F$, and for $L = 0.1H$, $C_4 = \frac{0.1}{10} \times 0.1 =$

1nF.

Alternatively, for a fixed C, where $\frac{1}{2\pi f C} = 10k\Omega$, $C_4 = \frac{1}{2\pi \times 10^4 \times 10^3} = .0159\mu F$. Use $0.01\mu F =$ **10nF**. Now select $R_1 = R_3 = R_2 = 10k\Omega$ in which case $L = C_4 R_1 R_3 R_5/R_2 = 10^4 C_4 R_5$. *For* $L =$ 10H, $R_5 = \frac{L}{10^4 C_4} = \frac{10H}{10^4 \times .01 \times 10^{-6}} =$ **100kΩ**. *For* $L = 0.1H$, $R_5 = \frac{0.1}{10^4 \times .01 \times 10^{-6}} =$ **1kΩ**.

11.24 From the solution to P11.23 above, for $L = 10H$, $R_1 = R_2 = R_3 = 10^4\Omega$, $R_5 = 10^5\Omega$, $C_4 = 10^{-8}F$, and for $L = 0.1H$, $R_1 = R_2 = R_3 = 10^4\Omega$, $R_5 = 10^3\Omega$, $C_4 = 10^{-8}F$. For $f_o = 1kHz$, $\omega_o = 2\pi \times 10^3 = 6.28 \times 10^3$ rad/s. Using $R_S = 20k\Omega$, with $\omega_o = (1/LC)^{1/2}$, or $C = \frac{1}{L\,\omega_o^2}$, and $Q = \omega_o CR$. *For* $L = 10H$, $C = \frac{1}{10 \times (6.28 \times 10^3)^2} = .00254\mu F$, and $Q = \omega_o CR = 6.28 \times 10^3 \times .00254 \times 10^{-6} \times 20 \times 10^3 = 0.319$. *For* $L = 0.1H$, $C = \frac{1}{0.1 \times (6.28 \times 10^3)^2} = .254\mu F$, and $Q = \omega_o CR = 6.28 \times 10^3 \times 0.254 \times 10^{-6} \times 20 \times 10^3 = 31.9$.

(a) Use $R_1 = R_2 = R_3 = 10^4\Omega$, $R_5 = 10^3\Omega$, $C_4 =$ **10nF**, and $C =$ **254nF**, to obtain a Q of **31.9**.

(b) Use $R_1 = R_2 = R_3 = 10^4\Omega$, $R_5 = 10^5\Omega$, $C_4 =$ **10nF**, and $C =$ **2.54nF**, to obtain a Q of **0.32**.

(c) For a design with equal-valued capacitors: For the simulated inductor $L = C_4 R_1 R_3 R_5/R_2$, and for $R_1 = R_3 = R_2 =$ **10kΩ**, $L = 10^4 C_4 R_5$. Now with $C = C_4$, the resonant frequency is $\omega_o = (\frac{1}{L C_4})^{1/2} = 2\pi \times 10^3$rad/s.

Thus $(C_4^2 R_5 \times 10^4)^{1/2} = \frac{1}{2\pi \times 10^3}$, and $R_5 = \frac{1}{(2\pi \times 10^3)^2} \times \frac{10^{-4}}{C_4^2} = \frac{2.533 \times 10^{-12}}{C_4^2}$, or $C_4 = (\frac{2.533 \times 10^{-12}}{R_5})^{1/2}$. Now, as $Q = \omega_o CR$, to raise Q, keep C_4 relatively large. Now, if R_5 is limited to $10^3\Omega$, $C_4 = (\frac{2.533 \times 10^{-12}}{10^3})^{1/2} = .0503\mu F$. (Use .05µF). Now for $C_4 = .05\mu F$, $R_5 = \frac{2.533 \times 10^{-12}}{(.05 \times 10^{-6})^2} =$ **1.013kΩ**. In which case, $Q = \omega_o C_4 R = 2\pi \times 10^3 \times .05 \times 10^{-6} \times 20 \times 10^3 =$ **6.28**. It is apparent that if a smaller value of R_5 were allowed, $C_4 = C$ could be higher, and Q would be raised. For example, with $C_4 = C = 0.1\mu F$, $R_5 = \frac{2.533 \times 10^{-12}}{(0.1 \times 10^{-6})^6} =$ **253Ω**, and $Q = 2\pi \times 10^3 \times .1 \times 10^{-6} \times 20 \times 10^3 =$ **12.6.**

11.25 For a 5th-order Butterworth with 3dB bandwidth of 10^4Hz, $\varepsilon = 1$, and $\omega_p = 2\pi \times 10^4$rad/s, with a pole radius $\omega_o = \omega_p (1/\varepsilon)^{1/N} = 2\pi \times 10^4$rad/s. Now poles are at $90° - \frac{\pi}{2(5)} = 82°$, and $82° - \frac{\pi}{5} = 46°$, and 0°. Thus the first complex pole pair has a Q such that $\cos 82° = \frac{\omega_o/2Q}{\omega_o} = \frac{1}{2Q}$, or $Q = \frac{1}{2\cos 82°} =$ 3.593. For the second complex pole pair, $Q = \frac{1}{2\cos 46°} = 0.7198$. For the 5th pole, $Q = 0.5$. Use a cascode of two circuits of the form shown in Fig. 11.22a with one of the form shown in Fig. 11.13a on the right. For a straightforward design, seven op amps would be needed. To achieve a low-frequency gain of 10, arrange a) a gain of 1 in one second-order section using a wire. b) a gain of 5 in the other second-order section using 2 series 33kΩ in the feedback path and 2 parallel 33kΩ to ground. c) a gain of 2 in the first-order section with one 33kΩ in the feedback and 2 parallel 33kΩ at the input.

Now for the first-order section (Fig. 11.13a), for which $\omega_o = \frac{1}{2\pi 10^4}$ and $Q = 0.5$, $R_1 =$ **33kΩ || 33kΩ**, $R_2 =$ **33kΩ**, $C = \frac{1}{2\pi \times 10^4 \times 33 \times 10^3} = 482.2pF$. Use $C =$ **400pF** || 80pF.

Now for one first second-order section (Fig. 11.22a), for which $\omega_o = \frac{1}{2\pi 10^4}$ and $Q = 3.593$, use $C_4 = C_6 = C =$ **400pF** || 80pF, and $R_1 = R_2 = R_3 = R_5 =$ **33kΩ**, with $R_6 = QR = 3.593\,(33) = 118.6k\Omega$.

Use **120kΩ.**

Now for the other second-order section (Fig. 11.22a), for which $\omega_o = \frac{1}{2\pi 10^4}$ and $Q = .7198$, use $C_4 = C_6 = C =$ **400pF** ∥ 80pF, and $R_1 = R_2 = R_3 = R_5 = 33k\Omega$, with $R_6 = QR = .7198\ (33) = 23.75k\Omega$. Use **24kΩ.**

11.26 From Table 11.1 for the BP filter, $T(s) = \dfrac{Ks/C_6 R_6}{s^2 + \dfrac{s}{C_6 R_6} + \dfrac{R_2}{C_4 C_6 R_1 R_3 R_5}}$. Now, $\frac{\omega_o}{Q} = \frac{1}{C_6 R_6}$, $\omega_o^2 = \dfrac{R_2}{C_4 C_6 R_1 R_3 R_5}$. Use $C_4 = C_6 = C$, with **R_6 to control Q** (and ω_o), and $R_1 = R_2 = R_3 = R$, with **R_5 to control ω_o**. Thus $\omega_o^2 = \dfrac{1}{C^2 R R_5}$, $\omega_o = \dfrac{1}{C\sqrt{R R_5}}$ --- (1), and $Q = \omega_o C R_6 = \dfrac{R_6}{\sqrt{R R_5}}$ --- (2). For capacitors using single digits 1, 2, 3, 5, the largest range of C to be accommodated by R_5 in establishing ω_o is 2/1 = 2. Since from (1), $R_5 = \dfrac{1}{\omega_o^2 C^2 R}$, the compensating range of R_5 will be $2^2 =$ **4**. Now from (2), $R_6 = \sqrt{R R_5}\, Q$, and for Q varying from 0.5 to 50 (that is by a factor of 100), and R_5 by a factor of 4, R_6 must vary by a factor of $\sqrt{4}$ (100) = **200.**

SECTION 11.7: SECOND-ORDER ACTIVE FILTERS BASED ON THE TWO-INTEGRATOR-LOOP TOPOLOGY

11.27 From Fig. 11.16c) for a BP filter, see that the 3dB bandwidth is $\frac{\omega_o}{Q}$. Here, $\omega_o = 2\pi \times 10^4$, and $\frac{\omega_o}{Q} = 500 \times 2\pi$, or $Q = \frac{10^4}{500} = \frac{100}{5} = 20$. Now for Fig. 11.24a, $CR = \frac{1}{\omega_o}$, $\frac{R_3}{R_2} = 2Q - 1$, and $\frac{R_f}{R_1} = 1$. For **C = 1nF**, $R = \dfrac{1}{2\pi \times 10^4 \times 1 \times 10^{-9}} =$ **15.92kΩ**. From Eq.11.59, $T_{bp}(s) = \dfrac{-K\,\omega_o\, s}{s^2 + s\,(\omega_o/Q) + \omega_o^2}$. Thus at ω_o, $\dfrac{V_{bp}}{V_i} = \dfrac{-K\,\omega_o\, j\omega_o}{-\omega_o^2 + j\omega_o\,(\omega_o/Q) + \omega_o^2} = -KQ$. Now, from Eq. 11.64, $K = 2 - (1/Q)$. Thus $\dfrac{V_{bp}}{V_i} = -Q\,(2 - 1/Q) = -2Q + 1 = -2\,(20) + 1 = -39$V/V. Now using Miller's theorem at resonance (ω_o), $R_{in} = \dfrac{(R_2 + R_3)}{1\text{-}gain} = \dfrac{R_2 + R_3}{1 - -39} = \dfrac{R_2 + R_3}{40}$. Now, $\dfrac{R_2 + R_3}{40} = 100k\Omega$ by specification. Thus $R_2 + R_3 = 4M\Omega$. Also $\frac{R_3}{R_2} = 2Q \times 1 = 39$ as well, and $R_3 = 39\,R_2$. Thus $R_2 + 39\,R_2 = 4M\Omega$, whence $R_2 = 100k\Omega$ and $R_3 = 3.9M\Omega$, unfortunately too large. To reduce R_3, replace the original $R_2\,R_3$ circuit by a network.

In the original situation, $V_x = V_i - \dfrac{R_2}{R_2 + R_3}(V_i - V_{bp})$

$= V_i \left\{ 1 - \dfrac{R_2}{R_2 + R_3}(1 - -39) \right\} = V_i \left\{ 1 - \dfrac{100k\Omega}{4M\Omega}(40) \right\} = 0$. Thus the positive input node of the leftmost amplifier is a virtual ground (as *might* have been obvious already). Thus use the network as shown:

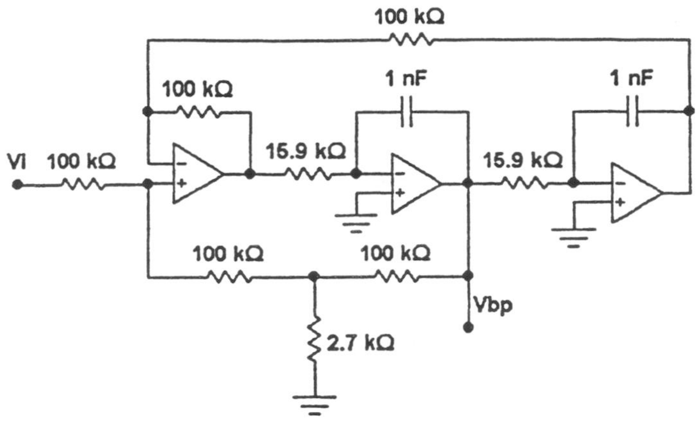

Note that since $V_x = 0$, the voltage $V_a = -V_i, \frac{R_x \parallel 100}{100 + R_x \parallel 100} = \frac{1}{39}$, $100 + R_x \parallel 100 = 39 (R_x \parallel 100)$, $\frac{38\, R_x\,(100)}{100 + R_x} = 100$, $38\, R_x = 100 + R_x$, $37\, R_x = 100$, $R_x = 100/37 =$ **2.7kΩ**. Use $R_2 = R_{3a} = R_{3b} = R_1 = R_f =$ **100kΩ**. The center-frequency gain is **–39V/V**.

11.28 From Ex. 11.22, for f_o = 5kHz, f_n = 8kHz, Q = 5, dc gain = 3V/V, and C = 1nF, one used R = 31.83kΩ, $R_1 = R_f = R_2$ = 10kΩ, R_3 = 90kΩ, R_H = 25.6kΩ, R_F = 42.7kΩ, $R_B = \infty$, R_L = 10kΩ. Here, f_o = 5kHz, f_n = 7.5kHz, Q = 10, and dc gain = 3V/V. From Eq. 11.67, $\frac{R_H}{R_L} = (\frac{\omega_n}{\omega_o})^2 = (\frac{7.5}{5})^2 = 2.25$. For R_L = **10kΩ**, R_H = **22.5kΩ**, $R_B = \infty$. From Eq. 11.66, dc gain = $\frac{-K\, R_F/R_L\, \omega_o^2}{\omega_o^2} = \frac{-K\, R_F}{R_L}$. Here $K = 2 - 1/Q$, with Q = 10. Thus K = 2 – 1/10 = 1.9, and $R_F = \frac{3R_L}{K} = \frac{3 \times 10k\Omega}{1.9}$ = **15.79kΩ**. Use **C = 1nF**, $R = \frac{1}{2\,\pi \times 5 \times 10^3 \times 10^{-9}}$ = **31.83kΩ**. Use $R_1 = R_f = R_2$ = **10kΩ**, and $R_3 = R_2\,(2Q - 1)$ = 10 (2 (10) – 1) = **190kΩ**.

11.29 Required to design a bandpass filter with f_o = 10kHz, f_{3dB} = 0.5kHz, R_{in} = 100kΩ using C = 1nF, with a center-frequency gain of –39V/V. Thus, for the Tow-Thomas circuit of Fig. 11.26, chose the negative bandpass version, for which $C_1 = 0$, $R_2 = R_3 = \infty$, and $R_1 = \frac{QR}{\text{center-frequency gain}}$. Here, from Fig. 11.16c, $f_{3dB} = \frac{f_o}{Q}$, whence $Q = \frac{f_o}{f_{3dB}} = \frac{10}{0.5} = 20$. Thus $R_1 = \frac{20R}{39}$, but the input resistance = $R_1 \geq$ 100kΩ. Thus $R = \frac{39}{20} R_1$ = 195kΩ. Use R = **200kΩ** for a slightly higher value of $R_{in} = R_1 = \frac{20}{39}$ (200) = **102.6kΩ**. Now $\omega_o = \frac{1}{RC}$, and $C = \frac{1}{2\,\pi \times 10^4 \times 20 \times 10^3}$ = **769pF**. As well, use r = **100kΩ**, and QR = 4MΩ.

SECTION 11.8: SINGLE-AMPLIFIER BIQUADRATIC ACTIVE FILTERS

11.30 Eq. 11.73 and 11.74 indicate that $\omega_o = (C_1\, C_2\, R_3\, R_4)^{-1/2}$ and $Q = \left\{ \frac{(C_1\, C_2\, R_3\, R_4)^{1/2}}{R_3} (1/C_1 + 1/C_2) \right\}^{-1}$. Try $R_3 = R_4$ = 1MΩ and $C_1 = C_2 = C$. Now for $\omega_o = 10^5$ rad/s, $C = \frac{1}{10^5 \times 10^6}$ = 10pF, $Q = 10^5 \times 10^6\, [1/10^{-11} + 1/10^{-11}]^{-1} = 10^{11} \times [2 \times 10^{11}]^{-1} = 1/2$ (as expected). Now since the required Q is 0.707, C must be raised to allow resistor values which are different, yet no larger than 1MΩ. Thus we could try C = 20pF and proceed to find $R_3\, R_4$, then each separately, as follows: See $R_3\, R_4 = (C_1\, C_2\, \omega_o^2)^{-1} = (20 \times 10^{-12} \times 10^5)^{-2} = (20 \times 10^{-7})^{-2} = \frac{10^{14}}{20 \times 20} = 0.25 \times 10^{12}$. Now for $Q = 1/\sqrt{2}$, $\sqrt{2} = \frac{(10^5)^{-1}}{R_3} (\frac{2}{20 \times 10^{-12}})$, or $R_3 = \frac{10^{-5}}{\sqrt{2}} \times 10^{11}$ = .707MΩ, for which $R_4 = \dfrac{.25 \times 10^{12}}{.707 \times 10^6} = .354\text{M}\Omega$. Thus, $C_1 = C_2 =$ **20pF**, $R_3 =$ **.707MΩ** and $R_4 = R_3/2 =$ **.354MΩ** is a possible solution. *Check:* $\omega_o = (20 \times 10^{-12} \times 20 \times 10^{-12} \times .707 \times 10^6 \times 0.354 \times 10^6)^{-1/2} = 10^5$. But this is possibly not the solution with the largest possible resistors!

Alternatively, try $C_1 = 10\text{pF}$, $C_2 = 20\text{pF}$, for which $R_3 R_4 = (10 \times 10^{-12} \times 20 \times 10^{-12} \times 10^5 \times 10^5)^{-1} = 0.5 \times 10^{-12}$, and for Q, $\sqrt{2} = \dfrac{(10^5)^{-1}}{R_3}\left[\dfrac{1}{20 \times 10^{-12}} + \dfrac{1}{10 \times 10^{-12}}\right]$, or $R_3 = \dfrac{10^{-5}}{\sqrt{2}} \times 10^{11}\ (1.5) = 1.061\text{M}\Omega$, for which $R_4 = \dfrac{0.5 \times 10^{12}}{1.061 \times 10^6} = .471\text{M}\Omega$. But this solution is not directly acceptable since R_3 is too large (although a series combination would clearly suffice). For example, one could use $C_1 =$ **10pF**, $C_2 =$ **20pF**, $R_3 =$ **1MΩ + 62kΩ** in series, $R_4 =$ **470kΩ**.

Alternatively, try $C_1 = 10\text{pF}$, $C_2 = 50\text{pF}$, for which $R_3 R_4 = (10 \times 50 \times 10^{-24} \times 10^{10}) = 0.2 \times 10^{12}$, and $\sqrt{2} = \dfrac{(10^5)^{-1}}{R_3}\left[\dfrac{1}{50 \times 10^{-12}} + \dfrac{1}{10 \times 10^{-12}}\right]$, or $R_3 = \dfrac{10^{-5}}{\sqrt{2}}\ (1.2 \times 10^{11}) = 0.849\text{M}\Omega$, for which $R_4 = \dfrac{0.2 \times 10^{12}}{.849 \times 10^6} = 0.236\text{M}\Omega$. Could use $C_1 =$ **10pF**, $C_2 =$ **50pF**, $R_3 =$ **.849MΩ**, $R_4 =$ **.236MΩ**. Of the three solutions, the first, with equal capacitors, is the most straightforward.

11.31 Now, for $\omega_o = 2\pi \times 10^4\text{rad/s}$, $Q = \dfrac{10^4}{5 \times 10^2} = 20$. Use $C_1 = C_2 = 1\text{nF} = 10^{-9}\text{F}$. From $\omega_o = (C_1 C_2 R_3 R_4)^{-1/2}$, $R_3 R_4 = \dfrac{1}{(2\pi \times 10^4 \times 10^{-9})^2} = 2.533 \times 10^8$. From $Q = \omega_o R_3 [1/C_1 + 1/C_2]^{-1}$, $R_3 = \dfrac{20}{10^4 \times 2\pi}\left(\dfrac{2}{10^{-9}}\right) = 6.366 \times 10^5 = 637\Omega$, and $R_4 = \dfrac{2.533 \times 10^8}{6.366 \times 10^5} = 398\Omega$. *Check:*

$$Q = \left[\frac{(C_1 C_2 R_3 R_4)^{1/2}}{R_3}(1/C_1 + 1/C_2)\right]^{-1} = \left[\frac{(10^{-9} \times 10^{-9} \times 6.37 \times 10^5 \times 398)^{1/2}}{6.36 \times 10^5}(1/10^{-9} + 1/10^{-9})\right]^{-1} = \left[\frac{159.2 \times 10^{-7}}{6.36 \times 10^5} \times \frac{2}{10^{-9}}\right]^{-1} = 19.97:$$

OK. Now, *for the center frequency gain:* For V_A (at the common node A of C_1, C_2, R_4/α and $R_4/(1-\alpha)$), with the voltage at the amplifier's negative input being zero volts, see $V_A = 0 - \dfrac{1}{sC_2}\dfrac{V_o}{R_3} = -\dfrac{V_o}{s C_2 R_3}$. Writing a node equation at node A yields: $\dfrac{V_o}{R_3} + s\, C_1\left[V_o + \dfrac{V_o}{s\, C_2 R_3}\right] - \dfrac{V_A}{R_4/(1-\alpha)} + \dfrac{V_i - V_A}{R_4/\alpha} = 0$, or $\dfrac{V_o}{R_3} + s\, C_1\, V_o + V_o\dfrac{C_1}{C_2 R_3} + \dfrac{(1-\alpha)\, V_o}{s\, C_2 R_3 R_4} + \dfrac{\alpha\, V_o}{s\, C_2 R_3 R_4} = \dfrac{-V_i}{R_4/\alpha}$, or $\dfrac{V_o}{V_i} =$

$$\frac{-\alpha/R_4}{s\, C_1 + \dfrac{1}{R_3} + \dfrac{C_1}{R_3 C_3} + \dfrac{1}{s\, C_2 R_3 R_4}} = \frac{-s\dfrac{\alpha}{C_1 R_4}}{s^2 + s\left(\dfrac{1}{C_1 R_3} + \dfrac{1}{C_2 R_3}\right) + \dfrac{1}{C_1 C_2 R_3 R_4}}$$

, for which the center-frequency gain is

$-\left[\dfrac{\alpha}{C_1 R_4}\right]\left[\dfrac{1}{C_1 R_3} + \dfrac{1}{C_2 R_3}\right] = \dfrac{-\alpha R_3 C_2}{R_4 (C_1 + C_2)}$. Here, $1 = \dfrac{\alpha R_3 C_2}{R_4 (C_1 + C_2)}$, whence $\alpha = \dfrac{R_4 (C_1 + C_2)}{R_3 C_2} = \dfrac{398\ (10^{-9} + 10^{-9})}{637 \times 10^3\ (10^{-9})} = 1.25 \times 10^{-3}$. Now, $\dfrac{R_4}{\alpha} = \dfrac{398}{1.25 \times 10^{-3}} =$ **318.4kΩ**, and $\dfrac{R_4}{1-\alpha} = \dfrac{398}{1 - 1.25 \times 10^{-3}} = 398.5\Omega$. As a result, at very high frequencies (where the capacitors are short circuits), $R_{in} \approx$ **318kΩ**, and at very low frequencies (where the capacitors are open), $R_{in} = 318k\Omega + 398\Omega \approx$ **318kΩ**, as well.

11.32 From the derivation in P11.31 above: $\dfrac{V_o}{V_i} = T(s) = \dfrac{-s\dfrac{\alpha}{C_1 R_4}}{s^2 + s\left[\dfrac{1}{C_1 R_3} + \dfrac{1}{C_2 R_3}\right] + \dfrac{1}{C_1 C_2 R_3 R_4}}$ – – (1), where the voltage at the join-point A of C_1, C_2, $\dfrac{R_4}{\alpha}$, and $\dfrac{R_4}{(1-\alpha)}$, is V_A. Note that the current in R_3 establishes the voltage across C_2, since their join-point is virtual-ground point with no current loss. Thus $V_A = \dfrac{-V_o}{s\,C_2 R_3}$. Now the input current is $I_i = \dfrac{V_i - V_A}{R_4/\alpha} = \dfrac{V_i + \dfrac{V_o}{s\,C_2 R_3}}{R_4/\alpha}$, and the input impedance is $Z_i(s) = \dfrac{V_i}{I_i} = \dfrac{R_4/\alpha}{1 + \dfrac{T(s)}{s\,C_2 R_3}}$. Thus $Y_i(s) = 1/Z_i(s)$

$$\left[1 - \frac{\alpha/(C_1C_2R_3R_4)}{s^2 + \dfrac{s}{R_3}\left[\dfrac{1}{C_1} + \dfrac{1}{C_2}\right] + \dfrac{1}{C_1C_2R_3R_4}}\right] = \left[\frac{s^2 + s\left[\dfrac{1}{R_3C_1} + \dfrac{1}{R_3C_2}\right] + \dfrac{1-\alpha}{C_1C_2R_3R_4}}{s^2 + \dfrac{s}{R_3}\left[\dfrac{1}{C_1} + \dfrac{1}{C_2}\right] + \dfrac{1}{C_1C_2R_3R_4}}\right]$$

Thus $Z_i(s) = \dfrac{R_4}{\alpha} \times \dfrac{s^2 + \dfrac{s}{R_3}\left[\dfrac{1}{C_1} + \dfrac{1}{C_2}\right] + \dfrac{1}{C_1C_2R_3R_4}}{s^2 + \dfrac{s}{R_3}\left[\dfrac{1}{C_1} + \dfrac{1}{C_2}\right] + \dfrac{1-\alpha}{C_1C_2R_3R_4}}$

Now (from Equations 11.73, 11.74), $\omega_o = 1/(C_1C_2R_3R_4)^{1/2}$, $Q = \left\{\dfrac{(C_1C_2R_3R_4)^{1/2}}{R_3}\left[\dfrac{1}{C_1} + \dfrac{1}{C_2}\right]\right\}^{-1}$,

or $1/Q = \dfrac{1}{\omega_o R_3}\left[\dfrac{1}{C_1} + \dfrac{1}{C_2}\right]$. Thus $Z_1(s) = \dfrac{R_4}{\alpha} \times \dfrac{s^2 + \dfrac{\omega_o}{Q}s + \omega_o^2}{s^2 + s\left(\dfrac{\omega_o}{Q}\right) + (1-\alpha)\omega_o^2}$. Now, at very low frequencies, as $s \to 0$, $Z_i = \dfrac{R_4}{\alpha} \times \dfrac{\omega_o^2}{(1-\alpha)\omega_o^2} = \dfrac{R_4}{\alpha\,(1-\alpha)}$, as can be seen quite directly with capacitors open, noting that $\dfrac{R_4}{\alpha} + \dfrac{R_4}{1-\alpha} = \dfrac{R_4}{\alpha\,(1-\alpha)}$. Now, at very high frequencies, as $s \to \infty$, $Z_i = \mathbf{R_4/\alpha}$, with capacitors shorted. Now at the center frequency, $s = -j\omega_o$, $Z_i = \frac{R_4}{\alpha} \times \frac{-\omega_o^2 + j\frac{\omega_o^2}{Q} + \omega_o^2}{-\omega_o^2 + j\frac{\omega_o^2}{Q} + (1-\alpha)\,\omega_o^2} =$ $\frac{R_4}{\alpha}\;\frac{j\,\omega_o^2/Q}{j\,\omega_o^2/Q - \alpha\,\omega_o^2} = \frac{R_4}{\alpha} \times \frac{1}{1 - j\,\alpha\,Q} = \frac{R_4\,(1 + j\,\alpha\,Q)}{\alpha\,(1 + \alpha^2 Q^2)}$, which, for $\alpha\,Q$ large, is $\frac{R_4}{\alpha}\left[\frac{1}{\alpha^2 Q^2} + \frac{j}{\alpha\,Q}\right]$, of relatively low magnitude, and with a 90° (inductor-like) phase.

11.33 For this Butterworth, $N = 7$, $A_{max} = 3\text{dB}$, $A_{dc} = 0\text{dB}$. Here $f_p = 5\text{kHz}$, and $\omega_p = 2\pi(5) = 10^5\pi$ rad/s, and $\varepsilon = (10^{A_{max}/10} - 1)^{½} = (10^{3/10} - 1)^{½} = 0.998$. (It should actually be 1.000. Why? Why is it not?). Thus for each filter stage, $\omega_o = \omega_p \left[\frac{1}{\varepsilon}\right]^{1/7}$, or $\omega_o = 10^5\pi \left[\frac{1}{.998}\right]^{1/7} = 3.14 \times 10^5$ rad/s. The 7 poles are located at 0°, ± 180/7 = ±25.7°, ±51.4°, and ±77.14°. For each pole pair, $\cos\Theta = \frac{\omega_o/2Q}{\omega_o} = 1/2Q$, whence $Q = 0.5/\cos\Theta = 0.5/\cos 0 = 0.5$, $0.5/\cos 25.7 = .555$, $0.5/\cos 51.4 = 0.801$, and $0.5/\cos 77.14 = 2.25$ respectively. From Ex. 11.28, see that for the Sallen and Key circuit of Fig. 11.34(c), that the dc gain is 1, and from Eq.11.77 and Eq. 11.78 that $\omega_o = (R_1R_2C_3C_4)^{-1/2}$, and $1/Q = \frac{1}{\omega_o C_4}\left[\frac{1}{R_1} + \frac{1}{R_2}\right]$. Use $C = 3.3\text{nF}$ for C_1, C_2 for all sections.

Now for the first-order section, $\omega_o = 1/RC$, and $R = \frac{1}{\omega_o C} = \frac{1}{3.14 \times 10^5 \times 3.3 \times 10^{-9}} = 965\Omega$ or **0.965kΩ.**

For the lowest-Q Sallen and Key Section, $1/Q = 1/.555 = \frac{1}{3.14 \times 10^5 \times 3.3 \times 10^{-9}}\left[\frac{1}{R_1} + \frac{1}{R_2}\right]$, or $1/R_1 + 1/R_2 = \frac{1}{.555} \times \frac{1}{965} = .001868$ – – – (1). Also $\omega_o = (R_1R_2C^2)^{-1/2}$ implies that $R_1\,R_2 = \frac{1}{\omega_o^2 C^2} = R^2 = 965^2$, or $R_1 = \frac{965^2}{R_2}$. Thus in (1) $\frac{R_2}{965^2} + \frac{1}{R_2} = \frac{1}{.555} \times \frac{1}{965}$. Therefore $R_2^2 + 965^2\,R_2 - \frac{965}{.555} = 0$ – – – (2),

or for R_2 in kilohm, $R_2^2 + .9312\,R_2 - 1.739 = 0$, $R_2 = \frac{-.9312 \pm \sqrt{.9312^2 + 4(1.739)}}{2} = \frac{.9312 \pm 2.797}{2}$ $=$ **0.933kΩ**, for which $R_1 = \frac{965^2}{.933} =$ **0.998kΩ.**

Now for the second Sallen and Key (by analogy from (2)), $R_2^2 + .965^2\,R_2 - \frac{.965}{.801} = 0$, $R_2^2 + .9312\,R_2 - 1.205 = 0$, $R_2 = \frac{-.9312 \pm \sqrt{.9312^2 + 4(1.205)}}{2} =$ **0.727kΩ**, for which $R_1 = \frac{.965^2}{.727} =$ **1.281kΩ.**

Now for the third Sallen and Key, $R_2^2 + .965^2\,R_2 - \frac{.965}{2.25} = 0$, $R_2^2 + .9312\,R_2 - .429 = 0$, $R_2 = \frac{-.931 \pm \sqrt{.931^2 + 4(.429)}}{2} =$ **0.348kΩ**, for which $R_1 = \frac{.965^2}{.348} =$ **2.68kΩ**, with all capacitors of 3.3nF value.

SECTION 11.9: SENSITIVITY

11.34 From Eq. 11.77 and 11.78, $\omega_o = (C_3\,C_4\,R_1\,R_2)^{-½}$, and $Q = \left[\frac{(C_3\,C_4\,R_1\,R_2)^{½}}{C_4}\left(\frac{1}{R_1} + \frac{1}{R_2}\right)\right]^{-1}$.

Now for ω_o, see $\frac{\partial\,\omega_o}{\partial\,C_3} = -½\,(C_4\,R_1\,R_2)^{-½}\,C_3^{-3/2} = -\frac{\omega_o}{2\,C_3}$, and $S_{C_3}^{\omega_o} = \frac{\partial\,\omega_o}{\partial\,C_3} \times \frac{C_3}{\omega_o} = -\frac{\omega_o}{2\,C_3} \times$

$\frac{C_3}{\omega_o} = -1/2$. Now $C_3 = C_{3a} + C_{3b}$ and $C_{3b} = k_3 C_{3a}$, nominally. However, note that C_{3a} and C_{3b} are independent from a sensitivity point of view. Thus $\frac{\partial C_3}{\partial C_{3a}} = 1$. Now $S_{C_{3a}}^{\omega_o} = \frac{\partial \omega_o}{\partial C_{3a}} \times \frac{C_{3a}}{\omega_o} = \frac{\partial \omega_o}{\partial C_3} \times \frac{\partial C_3}{\partial C_{3a}} \times \frac{C_{3a}}{\omega_o} = -\frac{\omega_o}{2 C_3} \times 1 \times \frac{C_{3a}}{\omega_o} = -\frac{1}{2}\frac{C_{3a}}{C_3} = -\frac{1}{2}\frac{C_{3a}}{(1+k_3) C_{3a}} = \frac{-1}{2(1+k_3)}$, and $S_{C_{3b}}^{\omega_o} = -\frac{1}{2}\frac{C_{3b}}{C_3} = -\frac{1}{2}\frac{k_3 C_{3a}}{(1+k_3) C_{3a}} = \frac{-k_3}{2(1+k_3)}$. Correspondingly, $S_{C_{4a}}^{\omega_o} = -\frac{1}{2(1+k_4)}$, and $S_{C_{4b}}^{\omega_o} = -\frac{k_4}{2(1+k_4)}$. When $k_3 = k_4 = 1$, all these sensitivities become **–1/4**.

Now for Q, see $\frac{\partial Q}{\partial C_3} = \left[\frac{(C_3 C_4 R_1 R_2)^{1/2}}{C_4}\left(\frac{1}{R_1} + \frac{1}{R_2}\right)\right]^{-1}\left[\frac{-1/2}{C_3}\right] = -1/2\frac{Q}{C_3}$. Thus $S_{C_3}^{Q} = -1/2$, where $S_{C_{3a}}^{Q} = -\frac{1}{2(1+k_3)}$, and $S_{C_{3b}}^{Q} = -\frac{k_3}{2(1+k_3)}$, both being **–1/4** for $k_3 = 1$. See also $\frac{\partial Q}{\partial C_4} = -\frac{-1/2\, Q}{C_4}$. Thus $S_{C_4}^{Q} = 1/2$, and $S_{C_{4a}}^{Q} = \frac{1}{2(1+k_4)}$, $S_{C_{4b}}^{Q} = \frac{k_4}{2(1+k_4)}$, both being **1/4** for $k_4 = 1$.

11.35 From Equations 11.77 and 11.78, $\omega_o = (C_3 C_4 R_1 R_2)^{-1/2}$, and $Q = \left\{\frac{(C_3 C_4 R_1 R_2)^{1/2}}{C_4}\left[\frac{1}{R_1} + \frac{1}{R_2}\right]\right\}^{-1}$. **Now for ω_o**, see $\frac{\partial \omega_o}{\partial R_1} = -1/2\,(C_3 C_4 R_2)^{-1/2} \times R_1^{-3/2} = -\frac{\omega_o}{2 R_1}$, and $S_{R_1}^{\omega_o} = \frac{\partial \omega_o}{\partial R_1} \times \frac{R_1}{\omega_o} = -\frac{\omega_o}{2 R_1} \times \frac{R_1}{\omega_o} = -1/2$. Likewise $S_{R_2}^{\omega_o} = -1/2$ as well.

Now for a fixed temperature, $R_1, R_2 = (1 \pm k)R$ with $\frac{\partial R_1, R_2}{\partial k} = \pm R$, and $S_k^{R_1, R_2} = \pm R \times \frac{k}{R} = \pm k$. Thus $S_k^{\omega_o} = S_{R_1}^{\omega_o} S_k^{R_1} + S_{R_2}^{\omega_o} S_k^{R_2} = -1/2\,(+k + (-k)) = 0$.

Now for small k, $R_1 \approx R_2 = R_0 (1 + a\,(T_o - T))$, whence $\frac{\partial R_1}{\partial T} = -a R_o$ and $S_T^{R_1} = \frac{\partial R_1}{\partial T} \times \frac{T}{R_1} = -a R_o \times \frac{T}{R_o (1 + a\,(T_o - T))} = -\frac{a\,T}{1 + a\,(T_o - T)} = S_T^{R_2}$. Thus $S_T^{\omega_o} = S_{R_1}^{\omega_o} \times S_T^{R_1} + S_{R_2}^{\omega_o} \times S_T^{R_2} = 2\left[-1/2 \times \frac{-a\,T}{1 + a\,(T_o - T)}\right]$, which around $T = T_o$, is **a T_o**.

Now for Q, see $Q = \left[\frac{C_4}{C_3 R_1 R_2}\right]^{1/2}\left[\frac{R_1 R_2}{R_1 + R_2}\right] = \left[\frac{C_4}{C_3}\right]^{1/2}\frac{(R_1 R_2)^{1/2}}{R_1 + R_2}$. Thus $\frac{\partial Q}{\partial R_1} = \left[\frac{C_4 R_2}{C_3}\right]^{1/2} \times \left[\frac{1/2}{R_1 + R_2} \times \frac{1}{R_1^{1/2}} + (-1) \times \frac{R_1^{1/2}}{(R_1 + R_2)^2}\right] = \left[\frac{C_4}{C_3}\right]^{1/2}\frac{(R_1 R_2)^{1/2}}{R_1 + R_2}\left[\frac{1}{2 R_1} - \frac{1}{R_1 + R_2}\right] = Q \times \frac{(R_2 - R_1)}{2 R_1 (R_1 + R_2)}$. Now for $R_2 = R_1$, $\frac{\partial Q}{\partial R_1} = 0$, and $S_{R_1}^{Q} = S_{R_2}^{Q} = 0$. But for $R_1, R_2 = (1 \pm k)R$ with $k << 1$, $\frac{\partial Q}{\partial R_1} \approx Q \times \frac{-2\,kR}{2R\,(2R)} = -\frac{k\,Q}{2R}$, with $S_{R_1}^{Q} = \frac{-k\,Q}{2R} \times \frac{R}{Q} = -k/2$. Likewise $\frac{\partial Q}{\partial R_1} = \frac{k\,Q}{2R}$, with $S_{R_2}^{Q} = +k/2$.

Now, $S_k^{Q} = S_{R_1}^{Q} S_k^{R_1} + S_{R_2}^{Q} S_k^{R_2}$, where here, $R_1, R_2 = (1 \pm k)R$, and $\frac{\partial R_1, R_2}{\partial k} = \pm R$, with $S_k^{R_1, R_2} = \pm R \times \frac{k}{R} = \pm k$. Thus $S_k^{Q} = -k/2(k) + k/2\,(-k) = -\mathbf{k^2}$. Also $S_T^{Q} = S_{R_1}^{Q} S_T^{R_1} + S_{R_2}^{Q} S_T^{R_2}$, where $R = R_o\,(1 + a\,(T_o - T)$, and $\frac{\partial R_1, R_2}{\partial T} = -(1 \pm k)\,R_o\,a$, with $S_T^{R_1, R_2} = -(1 \pm k)\,R_o\,a \times \frac{T}{(1 \pm k)\,R_o\,(1 + a\,(T_o - T)}$

$= -\dfrac{a\,T}{1 + a\,(T_o - T)}$. Thus $S_T^{f} = -k/2\left[\dfrac{-a\,T}{1 + a\,(T_o - T)}\right] + k/2\left[\dfrac{-a\,T}{1 + a\,(T_o - T)}\right] =$ **0.**

SECTION 11.10: SWITCHED-CAPACITOR FILTERS

11.36 For Φ_1 high, C_1 is discharged to zero (that is, virtual ground (VG) via the second switch. Then, for Φ_1 high, the first switch closes, charging C_1 to 1V, for a total charge transfer of $Q = CV = 0.1 \times 10^{-½} \times 1$ = **0.1pC**. For 1MHz operation, the corresponding average current is $I = 0.1 \times 10^{-12} \times 10^6 =$ **0.1μA.** Equivalent input resistance is $\dfrac{V}{I} = \dfrac{1V}{0.1\mu A} =$ **10MΩ**. For a 2pF feedback capacitance, 0.1pC charge produces an output change of $V = \dfrac{Q}{C} = \dfrac{0.1 \times 10^{-12}}{2 \times 10^{-12}} =$ **0.05V**. Thus the output change per cycle would be **50mV**. For $\upsilon_i = +1$V, C_1 charges positive when Φ_1 occurs. When Φ_2 occurs, VG tends to go positive, forcing the output negative to compensate. Thus the output changes in the negative direction. For saturation at ±10V, the maximum output change is 20 volts, requiring $\dfrac{20V}{50mV} =$ **400 cycles**. The average slope is –20V in 400 μs, or $-\dfrac{20}{400} = -.05$V/μs, or –50V/ms, or **–50,000V/s**. For a –0.1V input, the output slope becomes +5V/ms or **5000V/s**.

11.37 Require $f_{3dB} = f_o = 10^5$Hz, $Q = 1/\sqrt{2} = 0.707$, and $A_o = 1$V/V for $f_c = 1/T_c = 10^6$Hz, $C_1 = C_2 =$ 2pF.

Now $\omega_o = \dfrac{1}{T_c}\sqrt{\dfrac{C_3\,C_4}{C_2\,C_1}}$. Thus $\sqrt{C_3\,C_4} = \omega_o\,T_c\,\sqrt{C_2\,C_1} = 2\,\pi \times 10^5 \times 10^{-6} \times 2 \times 10^{-12} = 1.257 \times 10^{-12}$. Now for $\dfrac{T_c}{C_3}\,C_2 = \dfrac{T_c}{C_4}\,C_1$, with $C_1 = C_2$, then $C_3 = C_4 =$ **1.257pF**. Now $Q = \dfrac{C_4}{C_5}$. Thus $C_5 = \dfrac{C_4}{Q} = \dfrac{1.257 \times 10^{-12}}{.707} =$ **1.777pF**. *Check:* $C_5 = \omega_o\,T_c\,C/Q = 2\,\pi \times 10^5 \times 10^{-6} \times 2 \times 10^{-12} \times 1.414 = 1.7772 \times 10^{-12}$, as before. Now A_o of the low-pass function is $R_4/R_6 = C_6/C_5$. Thus $C_6 = 1 \times C_5 =$ **1.777pF**. In summary, $C_1 = C_2 =$ **2.000pF**, $C_3 = C_4 =$ **1.257pF**, $C_5 = C_6 =$ **1.777pF**. The output of the first integrator is the **bandpass output**. Its center frequency is $f_o = 10^5$**Hz**. Its 3dB bandwidth is $\dfrac{f_o}{Q} = \dfrac{10^5}{0.707} = 1.41 \times 10^5$**Hz**. Its maximum gain (from Eq. 11.100) is $\dfrac{C_6}{C_5} = \dfrac{1.777}{1.777} =$ **1V/V.**

SECTION 11.11: TUNED AMPLIFIERS

11.38 For $I_E = 1$mA, $r_e = \dfrac{25mV}{1mA} = 25\Omega$. *For* $R_E = r_e = 25\Omega$: $R_{in} = (\beta + 1)\,(r_e + R_E) = 201\,(25 + 25) =$ 10.05kΩ. Gain from base to collector, $A_{bc} = \dfrac{-R_L}{r_e + R_E} = \dfrac{-5 \times 10^3}{25 + 25} = -100$V/V. Gain from base to emitter, $A_{be} = \dfrac{R_E}{r_e + R_E} = \dfrac{25}{25 + 25} = 0.5$V/V. Equivalent input capacitance $C_{eg} = 10\,(1 - 0.5) + 1\,(1 - (-100)) = 5 + 101 = 106$pF. The total tuning capacitance is $C = 200 + 106 = 306$pF. Now $\omega_o = 1/\sqrt{LC} = (1 \times 10^{-6} \times 306 \times 10^{-12})^{-½} = 57.17 \times 10^6$rad/s. ≡ **9.1MHz.**

$B = \dfrac{1}{CR} = \dfrac{1}{306 \times 10^{-12}\,(10.05 \times 10^3 \parallel 10 \times 10^3)} =$ **652krad/s.** $Q = \dfrac{\omega_o}{B} = \dfrac{57.17 \times 10^6}{652 \times 10^3} =$ **87.7.**

Center-Frequency Gain $A = \dfrac{10.05}{10 + 10.05} \times (-100) =$ **–50.1V/V.**

For $R_E = 9\,r_e$: $R_{in} = 201\,(10)\,(25) = 50.25$kΩ, $A_{bc} = -\dfrac{5 \times 10^3}{10\,(25)} = -20$V/V, $A_{be} = \dfrac{9\,(25)}{10\,(25)} = 0.9$V/V, $C = 200 + 10\,(1 - 0.9) + 1\,(1 - -20) = 200 + 1 + 21 = 222$pF, $R = 10$kΩ ∥ 50.25kΩ = 8.34kΩ, $\omega_o = (1 \times 10^{-6} \times 222 \times 10^{-12})^{-½} =$ **67.1 × 10⁶**rad/s, $B = \dfrac{1}{222 \times 10^{-12} \times 8.34 \times 10^3} =$ **540 × 10³**rad/s, $Q = \dfrac{67.1}{0.54} =$ **124**, $A = \dfrac{50.25}{50.25 + 10} \times (-20) =$ **–16.7V/V.**

11.39 Here, I_E = 1mA. Thus r_e = 25Ω and r_π = 201 (25) = 5.025kΩ. Gain from base to collector = −5000/25 = −200V/V, C_{in} = 10pF + 1pF (1 − −200) = 211pF,

For direct connection (as in P11.72 of the Text): C = 200 + 211 = 411pF, R = 10kΩ ∥ 5.025kΩ = 3.34kΩ, $\omega_o = (LC)^{-1/2} = (1 \times 10^{-6} \times 411 \times 10^{-12})^{-1/2}$ = **49.3 × 10⁶rad/s**, $B = (CR)^{-1} = (411 \times 10^{-12} \times 3.34 \times 10^3)^{-1}$ = **0.728 × 10⁶rad/s**, $Q = \frac{\omega_o}{B} = \frac{49.3}{0.728}$ = **67.7**, $A = \frac{5.025}{5.025 + 10.0} \times (-200)$ = **−66.9V/V**.

For a tapped coil with k = 0.5: R_{in} = 5.025kΩ is transformed to $\frac{5.025 \times 10^3}{(0.5)^2}$ = 20.1kΩ, C_{in} = 211pF is transformed to 211×0.5^2 = 52.8pF, C = 200 + 52.8 = 252.8pF, R = 10kΩ ∥ 20.1kΩ = 6.68kΩ, $\omega_o = (1 \times 10^{-6} \times 252.8 \times 10^{-12})^{-1/2}$ = 62.9 × 10⁶**rad/s**, $B = (252.8 \times 10^{-12} \times 6.68 \times 10^3)^{-1}$ = **0.592** × 10⁶**rad/s**, $Q = \frac{62.8}{0.592}$ = **106**, $A = \frac{20.1}{10 + 20.1} \times 0.5 \times (-200)$ = **−66.8V/V**.

For a tapped coil with k = 0.1: $R_T = \frac{5.025 \times 10^3}{(0.1)^2}$ = 502kΩ, $C_T = 211 \times 10^{-12} \times (0.1)^2$ = 2.11pF, C = 200 + 2.1 = 202.1pF, R = 10kΩ ∥ 502kΩ = 9.8kΩ, $\omega_o = (1 \times 10^{-6} \times 202.1 \times 10^{-12})^{-1/2}$ = **70.3 ×** **10⁶**rad/s, $B = (202.1 \times 10^{-12} \times 9.80 \times 10^3)^{-1}$ = **0.505 × 10⁶**rad/s, $Q = \frac{70.3}{0.505}$ = **139.2**, $A = \frac{502}{10 + 502} \times 0.1 \times (-200)$ = **−19.6V/V**.

Comparison:

	ω_o Mrad/s.	Q	A V/V
Basic Circuit	49.3	68	−66.9
$R_E = r_e$	57.2	88	−50.1
$R_E = 9\ r_e$	67.1	124	−16.7
$k = 0.5$	62.9	106	−66.8
$k = 0.1$	70.3	139	−19.6

Thus the tapped coil gives the best results in general.

11.40 For the coil, $Q_o = \frac{\omega_o L}{r_s} \approx \frac{R_p}{\omega_o L}$. Here, at 10MHz, $\omega_o L = 2\pi \times 10^7 \times 2 \times 10^{-6}$ = 125.7Ω. For Q_o = 200, $r_s = \frac{125.7}{200}$ = 0.63Ω, and R_p = 200 (125.7) = **25.1kΩ**. For resonance at 10MHz, $\omega_o = (LC)^{-1/2}$, or $C = \frac{1}{\omega_o^2 L} = \frac{1}{(2\pi \times 10^7)^2 \times 2 \times 10^{-6}}$ = **126.7pF**. Bandwidth $B = \frac{1}{2\pi CR}$, for which $R = \frac{1}{2\pi \times 200 \times 10^3 \times 126.7 \times 10^{-12}}$ = 6.281kΩ. The resistor to be added is 6.281 ∥ (−25.1)kΩ = $\frac{6.281 \times 25.1}{25.1 - 6.28}$ = **8.38kΩ**.

11.41 For a single LC circuit, $Q = \frac{\omega_o}{B} = \frac{f_o}{f_b} = \frac{10^6}{10^5} =$ **10**. For synchronous tuning of N stages, from Eq. 11.110, $f_B = \frac{f_o}{Q}\sqrt{2^{1/N}-1}$. Here, $50 \times 10^3 = \frac{10^6}{10}\sqrt{2^{1/N}-1}$, or $1 = 2\sqrt{2^{1/N}-1}$, $2^{1/N}-1 = (\frac{1}{2})^2 = 0.25$, $2^{1/N} = 1.25$, $1/N \log 2 = \log 1.25$, $N = \frac{\log 2}{\log 1.25} = \frac{.301}{.0969} = 3.1$. Thus, use **3 stages**: *Check*: $f_B = \frac{10^6}{10}\sqrt{2^{1/3}-1} = 10^5\sqrt{1.2599-1} = 10^5\sqrt{.2599} = .51 \times 10^5$Hz. OK.

For the 30dB bandwidth, (from problem 11.77 on page 972 of the Text, part b)), for synchronous tuning: $|T(j\omega)| = \frac{|T(j\omega_o)|}{[1+4(2^{1/N}-1)(\delta f/f_b)^2]^{N/2}}$

For 1 stage: $N = 1$, $[1+4(2-1)(\delta f/f_b)^2]^{1/2} = \frac{|T_o|}{|T|} = 10^{+30/20} = 31.6$, or $1+4(\delta f/f_b)^2 = 1000$, $\delta f/f_b = (\frac{1000-1}{4})^{1/2} = 15.8$. Thus the skirt selectivity is $\frac{S}{B} = \frac{2\,\delta f}{f_b} = 2\,(15.8) =$ **31.6**, and the 30dB bandwidth is 31.6 (100kHz) = **3.16MHz**. For the synchronously-tuned cascade, with $N = 3$, $[1+4(2^{1/3}-1)(\delta f/f_b)^2]^{3/2} = 31.6$, $1+4(.2600)(\delta f/f_b)^2 = 9.995$, that is $\delta f/f_b = \left(\frac{9.995-1}{4\,(.26)}\right)^{1/2} = 2.94$. Thus the skirt selectivity factor is 2 (2.94) = **5.88**, and the 30dB bandwidth is 5.88 (50kHz) = **0.294MHz**.

11.42 Using Equations 11.115 and 11.116 of the Text: $f_{01} = f_o + \frac{B}{2\sqrt{2}} = 10.7 + \frac{0.1}{2\sqrt{2}} = 10.735$MHz, $f_{02} = f_o - \frac{B}{2\sqrt{2}} = 10.7 - \frac{0.1}{2\sqrt{2}} = 10.665$MHz, $B_1 = B_2 = \frac{B}{\sqrt{2}} = \frac{.100}{\sqrt{2}} = .0707$MHz, $Q_1 = Q_2 = \frac{\sqrt{2} \times 10.7}{0.1} = 151.3$. Now $f_o = \frac{1}{2\pi\sqrt{LC}}$, and $C_1 = \frac{1}{(2\pi f_{01})^2 L} = \frac{1}{(2\pi \times 10.735 \times 10^6)^2 \times 3 \times 10^{-6}} =$ **73.27pF**, and $C_2 = \frac{1}{(2\pi \times 10.665 \times 10^6)^2 \times 3 \times 10^{-6}} =$ **74.23pF**. Now $R = 2\pi f_o L Q$, and $R_1 = 2\pi \times 10.735 \times 10^6 \times 3 \times 10^{-6} \times 151.3 =$ **30.616kΩ**, $R_2 = 2\pi \times 10.665 \times 10^6 \times 3 \times 10^{-6} \times 151.3 =$ **30.416kΩ**. Since the voltage gain at resonance is proportional to R (see Ex. 11.37), the relative peak gain of each of the two is $\frac{R_1}{R_2} = \frac{30.616}{30.416} =$ **1.007**.

Chapter 12

SIGNAL GENERATORS AND WAVEFORM - SHAPING CIRCUITS

SECTION 12.1: BASIC PRINCIPLES OF SINUSOIDAL OSCILLATORS10

12.1 Look ahead to the diagram in the solution to P12.2 following only if you need to! Convert all resistive elements to equivalent values across the tank inductor L. Now, $Q = \frac{\omega_o L}{R_{ls}} \approx \frac{R_{lp}}{\omega_o L}$. Thus $R_{lp} = \frac{\omega_o^2 L^2}{R_{ls}}$ is the equivalent resistor across L due to inductor-wire resistance. Now the amplifier input resistance reflected through the turns ratio is $R_i = n^2 R_{in}$. Thus the total load on the tank is $R = R_o \parallel R_{lp} \parallel R_{cp} \parallel R_i = R_o \parallel \omega_o^2 L^2/R_{ls} \parallel R_{cp} \parallel n^2 R_{in}$. Loop gain (from the active end of the coil and back) is $G_m\ R/n$. Oscillation will occur when this is unity, that is when $\frac{G_m R}{n} = 1$, or $\frac{G_m}{n} = \frac{1}{R} = \frac{1}{R_o} + \frac{R_{ls}}{\omega_o^2 L^2} + \frac{1}{R_{cp}} + \frac{1}{n^2 R_{in}}$, at a frequency $\omega_o = \frac{1}{\sqrt{LC}}$. Now $\omega_o^2 = \frac{1}{LC}$, and $\frac{1}{\omega_o^2 L^2} = \frac{LC}{L^2} = \frac{C}{L}$, with oscillation occuring for $\frac{G_m}{n} = \frac{1}{R_o} + \frac{R_{ls}\ C}{L} + \frac{1}{R_{cp}} + \frac{1}{n^2 R_{in}}$, or for $\mathbf{G_m = \frac{n}{R_o} + \frac{n\ R_{ls}C}{L} + \frac{n}{R_{cp}} + \frac{1}{n\ R_{in}}}$.

12.2

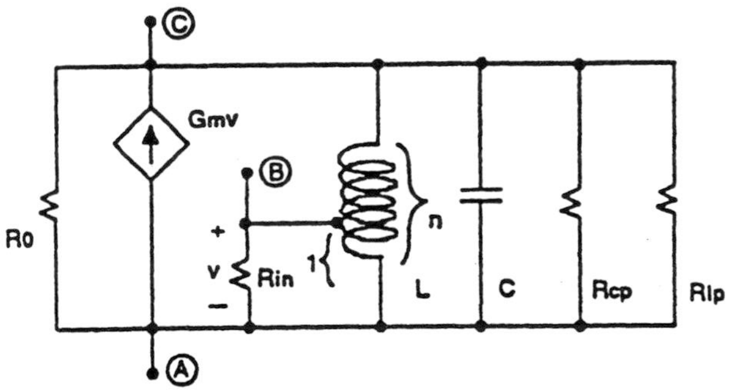

(a) Showing the inductor's parasitic resistance in its parallel form, the topology shown results for a non-inverting amplifier. Three classical circuit forms result depending on which of A, B, C is grounded in a practical implementation. Grounding A produces the design that is usually drawn from the description given.

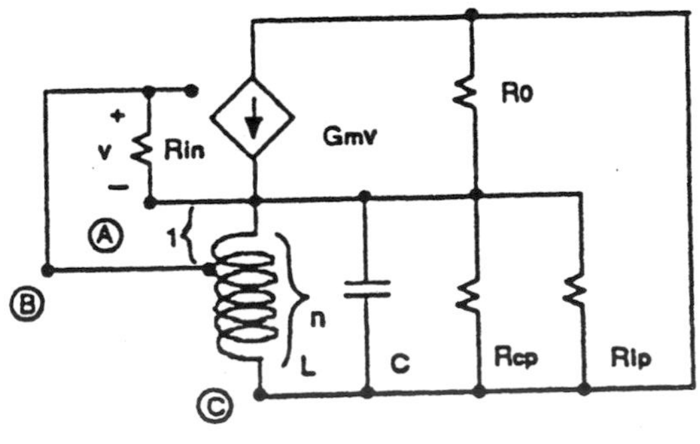

(b) For an inverting transconductance element, the topology shown results, again with three classical variations depending on where ground is connected. The one with C grounded, in which the device acts as a follower, is quite common in practice.

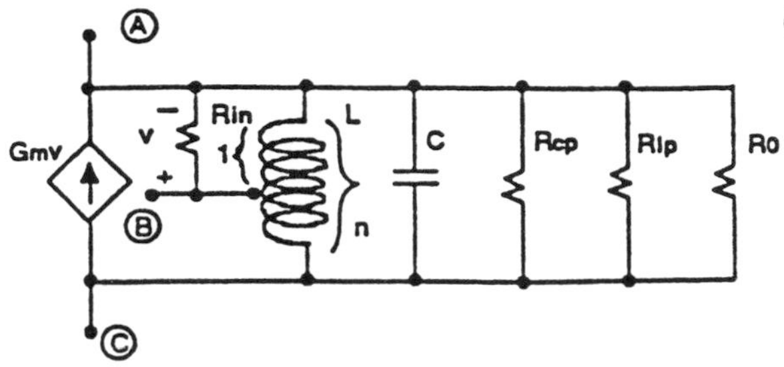

(c) This may be redrawn to resemble (a) more closely to illustrate that the change of amplifier sign is compensated simply by shifting the tap from one end of the auto-transformer to the other. For the negative-gain version, the loop-gain magnitude (as measured across L) is exactly as before, with the same conditions for oscillation. Note that the amplifier inversion is accounted for by the coil-tap-connection reversal.

12.3

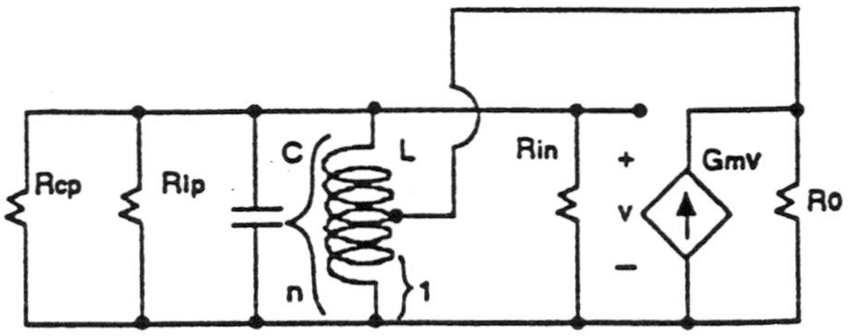

Here, at resonance, the load on the transconductor is $R = R_o \parallel \frac{1}{n^2}(R_{cp} \parallel R_{lp} \parallel R_{in})$, and the voltage across the tap is $G_m R$, with the loop gain equal to $n\, G_m R$. For oscillation to occur, $n\, G_m R = 1$, or $G_m = \frac{1}{n}\frac{1}{R} = \frac{1}{n}\left[\frac{1}{R_o} + \frac{n^2}{R_{cp}} + \frac{n^2}{R_{lp}} + \frac{n^2}{R_{in}}\right]$, or $G_{m2} = \mathbf{G_m = \frac{1}{n\,R_o} + \frac{n}{R_{cp}} + \frac{n}{R_{lp}} + \frac{n}{R_{in}}}$, with oscillation at a frequency $\boldsymbol{\omega_o = (LC)^{½}}$. Recall that for the connection in P.12.1 above, $G_{m1} = \frac{n}{R_o} + \frac{n}{R_{cp}} + \frac{n}{R_{lp}} + \frac{1}{n\,R_{in}}$.

Note that for a particular coil and capacitor, the relative value of the required G_m for each topology depends on the relative size of R_o and R_{in} of the amplifier. For $R_{in} = R_o$, the same G_m is required for a given n. Thus, for example, for $R_o = R_{in} = R_{cp} = R_{lp} = R$, $G_{m1} = G_{m2} = G_m = \frac{3n}{R} + \frac{1}{n\,R} = \frac{1}{R}(3n + 1/n)$, or $G_m\, R = 3n + 1/n$.

12.4 For an input resistance of 10kΩ, R_1 = **10kΩ**, and $R_f = 5\,(R_1)$ = **50kΩ**. Negative clamping occurs with D_1 conducting with 0.6V drop, the negative op-amp input at 0 volts, and V_A at –0.6V correspondingly. Thus at the edge of D_1 conduction, $V_{R2} = 10 - -0.6 = 10.6$V, and $V_{R3} = -0.6 - -2.5 = 1.9$V. Correspondingly, $R_2 = \frac{10.6}{1.9} R_3 = 5.58\, R_3$. For a limiting gain of 0.5, $\frac{R_f \parallel R_3}{R_1} = 0.5$, or $50\text{k} \parallel R_3 = 0.5\,(10\text{k}) = 5\text{k}\Omega$. Thus, $R_3 = 50\text{k} \parallel R_3 \parallel (-50\text{k}) = 5\text{k} \parallel (-50\text{k}) = \frac{5\,(-50)}{-50 + 5} = \frac{5\,(50)}{45}$ = **5.56kΩ**, and $R_2 = 5.58\,(5.56)$ = **31.0kΩ**. Overall, use R_1 = 10kΩ, R_f = 50kΩ, R_3 = 5.6kΩ, and R_2 = 31.0kΩ. Now, for the amplifier, $A = 1000$, $f_t = 10^6$Hz, and thus $f_{3dB} = \frac{10^6}{1000} = 10^3$Hz. For a gain of 5, $\beta = 1/5$, $A\beta = 1/5 \times 1000 = 200$, and $f_{3dB} \approx 200$kHz. *Alternatively*, from page 80 of the Text, $f_{3dB} = 10^6/(1 + R_f/R_1) = 10^6/(1 + 50/10) = 167$kHz. For a 3dB frequency of 167kHz, 2° phase shift occurs at $\tan^{-1}\frac{f}{167} = 2°$, or $f = 167 \tan 2°$ = **5.83kHz**.

12.5 For low input voltages, the gain is $-\frac{R_2}{R_1} = -\frac{10}{7.5} = \mathbf{-1.33V/V}$. For the zeners conducting, the gain is $-\frac{R_2 \parallel R_3}{R_1} = -\frac{10 \parallel 10}{7.5} = -\frac{5}{7.5} = \mathbf{-0.67V/V}$. Assume that the zener can be characterized by a linear resistor of value $R_2 = \frac{V_Z}{I_{ZK}}$ for voltages below the knee.

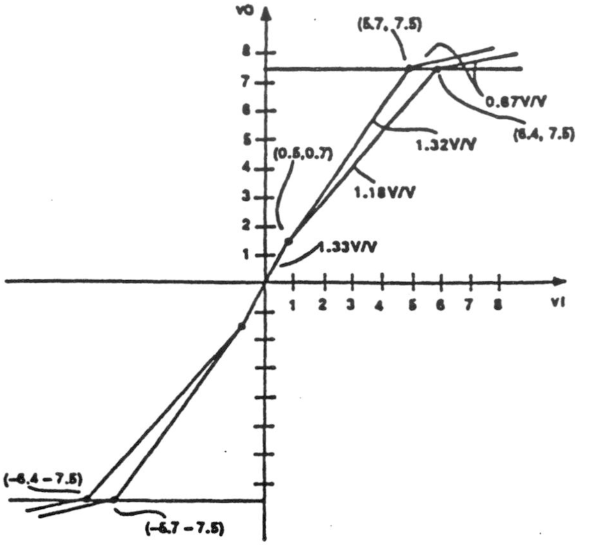

Here, $R_{Z1} = \frac{6.8}{100\mu A} = 68k\Omega$, and $R_{Z2} = \frac{6.8}{10\mu A} = 680k\Omega$. For the poorer zener, for output voltages beyond 0.7V (with corresponding inputs of $\frac{0.7}{1.33} = 0.53V$), $R_{Z1} = 68k\Omega$, and the gain is $-\frac{R_2 \parallel (R_3 + R_{Z1})}{R_1} = -\frac{10 \parallel (10+68)}{7.5} = -$ **1.18V/V**. For the better zener it is $-\frac{10 \parallel (10+680)}{7.5} = -$ **1.32V/V**. For outputs greater than 6.8 + 0.7 = 7.5V, the gain is –0.67V/V as calculated earlier. For the better zener, the corresponding input is about $\frac{7.5}{1.32} = \mathbf{5.68V}$, and for the poorer zener, it is about $\frac{7.5}{1.18} = \mathbf{6.36V}$.

SECTION 12.2: OP-AMP-RC OSCILLATORS

12.6 Now, $\omega_o = \frac{1}{CR}$. For C = 10nF, $R = \frac{1}{2\pi f_o C} = \frac{1}{2\pi \times 10^4 \times 10 \times 10^{-9}} = \mathbf{1.59k\Omega}$. For oscillation, $R_2 \geq 2R_1$. For 2V peak-to-peak output, peak output is 1V. At the threshold of oscillation the voltage at the positive op amp input must be 2/3V. Thus the voltage across the regulating diode plus its series resistor (R_2) must be 2/3V, as must be the voltage across R_2, with 1/3V across R_1. Thus the current in $R_1 = I = \frac{1/3}{R_1} = \frac{2/3}{R_2} + I_D$. Now $I_D = I_s\, e^{V_D/2V_T}$ where $V_D + I_D R_2 = 2/3V$. Now, $V_D = 0.70 + 0.05 \ln I_D/1\text{ mV}$, and $0.700 + .05 \ln I_D + I_D R_2 = 0.667$, or $I_D = \frac{1}{R_2}(-.05 \ln I_D - .033)$ – – – (1). Also $I_D = \frac{0.333}{R_1} - \frac{0.667}{R_2}$ – – – (2), with $R_2 > 2R_1$ – – – (3). With (1), try I_D = 0.1mA, whence $R_2 = \frac{1}{0.1}(-.05 \ln 0.1 - .0333) = .818k\Omega$. Try $I_D = 10\mu A$: $R_2 = \frac{1}{.01}(-.05 \ln .01 - .0333) = 19.7k\Omega$. Try $I_D = 20\mu A$: $R_2 = \frac{1}{.02}(-.05 \ln .02 - .0333) = 8.11k\Omega$. Use $R_2 = 8.2k\Omega$. From (2), $.01 = \frac{.3333}{R_1} - \frac{.6666}{8.2}$, or $R_1 = \frac{.3333}{.010 + .0813} = 3.65k\Omega$. Use 3.6kΩ. *Check*: from (3), $R_2 = 8.2 > 2R_1 = 2(3.65) = 7.3$ OK. *Check*: From (2), $I_D = \frac{.3333}{3.6} - \frac{.6666}{8.2} = .0926 - .0813 = .0113$mA. Now the drop across the diode plus the extra R_2 is 0.0113 (8.2) + .05 ln .0113 + 0.70 = .0927 – .2241 + 0.70 = .569V. Current in the feedback $R_2 = \frac{.569}{8.2} = .0693$mA. Voltage across R_1 = (.0693 + .0113) (3.6) = 0.200V, and the output voltage is .290 + .569 = .859V. Too low! Now, reduce R_1 slightly to R_1 = 3.3kΩ. From (2), $I_D = \frac{.3333}{3.3} - \frac{.6666}{8.2} = .1010 - .0823 = .0187$mA. Now the drop across the diode plus the extra R_2 is .0187 (8.2) + .05 ln .0187 + 0.70 = .1534 – .199 + 0.70 = .654V. Current in the feedback $R_2 = \frac{.654}{8.2} = .0800$mA. Voltage across R_1 = (.0800 + .0187) (3.6) = .355V. Thus the output voltage is .355 + .654

$= \mathbf{1.01V}$. Use $C = \mathbf{10nF}$, $R = \mathbf{1.59k\Omega}$, $R_1 = \mathbf{3.3k\Omega}$, $R_2 = \mathbf{8.2k\Omega}$.

12.7 From Eq. 12.11 of the Text, using an ideal amplifier, $L(j\omega) = \dfrac{1 + R_2/R_1}{3 + j(\omega CR - 1/(\omega CR))}$, where oscillation occurs when $\Phi_L = 0$, and $|L| = 1$ at $\omega_o = \dfrac{1}{CR}$. From Eq. 2.20 of the Text, for an amplifier with $\omega_t = 2\pi$rad/s and $A_o >> (1 + R_2/R_1)$, $\dfrac{V_o}{V_i}(j\omega) = \dfrac{1 + R_2/R_1}{1 + \dfrac{j\omega}{\omega_t/(1 + R_2/R_1)}}$. For oscillation at $\omega = 0.9\ \omega_o = 0.9/CR$ with this amplifier:

$$L(j\ 0.9\ \omega_o) = \frac{1 + R_2/R_1}{(3 + j\ (.9 - \frac{1}{0.9}))\ (1 + \frac{j\ 0.9\ \omega_o}{\omega_t\ (1 + R_2/R_1)})} =$$

$$\frac{1 + R_2/R_1}{3 - j\ (.211) + \frac{2.27\ j\ \omega_o}{\omega_t\ (1 + R_2/R_1)} + \frac{.1899\ \omega_o}{\omega_t\ (1 + R_2/R_1)}}.$$

Oscillation will occur when $0.211 = 2.27\omega_o/(\omega_t\ (1 + R_2/R_1))$, or $\omega_o = \omega_t\ (1 + R_2/R_1)\ (.09295)$. Now for a nominal frequency ω_o, oscillation will occur at $0.9\ \omega_o$, when $\dfrac{1 + R_2/R_1}{3 + \dfrac{.1899\ \omega_o}{\omega_t\ (1 + R_2/R_1)}} = 1$, or $1 + R_2/R_1 = 3 + .1899\ (.09295) = 3.018$, or $R_2/R_1 = 2.018$, or $R_2 = 2.018\ R_1$. Thus for $f_t = 1$MHz, the frequency of oscillation will be $0.9 \times 10^6\ (3.018)\ (.09295) = 0.252 \times 10^6$**Hz**, with nominal frequency, $f_o = \dfrac{1}{2\pi RC} = \dfrac{.252}{0.9} = \mathbf{0.280MHz}$. Thus, for operation at 0.252MHz, and $1 + R_2/R_1 = 1 + \dfrac{2.018\ R_1}{R_1} = 3.018$, the closed-loop op-amp gain must be $\dfrac{3.018}{1 + \dfrac{j\ .25 \times 10^6}{1 \times 10^6/3.018}} = \dfrac{3.02}{1 + 0.755\ \mathbf{j}}$, a value which seems quite reasonable.

12.8 From Eq. 2.20 of the Text, for an op amp with A_o large and a unity-gain frequency ω_t, the gain of the non-inverting topology is: $G(\omega) = \dfrac{1 + R_2/R_1}{1 + \dfrac{j\omega}{\omega_t\ (1 + R_2/R_1)}}$, and of the network is:

$T(\omega) = \dfrac{R}{R + \dfrac{1}{j\omega C}} = \dfrac{1}{1 + \dfrac{1}{j\omega RC}}$. Overall, $L(\omega) = \dfrac{1 + R_2/R_1}{(1 + \dfrac{j\omega}{\omega_t\ (1 + R_2/R_1)})\ (1 + \dfrac{1}{j\omega RC})} =$

$$\frac{1 + \mathbf{R_2/R_1}}{1 + \frac{\mathbf{j\omega}}{\omega_t\ (1 + \mathbf{R_2/R_1})} - \frac{\mathbf{j}}{\omega\ \mathbf{RC}} + \frac{1}{\mathbf{RC}\ \omega_t\ (1 + \mathbf{R_2/R_1})}}$$

Oscillation will occur when the net phase shift is 0°, when $\dfrac{\omega}{\omega_t\ (1 + R_2/R_1)} = \dfrac{1}{\omega RC}$ or $\omega = \left[\dfrac{\omega_t\ (1 + \mathbf{R_2/R_1})}{\mathbf{RC}}\right]^{1/2}$, provided the gain is at least one, that is $\dfrac{1 + R_2/R_1}{1 + \dfrac{1}{RC\ \omega_t\ (1 + R_2/R_1)}} \geq 1$, or

$\mathbf{R_2/R_1} = \dfrac{1}{\mathbf{RC}\ \omega_t\ (1 + \mathbf{R_2/R_1})}$, or $(R_2/R_1)^2 + (R_2/R_1) - 1/(RC\omega_t) = 0$. Now for $\omega = 4/RC = \left[\dfrac{\omega_t\ (1 + R_2/R_1)}{RC}\right]^{1/2}$, $\omega_t\ (1 + R_2/R) = 16/(RC)$, and $R_2/R_1 = \dfrac{1}{RC\ (16/RC)} = \dfrac{1}{16}$.

$\mathbf{R_2 = R_1/16} = .0625\, R_1$, for $\frac{1}{RC} = \omega_t$.

More generally, for $\frac{1}{RC} = \frac{\omega_t}{a}$, $\omega = \frac{\omega_t}{\sqrt{a}}\,(1 + R_2/R_1)^{1/2}$, with $(1 + R_2/R_1) = 1 + 1/(a\,(1 + R_2/R_1))$,

or $x = 1 + 1/(ax)$, $ax^2 - ax - 1 = 0$, $x^2 - x - 1/a = 0$,

for which $x = \frac{-1 \pm \sqrt{1 - 4(-1/a)}}{2}$, or $1 + R_2/R_1 = \frac{1 \pm \sqrt{4/a + 1}}{2}$. Now, for example: For $a = 1$, $1 + R_2/R_1 = \frac{1 + \sqrt{5}}{2} = 1.618$, $R_2 = .618\, R_1$ and $\omega = \omega_t\,(1.618)^{1/2} = 1.27\omega_t$. For $a = 2$, $1 + R_2/R_1 = \frac{1 \pm \sqrt{2 + 1}}{2} = 1.366$, $R_2 = .366\, R_1$ and $\omega = \omega_t\,(\frac{1.366}{2})^{1/2} = .826\omega_t$. For $a = 4$, $1 + R_2/R_1 = \frac{1 \pm \sqrt{2}}{2} = 1.2707$, $R_2 = .2707\, R_1$ and $\omega = \omega_t\,(\frac{1.2707}{4})^{1/2} = 0.564\omega_t$. For $a = 16$, $1 + R_2/R_1 = \frac{1 \pm \sqrt{1.25}}{2} = 1.059$, $R_2 = .059\, R_1$, and $\omega = \omega_t\,(\frac{1.059}{16})^{1/2} = 0.257\omega_t$.

Generalizing, we see that for frequencies significantly lower than ω_t for an amplifier with idealized single-pole rolloff, oscillation is possible for R_2 nearly zero, at a frequency which is the geometric mean of ω_t and $1/RC$. For excess phase shift due to additional poles near and above ω_t, more compensating phase shift will be required from the RC network, implying operation at a frequency lower than in the simple case. Note that this type of oscillator, operating at ω_o with a network for which $\omega = 1/RC$, allows an estimate of ω_t, and associated excess phase, if R_2 is adjusted to the maximum value for which oscillation is sustained. For $1/RC << \omega_o$, the oscillation frequency ω_o can become a sensitive function of various things, such as construction, but not of excess phase. For $1/RC$ very near ω_o, the frequency of oscillation is quite sensitive to excess phase. By varying R, find ω_o in the range two to four times $1/RC$ in order to evaluate ω_t. $\omega_o < 1/RC$ is an indication of excess phase, where operation also requires a higher value of R_2.

12.9 From the right, label the components C_1, R_1, C_2, R_2, C_3, R_3, with joining nodes N_1, N_2, N_3 respectively. Assume a virtual ground at the op-amp input into which a current flows from C_1. At N_1, $i_{C1} = i$, $v_1 = i/C_1 s$. At N_2, $v_2 = i_{C1} R_1 + v_1 = i\,(R_1 + 1/C_1 s)$, $i_{C2} = v_2/(1/C_2 s) = v_2 C_2 s = i\,(R_1 C_2 s + C_2/C_1)$. At N_3, $v_3 = (i_{C1} + i_{C2}) R_2 + v_2 = i\,(1 + R_1 C_2 s + C_2/C_1) R_2 + i\,(R_1 + 1/C_1 s) = i\,(R_1 R_2 C_2 s + R_2 C_2/C_1 + R_2 + R_1 + 1/C_1 s)$, $i_{C3} = v_3 C_3 s = i\,(R_1 R_2 C_2 C_3 s^2 + R_2 C_2 C_3 s/C_1 + R_2 C_3 s + R_1 C_3 s + C_3/C_1$. At x, $v_x = (i_{C1} + i_{C2} + i_{C3}) R_3 + v_3 = i\,(1 + R_1 C_2 s + C_2/C_1 + R_1 R_2 C_2 C_3 s^2 + R_2 C_2 C_3 s/C_1 + R_2 C_3 s + R_1 C_3 s + C_3/C_1) R_3 + i\,(R_1 R_2 C_2 s + R_2 C_2/C_1 + R_2 + R_1 + 1/C_1 s) = i\,(R_1 + R_2 + R_3 + R_3 C_2/C_1 + R_3 C_3/C_1 + R_2 C_2/C_1 + s\,(R_1 R_3 C_2 + R_1 R_3 C_3 + R_2 R_3 C_3 + R_1 R_2 C_2 + R_2 R_3 C_2 C_3/C_1) + R_1 R_2 R_3 C_2 C_3\, s^2 + 1/C_1\, s$. Now the loop gain $L = -i\, R_f/v_x$. Thus, for $C_1 = C_2 = C_3 = C$, $L(s) = -R_f/[R_1 + 2\,R_2 + 3\,R_3 + sC\,(R_1 R_2 + 2\,R_2 R_3 + 2\,R_1 R_3) + R_1 R_2\, R_3 C^2 s^2 + 1/(Cs)]$, and $L(s) =$ $\mathbf{-R_f Cs/[1 + (R_1 + 2\,R_2 + 3\,R_3)\,Cs + (R_1 R_2 + 2\,R_2 R_3 + 2\,R_1 R_3)\,C^2 s^2 + R_1 R_2\, R_3 C^3 s^3]}$. Now substituting $s = j\omega$ and multiplying top and bottom by j, $L(j\omega) = +R_f C\omega/[j - (R_1 + 2R_2 + 3R_3)\,C\omega - (R_1 R_2 + 2R_2 R_3 + 2R_1 R_3)\,C^2\omega^2 j + R_1 R_2\, R_3 C^3 \omega^3]$. For oscillation, the phase angle = zero, that is $1 = (R_1 R_2 + 2\;R_2 R_3 + 2\;R_1 R_3)\;C^2\;\omega^2$, where the frequency of oscillation is $\boldsymbol{\omega_o} = \frac{1}{\mathbf{C\sqrt{R_1\,R_2 + 2\,R_2\,R_3 + 2\,R_1\,R_3}}}$, which for $R_1 = R_2 = R_3 = R$ is $\boldsymbol{\omega_o} = \frac{1}{\mathbf{CR\sqrt{5}}}$, with $\frac{R_f\;C\;\omega_o}{(R_1 + 2R_2 + 3R_3)\;C\;\omega_o + R_1\;R_2\;R_3\;C^3\;\omega_o^3} = 1$, or $R_f = (R_1 + 2R_2 + 3R_3) + R_1 R_2 R_3 C^2 \omega_o^2 =$ $(R_1 + 2R_2 + 3R_3) + (R_1 R_2 R_3)\,(R_1 R_2 + 2R_2 R_3 + 2R_1 R_3)^{-1}$, which for $R_1 = R_2 = R_3 = R$ is $R_f = 6\,R + R/5 = \mathbf{6.2R}$.

Now for sensitivities (with $R_1 \approx R_2 \approx R_3 \approx R$):

$$\frac{\partial\,\omega_o}{\partial\,R_1} = -\frac{1}{2}\left[\frac{1}{C\,(R_1\,R + 2R^2 + 2R_1\,R)^{3/2}}\right](3R) = -\frac{3}{2}\,\omega_o\,\frac{R}{3R\;R_1 + 2R^2} = -\frac{3}{2}\,\omega_o\,\frac{1}{3R_1 + 2R},$$

and $\frac{\partial\,\omega_o}{\partial\,R_2} = -\frac{3}{2}\,\omega_o\,\frac{1}{3R_2 + 2R}$, and $\frac{\partial\,\omega_o}{\partial\,R_3} = -\frac{4}{2}\,\omega_o\,\frac{1}{4R}$. Now $S^{\omega_o}_{R_1, R_2} = \frac{\partial\,\omega_o}{\omega_o} \times \frac{R_1}{\partial\,R_1} = -\frac{3}{2}\,\omega_o$

$\frac{1}{3R_1 + 2R} \times \frac{R_1}{\omega_o}$, which, for $R_1 \simeq R$, is $-\frac{3}{2(5)} = \mathbf{-0.3}$. Similarly $S_{R_3}^{\omega_o} = \mathbf{-0.5}$. Also $\frac{\partial R_f}{\partial R_1} = 1 + R_2R_3\,(R_1R_2 + 2R_2R_3 + 2R_1R_3)^{-1} + (-1)\,R_1R_2R_3\,(R_1R_2 + 2R_2R_3 + 2R_1R_3)^{-2}\,(R_2 + 2R_3)$, which for $R_2 = R_3 = R$, is $\frac{\partial R_f}{\partial R_1} = 1 + R^2\,(R_1R + 2R^2 + 2R_1R)^{-1} - R_1R^2\,(R_1R + 2R^2 + 2R_1R)^{-2}\,(3R) = 1 + R\,(3R_1 + 2R)^{-1} - 3R_1R\,(3R_1 + 2R)^{-2} = \frac{9R_1^2 + 4R^2 + 12R_1R + 3RR_1 + 2R^2 - 3R_1R}{(3R_1 + 2R)^2} = \frac{9R_1^2 + 12R_1R + 6R^2}{(3R_1 + 2R)^2}$, and for $R_1 = R$, is 27/25 = 1.08. Thus, $S_{R_1}^{R_f} = \frac{\partial R_f}{\partial R_1} \times \frac{R_1}{R_f}$, for $R_1 \simeq R$, is $1.08 \times \frac{R}{6.2R} = \mathbf{0.174}$. Also with $R_1 \simeq R_3 \simeq R$, $\frac{\partial R_f}{\partial R_2} = 2 + R^2\,(RR_2 + 2RR_2 + 2R^2)^{-1} - R^2R_2\,(RR_2 + 2RR_2 + 2R^2)^{-2}\,(3R) = 2 + R\,(3R_2 + 2R)^{-1} - 3RR_2\,(3R_2 + 2R)^{-2}$, which for $R_2 = R$, is $2 + R\,(5R)^{-1} - 3R^2\,(5R)^{-2} = 2 + 1/5 - 3/25 = 2.08$. Thus $S_{R_2}^{R_f}$ is $2.08 \times \frac{R}{6.2R} = \mathbf{0.335}$. Similarly, $S_{R_3}^{R_f}$ is $3.08/6.2 = \mathbf{0.497}$.

12.10

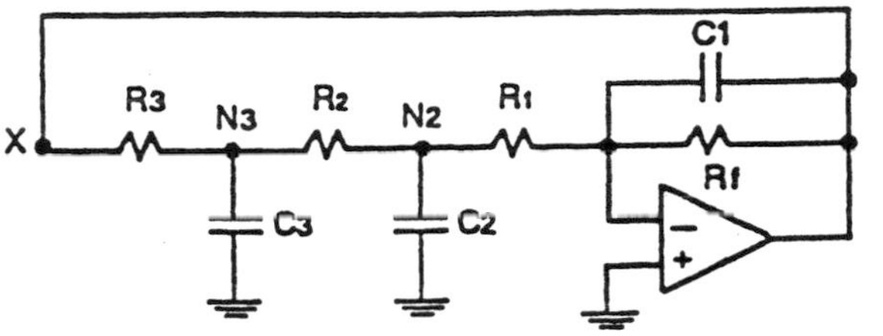

Here, $R_1 = R_2 = R_3 = R$, and $C_1 = C_2 = C_3 = C$. For a current i in R_1, at node N_2, $\upsilon_2 = i\,R_1 = iR$, $i_{C2} = \upsilon_2/(1/C_2s) = iRCs$. At node N_3, $i_{R2} = i_{R1} + i_{C2} = i\,(1 + RCs)$, $\upsilon_3 = \upsilon_2 + R_2\,(i_{R2}) = iR + R\,(i)(1 + RCs) = i\,(2R + R^2Cs)$, $i_{C3} = \upsilon_3C_2s = i\,(2RCs + R^2C^2s^2)$. At node x, $i_{R3} = i_{R2} + i_{C3} = i\,(1 + RCs) + i\,(2RCs + R^2C^2s^2) = i\,(1 + 3RCs + R^2C^2s^2)$, $\upsilon_x = \upsilon_3 + R_3\,i_{R3} = i\,(2R + R^2Cs) + Ri\,(1 + 3RCs + R^2C^2s^2) = iR\,(3 + 4RCs + R^2C^2s^2)$. Now, i flows in $R_f \,||\, C_1$. Thus the op-amp output voltage = $\upsilon_o = -i\,\frac{R_f\,(1/Cs)}{R_f + 1/Cs}$.

Loop gain $\mathbf{L(s)} = \frac{\upsilon_o}{\upsilon_x} = \frac{-\mathbf{R_f}}{\mathbf{R_f\,Cs + 1}} \times \frac{1}{\mathbf{R\,(3 + 4RCs + R^2C^2s^2)}}$.

Now $\mathbf{L(s)} = -(R_f/R)\,[3 + 4RCs + R^2C^2s^2 + 3R_fCs + 4RR_fC^2s^2 + R^2R_fC^3s^3]^{-1} = \mathbf{-(R_f/R)\,[3 + s\,(4RC + 3R_fC) + s^2\,(R^2C^2 + 4RR_fC^2) + R^2R_fC^3s^3]^{-1}}$. Substituting $s = j\omega_o$ and requiring the imaginary part to be zero, and $L(\omega_o) = 1$, see $(4RC + 3R_fC)\,\omega_o = R^2R_fC^3\omega_o^3$ – – – (1), and $1 = -(R_f/R)\,(3 - R^2C^2\omega_o^2\,4RR_fC^2\omega_o^2)^{-1}$ – – – (2). From (1), $\omega_o = \frac{1}{\mathbf{RC}}\left[\frac{\mathbf{4R + 3R_f}}{\mathbf{R_f}}\right]^{1/2}$. From (2), $R_f = R\,((R^2C^2 + 4RR_fC^2)\,\omega_o^2 - 3) = R\,(R^2C^2 + 4RR_fC^2)\,(\frac{4R + 3R_f}{R_fC^2R^2}) - 3R = (4R + 3R_f)\,(\frac{R}{R_f} + 4) - 3R$. Thus $R_f = \frac{4R^2}{R_f} + 3R + 16R + 12R_f - 3R$, $\frac{4R^2}{R_f} + 16R + 11R_f = 0$, $4R^2 + 16RR_f + 11R_f^2 = 0$, whence $R_f = \left[\frac{-16 \pm \sqrt{16^2 - 4\,(4)\,(11)}}{2\,(11)}\right]R = \left[\frac{-16 \pm 8.94}{22}\right]R$. Note that R_f/R is always negative, implying that oscillation is not possible. Why?

12.11 For each section, a maximum of 90° phaseshift is possible. Now with a positive-gain amplifier, the network must shift by 360°. Thus, realistically, for operation on a phase slope of more than 360/90 = 4%, **five sections** are needed. Label the sections 1 to 5 from right to left with the RC join nodes called N_1 through N_5 respectively. Now for $\upsilon_1 = \upsilon = \upsilon(1)$, $i_{C1} = \upsilon/R = \upsilon/R\,(1)$, $\upsilon_2 = i_{C1}/Cs + \upsilon_1 = \upsilon/RCs + \upsilon = \upsilon\,(1/RCs + 1)$, $i_{C2} = \upsilon_2/R + i_{C1} = (\upsilon/RCs + \upsilon)/R + \upsilon/R = \upsilon/R\,[1/RCs + 2]$, $\upsilon_3 = i_{C2}/Cs + \upsilon_2 = \upsilon\,[(\upsilon/RCs + 2)/RCs + 1/RCs + 1] = \upsilon\,(1/R^2C^2s^2 + 3/RCs + 1)$,

$i_{C3} = \upsilon_3/R + i_{C2} = \upsilon\,[(1/R^2C^2s^2 + 3/RCs + 1)/R + (1/RCs + 2)/R] = \upsilon/R\,[1/R^2C^2s^2 + 4/RCs + 3\,]$,
$\upsilon_4 = i_{C3}/Cs + \upsilon_3 = \upsilon\,[(1/R^2C^2s^2 + 4/RCs + 3)/RCs + 1/R^2C^2s^2 + 3/R\,Cs + 1] = \upsilon\,(1/R^3C^3s^3 + 5/R^2C^2s^2 + 6/R\,Cs + 1)$,
$i_{C4} = \upsilon_4/R + i_{C3} = \upsilon/R\,(1/R^3C^3s^3 + 5/R^2C^2s^2 + 6/RCs + 1 + 1/R^2C^2s^2 + 4/RCs + 3) = \upsilon/R\,(1/R^3C^3s^3 + 6/R^2C^2s^2 + 10/R\,Cs + 4)$,
$\upsilon_5 = i_{C4}/Cs + \upsilon_4 = \upsilon\,[1/R^4C^4s^4 + 6/R^3C^3s^3 + 10/R^2C^2s^2 + 4/RCs + 1/R^3C^3s^3 + 5/R^2C^2s^2 + 6/RCs + 1] = \upsilon\,(1/R^4C^4s^4 + 7/R^3C^3s^3 + 15/R^2C^2s^2 + 10/RCs + 1)$,
$i_{C5} = \upsilon_5/R + i_{C4} = \upsilon/R\,(1/R^4C^4s^4 + 7/R^3C^3s^3 + 15/R^2C^2s^2 + 10/RCs + 1 + 1/R^3C^3s^3 + 6/R^2C^2s^2 + 10/R\,Cs + 4) = \upsilon/R\,(1/R^4C^4s^4 + 8/R^3C^3s^3 + 21/R^2C^2s^2 + 20/RCs + 5)$,
$\upsilon_6 = i_{C5}/Cs + \upsilon_5 = \upsilon\,(1/R^5C^5s^5 + 8/R^4C^4s^4 + 21/R^3C^3s^3 + 20/R^2C^2s^2 + 5/RCs + 1/R^4C^4s^4 + 7/R^3C^3s^3 + 15/R^2C^2s^2 + 10/RCs + 1) = \upsilon\,(1/R^5C^5s^5 + 9/R^4C^4s^4 + 28/R^3C^3s^3 + 35/R^2C^2s^2 + 15/RCs + 1)$.

Now $L(s) = \dfrac{K\,\upsilon}{\upsilon_6} = K\,(1/R^5C^5s^5 + 9/R^4C^4s^4 + 28/R^3C^3s^3 + 35/R^2C^2s^2 + 15/RCs + 1)^{-1}$. Now substituting $s = j\omega$, the condition for zero phase angle is $1/R^5\,C^5\,\omega^5 - 28/R^3\,C^3\,\omega^3 + 15/R\,C\omega = 0$, or $1/R^4\,C^4\,\omega^4 - 28/R^2\,C^2\,\omega^2 + 15 = 0$. Now $1/R^2\,C^2\,\omega^2 = -\dfrac{-28 \pm \sqrt{28^2 - 4\,(15)}}{2} = \dfrac{28 \pm 26.91}{2} =$ 27.45 or .545. Now at $\omega = 1/RC$, the phase contribution of each section is 45°, and the total phase is about 225°. Thus operation must be at $\omega < 1/RC$. Thus choose the 27.45 solution, where $1/R^2\,C^2\,\omega^2 =$ 27.45, or $\omega_o = \dfrac{1}{RC} \times \dfrac{1}{\sqrt{27.45}} = \dfrac{0.191}{\mathbf{RC}}$. For this condition, $|L| = K\,(9/R^4\,C^4\,\omega^4 - 35/R^2\,C^2\,\omega^2 + 1)^{-1}$. Now for $|L| = 1$, $K = 9/(.191)^4 - 35/(.191)^2 + 1 = 6762.5 - 959.4 + 1 =$ **5804V/V.**

12.12 The modified amplifier has 4 sections, with all 4 capacitors of value C, and all 3 resistors of value R. The feedback resistor is R_f. Perhaps from the results of Exercise 12.5, we could reason that $L = \dfrac{\omega^2\,C^2\,R\,R_f}{5 + j\,(4\omega\,CR - 1/\omega\,CR)}$, (incorrectly!), but we must check: Label the network nodes, from the op-amp negative input toward the left, N_0, N_1, N_2, N_3, N_4. Note that $N_4 = x$. Now for a current i flowing into node N_0 from C_1, $\upsilon_o = 0\text{V}$, $i_{C1} = i$, $\upsilon_1 = i_{C1}/Cs = i/Cs$, $i_{C2} = \upsilon_1/R + i_{C1} = i\,(1/RCs + 1)$ $\upsilon_2 = i_{C2}/Cs + \upsilon_1 = i\,(1/R\,Cs + 1)/Cs + i/Cs$, $i_{C3} = \upsilon_2/R + i_{C2} = i\,[1/R^2C^2s^2 + 2/RCs + 1/RCs + 1] = i\,[1/R^2C^2s^2 + 3/RCs + 1]$, $\upsilon_3 = i_{C3}/Cs + \upsilon_2 = i\,(1/R^2C^2s^2 + 3/RCs + 1)/Cs + i\,(1/RCs + 2)/Cs = i\,(1/R^2C^2\,s^2 + 4/RCs + 3)/Cs$. Note that this corresponds to the result of Exercise 12.5. Now, $i_{C4} = \upsilon_3/R + i_{C3} = i\,[1/R^3C^3s^3 + 4/R^2C^2s^2 + 3/RCs + 1/R^2C^2s^2 + 3/RCs + 1] = i\,[1/R^3C^3s^3 + 5/R^2C^2s^2 + 6/RCs + 1]$, $\upsilon_4 = i_{C4}/Cs + \upsilon_3 = (i/Cs)\,[1/R^3C^3s^3 + 5/R^2C^2s^2 + 6/RCs + 1 + 1/R^2C^2s^2 + 4/RCs + 3] = (i/Cs)\,[1/R^3C^3s^3 + 6/R^2C^2s^2 + 10/RCs + 4]$. Now, $\upsilon_x = \upsilon_4$, and $\upsilon_o = -R_f\,i$. Thus $\mathbf{L(s)} = \upsilon_o/\upsilon_x = -R_f\,Cs\,[1/R^3C^3s^3 + 6/R^2C^2s^2 + 10/RCs + 4]$

$= \dfrac{-\mathbf{R_f\,C}}{\mathbf{1/R^3C^3s^4 + 6/R^2C^2s^3 + 10/RCs^2 + 4/s}}$. Now for $s = j\omega$, the phase of $L(j\omega) = 0$ when $6/R^2C^2\omega^3 = 4/\omega$, or $\omega = \left[\dfrac{3}{2}\dfrac{1}{RC}\right]^{1/2} = \dfrac{1.224}{\mathbf{RC}}$, where $|L|$ must be 1 (or more). For $1 = \dfrac{-R_f\,C}{1/R^3C^3\omega^4 - 10/RC\,\omega^2}$, $R_f = -1/R^3\,C^4\,\omega^4 + 10\,/\,R\,C^2\,\omega^2 = 10\,(2/3)\,R - 1\,(2/3)^2\,R = (6.666 - .444)\,R =$ **6.22R.**

Note that the result for υ_3 above allows the solution of the 3-capacitor circuit in Exercise 12.5 of the Text, namely $L(s) = \dfrac{\upsilon_o}{\upsilon_3} = \dfrac{-R_f\,i}{(i/Cs)\,[1/R^2C^2s^2 + 4/RCs + 3]} = \dfrac{-R_f\,s^2RC^2}{1/RCs + 4 + 3RCs}$, and $L(j\omega) = \dfrac{\omega^2RC^2R_f}{4 + j\,(3\omega RC - 1/\omega\,RC)}$, as stated there.

12.13 For the filter with $C_4 = C_6 = C$ and $R_1 = R_2 = R_3 = R_5 = R$, $f_o = \dfrac{1}{2\,\pi\,CR}$, with $Q = \dfrac{R_6}{R}$, from Eq. 11.53 and 11.54 on page 917 of the Text. Now for f_o = 10kHz with C = 10nF, R =

$\frac{1}{2\,\pi \times 10 \times 10^{-9} \times 10^4}$ = **1.59kΩ**. While the solution to satisfy the distortion specification is quite complex, we can simplify the process by assuming that the signal at υ_2 is a square wave of 0.7V peak amplitude, for which the fundamental is $\frac{4\,(0.7)}{\pi}$ = 0.89V and the 3rd harmonic is 1/3 (0.89) = 0.297V, (from page 4 of the Text). For this situation, since the gain from υ_2 to υ_1 is 2 times, the peak output at υ_1 will be 2 (0.89V) peak = **1.78V peak.** Now the 3rd harmonic component at υ_2 will be less than 1/3 of the fundamental there (since the wave at υ_2 is not very square. Thus a rejection of $\frac{1/3}{1/100} = 33.3 \equiv$ 20 log 33.30 = 30dB will be enough. Now for the second-order bandpass filter, $T(s) = \frac{(\omega_o/Q)s}{s^2 + (\omega_o/Q)\,s + \omega_o^2}$ with gain of 1 at ω_o. Now for gain of 1/33.3 at 3 ω_o $T(j\omega) = \frac{(\omega_o/Q)\,j\omega}{-\omega^2 + j\omega\,(\omega_o/Q) + \omega_o^2} = \frac{(\omega_o/Q)\,\omega}{\omega\,(\omega_o/Q) + j\,(\omega^2 - \omega_o^2)}$, that is,

$T(3\omega_o) = \frac{(\omega_o/Q)\,3\omega_o}{((3\omega_o\,\omega_o/Q)^2 + (9\omega_o^2 - \omega_o^2)^2)^{½}} = \frac{3}{100}$, whence $100/Q = ((3/Q)^2 + (9-1)^2)^{½}$, or $\frac{10^4}{Q^2} = 9/Q^2 + 64$, $\frac{10^4 - 9}{Q^2} = 64$, $Q^2 = \frac{10^4 - 9}{64} = 156$. Thus, $Q = 12.5$, for which $R_6 = 12.5\,R = 12.5\,(1.59)$ = **19.9kΩ**. Use $R_1 = \frac{1.78 - 0.7}{1mA}$ = **1.1kΩ.**

SECTION 12.3: LC AND CRYSTAL OSCILLATORS

12.14 For the FETs, operation is at I_{DSS} = 4mA, with $g_m = 2\,I_{DSS}/|V_p| = 2\,(4)/2 = 4$ mA/V. For each FET, $r_o = \frac{V_A}{I} = \frac{100V}{4mA} = 25k\Omega$. For L = 10μH and Q = 100 at 1MHz, the equivalent parallel resistance is $R_p = (2\,\pi \times 10^6 \times 10 \times 10^{-6}) \times 100 = 6.28k\Omega$. Now at resonance, R_p loads the tank consisting of L and C_1 with C_2 in series. Now the voltage across C_1, ie from the follower output to ground, is $\frac{C_2}{C_1 + C_2}$ of that across R_p. Since the power supplied to R_p comes via the capacitors, the current must be correspondingly increased.

Thus the corresponding load on R_p is $\left[\frac{C_2}{C_1 + C_2}\right]^2 \times R_p = \left[\frac{C_2}{C_1 + C_2}\right]^2 \times 6.28$. Thus the total load on Q_1 is $R_L = 25k\Omega \parallel 25k\Omega \parallel \left[6.28\left[\frac{C_2}{C_1 + C_2}\right]^2\right] \parallel 10k\Omega$, or $R_L = 5.56 \parallel \left[6.28\left[\frac{C_2}{C_1 + C_2}\right]^2\right]$, and the voltage gain of the amplifier is $\frac{R_L}{R_L + 1/g_m} = \frac{R_L}{R_L + 1/4}$ V/V. Now, looking back from the $C_1\,C_2$ node to the gate of Q_1, the Thevenin gain equivalent is a gain of $\frac{R_L}{R_L + 1/g_m} = \frac{5.56}{5.56 + .25} = 0.957$V/V, with a resistance of $\frac{1}{g_m} \parallel R_L = \frac{0.25\,(5.56)}{0.25 + 5.56} = .239k\Omega$. Now when loaded by the R_p L C_1 C_2 network, the gain to the gate end of C_2 becomes

$$\frac{(0.959) \times \left[\frac{C_2}{C_1 + C_2}\right]^2 6.28}{6.28\left[\frac{C_2}{C_1 + C_2}\right]^2 + 0.239} \times \frac{C_1 + C_2}{C_2}.$$

Now for oscillation, this must be unity, that is

$$0.957\,(6.28)\,\frac{1}{1 + C_1/C_2} = 6.28\left[\frac{1}{1 + C_1/C_2}\right]^2 + .239.$$

Now let $x = \frac{1}{1 + C_1/C_2}$. Thus $6.28\,x^2 - 6.01\,x + .239 = 0$, $x = \frac{6.01 \pm \sqrt{6.01^2 - 4\,(6.28)\,(.239)}}{2\,(6.28)} = \frac{6.01 \pm 5.49}{2\,(6.28)}$ = .915 or .0414 (unlikely). Therefore $\frac{1}{1 + C_1/C_2} = .915$, $1 + C_1/C_2 = 1.0929$, $C_1/C_2 =$

0.0929, or C_2/C_1 = **10.76**.

Alternatively, ignoring the effect of R_p and using Eq. 12.21, $C_2/C_1 = g_m R = 4 \times (25 \parallel 25 \parallel 10) = 22.22$. Now R_p reflected through the capacitor network becomes $\left[\frac{C_2}{C_1 + C_2}\right]^2 \times R_p = \left[\frac{22.22}{1 + 22.22}\right]^2 6.28 = 5.75\text{k}\Omega$, and $R = 25 \parallel 25 \parallel 10 \parallel 5.75 = 2.82\text{k}\Omega$, and $C_2/C_1 = g_m R = 4 (2.82) = 11.28$. Now R_p reflected becomes $\left[\frac{C_2/C_1}{1 + C_2/C_1}\right]^2 R_p = \left[\frac{11.28}{1 + 11.28}\right]^2 6.28 = 5.30\text{k}\Omega$, for which $R = 5.55 \parallel 5.30 = 2.71\text{k}\Omega$, and $C_2/C_1 = 4 (2.71) =$ **10.84**.

Alternatively, from Eq. 12.21, $C_2/C_1 = g_m R$, and here $R = 5.56 || 6.28 \left[\frac{C_2}{C_1 + C_2}\right]^2$. Now let $C_2/C_1 = x$. Thus $x = 4 \times 5.56 \parallel \left[6.28 \left[\frac{x}{1+x}\right]^2\right] = \dfrac{4 \times 5.56 \times 6.28 \times (\frac{x}{1+x})^2}{5.56 + 6.28 (\frac{x}{1+x})^2}$. Thus $5.56\, x + 6.28\, x (\frac{x}{1+x})^2 = 139.7 (\frac{x}{1+x})^2$, or $x = (\frac{x}{1+x})^2 \frac{(139.7 - 6.28x)}{5.56}$. As a means for solution, try $x = 10$: $x = (\frac{10}{11})^2 \frac{(139.7 - 62.8)}{5.56} = 11.34$. Try $x = 10.6$: $x = (\frac{10.6}{11.6})^2 \frac{(139.7 - 6.28\,(10.6))}{5.56} = 10.98$. Try $x = 10.8$: $x = (\frac{10.8}{11.8})^2 \frac{(139.7 - 6.28\,(10.8))}{5.56} = 10.83$. Thus $\mathbf{C_2 = 10.83\, C_1}$, and, from Eq. 12.20, $\omega_o = \left[L (\frac{C_1 C_2}{C_1 + C_2})\right]^{-1/2}$, $\frac{C_1 C_2}{C_1 + C_2} = \frac{1}{\omega_o^2 L} = \frac{C_1 (10.83\, C_1)}{C_1 + 10.83\, C_1}$, or $C_1 = \frac{11.83}{10.83} \times \frac{1}{(2\pi \times 10^6)^2 \times 10 \times 10^{-6}}$ = **2.767nF**, and $C_2 = 10.83\,(2.767\text{nF}) =$ **29.97nF**.

Check: for $C \approx 2.5\text{nF}$, $f = \frac{1}{2\pi} (LC)^{-1/2} = \frac{(2.5 \times 10^{-9} \times 10 \times 10^{-6})^{-1/2}}{2\pi} \approx 1.006\text{MHz}$.

For loop-gain > 1: With C_2 set 5% lower than calculated, the loop gain will be $\frac{C_1 + C_2}{C_2} = \frac{C_1 + 10.83\, C_1 (.95)}{10.83 C\, (.95)} = 1.097 = 1.1$, or 10% larger.

For diode limiting: There are several views of the limiting mechanism: One is to find the gate conducting-diode resistance, Miller-multiplied by the follower action which reduces the existing load at the gate (R_p) by 10%. This value is about 10 R_p. Now r_g at I_G is about $\frac{25mV}{I_G}$. Since the loaded follower gain is originally $\frac{C_2}{C_1 + C_2} = \frac{1}{1 + C_1/C_2} = (1 + 1/10.83)^{-1} = 0.915$. Thus, $10\,(6.28) = \frac{25 \times 10^{-3}}{I_G} \frac{1}{(1 - .915)}$, and $I_G = \frac{25 \times 10^{-3}}{10\,(6.28)} (11.76) = 4.68\mu\text{A}$. Now for $I_G = 4.68\mu\text{A}$, $V_G = 0.7 + 25 \ln \frac{4.68 \times 10^{-3}}{1} = 0.57\text{V}$. Now for v_{GS} raised by 0.57V, $i_D = 4 (1 - \frac{0.57}{-2})^2 = 6.60\text{mA}$, of which 4mA is absorbed by the current source.

Now for a load resistance of $5.56 \parallel (6.28 \parallel 6.28) (\frac{C_2}{C_1 + C_2})^2 = 5.56 \parallel \left[5.71 \left[\frac{1}{1 + C_1/C_2}\right]^2\right] = 5.56 \parallel \left[5.71 \left[\frac{1}{1 + \frac{1}{(10.84)\,(0.95)}}\right]^{-2}\right] = 5.56 \parallel (5.71\,(.831)) = 5.56 || 4.745 = 2.56\text{k}\Omega$, the peak output swing is $2.56\text{k}\Omega\,(6.60 - 4) =$ **6.66V**. A second, simpler, view of finding the peak output is simply to use $v_{GS} \le 0.7\text{V}$, say 0.6V, and proceed, as above, to find $i_D = 4 (1 - \frac{0.6}{-2})^2 = 6.76\text{mA}$, and a peak output of $(2.56)\,(6.76 - 4) =$ **7.1V**, or, using a much simpler view of the load as $5.56 \parallel 6.28 = 2.95\text{k}\Omega$, find a peak = $2.95\,(6.76 - 4) =$ **8.1V**.

For triode-region limiting: For $V_{GG} = 0$, $V_t = +3$V, and $\upsilon_{GS} \approx 0$, triode-mode operation for Q_1 begins for $\upsilon_{gd} = |V_p| = 2$V, that is for $\upsilon_{G1} = 3 - 2 = 1$V. Now for $\upsilon_{G1} = 1$V for a 2.56KΩ load, $i_d = 1/2.56 = 0.39$mA, $i_D = 4 + .39 = 4.39$mA, and $4.39 = 4\left(1 - \frac{\upsilon_{GS}}{-2}\right)^2$, $\upsilon_{GS1} = 2\left(\left(\frac{4.39}{4}\right)^{1/2} - 1\right) = 0.2$V, for which $\upsilon_{DS1} = 3 - 1 = 2$V, and $\upsilon_{DG1} = 3 - 1 - 0.2 = 1.8$V. Triode-mode operation begins for Q_2, for $\upsilon_{DG2} = \upsilon_{DS2} = 2$V, that is for $V_- = -3$V, when $\upsilon_{S1} = \upsilon_{D2} = \upsilon_o = -1$V. Now it is apparent that the worst case occurs for Q_1, although that for Q_2 is easier to calculate. For simplicity, use the $\upsilon_{GS} = 0$ characteristic. Now to reduce the loop gain from 1.1 to 1, the triode-region resistance must reduce the equivalent load resistance from $5.56 \parallel \left[(6.28)\left(\frac{C_2}{C_1 + C_2}\right)^2\right] = 5.56 \parallel (6.28\,(.831)) = 5.56 \parallel 5.22 = 2.69$kΩ to $2.69/1.1 = 2.45$kΩ to reduce the loop gain to 1. Now $R_T \parallel 2.69 = 2.45$, or $R_T = 2.45 \parallel -2.69 = 27.5$kΩ. Thus, the triode slope resistance must be about 27.5kΩ. Now, $i_D = K\,(2\,(\upsilon_{GS} - V_T)\,\upsilon_{DS} - \upsilon_{DS}^2)$, which for $I_{DSS} = K\,V_p^2$, and $V_T = -V_p$ is $i_D = I_{DSS}\left(2\left(1 - \frac{\upsilon_{GS}}{V_p}\right)\frac{\upsilon_{DS}}{-V_p} - \left(\frac{\upsilon_{DS}}{V_p}\right)^2\right)$. Now for $\upsilon_{GS} \approx 0$, $i_D = 4\left(2\,\frac{\upsilon_{DS}}{2} - \left(\frac{\upsilon_{DS}}{2}\right)^2\right)$. Now, $\frac{\partial i_D}{\partial \upsilon_{DS}} = 4\left(1 - \frac{2}{4}\,\upsilon_{DS}\right)$, and $R_T = \frac{\partial \upsilon_{DS}}{\partial i_D} = (4 - 2\,\upsilon_{DS})^{-1}$kΩ. [*Aside*: Note that $R_T = 0$ at $\upsilon_{DS} = 4/2 = 2 = |V_p|$, since V_A is assumed infinite.] Now $1/(4 - 2\,\upsilon_{DS}) = 27.5$kΩ, for $1 = 110 - 55\,\upsilon_{DS}$, or $\upsilon_{DS} = (110 - 1)/(55) = 1.98$V. Accordingly, for $\upsilon_{DS} = 1.98$V with 3V supplies, a peak signal output of about $3 - 2 =$ **1V** would result.

12.15 $\omega_s = (L\,C_s)^{-1/2} = 2\pi\,(2.015) \times 10^6 = 12.661 \times 10^6$ rad/s – – – (1). $\omega_p = \left[\frac{LC_s\,C_p}{C_s + C_P}\right]^{-1/2} = 2\pi\,(2.018) \times 10^6 = 12.679 \times 10^6$ rad/s – – – (2). $Q = \omega_o\,L/r = 50 \times 10^3$. Now $\omega_o \approx \omega_s$. Thus $r = \frac{12.66 \times 10^6}{50 \times 10^3} \times L = 253.2\,L$ – – – (3). From (2), $L = \frac{1}{\omega_p^2} \times \frac{C_s + C_p}{C_s\,C_p} = \frac{1}{\omega_p^2}\left(\frac{1}{C_s} + \frac{1}{C_p}\right)$, where $C_p = 4 \times 10^{-12}$F. Thus $L = (12.679 \times 10^6)^{-2}\left[\frac{1}{4 \times 10^{-12}} + \frac{1}{C_s}\right]$ – – – (4). Now, from (1), $L = \frac{1}{\omega_s^2}\left(\frac{1}{C_s}\right)$ or $1/C_s = \omega_s^2\,L$. From (4), $L = 1.555 \times 10^{-3} + \left(\frac{2.015}{2.018}\right)^2 L$, $L = \frac{1.555 \times 10^{-3}}{.00297} =$ **0.523H**, and $r = 253.2\,(.523) =$ **132.5Ω**, with $C_s = \frac{1}{\omega_s^2\,L} = \frac{1}{(12.661 \times 10^6)^2\,(.523)} =$ **0.0119pF**. Note, as suggested, that the topology of Fig. 12.16 is the same as that in Fig. 12.13, in which oscillation is possible with the crystal acting as an inductor. In both cases the amplifier input voltage across C_2 is C_1/C_2 times that across C_1. Now, for an amplifier with input resistance R_{in} and input voltage V_2 (across R_{in} and C_2), the voltage across C_1 is $V_1 = C_1/C_2\,V_2$, where, for no power loss, V^2/R must be the same in each case. Thus $\frac{V_2^2}{R_{in}} = \frac{V_1^2}{R_{pg}} = \frac{(C_1/C_2)^2\,V_2^2}{R_{eq}}$, whence $R_{eq} = (C_1/C_2)^2\,R_{in} \geq 1/100\,R_{in}$, for $C_2 = 10$pF and $C_1 \geq 1$pF. Now, the equivalent load on the amplifier is $R_L = (R_1 + R_{eq}) \parallel R_f$ and its gain is $-(g_{mp} + g_{mn})\,R_L = -2\,R_L$. For this gain, $R_{in} = \frac{R_f}{1 - -2R_L} = \frac{1000k\Omega}{1 + 2R_L}$, and $R_L = \left[R_1 + \frac{1000}{1 + 2R_L} \times (C_1/C_2)^2\right] \parallel 1000$ – – – (4). Now, the loop gain, $L = 2\,R_L\,C_1/C_2$, must exceed 1 for oscillation. Thus for $2\,R_L\,C_1/C_2 = 1$, $2\left[\left(R_1 + \frac{1000}{1 + 2R_L} \times \left(\frac{C_1}{C_2}\right)^2\right) \parallel 1000\right]\frac{C_1}{C_2} = 1$ – – – (5). Now, to solve (4) and (5), try appropriate values of C_2/C_1 as a measure of the required gain. *First*, for $C_2 = 10$pF and $C_1 = 1$pF, $C_2/C_1 = 10/1 = 10$. From (4) and (5), or directly, $2\left[\left(R_1 + \frac{1000}{1 + 2R_L} \times \frac{1}{10^2}\right) \parallel 1000\right] \times \frac{1}{10} = 1$, or $\frac{2R_L}{10} = 1$, or $R_L = 5$kΩ, whence from (4), $5 = \left[R_1 + \frac{1000}{100\,(1 + 2\,(5))}\right] \parallel 1000$, or $R_1 + \frac{10}{11} = 5$, $R_1 = 5 - 0.9 = 4.1$kΩ. Use **3.9kΩ**. *Second*, for larger C_1, say 2pF, $C_1/C_2 = 1/5$ and $2R_L \times 1/5 = 1$, or $R_L = 2.5$kΩ, whence $2.5 \approx R_1 + \frac{1000}{25\,(1 + 2\,(2.5))}$, or $R_1 = 2.5 - 40/6 = -4.17$kΩ. This implies that for $R_1 =$

0, a critical value of C_1/C_2 exists: Ignoring R_f, from (4), $R_L = \frac{1000}{1 + 2R_L} \times (\frac{C_1}{C_2})^2$, and from (5) $2R_L \frac{C_1}{C_2} = 1$, or $R_L = \frac{1}{2\, C_1/C_2} = \frac{C_2}{2C_1}$. Substituting, $\frac{C_2}{2C_1} = \frac{1000}{1 + 2\,\frac{C_2}{2C_1}} \times (\frac{C_1}{C_2})^2$, $\frac{C_2}{C_1} = \frac{2000}{1 + C_2/C_1} \times (\frac{C_1}{C_2})^2$, $\frac{C_2}{C_1} + (\frac{C_2}{C_1})^2 = \frac{2000}{(C_2/C_1)^2}$, or $(\frac{C_2}{C_1})^4 + (\frac{C_2}{C_1})^3 = 2000$. Thus, $C_2/C_1 < (2000)^{1/4} = 6.69$. Try $C_2/C_1 = (2000 - (C_2/C_1)^3)^{1/4}$. For $C_2/C_1 = 6$, $C_2/C_1 = (2000 - 6^3)^{1/4} = 6.5$. For $C_2/C_1 = 6.5$, $C_2/C_1 = (2000 - 6.5^3)^{1/4} = 6.44$. For $C_2/C_1 = 6.3$, $C_2/C_1 = (2000 - 6.3^3)^{1/4} = 6.46$. Thus $C_2/C_1 = \mathbf{6.4}$ for which $C_1 = \frac{10}{6.4} = 1.56$pF is an estimate of the critical value for which $R_1 = \mathbf{0k\Omega}$. Otherwise the largest value of R_1 occurs for the smallest value of C_1, that is for $C_1 = \mathbf{1pF}$, and is $R_1 \approx \mathbf{4.1k\Omega}$.

SECTION 12.4: BISTABLE MULTIVIBRATORS

12.16 As specified, there is at most a 0.1V drop in the output device(s) with a voltage of 4.9 – 0 = 5.0 – 0.1 = 4.9V across the series connection of R_1 and R_2. For corresponding operation in triode mode, $i_D = K\,[2\,(\upsilon_{GS} - V_t)\,\upsilon_{DS} - \upsilon_{DS}^2]$, or $\frac{4.9}{R_1 + R_2} = 1\,[2\,(5 - 1)\,0.1 - 0.1^2] = 0.8 - .01 = 0.79$, $R_1 + R_2 = \frac{4.9}{0.79} =$ 6.20kΩ (or more). Now assuming that the threshold of the symmetric input inverter is at 2.5V, V_{TL} and V_{TH} each lie 0.5V beyond the threshold. To establish this 0.5V offset, $\frac{4.9 - 2.5}{R_2} = \frac{0.5}{R_1}$, for example. Thus, $\frac{R_2}{R_1} = \frac{2.4}{0.5} = 4.8$, or $R_2 = 4.8\,R_1$. Now $R_1 + R_2 = (1 + 4.8)\,R_1 = 6.20$, for which $R_1 = \frac{6.20}{5.8} =$ **1.07kΩ**, and $R_2 = 4.8\,(1.07) =$ **5.13kΩ**. In practice, 5% resistors of higher values such as 2.7kΩ and 13kΩ would do the job, although 1% values of 5.11kΩ and 24.9kΩ would be better. In general, higher-valued resistors would save power, ensure a full output swing, and improve output transition times, while lower-valued ones would tend to reduce regeneration time.

Now for device variation, the highest value of V_{th} of the input device occurs for K_p = 20% high, K_n 20% low, V_{tn} 20% high, and V_{tp} 20% low. Now at V_{th}, $K_n\,(V_{th} - V_{tn})^2 = K_p\,(5 - V_{th} - V_{tp})^2$, or $0.8\,(V_{th} - 1.2)^2 = 1.2\,(5 - V_{th} - 0.8)^2$, or $V_{th} - 1.2 = (\frac{1.2}{0.8})^2\,(4.2 - V_{th}) = 9.45 - 2.25\,V_{th}$. Thus, $3.25\,V_{th} = 10.65$, or $V_{th} =$ **3.28V**. With V_{th} = 3.28V and input at V_{IH}, $\frac{V_{IH} - 3.28}{R_1} = \frac{3.28 - 0.1}{R_2}$, or $V_{IH} = \frac{1.07}{5.13}\,(3.27) + 3.28 =$ **3.96V**. Likewise $V_{IL} = \frac{-1.07}{5.13}\,(4.9 - 3.28) + 3.28 =$ **2.94V**. Now for the reversed set of extremes, by symmetry, $V_{th} = 2.5 - (3.28 - 2.5) =$ **1.72V**, for which $V_{IH} = 5 - 2.94 =$ **2.06V**, and $V_{IL} = 5 - 3.96 =$ **1.04V**. Notice that for the two extreme versions that the two threshold sets do not overlap, although both lie within the power-supply range (0 to 5V).

Concerning the rise and fall times, a complete calculation is somewhat complex, involving estimates of the transition times of the internal node, the V_{IL} to V_{IH} range of the second inverter and its gain, as well as switching of the output inverter. Alternatively, a quick estimate of output rise and fall times is defined by the current available from each output transistor as $K\,(5 - 1)^2 = 16K = 16$mA. Thus t_r, between 0.5 and 4.5V, for a 1pF load, exceeds $\frac{(4.5 - 0.5) \times 1 \times 10^{-12}}{16 \times 10^{-3}} =$ **0.25ns**. A closer estimate could be found by including the effect of current reduction due to triode operation over a large part of the switching range. As an approximation, the current for $\upsilon_{DS} = 0.5$V is $i_D = 1\,(2\,(5 - 1)\,0.5 - 0.5^2) =$ 3.75mA. Thus a better estimate of the average current would be $\frac{16 + 3.75}{2} = 9.875$mA, and of the transition time as $\frac{4.5 - 0.5 \times 1 \times 10^{-12}}{9.9 \times 10^{-3}} =$ **0.40ns**.

The delay requested includes a lot of factors, the most important of which are first the time for the first inverter to reach its threshold, and then for the second to do so. The former is dominated by the input itself which to rise (for example) from 0V to V_{IH} = 3V, by 3V, takes (3V/1V/μs) = 3μs!! Now the output of the first inverter changes at a rate defined by the input-signal rate and by its load capacitance. Likely the former dominates. Now, due to R_1, R_2 the signal at the gate of the first inverter pair is $\frac{R_2}{R_1 + R_2} = \frac{5.13}{1.07 + 5.13} = 0.83$ of the input. Now the gain of the first inverter at the middle of the switching region is $g_m\ r_o$ where, at the middle $i_D = K\ (\upsilon_{GS} - V_t)^2 = 1\ (2.5 - 1)^2 = 2.25\text{mA}$, $g_m = 2\ (2.5 - 1) = 3\text{mA/V}$, and $r_o = \frac{V_A}{i_D} = \frac{30}{2.25} = 13.3\text{k}\Omega$. Thus, the gain is $g_m\ r_o = 3 \times 13.3 = 40\text{V/V}$. Thus the rate of change of the output of the first inverter is limited to 1V/μs × .83 × 40 = 33.2V/μs, before regeneration. Thus the output of the first inverter will move from 0 or 5V to 2.5V in somewhat more than $\frac{2.5V}{33.2V/\mu s}$ = **75.3ns**. After regeneration, the remainder of the transition will-be faster, with the output current driving the 11pF load. Shortly after the transition begins as the circuit regenerates, the available current will be at least the 2.25mA middle bias current, for which the remaining transition time will be about $T = \frac{11 \times 10^{-12}\ (2.5 - 0.5)}{2.25 \times 10^{-3}} \simeq$ **10ns**. Now the final (isolated) conclusion about regeneration that can be easily made, is to estimate the time it takes for the input of the first inverter to cross the remaining half of its active region as driven through R_2 by the output changing by all (or part of) the output range. Now as implied earlier, the input active region is approximated by 5V/40 = .125V, and half is about 62mV. Now for a 5V change in output, the current change in R_2 is about 5V/5.13kΩ = 0.97mA, most of which supplies the 10pF input capacitance. For the input to move 62mV, takes about $\frac{10 \times 10^{-12} \times 62 \times 10^{-3}}{.97 \times 10^{-3}}$ = **0.64ns**. Thus, regeneration, once it begins as the output of the second gate begins to move, can be very fast.

12.17 Now for υ_O (at node B) high, υ_A = 1.4V. For I_{D1} = 1mA, $\frac{13.0 - 1.4}{R_2} - \frac{1.4}{R_1} = 1$, whence $\frac{11.6}{R_2} - \frac{1.4}{R_1} = 1$ --- (1). Now for υ_o low, $V_A = V_{D4} + V_{D3} - V_{D2}$ = **0.7V**, and for Φ_{D2} = 1mA, $\frac{0.7 - -13}{R_2} + \frac{0.7}{R_1} = 1$, or $\frac{13.7}{R_2} + \frac{0.7}{R_1} = 1$ --- (2). Now, adding (2) + (2) + (1) → $\frac{11.6}{R_2} + \frac{27.4}{R_2} + 0 = 3$, $R_2 = \frac{11.6 + 27.4}{3} = \frac{39}{3} = 13\text{k}\Omega$, and $\frac{11.6}{13} - \frac{1.4}{R_1} = 1$, and $R_1 = \frac{-1.4}{1 - .892} = -13\text{k}\Omega$. Now, the negative value of R_1, not available without active components, implies that the specifications are too tight. For example, allow the current in I_{D2} to increase, say to 2mA, for which change (2) becomes $\frac{13.7}{R_2} + \frac{0.7}{R_1} = 2$. Now combining the old (1) with the new (2) → $\frac{11.6}{R_2}$ - $\frac{27.4}{R_2} = 5$, $R_2 = \frac{39}{5}$ = **7.8kΩ**, and $R_1 = \frac{-1.4}{1 - 11.6/7.8} = \frac{-1.4}{1 - 1.487}$ = **2.87kΩ**. Note that while this is a solution, that the latter formulation indicates the possibility of a solution for which $R_1 = \infty$, where the term in the denominator reaches zero. Examining the circuit with R_1 removed, one sees this possibility directly. Thus, for $R_1 = \infty$, either $\frac{13.0 - 1.4}{R_2} = 1$ for which $R_2 = 11.6\text{k}\Omega$, or $\frac{0.7 - -13}{R_2} = 1$, for which $R_2 = 13.7\text{k}\Omega$. Now, to ensure that the associated current exceeds 1mA, the smaller value (or one even smaller) should be chosen. For convenience, make $R_1 = \infty$ and R_2 = **10kΩ**, for which $I_{D1} = \frac{13.0 - 1.4}{10} = 1.16\text{mA}$, and $I_{D2} = \frac{13 + 0.7}{10} = 1.37\text{mA}$. Now for D_2 conducting, the current in D_3 and D_4 must exceed 1mA. Thus $R_3 = \frac{15 - 0.7 - 0.7}{1 + 1.37} = 5.74\text{k}\Omega$. Use R_3 = **5.6kΩ**. Now, the maximum current in D_4 is $\frac{15 - 1.4}{5.6} + 1.16 = 3.6\text{mA}$ (< 4mA, as required). Finally, note that the input thresholds are defined by the possible voltages at node A, namely V_{TH} = **+1.4V** and V_{TL} = **+0.7V**. For inputs lower than 0.7V, the output is high, and υ_A = 1.4V. Now, as υ_I rises and just exceeds 1.4V, υ_O falls, and υ_A falls to 0.7V. For υ_I > 1.4V, υ_O = −13V. Now as υ_I falls, at υ_I = 0.7V, the output reverses again, and υ_O goes to +13V.

12.18

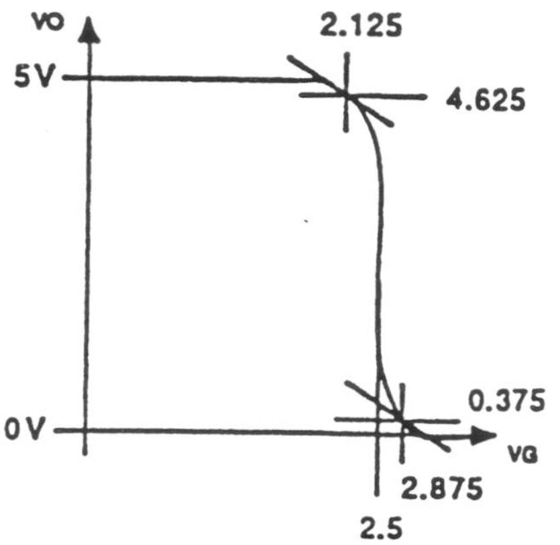

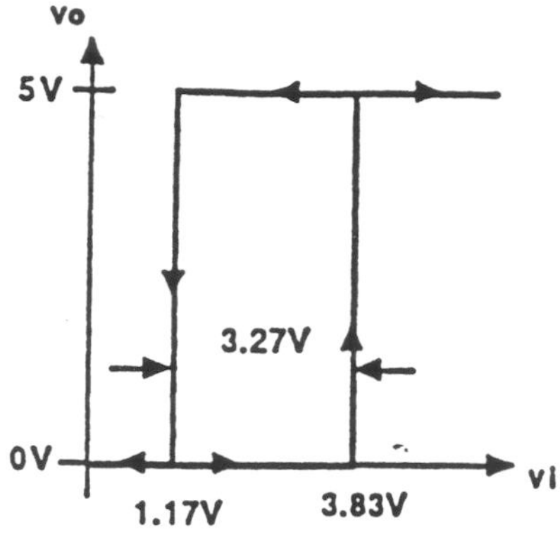

For the Q_3, Q_4 inverter, since $K_3 = K_4$ and $|V_t|$ are equal, $V_{th} = V_{DD}/2 = 5V$. As noted on page 934 of the Text, the transfer characteristic slope exceeds 1 in magnitude for υ_I between 5/8 $V_{DD} - V_t/4$, and 3/8 $V_{DD} + V_t/4$, or $V_{DD}/2 \pm (V_{DD}/8 - V_t/4) = 2.5 \pm (.625 - .25) = 2.5 \pm .375 = 2.875V$ and 2.125V, for which υ_O is $(V_{DD}/8 - V_t/4)$, and $V_{DD} - (V_{DD}/8 - V_t/4)$, or .375V and 4.625V respectively. While a relatively complete transfer characteristic is provided for interest, the value $V_{th} = 2.5V$ is the most essential feature. We will use this to estimate the overall characteristic υ_O/υ_I. Essentially, for the voltage at the internal node (A), $\upsilon_A < 2.5V$, υ_O is high, which for $\upsilon_A > 2.5$, υ_O is low, in both of which cases, positive feedback via Q_5, Q_6 forces node A away from 2.5V in the direction to which it tends. Correspondingly, the thresholds at the input are the voltages υ_I for which the voltage υ_A is 2.5V with $\upsilon_I = V_{IL}$ when Q_6 conducts (for υ_O high (5V)), and $\upsilon_I = V_{IH}$ when Q_5 conducts (for υ_O low (0V)). For $V_{IL} = \upsilon$, $i_1 = i_2 + i_6$ for all devices in saturation, $K_1 (5 - \upsilon - 1)^2 = K_2 (\upsilon - 1)^2 + K_6 (5 - 1)^2$. Now $K_1 = K_2 = 2 K_6$. Thus $2 (4 - \upsilon)^2 = 2 (\upsilon - 1)^2 + 4^2$, $32 - 16 \upsilon + 2 \upsilon^2 = 2 \upsilon^2 - 4 \upsilon + 2 + 16 \upsilon$ $(-16 + 4) = 16 + 2 - 32 = -14$, and $\upsilon = 14/12 = 1.167V$. Thus $V_{IL} =$ **1.167V** and, symmetrically, $V_{IH} = 5 - 1.167 =$ **3.83V**. *Check*: $2 (4 - 1.167)^2 = 2 (1.167 - 1)^2 + 1 (5 - 1)^2$, $16.052 = 0.056 + 16 \rightarrow$ **OK.**

SECTION 12.5: GENERATION OF SQUARE AND TRIANGULAR WAVEFORMS USING ASTABLE MULTIVIBRATOR

12.19 Here, for a 6.8V zener, the voltage values at υ_{02} are $\pm (6.8 + 2 (0.7)) = \pm 8.2V$. Now the υ_{03} output amplifier is a unity-gain follower. Thus the voltage at the RC common node is a triangle (approximately) of $\pm$ 1V peak amplitude. Now for the notation on pages 1003, 1004 of the Text, $L^+ = 8.2V$, $\beta = \dfrac{R_1}{R_1 + R_2}$ and β (8.2V) = 1V, whence $\beta = 0.12195$, and $R_1 = 0.12195 (R_1 + R_2)$ and $R_1 = .1389 R_2$. Now, from Eq. 12.31, with $L^+ = -L^-$, $T_1 = \dfrac{1/10kHz}{2} = RC \ln \dfrac{(1 - \beta(-1))}{(1 - \beta)} = RC \ln \dfrac{1 + \beta}{1 - \beta} = RC \ln \dfrac{1 + .12195}{1 - .12195} = .245RC$. Thus $R = \dfrac{50 \times 10^{-6}}{.245 \times 1000 \times 10^{-12}} = 0.204M\Omega$. Use $R =$ **200kΩ**, for which $RC = 200\mu s$, $R_2 = 200k\Omega$, $R_1 = 0.1389 (200) = 27.8k\Omega$, or, better, $R_2 =$ **240kΩ** and $R_1 = 33.3k\Omega$, for which use **33kΩ**. For R_3, the current exceeds $(1mA + 2 (\dfrac{8.2}{200k\Omega})) \approx 1.08mA$. Thus $R_3 = \dfrac{13 - 8.2}{1.08} = 4.4k\Omega$. Use **3.9kΩ**. Now, using the notation on page 1004 of the Text, $\upsilon^- = L^+ - (L^+ - \beta L^-)e^{-t/\tau}$, or $\upsilon^- = 8.2 - (8.2 - 1)e^{-t/RC}$, in general. Now $\dfrac{\partial \upsilon}{\partial t} = -7.2 (\dfrac{1}{RC})e^{-t/RC} = \dfrac{7.2}{RC} e^{-t/RC}$. Thus at $t = 0$, slope is $\dfrac{7.2V}{200 \times 10^{-6}s} =$ **0.036V/μs**, and at $t = 50$ μs, slope is $.036 e^{-\frac{50}{200}} = .036 (.779) =$ **0.028V/μs**. Thus the average slope is $\dfrac{0.036 + .028}{2} =$ **0.032V/μs**, and the slope change is $.036 - .028 = 0.008V/\mu s$. For the slope change reduced by half, reduce the voltage swing across C, so that $0.036 e^{-50/RC} = 0.036 - .008/2 = 0.032$, or $e^{-50/RC} = .8888$, $50/RC = -\ln .889 = 0.1178$. Thus $RC = 50/.1178 = 424/\mu s$, for which $R = \dfrac{424 \times 10^{-6}}{1000 \times 10^{-12}} =$ **424kΩ**. Now for this design, at $t = 0$, $\upsilon^- = \upsilon = L^+ - L^+ - \beta L^-) = \beta L^- =$

-8.2β, while at 50μs, $v^- = +8.2\beta$. Thus $8.2\beta = 8.2 - (8.2 + 8.2\beta)e^{-50/424}$, $\beta = 1 - (1 + \beta)(.8888)$, $\beta = 1 - .8888 - .888\beta$, $\beta = .1112/1.8888 = 0.0589$, and the peak voltages of the improved triangle wave are at $\pm 8.2\beta = 8.2\,(.0589) = \pm 0.483$V. Now to achieve a ± 1V output, two resistors must be added to the output amplifier to produce a gain of $1 + \dfrac{R_2'}{R_1'} = \dfrac{1}{.483} = 2.07$, or $R_2' = 1.07\,R_1'$. To arrange such values, one could make $R_1' =$ **10kΩ** and R_2', as **10kΩ in series with 680Ω.**

12.20 The required circuit is as shown:

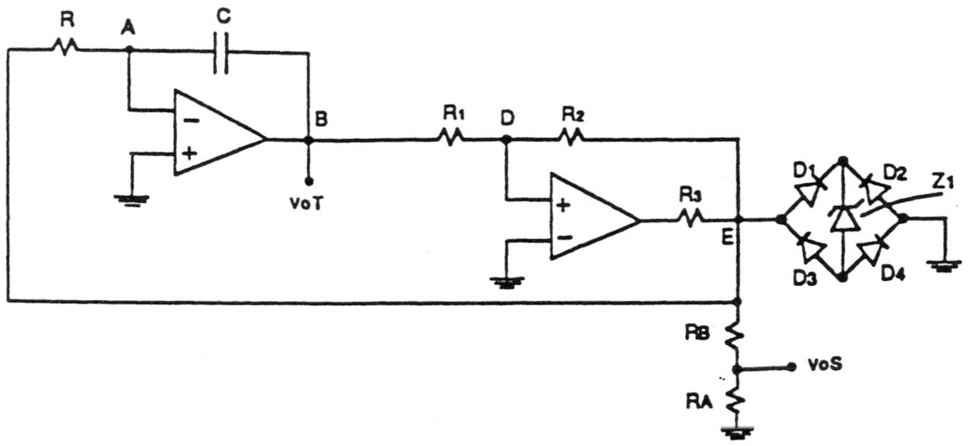

For a 6.8V zener and 0.7V diodes, the voltage at E is $\pm(6.8 + 2\,(0.7)) = \pm 8.2$V. For 1mA drain, $R_A + R_B = \dfrac{8.2}{1mA} = 8.2$kΩ. For ±1V square waves, $\dfrac{R_A}{R_A + R_B} \times 8.2 = 1$, that is $R_A =$ **1kΩ** and $R_B =$ **7.2kΩ.** For ±1V triangle waves, the thresholds at node B must be ±1V. Now, for $R_1 =$ **10kΩ**, $R_2 = \dfrac{8.2}{1} R_1 =$ **82kΩ.** Now for $C =$ **1000pF**, to charge to 1V in $\dfrac{1/10kHz}{2} = 50\mu s$, the current in R must be $I = \dfrac{1000 \times 10^{-12} \times 1}{50 \times 10^{-6}} = 20\mu A$. Thus $R = \dfrac{8.2V - 0V}{20\mu A} =$ **410kΩ.** Now the current in R_3 is 1mA for R_A and R_B, 1mA for the zener, 0.1mA for R_C, and 20μA for R. Thus $R_3 = \dfrac{13 - 8.2}{1 + 1 + 1 + .02} = 1.589$kΩ. Use $R_3 =$ **1.5kΩ.** Note how easy this circuit is to design! This is because it has a desirable direct relationship between particular components and specific functions!

SECTION 12.6: GENERATION OF A STANDARDIZED PULSE – THE MONOSTABLE MULTIVIBRATOR

12.21 See that for B more positive than A, $v_D = +10$V and v_A rises very slowly until $v_A > v_B$, at which point v_D goes to -10V, v_B falls to $-10 \times \dfrac{R_1 + R_5}{R_1 + R_5 + R_2}$, and A goes slowly to -0.7V.

12.21 (continued)

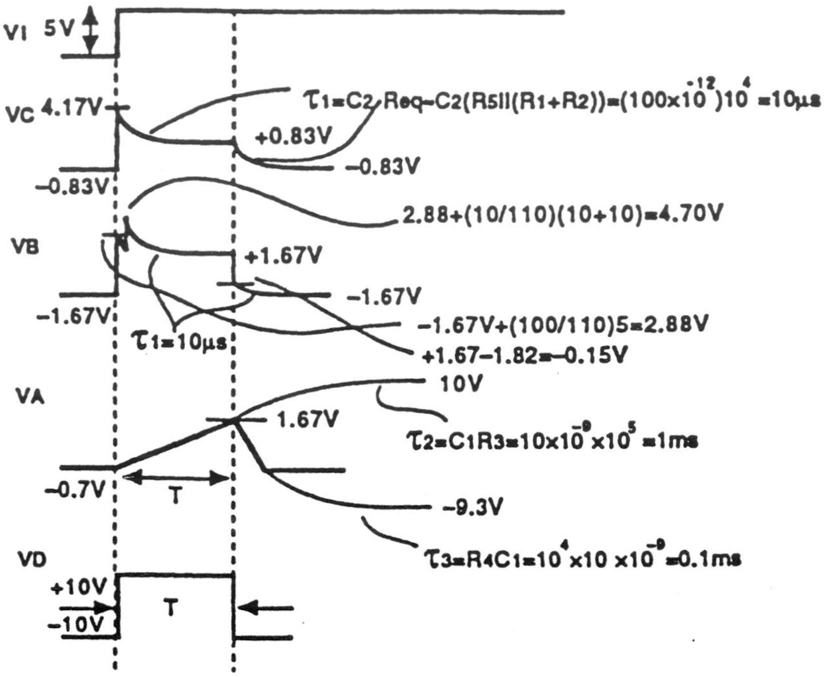

Since υ_A is limited to –0.7V, and υ_B can be made lower, the output remains at –10V in a stable state. In this state, for this design: $V_C = \dfrac{10R_5}{R_5 + R_1 + R_2} = \dfrac{-10\ (10)}{10 + 10 + 100} = \mathbf{-0.833V}$, $V_B = \dfrac{-10\ (R_5 + R_1)}{R_5 + R_1 + R_2}$ $= \dfrac{-10\ (10 + 10)}{120} = \mathbf{-1.67V}$, $V_A = \mathbf{-0.7V}$, $V_D = \mathbf{-10V}$. To trigger the circuit, υ_A must rise to –0.7 or by (1.67 – 0.7) = 0.97V, and υ_C to υ where $\upsilon \dfrac{100}{10 + 100} = 0.97$, or $\upsilon = +0.97\ (1.1) = 1.067$V, that is from –0.833 to +0.233V. Thus the positive step at υ_I required is $\upsilon_I = \mathbf{1.07V}$. From waveform υ_A, see that the pulse ends when $10 - (10 - -0.7)\ e^{-t/1ms} = 1.67$V, or $10.7\ e^{t/10^{-3}} = 8.33$, $e^{-t/10^{-3}} = 0.778$, or $t = -10^{-3}\ \ln .778 = 0.25$ms. Thus the pulse length produced is **0.25ms**. *For a rate-limited input*: $i = \dfrac{C\ d\upsilon}{dt}$ and $i\ (R_5 \parallel (R_1 + R_2)) = 1.07$V. Thus $1000 \times 10^{-9} \dfrac{d\upsilon}{dt} \times 10^4 \parallel (11 \times 10^4) = 1.07$, and $\dfrac{d\upsilon}{dt} = \dfrac{1.07}{0.91 \times 10^4 \times 1000 \times 10^{-9}} = 117.6$V/s = **118V/s**. *For recovery*: See that for a positive input pulse shorter than the output pulse, that the longest recovery time constant, controlled by R_4 is $\tau_3 = R_4\ C_1 = 0.1$ms, where $\upsilon_A = -9.3 + (9.3 + 1.67)e^{-t/10^{-4}}$, and recovery is complete when $\upsilon_A = 0.7 = -9.3 + (10.97)e^{-t/10^{-4}}$, or $e^{-t/10^{-4}} = \dfrac{10}{10.97}$, or $t = 10^{-4}\ (.093) = 9.3\mu$s.

Thus it appears that retriggering could occur 10μs or so after the output falls (assuming zero recovery time for the amplifier). In practice, recovery of the amplifier itself (from limiting) might require somewhat more time.

12.22

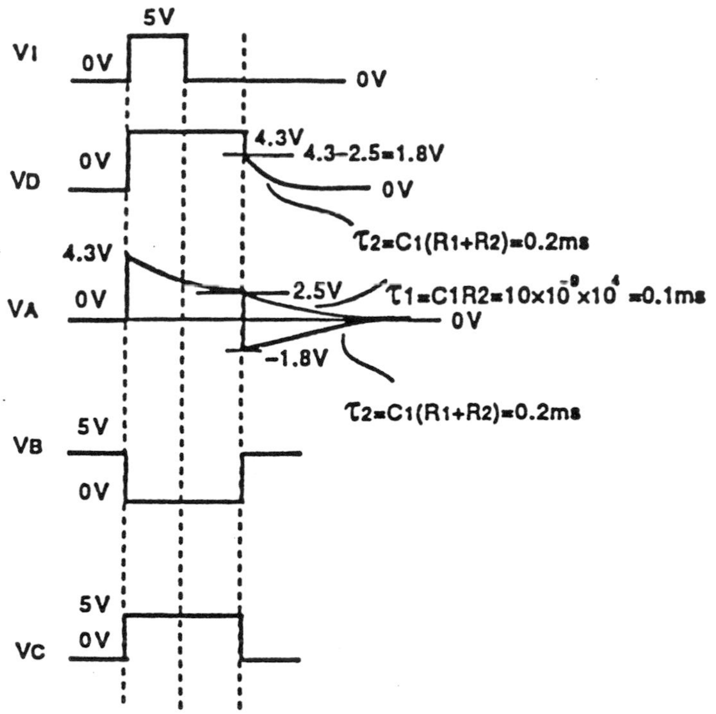

For node A, assuming the D_1, D_2 resistances to be zero, $\upsilon_A = 0 - (0 - 4.3\text{V})e^{-t/R_2 C_1}$. Now, $\upsilon_A = 2.5\text{V}$ where $e^{-t/10^{-4}} = \frac{2.5}{4.3}$, or $t = 10^{-4} \ln 1.72 = 54.2\mu\text{s}$. Thus the pulse is **54.2μs** long. Normally, υ_I would be very much shorter than that, a few tens or hundreds of ns, at most, the delay through the two inverters. For very long pulses at the input, the loop is held open, and the output fall time is an amplified version of that at node A. For gain of $40 \times 40 = 1600\text{V/V}$, and 5V output swing, the input active region is about $5/1600 = 3 \times 1 \times 10^{-3}\text{V}$. Now, $\upsilon_A = 4.3e^{-t/R_2 C_1} = 4.3e^{-t/10^{-4}}$, and $\frac{\partial \upsilon_A}{\partial t} = \frac{-4.3}{10^{-4}} e^{-t/10^{-4}}$.

At $t = 54.2\mu\text{s}$, $e^{-t/10^{-4}} = \frac{2.5}{4.3}$, $\frac{\partial \upsilon_A}{\partial t} = \frac{-4.3}{10^{-4}} \times \frac{2.5}{4.3} = \frac{2.5}{10^{-4}} = 2.5 \times 10^4\text{V/s}$. Thus the output fall time $\approx \frac{3.1 \times 10^{-3} V}{2.5 \times 10^4}$ = **124ns.**

SECTION 12.7: INTEGRATED-CIRCUIT TIMERS

12.23 From Eq. 12.39, $T = CR \ln 3 = 1.1 \times 10 \times 10^{-9} \times 10^4$ = **0.11ms.** The input pulse must be shorter than **0.11ms** by an amount which guarantees the relationship for component variation. For longer inputs, both comparator outputs are high, and the flip-flop is set and reset at the same time.

12.24 Extending Equation 12.38 for the case in which the capacitor voltage is not quite zero, but rather υ, $\upsilon_c = V_{CC} - (V_{CC} - \upsilon)\, e^{-t/RC} = 5 - (5 - \upsilon)e^{-t/RC}$. Now for $\upsilon_c = 2/3\ V_{CC} = 10/3\text{V}$, $10/3 = 5 - (5 - \upsilon)e^{-t/RC}$, $e^{-t/RC} = \frac{-3.333 + 5}{5 - \upsilon} = \frac{1.666}{5 - \upsilon}$, $t = RC \ln \frac{5 - \upsilon}{1.666} = RC \ln (3 - 0.6\upsilon)$, and $\frac{\partial t}{\partial \upsilon} = -0.6\ RC \frac{1}{(3 - 0.6\upsilon)}$. For υ small, $\frac{\partial t}{\partial \upsilon} = -\frac{0.6RC}{3} = -0.2\ RC$. Now for $T = 1.1\ RC$, for a 0.1V change in υ, the change in T is $-0.2\ RC\ (0.1)$, or $-\frac{0.2RC\ (0.1)}{1.1RC} \times 100$ = **−1.8%.**

12.25 From Eq. 12.41 on page 1013 of the Text, $T_H = 0.69\ C\ (R_A + R_B) = 0.69 \times 10 \times 10^{-9}\ (2 \times 10^4) =$ 138μs, and $T_L = 0.69\ C\ R_B = 0.69 \times 10 \times 10^{-9}\ (10^4) = \frac{138}{2} =$ 69μs. Thus the period is $T = 138 + 69$ = **207**μs, and the frequency $1/207 \times 10^{-6} =$ **4.83kHz**, with duty cycle $\frac{138}{207} \times 100 =$ **66.7%**. For R_A = 10kΩ and R_B = 1kΩ, $T_H = 0.69\ C\ (R_A + R_B) = 0.69 \times 10^{-8}\ (10 + 1) \times 10^3 = 75.9$μs, and $T_L = 0.69\ C\ (R_B) = 0.69 \times 10^{-8}\ (10^3) = 6.9$μs, for which the frequency is $\frac{1}{75.9 + 6.9} =$ **12.1kHz**, and duty cycle is $75.9/(75.9 + 6.9) \times 100 =$ **91.7%**. For the same frequency, $T = 207 \times 10^{-6} = 6.9 \times 10^{-6} + .69 \times 10^{-8}\ (R_A + 1)\ 10^3$, $R_A = \frac{200 \times 10^{-6}}{0.69 \times 10^{-5}} - 1 =$ **28kΩ**. For 10kHz, T = 100μs, $100 \times 10^{-6} = .69\ C\ (R_A + 2R_B)$, $R_A + 2\ R_B = \frac{100 \times 10^{-6}}{.69 \times 10^{-8}} = 1.45 \times 10^4 =$ **14.5kΩ**. There is no combination of resistors which will produce 10% duty cycle. Use 90% duty cycle and an inverter!

SECTION 12.8: NONLINEAR WAVEFORMPING CIRCUITS

12.26 For a sine-wave of peak output υ, ie $\upsilon \sin \omega t$, the zero-crossing | slope| is $\upsilon\omega$ volts per second. Thus the triangle wave reaches $\upsilon\omega \times T/4$, or $\upsilon\ (2\pi f) \times (1/4f) = 2\pi\upsilon/4 = \pi\upsilon/2 = 1.57\ \upsilon$, at the peak. Though the choice is arbitrary, let us assume a sine wave peak of 0.7V, with I_D peak = 1mA, such that the triangle input peak = 0.7 × 1.57 = **1.10V**, with the drop across R being (.57) (.7) = .400V. Thus R = 0.400/1mA = **400Ω**. Now, in general, $\upsilon_i = 1.10\ (\Theta/90)$, or $\Theta = 81.8\ \upsilon_i$, over the range 0° to 90°. Also $\upsilon = 700 + 50 \ln i/1$, and $i = e^{(\upsilon-700)/50}$. *For* υ_o = 0.7V, Θ = 90°, υ_i = 1.10V, $I = \frac{1.10 - 0.7}{400} =$ 1mA. *For* υ_o = 0.65V, $i = 1\ e^{(650-700)/50}$ = 0.368mA, υ_i = .65 + .368 (.4) = 0.79, Θ = 81.8 (.797) = 65.2°, and 0.7 sin 65.2° = 0.635V. *For* υ_o = 0.60V, $i = e^{(+600-700)/50}$ = .135mA, υ_i = .60 + .135 (.4) = 0.654, Θ = 81.8 (.654) = 53.5°, 0.7 sin 49.5° = .563V. *For* υ_o = 0.55V, $i = e^{(550-700)/50}$ = .050 mA, υ_i = 0.55 + .05 (.4) = 0.57V, Θ = 81.8 (.57) = 46.6°, 0.7 sin 46.6° = 0.509V. *For* υ_o = 0.50V, $i = e^{(500-700)/50}$ = .018mA, υ_i = 0.50 + .018 (.4) = .507, Θ = 81.8 (.507) = 41.5°, 0.7 sin 41.5° = .464V. *For* υ_o = 0.45V, $i = e^{(450-700)/50}$ = .007, υ_i = 0.45 + .007 (.4) = .453, Θ = 81.8 (.453) = 37.1°, 0.7 sin 41.5° = .422V. *For* υ_o = 0.40V, $i = e^{(400-700)/50}$ = .0025, υ_i = 0.40 + .0025 (.4) = .401, Θ = 81.8 (.401) = 32.8°, 0.7 sin 32.8° = .379V. *For* υ_o = 0.35V, $i = e^{(350-700)/50}$ = .0009, υ_i = .35 + .0009 (.4) = .350, Θ = 81.8 (.350) = 28.6°, 0.7 sin 28.6° = .335V. *For* υ_o = 0.3V = υ_i, Θ = 81.8 (.3) = 24.5°, 0.7 sin 24.5 = .291V. *For* υ_o = 0.2V = υ_i, Θ = 81.8 (.2) = 16.3°, 0.7 sin 16.3° = .197V. *For* υ_o = 0.1V = υ_i, Θ = 81.8 (.1) = 8.18°, 0.7 sin 8.18° = .099V. *For* υ_o = 0V = υ_i, Θ = 0, 0V. In summary:

Θ°	90	65.2	53.5	46.6	41.5	32.8	24.5	16.3	8.2	0
υ_o, V	0.7	0.65	0.60	0.55	0.50	0.40	0.30	0.2	0.1	0
0.7 sinΘV	0.7	0.635	0.563	0.509	0.464	0.379	0.291	0.197	0.099	0
ε, mV	0	15	37	41	36	21	9	3	1	0
ε%	0	2.4	6.6	8.1	7.8	6.4	3.1	1.5	1	0

Note that the output wave is generally "fatter" than the sine wave.

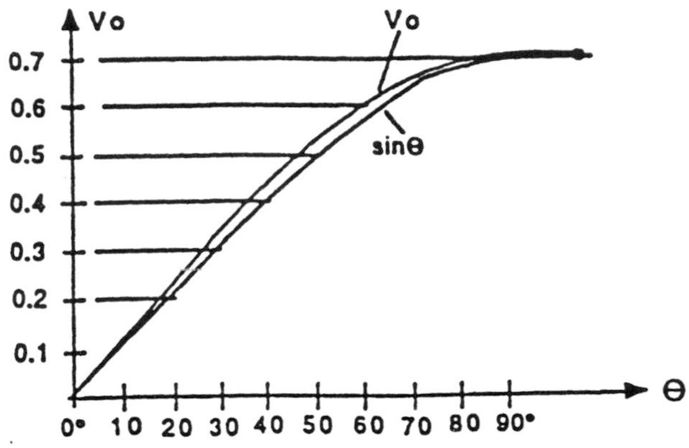

12.27 Here, $i = 0.1\upsilon^2$, with a match at υ = 2, 4, and 8V, using 0, 3, and 7V supplies. For υ = 2V, i = 0.4mA, R_1 = 2/0.4mA = **5kΩ** as before. Now chose $V_2 = 3 - 0.6 =$ **2.4V**, and $V_3 = 7 - 0.6 =$ **6.4V**. For υ = 4V, $i = 4/5 + \dfrac{4-0.6-2.4}{R_2+0.1} = 0.1(4)^2 = 1.6\text{mA}$. Thus, $\dfrac{1}{R_2+0.1} = 1.6 - .8 = .8\text{mA}$, $R_2 = \dfrac{1}{0.8} - 0.1 =$ **1.15kΩ.**

For υ = 8V, $i = \dfrac{8}{5} + \dfrac{8-3}{1.15+0.1} + \dfrac{8-0.6-6.4}{R_3+0.1} = 0.1(8)^2 = 6.4\text{mA}$. Thus $1/(R_3+0.1) = 6.4 - 1.6 - 4 = 0.8\text{mA}$, $R_3 = 1/0.8 - 0.1 =$ **1.15kΩ.**

Now for the errors: At 3V, $i = \dfrac{3}{5} = 0.6\text{mA}$, rather than $0.1\,(3^2) = 0.9\text{mA}$, with an error = **– 0.3mA.** At 5V, $i = \dfrac{5}{5} + \dfrac{5-3}{1.15+.1} = 1 + 1.6 = 2.6\text{mA}$, rather than $0.1\,(5)^2 = 2.5\text{mA}$, for an error of **+0.1mA.** Thus at 7V, the error is **–0.3mA**, and at 10V, it is **0mA**, just as in Ex. 12.22.

Now for 1mA diodes with n = 2, with V_2 = **2.4V**, V_3 = **6.4V**. Now at υ = 2, i = 0.4mA and R_1 = **5kΩ** as before. Now at υ = 4V, i = 1.6mA, with $I_{D2} = 1.6 - 0.8 = 0.8\text{mA}$. For 0.8mA, $\upsilon_{D2} = 700 + 50 \ln \dfrac{0.8}{1} = 689\text{mV}$. Thus $R_2 = \dfrac{4-2.4-.689}{0.8} =$ **1.139kΩ.** Now at υ = 8V, i = 6.4mA. Here $I_{R1} = \dfrac{8}{5} = 1.6\text{mA}$, and $I_{R2} \approx \dfrac{8-2.4-0.7}{1.139} = 4.30\text{mA}$, $\upsilon_{D2} = 700 + 50 \ln 4.3/1 = 773\text{mV}$, $I_{D2} = \dfrac{8-2.4-.773}{1.139} = 4.24\text{mA}$ (OK), Thus, $I_{D3} = 6.4 - 1.6 - 4.24 = 0.56\text{mA}$, $\upsilon_{D3} = 700 + 50 \ln .56 = 671\text{mV}$. Thus $R_3 = \dfrac{8-6.4-.671}{.56} =$ **1.659kΩ.**

Now for the errors: At 3V, $i = 3/5 + i_{D2}$, $\upsilon_D = 3 - 2.4 - I_{R2} = 0.6 - I_{R2}$. Try $i_D = .05\text{mA}$, $\upsilon_D = 700 + 50 \ln .05 = 550\text{mV}$, $\upsilon_{R2} = .05 \times 1.139 = .057\text{V}$, and $\upsilon_D + \upsilon_{R2} = .607\text{V}$ (rather than .600V) (OK). Thus $i = .6 + .05 = .65\text{mA}$, rather than 0.9mA, for an error of **–0.25mA.** At 5V, the required i is $0.1\,(5)^2 = 2.5\text{mA}$, and the actual $i = \dfrac{5}{5} + \dfrac{5-2.4-V_{D2}}{1.139} \approx 1 + \dfrac{5-2.4-0.7}{1.139} = 1 + \dfrac{1.9}{1.139} = 1 + 1.67 = 2.67\text{mA}$. Now for D_2, and $I_{D2} = 1.67\text{mA}$, $\upsilon_{D2} = 700 + 50 \ln 1.67 = 726\text{mV}$, $i_{D2} = \dfrac{5-2.4-.73}{1.139} = 1.64\text{mA}$. Overall, $i = 1 + 1.64 = 2.64\text{mA}$ with an error of **+0.14mA.** At 7V, required $i = (0.1)\,(7)^2 = 4.9\text{mA}$, and the actual $i = \dfrac{7}{5} + \dfrac{7-2.4-V_{D2}}{1.139} + \dfrac{7-6.4-V_{DS}}{1.659}$, which for $\upsilon_{D2} \approx .75\text{V}$, $i_{D2} = \dfrac{7-2.4-0.75}{1.139} = 3.38\text{mA}$. Now $\upsilon_{D2} \approx 700 + 50 \ln 3.38 = 761\text{mV}$, $i_{D2} = \dfrac{7-2.4-.76}{1.139} = 3.37\text{mA}$. Now $i_{D3} = \dfrac{7-6.4-.6}{1.659} \approx 0$. Thus $i = 1.4 + 3.37 = 4.77\text{mA}$, with an error of **– 0.13mA.** At 10V, required i = (0.1) 10^2 = 10mA. Actual $i = \frac{10}{5} + \frac{10 - 2.4 - V_{D2}}{1.139} + \frac{10 - 6.4 - V_{D3}}{1.659}$. Now $i_{D2} \approx \frac{10 - 2.4 - .78}{1.139}$ = 5.99mA, for which v_{D2} = 700 + 50 ln 5.99 = .79V, $i_{D2} = \frac{10 - 2.4 - .79}{1.139}$ = 5.99mA. Now $i_{D3} \approx \frac{10 - 6.4 - 0.72}{1.659}$ = 1.74mA, for which v_{D2} = 700 + 50 ln 1.74 = .73V, $i_{D2} = \frac{10 - 6.4 - .73}{1.659}$ = 1.73mA. Thus i = 2.00 + 5.99 + 1.73 = 9.72 with an error of **– 0.28mA.**

12.28 For each of the top two circuits to which v_1 and v_2 are applied as in Fig. 12.44, $i_D = \frac{v_I}{R} = I_S\, e^{-v_o/nV_T}$ where $v_O = -v_D$, $v_O = -nV_T \ln \frac{v_I}{R\, I_S}$, for $v_I > 0$. For the lower circuit to which we apply $-v_3$: $v_O = +nV_T \ln \frac{v_3}{R\, I_S}$. Thus $v_D = -\, nV_T\,(\ln \frac{v_3}{R\, I_S} - \ln \frac{v_1}{R\, I_S} - \ln \frac{v_2}{R\, I_S})\,(\frac{-R}{R}) = nV_T \ln \left[\frac{\frac{v_1 \times v_2}{(R\, I_S)^2}}{\frac{v_3}{R\, I_S}}\right] = nV_T \ln \left[\frac{v_1\, v_2}{v_3} \times \frac{1}{R\, I_S}\right]$. Now $i_{D4} = I_S\, e^{\frac{nV_T}{nV_T} \ln} \left[\frac{v_1\, v_2}{R\, I_S\, v_3}\right] = I_S \frac{v_1\, v_2}{R\, I_S\, v_3}$, and $v_O = -\frac{R\, I_S\, v_1\, v_2}{R\, I_S\, v_3} = -\frac{v_1\, v_2}{v_3}$. Thus to obtain $v_O = \frac{v_1\, v_2}{v_3}$, add one unity-gain inverter at the input of the lower logarithmic circuit (for v_3), and a second unity-gain inverter at the output of the antilog circuit. Now, as a check, for a 1mA diode with n = 2, 1kΩ inputs, and voltages of 0.5, 1, 2, 3V applied: at 1mA, v = 0.700V; at 0.5mA, v = 700 + 50 ln 0.5 = 665.3mV, at 2.0mA, v = 700 + 50 ln 2 = 734.6mV; at 3.0mA, v = 700 + 50 ln 3 = 754.9mV. Test $\frac{0.5 \times 0.5}{3}$: See v_D = –1/1 (– .6653 – .6653 + .7549) = .5757V. Now v_4 =
1 $e^{(-700+575.7)/50}$ = .0832mA, v_O = .0832 = 1/12 as required. Test $\frac{3 \times 3}{.5}$: See $v_D = \frac{-1}{1}$ (–754.9 – 754,9 + 665.3) = 844.5. Now $i_4 = e^{(-700+844.5)/50} \approx$ 17.993, v_O = 17.993V = 18 as required, provided of course the supply voltages are high enough! Test $\frac{1 \times 1}{1}$: see directly OK, with v_O = 1.

SECTION 12.9: PRECISION RECTIFIER CIRCUITS

12.29 This is called an absolute-value circuit.

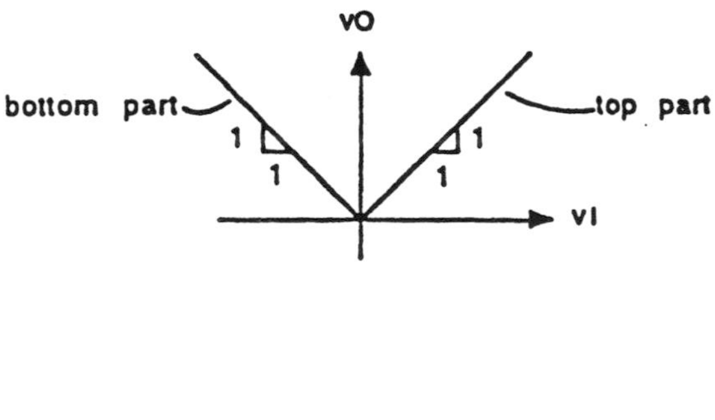

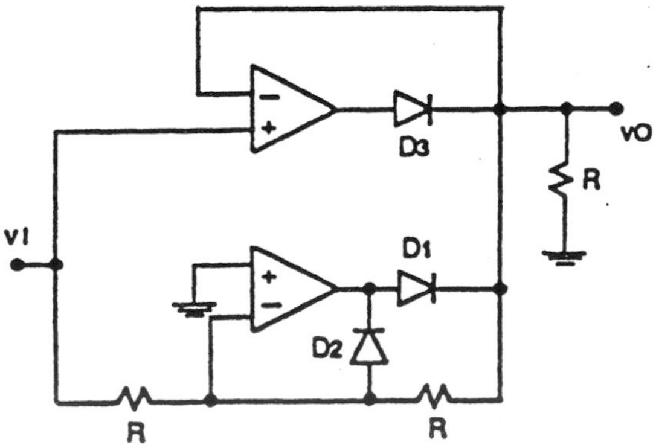

12.30

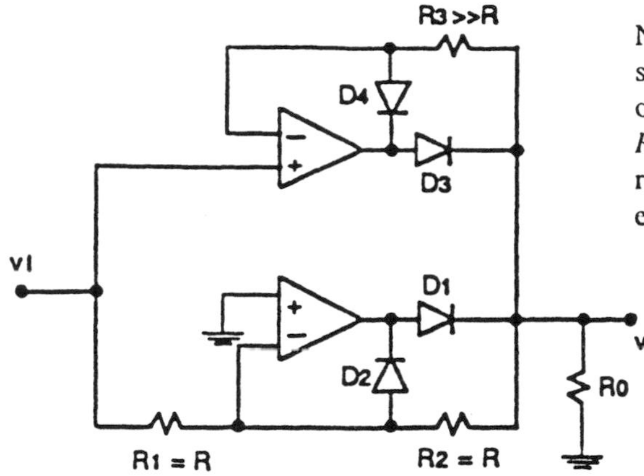

Note that while R_2 is a constant load on the output, since its leftmost end is always at ground, the non-output end of R_3 is connected to the ac input. Thus $R_3 >> R_2$ for least effect, and usually $R_2 > R$ for relative efficiency. For equivalent offset current effects, $R_3 = R_1 \parallel R_2 = R/2$ with $R >> R_0$.

12.31

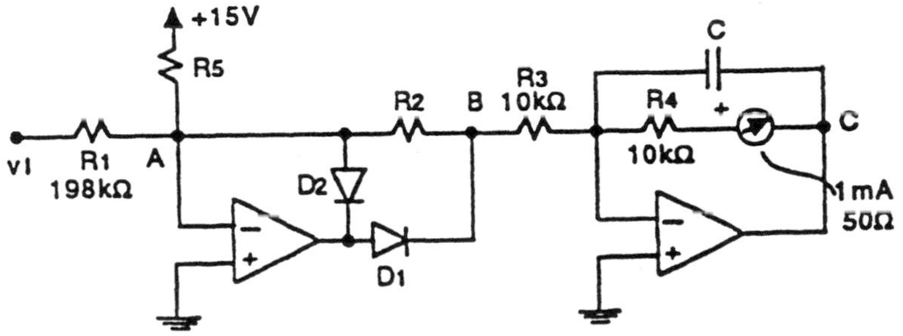

Here, 100Vrms = 141.4V peak, 140Vrms = 198.0V peak. Design for +10V at node B for 140Vrms input. Now R_5 supplies a current which cancels the effect of v_I as v_I goes negative, until it reaches 100V rms. Thus $\frac{R_5}{R_1} = \frac{15V}{141.4} \rightarrow R_5 = 0.1061\ R_1$. Chose R_1 so the change of input from 141.4V peak to 198V peak produces an extra 1mA. Thus $\frac{198 - 141.4}{R_1}$ = 1mA, or R_1 = 56.6kΩ – **use 56kΩ in series with 620Ω.** For v_B = 10V full scale with 1mA, R_2 = **10kΩ.** Also $R_5 = 0.1061 \times 56.6k\Omega$ = 6.00kΩ – **use two 12kΩ resistors in parallel.**

12.32 *For* v_I = +5V: v_O = **+10V**, and since v_B = +5V, v_C = 0V, $v_E = (1 + \frac{10}{10})\ 5$ = +10V, v_A = +10.7V, v_D = –0.7V. *For* v_I = 0V: v_O = **0V**, and since v_B = 0V, v_C = 0V, v_E = 0V. *For* v_I = –5V: v_O = **+10V**, and since v_B = –5V, v_C = 0V, $v_E = -\frac{20}{10}(-5)$ = +10V, v_A = –5.7V, v_D = +10.7V. The input resistance is (ideally) **infinite**. The circuit could be called a full-wave doubler rectifier.

12.33

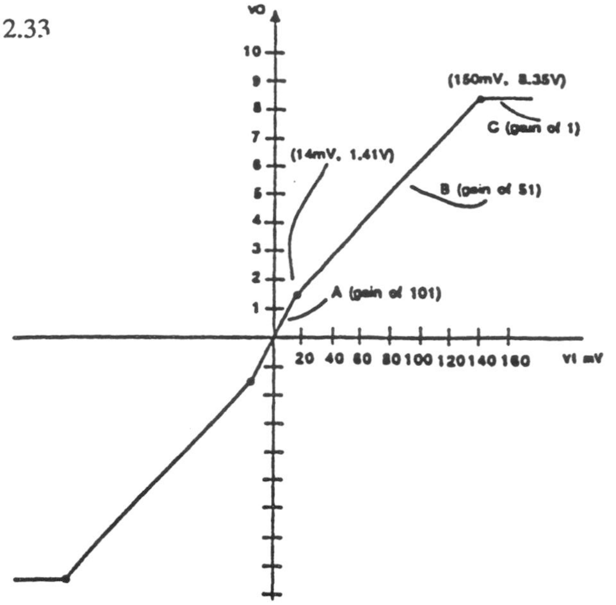

Note that the gain is generally $v_O/v_I = 1 + R_2/R_1$, for R_2/R_1 v_I < (+0.7 +0.7) = 1.4V, $1 + (R_2 \| R_3)/R_1$ for $((R_2 \| R_3)/R_1)$ v_I between 1.4V and 1.4 + 6.8 = 8.2V when the zener conducts, and $1 + (R_2 \| R_3 \| 0)/R_1$ = 1V/V beyond. In region A, for $0 < v_I < 1.4 \frac{R_1}{R_2} = 1.4 \frac{1}{100}$ = **14mV**, $\frac{v_O}{v_I} = 1 + \frac{R_2}{R_1} = 1 + \frac{100}{1}$ = **101V/V**, and v_O reaches $101 \times \frac{14}{1000}$ = **1.414V**. In region B, for 14mV < v_I < 14mV + 6.8V × $\frac{1}{100 \| 100}$ = **150mV**, $\frac{v_O}{v_I} = 1 + \frac{(R_2 \| R_3)}{R_1} = 1 + \frac{100 \| 100}{1}$ = **51V/V**, and v_O reaches $\frac{(150 - 14)}{1000} 51 + 1.414$ = **8.35V**. In region C, for 150mV < v_I < ∞, $\frac{v_O}{v_I}$ = **1V/V**. Due to the bridge connection, saturation is symmetric for positive and negative inputs.

12.34 The circuit is a dc restorer and rectifier, using the lower and upper amplifiers, respectively, to create ideal diodes. For a 100mV peak sine wave at the input, the voltage at the intermediate node (B) is a 100mV sine with lower peaks at 0V and upper peaks at 200mV. Correspondingly, the output v_o would be a dc level of 200mV, which would remain if the input is lowered or was removed. To return the output to zero from a peak-to-peak input v, one could use a resistor R_1 to ground at the output, or (better) to a negative supply connected to the output. For a ground connection, and a return to 0V in $10/f$, the time constant RC is such that $1/f < R_1 C < 10/f$. For 95% recovery, one must wait 3 time constants (since $e^{-3} = .05$). Thus, $3 R_1 C = 10/f$, and $R_1 = \frac{10}{3f\ C} = \frac{3.3}{fC}$. For input average drift at a low rate, the drift signal is coupled via the input C to node B where the resistance is infinite. Thus the average voltage at B would rise, and the output at C would rise to represent the maximum peak-to-peak value of the combined signal at f and $f/100$. To correct this, add a resistor R_2 from node B to ground (or to a negative supply). *Now*, from a filtering point of view, we want a high-pass filter with a pole at f_o, for which f is clearly in the passband and $f/100$ is rejected as much as possible. Now the transfer function from v_I to v_B is $T(s) = \frac{RCs}{1 + R\ Cs}$, and $T(j\omega) = \frac{jRC\omega}{1 + jw\ RC} = \frac{1}{1 + \frac{j}{RC\omega}}$, whence $|T| = \left[\frac{1}{1 + \frac{1}{(RC\omega)^2}}\right]^{½} = 0.95$. Now, at ω_o, $|T| = 1 + \frac{1}{(RC\omega_o)^2} = 1.108$, $\frac{1}{RC\omega_o} = (1.108 - 1)^{½} = 0.33$, or $\omega_o = \frac{3}{RC} = 2\pi f$, where $R_2 = \frac{3}{2\pi fC} = \frac{.48}{fC}$. Now at $\frac{f}{100}$, for $T(s) = \frac{1}{1 + \frac{1}{RCs}} = \frac{1}{1 + \frac{2\pi f}{3s}}$,

$$|T(\frac{f}{100})| = \left[\frac{1}{1 + \left[\frac{2\pi f}{3(2\pi f/100)}\right]^2}\right]^{½} = \left[\frac{1}{1 + (\frac{100}{3})^2}\right]^{½} \simeq \frac{1}{33.3} \equiv 20 \log .03 = -30.5\text{dB}.$$

This implies that a "volt or so" of drift will be reduced to "1/30 volts or so", or to around 30mV, still significant, but about as good as can be done.

12.34 (continued)

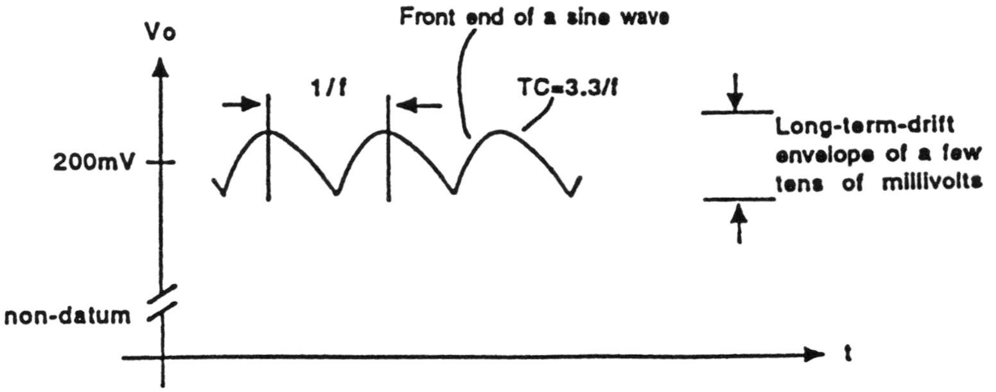

Chapter 13

MOS DIGITAL CIRCUITS

SECTION 13.1: DIGITAL CIRCUIT DESIGN – AN OVERVIEW

13.1

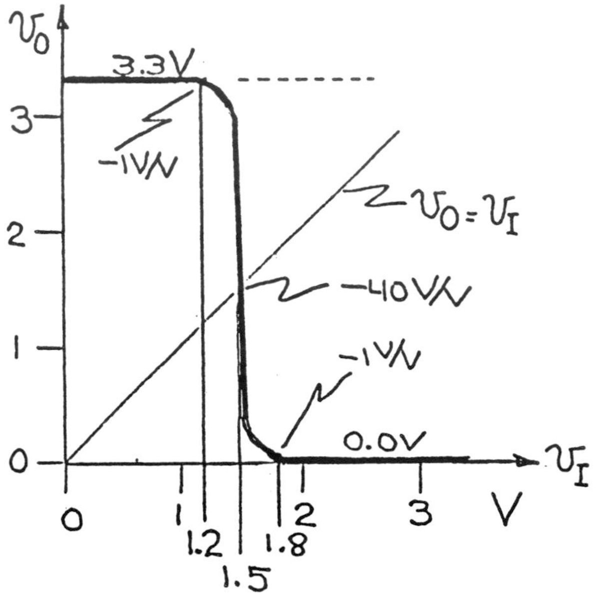

Here $V_{OL} =$ **0.0V**, $V_{OH} =$ **3.3V**, $V_{Il} =$ **1.2V**, $V_{IH} =$ **1.8V**, $V_{th} =$ **1.5V**, $V_M =$ **1.5V**. $NM_L = V_{Il} - V_{OL} = 1.2 - 0 =$ **1.2V**, $NM_H = V_{OH} - V_{IH} = 3.3 - 1.8 =$ **1.5V**.

13.2

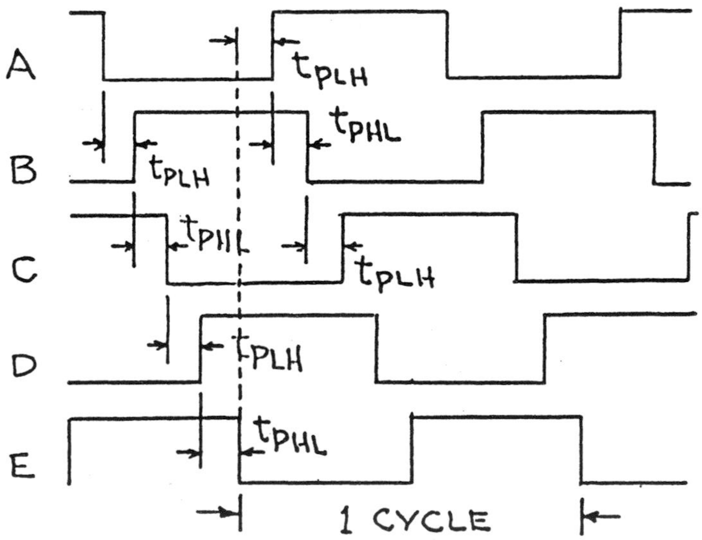

In one cycle, each inverter makes **2** transitions. There are **5** t_{PLH} and **5** t_{PHL} in one cycle. There are **10** transitions altogether in one cycle. [Count this from the diagram, or calculate as 5 × 2.]. At 100 MHz, the period is $1/(100 \times 10^6) = 10^{-8}s = 10ns$. Thus each of 10 transitions takes 10/10 = 1 ns on average. Thus $t_P =$ **1 ns**. If $t_{PLH} = 1.2t_{PHL}$ and $(t_{PHL} + t_{PLH})/2 = 1$, then $t_{PHL}(1 + 1.2) =$ 2, $t_{PHL} = 2/2.2 =$

0.909 ns and t_{PLH} = 1.2(0.909) = **1.091 ns.**

13.3 See that the static dissipation is zero. Thus the dynamic power/inverter is P_D = (300 × 10^{-6} × 3.3)/5 = 990 × 10^{-6}/5 = **198μW.**

Now $P_D = fCV^2$, and $C = P_D/(fV_2)$ = 198 × 10^{-6}/(100 × 10^6 × 3.3^2) = **0.182 pF.**

For this logic, t_P = (1/(100 × 10^6))/10 = 1*ns* and *DP* = 1 × 10^{-9} × 198 × 10^{-6} = **0.198 pJ**

13.4 For the gates, t_p = (30 + 10)/2 = 20ns. Total delay through 5 gates for 2 transitions each is T = 5 × 2 × 20 = 200ns. Frequency of oscillation = 1/*T* = 1/200ns = **5MHz.**

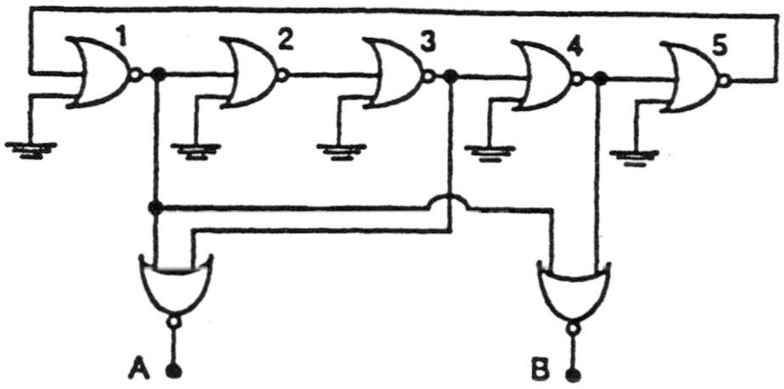

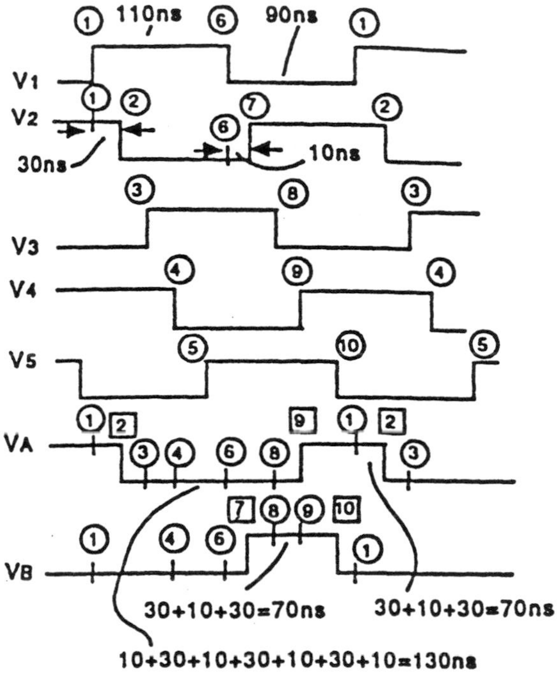

Note that [2]–(2) for matched gates.

SECTION 13.2: DESIGN AND PERFORMANCE ANALYSIS OF THE CMOS INVERTER

13.5 Since μ_n/μ_p = 100/40 = 2.5, then from Eq. 13.10, for a matched device $(W/L)_p = 2.5(W/L)_n$ = 2.5(1.2μm/0.8μm) = **(3.0 (μm/0.8μm).**

Since the generic process uses a supply of V_{DD} = 3.3V, $V_{OL} = 0V$ and $V_{OH} = 3.3V$.

From Eq. 13.8, with $V_{tn} = |V_{tp}|$ = 0.6V and $k_n = k_p$ for matching.

$$V_{tn} = \frac{V_{DD} - |V_{tp}| + \sqrt{k_n/k_p}\,V_{tn}}{1 + \sqrt{k_n/k_p}} = \frac{3.3 - 0.6 + \sqrt{1}(0.6)}{1 + \sqrt{1}} = 3.3/2 = \mathbf{1.65V.}$$

[Of course, this result could have been written down directly.]

From Eq. 5.94, $V_{IL} = (3V_{DD} + 2V_t)/8$ = (3(3.3) + 2(0.6))/8 = **1.3875V.**

From Eq. 5.93, $V_{IH} = (5V_{DD} - 2V_t)/8$ = [5(3.3) − 2(0.6)]/8 = **1.9125V.**

From Eq. 5.95, $NM_H = (3V_{DD} + 2V_t)/8$ = (3(3.3) + 2(0.6))/8 = **1.3875V.**

From Eq. 5.96, $NM_L = (3V_{DD} + 2V_t)/8$ = **1.3875V.**

Check:

Finally, $NM_L = V_{IL} - V_{OL} = 1.3875 - 0 =$ **1.3875V**, and $NM_H = V_{OH} - V_{IH} = 3.3 - 1.9125 =$ **1.3875V**

13.6 *For* V_{th}: Both transistors operate in saturation, sharing the same current i, with $\upsilon_I = \upsilon_O = V_{th}$.
Thus $i = 1/2 k_p (V_{DD} - V_{th} - |V_{tp}|)^2$, and $i = 1/2 k_n (V_{th} - V_{tn})^2$.
Equating and taking square roots: $V_{DD} - V_{th} - |V_{tp}| = \sqrt{k_n/k_p}(V_{th} - V_{tn})$.
Thus $\mathbf{V_{th} = [V_{DD} - |V_{tp}| + \sqrt{k_n/k_p} V_{tn}] / [1 + \sqrt{k_n/k_p}]}$, as provided in Eq. 13.8.

13.7 From part of the solution to P.36.6 above, for V_M, where
$V_M = [V_{DD} - |V_{tp}| + \sqrt{k_n/k_p}V_{tn}] / [1 + \sqrt{k_n/k_p}]$, where
$i = 1/2 k_p (V_{DD} - V_M - |V_{tp}|)^2$, and $i = 1/2 k_n (V_M - V_{tn})^2$.
Generally, $g_m = \partial i / \partial \upsilon$, where for $\upsilon = V_M$:
$g_{mp} = k_p (V_{DD} - V_M - |V_{tp}|) = k_p \sqrt{2 i / k_p} = \sqrt{2 i k_p}$, and $g_{mn} = k_n (V_M - V_{tn}) = \sqrt{2 i k_n}$.
Also $r_{op} = V_{Ap}/i$, and $r_{on} = V_{An}/i$.

$$\text{Gain} = -(g_{mn} + g_{mp})(r_{on} \parallel r_{op}) = \sqrt{2i}\,[\sqrt{k_n} + \sqrt{k_p}][\frac{V_{Ap} V_{An}}{V_{Ap} + V_{An}} \times 1/i]$$

$$= \sqrt{2/i}\,(\sqrt{k_n} + \sqrt{k_p})(V_{Ap} V_{An})/(V_{Ap} + V_{An})$$

$$= -[2/(V_{DD} - V_M - |V_{tp}|)][(\sqrt{k_n/k_p} + 1)(V_{Ap} V_{An})/(V_{Ap} + V_{An})].$$

Thus the transfer slope at V_M is $\mathbf{2(\sqrt{k_n/k_p} + 1)(V_{Ap} V_{An})/[(V_{Ap} + V_{An})(V_{DD} - V_M - |V_{tp}|)]}$
Now, for $V_{tn} = |V_{tp}| = 0.6$, $|V_A| = 20V$, $k_n = k_p = (1.2/0.8)(100) = 150\ \mu A/V^2$,
the gain is $-2(\sqrt{1} + 1)20 \times 20/(40 \times (3.3 - 1.65 - 0.6)) = \dfrac{-2(2)20(20)}{40(1.05)} =$ **− 38.1 V/V.**

Check: From first principles, numerically: $V_M = 1.65$, $i = 1/2(150 \times 10^{-6})(1.65 - 0.6)^2 = 82.7\mu A$, $r_o = 20/82.7 \times 10^{-6} = 0.242 M\Omega$, $g_m = 1/2(2)150 \times 10^{-6}(1.65 - 0.6) = 1575 \mu A/V$, and the *gain* $= -g_m r_o = -1575 \times 10^{-6} \times 0.242 \times 10^6 =$ **38.1 V/V**, OK.

13.8 See $k_n = (20) \times 8/2 = 80\mu A/V^2$, $k_p = (10) \times 16/2 = 80\mu A/V^2$. From Eq. 13.6 and 13.7 of the Text {or directly from the triode relation, that is, $i_D = k((\upsilon_{GS} - V_t)\upsilon_{DS} - \upsilon_{DS}^2/2) \approx k(\upsilon_{GS} - V_t)\upsilon_{DS}$, for small υ_{DS}}, see $r_{DS} = \upsilon_{DS}/i_D = \dfrac{1}{k(\upsilon_{GS} - V_t)}$. *For input high,* $r_{DS} = \dfrac{1}{80 \times 10^{-6}(5 - 1)} =$ **3.125kΩ**. *For input low,* $r_{DS} = \dfrac{1}{80 \times 10^{-6}(5 - 1)} =$ **3.125k Ω**, the same, since the inverter is matched.

13.9 Maximum currents are the same for both the p-channel and n-channel devices.

For the output connected to an opposing supply, $I_{MAX} = k/2(\upsilon_{GS} - V_t)^2 = 40 \times 10^{-6}(5 - 1)^2 =$ **640μA.**

For the output at $V_{DD}/2$, $I = k((\upsilon_{GS} - V_t)\upsilon_{DS} - \upsilon_{DS}^2/2) = 80 \times 10^{-6}((5 - 1)2.5 - 2.5^2/2) = 40 \times 10^{-6}(0 - 6.25/2) =$ **550μA.**

For the output $0.1V_{DD}$ *from the limit,* $I = 80 \times 10^{-6}((5 - 1)0.5 - 0.5^2/2) = 80 \times 10^{-6}(2 - 0.25/2) =$ **150 μA.**

13.10 For all the inverters, $V_{OL} = 0V$, $V_{OH} = V_{DD}$, as V_{DD} varies from 3.75 to 6.25V, and as $V_{tn} = V_{tp} = V_t$ varies from 0.75 to 1.25V, with $k_n = k_p$. From Eq. 5.94: $V_{IL} = 1/8(3V_{DD} + 2V_t)$. From Eq. 5.93: $V_{IH} = 1/8(5V_{DD} - 2V_t)$, or $V_{IH} = V_{DD} - V_{IL}$ generally (for symmetry). Now for V_{IL} to be largest, $V_{DD} = 6.25$ and $V_t = 1.25$, whence $V_{IL} = 1/8(3(6.25) + 2(1.25) =$ **2.66V**, for which $V_{IH} = 6.25 - 2.66 =$ **3.59V**. For V_{IL} smallest, $V_{DD} = 3.75$ and $V_t = 0.75$, $V_{IL} = 1/8(3(3.75) + 2(0.75)) =$ **1.59V**, for which $V_{IH} = 3.75 - 1.59 =$ **2.16V**. Now V_{OL} is always **0V**, and V_{OH} ranges from **3.75V** to **6.25V**, V_{IL} from **1.59V** to **2.66V**, V_{IH} from **2.16V** to **3.59V**. See that the noise margins between gates (with different supplies and different V_t) vary widely: Consider $NM_H = V_{OH} - V_{IH}$. Highest is $6.25 - 2.16 =$ **4.09V**. Lowest is $3.75 - 3.59 =$ **0.16V** (See that this is very bad). Consider $NM_L = V_{IL} - V_{OL}$. Highest is $2.66 - 0 =$ **2.66V**. Lowest is $1.59 - 0 =$ **1.59V**.

13.11 From Eq. 13.12, for an inverter with a fanout of 1,

$C = 2C_{gd1} + 2C_{gd2} + C_{db1} + C_{db2} + C_{g3} + C_{g4} + C_w$.

For the matched inverter, $(W/L)_n = (1.2/0.8)$, and $(W/L)_p = (100/40)(1.2/0.8) = (3.0/0.8)$.

Thus, $C_{gd1} = 0.5 \times 1.2 = 0.6fF$, $C_{gd2} = 0.5 \times 3.0 = 1.5fF$, $C_{db1} = 2.5 \times 1.2 = 3.0fF$, $C_{db2} = 2.5 \times 3.0 = 7.5fF$, $C_{g3} = 1.8 \times 0.8 \times 1.2 = 17.3fF$, $C_{g4} = 1.8 \times 0.8 \times 3.0 = 4.32fF$
$C_w = C_{g3} = 1.73fF$.

Thus $C = 2(0.6 + 1.5) + 3.0 + 7.5 + 1.73 + 4.32 + 1.73 =$ **22.5 fF**. Since the inverter is matched

$$t_{PLH} = t_{PHL} = t_P = \frac{1.7C}{k_n'(W/L)_n V_{DD}} = \frac{1.7 \times 22.5 \times 10^{-15}}{100 \times 10^{-6}} \times (1.2/0.8) \times 3.3 = \mathbf{77.3\ ps}$$

13.12 *For the assumption of constant current.* From Eq. 13.14, in saturation, $i_{DN}(0) = 1/2 k_n'(W/L)_n(V_{DD} - V_t)^2$.

From Eq. 13.17, $t_{PHL} = CV_{DD}/2/i_{DS}(O)$.

Now for $V_t = 0.2V_{DD}$, $i_{DS}(0) = 1/2 k_n'(W/L)_n(1 - 0.2)^2 V_{DD}^2 = k_n'(W/L)_n V_{DD}^2(0.32)$.

$$\text{Thus } t_{PHL} = \frac{CV_{DD}}{2(k_n')(W/L)_n V_{DD}^2(0.32)} \text{ or } \mathbf{t_{PHL}} = \frac{\mathbf{1.6C}}{\mathbf{k_n'(W/L)_nV_{DD}}}.$$

Alternatively, from Eq. 5.101,
$$t_{PHL} = \frac{2C}{k_n'(W/L)_n(V_{DD} - V_t)}\left[\frac{V_t}{V_{DD} - V_t} + 1/2 \ln\frac{3V_{DD} - 4V_t}{V_{DD}}\right].$$

For this case, in which $V_{DD} = 3.3V$ and $V_t = 0.6V$,

$$t_{PHL} = \frac{2C}{100 \times 10^{-6}(1.2/0.8)(3.3 - 0.6)}\left[\frac{0.6}{3.3 - 0.6} + 1/2 \ln\frac{3(3.3) - 4(0.6)}{3.3}\right]$$

$$= \frac{2C}{100 \times 10^{-6}(1.5)2.7} \times [0.22 + 0.41] = \mathbf{3111\ C\ s.}$$

Now, ignoring the fact that here $V_t = 0.6/3.3V_{DD} = 0.182V_{DD}$, (rather than 0.2 V_{DD}).

$$\text{From Eq. 13.18, } t_{PHL} = \frac{1.7C}{100 \times 10^{-6}(1.2/0.8)(3.3)} = \mathbf{3434\ C\ s.}$$

$$\text{From the constant-current calculation above, } t_{PHL} = \frac{1.6C}{100 \times 10^{-6}(1.2/0.8)(3.2)} = \mathbf{3232\ C\ s.}$$

Assuming the estimte from Eq. 5.101 to be the most accurate, it is interesting to see that of the simple approximations, the one found here, is best. It is certainly, the easiest to obtain from first principles.

13.13 For simplicity, use $k = k_n'(W/L)_n$, and substitute $V_t = 0.1V_{DD}$.

From Eq. 13.14, $i_D(N)i_D(0) = 1/2 k(V_{DD} - V_C)^2 = kV_{DD}^2(0.405)$ — — — (1)

From Eq. 1315, $i_D(AV)i_D(M) = k[(V_{DD} - V_t)V_{DD}/2 - (V_{DD}/2)^2/2] = kV_{DD}^2(0.9/2 - 1/8) = kV_{DD}^2(0.325)$

Thus, $i_D(av) = kV_{DD}^2(0.405 + 0.325)/2 = kV_{DD}^2(0.365)$

$$\text{From Eq. 13.18, } t_{PHL} = CV_{DD}/2/(0.365kV_{DD}^2) = \frac{\mathbf{1.37C}}{\mathbf{kV_{DD}}}.$$

Now for the current in (1) sustained for the half transition,

$t_{PHL} = CV_{DD}/2/(0.405kV_{DD}^2) = \dfrac{1.23\mathbf{C}}{\mathbf{kV_{DD}}}$. From Eq. 5.101 on page 434 of the Text,

$$t_{PHL} = \frac{2C}{k(V_{DD} - V_t)}\left[\frac{V_t}{V_{DD} - V_t} + 1/2\ln\frac{3V_{DD} - 4V_t}{V_{DD}}\right]$$

$$= \frac{2C}{kV_{DD}(0.9)}\left[\frac{0.1}{1-0.1} + 1/2\ln\left[\frac{3-4(0.1)}{1}\right]\right]\frac{2C}{kV_{DD}(0.9)}(0.111 + 0.478) = \frac{1.31\mathbf{C}}{\mathbf{kV_{DD}}}.$$

Obviously, the sustained saturation result is reasonably good and certainly simple to obtain from first principles. Notice, as well, the impact of change of V_t from 0.2 V_{Dd} to 0.1 V_{DD}, being a change in the coefficient from 1.6 to 1.2 in the simplest constant-current formula.

13.14 From P13.8 above, $k_n = k_p = 80\mu A/V$, $|V_t| = 1V$, $V_{DD} = 5V$. For $V_{in} = V_{DD}/2 = 2.5V$, $I_{peak} = i_D = 80/2\ (2.5 - 1)^2 = \mathbf{90\ \mu\ A}$. Assume (for the present purposes) that rise and fall times are measured from 0% to 100%. As the input goes from 1V to 4V, the current flow is triangular, with a peak value of 90μA. Over an interval of 3/5 × $T/4$, the average current is 90/2 = 45μA. This happens twice per cycle. Thus the average current per cycle from self conduction is $\dfrac{2\ (3/5) \times T/4 \times 45}{T} = \mathbf{13.5\mu A}$.

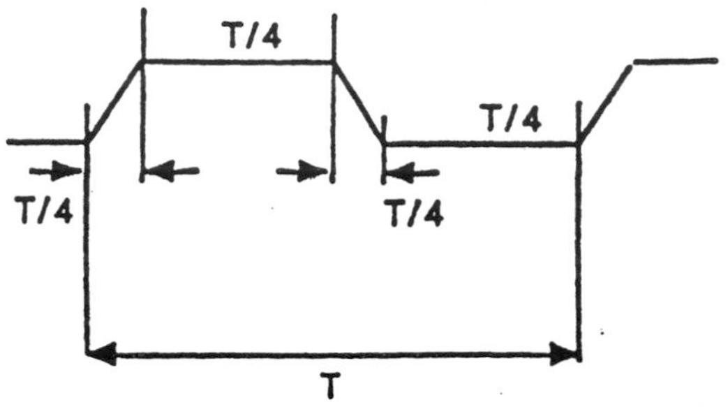

The average current due to capacitance load is $CV_{DD}\ f = 0.5 \times 10^{-12} \times 5 \times 20 \times 10^6 = \mathbf{50\ \mu A}$. Total average current = 50 + 13.5 = **63.5 μA**. With load, $P_D = 63.5 \times 10^{-6} \times 5 = \mathbf{317.5\ \mu W}$. Without load, $P_D = 13.5 \times 5 = \mathbf{67.5\ \mu W}$.

13.15 From Eqs.13.18 and 13.19, $t_{PHL} = \dfrac{1.7C}{k_n'(W/L)_n V_{DD}}$, and $t_{PLH} = \dfrac{1.7C}{k_p'(W/L)_p V_{DD}}$.

Here, $k_n' = 2k_p' = 20\mu A/V$, and $(W/L)_n = 1/2(W/L)_p = 8\mu m/2\mu m$.

For $V_{DD} = 5V$, $|V_t| = 0.2V_{DD} = 1V$, and $t_{PHL} = t_{PLH} = t_P. = \dfrac{1.7 \times 0.5 \times 10^{-12}}{20 \times 10^{-6}\ (8/2)(5)} = \mathbf{2.12\ ns.}$

For a 5-stage ring oscillator, $f = [10(2.12)]^{-1} = \mathbf{47.2\ MHz}$. Ignoring the transition-time peak-current flow per gate, $P_D = f\ CV_{DD}^2 = 47.2 \times 10^6 \times 0.5 \times 10^{-12}\ (5^2) = \mathbf{590\mu W}$.

The Delay-Power product $DP = 2.12 \times 10^{-9} \times 590 \times 10^{-6} = 1.25 \times 10^{-12}J = \mathbf{1.25pJ}$.

13.16 Here, $V_{DD} = 1.3V$, $|V_t| = 0.8V$, (a) Q_n conducts for v_I from 0.8 to 1.3V, and Q_p conducts for v_I from 0 to (1.3 − 0.8), or 0 to 0.5V. For $v_I = 1.3/2 = 0.65V$, neither transistor conducts and $i_D = 0$, (b) Output voltages range from 0 to 1.3V. $V_{OH} = \mathbf{1.3V}$; $V_{OL} = \mathbf{0.0V}$, (c) $V_{IL} = \mathbf{0.8V}$, and $V_{IH} = 1.3 - 0.8 = \mathbf{0.5V}$. [Note that $V_{IL} > V_{IH}$!] Between V_{IL} and V_{IH}, *no current* flows.

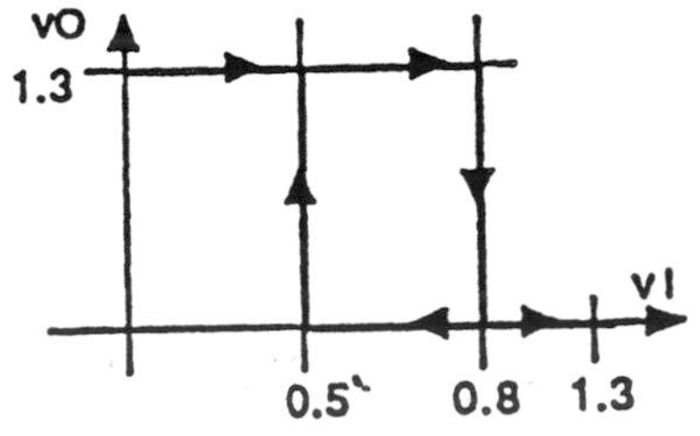

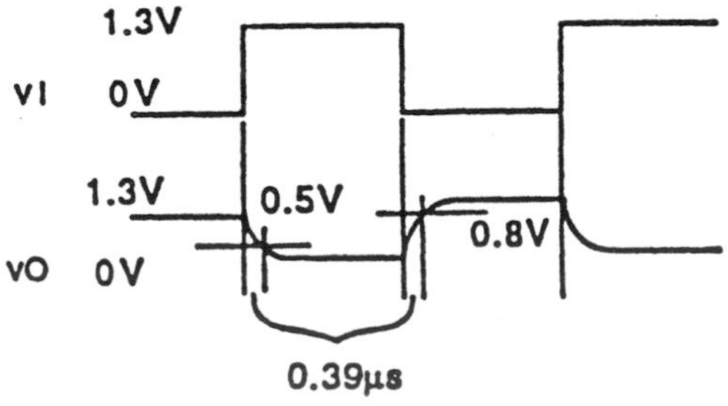

(d) (Note in considering the transfer characteristic, that the small capacitor at the output holds the output while neither transistor conducts. (e) $i_D = 1/2k\ (1.3 - 0.8)^2 = 1/2 \times 20 \times 10^{-6}\ (0.5)^2 = 2.5\mu A$ peak. For υ_O at 0.8V, $i_D = k\ [(\upsilon_{GS} - V_t)\ \upsilon_{DS} - \upsilon_{DS}^2/2] = 20 \times 10^{-6}\ [(1.3 - 0.8)\ 0.8 - 0.8^2/2] = 1.6\mu A$. Average current is $\frac{1.6 + 2.5}{2} = 2.05\mu A$, and the time for a 0.8V change (from 0 V to 0.8 V) is $\frac{CV}{I} = \frac{1 \times 10^{-12} \times 0.8}{2.05 \times 10^{-6}} = \mathbf{0.39}\ \mu$ s. Conclude that propagation delay is **more than** 0.39µs. Now, for the output moving from 0.8V to 1.3V, the average current available is (1.6+0)/2 = 0.8µA.

Thus, the time to reach 1.3V from 0.8V is about $1 \times 10^{-12} \times \frac{1.3 - 0.8}{0.8 \times 10^{-6}} = 0.625\mu s$, and the total transition time is about (0.39 + 0.625) = 1.02µs. Considering the driven stage with input at 0.8V, the available output current is 0µA. At an input of 1.3V, the available output current is 2.5µA. Thus, the average current $= \frac{0 + 2.5}{2} = 1.25\mu A$, and the propagation delay $\approx \frac{1.0 \times 10^{-12} \times 0.8}{1.25 \times 10^{-6}} = \mathbf{0.64}\ \mu s$.

(f) Frequency of oscillation of 5 gates is (at most) $= \frac{1}{10\ (0.64 \times 10^{-6})} = \mathbf{156kHz}$.

13.17 Here, $k_n = (20) \times 18/2 = 180\mu A/V^2$, $k_p = (20/2) \times 4/2 = 20\mu A/V^2$: $|V_t| = 1V$, $V_{DD} = 5V$. See $V_{OH} =$ **5V** $V_{OL} =$ **0V**. *For* $V_{th} = \upsilon$, $\upsilon_I = \upsilon_O = \upsilon$, both devices are in saturation, and $90\ (\upsilon - 1)^2 = 10\ (5 - \upsilon - 1)^2$. Taking the square root, $3(\upsilon - 1) = 4 - \upsilon$, $3\upsilon - 3 = 4 - \upsilon$, $4\upsilon = 7$. Thus, $V_{th} = \upsilon =$ **1.75V**. *For* $V_{IL} = \upsilon$, Q_p in triode, Q_n in saturation, $90\ (\upsilon - 1)^2 = 10\ (2\ (5 - \upsilon - 1)\ (5 - \upsilon_O) - (5 - \upsilon_O)^2)$, $9\ (\upsilon - 1)^2 = 2\ (4 - \upsilon)\ (5 - \upsilon_O) - (5 - \upsilon_O)^2$ – – – (1). Now, taking derivatives, $18\ (\upsilon - 1) = 2\ (4 - \upsilon)\ (-\frac{\partial \upsilon_O}{\partial \upsilon}) + 2\ (5 - \upsilon_O)\ (-1) - 2\ (5 - \upsilon_O)\ \frac{-\partial \upsilon_O}{\partial \upsilon}$. Now, with $\frac{\partial \upsilon_O}{\partial \upsilon} = -1 \rightarrow 18\ \upsilon - 18 = 18 - 2\ \upsilon - 10 + 2\ \upsilon_O - 10 + 2\ \upsilon_O$, $20\ \upsilon = 4\ \upsilon_O + 6$, $\upsilon = .2\ \upsilon_O + 0.3$ – – – (2). Substitute (2) in (1) $\rightarrow$ $9\ (.2\upsilon_O + 0.3 - 1)^2 = 2\ (4 - .2\upsilon_O - 0.3)\ (5 - \upsilon_O) - (5 - \upsilon_O)^2$, $9\ (.2\upsilon_O - 0.7)^2 = 2\ (3.7 - .2\upsilon_O - 0.3)\ (5 - \upsilon_O) - (5 - \upsilon_O)^2$, $0.36\ \upsilon_O^2 + 4.41 - 2.52\ \upsilon_O = 37 - 9.4\ \upsilon_O + .4\ \upsilon_O^2$ $-25 + 10\ \upsilon_O - \upsilon_O^2$, $0.96\ \upsilon_O^2 - 3.12\ \upsilon_O - 7.59 = 0$. Thus $\upsilon_O = \frac{3.12 \pm \sqrt{3.12^2 + 4\ (.96)\ (7.59)}}{2\ (.96)} =$ 4.87V, and $V_{IL} = \upsilon = .2\ (4.87 + 0.3) =$ **1.275V**. *For* $V_{IH} = \upsilon$, Q_n in triode, Q_p in saturation, $10\ (5 - \upsilon - 1)^2 = 90\ (2\ (\upsilon - 1)\ (\upsilon_O) - \upsilon_O^2)$, $(4 - \upsilon)^2 = 18\ (\upsilon - 1)\ (\upsilon_O) - 9\ \upsilon_O^2$ – – – (1). $\frac{\partial}{\partial \upsilon} \rightarrow 2\ (4 - \upsilon)\ (-1) = 18\ (\upsilon - 1)\ \frac{\partial \upsilon_O}{\partial \upsilon} + 18\ \upsilon_O - 18\ \upsilon_O\ \frac{\partial \upsilon_O}{\partial \upsilon}$. For $\frac{\partial \upsilon_O}{\partial \upsilon} = -1 \rightarrow 2\ \upsilon - 8 = 18 - 18\ \upsilon + 18\ \upsilon_O + 18\ \upsilon_O$, $20\ \upsilon = 36\ \upsilon_O + 26$, $\upsilon = 1.8\ \upsilon_o + 1.3$ – – – (2). Now, substituting (2) in (1), $(4 - 1.8\ \upsilon_O - 1.3)^2 = 18\ (1.8\ \upsilon_O + 1.3 - 1)\ \upsilon_O - 9\ \upsilon_O^2$, $(2.7 - 1.8\ \upsilon_O)^2 = 18\ (1.8\ \upsilon_O + .3)\ \upsilon_O - 9\ \upsilon_O^2$, $7.29 - 4.86\ \upsilon_O + 3.24\ \upsilon_O^2 = 32.4\ \upsilon_O^2 + 5.4\ \upsilon_O - 9\ \upsilon_O^2$, $\upsilon_O^2\ (20.16) + \upsilon_O\ (10.26) - 7.29 = 0$, $\upsilon_O^2 + .509\ \upsilon_O - .362 = 0$, $\upsilon_O = \frac{-.509 \pm \sqrt{.509^2 + 4\ (.362)}}{2} = 0.400V$, and $\upsilon = 1.8\ (.4) + 1.3 = 2.02V$. Thus $V_{IH} =$ **2.02V**.

For t_p: For $\upsilon_O \approx 5V$, $I_n = 90\ (5 - 1)^2 = 1.44mA$. For $\upsilon_O \approx 0V$, $I_p = 10\ (5 - 1)^2 = .160mA$. For $\upsilon_O = 1.75V$, $I_n = 90\ (2\ (5 - 1)\ (1.75) - 1.75^2) = 0.984mA$, and $I_p = 10\ (2\ (5 - 1)\ (5 - 1.75) - (5 - 1.75)^2) = 10\ (26 - 10.6) = 0.154mA$. *For discharging*, $t_{PHL} = \frac{CV}{I} = \frac{0.5 \times 10^{-12} \times (5 - 1.75)}{10^{-3}\ (1.44 + .984)/2} =$ **2.89ns**. *For charging*, $t_{PLH} = \frac{0.5 \times 10^{-12}\ (1.75)}{10^3\ (0.160 + 0.154)/2} =$ **5.57ns**.

SECTION 13.3: CMOS LOGIC-GATE CIRCUITS

13.18 $\overline{Y} = A\ (B + C)$. The corresponding pull-Down network (PDN) is shown as D_1.

$Y = \overline{A\ (B + C)} = \overline{A} + \overline{B + C} = \overline{A} + \overline{B}\ \overline{C}$. The corresponding pull-Up network (PUN) is shown as *Usbu* 1.

Now, the PUN dual to D_1 is U_2:

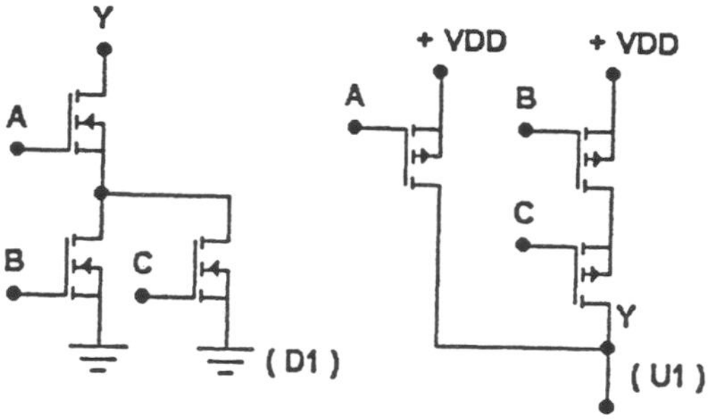

See that U_2 is similar to U_1, **but** *not identical, B* and C being interchanged with respect to the connection nearest to the power rail.

D_2 is a PDN obtained from U_1:

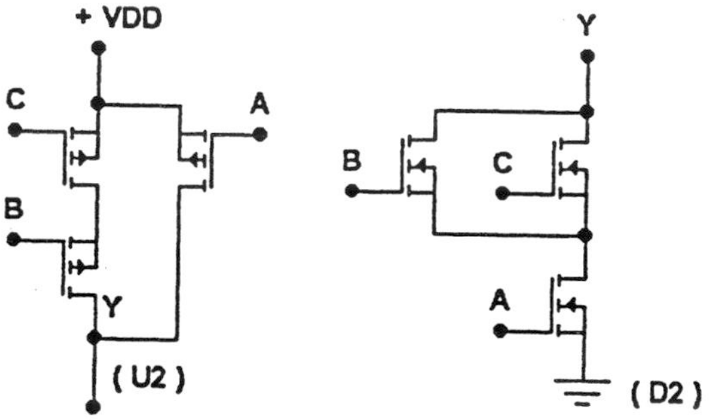

See that D_2 and D_1 are *not identical*, the transistor A (the one with input A) being near ground in D_2, but near the output in D_1. For this logic function, there are **2** PUN and **2** PDN which can form $2 \times 2 =$ **4** different gate topologies in all.

13.19 In Fig. 13.15, the PDN shown is to be called $\mathbf{D_l}$ and the PUN is to be called $\mathbf{U_l}$. Here, PDN D_2 is dual to U_1.

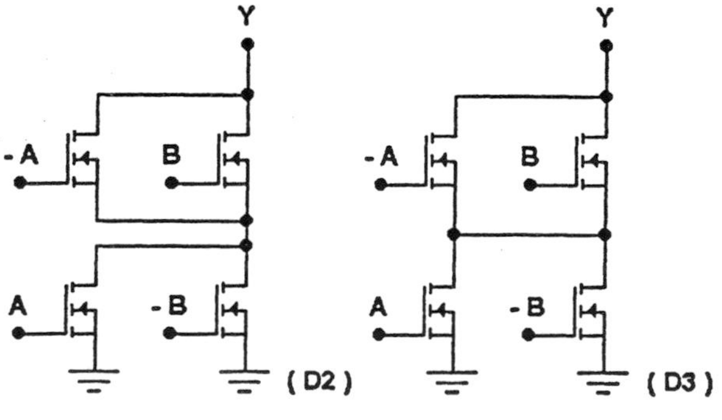

D_3 is D_2 redrawn more symmetrically. U_2 is the PUN dual to D_1, drawn directly. U_3 is U_2 redrawn.

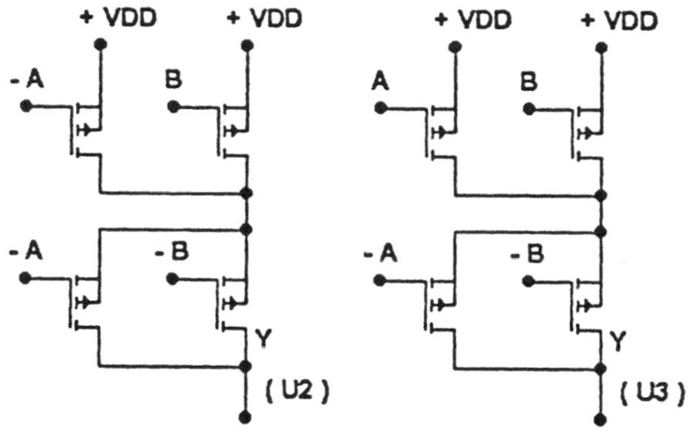

Using U_1, D_1 and U_3, D_3, there are 4 possible XOR circuits that can be constructed. When the relative placement of the inputs with respect to the supplies is considered, there are two versions of *each* of the perfectly square networks (like U_3, D_3) depending on proximity to the supply of each of the two series layers (of paralleled transistors). Thus for each of these U and D networks there are 2 variations. However for networks U_1 and D_2, in which the two series nodes are not joined (but, correspondingly, not for D_3 or U_3), there are 4 variations:

A B C D	A D C B	C B A D	C D A B

with respect to proximity to a supply.

Now, for all networks like U_1, D_1, there are $4^2 =$ 16 possible arrangements, but for half of these (that is, 8), there are twice as many combinations each (the diagonal exchanges in each group of 4 above). **Thus the number of combinations is 8 + 8(2) = 24**

13.20

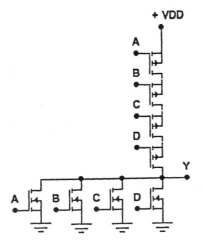

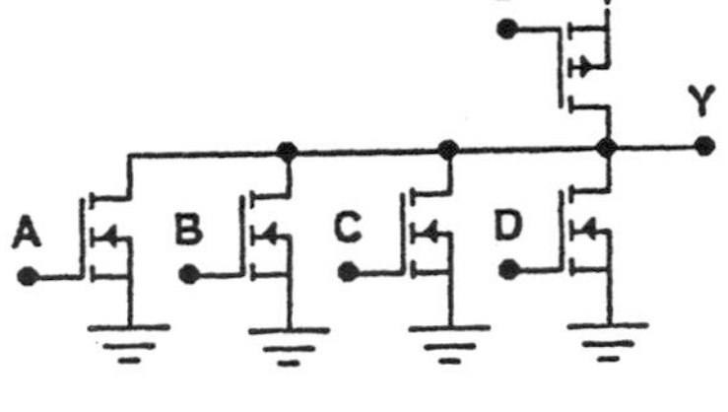

Refer to the transistors by their variable names P_A, N_A, etc. Now N_A, N_B, N_C, N_D are all of unit size, the same as N in the unit inverter, where $(W/L)_n = (1.2/0.8)$. Now for matching, P_A, P_B, P_C, P_D are all 4× larger than P in the inverter, which in turn is 2.5× the size of N, $4(W/L)_p = 4(3.0/0.8) = (12/0.8)$.

Total area of the NOR is $4[(1.2)0.8 + (12)0.8]$

$= 4(11)(0.8)(1.2) = 44(0.8 \times 1.2) =$ **42.24μm²**.

This is *44* × the area of a single NMOS. The area of an inverter is $(1 + 2.5)(0.8 \times 1.2) =$ **3.36μm²**.

Thus the NOR is $42.24/3.36 = 44/3.5 =$ **12.6 × larger** than a single inverter.

13.21 For a 4-input NAND gate, the N devices are in series and the P devices are in parallel. For mobility matching, each P device is the same size as P in the inverter, namely $(W/L)_p = 2.5(1.2/0.8) = (3.0/0.8)$.

For current-drive matching, each N of the NAND is 4 × larger than N in the inverter $= 4(1.2/0.8) = (4.8/0.8)$.

NAND gate area $= 4[(4(1) + 2.5)(0.8 \times 1.2)] = 4(6.5)(0.8)1.2 = 26(0.96) =$ **24.96μm²**.

This is 26× the area of one NMOS, and $26/3.5 =$ **7.43×** the area of the basic matched inverter, and $24.96/42.24 + 59\%$ of the area of a NOR?

13.22

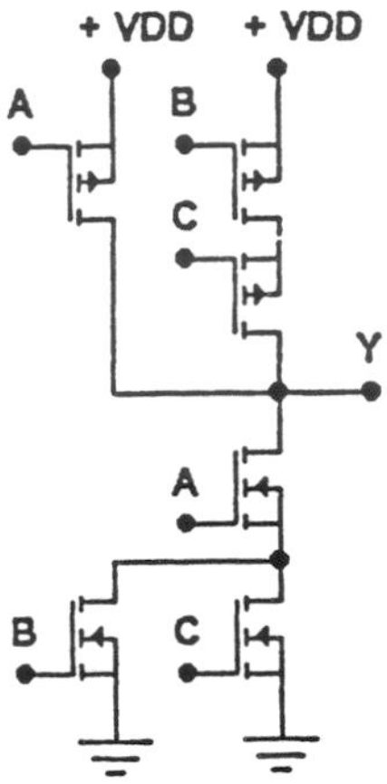

Consider one of the circuits resulting from the solution of P13.18 above for $\overline{Y} = A(B + C)$. For a matched inverter, the transistors are P, N where $(W/L)_p = 2.5(W/L)_n = 2.5(1.2/0.8)$. Thus, we can say $P = 2.5N$, referring to width or area (for all devices all of the same length). Here, $P_A = P$ and $P_B = P_C = 2A$. Also $N_A = N_B = N_C = 2N$. Total area

$= P(1 + 2 + 2) + N(2 + 2 + 2) = 2.5N(5) + N(6) = 18.5N =$ $18.5(1.2 \times 0.8) =$ **17.8 μm²**, or 18.5× the area of the basic NMOS, and $18.5/3.5 =$ **5.28×** the area of a basic matched inverter.

13.23 Consider Fig. 13.12 of the Text, with input A low and input B active at $\upsilon_I = 2.5\text{V}$. All devices share a current I with the upper PMOS which operates in triode mode with its drain voltage $= V_1$. Thus, *for one input active:* $I = k/2(2.5 - 1)^2 = k_p/2\,(V_1 - 2.5 - 1)^2$ and $I = k_p\,[(5 - 1)(5 - V_1) - (5 - V_1)^2/2]$. Thus, $(V_1 - 3.5)^2 = 8(5 - V_1) - (5 - V_1)^2$, $V_1^2 - 7V_1 + 12.25 = 40 - 8V_1 - 25 + 10V_1 - V_1^2$, $2V_1^2 - 9V_1 - 2.75 = 0$, and $V_1 = \dfrac{--9 \pm \sqrt{9^2 - 4(-2.75)(2)}}{2 \times 2} = \dfrac{9 \pm 10.15}{4} = 4.787\text{V}$,

whence $k_p = \dfrac{k\,(1.5^2)}{(4.787 - 3.5)^2} =$ **1.36k**.

Now, for two inputs active: $V_{th} = \upsilon$. Since the two NMOS operate in parallel, the current in the upper (series) transistors is $i = 2k/2\,(\upsilon - 1)^2 = k_p/2(V_1 - \upsilon - 1)^2$, and also $i = k_p\,[(5 - \upsilon - 1)(5 - V_1) - (5 - V_1)^2/2]$. From the first pair, $\upsilon - 1 = \left[\dfrac{k_p}{2k}\right]^{1/2} \times (V_1 - \upsilon - 1) = \left[\dfrac{1.36}{2}\right]^{1/2} \times (V_1 - \upsilon - 1) = 0.82\,V_1 - 0.82\,\upsilon - 0.82$. Thus $1.82\,\upsilon = 0.82\,V_1 + 0.18$, $\upsilon = 0.4505\,V_1 + 0.2195$, or $V_1 = 2.22\,\upsilon - 0.220$.

From the second pair: $(V_1 - \upsilon - 1)^2 = (4 - \upsilon)(10 - 2V_1) - (5 - V_1)^2$, $(2.22\,\upsilon - .220 - \upsilon - 1)^2 = (4 - \upsilon)(10 - 4.44\,\upsilon + .44) - (5 - 2.22\,\upsilon + .22)^2$, or $(1.22\,\upsilon - 1.22)^2 = (4 - \upsilon)(10.44 - 4.44\,\upsilon) - (5.22 - 2.22\,\upsilon)^2$, or $1.488\,\upsilon^2 - 2.977\,\upsilon + 1.488 = 41.76 - 17.76\,\upsilon - 10.44\,\upsilon + 4.44\,\upsilon^2 - 27.25 + 23.18\,\upsilon - 4.93\,\upsilon^2$, $1.98\,\upsilon^2 + 2.043\,\upsilon - 13.02 = 0$, $\upsilon^2 + 1.032\,\upsilon - 6.575 = 0$, $\upsilon = [-1.032 \pm \sqrt{1.032^2 - 4(-6.575)}]/2 = 2.10\text{V}$. Thus, the threshold is **2.10V**, for which $V_1 = 2.22(2.10) - .220 = 4.44\text{V}$.

Note that a simpler approach results from realizing that with both inputs active, the upper transistors behave as a single PMOS with twice the usual length, for which $(k_p)_{eq} = k_p/2 = 1.36k/2 = 0.68k$ while the lower pair operate with a combined $(k_n)_{eq} = 2\text{k}$.

Thus $I = ((2k)/2)(\upsilon - 1)^2 = (0.68k/2)(5 - \upsilon - 1)^2$, or $(\upsilon - 1)^2 = (0.34)(4 - \upsilon)^2$, or $\upsilon - 1 = 0.583(4 - \upsilon) = 2.33 - 0.583\upsilon$, or $1.583\upsilon = 3.33$, and $\upsilon = 3.33/1.583 =$ **2.10V**, as before.

13.24 For $\overline{Y} = AB + ACD$, obtain the PDN directly and the PUN as its dual:

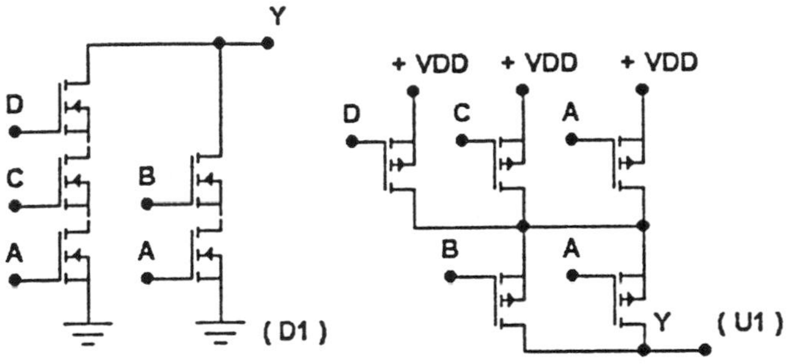

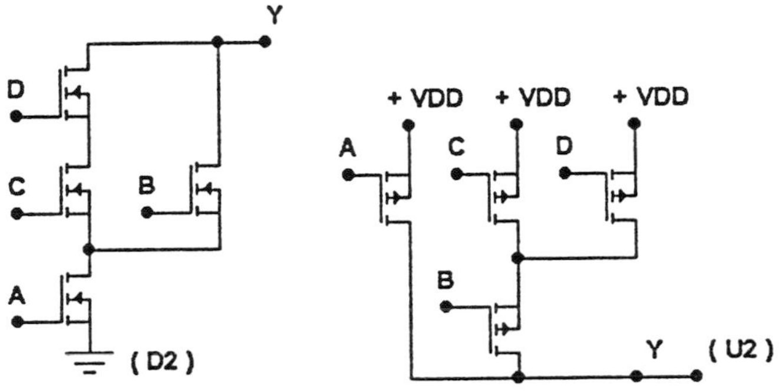

For $\overline{Y} = AB + ACD$: $W_n = [3(3) + 2(2)]1.213(1.2) = 15.6\mu m$, and $W_p = [5(2)](2.5)1.2 = 30\mu m$.
The total width is $W_T = 15.6 + 30 =$ **45.6μm.**
For $\overline{Y} = A(B + CD)$: $W_n [3(3) + 1(3/2)]1.2 = 10.5(1.2) = 12.6\mu m$, and $W_p [1 + 3(2)]2.5(1.2) = 17.5(1.2) = 21\mu m$. The total width is $W_T = 12.6 + 21 =$ **33.6μm,** a reduction of $\frac{45.6 - 33.6}{45.6} \times 100 = 26.3\%$!

13.25 *For a Buffered Inverter*: Total n-width is 1 + 3 + 9 = 13 units, while the total p-width is 2 + 2 (3) + 2 (9) = 26 units. Thus the total width (and area) is **39** units. *For a Buffered 4-Input NOR*: The input gate uses 4 n-channel devices of unit width, and 4 p-channel devices whose width is 4(2) units [since the input stage is "basic-inverter compatible"]. Thus, the total width (and area) of the input stage is 4 (1) + 4 (4)(2) = 36 units, and of the two buffer stages is 39 − 3 = 36. Thus, the total area of the buffered NOR is 36 + 36 = **72** units. *Input capacitance* is proportional to the area of the p and n devices connected to each input. For the buffered inverter, it is 1 + 2 = **3 units.** For the buffered 4-input NOR, it is 1 + 4 (2) = **9 units,** that is, **3** times greater. *For input inverters and an intermediate NAND*, the total area of the input inverters = 4 × (1 + 2) = 12. For the NAND, each n-channel device is 3 × 4 = 12 units wide, and each p-channel device is 3 × 2 = 6 units wide. Total area of the NAND is 4 (6 + 12) = 72. Total area of the output inverter is 9 (1 + 2) = 27 units. Overall area of the equivalent NOR is 12 + 72 + 27 = **111** units. For the equivalent NOR, the input capacitance is **3 units,** equal to that of an inverter and **1/3** that of the direct buffered NOR. It is interesting to note that an equivalent unbuffered NOR has an input capacitance of 9(3) = 324 units, 108× more than that of the inverter-input equivalent NOR!!

SECTION 13.4: PSEUDO-NMOS LOGIC CIRCUITS

13.26 Use a 3.3V supply with $(W/L)_n = (1.2/0.8)$, $(W/L)_p = (3.0/0.8)$, $k_n' = 2.5k_p' = 100\mu A/V^2$, and $|V_t| =$ 0.6V. To meet the capacitor current-drive specification, $i_{Dn} = 2i_{Dp}$ at $v_O = 3.3/2 =$ 1.65V.

Now, $i_{Dn} = k_n'(W/L)_n[(3.3 - 0.6)1.65 - 1.65^2/2] = 100(1.5)[1.65(2.7 - 1.65/2)] = 464\mu A$, and $i_{Dp} = 464 = 2 \times 40(W/L)_p[(3.3 - 0.6)1.65 - 1.65^2/2] = 40(W/L)_p(3.09)$. Now, noting that the terms [ ] are both the same (= 3.09), see that $(W/L)_p = 100(W/L)_n/(2 \times 40) = 1.25(W/L)_n$.

Here, $k_n = 100(1.2/0.8) = 150\mu A/V^2$ and $k_p = 40(1.25)(1.2/0.8) = 75\mu A/V^2$.

Thus $r = k_n/k_p = 150/75 = 2$.

$V_{OH} =$ **3.3 V**. From Eq. 13.36, $V_M = V_t + \dfrac{V_{DD} - V_t}{\sqrt{r} + 1} = 0.6 + \dfrac{3.3 - 0.6}{\sqrt{2} + 1} =$ **2.16V**.

From Eq. 13.38, $V_{IH} = V_t + 2/\sqrt{3r}\,(V_{DD} - V_t) = 0.6 + 2/\sqrt{6}(2.7) =$ **2.80V**.

From Eq. 13.35, $V_{IL} = V_t = \dfrac{V_{DD} - V_t}{\sqrt{r(r + 1)}} = 0.6 + \dfrac{2.7}{\sqrt{2(3)}} =$ **1.70V**.

From Eq. 13.39, $V_{OL} = (V_{DD} - V_t)[[1 - \sqrt{1 - 1/r}\,] = (2.7)[1 - \sqrt{1 - 1/2}] =$ **0.79V**.

Note that V_{OL} is very high, exceeding the threshold of the NMOS transistor of a succeeding inverter, whose current would reach $i_{DL} = V_2 \times 150 \times 10^{-6}(0.791 - 0.6)^2 = 2.75\mu A$.

Thus, this circuit might be appropriate for high-speed operation but leads to a supply current somewhat in excess of the usual 50% of the total PMOS current. Otherwise, $NM_L = V_{IL} - V_{OL} = 1.70 - 0.80 = 0.9V$, and $NM_H = V_{OH} - V_{IH} = 3.3 - 2.80 = 0.5V$ which are not large, but relatively balanced.

13.27 To meet this specification, the PMOS in the pseudo-NMOS gate must be identical to that in a matched minimum-size CMOS inverter. In this case, the most that a minimum-size NMOS driver could do is lower υ_O to $V_{DD}/2$! Correspondingly, V_{OL} would be $\simeq 3/2 =$ **1.5V**. $V_M = \upsilon$, would be very high:

Here, $i_{Dp} = k[(3 - 0.6)(3 - \upsilon) - (3 - \upsilon)^2/2]$, and $i_{Dn} = k/2[\upsilon - 0.6]^2$. For equality, $(\upsilon - 0.6)^2 = 2(2.4)(3 - \upsilon) - (3 - \upsilon)^2$, or $\upsilon^2 - 1.2\upsilon + 0.36 = 14.4 - 4.8\upsilon - 9 + 6\upsilon - \upsilon^2$, or $2\upsilon_2 - 2.4\upsilon - 5.04 = 0$, whence $\upsilon = [- \; - 2.4 \pm \sqrt{2.4^2 - 4(-5.04)(2)}]/4 = (2.4 + 6.79)/4 =$ **2.30V**

13.28

(a) *For minimum-size NMOS:* *For* $r = 4$, $k_p = k_n/4$. or $k_p'(W/L)_p = k_n'(W/L)_n/4$, or $(k_n'/2)(W/L)_p = k_n'(W/L)_n/4$ whence, $(W/L)_p = (W/L)_n/2$.

Since the width is a minimum of 1 unit, L_p must be raised to **2 units**.

Thus NMOS is 1×1 and PMOS is 1×2 and the total area is $1 \times 1 + 2 \times 1 =$ **3units²**.

For $r = 10$, $L_p = 10/2 =$ **5 units**, and the total area is $1 \times 1 + 5 \times 1 =$ **6units²**.

(b) *For minimum-size PMOS* (1×1*units*)

For $r = 4$, $W_n = 4/2 =$ 2 units, whence the total area is $(1 \times 2 + 1 \times 1) =$ **3units²**.

For $r = 10$, $W_n = 10/2 =$ 5 units, whence the total area is $(1 \times 1 + 5 \times 1) =$ **6units²**.

For design style (b), the output current drive (as measured by the PMOS current) is larger than that in (a) by a factor **r/2** although the areas are the same!

(c) In (a) and (b), arrange that the devices are (1×1) and $(r/2 \times 1)$ in size. Here, we try a design style where the devices are each $(1 \times \sqrt{r/2})$ and $(\sqrt{r/2} \times 1)$ in size.

For $r = 4$, the PMOS would have $W_p =$ 1 unit and $L_p =$ 1.414 units, with $W_n =$ 1.414 units and $L_n =$ 1 unit.

The total area is $2(1.414 \times 1) =$ **2.83 units** $\{ = 2\sqrt{2} = 2\sqrt{4/2} = 2\sqrt{r/2}\}$, in general.

For $r =$ 10,, $W_p =$ 1 unit, $L_p = \sqrt{10/2} =$ 2.236 units; $W_n =$ 2.236 units, $L_n =$ 1 unit.

The total area is $2(2.236 \times 1) = 4.47$**units²** $(= 2\sqrt{5} = 2\sqrt{10/2} = 2\sqrt{r/2})$, in general.

In general, the total area is $2\sqrt{r/2} = \sqrt{2r}$, as contrasted with $(1 + r/2)$ for styles (a), (b).

The output drive, proportional to the PMOS W/L ratio, is $1/\sqrt{r/2} = \sqrt{2/r}$ in (c), while its is proportional to $1/(r/2) = 2/r$ in style (a).

Thus style (c) produces a $\sqrt{r/2}$ improvement in current and a $\sqrt{2r}/(1 + r/2)$ reduction in area.

In summary:

Ratio	Current			Area		
r	a	c	ratio %	a	c	ratio%
4	0.5	0.707	141%	3.0	2.83	94%
10	0.2	0.447	224%	6.0	4.47	75%

13.29 For $V_{DD} = 3.3\text{V}$ and $V_t = 0\text{V}$:

From Eq. 13.41: $NM_L = 0.6 - 2.7\left[- \sqrt{1 - 1/r} - \frac{1}{\sqrt{r(+ 1)}} \right]$ - - - (1)

From Eq. 13.42: $NM_H = 2.7(1 - 1/\sqrt{3r})$.

From equality, evaluate by a process of "trial and success":

For $r = 1$: $NM_L = 0.6 - 2.7\left[- \sqrt{1 - 1/1} - \frac{1}{\sqrt{1(2)}} \right] = 0.6 - 2.7(1 - 0 - 0.707)$ negative, unworkable.

For $r = 2$: $NM_L = 06. - 2.7\left[- \sqrt{1 - 1/2} - \frac{1}{\sqrt{2(3)}} \right] = 0.910$;

$NM_H = 2.7(1 - 2/\sqrt{3 \times 2}) = 0.495$.

For $r = 3$: $NM_L = 0.6 - 2.7\left[- \sqrt{1 - 1/3} - \frac{1}{\sqrt{3(4)}} \right] = 0.884$;

$NM_H = 2.7(1 - 2/\sqrt{3(3)}) = 0.900$.

For $r = 2.9$: $NM_L = 0.6 - 2.7\left[1 - \sqrt{1 - 1/2.9} - \frac{1}{\sqrt{2.9(3.9)}} \right] = 0.8882$;

$NM_H = 2.7(1 - 2/\sqrt{3(2.)}) = 0.8692$.

For $r = 2.95$: $NM_L = 0.6 - 2.7\left[1 - \sqrt{1 - 1/2.95} - \frac{1}{\sqrt{2.95(3.95)}} \right] = 0.8860$;

$NM_H = 2.7(1 - 2/\sqrt{3(2.95)}) = 0.8848$.

For $r = 2.97$: $NM_L = 0.6 - 2.7\left[- \sqrt{1 - 1/2.97} - \frac{1}{\sqrt{2.97(3.97)}} \right] = 0.8853$;

$NM_H = 2.7(1 - 2/\sqrt{3(2.97)}) = 0.8909$.

For $r = 2.955$: $NM_L = 0.6 - 2.7\left[1 - \sqrt{1 - 1/2.955} - \frac{1}{\sqrt{2.955(3.955)}} \right] = 0.8859$;

$NM_H = 2.7(1 - 2/\sqrt{3(2.955)}) = 0.8863$.

Conclude that the noise margins are equal at $r =$ **2.954**, where $NM_L = NM_H = 0.886\text{V}$.

Now, for $r = 2.954$: $V_{OH} =$ **3.3V.**

From Eq. 13.39: $V_{OL} = (V_{DD} - V_t)[1 - \sqrt{1 - 1/r}\,] = 2.7(1 - \sqrt{1 - 1/2.954}) =$ **0.504V.**

From Eq. 13.38: $V_{IH} = V_t + 2(V_{DD} - V_t)/\sqrt{3r} = 0.6 + 2(2.7)/\sqrt{3(2.954)} =$ **2.414V.**

From Eq. 13.35: $V_{IL} = V_t + (V_{DD} - V_t)/\sqrt{r(r + 1)} = 0.6 + 2.7/\sqrt{2.954(3.954)} =$ **1.390V.**

From Eq. 13.36: $V_M = V_t + (V_{DD} - V_t)/\sqrt{r + 1} = 0.6 + 2.7/\sqrt{3.954} =$ **1.958V.**

From Eq. 13.40: $I_{stat} = 1/2k_p(V_{DD} - V_t)^2 = 1/2(100(1.2/0.8)/1.954)(2.7)^2 =$ **185μA** and $NM_H =$ **0.886V**, $NM_L =$ **0.886V**.

13.30 From Eq. 13.39: $V_{OL} = (V_{DD} - V_t)[1 - \sqrt{1 - 1/r} = \alpha V_t$, or $1 - \sqrt{1 - 1/r} = \alpha V_t/(V_{DD} - V_t)$.

Thus, $\sqrt{1 - 1/r} = 1 - \alpha V_t/(V_{DD} - V_t) = \dfrac{V_{DD} - V_t - \alpha V_t}{V_{DD} - V_t} = \left[\dfrac{V_{DD} - (1+\alpha)V_t}{V_{DD} - V_t}\right] = []$

Now, $1 - 1/r = []^2$, and $1/r = 1 - []^2 = \dfrac{(V_{DD} - V_t)^2 - [V_{DD} - (1+\alpha)V_t]^2}{(V_{DD} - V_t)^2}$, or

$$1/r = \frac{V_{DD}^2 - 2V_{DD}V_t + V_t^2 - V_{DD}^2 + 2V_{DD}V_t(1+\alpha) - (1+\alpha)^2V_t^2}{(V_{DD} - V_t)^2} = \frac{2 \times V_{DD}V_t^2[1 - (1+\alpha)^2]}{(V_{DD} - V_t)^2}.$$

Thus, in general, $r = \dfrac{(\mathbf{V_{DD} - V_t})^2}{\alpha \mathbf{V_t}[2\mathbf{V_{DD}} - (\alpha + 2)\mathbf{V_t}]}$.

For $V_{DD} = 3.3$ and $V_t = 0.6$: $r = \dfrac{(3.3 - 0.6)^2}{\alpha(0.6)[2(3.3) - (\alpha + 2)0.6]} = \dfrac{20.25}{\alpha(11 - \alpha + 2)}$, or $\mathbf{r} = \dfrac{20.25}{\alpha(9 - \alpha)}$.

Now, for $\alpha = 0.5$, $r = \dfrac{20.25}{0.5(9 - 0.5)} =$ **4.76.**

Check: From Eq. 13.39: $V_{OL} = (V_{DD} - V_t)[1 - \sqrt{1 - 1/r}] = 2.7[1 - \sqrt{1 - 1/4.76}] = 0.300 = V_t/2$, OK.

13.31

(a) Need $V_M = V_t + (V_{DD} - V_t)/\sqrt{r+1} = V_{DD}/2$: Thus $0.6 + 2.7/\sqrt{r+1} = 1.65$, and $\sqrt{r+1} = \dfrac{2.7}{1.65 - 0.6} = 2.57$, whence $r =$ **5.61.**

(b) Need $V_{OL} = (V_{DD} - V_t)[1 - \sqrt{1 - 1/r}] = V_t$: Thus $2.7[1 - \sqrt{1 - 1/r}] = 0.6$, and $\sqrt{1 - 1/r} = 1 - 0.222 = 0.777$, and $1/r = 1 - 0.777^2 = 0.395$, whence $r =$ **2.53.**

(c) Need $V_{OL} = (V_{DD} - V_t)[1 - \sqrt{1 - 1/r}] = 0.1V$: Thus, $2.7[1 - \sqrt{1 - 1/r}] = 0.1$, and $\sqrt{1 - 1/r} = 1 - .0370 = 0.963$, and $1/r = 1 - 0.963^2 = 0.0727$, whence $r =$ **13.75.**

(d) Need $V_{OL} = 0.01V$: Thus $2.7[1 - \sqrt{1 - 1/r} = 0.01$, and $\sqrt{1 - 1/r} = 1 - 3.7 \times 10^{-3} = 0.9963$, and $1/r = 1 - 0.9963^2 = 7.39 \times 10^{-3}$, whence $r =$ **135.2.**

(e) Need $V_{IL} = V_t + (V_{DD} - V_t)/\sqrt{r(r+1)} = 2V_t$: Thus, $2.7/\sqrt{r(r+1)} = 0.6$, and $\sqrt{r(r+1)} = 4.5$, and $r^2 + r = 20.25$, whence $r = (-1 \pm \sqrt{1 - 4(-20.25)})/2 = (- \pm 9.06)/2 =$ **4.02.** *For case a),* $r = 5.61$, and $V_{OL} = (V_{DD} - V_t)[1 - \sqrt{1 - 1/r}] = 2.7[1 - \sqrt{1 - 1/5.61}] =$ **0.252V.**

For case c), $r = 13.75$, and $V_{IL} = V_t + (V_{DD} - V_t)/\sqrt{r(r+1)} = 0.6 + 2.7/\sqrt{13.75(14.75)} =$ **0.789V**

For case d), $r = 135.2$, $V_{IL} = 0.6 + 2.7/\sqrt{135.2(136.2)} =$ **0.62V**

13.32 From Eq. 13.43, $t_{PLH} = \dfrac{1.7C}{k_p V_{DD}}$.

From Eq. 13.44, $t_{PHL} = \dfrac{1.7C}{k_n(1 - 0.46/r)V_{DD}}$, where $r = k_n/k_p$,

Thus, $t_{PLH} = \dfrac{\mathbf{1.7C}}{\mathbf{k_p V_{DD}}}$, and $t_{PLH} = \dfrac{\mathbf{1.7C}}{\mathbf{(k_n/r)V_{DD}}}$.

Also, $t_{PHL} = \dfrac{1.7C}{\mathbf{k_n(1 - 0.46/r)V_{DD}}}$, and $t_{PHL} = \dfrac{1.7C}{k_p(r - 0.46)V_{DD}}$

Now, see $t_{PLH}/t_{PHL} = \left(\dfrac{1.7C}{k_p V_{DD}}\right) \Big/ \left(\dfrac{1.7C}{k_p(r - 0.96)V_{DD}}\right) = (\mathbf{r - 0.46})$.

For $t_{PLH} = t_{PHL} = t_p$, $r - 0.46 = 1$ and $r =$ **1.46**

Correspondingly,

From Eq. 13.36, $V_M = V_t \dfrac{V_{DD} - V_t}{\sqrt{r+1}} = 0.6 + 2.7/\sqrt{1 + 1.46} =$ **2.05V**,

From Eq. 13.39, $V_{OL} = (V_{DD} - V_t)[1 - \sqrt{1 - 1/r}\,] = 2.7[1 - \sqrt{1 - 1/1.46} =$ **1.18V**.

From Eq. 13.41,

$$NM_L = V_t - (V_{DD} - V_t)\left[1 - \sqrt{1 - 1/r} - \frac{1}{\sqrt{r(r+1)}}\right] = 0.6 - 2.7\left[1 - \sqrt{1 - 1/1.46} - \frac{1}{\sqrt{1.46(2.46)}}\right]$$

$= -$ **0.089V**.

From Eq.13.42, $NM_H = (V_{DD} - V_t)[1 - 1/\sqrt{3r}\,] = 2.7[1 - 2/\sqrt{3(1.46)}] =$ **0.120 V**

Note that with these (small, negative) noise margins, this circuit would not normally be used in a string of similar circuits, but rather as a special solution to a particular problem:

For a matched inverter, $t_p = \dfrac{1.6C}{k_n'(W/L)_n V_{DD}} = \dfrac{1.6C}{k_n V_{DD}}$, where we assume that C is dominated by external load (as it is not otherwise the same as C for the pseudo-NMOS.

For the balanced pseudo-NMOS, $t_p = \dfrac{1.7C}{(k_n/r)V_{DD}}$ Ignoring the 1.6/1.7 coefficient difference see that the pseudo-NMOS has a delay of **1.46** of the CMOS, being therefore about 50% greater.

13.33 For all NMOS devices, $(W/L)_n = 1.2/0.8$, $k_n' = 100\mu A/V^2$, $V_{tn} = 0.6V$. For the PMOS, $|V_{tp}| = 0.6V$, $k_p' = 40\mu A/V^2$. Generally, $k_p = k_n/r$, and $(W/L)_p 40 = (W/L)_n 100/r$.

Thus $(W/L)_p = (1.2/0.8)(100/40)/r = 3.75/r$. Now, *for* $r =$, $(W/L)_p = 3.75/4 = 0.9375$. Thus, for $W_p = 1.2\mu m$, $L_p = 1.2/0.9375 = 1.28\mu m$.

Now, for $r = 10$, $(W/L)_p = 3.75/10 = 0.375$. Thus, for $W_p = 1.2\mu m$, $L_p = 1.2/.375 = 3.2\mu m$.

Concerning Capacitances: Follow the development on page 1053 of the Text:

Assume $C_w = 0$. For each NMOS: $C_{gd} = 0.5(1.2) = 0.6fF$, $C_{db} = 2.5(1.2) = 3.0fF$, $C_g = 1.81.2 \times 0.8 = 1.73fF$. For the PMOS with $r = 4$ or $r = 10$, assuming 1.2 μm width, and increased length: $C_{gd} = 0.5(1.2) = 0.6fF$, $C_{dB} = 2.5(1.2) = 3.0fF$.

Thus for the 8-input NOR with one input active, and an inverter load, $C_1 = 2(0.6) + 7(0.6) + 8(3.0) + 0.6 + 3.0 + 1.73 + 0 = 10(0.6) + 9(3.)) + 1.73 = 34.73fF$.

For 2 inputs active, $C_2 = 2(0.6) + 2(0.6) + 6(0.6) + 8(3.0) + 0.6 + 3.0 + 1.73 = 34.73 + 0.6 = 35.33fF$.

For a single input active: $t_{PHL} = 1.7C/[k_n(1 - 0.46/r)V_{DD}]$, and

$t_{PLH} = 1.7C/[k_p V_{DD}] = 1.7C/[(k_n/r)V_{DD}]$.

For $r = 4$: $t_{PLH} = 1.7(34.37 \times 10^{-15})/[(150 \times 10^{-6}/4)3.3] = 0.477 \times 10^{-9}s =$ **477 ps**, and

$t_{PHL} = 1.7(34.73 \times 10^{-15})/[150(1 - 0.46/4)3.3] =$ **135 ps**.

For $r = 10$: $t_{PLH} = 10/4(477) =$ **1.19 ns**, and $t_{PHL} = 135(1 - 0.46/4)/(1 - .46/10) =$ **125 ps**.

For 2 inputs active, k_n is effectively doubled as is r in the calculation for t_{PHL}. For t_{PLH}, the k_p and r used are the same. Why?

For $r = 4$: $t_{PLH} = 477 \times 35.33/34.73 =$ **485 ps**, and

$t_{PHL} = 1.7(35.33 \times 10^{-15})/[150 \times 2(1 - 0.46/8)3.3] =$ **64.5 ps** .

For $r = 10$: $t_{PLH} = 1.19 \times 35.33/34.73 =$ **1.21 ns**, and

$t_{PHL} = 1.7(35.33 \times 10^{-15})/[150 \times 2(1 - 0.46/20)3.3] =$ **62 ps**.

13.34 Assume all NMOS have unit length and width. For the matched complementary 8-input NOR, there are 8 unit-size NMOS and 8 PMOS which are each $8 \times 2.5 =$ 20 times larger, due to the fact that there are 8 in series and that $k_p' = k_n'/2.5$. Thus the area of the CMOS 8-input NOR is $8(1 \times 1 + 20 \times 1) = 8(21) = 168 units^2$. For the pseudo-NMOS with $r = 2$, all NMOS are unit-size and the PMOS is $(k_n'/k_p')/r = 2.5/2 = 1.25 \times$ wider. Thus the area of the pseudo-NMOS 8-input NOR is $8(1 \times 1) + (1.25 \times 1) = 9.25 units^2$. The complementary CMOS is $168/9.25 =$ **18.2 × larger!** Correspondingly, the pseudo-NMOS gate is only $9.25/168 \times 100 =$ **5.5% of the size** of the CMOS gate!

SECTION 13.5: PASS-TRANSISTOR LOGIC CIRCUITS

13.35 For this situation, $V_{OL} =$ **0.0V**, but $V_{OH} = 3.3 - V_t$, where V_t is increased by the body effect. From Eq. 5.30, and the data in the introductory NOTE:

$V_t = V_{t0} + \partial(\sqrt{V_{SB} + 2\Phi f} - \sqrt{2\Phi f}) = 0.6 + 0.5(\sqrt{V_{OH} + 0.6} - \sqrt{0.6})$,

or $V_t = 0.6 + 0.5\sqrt{3.3 - V_t + 0.6} - 0.5\sqrt{0.6}$ or $V_t = .2127 + 0.5\sqrt{3.9 - V_t}$, or

$\sqrt{3.9 - V_t} = 2(V_t - .2127) = (2V_t - 0.4254)$. Squaring, $3.9 - V_t = 4V_t^2 - 1.702V_t + 0.181$, or $4V_t^2 - 0.702V_t - 3.72 = 0$,

whence $V_t = (- - 0.702 \pm \sqrt{0.702^2 - 4(4)(-3.72)})/2(4) = (0.702 \pm 7.747)/8 = 1.06V$

Thus $V_{OH} = 3.3 - 1.06 =$ **2.24V**.

Assuming that the specification "driven" means that the two gates are joined drain to source (*not* drain to gate), they behave as a single transistor with twice the length, but with the same V_t. Thus for no dc load on the gate, $V_{OH} =$ **2.24V**, as before. However, dynamically, they are different: For a matched inverter using the standard devices for which $|V_t| =$ 0.6V, $k_n' = 2.5k_p' = 100\mu A/V^2$, and $(W/L)_n = (1.2/0.8)$, $k_n = k_p = 100(1.2/0.8) = 150\mu A/V^2$.

For a 2.24V input, the current in the connected inverter, $i_D = 1/2(150)(3.3 - 2.24 - 0.6)^2 =$ **15.87μA**

For the NMOS with $V_O = \upsilon$, $15.47 = 150[(2.24 - 0.6)\upsilon - \upsilon^2/2]$, and $3.28\upsilon - \upsilon^2 = 0.2063$, or $\upsilon^2 - 3.28\upsilon + .2063 = 0$, whence $\upsilon = (- - 3.28 \pm \sqrt{3.28^2 - 4(.2063)})/2 = (3.28 \pm 3.152)/2 =$ **0.064V**.

For the capacitance at the inverter input including the input pass gate: and a grounded pass gate.

$C_{gn} = 1.2 \times 0.8 \times 1.8 \times 10^{-15} =$ 1.728 fF, $C_{gp} = 2.5 \times 1.2 \times 0.8 \times 1.8 \times 10^{-15} =$ 4.32 fF, $C_{sb} = C_{db} = 2.5 \times 10^{-15} \times 1.2 =$ 3.0 fF. Now, assuming $C_w = 0$. Thus, $C = 1.728 + 4.32 + 3.0 + 3.0 + 0 =$ **12.04 fF.**

For propagation times: Consider the gate of the pass transistor to be always at + 3.3V.

At $\upsilon_S = 3.3/2 = 1.65V$, $V_t = 0.6 + \gamma[\sqrt{V_{SB} + 2\Phi f} - \sqrt{2\Phi f}] = 0.6 + 0.5\sqrt{1.65 + 0.6} - 0.5\sqrt{0.6} =$ 0.962V

For t_{PLH} $(\upsilon_I = 3.3V)$: $i_D(L) = 1/2(150)(3.3 - 0.6)^2 = 547\mu A$, and

$i_D(M) = 1/2(150)(3.3 - 1.65 - 0.962)^2 = 35.5\mu A$, for which $i_D(a\upsilon) = (547 = 35.5)/2 = 291\mu A$. Correspondingly, $t_{PLH} = (12.04 \times 10^{-15} \times 1.65)/(291 \times 10^{-6}) =$ **68.3 ps.**

For t_{PHL} $(\upsilon_I = 0V)$. Assuming the output starts from 3.3V! $i_D(H) = 1/2(150)(3.3 - 0.6)^2 = 547\mu A$ (as before), and $i_D(M) = 150[(3.3 - 0.6)1.65 - 1.65^2/2] = 463\mu A$, for which $i_D(a\upsilon) = (547 + 463)/2 = 505\mu A$.

Thus $t_{PHL} = 12.04 \times 10^{-15} \times (3.3 - 1.65)/(505 \times 10^{-6}) =$ **39.3 ps**

For an imperfect initial signal (2.24V rather than 3.5V).

$t_{PHL} = 12.04 \times 10^{-15}(3.3 - 2.24)/505 \times 10^{-6} = 4$ **25.3 ps.**

13.36 For $(W/L)_p = 0.1$, with $W_p = 1.2\mu m$, $L_p = 1.2/0.1 = 12\mu m$. Now, Q_R begins to conduct when V_{O2} reaches $3.3 - 0.6 = 2.7V$. For $\upsilon_{01} = \upsilon$. For the inverter, $i_{Dn} = 1/2(150)(\upsilon - 0.6)^2 = 75(\upsilon - 0.6)^2$, $i_{Dp} = 150[(3.3 - \upsilon)0.6 - 0.6^2/2] = 90(3.3 - \upsilon) - 27$.

Now $i_{Dn} = i_{Dp} \rightarrow 75(\upsilon - 0.6)^2 = 90(3.3 - \upsilon) - 27$, $75\upsilon^2 - 90\upsilon + 0.36 = 297 - 90\upsilon - 27$, $75\upsilon^2 = 297 - 27 - 0.36 = 269.6$, $\upsilon^2 = 269.6/75 = 3.595$, whence $\upsilon = 1.896V$. Thus Q_R begins to conduct with $\upsilon_{01} =$ **1.896 V.**

As Q_R begins to conduct it assists in the LH transition pulling υ_{O1} to $V_{OH1} =$ **3.3V**, and reducing t_{PLH} by speeding up the upper end of the transition.

Now for t_{PHL} at υ_{O1} with Q_R: Initially, $i_{D1}(H) = 1/2(150)(3.3 - 0.6)^2 = 547\mu A$. Then $i_{D1}(M) = 150[(3.3 - 0.6)(1.65 - 1.65^2/2)] = 463\mu A$.

Also, at $\upsilon_{O1} = V_{DD}/2 = V_M$, $\upsilon_{O1} = V_{DD}/2 = 1.65V$ eventually and the minimum size Q_R conducts $i_{DR}(M) = 1/2(150/2.5)(3.3 - 1.65 - 0.6)^2 = 33.1\mu A$ but before υ_{O2} responds (and is still at 0V), $i_{DR}(x) = 1/2(150/2.5)(3.3 - 0.6)^2 = 218.7\mu A$.

Thus there are two choices represented: First, there is the case in which turnoff is regenerative, and Q_R helps, correspondingly, the average capacitor current is

$((547 - 0) + (463 - 33))/2 = 489\mu A$.

Now for $C =$ 12 fF, from P13.35 above, $t_{PHL} = 12 \times 10^{-15} \times (3.3 - 1.65)/489 \times 10^{-6} = 40.5ps$.

In the second case, regeneration is assumed to be delayed and Q_R conduction reduces the current available to the capacitor, namely, $i_{Q(max)} = 150/2.5[(3.3 - 0.6)(1.65) - 1.65^2/2] = 463/2.5 = 185\mu A$.

If this is assumed to flow all the time (it is actually smaller for high υ_{O1}) then the average capacitor current is $[(547 - 185) + (463 - 185)]/2 = 320\mu A$, for which $t_{PHL} = 12 \times 10^{-15} \times 1.65/320 \times 10^{-6} =$ **56.25 ps.**

13.37 For this situation, the pass transistors, the grounding switch and the inverter NMOS are all minimum size ($1.2\mu m \times 0.8\mu m$), while the inverter PMOS is 2.5 × wider.

Using the data for the standard devices (in the introductory NOTE above), $C_{db} = Csb = 2.5 \times 1.2 = 3.0fF$, $C_g = 1.8 \times 1.2 \times 10.8 = 1.73fF$, $C_{gs} = 0.5 \times 1.2 = 0.6fF$, while C_w is assumed to be zero.

At the Inverter input:

Thus $C = 3(3.0) + (1 + 2.5)(1.73) + 3(0.6) + (1 + 2.5)(0.6) = 9.0 + 6.06 + 1.8 = 2.1 =$ **18.9 fF.** Also $V_{OL} =$ **0.0V** and $V_{OH} =$ **3.3V.** For propagation calculations, the gate signals of the pass device are assumed to keep the devices on while the input (to the drain source connection) rises and falls.

For t_{PLH}: At 0V, $i_{Dn}(L) = 1/2(150)(3.3 - 0.6)^2 = 547\mu A$, and

$i_{Dp}(L) = 1.2(150/2.5)(3.3 - 0.6)^2 = 219\mu A$.

At 1.65V, $V_t = V_{t0} + \partial(\sqrt{V_{SB} + 2\Phi f} - \sqrt{2\Phi f}) = 0.6 + 0.5(\sqrt{1.65 + 0.6} - \sqrt{0.6}) =$ 0.96V.

Thus $i_{Dn}(M) = 1/2(150)(3.3 - 1.65 - 0.96)^2 = 35.7\mu A$.

$i_{Dp}(M) = 150/2.5[(3.3 - 0.6)1.65 - 1.65^2/2] = 185\mu A$.

Thus the current available to charge the output is $i_C(Av) = (547 + 219 + 35.7 + 185)/2 = 493\mu A$.

Correspondingly, $t_{PLH} = 18.9 \times 10^{-15} \times 1.65/493 \times 10^{-6} =$ **62.3 ps.** *For t_{PHL}*: Roles interchange: p-channel currents are 1/2.5 those available from the n-channel.

$i_C(Av) = (547/2.5 + 219(2.5) + 35.7/2.5 + 185(2.5))/2 = (219 + 547 + 14.3 + 463)/2 = 622\mu A$.

Thus $t_{PHL} = 18.9 \times 10^{-5} \times 1.65/633 \times 10^{-6} =$ **50.1 ps.**

13.38

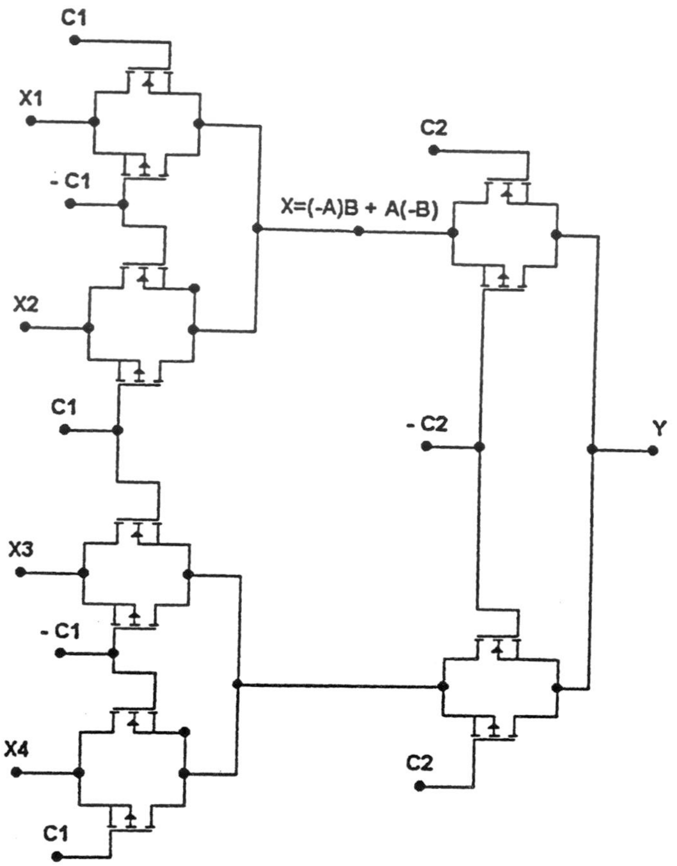

13.39

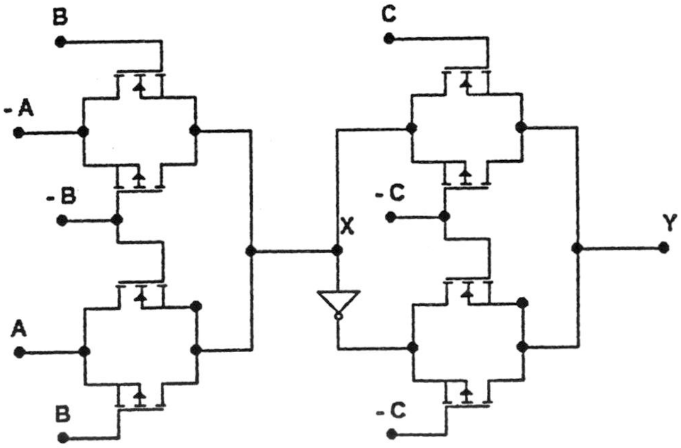

13.39 (continued)

Another implementation uses a pass network to create $\overline{Y}$ rather than using an inverter. Note that $Y = \overline{A}\ B + A\ \overline{B}$, $\overline{Y} = \overline{\overline{A}\ B + A\ \overline{B}} = \overline{\overline{A}\ B} \cdot \overline{A\ \overline{B}} = (A + \overline{B})\ (\overline{A} + B) = A\ B + \overline{A}\ \overline{B}$ (See that to obtain the inverse, one exchanges A and $\overline{A}$, or B and $\overline{B}$.

For 3 variables, $X = \overline{A}\ B + A\ \overline{B}$ and

$Y = \overline{C}\ X + C\ \overline{X} = \overline{C}\ (\overline{A}\ B + A\ \overline{B}) + C\ (A\ B + \overline{A}\ \overline{B}\ ) = \overline{A}\ B\ \overline{C} + A\ \overline{B}\ \overline{C} + A\ B\ B + \overline{A}\ \overline{B}\ C$:

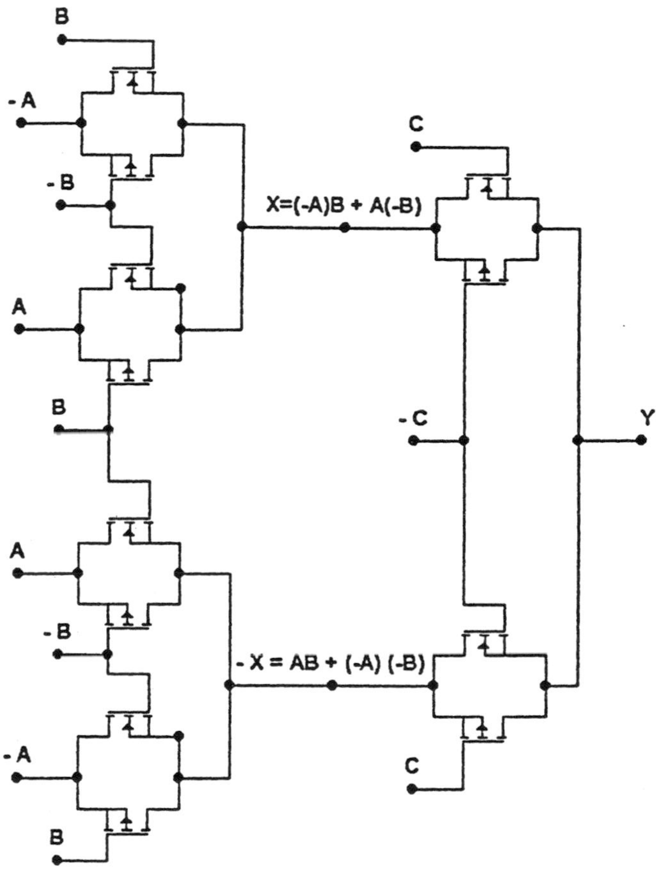

13.40

C	X	Y	Z	F
0	0	0	0	0
0	0	0	1	0
0	0	1	0	0
0	0	1	1	0
0	1	0	0	0
0	1	0	1	0
0	1	1	0	1
0	1	1	1	0
1	0	0	0	0
1	0	0	1	0
1	0	1	0	0
1	0	1	1	0
1	1	0	0	0
1	1	0	1	0
1	1	1	0	0
1	1	1	1	0

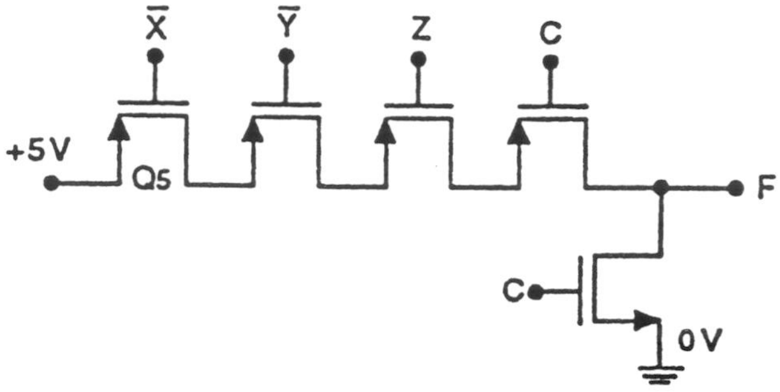

See that F goes high when X is high, $\overline{Y}$ is low, Z is low, as C goes low. That is, $F = C\ X\ Y\ \overline{Z}$. If $\overline{X}$ is available, but X is not, and Y is available, replace X (as shown) by Y, and $\overline{Y}$ (as shown) by X. If $\overline{X}$ is available, but X is not and Y is not, use an additional p-channel MOSFET at the left connected to the supply and $\overline{X}$. For the basic symmetric inverter $(W/L)_p$ = 20/5. Thus $(W/L)_4$ = 10/5, $(W/L)_{1,2,3}$ = 10/5 × 2 × 3 = 60/5, to account for hole mobility and the fact that there are 3 transistors in series. In many applications in which the load is an inverting input, and the goal is circuit simplicity, all devices can be of minimum size.

13.41

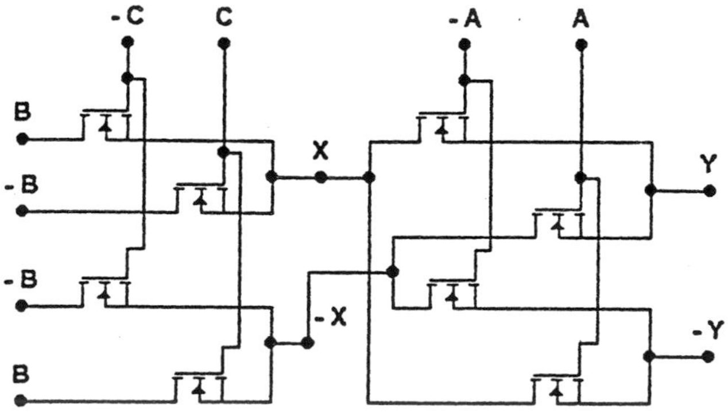

See $X = B\ \overline{C} + \overline{B}\ C$ and $\overline{X} = \overline{B}\ \overline{C} + B\ C$

See $Y = X\ \overline{A} + \overline{X}\ A$ and $\overline{Y} = \overline{X}\ \overline{A} + X\ A$

whence, $Y = \overline{A}\ (B\ \overline{C} + \overline{B}\ C\ ) + A\ (\overline{B}\ \overline{C} + B\ C\ ) = \overline{A}\ B\ \overline{C} + A\ \overline{B}\ C + A\ \overline{B}\ \overline{C} + A\ B\ C$

and $\overline{Y} = \overline{A}\ (\overline{B}\ \overline{C} + B\ C\ ) + A\ (B\ \overline{C} + \overline{B}\ C\ ) = \overline{A}\ \overline{B}\ \overline{C} + \overline{A}\ B\ C\ + A\ B\ \overline{C} + A\ \overline{B}\ C$

See that for the XOR, the output is 1 for all cases of an *odd* number of inputs, that is (1 or 3).

See that for the XNOR, the output is 1 for all cases of an *even* number of inputs, that is (0 or 2).

Concerning dynamic operation, there are several principles to observe:

(1) Most active input should be closest to the output.

(2) Most active input should drive the least capacitance.

(3) Most active input should activate the least capacitance charging activity.

(4) Source-drain paths are fastest (since there are no thresholds to overcome).

(5) Mutual conflict between principles will exist.

For this situation, placement of input A follows rule (1), rule (2), and rule (3), but is in conflict with rule (4). Placement of inputs B and C follows rules (4) and (1), but is in conflict with rules (2), (3).

SECTION 13.6: DYNAMIC LOGIC CIRCUITS

13.42 For these devices, $k_n = 2.5k_p = 150\mu A/V^2$, $|V_t| = 0.6$ V. In general, use the techniques suggested in Excercises 13.10 and 13.11. Note in particular that the usual definition of t_{PLH} (to 50%) is not particularly relevant, since it is important for a dynamic gate that the output be at V_{OH} before evaluation takes place.

For t_{PLH}: Use a 0% to 90% rise-time estimate.

For v_O low, $i_{Dp}(L) = 1/2(150/2.5)(3.3 - 0.6)^2 = 21.87\mu A$.

For v_O high, $i_{Dp}(H) = (150/2.5)[(3.3 - 0.6)(0.1 \times 3.3) - (0.1 \times 3.3)^2/2] = 50.2\mu A$.

Thus, $i_{Dp}(Av) = (218.7 + 50.2)/2 = 134\mu A$, whence $t_{PLH} \approx 100 \times 10^{-15}(0.9 \times 3.3)/(134 \times 10^{-6}) =$ **2.22 ns**.

For t_{PHL}: Consider 4 NMOS in series and t_{PLH} measured from $v_O = 3.3$V to $v_O = 1.65$V:

For v_O high, $i_{Dn}(H) = 1/2(150/4)(3.3 - 0.6)^2 = 136.7\mu A$.

For v_O middle, $i_{Dn}(M) = (150/4)[(3.3 - 0.6)1.65 - 1.65^2/2] = 115.7\mu A$.

Thus, $i_{Dn}(Av) = (136.7 + 115.7)/2 = 126\mu A$, whence $t_{PHL} = 100 \times 10^{-15} \times 1.65/(126 \times 10^{-6}) =$ **1.31 ns**.

For a 3-input NOR; t_{PLH} is about the same, namely $t_{PLH} = 2.2ns$. However t_{PHL} is reduced by a factor of 2, since only 2 transistors are in series during evaluation. Thus $t_{PHL} = 1.31/2 =$ **0.65 ns**.

13.43 For standard minimum-size devices, $k_n = 150\mu A/V^2$, $|V_t| = 0.6$V, $(W/L) = (1.2/0.8)$, $C_{ox} = 1.8fF/\mu m^2$, $C_{gd} = C_{gs} = 0.5fF/\mu m$, and $C_{db} = C_{sb} = 2.5fF/\mu m$.

For a 3-input NAND, the equivalent capacitance includes 5 drains, 3 sources, 3 gates and 2 overlaps:

$C_{eq} = 5(2.5 \times 1.2) + 3(2.5 \times 1.2) + 3(1.2 \times 0.8 \times 1.8) + 2(0.5 \times 1.2)$

$= 5(3) + 3(3) + 3(1.73) + 2(0.6) = 15 + 9 + 5.2 + 1.2 =$ **30.4 fF**.

For 2 inverters, $C_L = 2[1.2 \times 0.8 \times 1.8 \times (1 = 2.5)] = 7(1.73) = 12.1fF$. Thus $C_{OL} = 30.4 + 12.1 =$ **42.5 fF**.

Initially at t_{0+}, Q_A is cut off, node 1 is precharged to 3.3V, and nodes 2, 3, 4, are pulled to 0V through transistors driven by Φ, C, B. Now at t_{1+}, C goes low, isolating node 3 (at 0V). Now at t_{2+}, A goes high, joining nodes 2, 3 to the output node 1.

Capacitance of each of the nodes 2 and 3 is

$C = 2(2.5 \times 1.2) = 1(1.2 \times 0.8 \times 1.8) = 2(3) = 1(1.73) = 6 + 1.73 = 7.73fF$.

Capacitance of both nodes 2.3 together $= 2(7.73) = 15.5fF$.

Time	Node 1	Node 2	Node 3	Node 4	Φ	A	B	C
t_0	–	–	–	–	L	L	H	H
t_{0+}	3.3(F)	0	0	0	H	L	H	H
t_{1+}	3.3(F)	0(F)	0(F)	0	H	L	H	L
t_{2+}	1.83(F)	1.83(F)	1.83(F)	0	H	H	H	L

Note: Here, (F) indicates a floating voltage stored on the node capacitance.

Notice that the capacitance of node 1, when isolated by A low, is $C_{dbp} + C_{dbn} + 2(C_{ov}) + 2(C_{gp} + C_{gn}) = 2.5 \times 1.2 + 2.5 \times 1.2 + 2(0.5 \times 1.2) + 2(3.5)(1.73)$

$= 3.0 + 3.0 + 1.2 + 12.1 = 19.3fF$.

Now at t_{2+}, nodes 1, 2, 3 are joined with a common voltage υ. Charge is conserved.

Thus $Q = CV$, and $19.3(3.3) + (15.5)0 = \upsilon(19.3 + 15.5) = 34.8\upsilon$.

Thus $\upsilon = (19.3/34.8)3.3 =$ **1.83V.**

Note the dramatic loss in voltage! In practice addtional wiring capacitance C_w at the output node would help.

13.44 For CfF to discharge V_t volts in t_E ns, the leakage is $i_L = CV_t/t_E = (50 + 10n)(0.6)/t_E\ \mu A$.

For a single device for which $V_{tn} = 0.6V$ and $k_n = 150\mu A/V^2$, with an input voltage υ

$(50 + 10n)(0.6)/t_E = 150/2(\upsilon - 0.6)^2$, whence $\upsilon = 0.6 + (0.4 + 0.08\mathbf{n})/(t_E)^{1/2}$.

For $n = 5$, and $t_E = 10ns$, the voltage needed at one input is υ_1 where

$\upsilon_1 = 0.6 + ([0.4 + (0.08)^5]/10)^{1/2} = 0.6 + 0.283 =$ **0.883V**

For all 5 inputs acting together, $\upsilon_5 = 0.6 + (1/5)^{1/2}(0.283) =$ **0.727V**

13.45 For these transistors, $(W/L) = (1.2/0.8)$, $k_n = 2.5k_p = 150\mu A/V^2$.

From Q_1, $i_D(H) = 1/2(150)(3.3 - 0.6)^2 = 547\ \mu A$.

Time taken for V_1 to fall from 3.3V to 0.6V is $t = 50 \times 10^{-15}(3.3 - 0.6)/(547 \times 10^{-6}) =$ **247 ps.**

Now, for the leakage in Q_2!

$i_{D2}(3.3) = 1/2(150)(3.3 - 0.6)^2 = 547\mu A$, and $i_{D2}(0.6) = 0\mu A$, such that $i_{D2}(A\upsilon) = (547 + 0) =$ **273μA.**

Thus, the change of voltage on C_{L2} is $\Delta\upsilon_2 \approx 273 \times 10^{-6} \times 247 \times 10^{-12}/(50 \times 10^{-15}) =$ **1.35V**! Clearly, this circuit has a problem! It is this problem which motivated the invention of *Domino CMOS.*

13.46 The inverter uses $(W/L) = 1.2/0.8$ for *both* devices.

At node X_1, $C_{X1} = 2C_{db} + 2C_{ov} = 2C_g = 2 \times 2.5 \times 1.2 + 2 \times 0.5 \times 1.2 + 2 \times 1.8 \times 1.2 \times 0.8$

$= 2(3) + 2(0.6) + 2(1.73) = 6.0 + 1.2 + 3.46 = 10.66fF$.

Now for the simple inverter, $V_M = \upsilon$ is reached when

$i = 1/2(150)(\upsilon - 0.6^2) = 1/2(150/2.5)(3.3 - \upsilon - 0.6)^2$, or $2.7 - \upsilon = \sqrt{2.5}(\upsilon - 0.6)$, or

$2.7 - \upsilon = 1.58\upsilon = 0.949$ or $2.58\upsilon = 3.65$, or $\upsilon = 1.41V$.

Now for A rising to 3.3V, Q_1 and Q_{e1} are in series, and the current in Q_1 varies from $i_{Dn}(H) = 1/2(150/2)(3.3 - 0.6)^2 = 273\mu A$ to $i_{Dn}(M) = 150/2[(3.3 - 0.6)1.41 - 1.41^2/2] = 211\mu A$, as X_1

falls, where the average is $i_{Dn}(Av) = (273 + 211)/2 = 242\mu A$.

Thus the time taken for X_1 to fall to $V_M = 1.41V$ is

$t = 10.66 \times 10^{-15} \times (3.3 - 1.41)/(242 \times 10^{-6}) = 83.2ps$.

For the inverter loaded by Q_2, the output capacitance is

$C_{Y1} = 2C_{db} + 2(2C_{ov}) + C_g = 2(3) + 4(0.6) + 1(1.73) = 6 + 2.4 + 1.73 = 10.13fF$.

For the p device in the simple inverter, at $X_1 = 0V$, $i_{Dp}(L) = 1/2(150/2.5)(3.3 - 0.6)^2 = 219\mu A$, and $i_{Dp}(M) = (150/2.5)[(3.3 - 0.6)1.65 - 1.65^2/2] = 80\mu A$, with $i_{Dp}(Av) = (219 + 80)/2 = 150\mu A$, for which $t_p = 10.13 \times 10^{-15}(1.65)/150 \times 10^{-6} = 111ps$.

Thus t_{PLH} from X_1 to Y_1 is approximately (83.2 + 111) = **194 ps.**

For 10 such gates in cascade, Φ must be high for at least 10(194) ≃ **1.94 ns**

SECTION 13.7: LATCHES AND FLIP FLOPS

13.47 For this technology, $|V_t| =$ 0.0V, $k_n = k_p = 150\mu A/V^2$, $V_A = 20V$, and $V_{DD} = 3.3V$.

From Eq. 5.94, $V_{IL} = 1/8(3V_{DD} + 2V_t) = (3(3.3) + 2(0.6))/8 =$ **1.3875V.**

From Eq. 5.93, $V_{IH} = 1/8(tV_{DD} - 2V_t) = (5(3.3) - 2(0.6))/8 =$ **1.9125V.**

and $V_M = V_{DD}/2 = 3.3/2 =$ **1.65V.**

Now at V_M, $i_{Dn} = 150/2(1.65 - 0.6)^2 = 82.7\mu A$, for which $g_m = 150 \times 10^{-6}(1.65 - 0.6) = 157\mu A/V$, and $r_o = V_A/i_D = 20/82.7\mu A = 242k\Omega$.

Thus, for each inverter, the voltage gain at $v_I = V_M$ is $-(g_n + g_m)(r_o \parallel r_o) = -g_m r_o = -157 \times 10^{-6} \times 242 \times 10^3 =$ **− 38.0 V/V.**

Thus the maximum loop gain is $38.0^2 =$ **1444 V/V.**

13.48

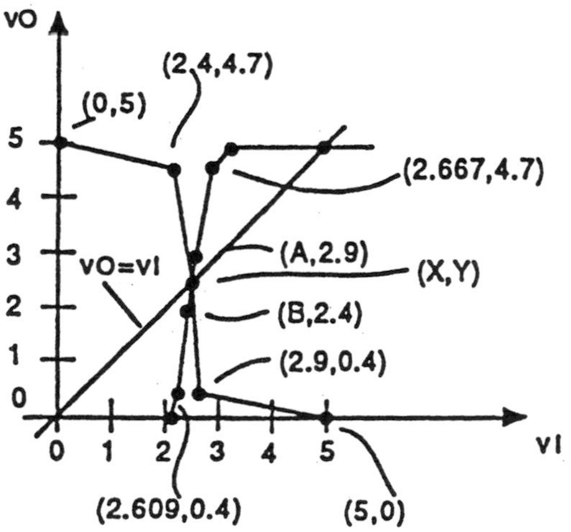

For the input coordinates corresponding to the 2.9V and 2.4V output levels (ie nodes A, B):

$$v_A = 2.4 + \frac{4.7 - 2.9}{4.7 - 0.4} \times (2.9 - 2.4) =$$

$$2.4 + \frac{1.8}{4.3}(0.5) = 2.609V.$$

$$v_B = 2.4 + \frac{4.7 - 2.4}{4.7 - 0.4} \times (2.9 - 2.4) =$$

$$2.4 + \frac{2.3}{4.3}(0.5) = 2.667V.$$

There are 3 points for which input and output are at equal levels: Two of these are at **(0V, 0V)** and **(5V, 5V)** where the loop gain is 0, and the other (in the middle, more or less) is where, $v_O \simeq \frac{2.609 + 2.667}{2} \simeq$ **2.64V** $= v_I$, that is at **(2.64, 2.64)**. For each inverter, the gain there exceeds $\frac{4.7 - 0.4}{2.4 - 2.9} = -\frac{4.3}{0.5} = -8.6V/V$. Thus the overall loop gain exceeds 8.6^2 or **74 V/V** at the (middle) unstable point, and is 0 V/V at the other two points of equality, which are stable.

13.49 For transistors for which $|V_t| = 0.6\text{V}$ and $(W/L)_n = (W/L)_p$, and $k_n' = 2.5k_p' = 150\mu A/V^2$, $V_M = \upsilon$ is found from $1/2(150)(\upsilon - 0.6)^2 = 1/2(150/2.5)(3.3 - \upsilon - 0.6)^2$, or $\sqrt{2.5}(\upsilon - 0.6) = 2.7 - \upsilon$, or $1.58\upsilon - .949 = 2.7 - \upsilon$, $2.58\upsilon = 3.649$, or $\upsilon = 1.414V$. Thus $V_{th} = V_M =$ **1.41V**

Now to lower Q with R high to $\upsilon_Q = V_M$ requires that the current in Q_8 and Q_7 equal that in Q_4. Now $i = i_{D4} = (150/2.5)[(3.3 - 0.6)1.65 - 1.65^2/2] = 185\mu A$.

Now for Q_7, $\upsilon_{DS} = \upsilon$ and $185 = 150[(3.3 - 0.6)\upsilon - \upsilon^2/2]$, or $1.23 = 2.7\upsilon + \upsilon^2/2$, or $\upsilon_2 - 5.4\upsilon + 2.46 = 0$, whence $\upsilon = [- - 5.4 \pm (5.4^2 - 4(2.46))^{1/2}]/2 = (5.4 \pm 4.395)/2 = 0.302V$.

Now the voltage needed on Φ to lower υ_Q to V_M is $\upsilon_\Phi = \upsilon$ where $185 = 150/2(\upsilon - 0.6 - 0.302)^2$, or $\upsilon - 0.902 = (2.46)^{1/2} = 1.571$. Thus $\upsilon = 1.571 + 0.902 =$ **2.47V**, relatively high, but workable in a 3.3V system.

13.50

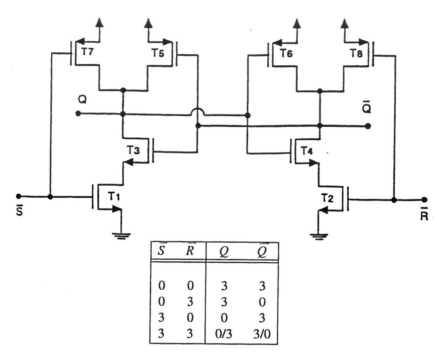

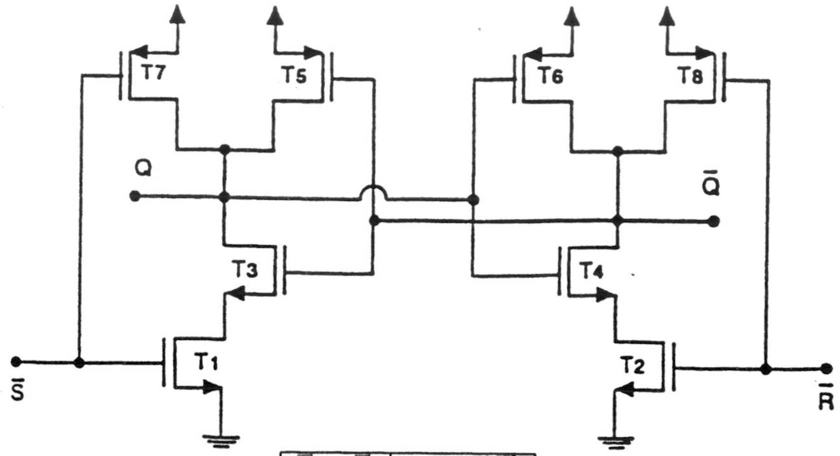

$\overline{S}$	$\overline{R}$	Q	$\overline{Q}$
0	0	3	3
0	3	3	0
3	0	0	3
3	3	0/3	3/0

Note that $|V_t| < 3/2\text{V}$ is necessary in order to ensure that at least one transistor conducts for all input voltages in the range 0 to 3V.

13.51 The flip flop inverters use minimum-size devices, but the load inverters are matched. For the flip flop inverters:

$C = 3C_{db} + 3C_{ov} + 2C_g + 2(1 + 2.5)C_g = 3(2.5 \times 1.2) + 3(0.5 \times 1.2) + 9(1.2 \times 0.8 \times 1.8)$

$= 3(3) + 3(0.6) = 9(1.73) = 9 = 1.8 + 15.6 =$ **26.4 fF**

Note from the solution of P13.49 above, that for the internal inverters, $V_M = 1.41\text{V}$. We will use this rather than $V_{DD}/2$ for the t_p calculations.

Now with Φ high (at 3.3V) the current in Q_7, Q_8 is that created by a double-length device:

$i_{D8}(H) = 1/2(150/2)(3.3 - 0.6)^2 = 273\mu A$.

As Q falls, the competing current in Q_4 is initially zero, increasing to $i_{D4}(M) = 1/2(150/2.5)(3.3 - 0.6)^2 = 219\mu A$ as Q falls. Thus the average discharging current is

$[(273 - 0) + (273 - 219)]/2 = 164\mu A$ and t_{PHL} at Q is $26.4 \times 10^{-15}(3.3 - 1.41)/164 \times 10^{-6} = 304ps$

For the second internal inverter, one can estimate its propagation time roughly on the basis of the maximum available p-channel current, just calculated as $219\mu A$.

Thus t_p to $\overline{Q}$ is $26.4 \times 10^{-15}(1.41)/219 \times 10^{-6} = 170ps$.

Thus the propagation delay from Φ rising with R high is approximately **304 ps** to Q falling and $304 + 170 =$ **474 ps** to $\overline{Q}$ rising.

13.52 For the internal inverters, $V_M = V_{DD}/2$. Correspondingly, $k_n = k_p = 150\mu A/V_2$, and $W_p = 2.5W_n$.

For gating, the maximum current from Q_4 is $i_{D4} = 1/2(150)(3.3 - 0.6)^2 = 547\mu A$. Now for Q_6, whose source may be as high as $V_t = 0.6V$m conducting this with $v_Q = 1.65V$:

$547 = 100 \times (W_6/0.8)[(3.3 - 0.6 - 0.6)(1.65 - 0.6) - (1.65 - 0.6)^2/2]$,

$W_6 = (547/100)0.8/1.65$, whence $W_6 =$ **2.65μm.**

Now, for the bit-line driver, whose width is W_D, as a worst case operating at $v_{DS} \leq 0.6$V with

$547 = 100(W_D/0.8)[(3.3 - 0.6)0.6 - 0.6^2/2]$,

whence $W_D = (547/100)(0.8/1.44 =$ **3.04μm.**

Check the current from a transistor representing the series connection of Q_6, Q_D, whose width is 3 μm and length is $2(0.8) = 1.6\mu m$: $i_D = 1/2(100(3.0/1.6)(3.3 - 0.6)^2 = 683\mu A$,

which exceeds 547 due to effectively enlarging Q_6 from 2.65 to $3.04\mu m$ (OK)

13.53 For the internal inverters, using devices of minimum size, $k_n = 2.5k_p = 150\mu A/V^2$, and $|V_t| = 0.6$:

Thus $V_{th} = v$, where $i = 1/2(150)(v - 0.6)^2 = 1/2(150/2.5)(3.3 - v - 0.6)^2$, $\sqrt{2.5}(v - 0.6) = 2.7 - v$, $1.58v - 0.949 = 2.7 - v$, $2.58v = 3.649$, whence $V_{th} = v =$ **1.414V.**

For D high: The threshold of the input pass NMOS at the switching point (1.414V) will be

$V_t = V_{t0} + \gamma(\sqrt{V_{SB} + 2\Phi f} - \sqrt{2\Phi f} = 0.6 + 0.5(\sqrt{1.414 + 0.6} - \sqrt{0.6} = 0.922V$.

Now, $i_{DI}(M) = 1/2(150)(3.3 - 1.414 - 0.922)^{1/2} = 69.7\mu A$, and $i_{DI}(L) = 1/2(50)(3.3 - 0.6)^2 = 547\mu A$, for which $i_{DI}(A_v) = (69.7 + 547)/2 = 308\mu A$, and $t_{PLH} = t_{PLH1} = 50 \times 10^{-15}(1.414)/308 \times 10^{-6} = 230ps$.

For $\overline{Q}$ falling, $i_{DN} \approx 1/2(150)(3.3 - 0.6)^2 = 547\mu A$, and

$t_{PHL2} \approx 50 \times 10^{-15}(3.3 - 1.414)/547 \times 10^{-6} = 172ps$.

For $\overline{Q}$ rising to regeneration, $i_{DD} \approx 1/2(150/2.5)(3.3 - 0.6)^2 = 547.2.5 = 219\mu A$, and

$t_{PLH3} \approx 50 \times 10^{-15} \times 1.414/219 \times 10^{-6} = 323ps$.

Thus for Φ *rising with D* held high until Q rises, $t_{Pu} = 230 + 172 + 323 =$ **725 ps.**

For D low: $i_{Di} = 1/2(150)(3.3 - 0.6)^2 = 547\mu A$, and $t_{PHL1} = 50(3.3 - 1.414)/547 \times 10^{-6} = 172ns$.

Using the earlier inverter calculations, $t_{PD} = 172 + 323 + 172 =$ **667 ps.**

To ensure correct data flow, Φ must be high for the longer of t_{Pu} or t_{Pd}, namely **725 ps.**

For 5% overlap and equal phases, the maximum clock frequency will be $\dfrac{1}{2(1.05)725 \times 10^{-12}} = 657$ MHz.

Since $\overline{\Phi}$ ensures the latched state, it is normally relatively long, but can be as short as one wants or needs. There is generally **no restriction.** Overlap at the falling edge of Φ does not matter in general provided D is stable. Moreover if D changes while Φ is high, there is trouble in any case, since this is a different mode of operation than we consider here. Overlap at the falling edge of $\overline{\Phi}$ is a bit more of a problem, since during overlap a short-circuit develops from Q to D which will load the source and will induce reduced logic levels at Q. It is not generally obvious (without detailed calculations) whether the new state will be established at the rise of Φ or at the fall of $\overline{\Phi}$. It depends · · · !

Aside: Notice that this circuit can be interestingly improved by replacing the feedback NMOS by a PMOS and driving its gate with the same signal as that of the input gate. There are several advantages:

(1) Only a single clock signal is used.

(2) Non-overlapping is not an issue.

(3) The weak upper level at the first inverter input is restored by the PMOS feedback circuit.

Note, however, that the input inverter low level is brought down to zero, but not held there by the PMOS. In practice, it is likely to stay low. Discuss these points and do some appropriate analysis.

13.54 Compared with the situation in P13.53 above, there are several differences:

(1) The upper voltage level at the input to the first inverter will be a regular 3.3V value.

(2) The leakage current in the first inverter will be reduced essentially to zero.

(3) The delay in the input switch will be reduced due to the current-drive contribution of the PMOS.

(4) Φ and $\overline{\Phi}$ will be used in both the input and feedback switches.

(5) Clock overlap will be more complex, but OK at the falling edge of Φ. The situation is trickier at the falling edge of $\overline{\Phi}$.

From P13.53, we know that the minimum-size inverter, for which $(W/L)_n = (W/L)_p$, has $V_M = 1.414V$.

For D high, the PMOS current for output low is $i_{Dp}(L) = 1/2(150/2.5)(3.3 - 0.6)^2 = 219\mu A$.

For output at the middle ($1.414V$), the current is essentially the same,

$i_{Dp}(M) = (150/2.5)[(3.3 - 0.6)(3.3 = 1.414) - (3.3 - 1.414)^2/2] = 199\mu A$.

For output low, the NMOS current is $i_{DN}(L) = 1/2(150)(3.3 - 0.6)^2 = 547\mu A$

For output at the middle (where $V_{tn} \approx 1.0V$), $i_{Dn}(M) = 1/2(150)(3.3 - 1.414 - 1.0)^2 = 59\mu A$.

Thus the average current $i_{Dn}(Av) = (219 + 199 + 547 + 59)/2 = 512\mu A$ and $t_{PLH1} = 50 \times 10^{-15} \times 1.414/522 \times 10^{-6} = 135ps$.

For D low, $i_{Dn}(H) = 1/2(150)(3.3 - 0.6)^2 = 547\mu A$,

$i_{Dp}(H) = 1/2(150/2.5)(3.3 - 0.6)^2 = 547/2.5 = 219\mu A$,

$i_{Dn}(M) = 150[(3.3 - 0.6)1.414 - 1.414^2/2] = 423\mu A$.

Now assuming $V_{tp} \approx 1.1V$ with back bias at $\upsilon = 1.414V$,

$i_{Dp}(M) = 1/2(150/2.5)(1.414 - 1.1)^2 = 3\mu A$.

Thus $i(Av) = (547 + 219 + 423 = 3)/2 = 596\mu A$,

and $t_{PHL1} = 50 \times 10^{-15}(3.3 - 1.414)/596 \times 10^{-6} = 158ps$.

For the inverters:

For t_{PLH}, $i_{Dp}(L) = 1/2(150/2.5)(3.3 - 0.6)^2 = 219\mu A$. Assuming this current is nearly constant, $t_{PLH2} = 50(1.414)/219 = 323$ *ps.*

For t_{PHL}, $i_{Dn}(H) = 1/2(150)(3.3 - 0.6)^2 = 547\mu A$ and

$t_{PHL2} = 50 \times (3.3 - 1.414)/547 = 172ps$.

For loop delay

For *D* high: $t_p = 135 = 172 + 313 =$ **630 ps**

For *D* low: $t_p = 158 + 323 + 172 =$ **653 ps**

[Note that doubling the size of the PMOS in the inverters would reduce these delays by 100 ps or so (Check this yourself, but be careful to correct, everywhere(!), for the new value of V_M which this produces.)]

It is apparent that the minimum phase duration should be about **650 ps** as it affects the input gate. If Φ and $\overline{\Phi}$ are used and implimented (with an inverter) the clock period would be 1.3 ns.

For this situation, the maximum clock frequency would be 1/1.3 = **769 MHz.** If 2 inverters are used to produce Φ_1, $\overline{\Phi}_1$, Φ_2, $\overline{\Phi}_2$, the length of Φ can be much much reduced (if that is useful). However, in most systems the holding time of the flip flop would be as long or longer than the gating time.

See that the added PMOS increase the speed slightly, the reliability a lot, and the power efficiency a lot, the latter by reducing the static leakage current to near zero.

13.55 The circuit of Fig. 13.45 in the Text needs **16** transistors for full complementary operation. For all devices of minimum size with $W = 1.2\mu m$ a total of $16W =$ **19.2μm** is needed. If devices are matched for $k_n = k_p$ (while $k_n' = 2.5k_p'$), the total width needed is 1.2(8 + 8(2.5)) = **33.6μm.** Matching increases the cost by 33.6 − 19.2/19.2 × 100 = **75%.**

Suggestions:

(1) Matching of the D-input gate is important for speed and noise reduction.

(2) Matching of G_1 centers V_M and balances the noise margins.

(3) Making the Φ_1 gate NMOS alone and the Q_2 gate PMOS alone reduces the leakage current problem while maintaining low cost.

(4) There is little need to match the feedback switch.

(5) Matching the slave input switch is of some important for speed and to reduce leakage, although approach (3) would suffice at much lower cost.

(6) Matching the inverters would reduce the latching time.

(7) Matching G_4 and making it large would improve output symmetry and drive capability.

(8) Making the input switch large and matched and the feedback switch very small (and one device) would be useful.

13.56

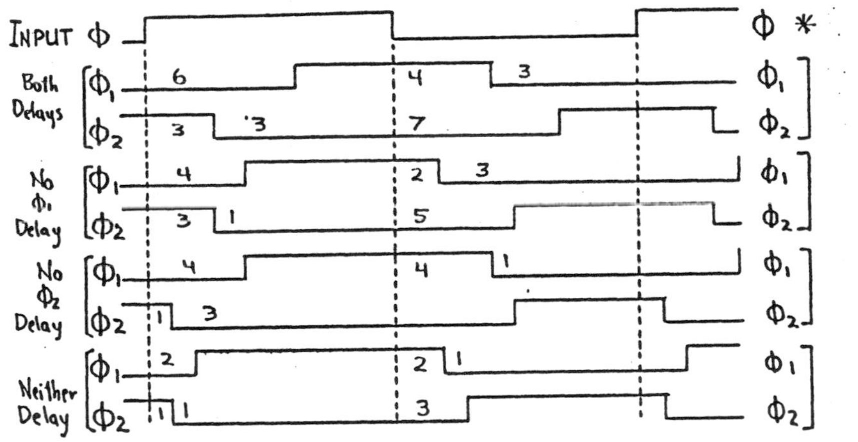

Note from the figure that the non-overlap periods (between Φ_1 and Φ_2) are:

3t, 3t for both Φ_1 and Φ_2 paired-inverter delays installed, **3t, 1t** for no Φ_1 delay, **1t, 3t** for no Φ_2 delay, and **1t, 1t** for no delays installed

To increase the gap between Φ_1 falling and Φ_2 rising, increase the number of inverters in the Φ_2 inverter string (the one on *top*). The gap is originally 3 units. To increase it to 5 units, add 2 inverters; To 7 units, add 4 inverters. Generally speaking, the reason for at least 2 added inverters, is that gate-to-gate variability makes the actual length of the nominal 1t non-overlaps relatively uncertain.

13.57

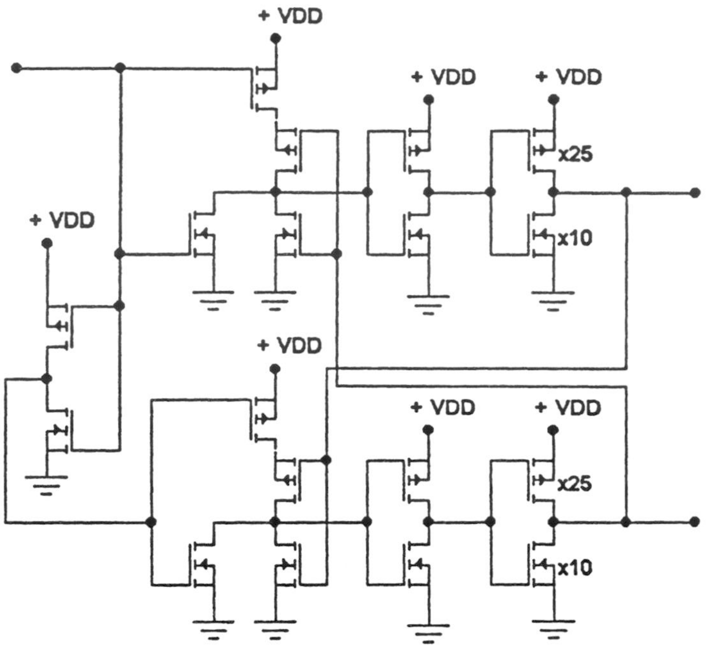

All transistors are minimum-size except those marked specifically (by $\times k$).

Total device width required is $2 + 2[4 + 2 + 25 + 10] =$ 84 units or $84(1.2\mu m) =$ **100.8μm**

SECTION 13.8: MULTIVIBRATOR CIRCUITS

13.58 From the Equation just preceding Exercise 13.15,

$$T = C(R + R_{on})\ln\left[\frac{R}{R + R_{on}} - \frac{V_{DD}}{V_{DD} - V_{th}}\right],$$

whence $200 \times 10^{-9} = 20 \times 10^{-12}(R + 200)\ln\left[\frac{R}{R + 200} \times \frac{5}{5 - 5/2}\right]$, and for R in kΩ,

$$10 = (R + 0.2)\ln\left[\frac{2R}{R + 0.2}\right]$$

Now, for R assumed to be very large ($R >> 0.2k\Omega$), $R = 10/\ln 2 = 10/0.693 = 14.43k\Omega$

Noting that $\ln\left[\frac{2R}{R + 0.2}\right] = \ln\left[\frac{2(14.43)}{14.43 + 0.2}\right] = 0.6794$, we see that $(R + 0.2)$ should be changed to $14.63 \times 0.6794/0.6931 = 14.34k\Omega$ and R to 14.14 kΩ.

[Note that, now, $\ln[2(14.14/14.34)] = 6.791$, unchanged essentially.] Thus, use $R =$ **14.14 kΩ.**

To ensure correct operation, the triggering input pulse must be long enough to allow the positive feedback path through G_1 and G_2 to close before the input goes away. Thus v_i must be positive for at least $2(15) =$ **30 ns.**

If the input pulse is longer than the output pulse, it inhibits the regenerative action which normally occurs at the pulse end, by keeping v_{O1} low when v_{i2} reaches V_{th} of G_2. At this point, the current in R is $2.5/14.1 \times 10^3 = 177.3\mu A$ and the rate of rise of v_{i2} is $177.3 \times 10^{-6}/20 \times 10^{12} = 8.86 \times 10^6 V/s$.

If G_2 has a | gain | of 20 V/V and an output swing of 5 V, its input transition region is $5/20 = 0.25V$ wide. This region is crossed by v_{I1} in $0.25/(8.86 \times 10^6) = 2.82 \times 10^{-8}s$ or $28.2ns$. Thus the fall time of υ_{O2} would be **28.2 ns**, that is, somewhat long. Thus, to sustain an ideal 200 ns output, the input pulse should be no longer than **215 ns**. Transition times for SSI CMOS gates are likely to be about equal to the propagation delay, and about **15 ns** here.

13.59 Here, $T = C\ (R + R_{on})\ \ln\left[\dfrac{R}{R + R_{on}}\ \dfrac{V_{DD}}{V_{DD} - V_{th}}\right]$, or $T = 1500 \times 10^{-12}\ (22 + 0.18) \times 10^3\ \ln$

$\left[\dfrac{22}{22 + 0.18}\ \dfrac{5}{5 - 0.6\,(5)}\right] = 1.5 \times 10^{-6}\ (22.18)\ \ln\left[\dfrac{22}{22.18} \times \dfrac{1}{0.4}\right] = 30.2\mu s$. Now,

$\Delta V_1 = V_{DD}\ \dfrac{R}{R + R_{on}} = 5\ \dfrac{22}{22 + 0.18} = \mathbf{4.96V}$, $\Delta V_2 = V_{DD} + V_{D1} - V_{th} = 5 + 0.7 - 0.6\ (5) = \mathbf{2.7V}$.
During the interval T, υ_{i2} changes by $V_{th} - (V_{DD} - \Delta V_1) = 0.6\ (5) - (5 - 4.96) = 2.96V$. The current changes by $\dfrac{2.96V}{22k\Omega}$, and υ_{O1} changes by $0.18 \times \dfrac{2.96}{22} = \mathbf{0.024V}$. The peak sink current for G_1, occurring as υ_{O1} falls at the beginning of the timing interval, is $i_1 = \dfrac{\Delta V_1}{R} = \dfrac{4.96}{22k\Omega} = \mathbf{0.225mA}$. The peak source current occurs at the end of the timing interval, where D_1 conducts, and υ_{O1} (at the interior end of R_{on}) changes from 0 to +5V, while the gate of G_2 changes from $V_{th} = 3V$ to $V_{DD} + V_{D1} = 5.7V$, for which, $i_2 = \dfrac{5 - (5.7 - 3)}{0.18k\Omega} = \mathbf{12.8mA}$.

13.60

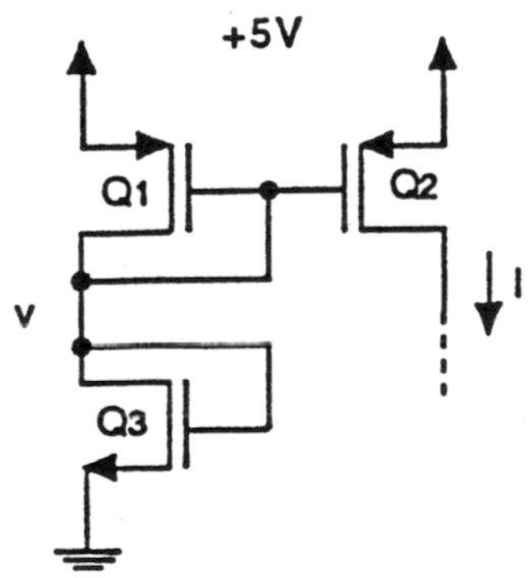

In the mirror, for Q_1, Q_2, $K = 1/2k = 1/2 \times 10 \times 2 = 10\mu A/V^2$, and for Q_3, $K = 1/2 \times 20 \times 2 \times 1/10 = 1\mu A/V^2$. For node voltage υ, $10\ (5 - \upsilon - 1)^2 = 1\ (\upsilon - 1)^2$, $\upsilon - 1 = \sqrt{10}\ (4 - \upsilon)$, $\upsilon - 1 = -3.16\ \upsilon + 12.65$, $7.16\ \upsilon = 13.65$, $\upsilon = 1.906V$, and $I = 1\ (1.906 - 1)^2 = 0.822\mu A$. For G_1 loaded by I, at the beginning of the timing interval, the output voltage is υ, where: $I = 0.822 = 1/2 \times 20 \times 2\ [2\ (5 - 1)\ \upsilon - \upsilon^2] = 20\ [8\ \upsilon - \upsilon^2]$, and $\upsilon^2 - 8\ \upsilon + .0411 = 0$, and $\upsilon = \dfrac{--8 \pm \sqrt{8^2 - 4\ (.0411)}}{2}$, whence $\upsilon_{O1} = 5mV$. *For the Inverter Threshold*: at $V_{th} = \upsilon$, $\upsilon_O = \upsilon_I = \upsilon$, with $K_r = K_n/K_p = 1/2$, $I = K_p\ (5 - \upsilon - 1)^2 = K_n\ (\upsilon - 1)^2$, $(4 - \upsilon) = \sqrt{2}\ (\upsilon - 1)$, $4 - \upsilon = 1.414\upsilon - 1.414$, or $2.414\upsilon = 5.414$, whence $V_{th} = \upsilon =$ **2.24V**.

Now, the time for υ_{I2} to rise to V_{th} from υ_{O1} is $\dfrac{(2.24 - 0.005) \times C}{0.822 \times 10^{-6}} = 10 \times 10^{-6}$ s, and $C = \dfrac{8.22 \times 10^{-12}}{0.24} = \mathbf{3.67pF}$.

13.61 The maximum current supplied by Q_6 is $i_{D6} = 1/2(100/2.5)(1.2/0.8)[3.3 - 0.6]^2 = 218.7\mu A$.

The maximum current in Q_5 is $L5/L6 = 1/10$ of this. Correspondingly, $i_{D5}(L) = 21.87\mu A$.

At $\upsilon_{I2} = V_{DD}/2 = 1.65V$, $i_{D5}(M) = (100/2.5)(1.2/8.0)[(3.3 - 0.6)1.65 - 1.65^2/2] = 18.5\mu A$.

Note that $V_{DD}/2$ is the threshold of the Q_3, Q_4 inverter since the devices are matched. The low output of G_1, namely $\upsilon_{DS1} = \upsilon$ is such that

$218.7 \times 10^{-6} = 100 \times 100^{-6}(12/0.8)[(3.3 - 0.6)\upsilon - \upsilon^2/2]0.2916 = 5.4\upsilon - \upsilon^2$ or

$\upsilon^2 - 5.4\upsilon + 0.2916 = 0$, and $\upsilon = [--5.4 \pm (5.4^2 - 4(1)0.2916)^{1/2}]/2 = (5.4 \pm 5.291)/2 = 54.5mV$, or 60 mV is i_{D5} if included).

Now for the circuit at rest, with input low, $\upsilon_{D6} = 3.3V$, $\upsilon_{D5} = 3.3V$, $\upsilon_{O2} = 0V$. For υ_I rising to 3.3V, υ_{D6} falls to 60 mV (or so), and υ_{D5} falls to the same value.

Now the average current flow into C is $(21.9 + 18.5)/2 = 20.2\mu A$, and the time for v_{D4} to reach $V_{th} = V_{DD}/2 = 1.65$ is $T = 10 \times 10^{-12}(1.65 - 0.060)/20.2 \times 10^{-6} = 0.795\mu s = 795ns$.

Thus the negative output pulse is about $T =$ **795 ns** long.

For capacitances: At the internal node:

$C_i = 2(C_{dbn}) + 2C_{dbp} + 2C_{ovn} + 2C_{ovp} + C_{g4} + C_{g5} = 2(12 \times 2.5) + 2(1.2 \times 2.5) + 2(12 \times 0.5) + 2(1.2 \times 0.5) + (1 + 2.5)(10 \times 1.2 \times 0.8 \times 1.8) = 60 + 6 + 12 + 1.2 + 60.5$, or $C_i = 140fF$.

At the output:

$C_o = C_{dbn} + C_{dbp} + C_{ovn} + C_{ovp} + C_{gn} = 12 \times 2.5 + 2.5(12) \times 2.5 + 12 \times 0.5 + 2.5(12) \times 0.5 + 12 \times 0.8 \times 1.8 = 30 + 75 + 6 + 15 + 17.3$ or $C_o = 143fF$.

For transition and propagation time: At the output of G_1:

$i_{D6}(H) \approx 1/2(150 \times 10)(3.3 - 0.6)^2 = 5.467mA$; $t_{THL1} = 140 \times 10^{-15} \times 3.3/5.467 \times 10^{-3} =$ **84.5 ps;**

t_{PHL1} is half this, namely **42.2 ps.** [For more on t_{TLH} and t_{PLH}, see the later discussion on timing-capacitor recovery.]

At the output of G_2: $i_{D3}(M) \approx 150 \times 10[(3.3 - 0.6)(1.65 - 1.65^2/2)] = 4.63mA$.

Thus $t_{PHL2} = t_{PLH2} \approx 143 \times 10^{-15} \times 1.65/4.63 \times 10^{-3} =$ **51.0 ps,** and t_T may be about 3x larger, at **150 ps.** {Aside: Why is the factor 3 reasonable? [Hint: Consider the current as constant until $V_{DD}/2$, then reducing to zero as the transition is completed.]}

The minimum-length trigger pulse is $t_{min} = t_{PHL1} + t_{PLH2} = 42.2 + 51.0 =$ **92 ps.**

The maximum length of the trigger pulse that allows regeneration is $t_{max} = T + t_{PHL1} \approx$ **795 ns.**

Between pulse inputs, the power-supply current is **0 µA.** Immediately after triggering, the current is $(218.7 + 21.87)\mu a =$ **241µA.**

During the period between inputs, the voltage across the timing capacitor C is **0V.**

For the Recovery Time: At the pulse end, v_{D6} rises from near 0V to 3.3V, and v_{D5} rises from 3.3/2 to nearly $3.3/2 + 3.3 = 1.5(3.3) = 4.95V$. Meanwhile, Q_5, operating with drain and source interchanged, conducts a drain current of i_{D5}, nearly $i_{D5} = (150/2.5)/10[(4.95 - 0.6)(4.95 - 3.3) - (4.95 - 3.3)^2/2]$, or $i_{D5} = 17.9\mu A$.

This current causes a small drop in v_{D6} below 3.3V by an amount

$v = 17.9 \times 10^{-6}/(150/2.5(3.3 - 0.6)) = 0.1V$, or so.

During recovery, the average recovery current is about 17.9/2 or $9\mu A$. Recovery takes about $10 \times 10^{-12} \times 1.65/9 \times 10^{-6}$, or about **1.83µs, quite long!**

Very fast recovery can be arranged using a circuit addition which clamps v_{D5} to $+3.3V$ when both v_{O2} and v_I are low. Such a circuit, which can produce any desired recovery time, is shown.

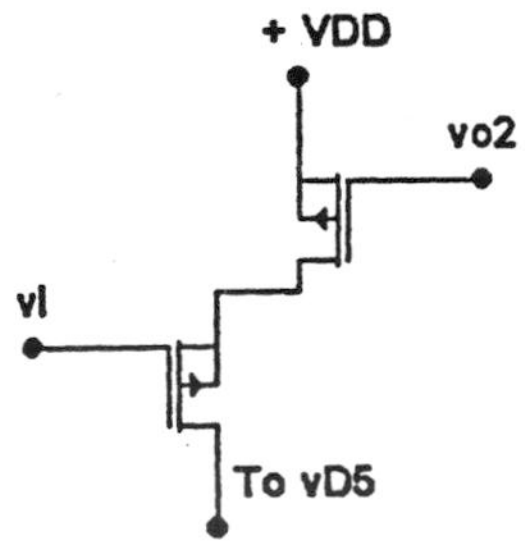

13.62 From P13.77 in the Text, $T = CR \ln \left[\frac{2V_{DD} - V_{th}}{V_{DD} - V_{th}} \times \frac{V_{DD} + V_{th}}{V_{th}} \right]$, for an added gate resistor of value 10 R or more. Thus, $10^{-6} = 100 \times 10^{-12} R \ln \left[\frac{2(5) - 0.44(5)}{5 - 0.44(5)} \times \frac{5 + 0.44(5)}{0.44(5)} \right]$, $R = 10^4 \ln \left[\frac{10 - 2.2}{5 - 2.2} \times \frac{5 + 2.2}{2.2} \right] = 10^4 \ln \left[\frac{7.8}{2.8} \times \frac{7.2}{2.2} \right]$ = **22kΩ**, with the gate resistor used being **220kΩ**, (or more).

13.63 In the circuit, Q_5 tends to keep Q_1, Q_2 in the active region (near V_{th}): For example, if $\upsilon_I > V_{th}$, υ_O is lower, and Q_5 will conduct to tend to equalize them. As well, if $\upsilon_I < V_{th}$, υ_O is higher. In no case can both υ_I and υ_O be low enough to turn off Q_5. Now if Q_1, Q_2 is active, so also is Q_3, Q_4. Two inverters in a loop makes a bistable. A bistable with a time-dependent loop is an astable.

Now, when υ_{O2} goes high, υ_I goes high, and υ_{O1} goes low to establish one astable state. While υ_{O2} is high, Q_5 operates (as a follower) to charge C and pull υ_{I1} low. When υ_{I1} reaches V_{th}, υ_{OO1} rises to V_{th} of Q_3, Q_4 and υ_{O2} falls, regenerating with Q_1, Q_2. With υ_{O1} high, Q_5 conducts current to C causing υ_{I1} to rise, eventually to reach V_{th} for a new cycle to begin.

The voltage change on υ_{01} and υ_{02} is 0 to 3.3V and on υ_{I1}, from $V_{th} = 1.65V$ to $V_{th} + V_{DD} = 4.95V$ to $V_{th} - V_{DD} = -1.65V$.

For υ_{O1} high, $\upsilon_{SG5} = 3.3V$, υ_{I1} varies from $-1.65V$ to $1.65V$, and υ_{SD} varies from 4.95V to 1.65V for which $i_{D5} \approx (150/2.5)/100(3.3 - 0.6)^2 = 4.37\mu A$, reducing to

$(150/2.5)/100[(3.3 - 0.6)1.65 - 1.65^2/2] = 1.85\mu A$, for an average value of $(4.37 + 1.85)/2 = 3.1\mu A$.

For υ_{O1} low, drain and source roles interchange, and υ_{I1} varies from 4.95 to 1.65, υ_{SG1} varies from 4.95V to 1.65V, and υ_{SD1} varies from 4.95V to 1.65V, for which i_{D5} varies from $i_{D5} \approx (150/2.5)/100(4.95 - 0.6)^2 = 11.35\mu A$ to $(150/2.5)/100(1.65 - 0.6)^2 = 0.66\mu A$ for an average value of $(0.66 + 11.35)/2 = 6.0\mu A$.

Thus the waveform is asymmetric, with υ_{O2} low for about twice as long as it is high; that is

$t_{high} = 10 \times 10^{-12}(4.95 - 1.65)/6.0 \times 10^{-6} =$ **5.5μs**

$t_{low} = 10 \times 10^{-12}(4.95 - 1.65)/3.1 \times 10^{-6} =$ **10.6μs**

To achieve a 50% duty cycle, (that is, equal high and low times), there are several approaches:

(1) To maintain symmetry, add an NMOS Q_6 with gate connected to 3.3V, and drain and source connected to υ_{O1} and υ_{I1} (just like Q_5). Unfortunately its length should be 250× regular to achieve symmetry in this case as specified.

(2) Another approach is to vary V_{th} of G_1, by one of several means: For example, by increasing the size of Q_1 to lower V_{th}. One might also add a fixed bias current to node υ_{O1} to change both V_{th} and the υ_{O1} output levels.

(3) Yet another approach is to use 2 capacitors with switches to allow the two parts of the cycle to be adjusted separately. One approach is shown. Unfortunately, there is some interaction between the two parts of the circuit (Explore this!). Q_p and Q_n can be minimum size, but matched.

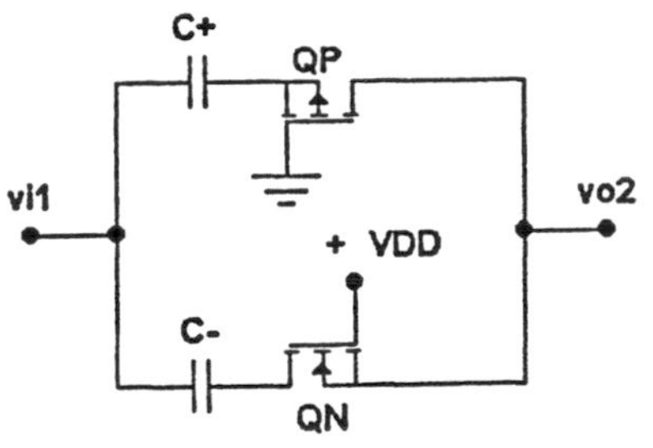

13.64 For each matched inverter, $V_{th} = V_{DD}/2 = 1.65V$ and $i_D = 1/2(100)(1.2/0.8)(3.3 - 0.6)^2 = 547\mu A$. Also $C_{eq} = C_{dbn} + C_{dbp} + C_{ovn} + C_{ovp} + C_{gn} + C_{gp}$. Now, since $W_p = 2.5W_n$,
$C_{eq} = (1 + 2.5)(1.2 \times 2.5) + (1 \times 2.5)(1.2 \times 0.5) + (1 + 2.5)(1.2 \times 0.8 \times 1.8)$
$= 3.5 \times [3.0 + 0.6 + 1.73] = 3.5(5.33) =$ **18.7 fF**, whence $t_P \approx 18.7 \times 10^{-15}(1.65)/547 \times 10^{-6} =$ **56.4 ps.**

The oscillation frequency is $f = 1/10t_p = 1/(10(56.4) \times 10^{-12}) =$ **1.77 GHz.** For changes:

(a) For an extra inverter load on each stage C_{eq} increases by $3.5(1.73) = 6.06fF$, or $6.06/18.7 \times 100 = 32.4\%$. Thus the new frequency will lower from $f_1 = 1.77GHz$ to $f_2 = 1.77/(1.324) =$ **1.34 GHz.**

(b) If, separately, the supply is reduced to 2.0V, i_D becomes $i_D = 1/2(150)(2 - 0.6)^2 = 105\mu A$, a reduction to $105/547 = 0.192$ of the previous. Thus $f_3 = 1.77(0.192) =$ **340 MHz.**

SECTION 13.9: SEMICONDUCTOR MEMORIES: TYPES AND ARCHITECTURES

13.65

	Address Bits				Structure						
#	Block	Row	Col	Total	Blocks	Rows	Columns	Words	Bits/Word	Bits/Block	Total Bits
a	5	10	7	22	32	1024	128	4M	1	128K	4M
b	0	8	8	16	1	256	256	64K	16	1M	1M
c	4	11	10	25	16	2048	1024	32M	8	16M	256M
d	4	10	10	24	16	1024	1024	16M	16	16M	256M
e	3	12	11	26	8	4096	2048	64M	1	8M	64M
f	3	11	10	24	8	2048	1024	16M	4	8M	64M

13.66 Each decoder handles 1M/4 = 256K bits in an array of size $\sqrt{256K}$, or 512 by 512. Now $512 = 2^9$. Thus each decoder uses **9 bits**, such that the required address is 2 + 9 + 9 = 20 bits where $2^{20} = (1K)^2 =$ 1M. For the address being two array-bits, nine row-bits, and nine column-bits, the address 102476 is in the zeroth quadrant. Working from the top (bit 19), the bits are found, by successive trial subtractions, as follows: Thus $102476 - 2^{16} = 102476 - 65536 = 36940$, $36940 - 2^{15} = 36940 - 32768 = 4172$, $4172 - 2^{12} = 4172 - 4096 = 76$, $76 - 2^6 = 76 - 64 = 12$, $12 - 2^3 = 12 - 8 = 4$, $4 - 2^2 = 0$.

Thus the total address is: 0 0 0 1 1 0 0 1 0 0 0 0 0 1 0 0 1 1 0 0, where 4 + 8 + 64 + 4096 + 32768 + 65536 = 102476, OK. Thus, the corresponding column address (the rightmost 9 bits) is: **0 0 1 0 0 1 1 0 0** $\equiv 76_{10}$.

SECTION 13.10: RANDOM-ACCESS MEMORY (RAM) CELLS

13.67 Total area of all the devices in the cell is

$(1.2 \times 0.8)[2(1) + 2(2.5) + 2(3)] = (0.96)[2 + 5 + 6] = 13(0.96) =$ **12.48µm²**. Expected area of the wired cell $= 2(12.48) = 24.96\mu m^2$. The side length of a square cell $= \sqrt{24.96} =$ **5.0µm.**

For the connection to the 0V side of the cell: The available current is:

$i_D = 3(150)[(3.3 - 0.6)(1.65 - 1.65^2/2) =$ **1.389 mA** directed to pulling down the bit line.

For the connection to the 3.3V side of the cell:

$V_t = V_{t0} + \gamma(\sqrt{V_{SB} + 2\Phi f} - \sqrt{2\Phi f} = 0.6 + 0.5((1.65 + 0.6)^{½} - (0.6)^{½}) = 0.96V$.

and $i_D = 3(150)(1.65 - 0.96 - 0.6)^2 =$ **3.6 µA**, directed to pulling up the bit line.

Bit-line capacitance for the $5\mu m \times 5\mu m$ cell is $C = C_{ox} \times Area \times n = 1.8 \times 10^{-15} \times 1 \times 5 \times 128 =$ **1.152 pF.**

Clearly, the majority of the action will be to pull down one of the bit-lines, the other going up only infinitesimally. However, including its effect, the differential current is $1.389 + 0.004 = 1.393mA$.

For a 0.2V change, $t = 1.152 \times 10^{-12} \times 0.2/1.393 \times 10^{-3} =$ **165 ps.**

13.68 For the standard minimum-size matched inverter, the current required to hold its output at $V_{DD}/2$ in either direction is $i = 150[(3.3 - 0.6)1.65 - 1.65^2/2] = 463\mu A$. Current from the gate NMOS whose source is grounded is $i_G = 3(150)[(3.3 - 0.6)1.65 - 1.65^2/2] = 3(0.463) = 1.388mA$.

Here the net current for capacitor drive is $i_C = (3 - 1)0.463 = 2(0.463) =$ **0.926 mA** pulled from the cell.

Now, for the connection to the 3.3V input side, and the cell voltage pulled to 1.65V: V_t for the gate transistor is $V_t = V_{t0} + \gamma((V_{SB} + 0.6)^{½} - (0.6)^{½}) = 0.6 + 0.5(1.65 + 0.6)^{½} - 0.6^{½} = 0.963V$, for which $i_D = 3(150)[3.3 - 1.65 - 0.96 - 0.6]^2 =$ **3.64 μA**! Very low!

Thus lifting the low side up is not possible! However switching is still possible, with a one-sided drive forcing the high side of the cell (ie Q) toward 0V.

For the cell output capacitance:

$C_{eq} = C_{dbn} + C_{dbp} + C_{dbg} + C_{ovn} + C_{ovp} + C_{ovg} + C_{gp} + C_{gn} = (1 + 2.5 + 3) \times 1.2 \times 2.5 + (1 + 2.5 + 3) \times 1.2 \times 0.5 + (1 + 2.5)(1.2 \times 0.8 \times 1.8)$, or

$C_{eq} = 6.5 \times 3.0 + 6.5(0.6) + 3.5(1.73) = 6.5(3.6) + 3.5(1.73) = 29.5fF$.

Returning to the current drive, the excess drive current at $V_{DD}/2$ is 0.926 mA. But at V_{DD}, the cell output is 0 mA, while the gate drive is $i_G = 3(150)(3.3 - 0.6)^2 = 3.28mA$. Thus the average switching drive is $i_D(Av)(3.28 + 0.926)/2 =$ **2.10 mA.**

Time for regeneration to begin is $t_{reg} = 29.5 \times 10^{-15} \times 1.65/3.28 \times 10^{-3} =$ **14.8 ps**

13.69 For each inverter, $V_{th} = \upsilon$ is the voltage at which $\upsilon_O = \upsilon_I = \upsilon$. Now $K_n = 1/2k_n' = 1/2 \times 25\ (2/3) = 8.33\ \mu A/V^2$, $K_p = 1/2 \times 10 \times 2/3 = 3.33\mu A/V^2$. For both devices in pinchoff, $8.33\ (\upsilon - 1)^2 = 3.3\dot{3}\ (5 - \upsilon - 1)^2$, $\upsilon - 1 = (4 - \upsilon)\ (0.632)$, $\upsilon - 1 = 2.53 - 0.632\ \upsilon$, $1.632\ \upsilon = 3.53$, and $\upsilon =$ **2.16V**. For input current i and $\upsilon_O = 2.16/2 = 1.08V$, with the n-channel device in triode mode with $\upsilon_{GS} = 5V$, $i = 8.33\ (2\ (5 - 1)\ (1.08) - 1.08^2) = 8.3\dot{3}\ (8\ (1.08 - 1.08^2))$. Thus the current to the cell = **62.3 μ A**. For the p-channel device, $i = 3.3\dot{3}\ [2(5-1)\left(\frac{5 - 2.16}{2}\right) - \left(\frac{5 - 2.16}{2}\right)^2] = 3.3\dot{3}\ (8(1.42) - 1.42^2) = 31.1\mu A$. Thus, the current from the cell = **31.1 μA**. Now, for reading, the digit line is assumed held at $V_{DD}/2 = 2.5V$ by its large capacitance, with a noise margin of half the threshold. Assume Q_5 and Q_6 are n-channel FETs and that the p-channel output connection is the most sensitive. Thus, arrange the size of Q_5 (or Q_6) so that the current is 31.1μA (or less), with $\upsilon_{S5} = 2.5V$, $\upsilon_{D5} = 3.58V$, and $\upsilon_{G5} = 5V$. Thus, $31.1 = K\ (2\ (5 - 2.5 - 1)\ (3.58 - 2.5) - (3.58 - 2.5)^2) = K\ (2\ (1.5)\ (1.08) - 1.08^2) = 2.074K$, $K = 31.1/2.074 = 15.00\mu A/V^2$. Since $K = 1/2\ (25) \times W/3 = 15$, $W = W_5 = \frac{3 \times 15 \times 2}{25} =$ **3.6 μ m**. For writing, the worst case is likely to be raising the output against the n-channel device. Here $\upsilon_{G5} = +5V$, $\upsilon_{D5} = +5V$ and $\upsilon_{S5} \geq 2.16V$, while $\upsilon_{G1} = +5V$, $\upsilon_{D1} \geq 2.16$, $\upsilon_{S1} = 0V$. Correspondingly, $i_5 = (1/2 \times 25 \times \frac{3.6}{3})\ (5 - 2.16 - 1)^2 =$ **27.6 μ A**, $i_1 = (1/2 \times 25 \times \frac{2}{3})\ [2\ (5 - 1)\ (2.16) - 2.16^2] =$ **105.1 μ A**. Thus writing cannot occur by Q_5 overpowering Q_1. *However*, the other digit line at 0V may succeed. For this case, Q_4 and Q_6 compete, where $i_6 \approx (1/2 \times 25 \times 3.6/3)\ (5 - 1)^2 =$ **240 μ A**, while $i_4 \approx (3.3\dot{3})\ (5 - 1)^2 =$ **53.35 μ A**. Thus i_6 clearly exceeds i_4, and the design is viable. Therefore writing will occur, with a major role for one of inputs in pulling down, and a secondary role for the other input in pulling up.

13.70 *For the bit-line capacitance:*

Bit-line length $= 256 \times 2.5\mu m = 640\mu m = 0.64mm$.

Bit -line capacitance $= 640 \times 1 \times 1.8 = 1.152pF$.

Total bit-line capacitance $= 1.152 + 0.070 =$ **1.222 pF**

For the stored signals:

For Q with the gate at 3.3V, the cell voltage is $3.3 - V_t$ where $V_t = V_{t0} = \gamma((V_{SB} + 0.6)^{1/2} - (0.6)^{1/2})$ Try $V_t = 1.1V$, whence $V_{SB} = 2.2V$. $V_t = 0.6 + 0.5((2.2 + 0.6)^{1/2} - 0.6^{1/2}) = 1.049$

For $V_t = 1.05$, $V_{SB} = 2.25$, $V_t = 0.6 + 0.5((2.25 + 0.6)^{1/2} - 0.6^{1/2}) = 1.06$.

Use $V_t = 1.06V$ in which case the cell voltages are **0V** and $3.3 - 1.06 =$ **2.24V**.

Thus the total stored voltage is $2.24 - 0 = 2.24V$ of which 20% is 0.448 V. Thus the stored signals could be a high as $0.45V$ or as low as $2.25 - 0.45 = 1.79V$.

Now for the bit lines charged nominally to $(V_{DD}/2 - V_t) = (3.3/2 - 0.6) = 1.05V$,

the positive signal (from Eq. 13.58) is $\frac{40}{40 + 1222}(1.79 - 1.05) =$ **23.5 mV**,

and the negative signal is $\frac{40}{40 + 1222}(1.05 - 0.45) =$ **19.0 mV**.

13.71 Where the body effect raises V_t to about 1V, the initial stored cell voltage is $3.3 - 1.0 = 2.3V$. Thus a 20% change is $0.2(2.3) = 0.46V$. For this to occur in 10 ms across a 40 fF capacitor, the leakage current must be $I_L = 40 \times 10^{-15} \times 0.46/10 \times 10^{-3} = 1.84pA$.

For an average voltage of $2.3 - 0.46/2 = 2.07V$, the equivalent leakage resistance must be

$R_L = \frac{2.07}{1.84 \times 10^{-12}} = 1.125 \times 10^{12} =$ **1.12Ω**

13.72 Here, $Q = CV = IT$. Thus, $C = \frac{10 \times 10^{-15} \times 4 \times 10^{-3}}{1.5} = 26.7 \times 10^{-18} \equiv$ **0.027fF.**

13.73 Refresh is done 1 row (or word) at a time, in each block. Thus a total refresh takes 1024 cycles. In 10 ms, there are $10 \times 10^{-3}/30 \times 10^{-9} = 3.33 \times 10^5$ read/write cycles. Thus of the available cycles, $1024/3.33 \times 10^5 = 3.07 \times 10^{-3}$, or 1 in 325, or 0.3% are spent on refresh. Such a design needs 1024 sense amps in each block, or $16 \times 1024 =$ **16.384** altogether. If only 1024 amplifiers must be used, the overhead increases to 16×0.307 or **4.9%**, corresponding to 1 in 20.4 cycles!

SECTION 13.11: SENSE AMPLIFIERS AND ADDRESS DECODERS

13.74 For the body effect on Q_7, $V_t = V_{t0} + \gamma((V_{SB} + 0.6)^{1/2} - 0.6^{1/2}) = 0.6 + 0.5((1.65 - 0.6)^{1/2} - 0.6^{1/2}) = 0.96V$

Assume the difference is quite small, and that Q_7 operates in the triode mode with resistance r where

$1/r = 150(3.3 - 1.65 - 0.96) = 103.5\mu A/V$, or $r = 9.66k\Omega$.

With respect to this resistor, the 1 pF line capacitances are in series, corresponding to an equivalent capacitance of 0.5 pF. Thus the time constant is $0.5 \times 10^{-12} \times 9.66 \times 10^3 = 4.83ns$.

The voltage across r is $\upsilon = \Delta\upsilon e^{-t/4.83}$. Now $\upsilon/\Delta\upsilon = 1/100$, when $t = -4.83 \ln 1/100 =$ **22.2 ns.**

To reduce this to 1 ns, Q_7 must be 22.2 times wider, that is $W_7 = 22.2 \times 1.2 =$ **26.6μm**

13.75 The regeneration time constant is C/G_m where C is the line capacitance and $G_m = (g_{mp} + g_{mn})$ for each inverter. Here, for a matched minimal inverter, at $V_{tn} = V_{DD}/2 = 1.65V$, $i_{Dn} = 1/2(150)(1.65 - 0.6)^{1/2}$ and $g_m = 2(1/2)(150)(1.65 - 0.6) = 157.5\mu A/V$, with $G_m = 2(157.5) = 315\mu A/V$.

Now for an inverter $n\times$ minimum size needed to achieve a 2 ns time constant, $2 \times 10^{-9} = 2 \times 10^{-12}/n(315 \times 10^{-6})$, and $n = 2 \times 10^{-12}/(315 \times 10^{-6} \times 2 \times 10^{-9})$, or $n =$ **3.17.**

Now, generally $\upsilon = V_{DD}/2 - \Delta\upsilon e^{-t/2}$, and $t = 2\ln\left[\dfrac{\upsilon - V_{DD}/2}{\Delta\upsilon}\right]$.

Time to rise from 0.5 V_{DD} to 0.9 V_{DD} for by 50 mV input is $t = 2\ln\left[\dfrac{(0.9 - 0.5)3.3}{50 \times 10^{-3}}\right] =$

6.55 ns.

Time to fall from 0.5 V_{DD} to 0.1 V_{DD} initiated by 25 mV is $t = 2\ln\left[\dfrac{(0.5 - 0.1)3.3}{25 \times 10^{-3}}\right] =$ **7.93 ns**

13.76 For a so-called 1 Mb memory, the number of cells is $2^{20} =$ 1,048,576. For a square array of 1024 × 1024, there are 1024 rows (words) requiring a decoder with **10 input bits** and **1024 output lines.** Since bits are grouped in 4-bit packs for readout, only 10 – 2 = 8 bits of column addressing are needed. The column decoder would have **8 input bits** and 1024/4 = **256 output lines.**

For a 1024 bit row decoder, needing n bits, $2^n = 1024$, and $\log 1024 = n \log 2$. Thus $n = \log 1024/\log 2 =$ **10** bits.

For a 10-bit 1024 line output decoder as in Fig. 13.63 of the Text, there are 1024 output lines and 10 pairs of input lines with 1 NMOS/pair/row. Thus there are 1024 × 10 = **10,240** decoder NMOS, with **1024** dynamic PMOS and **10** address inverters.

13.77 Now a 256-line decoder rquires $n = \log 256/\log 2 = 8 bits$ of input. Thus a corresponding tree decoder has **8** input layers. Since $2^{10} = 1024$, there are **10** layers in a 1024-line tree decoder. For n bits input, the number of tree transistors is $N = 2 + 4 + 8 + \cdots + 2^n$. Now, see that $N = 1/2(N + 2^{n+1}) - 1$, $2N = N + 2^{n+1}2$, and $N = 2^{n+1} - 2$.

Now for 256 lines, $n = 8$ and $N = 2^9 - 2 = 512 - 2 =$ **510** transistors. For 1024 lines, $n = 10$ and $N = 2^{11} - 2 = 2048 - 2 =$ **2046** transistors.

For a standard minimum-size NMOS at low υ_{DS}, the series resistance is r, where $1/r = 150 \times 10^{-6}(3.3 - 0.6)$ and $r = 1/(150 \times 10^{-6} \times 2.7) =$ **2.469 kΩ**

Thus a resistance of almost 2.5 kΩ is associated with each switch. For n layers, the associated time constant is $n(2.469) \times 10^3 \times 1 \times 10^{-12} = 2.47\ ns$.

For the line to fall from V_{DD} to $V_{DD}/10$ in 7 ns, $0.1 = 1e^{-7/2.47n}$ $-7/2.47n = \ln 0.1 = -2.302$, and $n = 7(2.47 \times 2.3) = 1.23$.

Thus only one layer is possible! Clearly this is quite unsatisfactory! To resolve this problem, two possible solutions exist:

(a) Increase the width of all NMOS by a factor m, allowing n to become 1.23 m.

(b) Buffer the decoder output with a buffer consisting of 2 cascaded inverters.

For situation (b), a single NMOS provides a current of about $i_D = 150(3.30 - 0.6)^2 = 1.09 mA$. This will discharge 1 pF from 3.3 to 0.33 in $t = 1 \times 10^{-12} \times (0.9 \times 3.3)/1.09 \times 10^{-3} = 2.7 ns$.

Thus the buffer could use 3 minimum-size NMOS (2 for the inverter, and one output transistor). For n layers, this buffered solution would increase the number of minimum-size transistors to $3(2^n) + 2^{n+1} - 2 = 5 \times 2^n - 2$ for a total device width of $(5 \times 2^2 - 2)W_n$ For the direct solution (a) with wide transistors, and the approximation that a minimum-size device can serve only a single layer, we need for n layers, $(2 \times 2^n - 2)$ transistors of width n, for a total width of $(2 \times 2^n - 2)n$.

For example, for 10 layers and $n = 10$: (a) gives $10(2^{22} - 2) = 20.5 \times 10^3 W_n$, while (b) gives $5 \times 2^{10} - 2 =$ $(5.2 \times 10^3)W_n$. Obviously a buffered solution is attactive, economic, and potentially faster!

13.78

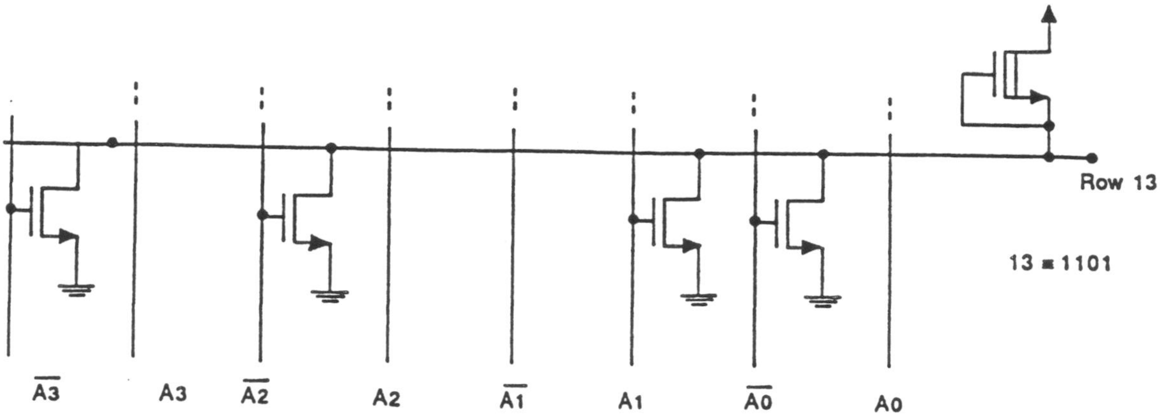

Here, $13_{10} \equiv (1101)_2$. Thus the decoding pattern is as shown. For a 256K-bit square array, there are two 512-line decoders, each with a 9-bit input requiring 9 + 1 = **10** transistors per row.

SECTION 13.12: READ-ONLY MEMORY (ROM)

13.79 Basic NMOS are $1.2\mu m \times 0.8\mu m$. With 30% overhead in each dimension, the cell size may be $1.2 \times 1.3 = 1.56\mu m$ by $0.8 \times 1.3 = 1.04\mu m$. Perhaps a **1.6 μm × 1 μm** cell design is possible. For a chip $1mm^2$ in area, 85% can be used for the array whose area would be $0.85 \times 10^{-6} m^2$. The number of cells that can be accommodated is $0.85 \times 10^{-6}/(1.56 \times 10^{-6} \times 1.04 \times 10^{-6}) = 5.24 \times 10^5$. Thus it seems that the ROM could accommodate 2^{19} cells where 2^{19} = **5.243** $\times 10^5$. For 32-bit words, where $32 = 2^5$, the ROM capacity is $2^{19-5} = 2^{14}$ = 16,384, or **16K** words.

13.80 For the computation of X/Y as $F + Q + R$, where "1" in the 5 output columns represents the location of a ROM transistor:

x_1	x_0	y_1	y_0	f_1	q_1	q_0	r_1	r_0
0	0	0	0	1	0	0	0	0
0	0	0	1	0	0	0	0	0
0	0	1	0	0	0	0	0	0
0	0	1	1	0	0	0	0	1
0	1	0	0	1	0	0	0	0
0	1	0	1	0	0	1	0	0
0	1	1	0	0	0	0	0	1
0	1	1	1	0	0	0	1	1
1	0	0	0	1	0	0	1	0
1	0	0	1	0	1	0	0	0
1	0	1	0	0	0	1	0	0
1	0	1	1	0	0	0	1	0
1	1	0	0	1	0	0	1	1
1	1	0	1	0	1	1	0	0
1	1	1	0	0	0	1	0	1
1	1	1	1	0	0	1	0	0

Now, count the ones in the output columns of the table (the rightmost 5 columns). See that for the 5 column outputs, there are 20 transistors in the array itself, plus 5 loads, plus 5 inverters of 2 transistors each. Thus the design needs 20 + 5 = **25** transistors in the array, and 5(2) = **10** for the inverters for a total of 25 + 10 = **35**. Without the inverters (and using transistors to represent logic zero), the design would use 5 (16) −20 +5 = 65 transistors total. For the 4-bit decoder, one needs 4 + 1 transistors per row (including the load), for 16 rows = **80** transistors in total.

13.81

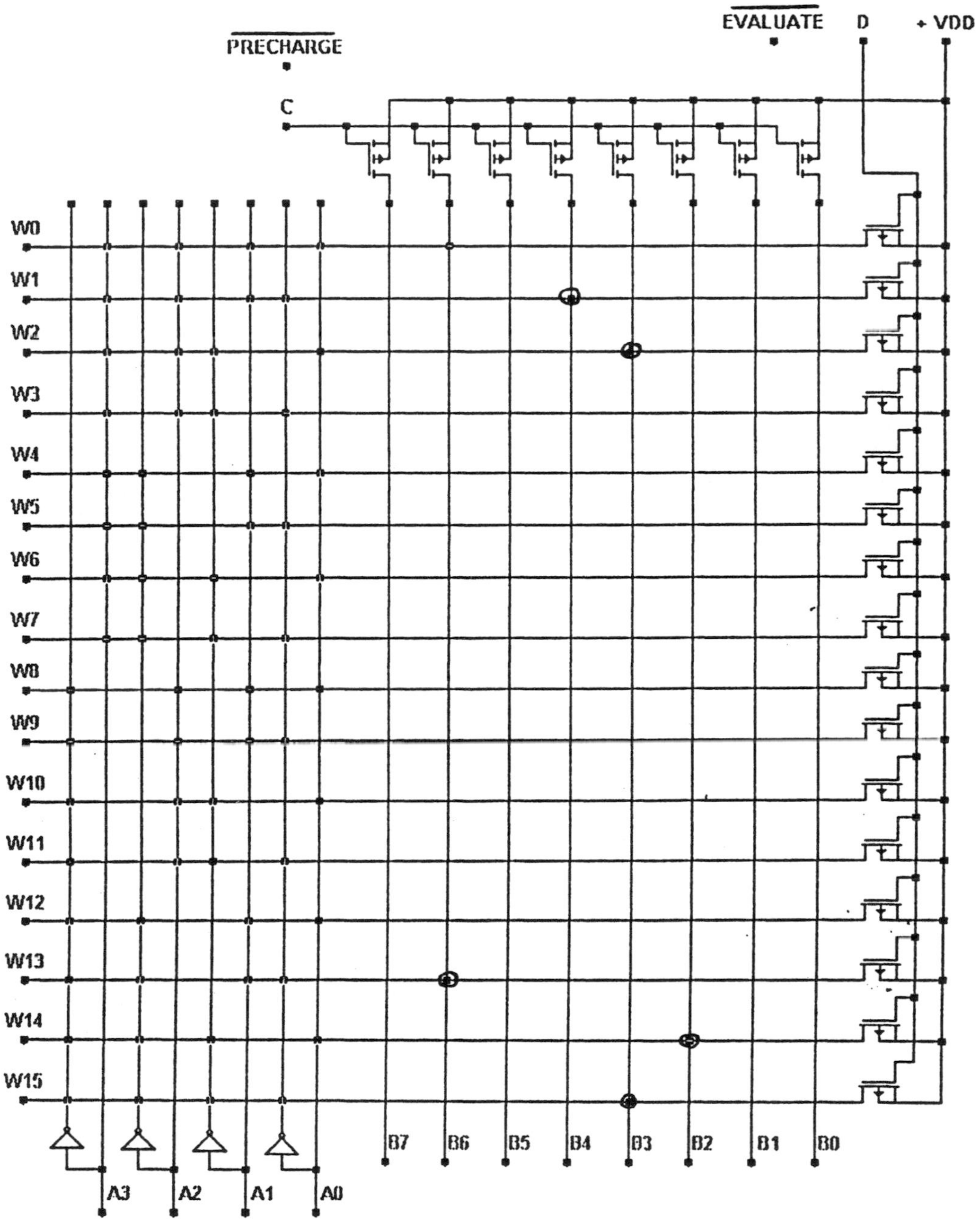
PRECHARGE
EVALUATE
D
+ VDD
C
W0
W1
W2
W3
W4
W5
W6
W7
W8
W9
W10
W11
W12
W13
W14
W15
B7
B6
B5
B4
B3
B2
B1
B0
A3
A2
A1
A0

(a)

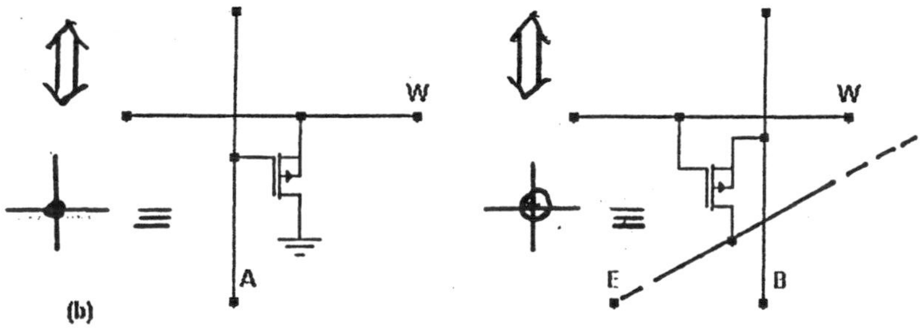
W
A
W
E
B

(b)

Here, E is a global signal input to the ROM array which can be grounded (or earthed), or driven by a single (large) NMOS by signal F for evaluation control. In the latter case, the design is totally dynamic and nonOratioed.

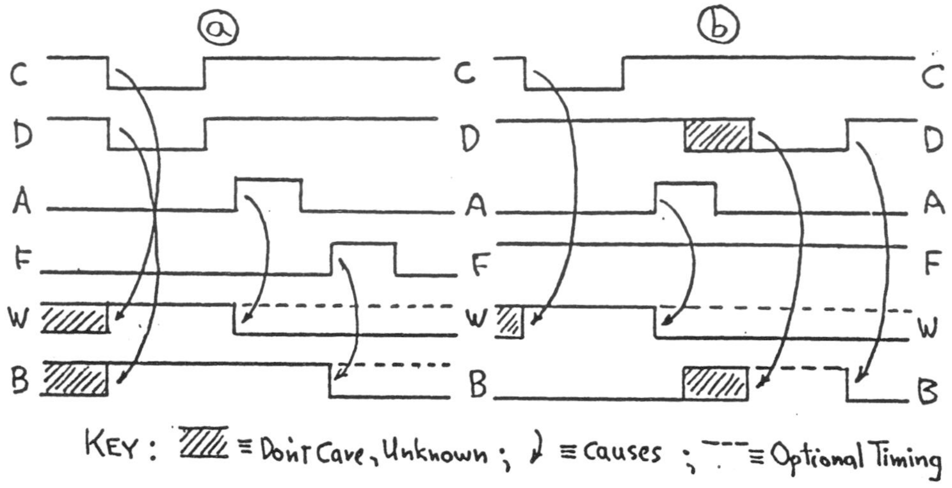

There are at least two possible timing modes, depending on the use of line E (and signal F):

Case a): Precharge is done early by lowering C, D. {They can be connected together.} Evaluation does not occur until F is raised, lowering the common line E and allowing some bits B to fall. This design is unratioed.

Case b): F is always high (or E is grounded). C precharges the address decoder. A allows only one word line to be high. D falls to raise the bit-line B of only the selected word. The design is ratioed.

13.82 To reduce the initial cost of programming the ROM:

(i)

(a) Use a high value for a '1' so that few transistors need be "removed".

(b) Use a low value for a '1' so that few transistors must be used.

(c) Use a low value for '1' so that few fuses need to be blown.

(d) Use a high threshold value for '1' so that little threshold-raising is needed.

But there are other costs, for example in:

(ii) Creating the basic ROM IC which is cheaply programmable.

(iii) Providing the operating power.

For example, in the case of operating power, such power can increase if larger currents flow in one signal convention than in another. For this reason, one might choose for the 4 cases above:

(a) a low '1',

(b) a low '1',

(c) either choice,

(d) a low threshold for '1'.

NOTES

Chapter 14

BIPOLAR DIGITAL CIRCUITS

SECTION 14.1: THE BJT AS A DIGITAL CIRCUIT ELEMENT

14.1 Generally, $\alpha = \beta/(\beta + 1)$. Thus $\alpha_F = 100/101 = \mathbf{0.990}$, $\alpha_R = 0.25/1.25 = \mathbf{0.200}$.

From Eq. 4.100, $i_{DE} = I_{SE}(e^{\upsilon_{BE}/nV_T} - 1)$.

Here, $1/0.99 = I_{SE}e^{700/1(25)}$ or $I_{SE} = e^{-28/0.99} = \mathbf{6.98 \times 10^{-13}A}$.

From Eq. 4.102, $\alpha_F I_{SE} = \alpha_R I_{SC} = I_S$. Thus $I_S = 0.99 \times 6.98 \times 10^{-13} = \mathbf{6.91 \times 10^{-13}A}$,

and $I_{SC} = I_S/\alpha_R = 6.91 \times 10^{-13}/0.2 = \mathbf{3.46 \times 10^{-12}A}$.

For saturation, $I_{Bsat} \approx (4.3 - 0.7)/10k\Omega = 0.36$ mA,

and, assuming $V_{CEsat} \approx 0.2$ V, $I_{Csat} \approx (5.0 - 0.2)/(2k\Omega = \mathbf{2.4\ mA}$.

From Eq. 4.113 simplifed, $i_B \approx I_S/\beta_F e^{\upsilon_{BE}/nV_T}$ or $0.36 \times 10^{-3} = 6.91 \times 10^{-13}/100e^{V_{BE}/25}$, and

$\upsilon_{BE} = 25\ln(0.36 \times 10^{-3} \times 100/6.91 \times 10^{-13}$, or $\upsilon_{BEsat} = \mathbf{616.9\ mV}$.

now, a better version of I_{Bsat} is $4.3 - 0.617)/10k\Omega = \mathbf{0.368\ mA}$ for which $\upsilon_{BEsat} = 25\ln(0.368 \times 10^{-3} \times 100/6.91 \times 10^{-13}) = \mathbf{617.5\ mV}$.

From Eq. 4.114 $V_{CEsat} = V_t \ln \dfrac{1 + (\beta_{forced} + 1)/\beta_R}{1 - \beta_{forced}/\beta_F}$, where $\beta_{foreced} \approx 2.4/0.368 = 6.52$.

$$\text{Thus } V_{CEsat} = 25\left[\frac{1 + (6.52 + 1)/0.25}{1 - 6.52/100}\right] = 25\ln\left[\frac{31.08}{0.935}\right] = 87.6 \text{ mV}.$$

Now $i_{Csat} = (5 - 0.0876)/2k\Omega = \mathbf{2.46\ mA}$, and $\beta_{forced} = 2.46/0.368 = \mathbf{6.68}$,

$$\text{whence } V_{CEsat} = 25\ln\left[\frac{1 + (6.68 + 1)/0.25}{1 - 6.68/100}\right] = 25\ln[31.72/0.933] = \mathbf{88.2\ mV}$$

14.2 *For turnon:*

For the turnon delay: The total capacitance at the base, $C = C_{je} + C_\mu = 0.5 + 0.5 = 1.0$ pF, must charge from 0 to 0.7 V through $R_B = 10$ kΩ, with $\upsilon_I = 4.3$ V and V_B between 0 and 0.7 V, providing an average current $I_{B3} = \dfrac{4.3 - (0.7 + 0)/2}{10k\Omega} = 0.395$ mA.

Thus, using $CV = IT$, $t_d = 1 \times 10^{-12} \times 0.7/(0.395 \times 10^{-3}) = \mathbf{1.77\ ns}$.

{*Aside:* Note that from Exercise 14.1, considering the early part of the exponential:

$t_d = (R_B + r_x)(C_{je} + C_\mu)\ln[(V_2 - V_1)/V_2 - 0.7)] = (10^4 + 0)(0.5 + 0.5)^{-12}10\ln[(4.3 - 0)/(4.3 - 0.7)] =$ **1.777 ns**, which is nearly the same.}

For the turnon rise time: The base current charges the active base capacitances. The equivalent resistance is $R_B \parallel r_\pi$ as r_π falls from ∞ to the value just before saturation. Use the value that applies when $V_C = 2.5$ V, where $I_C = (5 - 2.5)/2 = 1.25$ mA, $I_B = I_C/\beta = 125$ µA, and

$r_\pi = V_T/I_B = 25 \times 10^{-3}/12.5 \times 10^{-6} = \mathbf{2\ k\Omega}$. Thus, at the base,

$R_{eq} = 10k\Omega \parallel 2k\Omega = 10k\Omega/6 = 1.67k\Omega$.

LIkewise, use C_π at 2.5 mA, where $g_m = \beta/r_\pi = 100/2 \times 10^3 = 50$ mA/V.

Now, from Eq. 4.130, $C_{eq} = C_\pi + C_\mu = \dfrac{g_m}{2\pi f_T}$ or $C_\pi + C_\mu = \dfrac{50 \times 10^{-3}}{2\pi(10^9)} = 7.96\ pF$.

Ignoring the Miller effect, the base time constant becomes

$\tau = R_{eq}C_{eq} = R_{eq}(C_\pi + C_\mu) = 1.67 \times 10^3 \times 7.96 \times 10^{-12} =$ *13.3 ns.*

If the Miller effect is included, with an average gain equal that at $v_O =$ 2.5 V and $i_C =$ 1.25 mA, the corresponding gain is $-g_m R_L = -50 \times 10^{-3}(2 \times 10^3) = -$ 100 V/V,

for which $C'_{eq} = C_\pi + C_\mu + 100(C_\mu) = 7.96 + 100(0.5) =$ 57.96 pF, and the base time constant is $\tau' = R_{eq}C'_{eq} = 1.67 \times 10^3(57.96 \times 10^{-12} = 96.8ns$.

Now $i_C = \beta I_{B2}(1 - e^{-t/\tau})$, and t_r is the time it takes from essentially zero to when $i_C =$ 0.9 (I_{sat}) $\simeq 0.9(2.5) =$ 2.25 mA, where $I_{B2} = (4.3 - 0.7)/10 =$ 300 μA.

Thus at 90% of the current rise, $2.25 \times 10^{-3} = 100(360 \times 10^{-6})(1 - e^{-t/\tau})$, or

$-e^{-t/\tau} = (2.25 \times 10^{-3})/(3 \times 10^{-2}) - 1 = -0.325$, $-t_r/\tau = -1.124$, whence $t_r = 1.124\tau$.

Here, $t_r = 1.124(13.3) =$ **14.9 ns**, or with the Miller effect, $t_r = 1.124(96.8) =$ **109 ns**.

Of these, the first applies to the collector current if the load is shorted. For the RTL inverter, as shown, $t_r =$ 109 ns is a better estimate.

Now, overall, $t_{on} = t_d + t_r/2 = 1.77 + 109/2 =$ **56.3 ns**, or for a shorted load, $t_{on} = 1.77 + 14.9/2 =$ **9.2 ns.**

For turnoff: For the turnoff delay: From Eq. 14.3,

$t_s = \tau_s \dfrac{I_{B2} - I_{Csat}/\beta}{I_{B1} + I_{Csat}/\beta}$, where $I_{Csat} = (5 - 0.1)/2 \times 10^3 =$ 2.45 mA.

Here, $I_{B2} = (4.3 - 0.7)/10 \times 10^3 =$ 360 μA, and $I_{B1} = 0.7/10 \times 10^3 =$ 70 μA, $\beta =$ 100, and $\tau_s =$ 1.5 ns.

Thus, $t_s = 1.5 \times 10^{-9}(0.360 - 2.45/100)/(0.07 + 2.45/100) = 1.5 \times 10^{-9}(0.3355)/(0.0945) =$ **5.325 ns.**

For the (turnoff) fall time: (i) Using the Miller-Effect approach introduced earlier for the rise time, and the corresponding average resistance and capacitance data: $R_{eq} = 1.67k\Omega$, and $C'_{eq} = 57.96pF$, for which $\tau' = 96.8ns$, t_f is calculated for the current level falling from I_{Csat} to 0.1 I_{Csat},

where $i_C = I_{Csat}e^{-t/\tau}$, $0.1I_{Csat} = I_{Csat}e^{-t/\tau}$, and $t_f = -\tau\ln 0.1 = -96.8\ln 0.1 = 223ns$, a very long time. *Note,* just as in the case of t_r above, an estimate of t_f for the collector shorted is much smaller, $t_f = 13.3(2.3) =$ **30.6 ns.**

Alternatively: (ii) One might consider using $R_{eq} = R_B =$ 10 kΩ, with a) $C_{eq} = C_{je} + C_\mu =$ 1 pF, for which $\tau =$ 10 ns, and $t_f = -10\ln(0.1) = 23ns$. or b) $C_{eq} = C_\pi + C_\mu =$ 7.96 pF, for which the time constant is $7.96pF \times 10k\Omega =$ 79.6 ns, and $t_f = 79.6(2.3) = 183ns$.

A better approximation could be obtained by using the first approach for (say) the initial half of the current fall, and the basic second approach for the final part. In that event, the current fall time would be $t_f = [-96.8\ln(0.5) - 10\ln(0.2)] = 96.8(0.693) + 10(1.61) = 67.1 + 16.1 = 83.2ns$.

Yet another estimate can be made by considering R_B as a source of current drawn from the base lead, with most current flowing in C_μ, as V_C rises while i_C falls. Time for C_μ to charge to 5 V is $t_f = 0.5 \times 10^{-12} \times 5/[(0.7 - 0)/(10 \times 10^3)] = 35.7ns$.

While no firm conclusion can be reached simply, t_f is likely to be quite large, perhaps in the vicinity of *100 ns*. This situation is very appropriate for SPICE simulation!

Now $t_{off} = t_3 + t_f/2 = 5.3 + t_f/2$. For the load shorted, this becomes $t_{off} = 5.3 + 30.6/2 =$ **20.6 ns.** Otherwise, t_f dominates. For the worst-case Miller-Effect calculation, $t_{off} = 5.3 + 223/2 =$ **117 ns.**

Finally, the base will stay essentially in the conduction region until conduction ceases. From the Miller-Effect viewpoint, this is for most (but not all) of t_f, up to **223 ns.** Certainly for no Miller Effect, the base voltage will hold up until the excess stored charge is removed, namely, for the period $t_s =$ **5.3 ns.**

14.3 Here, $t_s = \tau_s \dfrac{I_{B2} - I_{C\,sat}/\beta}{I_{B1} + I_{C\,sat}/\beta}$. Now, for $I_{B1} = 0$, $t_s = 20 \times 10^{-9}\dfrac{1 - 10/200}{0 + 10/200} = 20\left[\dfrac{1 - .05}{.05}\right] = 20(19) =$

380ns. For $I_{B1} = 1$ mA: $t_s = 20\,\frac{(1-.05)}{1+.05}$ = **18.1ns**; and for $I_{B1} = 10$ mA: $t_s = 20\,\frac{(1-.05)}{10+.05}$ = **1.89ns**.

14.4 Here, for $\upsilon_{BE} = 0.7$ V, $\upsilon_{CEsat} = 0.2$ V, $I_{B2} = \frac{5-0.7}{1k} = 4.3$ mA, $I_{B1} = \frac{0.7-0.2}{1k} = 0.5$ mA, and $I_{Csat} = \frac{5-0.2}{0.5k} = 9.6$ mA. Now, the original stored base charge = $\tau_s\,(4.3 - \frac{9.6}{\beta})$.

Without a Capacitor: charge removed = 80 ns $(0.5 + \frac{9.6}{\beta})$ mA = 80 (0.5 + 9.6/β) pC.

With a Capacitor: charge removed by capacitor = CV = 8 pF × (5 − 0.2) V = 38.4 pC. Subsequent charge removed = 30 (0.5 + 9.6/β). For a charge balance, 80(0.5 + 9.6/β) −38.4 = 30 (0.5 + 9.6/β). Thus 50(0.5 + 9.6/β) = 38.4, 0.5 + 9.6/β = 38.4/50 = 0.768, 9.6/β = 0.268, and β = 9.6/0.268 = **35.8**.

Also $\tau_s(4.3 - 9.6/\beta) = 80(0.5 + 9.6/\beta)$. Thus, $\tau_s = 80\,\frac{0.5+0.268}{4.3-0.268}$ = **15.2 ns**. Now, $t_s = 0$ when the charge provided via C just cancels the internal charge: Thus $CV = \tau_s\,(I_{B2} - \frac{I_{C\,sat}}{\beta})$, or

$$C = \frac{15.2}{5-0.2}\,(4.3 - \frac{9.6}{35.8}) = \frac{15.2}{4.8}\,(4.032) = 12.77\text{pF or } \mathbf{12.8pF}.$$

Now for C = 12.8 pF, and β = 2 (35.8) = 71.6,

$$t_s = \frac{15.2\,(4.3 - \frac{9.6}{71.6}) - 12.8\,(5.0 - 0.2)}{0.5 + \frac{9.6}{71.6}}$$

$$= \frac{15.2\,(4.166) - 12.8\,(4.8)}{.634} = 2.97, \text{ or } \mathbf{3ns}.$$

Now for $C = 0$, $t_s = \dfrac{15.2\,(4.3 - \frac{9.6}{71.6})}{0.5 + \frac{9.6}{71.6}} = \dfrac{15.2\,(4.166)}{.634}$ = **99.9ns!** (compared with 80 ns originally).

Check: $t_s = \dfrac{15.2\,(4.3 - \frac{9.6}{35.8})}{0.5 + \frac{9.6}{35.8}} = 79.8$ ns.

SECTION 14.2 EARLY FORMS OF BJT CIRCUITS

14.5

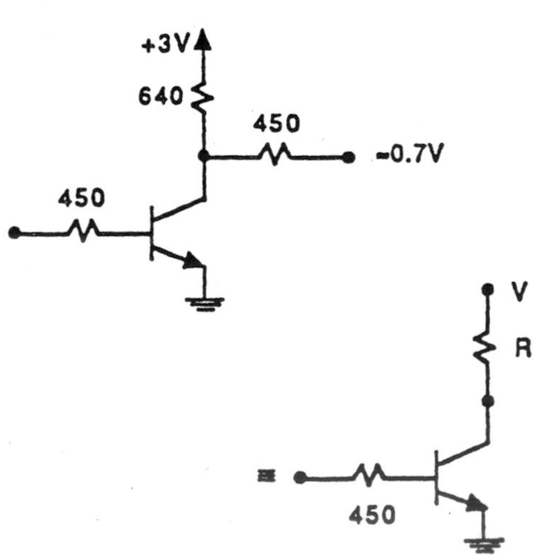

$R = 450 \parallel 640 = 264\ \Omega$, $V = 0.7 + \frac{450}{1090}\,(3-0.7) = 1.65$ V.

For gain = −1: Gain $= -\frac{\beta R_L}{R_S + r_\pi} = \frac{-50(264)}{450 + r_\pi} = -1$, $450 + r_\pi = 13200$, and $r_\pi = 12750$. Thus $r_e = \frac{12750}{50+1} = 250\ \Omega$, and $I_E = \frac{25mV}{250\Omega} = 0.1$ mA, for which $\upsilon_{BE} = 700 + V_t \ln \frac{0.1}{1} = 642$ mV. Thus $V_{IL} = 0.642 + \frac{0.1}{51} \times 0.45 =$ **0.643V** (rather than 0.60V, as assumed).

Now for two inputs, at a gain of −1, equivalent r_π is the same; total current is the same; current in each transistor is half; and $\upsilon_{BE} = 700 + V_t \ln \frac{0.05}{1}$

$= 625$ mV. Thus $V_{IL} = \mathbf{0.625V}$.

14.6

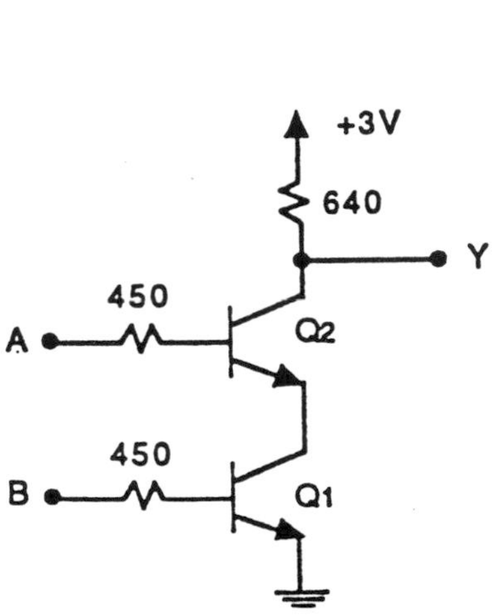

If for input A, $\beta_f = \frac{\beta_F}{2} = 25$, or $\frac{I_{C2}}{I_{B2}} = 25$, then $I_{C1} = 26$ $I_{B2} = \frac{26}{25} I_{C2}$. For simplicity, let $I_{C1} = I_{C2}$, and $V_{CEsat} = 25 \ln \frac{1 + (\beta_f + 1)/\beta_R}{1 - \beta_f/\beta_F} = 25 \ln \left[\frac{1 + (25+1)/0.1}{1 - 25/50} \right] = 25 \ln$ $(261/0.5) = 156$mV. Thus $V_Y = 0.156 + 0.156 = 0.313$ V, $I_{C2} = \frac{3 - 0.31}{640} = 4.20$ mA,

$I_{E2} = \frac{26}{25}(4.20) = 4.37$ mA, and $I_{E1} \approx (\frac{26}{25})^2 (4.20) = 4.54$ mA. Thus $v_{BE1} = 700 + 25 \ln 4.54/1 = 738$ mV, and $v_{BE2} = 700 + 25 \ln 4.37/1 = 737$ mV. Now, $V_{IH1} = 0.738 + 0.45 \times 4.54/25 = \mathbf{0.820V}$, $V_{IH2} = 0.737 + 0.156 + 0.45 \times 4.37/25 = \mathbf{0.972V}$.

14.7

(a) $I_C = \frac{4V - 2V}{4k\Omega} = 0.6$ mA; $I_B = \frac{0.5}{30} = \mathbf{16.7\mu A}$.

(b) $V_{BE} = 700 + 25 \ln 0.517/1 = 0.684$V, $I_{R2} = \frac{2 + .684}{5} = 0.537$ mA, $I_{D3,4} = 0.537 + .017 = 0.554$ mA, $V_{D3,4} = 700 + 25 \ln .554/1 = 0.685$V. Thus $V_X = 0.684 + 2\ (.685) = 2.054$ V, $I_{R1} = \frac{4 - 2.054}{2} = 0.973$ mA, $I_{D1} = 0.973 - 0.554 = 0.419$ mA, and $V_{D1} = 700 + 25 \ln 0.419/1 = 0.678$V. Thus $V_{th} = V_X - V_{D1} = 2.054 - 0.678 = \mathbf{1.376V}$.

(c) For diode drops as calculated, $V_X = 2.054$V, and $I_{R1} = 0.973$ mA. For A high, $I_{D3,4} = 0.973$mA, $V_{D3,4} \approx 700 + 25 \ln 0.973/1 = 699$mV, and $I_B \approx 0.973 - 0.537 = 0.436$ mA, for which Q is saturated with $I_C \approx \frac{4-0}{4} = 1$ mA, and $I_E \approx 1.44$ mA, with $V_B \approx 700 + 25 \ln 1.44/1 = 0.709$V. Thus the corrected value of $V_X = 0.709 + 2\ (.699) = 2.107$ V, for which $I_{R1} = \frac{4-2.107}{2} = 0.946$ mA, and $I_B \approx 0.946 - 0.537 = \mathbf{0.409mA}$.

(d) $I_{in} \approx \frac{4 - 0.7}{2} = 1.65$ mA, for which $v_D = 700 + 25 \ln 1.65/1 = 713$mV, and $I_{in} = \frac{4 - 0.713}{2} = \mathbf{1.64mA}$.

(e) Add another *diode in series with* D_3, D_4, which (for the same base drive) has a 0.678 V drop. Thus V_{th} is raised by 0.678 V, provided that the current levels remain the same. Now, for same maximum base drive, $V_X = 0.709 + 3\ (0.699) = 2.806$V, and $R_1 = \frac{4 - 2.806}{0.946} = \mathbf{1.26k\Omega}$. Now, for $V_A = 0$V, $I_{in} \approx \frac{4 - 0.7}{1.26} = 2.62$ mA, for which $V_D = 700 + 25 \ln 2.62/1 = 724$mV, and $I_{in} = \frac{4 - .724}{1.26} = \mathbf{2.60mA}$.

For fanout, I_B max $= 0.436$ mA, and with $\beta_{forced} = \beta/2 = 30/2 = 15$, $I_{C\ sat} = 15\ (0.436) = 6.54$ mA. Thus the maximum load current $= 6.54 - (4-0)/4 = 5.54$ mA. Thus the maximum fanout $= 5.54/2.60 = 2.13$, or 2, conservatively. For a fanout $= 3$, $\beta_{forced} = \frac{3(2.60)+1}{0.436} = 20.2$. Thus $\beta_{forced} < \beta$, as required.

14.8

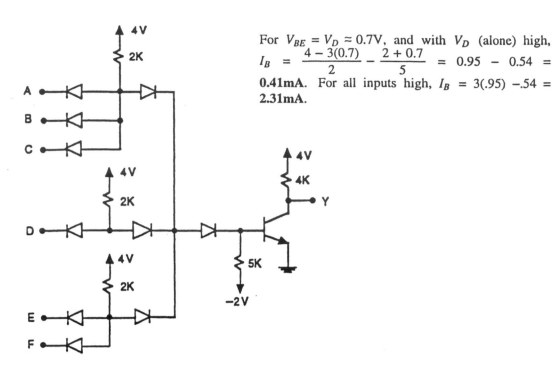

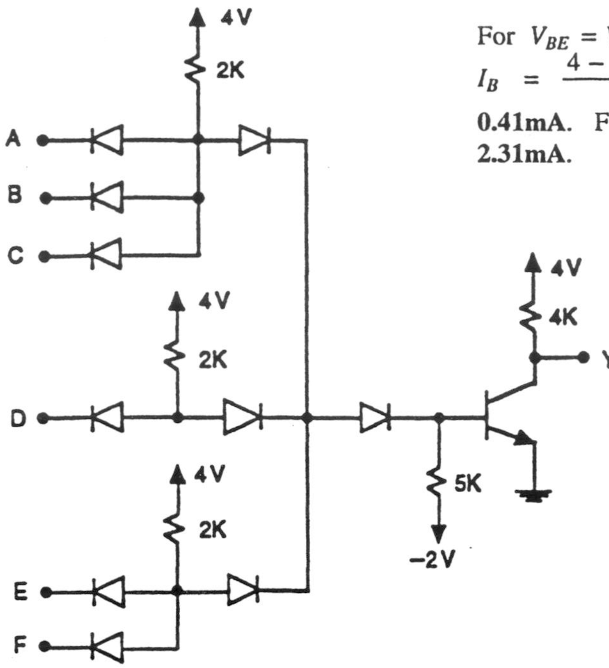

For $V_{BE} = V_D \approx 0.7$V, and with V_D (alone) high, $I_B = \frac{4 - 3(0.7)}{2} - \frac{2 + 0.7}{5} = 0.95 - 0.54 =$ **0.41mA**. For all inputs high, $I_B = 3(.95) - .54 =$ **2.31mA.**

SECTION 14.3: TRANSISTOR-TRANSISTOR LOGIC

14.9 Reduce R_B to 5/2 = 2.5 kΩ. From Exercise 14.4, the turn-on base current was 1.6 mA. Now, it is 1.6 − 0.7/5 = **1.46mA**. Now, for β = 50, $I_{C\ max}$ = 1.46(50) = 73 mA. The load current = 73 − 5/2 = 70.5 mA. Thus, for I_{in} = 1.1 mA, the maximum fanout is $\frac{70.5}{1.1}$ = 64.1, ie N = **64**. Now for N = 64/4 = 16, I_C = 16(1.1) + 2.5 = 20.1 mA, and β_{forced} = 20.1/1.46 = 13.8. Now, for β_F = 50, β_R = 50/100 = 0.5, $V_{ECsat} = 25 \ln \frac{1 + (\beta_f + 1)/\beta_R}{1 - \beta_f/\beta_F} = 25 \ln \frac{1 + (13.8+1)/0.5}{1 - 13.8/50} =$ **95.6mV** $= V_{OL}$.

14.10 Generally, I_{B2} = 1.46 mA, $I_{B1} = \frac{0.7V}{2.5k\Omega}$ = 0.28 mA, $I_C \approx 5/2$ = 2.5 mA, and $t_s = \tau_s \frac{I_{B2} - I_C/\beta}{I_{B1} + I_C/\beta}$.

For β = 50 *and,* I_C = 2.5 mA, becomes $t_s = 10 \frac{1.46 - \frac{2.5}{50}}{.28 + \frac{2.5}{50}} = \frac{10(1.41)}{.33} =$ **42.7ns.**

For N = 16, I_C = 5/2 + 16(1.1) = 20.1 mA, and $t_s = 10 \frac{1.46 - \frac{20.1}{50}}{0.28 + \frac{20.1}{50}} =$ **15.5ns.**

For β = 100, *and* N = 0, $t_s = 10 \frac{1.46 - \frac{2.5}{100}}{.28 + \frac{2.5}{100}} =$ **47.0ns.**

For N = 16, $t_s = 10 \frac{1.46 - \frac{20.1}{100}}{0.28 + \frac{20.1}{100}} =$ **26.2ns.**

14.11 For a load of 2 kΩ, $I_C = \frac{5-0.2}{2k} = 2.4$ mA. For $\beta_{forced} = 10$, $I_B = 2.4/10 = 0.24$ mA. Thus, the maximum collector current available = 50 × 0.24 = 12 mA. Thus, the additional load current allowed = 12 – 2.5 = 9.5 mA. Thus, the minimum external load = $R_L = \frac{5-0.2}{9.5}$ = **505Ω**. Now, consider a load of 2(505 Ω) = 1.01 kΩ, for which the total equivalent collector resistance = 2k ‖ 1.01 k = 0.67 kΩ.

For the rise time: Here $\upsilon_O = 5.0 - (5-0.2)e^{-t/R_L C}$. Now the time for 10% = time for 0.2 V ≈ 0 ns. Now, for the time for 90%: $0.9(4.8) + 0.2 = 5.0 - 4.8\,e^{-t/R_L C}$, $4.32 + 0.2 = 5.0 - 4.8\,e^{-t/R_L C}$, $t = -R_L\,C \ln \frac{5-4.52}{4.8} = -0.67 \times 10^3 \times 10 \times 10^{-12} \times (-2.30) = 15.4$ns. Thus the rise time ≈ **15.4ns**.

For the fall time: Here the output starts at 5.0 V and heads toward 5 – 12 mA(0.67k) = –3.04 V. Thus $\upsilon_O = -3.04 + (5+3.04)e^{-t/R_L C}$. Now, the fall time is complete when $-3.04 + 8.04e^{-t/R_L C} = 0.2 + 0.1(4.8)$, or $t = -6.7$ ns × ln 3.72/8.04 = **5.16ns**. Because of the load, the rise and fall times have become more equal than they are with very light loads. Note that if an external load of 505 Ω is used, t_r and t_f become equal. Why?

14.12 Here, (a) For I_C = 20 mA and $\beta_f = 40/2 = 20$, $I_B = 20/20 = 1$ mA, (b) $R_C = R$, (c) Supply = V is small, (d) $V_{OL} \approx 0.2$ V. Now, $V_{IL} = 0.6 - 0.2 = 0.4$V, and assuming Q_1 conducting in inverted saturation, $V_{IH} = 0.7 + 0.2 = 0.9$ V. Now, *for loading*:

For υ_O high: For reverse-current flow in each input transistor and fanout N, the total outward-directed load is $N(I_{B3})\beta_R$. For $N = 10$, $I_{LH} = 10(1)(0.1) = 1$ mA. Thus $V_{OH} = V - R(1)$.

For υ_O low (=0.2V): The current required by each input is $\frac{V-0.7-0.2}{R}$. For $N = 10$, $I_L = 10(\frac{V-0.9}{R}) \le \beta_f\, I_B = 20$. Thus, $V \le 2R + 0.9$ – – – (1). Now, $NM_L = V_{IL} - V_{OL} = 0.4 - 0.2 = 0.2$V, and $NM_H \ge 1.5\,NM_L = 1.5(0.2) = 0.3$V. Now, since $V_{IH} = 0.9$V, $V_{OH} \ge 0.9 + 0.3 \ge 1.2$V. But, $V_{OH} = V - R(1) \ge 1.2$V. Thus $V \ge 1.2 + R$ – – – (2). Now from (a), for $I_B \ge 1$ mA, and for Q_3 saturated with Q_1 operating in reverse mode with $\beta_R = 0.1$, the current in the base of $Q_3 = \left(\frac{V-0.7-0.7}{R}\right)(1+0.1) \ge 1$mA. Thus $V - 1.4 \ge R/1.1$, or $V \ge 1.4 + 0.91R$ – – – (3). Now, overall, there are three conditions: (1) $V \le 2R + 0.9$, (2) $V \ge 1.2 + R$, and (3) $V \ge 1.4 + 0.91\,R$.

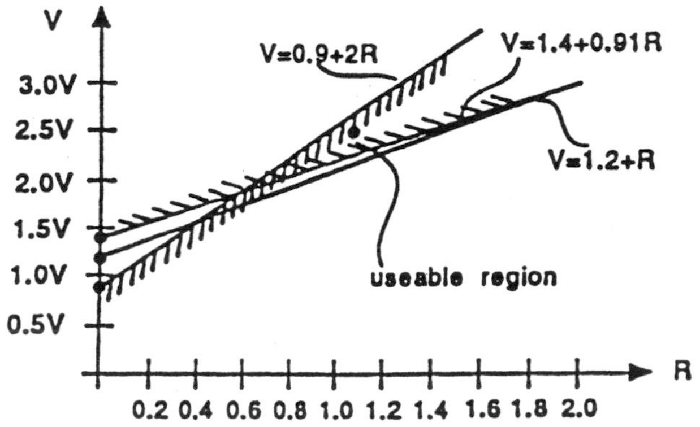

See that the minimum V occurs for the intersection of conditions (1) and (3), where $1.4 + 0.91\,R = 2R + 0.9$, or $1.09R = 0.5$, whence $R = 0.459$kΩ, and $V = 0.9 + 2(0.459) = 1.817$V. Use (conservatively) V = **2.0V**, and (from (1)) $2R \ge 2 - 0.9$, $R > 0.55$ kΩ, and (from (3)) $0.91R \le 2.0 - 1.4 = 0.6$, or $R \le 0.66$K. Use R = 600Ω. Here, it is apparent that the higher the voltage chosen (and the resistor chosen), the greater is the range of adequate operation.

Thus for V = 2.5V, $R \ge \frac{2.5-0.9}{2} = 0.8$ kΩ, and $R \le \frac{2.5-1.4}{0.91} = 1.21$ kΩ. Thus use **1.0kΩ**.

For V = 2.0, R = 0.6 kΩ: For υ_I high, $I_{C3} \le 40\left(\frac{2-0.7-0.7}{0.6}\right) = 40$ mA. For υ_I

low, $I_{in} = \frac{2-0.2-0.7}{0.6} = 1.83$mA. Thus, for the edge of saturation, $N \le 40/1.83 = 21.8$, say **21**. *For V* $= 2.5$, $R = 1.0$ kΩ: For υ_I high, $I_{C3} \le 40 \left[\frac{2.5-0.7-0.7}{1}\right] = 44$ mA. For υ_I low, $I_{in} = \frac{2.5-0.2-0.7}{1} =$ 1.6 mA. Thus $N \le 44/1.6 = 27.5$, say **27**.

14.13

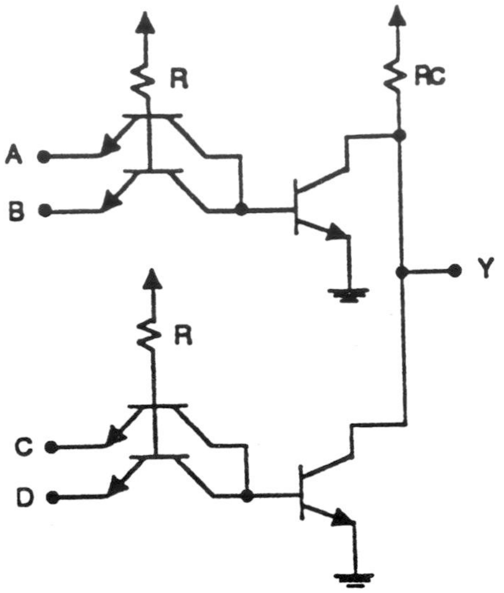

14.14 *For input high,* assuming $V_{BE} = 0.70$ V and $V_{CE\,sat} = 0.2$ V, see $V_{B3} =$ **0.7V**, $I_{1K} = 0.7$ mA, $V_{C2} =$ **0.9V**, $I_{C2} = \frac{5-0.9}{1.6} =$ **2.56mA**, $V_{B2} =$ **1.4V**, $V_{B1} \approx$ **2.1V**, $I_{B1} = \frac{5-2.1}{4} =$ **0.725mA**, $I_{E1} = 0.725(.05) =$ **36 μA**. Thus $I_{E2} = 0.725 + 2.56 + .036 =$ **3.33mA**, and $I_{B3} = 3.33 - .70 =$ **2.63mA**.

Check Saturation: For Q_2, $\beta_f = \frac{2.56}{.725+.036} = 3.36$, $\beta_F = 9$, $\beta_R = .05$. Thus $V_{CE\,sat} = 25 \ln \frac{1+(\beta_f+1)/\beta_R}{1-\beta_f/\beta_F} = 25 \ln \frac{1 + (3.36+1)/.05}{1 - 3.36/9} =$ **0.124 V**.

For Q_3, $V_{C3} \approx 0.2$V, $I_{C3} = \frac{5-0.2}{1k\Omega} = 4.8$ mA, $\beta_f = 4.8/2.63 = 1.83$. Thus $V_{CE\,sat} = 25 \ln \frac{1 + (1.83+1)/.05}{1 - 1.83/9} =$ **107mV**. Now for input low, $I_{in} \approx \frac{5 - 0.7 - 0.2}{4} = 1.025$mA. Thus the maximum fanout $N = \frac{2.63 \times 9}{1.025} = 23.1$, say **23**.

14.15 *For input low* at 0.3 V, $V_{E1} =$ **0.3V**, $V_{B1} = 0.3 + 0.7 =$ **1.0V**, $V_{C1} \approx 0.3 + 0.2 =$ **0.5V**. Now assuming $V_{out} \approx 3$V, $I_{out} \approx 3$ mA, $I_{B4} \approx \frac{3}{9+1} =$ **0.3mA**. Thus $V_{B4} \approx 5 - 1.6(0.3) =$ **4.5V**, $V_{E4} =$ **3.8V**, $V_{out} =$ **3.1V**. Thus $I_{out} =$ **3.1mA**, $I_{B4} =$ **0.31mA**, and $V_{B4} = 5 - 1.6(0.31) =$ **4.5V**, with $V_{C4} = 5 - .13 \times 3.1 \times \frac{9}{9+1} =$ **4.64V**.

14.16 Here, $V_{B3} =$ **0.7V**, $V_{B2} =$ **1.4V**, $V_{B1} \approx$ **2.1V**, $V_{E1} \approx$ **1.4V**, $I_{E1} \approx \frac{5-2.1}{4} = 0.725$mA, $V_O \approx 1.4 - 0.725(.200) = 1.26\text{V} = V_{C3}$, $V_{E4} = 1.26 + 0.7 = 1.96$V, $V_{B4} = 1.96 + 0.7 = 2.66$V, $I_{1.6k} = \frac{5-2.66}{1.6} =$ 1.46mA. Now Q_3 and Q_4 will both conduct at approximately the same levels (except for I_{200}). Thus

$I_{B4} = I_{B3} = I$, and $(1.46\text{ mA} - I)\,\frac{10}{9} - \frac{0.7}{1k} = I$, $1.62 - 1.11I - 0.7 = I$, $2.11\,I = 0.92$. Thus $I = I_{B3}$ $= I_{B4} = 0.44$mA, and $I_{C3} \approx 9(0.44) = 3.92$mA. Also $I_{E2} = .44 + 0.7 = 1.14$mA, $I_{B2} = \frac{1.14}{9+1} = 0.114$mA. Thus $I_{E1} = 0.725 - .114 = 0.61$mA, with $V_O \approx 1.4 - 0.2(.61) = 1.29\text{V} = V_{C3}$. Thus $I_{E4} = 3.92 - .61 =$ 3.31mA, $I_{B4} = 3.31/9{+}1 = 0.33$mA, $V_{B4} = 1.29 + 0.7 + 0.7 = 2.69$V, $I_{1.6k} = \frac{5-2.69}{1.6} = 1.44$mA, $I_{C2} =$ $1.44 - .33 = 1.11$mA, $I_{E2} = \frac{10}{9}(1.11) = 1.23$mA, $I_{B3} = 1.23 - 0.7 = 0.53$mA, $I_{C3} = 9(.53) = 4.77$mA, $I_{E4} = 4.77 - .61 = 4.16$mA, $I_{B4} = \frac{4.16}{9+1} = 0.416$mA, $I_{C2} = 1.44 - .416 = 1.024$mA, $I_{E2} = (10/9)(1.024)$ $= 1.14$mA, $I_{B3} = 1.14 - 0.7 = 0.44$mA, $I_{C3} = 9(.44) = 3.96$mA, $I_{B2} = 1.14/(9+1) = 0.114$mA, $I_{E1} = .725$ $- .114 = 0.61$mA, $I_{E4} = 3.96 - .61 = 3.35$mA, $I_{B4} = 3.35/(9+1) = 0.335$mA, $I_{C2} = 1.444 - .335 =$ 1.11mA, $I_{E2} = 1.233$mA, $I_{B3} = 1.233 - 0.7 = 0.533$mA, $I_{C3} = 9(.533) = 4.80$mA. Conclude $I_{C3} \approx \frac{4.80 + 3.96}{2}$ = **4.4mA**, $I_{B3} = 4.4/9 =$ **0.49mA**, $I_{E3} = 0.49 + 4.4 =$ **4.9mA**, $I_{E2} = 0.49 +$ $0.7 =$ **1.19mA**, $I_{E1} \approx$ **0.61mA**, and $V_O =$ **1.29V**.

For $V_{CE\,sat1}$: $I_{E1} \approx 0.61$mA, $I_{B1} \approx 0.725$mA, $I_{C1} \approx -0.114$mA. Thus $\beta_f = -0.114/0.725 = -0.157$, whence $V_{CE\,sat} = 25\ln\frac{1 + (\beta_f + 1)/\beta_R}{1 - \beta_f/\beta_F} = 25\ln\frac{1 + .843/0.5}{1 + .157/9} =$ **72mV**.

14.17 From the solution of P14.15, for no load, $V_O = V_{C3} \approx 1.29$ V, $I_{E1} = I_{200} = 0.61$ mA, $V_{B4} = 2.69$ V, $I_{1.6k}$ $= 1.44$ mA, $I_{C3} \approx 4.4$ mA, with $\beta_F = 9$, $V_{BE} = 0.7$V, and $\beta_F + 1 = 10$. *Now iterate*:

(a) *With* $R_L = 200\Omega$ *to ground*, $I_L = \frac{1.29}{0.2} = 6.45$ mA. Assume I_{C3} reduces, say to 2 mA. Thus I_{E4} $= 6.45 + 2 = 8.5$ mA, $I_{B4} = 8.5/10 = 0.85$mA. Thus $I_{C2} = I_{1.6k} - I_{B4} \approx 1.44 - 0.85 = 0.59$ mA, $I_{E2} = 10/9(0.59) = 0.66$ mA, $I_{1k} \approx 0.7$ mA. Thus $I_{B3} \approx 0$mA. Thus $I_{C3} \approx 0$. Thus $I_{E4} \approx I_L = 6.5$ mA, $I_{B4} = 6.5/10 = .65$mA – – – (1), $I_{C2} = 1.44 - .65 = 0.79$ mA, $I_{E2} = 10/9(0.79) = 0.88$ mA, $I_{B3} = 0.88 - 0.7 = 0.18$ mA, $I_{C3} = 9(0.18) = 1.62$ mA, $I_{E4} = I_L + I_{C3} - I_{200} = 6.5 + 1.62 - 0.61$ $= 7.51$ mA, $I_{B4} = 7.51/10 = .75$mA, $I_{C2} = 1.44 - .75 = 0.69$mA, $I_{E2} = 10/9(.69) = 0.766$mA, $I_{B3} =$ $0.766 - 0.70 = .066$mA, $I_{C3} = 9(.066) = 0.6$mA, $I_{E4} = 6.5 + 0.6 - 0.61 = 6.5$mA, $I_{B4} = 6.5/10 =$ .65mA, $I_{C2} = 1.44 - .65 = .79$mA, $I_{E2} = 10/9\,(.79) = .88$mA, and $I_{B3} = .88 - .7 = .18$, etc.

See that we begin a cycle initiated in (1) above, presumed to converge {with a more detailed analysis (eg exponential junctions and consideration of the effect of I_{B2} on I_{E1}, V_O, etc)} on the intermediate value. Thus $I_{C3} = \frac{0.60 + 1.62}{2}$ = **1.1mA**, $I_{E4} = 6.5 + 1.1 - .6 =$ **7.0mA**, $I_{B4} = 7.0/10$ = **0.7mA**, $I_{C2} = 1.44 - .7 =$ **0.74mA**, $I_{E2} = (10/9)(.74) =$ **0.82mA**, $I_{B3} = 0.82 - 0.7 =$ **0.12mA**, I_{C3} $= 0.12 \times 9 = 1.08 =$ **1.1mA**, as conjectured, where $V_O = V_{C3} =$ **1.29V**, and $I_L =$ **6.5mA**.

(b) *With* $R_L = 200\Omega$ *to* +5V: For $V_O \approx 1.29$V, $I_{C3} = I_L + I_{E1} = \frac{5-1.29}{.200} + 0.61 = 19.2$ mA. Likely Q_4 is nearly cutoff $\rightarrow I_{B4} \approx 0$mA. Thus $I_{C2} = I_{1.6k} = 1.44$ mA, $I_{E2} = (10/9)(1.44) = 1.6$mA, I_{B3} $= 1.6 - 0.7 = 0.9$mA, $I_{C3} = 9(0.9) = 8.$mA (too small). Conclude that as V_O goes up, I_L reduces, I_{E1} reduces, I_{B2} increases, possibly Q_2 saturates.

For Q_2 saturated: $I_{C2} \approx \frac{5 - 0.7 - 0.2}{1.6k} = 2.56$ mA, $I_{B1} \approx \frac{5 - 0.7 - 0.7 - .7}{4} = 0.725$mA, $I_{E2} \le 2.56 + .73 = 3.29$mA, $I_{B3} \le 3.29 - 0.7 = 2.59$mA, $I_{C3} \le 9(2.59) = 23.3$ mA. Conclude that Q_2 is *not saturated*, but I_{E1} is small, $V_O \approx 1.4$V, for which $I_{C3} \approx \frac{5 - 1.4}{200}$ = **18mA**, $I_{B3} = 18/9 =$ **2mA**, $I_{E2} = 2 + .7 =$ **2.7mA**, $I_{C2} \approx 9/10 \times 2.7 =$ **2.43mA**, $V_{C2} = 5 - 2.43(1.6) =$ **1.11V**, not quite saturated, as assumed. Thus $I_{B2} = 2.7/10 =$ **0.27mA**, $I_{E1} = .725 - .27 =$ **0.46mA**, and $V_O = 0.7 +$ $0.7 + 0.7 - 0.7 - 0.46(0.2) =$ **1.3V**, with Q_4 cut off.

SECTION 14.4: CHARACTERISTICS OF STANDARD TTL

14.18 For $v_O(C) = 3.0$ V, $v_{C2}(C) = 3 + 0.65 + 0.65 = 4.3$ V. Now at C, $V_{BE3} \approx 0.6$V and $I_{E2} \approx I_{C2} = \frac{0.6}{R_2}$. Thus $v_{C2} = 5 - 1.6\left[\frac{0.6}{R_2}\right] = 4.3$V. Thus $R_2 = \frac{1.6(0.6)}{0.7} =$ **1.37kΩ**. For this change, $v_I(C) = 1.2$V, $v_O(C) = 3.0$V, $v_I(B) = 0.5$V, $v_O(B) = 3.7$V. Thus gross slope $BC = -\frac{3.7 - 3.0}{1.2 - 0.5} = -\frac{0.7}{0.7} =$ **−1 V/V.**
For Incremental gain: I_{E2} varies from 0 to 0.6/1.37 = 0.44mA, and $I_{E2\,av} = 0.22$ mA. Thus $r_{e2} \approx 25/.22 = 114\Omega$, and gain $= -0.98 \times \frac{1.6k}{.114k + 1.37k} = -1.06$ V/V. For $R_2 = 1\text{k}\Omega$ and $v_O(C) = 3.0$V, $v_{C2} = 5 - R_1\,(0.6/1) = 4.3$V, and $R_1 = \frac{0.7}{0.6} =$ **1.17kΩ.**
For turnon of Q_3: Previously: $I_{B3} = I_{B2} + I_{C2} - I_{R2} = \frac{5 - 0.7 - 0.7 - 0.7}{4} + \frac{5 - 0.7 - 0.2}{1.6} - \frac{0.7}{1} = .725 + 2.563 - 0.7 = 2.59$ mA. Now: $I_{B3} = .725 + \frac{4.1}{1.17} - 0.7 = 3.53$ mA. Thus the increase = 3.53 − 2.59 = **0.94mA** ≈ **36%.** Now the Storage Delay will increase even more, by a factor due to the $I_{B2} - \frac{I_C}{\beta_F}$ difference. For example, with $\beta_F = 50$, and $I_C = 10$, the increase is by $\frac{3.53 - 0.2}{2.59 - 0.2} = 1.39$, or by39%!

14.19 Use coordinates of points C, D:

Temp.	V_{OL}	V_{IL}	V_{OH}	V_{IH}
−55°C	0.1	1.52	2.16	1.72
125°C	0.1	0.80	3.46	1.0

For Source at −55°C, load at 125°C:

$NM_L = V_{IL}(125) - V_{OL}(-55) = 0.80 - 0.1 =$ **0.7V,**

$NM_H = V_{OH}(-55) - V_{IH}(125) = 2.16 - 1.0 =$ **1.16V.**

For Source at 125°C, load at −55°C:

$NM_L = V_{IL}(-55) - V_{OL}(125) = 1.52 - 0.1 =$ **1.42V,**

$NM_H = V_{OH}(125) - V_{IH}(-55) = 3.46 - 1.72 =$ **1.74V.**

For comparison, for the nominal circuit with source and load both at 25°C:

$NM_L = 1.2 - 0.1 = 1.1$V,

$NM_H = 2.7 - 1.4 = 1.3$V.

14.20 The maximum base current $I_{B3} = 4\left[\frac{5 - 2.1}{4}\right] + \frac{5 - 0.7 - 0.2}{1.6} - \frac{0.7}{1} = 2.9 + 2.56 - 0.7 =$ **4.76mA.**
For three inputs low, $I_{B3} = .725 + 2.56 - 0.7 = 2.585$ mA. Now $t_s = \tau_s \frac{I_{B2} - I_C/\beta}{I_{B1} + I_C/\beta} = \frac{I_{B2}}{I_{B1}}$, for I_C = 0mA. Thus the delay for four inputs dropping together is $t_s = 10 \times \frac{4.76}{2.585} =$ **17.8ns.**

14.21 For $I_{C6} = 1$ mA, the voltage drop in 4kΩ = 4V. Thus $V_{C6} = 5 - 4 = 1$ V. Now $I_{D2} \approx 1$ mA, $V_{E6} \approx 0.7$V. Thus Q_6 is barely saturated (if at all). Assume linear → $I_{B6} = \frac{1mA}{50} = 0.02$ mA. Thus I_{E6} = 1.02 mA. Thus $v_{R6} \approx 2\left[(25)\ln\frac{1.02}{1} + 700\right] = 1.401$V. *Check* $V_{CE\,sat}$ for $\beta_f = 49$ (say): $V_{CE\,sat}$ =

$25 \ln \frac{(1 + (49+1)/0.1)}{1 - 49/50} = 253$ mV. Thus Q_6 operates linearly, and $I_{B5} \approx \frac{5-2.1}{4} = 0.725$ mA, $I_{C5} = -0.02$ mA, $I_{E2} = 0.705$ mA, for which $\beta_f = -\frac{0.02}{.725} = -.028 \rightarrow V_{CE\,sat} = 25 \ln \frac{(1 + (.028+1)/0.1)}{1 + .028/50} = 59$ mV, and $v_I = 1.401 - .059 =$ **1.34V**.

14.22 $I_{B1} = \frac{5 - 3(0.7)}{4} = 0.725$ mA. Via Q_7, $V_{C2} = 0.7 + 0.2 = 0.9$V. Now, from the symmetry of Q_7, D_2 and 4kΩ, with Q_2, V_{BE3} and 4kΩ, it is likely that $I_{C2} \approx \frac{I_{1.6k}}{2} = \frac{5-0.9}{2(1.6k)} = 1.28$ mA. Thus $I_{B3} \approx 0.725 + 1.28 - 0.7 =$ **1.31mA**. That is, Q_3 is distinctly turned on. Thus we see the need for the connection from the tristate input to Q_1!

SECTION 14.5: TTL FAMILIES WITH IMPROVED PERFORMANCE

14.23

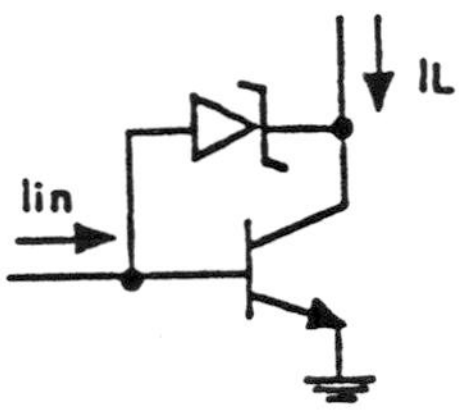

$I_{in} = 1$ mA, $I_L = 0$ mA. Thus $I_E =$ **1mA**, $I_B = \frac{1}{\beta+1} = \frac{1}{51} =$ **19.6 μA**, $I_D = 1$ mA − 19.6μA = **0.98mA**. Thus $V_B = 750 + 25 \ln 1/1 = 750$ mV = **0.750V**, and $V_C = 750 - (500 + 25 \ln .98/1) =$ **0.25V**.

$I_{in} = 1$ mA, $I_L = 1$ mA. Thus $I_E =$ **2mA**, $I_B =$ **0.04mA**, $I_D =$ **0.96mA**. Thus $V_B = 750 + 25 \ln 2 = 767$ mV = **0.767V**, $V_C = 767 - (500 + 25 \ln \frac{0.96}{1}) =$ **0.266V**.

(c) $I_{in} = 1$ mA, $I_L = 10$ mA. Thus $I_E =$ **11mA**, $I_B = 11/51 =$ **0.216mA**, $I_D = 1 - .216 -$ **0.784mA**. Thus $V_B = 750 + 25 \ln 11/1 =$ **0.810V**, $V_C = 810 - (500 + 25 \ln .784) =$ **0.316V**.

(d) $I_{in} = 10$ mA, $I_L = 10$ mA. Thus $I_E =$ **20mA**, $I_B = 20/51 =$ **0.392mA**, $I_D = 10 - .392 =$ **9.61mA**. Thus $V_B = 750 + 25 \ln 20 =$ **0.825V**, $V_C = 825 - (500 + 25 \ln 9.61) =$ **0.268V**.

(e) $I_{in} = 10$ mA, $I_L = 1$ mA. Thus $I_E =$ **11mA**, $I_B = 11/51 =$ **0.216mA**, $I_D = 10 - .46 =$ **9.78mA**. Thus $V_B = 750 + 25 \ln 11 =$ **0.810V**, $V_C = 810 - (500 + 25 \ln 9.78) =$ **0.256V**.

14.24 For the transistor: $I_E = 1$ mA at $V_{BE} = 0.75$ V, with $n = 1$, and $\beta = 50$. For the SBD, $I_D = 1$ mA at $V_D = 0.5$ V, with $n = 1$.

Consider the problem in three parts: *First,* use a constant-voltage junction approximation to find the approximate voltage values. *Second,* estimate device currents. *Third,* refine the junction voltages based on the current levels found.

(a) *Voltages for $V_{BE} = 0.75V$, and $V_D = 0.5V$:* Now, for inputs A and B both high, Q_1 is cutoff, Q_2, Q_3, Q_6 all conduct and Q_4, Q_5 are cutoff. See $V_{B3} =$ **0.75 V**, $V_{C3} = 0.75 - 0.5 =$ **0.25 V**, $V_{B2} = 0.75 + 0.75 =$ **1.5 V**, $V_{C2} = 0.75 + 0.25 =$ **1.00 V**, $V_{C1} = V_{B2} =$ **1.5 V**,

$V_{B1} = 1.5 + 0.5 =$ **2.00 V**. For Q_6, $V_{B6} \approx$ **0.75 V**, $V_{C7} \geq (0.75 - 0.5) =$ **0.25 V**.

(b) *Device Current Estimates:*

$I_R = (5 - 2)/2.8 = 1.07mA = I_{D1}$, $I_{R1} = (5 -)/0.9 = 4.44mA = I_{C2}$

Now, $I_{B2} = I_{C2}/\beta = 4.44/50 = 0.09mA$. Thus $I_{D2} = I_{D1} - I_{B2} = 1.07 - .09 = 0.98mA$, and $I_{E1} = I_{R1} + I_R = 4.44 + 1.07 = 5.51mA$

Now the load on the output is from resistor R of 2 similar gates, with input at 0.25 V, $V_{BE1} = 0.75$ and $I_L = 2(5 - 0.75 - 0.25)/2.8 =$ 2.86 mA.

Thus $I_{C3} = 2.86mA$, $I_{B3} = 2.86/50 = 0.06mA$.

Now I_{D3} is relatively large depending on the fraction of current from Q_2 that is conducted away by Q_6. However its value does not matter, since its voltage-drop variation is absorbed by V_{C3} changing slightly, leaving V_{B3} as is.

Now for the BJT, V_{BE} = 0.75 V for $I_E \simeq I_C$ = 1 mA.

Since $i = I_S e^{v/V_T}$, $v_1 - v_2 = V_T \ln(i_1/i_2)$, and $v_2 = v_1 + V_T \ln(i_2/i_1)$ – – – (1)

and, as well, $i_2 = i_1 e^{(v_2 - v_1)/V_T}$ – – – (2)

Thus for I_{C3} = 2.86 mA, $V_{B3} = 750 + 25\ln(2.86/1) = 776.3mV$.

Now for V_{B6} = 776 mV, from (2), $I_{C6} = 1e^{(776-750)/25} = 2.83mA$, for which $I_{B6} = 2.83/50 = 0.057mA$, and

$V_{R2} = 500 \times 50 \times 10^{-6}$ = 0.0285 V.

Thus V_{B6} is lower than expected; but *not* 776.3 – 28.5 = 748 mV.

Iterate: To start, for V_{BE6} = 750 mV, I_{C6} = 1 mA and I_{B6} = 20μA, for which $V_{R2} = 500 \times 20 \times 10^{-6}$ = 10 mV, and $V_{B3} = 750 + 10 = 760mV$.

For $V_{BE6} = 760mV$, $I = 1e^{(760-750)/25}$ = 1.49 mA, and $V_{R2} = 500 \times 1.49 \times 10^{-3}/50$ = 14.9 mV, and V_{B3} = 760 + = 775 mV.

For V_{BE6} = 761 mV, $I = 1e^{(761-750)/25)}$ = 1.82 mA, and $V_{R2} = 500 \times 1.82 \times 10^{-3}/50$ = 1.55 mV, and V_{B3} = 761 + 15.5 = 776.5 mV more or less as needed.

Conclude that V_{B6} = **761 mV** with V_{B3} = **776 mV**, $I_{B6} = 36\mu A$, and $I_{C6} = 1.82mA$

For this condition $V_{R5} = 250 \times 1.82 \times 10^{-3}$ = 0.455 V, and V_{C6} = 776 – 455 ≡ **321 mV**, implying that D_6 is not conducting.

Now the current supplied by Q_2 is $I_{E2} = I_{R1} + I_R = 5.51mA$ and that needed by I_{B3}, is 0.09 mA.

Thus the excess current is 5.51 – 1.82 – .09 – .04 = 3.56 mA. This current will flow in D_3.

For Eq. (1), the voltage drop in D_3 will be v = 500 + 25ln3.56/1 = 532 mV.

Thus V_{C3} = 776 – 532 = **244 mV**.

Now, returning to Q_2 and Q_1, I_{C2} = 4.44 mA and V_{BE2} = 750 + 25ln4.44 = **787 mV**.

Now I_{D2} = 0.98 mA and $V_{D1} = 500 + 25\ln 0.98 = 499mV$, in which case,

$V_{B2} = V_{B3} + V_{BE2}$ = 776 + 787 = **1563 mV**, and $V_{C2} = V_{B3} + V_{BE2} - V_{D2}$ = 776 + 787 – 499 = **1064 mV**.

Thus V_{C1} = **1064 mV**. Now I_{D1} = 1.07 mA, and $V_{D1} = 500 + 25\ln 1.07 = 502mV$. Thus V_{B1} = 1064 + 502 = **1566 mV**.

14.25 Using some of the results from the solution of P14.24 above, the emitter current of Q_2 is I_{E1} = 5.51 mA. For half this current flowing in Q_6, I_{E6} = 5.51/2 = 2.75 mA, and I_{B3} = 2.75 mA. For 1 mA required to operate Q_3, and β = 50, then I_{C3} = 50 mA and $V_{B3} = 750 + 25\ln 50/1 = 848mV$.

For $I_{C6} \simeq 50/51 \times 2.75$ = 2.70 mA

$V_{BE6} = 750 + 25\ln 2.70 = 775mV$. Thus I_{B6} = (848 – 775)/0.5 = 146 μA, flowing in R_2 = 500 Ω.

But the base current needed to support I_{E6} = 2.75 mA is 2.75/51 = 54 μA. Thus D_6 conducts a current of 146 – 54 – 92 μA for which V_D = 500 + 25ln0.092 = 440 mV. Thus R_5 = (848 – 440)/(2.750 – 0.146) = **157Ω.**

For larger values of R_5 (like the 250 Ω in the current design, more current would enter the base of Q_3. Further, as the load on Y and Q_3 is reduced, and less current is actually needed by the base of Q_3, V_{B3} will lower, and the current in Q_6 will also diminish, with extra current (ie > 1/2) flowing into the base lead of Q_3, but then via D_3 and the collector junction to ground. Of course, the minimum emitter current in Q_3 is the fraction of I_{E2} that does not flow in Q_6 when the output is open-circuited, and its collector current is nearly as large as its base current. It is interesting to calculate R_2 for equal currents in Q_6 and the base of Q_3, while R_6 is 250 Ω.

14.26 To establish the base drive for Q_3, assume that the conducting transistors, Q_3, Q_2 and Q_6 have V_{BE} = 0.75 V and V_{BC} = 0.5 V, and that for Q_1, V_{BC} = 0.5 V while the *EB* junction is cutoff.

Correspondingly, $V_{B3} = 0.75$ V, $V_{B2} = 0.75 + 0.75 = 1.5$ V, $V_{B1} = 1.5 + 0.5 = 2.0$ V, and $V_{C2} = 1.5 - 0.5 = 1$ V.

Thus $I_R = (5 - 2)/2.8k\Omega = 1.07$ mA, $I_{R1} = (5 - 1)/0.9 = 4.44$ mA, $I_{R5} = (0.75 - 0.25)/0.25 = 2.0$ mA, whence $I_{E2} = 1.07 + 4.44 = 5.51$ mA and $I_{B3} = 5.51 - 2.0 = 3.51$ mA, neglecting I_{B6}.

(a) Now for load current $I = 1$ mA, $I_{E3} = 3.51 + 1.0 = 4.51mA$, the internal base current of $Q_3 = I_{b3} = 4.51/51 = 88.4\mu A$, and the SBD current in $Q_3 = I_{D3} = 3.51 - .0884 = 3.42mA$.

For these currents: $r_d = nV_T/I_D = 25/3.42 = 7.3\Omega$

$r_{\pi 3} = 25 \times 10^{-3}/(88.4 \times 10^{-6}) = 283\Omega$. and $g_{m3} = \beta_3/r_{\pi 3} = 50/283 = 177$ mA/V.

Assuming that the base source of Q_3 can be represented as R_6, the equivalent circuit is as shown. Here v is a test voltage and i the response current.

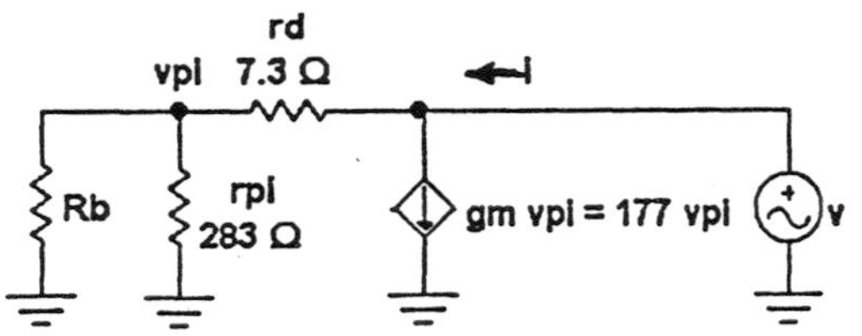

Now for R_b assumed to be large enough to igore, $i = g_m v_\pi = g_m \dfrac{r_\pi}{r_\pi + r_d} v$

and the output resistance

$$r = v/i = \frac{v}{g_m v r_\pi/(r_\pi + r_d)} = \frac{r_\pi + r_d}{g_m r_\pi} = \frac{r_\pi + r_d}{\beta},$$

or $r = r_e + r_d/\beta = (283 + 7.3)/50 =$ **5.8Ω.**

Note that this is essentially $r_e = r_\pi/(\beta + 1) = 283/51 = 5.5\Omega$.

Now to include R_6: $r_{e6} = nV_T/I_{E6} \simeq nV_T/I_{R5} = 25/2 = 12.5\ \Omega$. Thus,

$R_b \simeq R \parallel R_1 \parallel (R_5 + r_{e6}) = 2.8 \parallel 0.9 \parallel [(250 = 12.5)10^{-3}] = 2.8 \parallel 0.9 \parallel 0.263 = 0.195k\Omega = 195\Omega$.

Clearly, this may be too small to be ignored.

Now, $i = g_m v \dfrac{r_\pi \parallel R_b}{r_d + r_\pi \parallel R_b} = 177v \dfrac{283 \parallel 195}{7.3 + 283 \parallel 195} = 177v \times \dfrac{115.4}{7.3 + 115.4} = 166.5\ v$ mA and $r = v/i = 1/166 \times 10^{-3} =$ **6.02Ω.**

Thus, in fact, R_b does not have much effect, since r_d is so small.

(b) Now for a load current $I +$ 10 mA, $I_{E3} = 3.51 + 10 = 13.51$, and $I_{b3} = 13.51/51 = 265\ \mu A$, $I_{D3} = 3.51 - .265 = 3.24$ mA, with $r_d = 25/3.24 = 7.7\Omega$, $r_{\pi 3} = 25/0.265 = 94.3\Omega$, $g_{m3} = \beta_3/r_{\pi 3} = 50/94.3 = 530$ mA/V.

Thus, ignoring R_b, $r = (r_\pi + r_d)/\beta = (94.3 + 7.7)50 =$ **2.04Ω.**

Note that this is small due to shunt-shunt negative feedback provided by the SBD.

14.27 Use $V_{BE} = 0.75$ and $V_D = 0.5$V.

(a) Input is high: $I_{20k} = \dfrac{5 - .75 - .75}{20} = 0.175$ mA, $I_{8k} = \dfrac{5 - 0.75 - 0.75 + 0.5}{8} = 0.5$ mA. Thus $I_{CC} = 0.175 + 0.5 =$ **0.675mA**, for output open or shorted to ground.

(b) Input is low; at $v_I = 0.75 - 0.5 = 0.25$V. $I_{20k} = \dfrac{5 - 0.5 - 0.25}{20} = .2125$ mA. Output shorted: Q_4 and Q_5 conducting with drop of $0.75 + 0.75 - 0.5 = 1$V. Thus $I_{SC} = \dfrac{5 - 1}{.120} = 33.3$mA. Thus $I_{CC} =$ **0.213mA** for output open, or $I_{CC} =$ **33.5mA** for output grounded.

Now, Power Loss is 5 × .675 = **3.37 mW** for inputs high, and 5 × .213 = **1.07mW** for inputs low, output open, or 5 × 33.5 = **167.5mW** for inputs low, output grounded. The average dc power loss with no load = $\dfrac{3.37 + 1.07}{2} = 2.22$ mW. Dynamic power with a 10 pF load at 30 MHz = $fCV^2 = 30 \times 10^6 \times$

$10 \times 10^{-12} \times (5 - 1.4 - 0.3)^2 = 3.27$mW. Thus total power = 3.27 + 2.22 = 5.49 mW. Now, DP product = $5.49 \times 10^{-3} \times 10 \times 10^{-9}$ = **54.9 pJ.**

SECTION 14.6: EMITTER-COUPLED LOGIC (ECL)

14.28 (a) $V_{OH} = 0 - 0.75 = -0.75$V. $V_{OL} = 0 - RI - 0.75 = -$ **(RI + 0.75)V**.

(b) $V_{th\ A} = -IR/2 - 0.75 = -$**(RI/2 + 0.75)V**, for which $V_{BE} = 750 + 25 \ln i/I = 750 + 25 \ln \frac{I/2}{I} =$ **682.7mV**.

(c) For $i = 0.95I$: $V_{BE} = 750 + 25 \ln 0.95 = 750 - 1.3 =$ **748.7mV**.

(d) For $i = 0.05I$: $V_{BE} = 750 + 25 \ln .05 = 750 - 74.9 =$ **675.1mV**, and $\Delta V_{BE} = 748.7 - 675.1 =$ **73.6mV**.

(e) Thus $V_{IL} = -RI/2 - 0.750 - .0736 = -$ **(RI/2 + 0.824)V**, and $V_{IH} = -RI/2 - 0.750 + .0736 = -$ **(R I / 2** + 0.676)V.

(f) $NM_H = V_{OH} - V_{IH} = -0.750 + RI/2 + 0.676 =$ **(RI/2 −0.074)V**, and $NM_L = V_{IL} - V_{OL} = -RI/2 - .824 + RI + .750 =$ **(RI/2 −0.074)V**.

(g) Transition region = $2\,\Delta V_{BE}$ = 2(73.6) = 147.2 mV. Thus $V_{IH} - V_{IL}$ = 147 mV. Now $RI/2 - .074 = .147\text{V} \rightarrow RI = 2(.147 + .074) =$ **0.442V** = IR.

(h) Now for $IR = 0.442$V, $V_{OH} =$ **−0.750V**, $V_{OL} =$ **−1.192V**, $V_{IH} =$ **−0.897V**, $V_{IL} =$ **−1.045V**, $V_R = (-.897 - 1.045)/2 =$ **−0.971V**.

14.29 Here, $V_{BE} = 0.75$ @ 4 mA, and $V_{th} = -R/2 - 0.75 = -1.32$V. Thus $R = \frac{1.32 - .75}{2} = 0.285 = 285\Omega$. Also, $V_{OH} = 0 - 0.75 =$ **−0.750V**, and $V_{OL} = -0.750 - 0.285(4) =$ **−1.89V**. For a 1000-to-1 current split, $\Delta V_{BE} = 25 \ln 10^3 = 173$ mV. Thus $V_{IL} = -1.32 - .173 =$ **−1.493V**, and $V_{IH} = -1.32 + .173 =$ **−1.147V**. Thus $NM_L = V_{IL} - V_{OL} = -1.493 + 1.89$ V = **0.397V**, and $NM_H = V_{OH} - V_{IH} = -0.750 + 1.147$ V = **0.397V**.

14.30

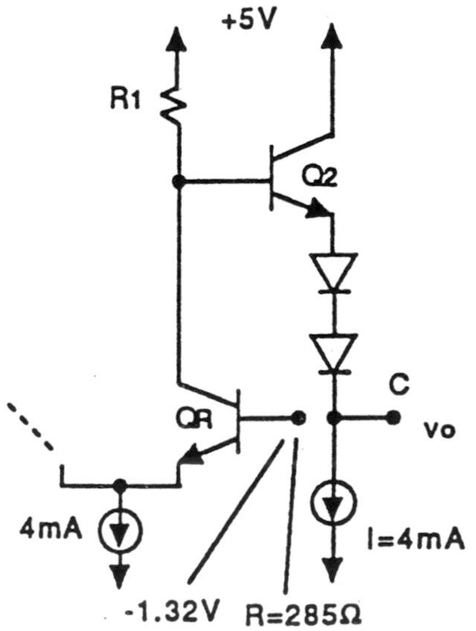

Here, I = 4 mA, $V_{th} = -(I_R/2 + 0.75) = -1.32$ V. Thus $R = \frac{1.32 - .75}{2} =$ **285Ω**. For N diodes, $V_{OH} = 5 - 0.75 - N(0.75) \geq 2.7$ V − − − (1), and $V_{OL} = 5 - 0.75 - N(0.75) - R_1 I \leq 0.5$ − − − (2).

From (1), $N \leq \frac{-2.7 + 5 - 0.75}{0.75} = 2.07$. Use **two diodes** with $V_{OH} = 5 - 3(0.75) =$ **2.75V**. *From* (2), $5 - 3(0.75) - R_1(4) \leq 0.5$, $R_1 > \frac{5 - 2.25 - .50}{4} =$ 0.5625 kΩ. Use $R_1 = 570\Omega$, for which, $V_{OL} = 5 - 2.25 - .570(4) =$ **0.47V**.

14.31

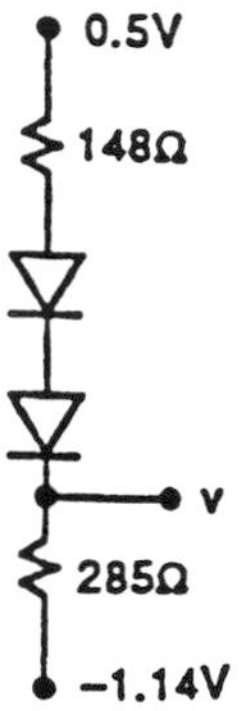

Use I = 4 mA, and R = 285Ω. See with υ_I high, D_3 operates at 8 mA with V_D = 750 + 25 ln 8/4 = 0.767V, while D_1 and D_2 operate at 4 mA. Thus $R_2 = \dfrac{2.70 - .767 - .75}{8}$ = 148Ω. For T^2L output low enough, I flows in R and diodes cut off: V_{Base} = –4(.285) = –1.14V. For V_{OL} (of T^2L) = 0.5V, voltage across two diodes (D_3, D_2) and R_2 is 1.14 + 0.5 = 1.64 V. Thus $I_{R2} \approx \dfrac{1.64 - 0.75 - 0.75}{.148} = \approx$ **0.95mA**, for which V_{Base} is higher. For 0.75 V diodes, $\upsilon = -1.14 + \dfrac{285}{148 + 285}(0.5 + 1.14 - 2(0.75)) = -1.04$ V $= V_{OL}$. This is not ideal, but fairly good, and certainly OK. Now for $V_{O\,TTL} > V_{OH\,TTL}$, D_1 conducts excess current, causing the base of Q_2 to rise above ground, and V_{OH} of the converter to rise above –0.75 V. In summary, for the converter, V_{OH} = **–0.75V**, or slightly more positive, and V_{OL} = **–1.04V** to **–1.14V**.

14.32 For this smaller transistor, V_{BE} = 0.75 at I_E = 0.5 mA, with β = 30, and V_{OH} lowers.

Iterate: Normally V_{OH} = –0.88V. Try V_{OH} = –0.90V. Thus $I_{50} = \dfrac{2\text{–}0.90}{50}$ = 22 mA. Thus V_{BE} = $750 + 25 \ln \dfrac{22}{0.5}$ = 0.845V, $I_B = \dfrac{22}{31}$ = 0.710 mA, and V_B = 0.710 × 245 = 0.174V. Thus V_{OH} = 0 –.174 –.845 = –1.02V, for which $I_{50} = \dfrac{2\text{–}1}{50}$ = 20 mA, and $V_{BE} = 750 + 25 \ln \dfrac{20}{0.5}$ = .842V, and V_B = $\dfrac{20}{31}$ × .245 = 0.158V. Thus V_{OH} = –0.158 –0.842 = –1.000 V, as conjectured. Now $NM_H = V_{OH} - V_{IH}$ = –1.000 –(–1.205) = **0.205V**.

14.33 Assume υ_{BE} = 0.75 at 1 mA, n = 1, and β = 100. For $\upsilon_I = \upsilon_{R1}$, I_E splits equally between Q_A and Q_R. Using $\upsilon_{BE} \approx$ 0.75 V, I_E = (– 1.32 – 0.75 – – 5.2)/779 = 4.02 mA.

Thus $I_{EA} = I_{ER}$ = 2.01 mA, and $r_{eA} = r_{eR}$ = 25/2.01 = 12.4Ω.

Iterate for the current in Q_2:

Now, the voltage at the base of Q_2 is approximately υ_{B2} = – 245(2.01) = – 0.492 V.

Assuming υ_{BE} = 0.75 V, υ_{E2} = – .492 – .75 = – 1.242 and $i_{E2} \approx$ (– 1.242 – – 2)/50 = 15.2mA.

For i_E = 15.2 mA, υ_{BE} = 750 = 25ln(15.2/1) = 818 mV and i_B = 15.2/101 = 0.15 mA.

Thus $\upsilon_{E2} \approx$ 0 – 245(2.01 + 0.15) – 0.818 = – 1.35V, for which i_{E2} = (2 – 1.35)/50 = 130 mA, and for which υ_{BE2} = 750 = 25ln13.0 = 814 mV, and i_{B2} = 13.0/101 = 0.129 mA in which case

υ_{E1} – 0 – 245(2.01 + 0.129) – .814 = – 0.524 – .814 = – 1.34 V.

Thus, i_{E1} = 13.0 mA and r_{e2} = 25/13 = 1.92 Ω

Now the load reflected to the base of Q_2 is (β + 1)(r_{e2} + 50) = 101(1.92 + 50) = 5.24 kΩ.

Thus the voltage gain from υ_I to υ_{OR} is + (245 ll 5240)/(2(12.4)) = 234/24.8 = **9.44 V/V.**

14.34 *For* t_{PHL}, operation is likely dominated by R_T and the load capacitance as Q_2 cuts off. For this situation, $\tau = R_T C_L = 50(3 + 2) \times 10^{-12}$ = 250ps and the propagation delay to 50% of the signal swing is t_{PHL} – 250ln0.5 = **173 ps.**

For t_{PLH}, the situation is more complex, since Q_2 conducts. At the low end, the current is (V_{OI} – – 2)/50 = (– 1.77 + 2)/50 = 4.6 mA.

At the middle ($\approx V_R$), the current is (V_R – – 2)/50 = (– 1.32 + 2)/50 = 13.6 mA.

Use an averge current $= (13.6 + 4.6)/2 = 49.1mA$ for which $g_m \approx 9.1 \times 10^{-3}/(25 \times 10^{-3}) = 364$ mA/V.

Now $f_T = \dfrac{g_m}{2\pi(\pi + C_\mu)}$ and $(C_\pi + C_\mu) = g_m/(2_\pi f_T)$, for which

$C_\pi + C_\mu = 364 \times 10^{-3}/(2\pi \times 5 \times 10^9) = 11.6$ pF, and $C_\pi = 11.6 - 0.1 = 11.5$ pF.

As well, for $\beta = 100$, $r_\pi = \beta/g_m = 100/364 \times 10^{-3} = 275\ \Omega$.

The equivalent circuit is as shown:

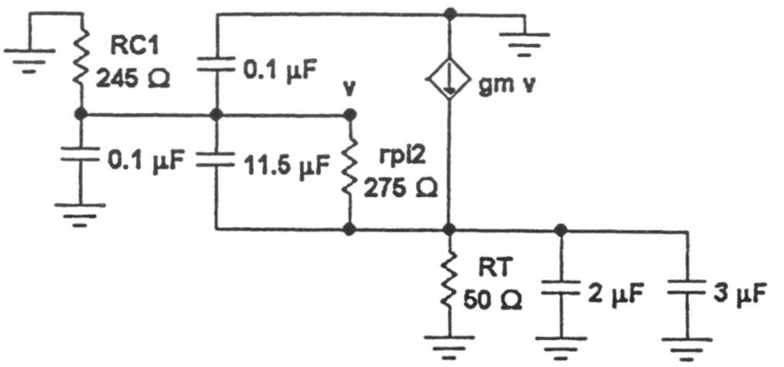

Now follow the general approach to follower frequency response to Section 7.6 of the Text, working with a basic-equivalent circuit and using the Miller approach to handling r_π and C_π, with the follower transformation to handle r_T and C_L. Here the follower voltage gain is $G = r_E/(1/g_m + r_E) = g_m r_E/(1 + g_m r_E)$.

The follower current gain is $(\beta + 1)$. The Miller multiplier for the base-to-emitter circuit is $(1 - G = 1 - g_m r_E)/(1 + g_m r_E) = \dfrac{1}{1 + g_m r_E}$.

Here, $1 + g_m r_E = 1 + 364 \times 10^{-3} \times 50 = 19.2$ and $\beta + 1 = 101$. Thus, in the base, we see a resistance $R_{eq} = 245 \parallel (275 \times 19.2) \parallel (50 \times 101) = 245 \parallel 5280 \parallel 5050 = 224\Omega$,

and a capacitance $C_{eq} = 0.1 + 0.1 + 11.5/19.2 + (2 + 3)/101 = 0.85pF$.

for which the time constant is $224 \times 0.85 \times 10^{-12} = 190ps$ and $t_{PLH} = -190\ln(0.5) =$ **132 ps.**

Thus $t_P = (132 + 173)/2 =$ **152 ps.**

For a fanout of 10, $C_L = 3/10 = 30$ pF. For t_{PHL}, the load capacitance enters the calculation directly. Thus $t_{PHL} = 17330 + 2)/(3 + 2) =$ **1.11 ns.**

For t_{PLH}, an additional capacitance of $9(3)/101 = 0.267$ pF is added to 0.85 pF to make *1.12 pF* for which $t_{PLH} = 132 \times 1/12/0.85 =$ **174 ps.**

Thus $t_P = (174 + 1110)/2 =$ **642 ps**

14.35 Here, the signal velocity = 2/3 × 0.3 = 0.2 mm/ps. For a rise time of 1.2 ns, and rise/return ratio of 6, wire length = 1/2 × 1.2/6 = 0.1ns, or 100ps. A line whose end-to-end delay is 100 ps is 0.2 × 100 = 20 mm long. Thus the longest allowed interconnect is **2 cm!**

14.36 Here, $I_{RE} = (5.2 - 1.32 - 0.75)/779 = 4.02$ mA

$I_{R3} = (5.2 - 1.32)/6100 = 0.656$ mA

$I_{R2} = (5.2 - 1.32 + 0.75 - 2(0.75))/4.98 = 0.653$ mA.

Use values of about **4, 0.66** and **0.66** mA to replace R_E, R_2, R_3.

Now, since I_E is constant, there is no need to make R_{C1} and R_{C2} different. R_{C1} should be raised to 245 Ω since the current in R_{C1} and R_{C2} is always 4 mA at its maximum. (Note that it was higher in $Q_{A,B}$ formerly because of R_E.)

As well, since I_{R2} is constant, D_1 and D_2 can be removed and replaced by a direct connection.

For the temperature effects:

Since the bias currents are assumed to be temperature-indpendent, most changes are due directly to the base-emitter variation, $\delta = -2$ mV/°C.

Following Example 14.3: Directly $\Delta V_R = -\delta =$ **2mV /°C.**

For V_{OL}, only the Q_2 variation is observed, through a voltage divider comprising re_2 and R_T.

Thus $\Delta V_{OL} = -\delta(R_T/(r_{e2} + R_T)$.

Now from page 1202, $I_{E2} = (-0.98 - 0.79 - -2)/50 = 4.6$ mA, for which $r_{e2} = 25/4.6 = 5.43\ \Omega$.

Thus $\Delta V_{OL} = -\delta(50/(50 + 5.43) = -0.90\delta =$ **1.8 mV /°C.**

For V_{OH}, $I_{E2} = 22.4$ mA, $r_{e2} = 25/22.4 = 1.1\ \Omega$

and $\Delta V_{OH} = -\delta 50/(50 + 1.1) = -0.98\delta =$ **2 mV/°C.**

Variation of the midpoint is $1/2(\Delta V_{OL} + \Delta V_{OH}) = \delta(-0.90 + 0.98)/2 = -0.94\delta \equiv$ **1.9 mV/°C**

Other changes:

- Of course, D_1 and D_2 are eliminated.
- R_A and R_B could be replaced by current sources of about $(5.2 - 1.32)/50k \approx 80\ \mu A$. This may save space, although R_A and R_B can be made with very imprecise processes that may be physically more compact than regular resistor designs (eg. using very narrow devices, or FETs).
- Constant currents could be supplied to replace some aspects of R_T. One might consider a current for which $(r_e + R_C/\beta) \approx 50\ \Omega$ for which $r_e = 5 - -2.45/100 = 47.5\ \Omega$ and $I_E = 25/47.5 = 0.53$ mA. Larger currents are a problem; certainly 22 mA is too power consuming; perhaps 4 mA current would be a compromise..RE

12SECTION 14.7: BiCMOS DIGITAL CIRCUITS

14.37 For Q_P in triode-mode operation:

$i_D = k_P((\upsilon_{SG} - V_t)\ \upsilon_{SG} - \upsilon_{SG}^2/2)$, and for small υ_{DS}, $i_D \approx k_P\ (\upsilon_{SG} - V_t)\ \upsilon_{SG}$, which for $\upsilon_{SGp} = V_{DD} =$ 5 V, results in $r_{SDp} = \dfrac{\upsilon_{SDp}}{i_{Dp}} = \dfrac{1}{k_P\ (5-1)} = \dfrac{1}{4k_P} = \dfrac{1}{2k_n} = r = 1/(2 \times 400 \times 10^{-6}) = 1.25k\Omega$.

For no load: $V_{OH} =$ **5.0V**. $V_{OL} =$ **0.0V**. For V_{th}: with R_2 small, so that Q_2 does not conduct, V_{th} is reached with Q_P, Q_N both in pinchoff and sharing the same current, ie $i_D = 1/2k_P(5 - V_{th} - 1)^2 = 1/2k_n(V_{th} - 1)^2$, or $k_n/2(4 - V_{th})^2 = k_n\ (V_{th} - 1)^2$, $4 - V_{th} = \sqrt{2}\ (V_{th} - 1)$, $2.414\ V_{th} = 5.414$, whence $V_{th} = \dfrac{5.414}{2.414} =$ **2.24V**, for which $i_D = k_n/2\ (2.24 - 1)^2 = 0.77\ k_n$. Now, the drop across $R_2 = r = \dfrac{1}{2k_n} = 1.25k\Omega$ is $\dfrac{1}{2k_n} \times 0.77k_n = 0.39$ V, and Q_2 does not turn on, as desired.

For a 5 kΩ load to 2.5 V: Assume, initially, that $V_{OH} = 5 - 0.7 = 4.3$V, and $V_{OL} = 0.7$V, for which $I_L = \dfrac{2.5 - 0.7}{5k\Omega} = 0.36$mA, directed either in or out.

For V_{OH} in detail: Q_1 conducts: $\upsilon_{SDp} = \upsilon$ is assumed small. Now, $i_{Dp} = k_p\ ((5 - 0 - 1)\ \upsilon - \upsilon^2/2)$ $\approx k_n/2(4\ \upsilon) = \dfrac{400}{2} \times 10^{-6} \times 4\upsilon = 800 \times 10^{-6}\ \upsilon$, and $r_{DSp} = \dfrac{\upsilon}{i_{Dp}} = \dfrac{1}{800 \times 10^{-6}} = 1.25$ kΩ. Now $R_1 =$ 1.25 kΩ, as well. Now for $\upsilon_O = +4.3$ V and $V_{B1} \approx 5$V, $I_L = 360\ \mu A$, $I_{B1} \approx \dfrac{360}{100} = 3.6\mu A$, and $i_{Dp} \approx \dfrac{V_{BEo}}{R_1} + I_{Bp} \approx \dfrac{0.7}{1.25k} + 3.6\mu A = 0.564$ mA. Now it is apparent that Q_3 will not conduct, but rather that Q_1 and R_1 in series will support the load, in which case $V_{OH} \approx 2.5 + \dfrac{5}{5 + 1.25 + 1.25}(5 - 2.5) =$ **4.17V**, with $V_{BE} \approx \dfrac{5 - 4.17}{2} = 0.42$V.

For V_{OL} in detail: Assume Q_2 does not conduct, but that V_{OL} is sustained by r_{DSN} and R_2 in series. Now for small υ_{DSN}, $r_{DSN} \approx \frac{r}{2} = 0.625$ kΩ. Thus $V_{OL} = 2.5 - 2.5 \left[\frac{5}{5 + 1.25 + 0.625}\right] =$ **0.68V**, with $V_{BE2} \approx \frac{1.25}{1.25 + 0.625} \times 0.68 = 0.45$ V. Now with this centred load, V_{th} will be that voltage for which the output voltage is centred, that is where Q_1 and Q_2 conduct equally. Thus V_{th} will be essentially as before, ≈ **2.24V**.

14.38 See that Q_{PX} and Q_{NX} below, form an output inverter connected in parallel with the main BiCMOS inverter.

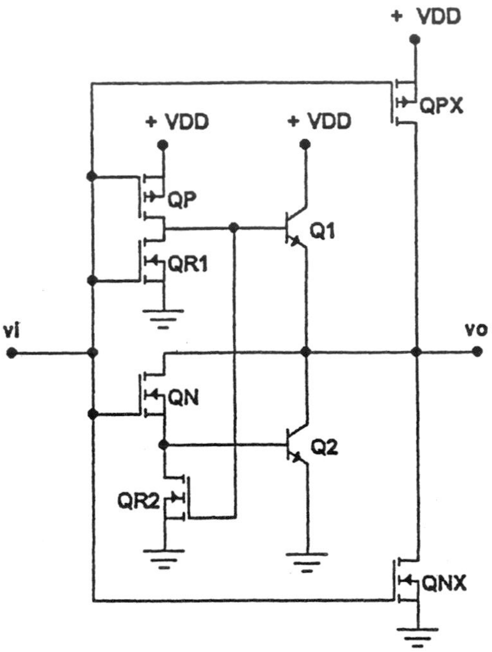

For the devices specified in P14.37, $k_n = 2k_p = 400\mu A/V^2$, $|V_t| = 1$ V, and $V_{DD} =$ 5V. Use such devices for Q_{PX} and Q_{NX}.

For Q_{PX} operating in the triode region, $i_D = k_p[(\upsilon_{SG} - V_t)\upsilon_{SD} - \upsilon_{SD}^2/2]$

For very small υ_{SD}, $i_D = k_p(\upsilon_{GS} - V_t)\upsilon_{SD}$ and $r_{SD} = -\upsilon_{SD}/i_D = 1/(k_p(\upsilon_{GS} - V_t))$

Here $r_{SDP} = 1/[200 \times 10^{-6}(5 - 1)] =$ **1.25 kΩ.**

Check: For $\upsilon_{SD} =$ 0.7V, $i_D = 200 \times 10^{-6}[(5 - 1)0.7 - 0.7^2/2] = 511\ \mu A$

$r_{eq} = 0.7/511 \times 10^{-6} = 1.37k\Omega$, which is not too much larger!

For Q_{NX} operating in the triode region (with $k_n = 2k_p$), $r_{DSN} = 1.25/2 =$ **0.625 kΩ.**

Thus the output resistances are **1.25 kΩ** with output high, and **625Ω** with output low.

14.39 From the introductory NOTE to Chapter 13, the corresponding minimum-size matched CMOS inverter has $V_{tn} = -V_{tp} = 0.6$ V, $\mu_n C_{ox} = 100\ \mu A/V^2$, $\mu_p C_{ox} = 40\mu A/V$ with $(W/L)_n = 1.2\mu m/0.8\mu m$ and $(W/L)_p = 3.0\mu m/0.8\mu m$. For this design, $k_n = k_p = 100 \times 10^{-6} \times 1.2/0.8 = 150\mu A/V^2$. The body-effect parameters are $\gamma = 0.5V^{1/2}$, and $2\Phi f = 0.6$ V.

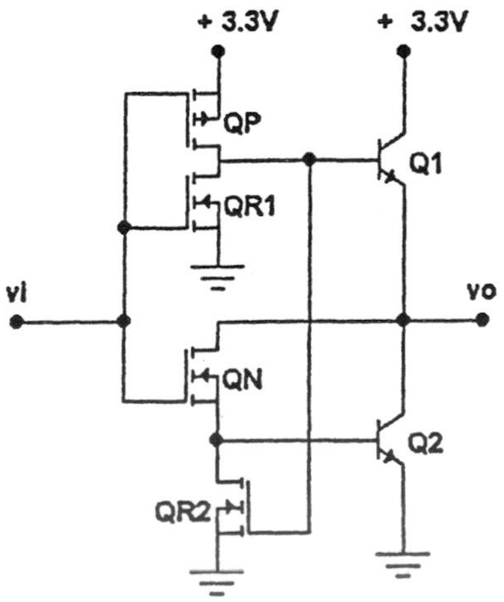

As implied in the specifications, all NMOS are minimum size and the PMOS 2.5 × wider. All have $k = 150\mu A/V^2$. For the BJTs, $V_{BE} = 0.5$ at the edge of conduction and 0.7 in conduction, with $\beta = 50$. The supply is 3.3 V.

For V_{OH}, V_{OL}: For very short-term signals V_{OH} and V_{OL} are $3.3 - 0.7 =$ **2.6 V** and **0.7 V** respectively. On a somewhat longer term, as the BJTs continue to conduct small currents, $V_{OH} =$ **+ 2.8V** and $V_{OL} =$ **0.5 V.**

Note that as the input rises and as Q_N begins to conduct, its source raises to 0.7V, for which $V_{tn} = Vsutt0 + \gamma[\sqrt{V_{SB} + 2\Phi f} - \sqrt{2\Phi f}$ (from Eq. 5.30).

Here $V_{tn} = 0.6 + 0.5[\sqrt{0.7 + 0.6} - \sqrt{0.6}] =$ $0.78V$. Earlier, for the source at 0.5V, $V_{tn} = 0.6 = 0.5[\sqrt{0.5 + 0.6} - \sqrt{0.6} = 0.74V$.

For V_{IH} and V_{IL}:

V_{IL} will occur approximately at the point when V_{B2} reaches 0.5V, as Q_N conducts in competition with Q_{R2}. First assume that V_{B1} does not change much in this interval.

At $V_{B2} =$ 0.5V,

$i_{DR2} = k[(\upsilon_{GS} - V_T)\upsilon_{DS} - \upsilon_{DS}^2/2] = 150 \times 10^{-6}[(3.3 - 0.6)0.5 - 0.5^2/2] = 184\mu A$.

For $V_{IL} = \upsilon$, $184 = 150/2(\upsilon - 0.5 - 0.78)^2$, $\upsilon - 1.28 = [2(184/150)]^{1/2} =$ 1.57, $\upsilon = 2.85V$, clearly very high.

Now V_{th} for the Q_P, Q_{R1} matched inverter is $3.3/2 =$ 1.65 V where $\upsilon_B =$ 1.65 and $i_{DR2} = 150 \times 10^{-6}[(1.65 - 0.6)0.5 - {}^2/2] =$ 60 μA.

For $V_{IL} = \upsilon$,

$60 = 150/2(\upsilon - 0.5 - 0.98)^2$ or $\upsilon - 1.28 = (120/150)^{1/2} =$ 0.89 $\upsilon =$ 2.17 V > 1.65 V.

Thus it is apparent that V_{IL} is too high, being controlled by the matched inverter.

Now modify the design to increase the width of Q_n to comparator. Use a 4× minimum device. Thus, repeating the last calculation but with $V_t =$ 0.74V: $60 = 4(150)/2(\upsilon - 0.74 - 0.5)^2$, or $\upsilon - 1.24 = (120/(4(150)))^{1/2} =$ 0.45, whence $\upsilon =$ 1.69V.

Though this may be acceptable, try **6× minimum Q_N**:

$60 = 6(150)2(\upsilon - 1.24)^2$, or $\upsilon - 1.24 = \frac{(120}{6/150})^{1/2} =$ .365, whence $\upsilon = 1.61V$

Thus $V_{IL} \approx$ **1.61V**, with Q_n with 6× minimum width.

For V_{IH}:

As estimate is when the output of the matched inverter reaches a level at which Q_1 and Q_{B2} begin to turn on . Since at the time $\upsilon_O =$ 0.5V to 0.7V, V_{B1} is at least 1.0V.

For the matched inverter with input υ

$1/2150(3.3 - \upsilon - 0.6)^2 = 150[(\upsilon - 0.6)1 - 1^2/2]$, or $(2.7 - \upsilon)^2 = 2(\upsilon - 0.6) - 1$, or $7.29 - 5.4\upsilon + \upsilon^2 = 2\upsilon - 1.2 - 2$, or $\upsilon^2 - 7.4\upsilon + 10.49 =$ 0, or

$\upsilon = (- \ - 7.4 \pm \sqrt{7.4^2 - 4(10.40})/2 = (7.4 \pm 3.58)/2 = 0.6$, and $V_{OH} =$ **1.91V**

For propagation delays:

For the *output rising from near zero*, the peak current from Q_P is nearly $i_{DP}/2)1/2(150)(3.3 - 0.6)^2 = 547\mu A$ (actually it will be slightly lower because the drain voltage must be high enough to cause Q to conduct, puring Q_P somewhat into the triode region). Correspondingly the peak output current from Q_1 will be $(50 + 1)(547\mu A) =$ **27.9 mA.**

For υ_O at 1.65V and $\upsilon_{B1} = 1.65 + 0.7 =$ 2.35, $\upsilon_{SDp} = 3.3 - 2.35 =$ 0.95V, in which case, $i_{Dp}(M) = 150[(3.3 - 0.6)(0.95) - 0.95^2/2] = 318\mu A$.

Corresponding, the available output current from Q_1 is 51(318) = **16.2 mA.**

Thus the average current causing the output to rise is $(16.2 + 27.9)/2 = 22.05mA$ and t_{PLH} (from $\upsilon_O =$ 0.5V to 1.65V) is $t_{PLH1} = 10 \times 10^{-12}(1.65 - 0.5)/22.05 \times 10^{-3} =$ **0.52 ns.**

For the output falling from 2.7V, the peak current from Q_n is

$i_{Dn}(H) = 1/2(6)150[3.3 - 0.7 - 0.78]^2 = 1490\mu A$.

Correspondingly, the current drain into Q_2 is 51(1490) = **76.0 mA.**

For $\upsilon_O =$ *1.65V*, $i_{DN}(M) = 6(150)[(3.3 - 0.7 - 0.78)(1.65 - 0.7) - (1.65 - 0.7)^2/2] = 1.154mA$.

Correspondingly, the current drawn from the load by Q_2 is 51(1.154) = **58.9 mA.**

Thus the average available current is $(76.0 + 58.9)/2 = 67.5mA$, and t_{PHL} (from v_O = 2.8V to 1.65V) is $t_{PHL1} = 10 \times 10^{-12}(2.8 - 1.65)/67.5 \times 10^{-3} =$ **0.170 ns.**

Thus $t_p = (0.170 + 0.52)/2 =$ **0.345 ns**

Augmented Circuit:

For a minimum-size matched inverter connected in parallel with this circuit, input to input, and output to output, the long-term values of V_{OH} and V_{OL} become – **3.3V** and **0.0V** respectively. On the short term, values of **2.6V** and **0.7V** would be expected.

For full-swing output; the BJTs bring the output to within 0.7V of the supplies, and the matched inverter drives it to the full swing. At the very end of the range over which the BJT operates, (v_{SD}) of Q_P is zero and the drive current is zero. Thus for the last half of the rising edge, the average current is $(0 + 16.2)/2 =$ 8.1 mA and the rise from 1.65 to $(3.3 - 0.7) =$ 2.6, (by 0.95 V) takes about

$t_{TLH2} = 10 \times 10^{-12}(0.95)/8.1 \times 10^{-3} =$ 1.17 ns. Finally, the current from the inverter *at 0.7V from the 3.3 V supply* is $150[(3.3 - 0.6)(0.7) - 0.7^2/2] = 247\mu A$, *and at 0.2V from the supply is* $150[(3.3 - 0.6)0.2 - 0.2^2/2] = 78\mu A$.

Thus the average current is $(247 + 78)/2 = 163\mu A$. Time taken is

$t_{TLH3} = 10 \times 10^{-12}(0.7 - 0.2)/163 \times 10^{-6} =$ 123 ns.

Now, the first part of the transition from 0.5 V to 1.65 V, t_{TLH1} is calculated above as $t_{PLH} = 0.52ns$. Thus, $t_{TLH} = t_{TLH1} + t_{TLH2} + t_{TLH3} = 0.52 + 1.17 + 123 =$ **125 ns.**

Correspondingly, the fall time is dominated by the final half-volt change. Thus $t_{THL} \approx$ **125 ns**, as well.

14.40 For short-duration signals, $V_{OH} = 3.3 - 0.7 =$ **2.6V**. For longer signals $V_{OH} =$ **3.3V**. For all signals $V_{OL} =$ **0V**.

For t_{PLH}:

For $v_O = 0$, $i_{D2}(L) = 1/2(100/2.5)(2.5)(3.3 - 0.6)^2(1.2/0.8) = 547\mu A$.

For $v_O =$ 1.65V, $i_{D2}(M) = 150[(3.3 - 0.6)(1.65) - \frac{1.65^2}{2}] = 469\mu A$.

Thus $i_{D2}(AV) = (547 + 469)/2 = 508\mu A$, and the average current available to the load capacitor is $51(508) = 25.9mA$.

Thus $t_{PLH} = 10 \times 10^{-12}(1.65)/25.9 \times 10^{-3} =$ **0.64 ns.**

For t_{PHL}:

For $v_O =$ 3.3V or 2.6V, $i_{D4}(N) = 1/2(100)(1.2/0.8)(\beta)(3.3 - 0.6)^2 = 1/2(150)(50)(2.7^2) = 27.3mA$.

For $v_O =$ 1.65, $i_{D4}(M) = 150(50)[(3.3 - 0.6)(1.65) - 1.65^2/2] = 23.1mA$.

Now, $i_{D4}(AV) = (27.3 + 23.1)/2 = 25.2mA$, and for a fall from 3.3V,

$t_{PHL} = 10 \times 10^{-12}(3.3 - 1.65)/25.2 \times 10^{-3} =$ **0.65 ns.**

Thus we see that the propagation delays are well-matched through the use of a *very large* MOS device.

For the supply current at $v_O = V_{DD}/2$, current I in Q_4 in $i_{B5} = (I - i_{D3})/51$, $i_{O1} = I/50$, and $i_{D2} = I/50 + (I - i_{D3})/51$.

But $i_{D3} = i_{D2}$. Thus $i_{D2} = I/50 + I/51 - i_{D2}/51$, $i_{D2}(52/51) = I(\frac{50 + 51}{50 \times 51})$, whence

$i_{D2} = \frac{50 + 51}{50 \times 52}I =$ 0.039I, $i_{D1} = I/50 =$ 0.020I and $i_{D2}/i_{D1} = 0.039/0.020 =$ 1.94, and $i_{D2} = 1.94 i_{D1}$.

Assuming Q_1 and Q_2 are in saturation, and $v_I = v$, where

$i_{D2} = 1/2(150)(3.3 - v - 0.6)^2 = 1.94 i_{D1} = 1.94(1/2)(150)(v - 0.6)^2$.

Thus, $1.94(2.7 - v)^2 = (v - 0.6)^2$, $\sqrt{1.94}(2.7 - v) = \pm(v - 0.6)$, and $3.76 - 1.39v = \pm(v - 0.6)$.

For the + choice: Now $2.39\upsilon = 4.36$ (the other choice leading to a non-physical result).

Thus $\upsilon = 4.36/2.39 = 1.82$V.

Now for $\upsilon_I = 1.82$V, $i_{D4} = 1/2(150)(50)(1.82 - 0.6)^2 = 5.6$ mA.

Thus the supply current is **5.6 mA**

14.41 A two-input BiCMOS NOR is as shown:

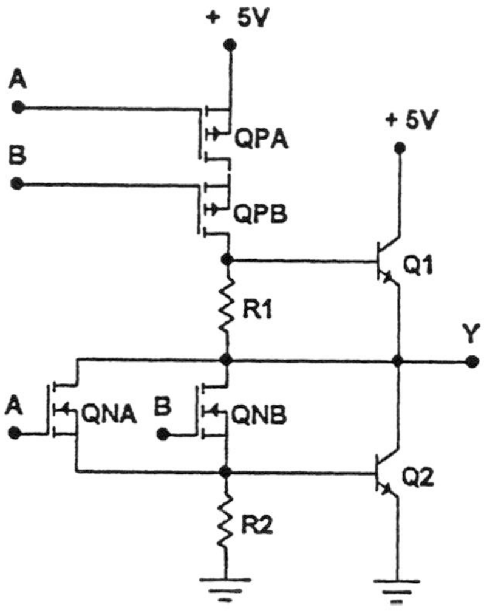

For $(W/L)_N = (W/L)_P$ in the prototype inverter, here $(W/L)_{PA} = (W/L)_{PB} = 2(W/L)_P$ and $(W/L)_{NA} = (W/L)_{NB} = (W/L)_N$. Since the inverter prototype uses equal-sized devices and $\mu_n = 2\mu_p$, in the NOR circuit, the NMOS have $k_n = k$, and the PMOS, $k_p = 2k/2 = k$.

Chose A low for simplicity (keeping Q_{PA} in tje triode mode).

At V_{th}, $V_B = V_Y = V_{th} = \upsilon$ and the voltage across Q_{PA} is υ_a. The current in all 3 devices Q_{PA}, Q_{PB} ad Q_{NB}, is the same, i. Assume the voltage drops across R_1 and R_2 are each 0.7V, raising the lower end of Q_{NB}, but otherwise having no effect.

For Q_{PA}, $i = k[(5-1)\upsilon_a - \upsilon_a^2/2]$ – – – (1).

For Q_{PB}, $i = k/2(5 - \upsilon - \upsilon_a - 1)^2$ – – – (2).

For Q_{NB}, $i = k/2(\upsilon - 0.7 - 1)^2$ – – – (3).

From (2), (3), $(\upsilon - 1.7) = (4 - \upsilon - \upsilon_a)$, whence $\upsilon_a = 5.7 - 2\upsilon$ – – – (4).

{*Check:* $\upsilon \le 5.7/2 = 2.85$ (seems OK)}

Substitute (4) in (1), and use (3) for i:

$i = k/2(\upsilon - 1.7)^2 = k[4(5.7 - 2\upsilon) - (5.7 - 2\upsilon)^2/2]$,

$\upsilon^2 - 3.4\upsilon = 2.89 = 45.6 - 16\upsilon - 32.49 = 22.8\upsilon - 4\upsilon^2$,

$5\upsilon^2 - 10.2\upsilon - 9.81 = 0$

whence $\upsilon = (+10.2 \pm \sqrt{10.s^2 - 4(5)(-9.81)}/25) = (10.2 + 17.3)/10 =$ **2.75V**

and from (4), $\upsilon_a = 5.7 - 2(2.75) = 0.20V$.

One (reasonable approach to designing this circuit, now, would be to first establish the value for k (possibly using minimum-size NMOS, as the original inverter specification implied), then finding i, then selecting $R_1 - R_2$ to remove a suitable fraction (say 10 to 20%) from each transistor base. You might try this for interest!!

SECTION 14.8: GALLIUM ARSENIDE DIGITAL CIRCUITS

14.42 For the load MESFET reduced in width from 6μm to 5μm: V_{DD} = 1.5V, V_{tD} = –1V, V_{tE} = 0.2V, β = 10^{-4}A/V^2 per μm, λ = 0.1V^{-1}. For the input device, W = 50μm and β_I = 5mA/V^2. For the load device, W = 5μm and β_L = 0.5mA/V^2. V_{OH} = **0.7V**.

For V_{OL} = υ: 5×10^{-3} [2 (0.7 – 0.2) υ – υ^2] (1 + 0.1 υ) = 0.5×10^{-3} [0 – 1 (–1)]^2 (1 + 0.1 (1.5 – υ), 10 (2 (0.5) υ – υ^2) (1 + 0.1 υ) = (1) (1 + 0.15 – 0.1 υ), or (10 υ – υ^2) (1 + 0.1 υ) = 1.15 –0.1 υ, 10 υ – 1υ^2 + 1 υ^2 – 0.1 υ^3 = 1.15 – 0.1 υ, 0.1 υ^3 – 10.1 υ + 1.15 = 0, and υ^3 – 101 υ + 11.5 = 0. For υ small, υ ≈ 11.5/101 = 0.114V. Thus V_{OL} ≈ **0.114V**.

For V_{IL} = υ: 5×10^{-3} (υ – 0.2)^2 (1 + 0.1 (0.7)) = 0.5×10^{-3} [2 (1) (1.5 – 0.7) – (1.5 – 0.7)^2] (1 + 0.1 (1.5 – 0.7)), or 10 (υ – 0.2)^2 (1 + .07) = 1 [2 (0.8) – .8^2] [1.08], υ^2 – .4 υ + .04 = .0969, υ^2 – .4 υ – .0569 = 0, $υ = \frac{- -.4 \pm \sqrt{.4^2 - 4(-.0569)}}{2} = \frac{.4 \pm .6226}{2}$ = 0.511, or –0.113V. Conclude V_{IL} = **0.51V**.

For V_{IH} = υ: Q_I in triode, Q_L in saturation, 5×10^{-3} [2 (υ – 0.2) $υ_O$ – $υ_O^2$] [1 + 0.1 $υ_O$] = 0.5×10^{-3} (0 – – 1)^2 (1 + 0.1 (1.5 – $υ_O$)). Now, neglecting terms in 0.1 $υ_O$, 2 (υ – 0.2) $υ_O$ – $υ_O^2$) = 0.1 (1 + .15), $υ_O^2$ – 2 υ $υ_O$ + 0.4 $υ_O$ + .115 = 0 – – – (1). Now taking $\frac{\partial}{\partial υ}$ → $2υ_O \frac{\partial υ_O}{\partial υ} - 2υ\frac{\partial υ_O}{\partial υ} - 2υ_O + 0.4\frac{\partial υ_P}{\partial υ}$ + .115 = 0, which for $\frac{\partial υ_O}{\partial υ}$ = – 1, becomes –2 $υ_O$ + 2υ – 2$υ_O$ –.4 + .115 = 0, 2υ = 4$υ_O$ + 0.285, υ = 2$υ_O$ + 0.1425 – – – (2).

Substituting (2) in (1) → $υ_O^2$ – 2$υ_O$ (2$υ_O$ + 0.1425) + 0.4 $υ_O$ + 0.115 = 0, $υ_O^2$ – 4 $υ_O^2$ – = 0, – 3 $υ_O^2$ – .115 $υ_O$ + .115 = 0, $υ_O^2$ + .0383 $υ_O$ – .0383 = 0, $υ_O = \frac{-.0383 \pm \sqrt{.0383^2 - 4(-.0383)}}{2} = \frac{-.0383 + .3933}{2}$ = 0.177V, whence V_{IH} = υ = 2 (.177) + .1425 = **0.497V**. Now NM_H = V_{OH} – V_{IH} = 0.700 –.497 = **0.203V**, and NM_L = V_{IL} – V_{OL} = 0.510 – .114 = **0.396V**.

14.43 From P13.45: β_I = 5mA/V^2; β_L = 0.5mA/V^2.

Output High: I_{DD} = 0.5mA/V^2 [2 (0 – – 1) (1.5 – 0.7) – (1.5 – 0.7)^2] [1 + 0.1 (1.5 – 0.7)] = 0.5 [2 (0.8) – 0.8^2] [1 + .08] = 0.5184mA.

Output Low: (V_{OL} = 0.114V) Q_L in saturation, I_{DD} = 0.5×10^{-3} (0 – – 1)^2 (1 + 0.1 (1.5 – .114) = 0.5693mA. Thus average current = $\frac{.5184 + .5693}{2}$ = **0.544mA**. Average static power + 1.5 (.574) = **0.816mW**.

For t_{pHL}: Need to calculate the time to fall from V_{OH} to $\frac{V_{OH} + V_{OL}}{2}$, that is, from 0.7 to $\frac{0.7 + .11}{2}$ = 0.407V. For i_I at $υ_O$ = 0.7, Q_I is in saturation (with $υ_{GS}$ = 0.7V), i_I = 5×10^{-3} (0.7 – 0.2)^2 (1 + 0.1 (0.7)) = 1.38mA. For i_I at $υ_O$ = 0.407, Q_I is nearly in saturation (with $υ_{GS}$ = 0.7V), i_I = 5×10^{-3} (0.70 – 0.2)^2 (1 + 0.1 (0.41)) = 1.30mA. Also i_L at $υ_O$ = 0.7V is 0.518mA, and i_L at $υ_O$ = .114V is 0.569mA, and (probably) i_L = .518 + $\frac{.569 - .518}{2}$ = 0.544mA, at $υ_O$ = 0.407. The discharge current = $\frac{1.38 - .518 + 1.30 - .544}{2}$ = 0.809mA, and $t_{pHL} = \frac{CV}{I}\ \frac{30 \times 10^{-15} \times (0.70 - .407)}{.809 \times 10^{-3}}$ = **10.9ps.**

For t_{pHL}: At $υ_O$ = .114V, i_L = .569mA. At $υ_O$ = 0.407V, i_L ≈ 0.544mA. Thus the charging current = $\frac{.544 + .569}{2}$ = 0.556mA, and $t_{pLH} = \frac{30 \times 10^{-15} (.407 - .114)}{.556 \times 10^{-3}}$ = **15.8ps**. Thus, overall, $t_p = \frac{15.8 + 10.9}{2}$ = **13.4ps.**

Dynamic Power at 2GHz: P_D = 30×10^{-15} × (0.70 – .114) (2 × 10^9) = **35.2μW**. Total power = 35.2μW + 816μW = **851μW**. Delay-power product DP = 851 × 10^{-6} × 13.4 × 10^{-12} = **0.011pJ.**

14.44 For the nominal design, $\beta_S = 10^{-4} \times 20 = 2\text{mA/V}^2$, $\beta_L = 2\text{mA/V}^2$, $\beta_{PD} = 10^{-4} \times 10 = 1\text{mA/V}^2$, $V_{tD} = -0.9\text{V}$, $\lambda = 0\text{V}^{-1}$, $V_{DD} = 3\text{V}$, $V_{SS} = 2\text{V}$, $V_D = 0.7\text{V}$. From Fig. 13.54 of the Text, for the nominal design, $V_{OH} = 0.7\text{V}$, $V_{OL} = -1.27\text{V}$, $V_{IL} = -0.26\text{V}$, $V_{IH} = -0.16\text{V}$, $NM_H = 0.86\text{V}$, $NM_L = 1\text{V}$.

(a) *For* $V_{tD} = -0.8\text{V}$, $V_{OH} = \mathbf{0.7V}$.

For V_{IL}, $i_S \simeq i_L - i_{PD}$, $i_{PD} = 1\,(0 - -0.8)^2 = .64\text{mA}$, $i_L = 2\,(0 - -0.8)^2 = 1.28\text{mA}$, $i_S = 1.28 - 0.64 = 0.64$ mA. Thus, $0.64 = 2\,(\upsilon - -0.8)^2$, $\upsilon + 0.8 = \sqrt{.64/2} = .566$, $\upsilon = .566 - 0.8 = \mathbf{-0.234V}$. Thus, $V_{IL} = \mathbf{-0.234V}$.

For V_{OL}, $i_{PD} = .64\text{mA}$, $i_L = 1.28\text{mA}$, $i_S = 1.28 - 0.64 = 0.64\text{mA}$. But $\upsilon_I = 0.7\text{V}$ with Q_S in triode mode, with $\upsilon_{DSS} = \upsilon_O$, $0.64 = 2\,[2\,(0.7 - -0.8)\,(\upsilon_O) - \upsilon_O^2]$, $0.32 = 3\upsilon_O - \upsilon_O^2$, $\upsilon_O^2 - 3\upsilon_O + 0.32 = 0$,

$$\upsilon_O = -\frac{-3 \pm \sqrt{3^2 - 4\,(.32)}}{2} = 0.111\text{V}. \text{ Thus, } V_{OL} = 0.111 - 0.7 - 0.7 = \mathbf{-1.29V}.$$

For $V_{IH} = \upsilon$: As above, but with input υ (rather than 0.7V), see $0.64 = 2\,[2\,(\upsilon - -0.8)\,\upsilon_O - \upsilon_O^2]$, $0.32 = 2\,(\upsilon + 0.8)\,\upsilon_O - \upsilon_O^2$, or $\upsilon_O^2 - 2\,(\upsilon + 0.8)\,\upsilon_O + 0.32 = 0$ — — — (1). $\frac{\partial}{\partial \upsilon} \rightarrow 2\upsilon_O \frac{\partial \upsilon_O}{\partial \upsilon} - 2\,(\upsilon + 0.8)\frac{\partial \upsilon_O}{\partial \upsilon} - 2\upsilon_O = 0$. Now, for $\frac{\partial \upsilon_O}{\partial \upsilon} = 1$, $-2\upsilon_O + 2\,(\upsilon + 0.8) - 2\upsilon_O = 0$, $2\upsilon = 4\upsilon_O - 1.6$, $\upsilon = 2\upsilon_O - 0.8$ — — — (2).

Substituting (2) in (1), $\upsilon_O^2 - 2\,(2\upsilon_O^2) + .32 = 0$, $3\upsilon_O^2 = .32$, $\upsilon_O = (.32/3)^{1/2} = 0.327\text{V}$, $\upsilon = 2\,(.327) - 0.8 = -0.146\text{V}$. Thus, $V_{IH} = \mathbf{-.146V}$. Now $NM_H = V_{OH} - V_{IH} = 0.7 - -.146 = \mathbf{0.85V}$, $NM_L = V_{IL} - V_{OL} = -.234 - -1.29\text{V} = \mathbf{1.06V}$.

(b) *For all* $V_{tD} = -1.0\text{V}$, $V_{OH} = \mathbf{0.7V}$.

For V_{IL}: $i_{PD} = 1\,(0 - -1.0)^2 = 1\text{mA}$, $i_L = 2\,(0 - -1.0)^2 = 2\text{mA}$, $i_S = 2 - 1 = 1\text{mA}$. Now $1 = 2\,(\upsilon - -1.0)^2$, $\upsilon = \sqrt{1/2} - 1 = -0.293\text{V}$, and $V_{IL} = \mathbf{-0.293V}$.

For $V_{OL} = \upsilon_O - 1.4$: $1.0 = 2\,[2\,(0.7 - -1)\,(\upsilon_O) - \upsilon_O^2]$, $0.5 = 3.4\,\upsilon_O - \upsilon_O^2$, $\upsilon_O^2 - 3.4\,\upsilon_O + 0.5 = 0$, and $\upsilon_O = -\frac{-3.4 \pm \sqrt{3.4^2 - 4\,(0.5)}}{2} = 0.308\text{V}$, $V_{OL} = 0.308 - 1.4 = \mathbf{-1.09V}$.

For $V_{IH} = \upsilon$, $1 = 2\,[2\,(\upsilon - -1)\,\upsilon_O - \upsilon_O^2]$, $0.5 = (2\upsilon + 2)\,\upsilon_O - \upsilon_O^2$, $\upsilon_O^2 - 2\,(\upsilon + 1)\,\upsilon_O + 0.5 = 0$ — — — (1), $\frac{\partial}{\partial \upsilon} \rightarrow 2\,\upsilon_O \frac{\partial \upsilon_O}{\partial \upsilon} - 2\,(\upsilon + 1)\frac{\partial \upsilon_O}{\partial \upsilon} - 2\,(\upsilon_O) = 0$. Now, for $\frac{\partial \upsilon_O}{\partial \upsilon} = -1$, $-2\upsilon_O + 2\upsilon + 2 - 2\upsilon_O = 0$, $2\upsilon = 4\upsilon_O - 2$, $\upsilon = 2\upsilon_O - 1$ — — — (2).

Substituting (2) in (1), $\upsilon_O^2 - 2\,(2\upsilon_O)\,\upsilon_O + 0.5 = 0$, $\upsilon_O = (0.5/3)^{1/2} = .408\text{V}$, $\upsilon = 2\,(.408) - 1 = -.184\text{V}$, $V_{IH} = \mathbf{-0.184V}$. Now, $NM_H = 0.7 - -.184 = \mathbf{0.88V}$, and $NM_L = -.293 - -1.09 = \mathbf{0.80V}$.

(c) *For* $W_L = 2W_{PD} = 5\mu\text{m}$, $\beta_S = 10^{-4} \times 20 = 2\text{mA/V}^2$, $\beta_L = 10^{-4} \times 5 = 0.5\text{mA/V}^2$, $\beta_{PD} = 10^{-4} \times 5/2 = .25\text{mA/V}^2$, $V_{tD} = -0.9\text{V}$, $\lambda = 0\text{V}^{-1}$, $V_{DD} = 3\text{V}$, $V_{SS} = 2\text{V}$. See $V_{OH} = \mathbf{0.7V}$.

For $V_{IL} = \upsilon$: $i_{PD} = 0.25\,(0 - -.9)^2 = .2025\text{mA}$, $i_L = 0.5\,(0 - -.9)^2 = .4056\text{mA}$, $i_S = .405 - .2025 = .2025\text{mA}$. Now $0.2025 = 2\,(\upsilon - -0.9)^2$, $\upsilon = (\frac{.2025}{2})^{1/2} - 0.9 = -.582\text{V}$, $V_{IL} = \mathbf{-0.582V}$.

For $V_{OL} = \upsilon_O$: $\upsilon_O^2 - 3.2\upsilon_O + .10125 = 0$, $\upsilon_O = \frac{3.2 \pm \sqrt{3.2^2 - 4\,(.10125)}}{2} = \mathbf{0.032V}$, $V_{OL} = .032 - 1.4 = \mathbf{-1.37V}$.

For $V_{IH} = \upsilon$, $0.2025 = 2\,[2\,(\upsilon - -0.9)\,\upsilon_O - \upsilon_O^2$, $\upsilon_O^2 - 2\,(\upsilon + 0.9)\,\upsilon_O + .10125 = 0$ — — — (1). $\frac{\partial}{\partial \upsilon} \rightarrow 2\upsilon_O \frac{\partial \upsilon_O}{\partial \upsilon} - 2\,(\upsilon + 0.9)\frac{\partial \upsilon_O}{\partial \upsilon} - 2\upsilon_O = 0$. Now for $\frac{\partial \upsilon_O}{\partial \upsilon} = -1 \rightarrow -2\upsilon_O + 2\upsilon + 1.8 - 2\upsilon_O = 0$, $\upsilon = 2\upsilon_O - 0.9$ — — — (2).

Substituting (2) in (1), $\upsilon_O^2 - 2\,(2\upsilon_O)\,\upsilon_O + .10125 = 0$, $\upsilon_O = (\frac{.10125}{2})^{1/2} = 0.184\text{V}$, $\upsilon = 2\,(.184 - 0.9) = -.532\text{V}$, $V_{IH} = \mathbf{-0.532V}$. Now, $NM_H = V_{OH} - V_{IH} = 0.7 - -.532 = \mathbf{1.23V}$, and $NM_L =$

$V_{IL} - V_{OL} = -.582 - -1.37 =$ **0.79V**.

13.45 Use $V_{tD} = -0.9$V for all devices, and use $W_S = W_L = 20\mu$m, and $W_{PD} = 10\mu$m for easy equivalence to the *FL* design. However, actually, since Q_{PD} is the load presented by each gate, and since the loading of the Q_{PD} drain is very light (the gate of Q_S), then Q_{PD} can be made much smaller. For a fanout of 4, reduce the width of Q_{PD} to 2.5 μm or less. $\beta_S = \beta_L = 20 \times 10^{-4} = 2$mA/V^2, $\beta_{PD} = 10 \times 10^{-4} = 1$mA/V^2, $V_D = 0.7$V. For the *FL* gate, $V_{OH} = 0.7$V, $V_{IL} = -0.26$V, $V_{IH} = -0.16$V, $V_{OL} = -1.27$V. At the gate of Q_S of the connected circuit, $\upsilon_O = 0.7$V, $V_{OH} = 0.7 + 0.7 + 0.7 =$ **2.1V**, with a fanout of 1! $V_{OL} = -1.27 + 2\ (0.7) =$ **0.13V**, or lower with a fanout > 1. $V_{IL} = -0.26\text{V} + 1.4\text{V} =$ **1.14V**, $V_{IH} = -.16\text{V} + 1.4\text{V} =$ **1.24V**. Thus, $NM_H = 2.10 - 1.24 =$ **0.86V**, $NM_L = 1.14 - .13 =$ **1.01V**. For a design in which W_{PD} is reduced to 2.5μm, these results are an approximation for a fanout of 4. For a fanout of 1, V_{OL} increases, as do V_{IH} and V_{IL} slightly.

PART III
ANSWERS

pages 427 to 458

Chapter 1

INTRODUCTION TO ELECTRONICS

1.1 See pages 132 and 134 for sketches.

1.2 377 rad/s, 120 Hz, 400 Hz, 6.35×10^6 rad/s, 611×10^6 rad/s, 6.28 rad/s, 60 Hz, 0.159 Hz, 62.8×10^9 rad/s, 25.1×10^{11} rad/s.

1.3 1.67×10^{-2}s, 16.7 ms; 8.33×10^{-3}s, 8.33 ms; 2.50×10^{-3}s, 2.50 ms; 9.90×10^{-7}s, 990 ns; 1.03×10^{-8}s, 10.3 ns; 1.00 s, 1.00 s; 1.67×10^{-2}s 1.67 ms; 6.29 s, 6.29 s; 1.00×10^{-9}s, 1.00 ns; 2.50×10^{-12}s, 2.50 ps.

1.4 7.01 ps.

1.5 a) 1 times, 1 times; b) 30 times, 50 times; c) 10 times, 1/500 times.

1.6 4.05%; 1.89%. See page 200 for square-wave recompositions.

1.7 Between 5 kHz and 6 kHz.

1.8 a) 2.82 Vpp square wave; b) 2.82 Vpp square wave; c) 2 Vpp square wave; d) sequence of positive and negative pulses of amplitude 1.41 V, width 1/4 f, spaced 1/4 f apart; dc level of 1.41 V.

1.9 6 bits; 000000, 000111, 001111, 011111, 100001; 63.

1.10 4 bits; 0000, 0100, 0111, 1110; 30.

1.11 a) 180, 52; b) – 76, 52.

1.12 a) 45, 173; b) 45, – 83; c) 45, – 45.

1.13 $1111_2 = 31_{10}$, $0000_2 = 0_{10}$; 13; 1.219 V; 2.906 V; 0.09375 V; 1.03125; 01011_2.

1.14 b) 1 mA, 20 mW, 2×10^3μA, 20 μW, 1 mA, 0.5 mW, 0.05 V/mV, 34 dB, 0.5×10^{-3} mA/μA, –6 dB, 0.025 mW/μW, 14 dB, 2.5%; c) 0.05 mA, 0.05 mA, 1 mW, 100 mV, 50 μW, 2 V, 0.2 kΩ, 0.02 V/mV, 26 dB, 0.01 mA/μA, 20 dB, 0.2 mW/μW, 23 dB; d) 10 mA, 10 mA, 14.1 mV, 1.41×10^3 μA, 2.82 V, 28.2 mA, 0.1 kΩ, 46 dB, 0.02 mA/μA, 26 dB, 4 mW/μW, 36 dB, 20%; e) 3.1×10^{-3} mA, 3.1×10^{-3} mA, 0.01 V/mV, 20 dB, 0.01 mA/μA, 20 dB, 20 dB.

1.15 140 mVp.

1.16 5.66 Vrms, –1 V, – 20 mV.

1.17 3.125%.

1.18 0.598 V, 5 V, 0.613 V, 1.89 Vp, –97.8 V/V.

1.19 430 Ω, 53.8 V/V.

1.20 2.0 kΩ.

1.21 $A_{\upsilon 0} = 1 + R_o/R_i$.

1.22 A_1A_2, 12.5 V/V, 0.05 V/V, 0.5 V/V, double the gain.

1.23 0.99×10^4 V/V, 0.5×10^4 V/V, 9.99 V/V, 1 A/V.

1.24 a) 0.066 V/V, 0.178 V/V, 0.016 V/V; b) 1, 2 or 3, 3 or 2; 3, 2, 1; 2, 1, 3; c) A_1A_2, A_1A_3; d) 30.3 V/V for A_1A_2.

1.25 9 amplifier pairs: (2,1): 24752 V/V; (1,1): 4901 V/V; (1,2), (1,3): 4541 V/V; (2,2), (2,3), (3,2), (3,3): 4132 V/V; (3,1): 2500 V/V.

1.26 A_1, A_3, A_2; 4132 A/A.

1.27 10 A/V, 10^{-1} A/V, 10^{-2} A/V; $FM2 = g_m R_o R_i^2$; A_3, A_2, A_1.

1.28 200 mA/mA, 40 mA/V.

1.29 $v_e/v_b = 1/(1 + r_\pi/[(\beta + 1)R_e] = r_e/(r_e + R_e)$, $R_{in} = r_\pi/(\beta + 1) = r_e$.

1.30 $v_c/v_b = -\alpha R_L/(r_e + R_e)$, where $r_e = r_\pi/(\beta + 1)$; $R_{out} = \infty$.

1.31 $v_e/v_s = [(r_\pi + (\beta + 1)R_e)/(R_S + r_\pi + (\beta + 1)R_e)] \times [(\beta + 1)R_e/(r_\pi + (\beta + 1)r_e)]$; $R_S = r_\pi + (\beta + 1)R_e$.

1.32 2000 V/V, 72 ° lagging.

1.33 100 kHz, 80 kHz.

1.34 high-pass output, low-pass output.

1.35 7.27 kHz, 80 MHz, 12.6 kHz.

1.36 1/(RC) rad/s, 0.0644/(RC) rad/s, k = 0.25.

1.37 60 dB, 0 dB, –20 dB, 40 dB; 60 dB, 0 dB, –3 dB, 57 dB; 60 dB, 0 dB, 0 dB, 60 dB; 60 dB, 0 dB, 0 dB, 60 dB; 60 dB, –3 dB, 0 dB, 57 dB; 60 dB, –20 dB, 0 dB, 40 dB; 10^5 Hz; from 100 Hz to 10^4 Hz.

1.38 10 Hz, 10^5 Hz, 10^3 V/V; The standard form is more straightforward.

1.39 $A_v = R_i/[R_s + R_i] \times G_m R_o$, ω_H = $1/(R_o C)$; 155 V/V; $GB = 100/(R_s C)$ for large I, independent of I; 15.9 V/V, 126 mA.

1.40 59.5 V/V; 119 pf, 6.8 nF; 47.6 V/V; 39.304 kHz to 40.708 kHz; 1.404 kHz.

1.41 $[RCs-1] / [RCs + 1]$.

1.42 0.5 V, 3.5 V, 2.35 V, 2.65 V, 0.85 V, 1.85 V; 0.3 V; 0.55 V, 1.55 V; 2.0 V.

1.43 5.0 V, 0.286 V, 23.6 mW, 0 mW, 4.17 V, 13.9 mW.

1.44 0.0 V, 5.0 V, 0 mW; 49.5 mV, 4.95 V, 4.95 mW.

1.45 250 Ω, 250 Ω; 0.0 V, – 1.0 V; 0.5 V; 20 mW; No.

1.46 25 mW; 1.15 ns, 1.15 ns; 0.35 ns.

1.47 0.0 V, 3.0 V, 0 mW; 29.7 mV, 2.97 V, 1.78 mW.

1.48 – 1 V, 0 V, 1.73 ns, 1.73 ns, 0.52, ns, 0.52 ns, 20 mW, 1.2 mW, 21.2 mW.

Chapter 2
OPERATIONAL AMPLIFIERS

2.1 2 op amps, no pins unused; 4 op amps, no pins unused.

2.2 – 300 μV, 100.3 mV.

2.3 0.35 V.

2.4 – 10 V/V, – 0.1 V/V.

2.5 – 1 V, $10^{-7}A$ (from ground), – 1.002 V.

2.6 2 solutions: a) 100 kΩ; b) 50 kΩ.

2.7 $R_1 =$ 22 kΩ.

2.8 Use $R_1 =$ 100 kΩ with $R_2 =$ 2 MΩ, using two 1 MΩ resistors in series. Other options are discussed.

2.9 a) – 10 V/V, – 0.1 V/V; b) – 9.009 V/V, – 0.0989 V/V.

2.10 909 V/V, 10^4 V/V.

2.11 $R_i = R_2/(1 + A)$; $G = AR_i/(R_1 + R_i)$.

2.12 1.010 kΩ, – 96.8 V/V, 98.8 kΩ, 100 kΩ, – 1.2%, 8.77 kΩ.

2.13 $R_1 =$ 100kΩ, $R_2 =$ 1MΩ, $R_3 =$ 50kΩ, $R_4 =$ 1MΩ.

2.14 $R_1 =$ 1MΩ in series with 1MΩ, $R_2 =$ 1MΩ, $R_3 =$ 2.5kΩ, $R_4 =$ 1MΩ.

2.15 $R_1 = R_2 = R_4 =$ 1MΩ, $R_3 = 100k\Omega \| 100k\Omega$.

2.16 $v_O = -(R_4/R_3)v_2$, $v_O = -(R_2/R_2)(1 + R_4/R_2 + R_4/R_3)v_1 - (R_4/R_3)v_2$.

2.17 $v_o(s)/v_i(s) = -(R_2/R_1)(1 + R_1C_1s)/(1 + R_2C_2s)$, independent of frequency if $R_1C_1 = R_2C_2$: a) – 10 V/V, independent of frequency; b) – 100 (1 + s / 100) / (1 + s / 10); c) – 1 (1 + s / 100) / (1 + s / 1000).

2.18 Negative-going ramp of slope 1000 V/s, falling from 10 V to 0 V in 10 ms.

2.19 An inverted sine-wave of 26.5 mV peak lagging by 90°.

2.20 Input falls at a rate of 200 V/s; for the rise, $v_O =$ – 5 V; for the fall, $v_O =$ +5 V.

2.21 See pages 163 and 164: a) – 10 V, 0.1 ms, +1 V; b) 5 V, 0.1 ms, 0 V.

2.22 Use $R_1 =$ 30kΩ, $R_2 =$ 15kΩ, $R_3 =$ 10 kΩ, $R_f =$ 30kΩ.

2.23 See pages 164 and 165: two op amps with $R_1 =$ 20kΩ, $R_2 =$ 10kΩ, $R_3 =$ 15kΩ, $R_4 =$ 30kΩ, $R_5 =$ 10kΩ, $R_8 =$ 10kΩ, with a total resistance of 95kΩ.

2.24 $v_O = V_O - 1000\int_0^t \left[v_1(t) + 2v_2(t)\right]dt$.

2.25 11 V/V, 1.10 V/V.

2.26 See page 166.

2.27 $R_1 = R_4 =$ 20kΩ, $R_2 = R_3$ 10kΩ, $R_f =$ 30kΩ.

2.28 $R_1 = R_2 =$ 10kΩ, $R_3 = R_f =$ 100 kΩ. This configuration, with $R_f/R_2 = R_3/R_1$, is called a difference amplifier.

2.29 $G =$ 0.909 V/V; $R_1 =$ 90kΩ, $R_2 =$ 10kΩ.

2.30 20kΩ to the – 10 V supply from the op amp negative input, with $R =$ 2.22kΩ.

2.31 See page 169: $R_1 =$ 10kΩ, $R_2 =$ 20kΩ; total power = 367.5 mW.

2.32 Gain = – 10 V/V.

2.33 Gain = – 5 V/V; Remove R_1 and R_2 and connect sources directly, or make $R_2 = R_4 =$ 200kΩ; Add an additional 2kΩ resistor in series with R_{S2} and R_3, or change R_4 to 180kΩ.

2.34 $v_O = -v_1(R_2/R_1)+v_2(1+R_2/R_1)/(1+R_3/R_4)+v_3(1+R_2/R_1)/(1+R_4/R_3)$; $v_O = (R_2/R_1)(v_2-v_1) + v_3$.

2.35 a) 5 V, b) 5 V, c) 0 V; a) 5 V, b) 5 v, c) 0 V.

2.36 For $R_1 =$ 10kΩ, $R_2 =$ 45kΩ, and for $R_3 =$ 10kΩ, $R_4 =$ 100kΩ; $v_{O1} =$ 5.4 V, $v_{O2} =$ 4.5 V.

2.37 $R_i = R_4 - R_1R_3/R_2$; $v_X/v_W = 1/(1-(R_2/R_1)(R_4/R_3))$, $v_Y/v_W = (1+R_2/R_1)/(1+(R_2/R_1)/(R_4/R_3))$;
a) R_3, – 1 V/V, – 2 V/V; b) 0 Ω, ∞, ∞ ; c) $-R_3/2$, 2 V/V, 4 V/V.

2.38 $I_N = v_I/R$, $R_N = \infty$; $i_2 = v_I/R$; Z/R, 1/sCR; RC, 1/RC.

2.39 30 V/V, 0.075 V/V; CMRR = 400 V/V, or 52 dB.

2.40 10 Vpp.

2.41 10 Hz, 100 V/V.

2.42 0.909 MHz, 1.00 MHz, 90.9 kHz, 100 kHz.

2.43 10^3 V/V for a single amplifier; 4.14×10^5 V/V for two stages in cascade.

2.44 – 100 V/V, 30 kHz, 6 MHz.

2.45 4.55 MHz.

2.46 1.6 V.

2.47 0.83 MHz.

2.48 $V_o = SR/(2\pi f_b)$, 0.64 Vpeak.

2.49 0.40 mV.

2.50 0.40 V; Use $R_3 =$ 100 kΩ for which $v_O =$ 0.13 V; For case a), bias current dominates; For case b), offset voltage dominates. For each effect halved, the output offset becomes 0.25 V and 0.08 V respectively.

2.51 a) 0.3 V; b) 0.03 V, with 10 MΩ to compensate.

2.52 $R =$ 15 kΩ, $R_{in} =$ 15 kΩ.

2.53 a) 3.22 s, b) 9.09 s; 10 kΩ, longer by perhaps 10×, nothing happens; 2.1 V.

2.54 101 V/V, 1.19 μA; $V_{OS} =$ 3.7 mV, $I_{OS} =$ 0.22 μA; 1.41 V; Reduce resistors to 1 kΩ and 100 kΩ, and use 10 kΩ in series with the positive input; 22 mV.

Chapter 3
DIODES

3.1 a) 0 V, 5 mA; b) – 5 V, 10 mA; c) – 10 V, 10 mA; d) 5 V, 5 mA; e) 5 V, 0 mA; f) – 5 V, 0 mA.

3.2 a) 5 V, logic OR (in positive logic), logic AND (in negative logic); b) 0 V, OR (in positive logic), AND (in negative logic); c) 5 V, AND (in positive logic), OR (in negative logic); d) 0 V, AND (in positive logic), OR (in negative logic); e) 0 V, AND (in positive logic), OR (in negative logic).

3.3 $Y = AE + BC + D$ in positive logic, 5 V.

3.4 82.8 mA, 13.5 mA, 49.5 mA, 6.2 mA.

3.5 5 mA, 3 mA, 1 mA, 1 mA, 0 V.

3.6 $n = 2.00$, $I_S = 8.32 \times 10^{-17} A$.

3.7 0.758 V, 9.95 mA.

3.8 0.355 V, 9.2 mV.

3.9 0.445 V.

3.10 128 nA, 181 nA.

3.11 $p_{po} = 10^{-m}$, $n_{po} = 10^{-n}$.

3.12 $1.6 \times 10^5 carriers/cm^2$ at 200 °C, $1.5 \times 10^{10} carriers/cm^3$ at 300 °C, $5.2 \times 10^{12} carriers/cm^2$ at 400 °C; (3.5×10^4)%; 1 in 10^{10}.

3.13 a) $2.3 \times 10^5 \Omega$ cm and 6.59 Ω cm; b) 1.73 Ω cm and 1.73 Ω cm.

3.14 Larger in the lighter-doped p region, by ten times.

3.15 a) 0.307 μm, 154 nm, 154 nm; 0.665 pC, 0.665 pC; b) 0.405 μm, 0.368 μm, 0.037 μm; 0.504 pC, 0.504 pC.

3.16 5 nA.

3.17 $q_J = A\left[2\varepsilon_s q \frac{N_A N_D}{N_A + N_D}(V_0 + V_R)\right]^{1/2}$; a) 0.67 pC at 0 V, 2.60 pC at 10 V, 2.72 pC at 11 V; 119 fF, 119 fF; 1.89 fF at 10.5 V, 90.6 fF at 100 V; b) 0.50 pC at 0 V, 1.97 pC at 10 V, 1.97 at 11 V; 90.1 fF, 90.0 fF; 143 fF at 10.5 V, 68.7 fF at 100 V.

3.18 0.1 pF.

3.19 $V_O = 0.71$ V, $C_{j0} = 15.4$ pF, $C_j = 15.4$ pF.

3.20 0.42 mA, 4.2 mA.

3.21 1 μm, 2.3 μm.

3.22 $I_S = 6.0 \times 10^{-17} A$.

3.23 + 19.4% / °C.

3.24 holes: 85.%, 0.895 pC; electrons: 10.5%, 0.210 pC; 1.1 ns, 4 pF.

3.25 50 ps; same; 50 fC, 0.5 pC.

3.26 $C_d = \tau I / (nV_T)$.

3.27 a) 0.75 V, 2.5 mA; b) 0.73 V, 1.7 mA; c) 0.74 V, 1.8 mA.

3.28 0.741 V, 2.59 mA.

3.29 0.708 V, 2.92 mA.

3.30 0.723 V, 2.73 mA; 0.750 V, 2.5 mA.

3.31 a) 0.7 V, 1 mA; b) 0.675 V, 1.25 mA; c) 0.701 V, 0.91 mA; d) 0.750 V, 0.50 mA.

3.32 600Ω

3.33 4.04 V.

3.34 0.752 V.

3.35 At 0.1 mA, $r =$ 500Ω; At 10 mA, $r =$ 5Ω; The geometric mean is likely to be best; $(r_1 + r_2)/2 =$ 252.5Ω, $(r_1 r_2)^{1/2} =$ 50Ω; At 5.05 mA, $r =$ 9.9Ω. At 1 mA, $r =$ 50Ω. The arithmetic mean is clearly not very relevant!

3.36 25Ω; 1 mA in each; 50 Ω; 25 Ω; This demonstrates that diode incremental resistance is independent of diode junction size.

3.37 0.005 V/V, 0.05 V/V, 0.33 V/V, 0.83 V/V.

3.38 Currents split equally; Thus all diode currents are equal; $R_T =$ 100/I; At 10 mA, $\upsilon_O/\upsilon_S =$, 0.999 V/V; At 1 μA, $\upsilon_O/\upsilon_S =$ 0.0909 V/V; Linearity is critical for small currents; υ_S is limited to 22 mVpeak for $I =$ 1 μA.

3.39 ± 40 mV, or ± 1%; – 50 mV, or – 1.25%; Combined as – 90 mV, or – 2.25%; 3.91 V.

3.40 $C_{j0} =$ 1.99 pF, $V_0 =$ 0.70 V, $m =$ 1.10, $n =$ 2.0, $\tau_T =$ 300 ps, $C_T =$ 64 pF.

3.41 10 Ω, 30 pF, 40 pF, 70 pF, 1 pF.

3.42 6.70 V, 6.7 V, 11.3 mA, 7.9 V.

3.43 90.9 mV/V, – 18.2 mV/mA.

3.44 150Ω, 6.57 V, 7.19 V.

3.45 270Ω, 15.39 V, 296 mW.

3.46 10.6 V; Conduction for one-half cycle, or more precisely, 48% of a cycle; 3.60 V, 11.3 V; 9.97 V, 3.06 V.

3.47 See page 188: 2.40 V.

3.48 See page 188: 152.7 mA.

3.49 See page 189: 15.57 V, 0.876 ms, 9.40 V, 16.3 V.

3.50 16.3 V dc output; 33.2 V PIV.

3.51 41.7 μF, 167 μF, 28.8 mA, 81.8 mA.

3.52 20.9 μF, 83.4 μF, 14.4 mA, 40.9 mA, 33.2 V.

3.53 Use an 18.2 V rms centre-tapped secondary, 2500 μF, 25.1 V, 2.78 A.

3.54 $i_D(Av) =$, $1.05A$, $\upsilon_O =$ 18.2 V; With a source resistance, the output drops by 0.3 V or 1.6% to 17.9 V on average.

3.55 +3.0 V for 6.0 V input; – 3.0 V for – 6.0 V input; $K =$ 0.5 V/V; 0.9 mA.

3.56 See page 192.

3.57 9.1 Vpp, 313Ω.

3.58 For light load, the output is a square wave of period T going from + 0.5 V to – 89.5 V; As R reduces, the negative side rises toward ground. For $RC =$ 2T, the waveform rises initially to 0.7 V, then drops to 0.55 V, then falls to – 89.5 V, then reduces to – 69.7 V, then rises to 0.7 V to begin a new cycle.

3.59 198.6 V, 193.6 V, 19.9 mA.

3.60 See page 194: 50 V, 75 V, 87.5 V, 96.9 V, 99.8 V.

3.61 88.4 V using 0.7 V diodes.

3.62 $n =$ 1.216, $5.18 \times 10^{-6}A$, 17 mΩ.

3.63 334 mV, 880 mV.

3.64 696 mV, 823 mV.

3.65 59.1 pF, 23.6 pF, 7.1 pF, 59.1 pF.

3.66 70 μA, 98.7 μA; 1.5 nA, 1.54 μA.

3.67 1.40 V, 0.35 V; See page 195.

3.68 0 V; v_O rises' 70 mV.

3.69 1.59 W; 36 series diodes (cells); $n \approx 2$; 0.6 mA.

3.70 19 mW; 15.7 mA, 1.91 V; 453 Ω, 287 Ω.

3.71 Solar-cell mode; 30 μA to 80 μA; See page 198.

Chapter 4

BIPOLAR JUNCTION TRANSISTORS (BJTs)

4.1 (1) npn, active; (2) npn, cutoff; (3) pnp, cutoff; (4) pnp, saturated; (5) pnp, cutoff; (6) npn, saturated.

4.2 4 modes; EBJ Reverse, CBJ Forward.

4.3 $27.5 cm^2/S$, $1.39 \times 10^4 W \mu m^2$ (for W is μm); $2.78 \times 10^4 \mu m^2$.

4.4 $1.38 \times 10^{-13} A$, 200 mA, 578 mV, $1.66 \times 10^{-8} A$, 275 mV.

4.5 $n_{p0} = 2250/cm^3$, $n_p(0) = 3.25 \times 10^{15}/cm^3$; For $W = 1$ μm: $I_n = -$ 0.443 mA, $I_S = 3.07 \times 10^{-16} A$, $i_E =$ 0.44 mA, $\beta =$ 368, $\alpha =$ 0.997; For $W = 0.1 \mu m$: $I_n = -$ 4.43 mA, $I_S = 3.07 \times 10^{-16} A$, $i_E =$ 4.44 mA, $\beta =$ 6807, $\alpha =$ 0.9999.

4.6 0.469 μm, 0.288 μm.

4.7 $5.11 \times 10^{-17} A$, 403×, 4.03 mA, 29.8 mA.

4.8 $I_S = 1.03 \times 10^{-14} A$, $I_S/\alpha = 1.04 \times 10^{-14} A$, $I_S/\beta = 7.73 \times 10^{-17} A$; $\alpha =$ 0.993, $\beta =$ 133.3.

4.9 β from 125 to 375; α from 0.9920 to 09973.

4.10 1.46 pC, 87.3 ns.

4.11 $\upsilon_{CB} \geq 0$ V, $\upsilon_{CE} \geq 700$ mV; $I_S =$ 6.91×10^{-15} A, $\beta =$ 100, $i_E =$ 10.1 mA.

4.12 b) 20 μA, 1.02 mA, 0.980; c) 1.96 mA, 40 μA, 49; d) 1.99 mA, 2.00 mA, 199; e) 100 mA, 10 mA, 0.909; f) 1 mA, 1.001 mA, 0.999 mA.

4.13 See page 204: αi_E or βi_B.

4.14 0.47 V, 0.24 V.

4.15 10 nA, 1.28 μA.

4.16 0.01.

4.17 0.390 mA, 0.429 mA, 676 mV.

4.18 99, 1.46 A.

4.19 a) I_E at 1 mA; V_E from – 0.6 V to – 0.8 V; I_C from 0.909 mA to 0.997 mA; b) V_C from 5.45 V to 5.02 V; V_E has no effect on I_C or V_C; c) $R_C =$ 10 kΩ.

4.20 $i_C = \alpha(I + i_e) = 1.1$ mA for high beta, and 1 mA for $\beta =$ 10; 9.90 kΩ, 1.82 Vpp; 1.65 Vpp.

4.21 9.3 kΩ, 1.021 mA, 9.7 kΩ, – 2.8 mV.

4.22 0.055 mA, 0.407 mA, 3.012 mA.

4.23 77.8 kΩ, 167 V; 1.7 MΩ; 17 KΩ.

4.24 122.5 μA.

4.25 a) – 3.3 V, – 5.39 V, 1 mA, 0.98 mA, 19.6 μA; b) – 5.3 V, – 5.5 V, 1.606 mA, 0.957 mA, 0.648 mA; c) – 1.3 V, – 3.39 V, 1 mA, 0.98 mA, 19.6 μA; d) 0 V, – 10 V, 0 mA; e) – 4.7 V, – 3.67 V, 1.128 mA, 1.105 mA, 22.1 μA; f) – 6.7 V, – 2.7 V, 0.702 mA, 0.688 mA, 13.8 μA.

4.26 a) 6.6 kΩ, 12 kΩ; b) 10.6 kΩ, 8 kΩ.

4.27 a) 0.930 mA, 1.023 mA, 93 μA, 8.14 V, 0 V, 0.7 V; b) 0.930 mA, 1.023 mA, 93 μA, 1.86 V, 10 V, 9.3 V; c) 0.230 mA, 0.253 mA, 23 μA, 9.54 V, 0 V, 0.7 V; d) 0 mA, 0 mA, 0 μA, – 10 V, 0 V, 0 V.

4.28 a) 1.55 V, – 0.423 V, – 1.123 V, 0.845 mA, 42.27 μA, 0.888 mA; b) 3.55 V, – 3.23 V, – 3.93 V, 0.645 mA, 32.26 μA, 0.677 mA; c) – 8.34 V, – 7.56 V, 8.26 V, 0.166 mA, 8.29 μA, 0.174 mA.

4.29 3 V, 2.3 V, 7.7 V; 2.465 V, 1.765 V, 8.23 V; 1.313 V, 0.613 V, 9.44 V.

4.30 120.

4.31 a) 2.3 mA, 2.1 V; b) 2.156 mA, 2.575 V; c) 1.432 mA, 4.964 V.

4.32 a) 0.582 mA, 3.86 V; b) 0.590 mA, 3.79 V; c) 0.571 mA, 3.96 V; d) 0.55 mA, 3.69 V.

4.33 5 V, 4.3 V, 10.7 V, 11.4 V, 3.6 V; 4.73 V, 4.03 V, 11.31 V, 12.07 V, 2.90 V.

4.34 40 μA/V, 4 mA/V, 40 mA/V, 4 A/V.

4.35 25 kΩ, 2.5 MΩ; 250 Ω, 25 kΩ; 25 Ω, 2.5 kΩ; 0.25 Ω, 250 Ω.

4.36 Gain = K/V_T, a constant.

4.37 37.8 kΩ, – 39.7 V/V.

4.38 – 4000 V/V, – 1000 V/V.

4.39 $g_m' = g_m r_e / (r_e + r_E)$, $r_\pi' = (\beta+1)(r_e + r_E)$, 9.9 mA/V, 10.1 kΩ.

4.40 a) 0.976 V/V; b) – 8 V/V; c) – 40 V/V; d) – 7.92 V/V; e) – 4.44 V/V.

4.41 – 297 V/V, 2.575 Vp, 8.67 mVp.

4.42 Increases by 49%, or reduces by 33%; 6.7 kΩ.

4.43 – 4000 V/V.

4.44 $r = \left[R_1 R_2/(\beta+1) + r_e(r_1 + R_2)\right] / (r_e + R_1)$: a) r_e; b) $2r_e$; c) $3r_\pi / (\beta+2) \simeq 3r_e$.

4.45 $Gain = -g_m(r_o \| R_f)$, $R_i = r_\pi \| (R_f/(1 + g_m(r_o \| R_f)))$; $Gain = -g_m r_o/2$, $R_i = r_\pi \,||\, (2/g_m) \simeq 2r_e$.

4.46 See page 213: $r_\pi' = 2(\beta_1+1) r_{\pi_1}$, $g_m' = g_m/2$.

4.47 See page 213: $i_B = 5$ μA, $i_C = 1050$ μA, $v_{EC} = 4.75$ V; 2.9 V, 0.582 mA; 3.09 V, 0.618 mA; clipping for 50% of the cycle.

4.48 49.

4.49 30 Ω, 400 Ω, 200 Ω, 40 Ω; 101.2 mA, 4.995 V.

4.50 $R_E = 1.00$ kΩ, $R_B = 68$ kΩ.

4.51 For $R_B = 95.2$ kΩ, I_E varies from 1.056 mA to 0.787 mA, and V_{CB} from 0.5 V to 1.47 V. For $R_B = 100$ kΩ, I_E varies from 1.049 mA to 0.773 mA, and V_{CB} from 0.522 V to 1.516 V.

4.52 $R_\beta = 120$ kΩ, $R_B = 91$ kΩ; I_E varies from 1.047 mA to 0.700 mA and V_{BC} varies from 0.530 V to 1.78 V.

4.53 $R_\beta = 68$ kΩ, $R_B = 47$ kΩ; 0.484 V, 1.06 mA; 1.26 V, 0.844 mA.

4.54 176 kΩ; – 4.29 V
to – 0.876 V; – 2.44 V to – 0.50 V.

4.55 9.3 kΩ; from + 5 V to – 4.8 V, or so.

4.56 – 0.84 V, – 1.54 V, 8.4 mA, +1.6 V, 336 mA/V, 2.95 Ω, 298 Ω, 11.9 kΩ; Note constant voltages, inversely-scaled currents, and directly-scaled parameters.

4.57 289 Ω, – 336 mA/V, 0.922 kΩ, – 310 V/V, – 97.1 A/A, 36.2 V/V, 46.7 A/A; Note that for resistance-scaled designs, parameters scale correspondingly and gains are constant.

4.58 Use $R_E = 10$ kΩ: For $\beta = \infty$, – 0.7 V, +0.7 V, 0.83 mA, – 307 V/V, 3.26 mV; For $\beta = 90$, – 1.52 V, 1.60 V, 0.74 mA, – 64.1 V/V, 3.66 mV, with $v_s = 15.6$ mV.

4.59 – 37.5 V/V to – 79.0 V/V.

4.60 – 20.8 V/V to – 29.3 V/V.

4.61 – 199 V/V, 3.78 kΩ, – 109 V/V, 3.78 kΩ, 0.274 V/V, 5934 V/V.

4.62 71 Ω, – 4.55 V/V.

4.63 21.4 Ω to 8.3 Ω; 6.14 V/V to 6.92 V/V.

4.64 5.216 V, 4.516 V, 3.784 V, 0.153 mA; a) Source coupled to B, load to E, ground to C; b) Source to B, 10 kΩ from E to ground or 10/3 kΩ from E, with load to C; c) Source to B, ground to E, load to C; d)

Source to E, ground to B, load to C.

4.65 a) 0.968 V/V; b) – 0.5 V/V or – 1.0 V/V for an added 10/3 kΩ resistor; c) – 20.4 V/V; d) 20.4 V/V.

4.66 0.685 V/V.

4.67 a) – 30.8 V/V, 193 kΩ; b) – 27.4 V/V, 221 kΩ; c) 27 kΩ.

4.68 2.7 kΩ, $R_B \leq 0.448\beta$ kΩ.

4.69 a) Q_1 cut off and Q_2 saturated; b) Q_1 saturated and Q_2 cut off, with $\beta_{forced} = 1.12$.

4.70 a) 2.05 V, 2.75 V, 2.25 V, 3.44, 0.561 mA; b) 4.8 V, 5.5 V, 5.0 V, 9, 10.1 mA.

4.71 $2 \times 10^{-13}A$, $8 \times 10^{-12}A$; 0.9934, 0.0248; 0.0254.

4.72 673 mV.

4.73 Flow in the forward BCJ direction with (text) i_C negative (as defined); Reverse Active mode; 581 mV.

4.74 0.0203, 97.9 mV.

4.75 $\beta_{forced} = 0$, 58.0 mV to 57.7 mV.

4.76 49.3 mV, 48.7 mV, 12 Ω.

4.77 $\beta_{forced} = 0.1$; V_{ECsat} (mV): ∞, 58.1, 17.8, 6.1, 3.1, 0.75, 0.50.

4.78 199; V_{ECsat} (mV): ∞, 223, 168, 133, 114, 64, 40; In normal mode, 42.2 mV; In inverted mode, 12.9 mV.

4.79 a) $i_{DC} = 2.80$ mA, $i_{DE} = 12.87$ mA; b) 697 mV, 649 mV, 48 mV; c) 45.6 mV.

4.80 432 μA.

4.81 a) 0.85 V, 0.3 V, 0.76 V, 0.5 V; 0.088 V, 0.20 V; – 5.09 V/V; b) 1.65 V, 0.3 V, 0.76 V, 0.5 V; 0.89 V, 0.20 V; – 32 V/V.

4.82 1.67 V, 66.8 mV, 1.10 V, 0.64 V, 0.57 V, 0.57 V; – 3.43 V/V, – 28.8 V/V.

4.83 2 MΩ, 1.2 GΩ.

4.84 a) 50 V; b) 30 V; c) 7 V.

4.85 50 Ω, 0.05 V.

4.86 109, 90, 60 V.

4.87 4.71 V, 4.85 V.

4.88 1.15 % / °C, 0.63 %°C.

4.89 $4.69 \times 10^{-13}C$, 0.47 ns, 1.88 pF; $1.17 \times 10^{-11}C$, 11.8 ns, 47 pF.

4.90 0.208 pF, 90 fF; 2.09 pF, 2.92 GHz.

4.91 0.302 pF; 6.07 pF, 6.1 fF; 206 MHz.

4.92 20 ps, 80 fF, 80 fF, 0.9 V, 0.33, 3.2 pF, 160 fF, 3.36 pF, 48 fF, 7.5 GHz; 6.3 GHz.

4.93 15.6 MHz, 1.59 MHz; 501 MHz, 131 MHz.

Chapter 5

FIELD-EFFECT TRANSISTORS (FETs)

5.1 $\upsilon_{GS} \geq 5$ V, $\upsilon_{DS} \geq 1.5$ V, $\upsilon_D \geq 1.5$ V, $\upsilon_D \leq 1.5$ V.

5.2 a) Saturated mode; b) Triode mode; c) Cut-off mode; d) Saturated mode; e) N-channel, $\upsilon_D \geq -2$ V; f) P-channel, saturated mode; g) P-channel, $\upsilon_G \geq 1$ V.

5.3 9 mA, 8 V, 8 V, 4 mA, 4 mA, 3 mA, 10 V, 4 V.

5.4 See page 230.

5.5 $V_D \leq 2$ V, 400 μA, 300 μA, 175 μA, 2.5 kΩ, 0.04 V, 0.36 V.

5.6 6 V to 1.005 V, 826 μA, 0.909 V, 0.43 nA, 0.45 mV.

5.7 $\upsilon_{GS}-V_t$, 0.859 $(\upsilon_{GS}-V_t)$, 0.564 $(\upsilon_{GS}-V_t)$, 0.134 $(\upsilon_{GS}-V_t)$; 2 V, 1.72 V, 1.13 V, 0.26 V.

5.8 50 kΩ, 0.0093 V^{-1}, 107.5 V.

5.9 1.575 mA.

5.10 – 2 V, – 2.96 V.

5.11 a) 0.9 mA; b) 4 V; c) 0 mA; d) – 3 V.

5.12 a) – 4 V, 5 V, Cutoff, 0 mA; b) – 2 V, 3 V, Saturation, 4 mA; c) 0 V, 5 V, Saturation, 16 mA; d) 0 V, 2 V, Triode, 12 mA; e) – 1 V, 4 V, Edge of saturation, 25 mA; f) 2 V, 3 V, Triode, 35 mA; g) 2 V, 0 V, Triode, 0 mA; h) – 2 V, 2 V, Triode, 20 mA.

5.13 a) 0.4 mA; b) 0 V; c) 1 V; d) 0.9 mA, +1 V; e) 4.172 V.

5.14 0.296 mA, – 1.075 V.

5.15 0.4 mA, – 2 V, 7.5 kΩ.

5.16 0.4 mA/V^2, 2.43 V, 21.5%.

5.17 20 kΩ.

5.18 200 kΩ, 2.15 V.

5.19 2 V, V_t raised by 10.1%, K lowered by 18.4%.

5.20 4 mA, 11 V, 8 V, ≥ 14 V.

5.21 – 1.60 V.

5.22 20 kΩ; $M3 \parallel M2$; 10 kΩ.

5.23 9 mA, 7.5 V; 6.0 mA, 3.0 V.

5.24 0.076 kΩ, 3.12 to 1.

5.25 9 mA, 7.5 V, 6 mA/V, 6 mA/V, – 3.0 V/V, ± 1.5 V, + 5.875 V and 8.875 V, versus 6.0 V and 9.0 V.

5.26 2.83 mA/V, 3.53 kΩ; Linear for $\upsilon_{gs} << 5.66$ V, for 1%, 0.06 V peak, or for 10%, 0.6 V peak.

5.27 0.240 mA/V^2, 127 V, 2.5 V.

5.28 $(I+1)^{1/2}$ V, I mA, $2\sqrt{I}$ mA/V, 50/I kΩ, $-2R\sqrt{I}$ V/V, $10^4/(1+2R\sqrt{I})$ kΩ; – 100 V/V, 99 kΩ; – 50 V/V, 196 kΩ; – 73.0 V/V, 135 kΩ.

5.29 – 2.38 V/V, – 4.76 V/V, – 1 V/V.

5.30 1.34 kΩ; 0.971 V/V; > 63 kΩ, > 17 kΩ.

5.31 No; 1.12 kΩ; 0.814 V/V; > 1.78 kΩ.

5.32 0.189 mA, 5.22 V, 4.61 V, 1.89 V for cutoff, 0.276 mA, 3.48 V, 2.74 V, 2.76 V for cutoff.

5.33 10 MΩ, 8.6 MΩ, 1.75 kΩ, 5.33 kΩ.

5.34 0.311 mA, 2.79 V, 2 Vp; 0.358 mA, 1.85 V, 1 Vp.

5.35 0.189 mA, 5.22 V, 4.61 V, 0.276 mA, 3.48 V, 3.74 V.

5.36 See page 237: 3.0 kΩ.

5.37 10 MΩ, 2 MΩ, 2.1 kΩ.

5.38 0.25 kΩ, 0.375 kΩ, ∞ Ω, 10 MΩ.

5.39 45 μA, 55.6 kΩ; ≥ 1.5 V; 0.439 V.

5.40 10 μm, 100 μm, 400 μm; 1.5 V above the negative supply; 103.5 μA.

5.41 See pages 239 and 240; Topology A: 8, 66 μm, 14 μm or 28 μm; Topology B: 7, 62 μm, 14 μm or 28 μm; Topology C: 9, 36 μm, 16 μm.

5.42 a) 2 V, 1 V to + 4.5 V; b) – 2 V, – 1 V to 4.5 V; c) – 2.5 V to 6 V, – 4.5 V to 4 V.

5.43 – 63.2 V/V, – 632 V/V.

5.44 225 μA, 1.75 V, – 66.7 V/V; 3.5 V, 0.75 V.

5.45 1.707 V, 85.2 V/V, 11.8 kΩ.

5.46 v_O from 4.29 V to – 4.29 V, with v_I from + 8.41 V to – 3.12 V; 0.832 V/V, 5.88 kΩ; 5.88 kΩ.

5.47 1.002 mA/V, 20.62 kΩ, 20.6 V, ≥ 8.57 kΩ.

5.48 See page 242. There are 2 enhancement and 4 depletion configurations, with 6 in total, for which current flows. Of these, four allow saturation operation.

5.49 a) 2.5 V, 0.25 mA; b) 2.414 V, 0.343 mA; c) 3.82 V, 3.31 mA; d) 2.5 V, 0.25 mA; e) 2.5 V, 16.25 mA.

5.50 – 3 V/V, 3.5 V, 1 V, for v_I = 4 V, and 2.5 V for v_I = 3.5 V.

5.51 3.48 V, for V_t = 1.25 V, and χ = 0.172 and g_m = 0 μA/V; 0 V at V_t = 0.9 V, and χ = 0.323 and g_m = 82 μA/V.

5.52 – 2.62 V/V.

5.53 – 8.34 V/V; The output range is from 3.0 V to about 1.2 V.

5.54 3.3 V, 0.0 V, 1.86 V, 1.44 V, 1.44 V, 1.44 V, 1.65 V; 72.25 μA; 1.40 V, 1.90 V; 1.07 V, 2.23 V; 2 kΩ, 2 kΩ; 0.15 V, 3.15 V.

5.55 121 ps or 142 ps; a) 1.125 mW, b) 1.19 mW; 0.17 pJ; The match is poor since f is defined on a different basis!

5.56 1963 μm.

5.57 2.4%, 3.2%, 2.4%.

5.58 23.9 μm, 53.5 μm; 198 T, 994 T.

5.59 a) 26.25 fF, 306 fF, 26 fF, 214 fF, 214 fF; b) 2.62 fF, 282.6 fF, 2.6 fF, 21.4 fF, 21.4 fF.

5.60 a) 437 MHz and 138 MHz; b) 50.9 MHz and 29.9 MHz; for 100 μA and 10 μA, respectively.

5.61 3.47 kΩ, capacitive; 30.0 kΩ, capacitive.

5.62 10 mA, 10 mA, 7.5 mA, 0.586 V.

5.63 – 0.5 V, – 1.3 V.

5.64 100 Ω, 200 Ω, ∞Ω.

5.65 – 0.586 V, 15 kΩ, 0.004 V^{-1}, 250 V; or, alternatively 50.02 kΩ, 0.00403 V^{-1}, 248 V.

5.66 a) 4 mA; b) 1 V; c) 0.268 V; d) – 0.236 V.

5.67 1 mA, 0 V; 0.61 mA, 0 V.

5.68 10 mA, 5 V, 5 kΩ, – 50 V/V, 19.2 kΩ, – 25 V/V, 37 kΩ.

5.69 38.4 mA/V to 25.6 mA/V, 347 Ω to 781 Ω, 13.3 V/V to 20.0 V/V.

5.70 1.21 V to 2.09 V, – 2.2 V/V.

5.71 15 mA, 0 V, 30 mA/V, – 15 V/V.

5.72 0.144 V, 29.7 mA/V, – 12.9 V/V.

Chapter 6

DIFFERENTIAL AND MULTISTAGE AMPLIFIERS

6.1 a) 115 mV; b) 73.6 mV; c) 54.9 mV.

6.2 a) 0 V, 6 V; b) 1.3 V, 6 V; c) 2.0 V, 2.0 V, 10 V; d) 1 V, 2 V; e) 3.5 V, 10 V; f) – 4.7 V, 4 V; g) 4.0 V, 3.5 V, 10 V; h) 2.8 V, 10 V, 3 V.

6.3 a) 0.0 V, 9.6 V; b) – 0.695 V, 9.52 V, 9.68 V; c) – 0.664 V, 9.90 V, 9.30 V; d) 0.037 V, 9.35 V, 9.85 V; e) – 1.05 V, – 1.714 V, 9.30 V; f) – 1.758 V, 6.00 V, 6.00 V; g) – 0.752 V, 5.21 V, 6.79 V; h) – 0.722 V, 9.05 V, 2.95 V; i) 0.951 V, 0.228 V, 9.00 V.

6.4 $i_{C1} = \alpha I/2 + \alpha I/(2V_T)(\upsilon_d/2)[(1-(1/2)(\upsilon_d/(2V_T))^2-(1/4)(\upsilon_d/(2V_T)^2))]$, 10% (or 5%); 11 mV, 7.9 mV, 3.5 mV.

6.5 40 V/V, 20 V/V, 2.4 V.

6.6 400 V/V, 75.5 kΩ, 200 V/V.

6.7 50.25 kΩ; 56.94 V/V, 100 MΩ, 0 V/V, ∞ V/V, ∞ dB; 28.5 V/V, – 0.00995 V/V, 2864 V/V, 69.1 dB.

6.8 2×10^{-5} V/V, – 94 dB, 2.85×10^{6} V/V, 129 dB; 2×10^{-4} V/V, – 74 dB, 2.85×10^{5} V/V, 109 dB.

6.9 79.7 V/V, 12×10^{-4} V/V, 66.4×10^{3} V/V, 96.4 dB.

6.10 79.9 V/V, 1.5×10^{-4} V/V, 114.5 dB.

6.11 2.5 mV, 25 mV.

6.12 47.5 mV, 33.6 mV, 22.5 mV.

6.13 4 mV, 4.61 mV.

6.14 9.9 mV.

6.15 2.7 V.

6.16 250 Ω, 250 Ω, 500 Ω.

6.17 198, 1998.

6.18 60 Ω, 58.15 Ω; 0.990 A/A, 1.005 A/A; 0.982 A/A, 0.991 A/A, 0.996 A/A.

6.19 1.974 V, 0.474 V to 3.474 V.

6.20 20 kΩ, 2.7 V.

6.21 a) See pages 257 and 258: 9 BJTs; b) See page 215: 10 BJTs; c) 9 BJTs.

6.22 $I_O/I_R = 1/(1+3/(\beta^2+\beta))$.

6.23 $I_O/I_R = 1/((\beta_1/\beta_2)(\beta_2+1)/(\beta_1+1)+((\beta+1)/(\beta_1+1)+1)/(\beta_2(\beta_3+1)))$; $\beta_1 = (1-k)\beta$, $\beta_2 = (1+k)\beta$, $\beta_3 = \beta$;
$I_O/I_R = (\beta^2+\beta+k\beta-k^2\beta^2)/(\beta^2+\beta-k\beta-k^2\beta^2+2 = 1$, for $k = 1/\beta$.

6.24 $I_O/I_R = 1/(1+2(\beta_2-\beta_3+1)/(\beta_2\beta_3+2\beta_3))$; Use $\beta_1 = \beta$, $\beta_2 = (1-k)\beta$, $\beta_3 = (1+k)\beta$;
$I_O/I_R = (\beta^2-k^2\beta^2+2k\beta+2\beta)/(\beta^2-k^2\beta^2-2k\beta+2\beta+2) = 1$ for $k = 1/(2\beta)$.

6.25 For 1 μA, $R_E =$ 115 kΩ; For 10 μA, $R_E =$ 5.75 kΩ.

6.26 2 mA/V, – 3000 V/V, 1.5 MΩ, 76 kΩ, – 144.7 V/V.

6.27 1 mA/V, 3 MΩ, – 3000 V/V.

6.28 2 mA/V, 51.8 MΩ, 103.5×10^{3} V/V.

6.29 ± 0.2 V/V, ± 0.1 V/V, ± 0.02 V/V, for 10%, 5%, 1% drop, respectively.

6.30 1.35 V, 70.8 μA/V, 283.2 V/V, 141.6 V/V.

6.31 36 mV.

6.32 ± 1.25 mV, ± 1.25 mV, ± 0.6 mV; ± 3.1 mV, ± 1.87 mV.

6.33 1.95 V, 100 μA, 101 μA, 8 MΩ.

6.34 1.92 V, 91.9 μA, 92.9 μA, 8 MΩ.

6.35 $I_{O1} = I_{O2} =$ 50.8 μA, 5.6 kΩ, 800 kΩ; To improve, split both Q_3 and Q_2 into two parts, requiring a total device width of 4 W as in the original case, rather than 5 W as required by the initial modification.

6.36 – 20 V/V, 100 kΩ.

6.37 a) 400 μA/V, 250 kΩ, 10 MΩ, – 4000 V/V; b) 63.2 μA/V, ∞ Ω, 2 MΩ, – 126.4 V/V.

6.38 – 55.6×10^3 V/V.

6.39 – 17.6×10^3 V/V.

6.40 Normally, $R_{out} = 115 \times 10^9 \Omega$; $0.889 \times 10^9 \Omega$ with no Q_3, Q_6; $1.264 \times 10^9 \Omega$ with no Q_5, Q_2.

6.41 a) 1.25 mA, 2.5 V; b) 2.95 mA, – 0.13 V; c) 3.0 mA, 1.5 mA, – 0.12 V, 2 V; d) 0.12 mA, 4.94 V; e) 1.5 mA, 0.37 V.

6.42 a) – 4.5 V; b) – 3 V; c) 0.55 mA; d) 880 kΩ; e) 4.5 μA.

6.43 a) – 4.3 V; b) – 3.3 V; c) 1.1 mA; d) 8.16 MΩ; e) 0.49 μA.

6.44 $W_1 = W_3 =$ 43.5 μm, $W_2 =$ 21.7 μm; Diode drop is 0.75 V at 2.5 mA; 0.926 V/V; 31.1 kΩ or 26.5 kΩ.

6.45 $W_1 =$ 4.67 μm, $W_2 =$ 45.2 μm; Use $i_D =$ 0.502 mA; 1.76 MΩ; 2 μA change is consistent.

6.46 0 V, 0.88 MΩ, – 822 V/V.

6.47 $W_1 = W_2 = W_3 =$ 16.3 μm; Use $I =$ 1.0 mA; Needs an 11 mV offset to keep v_O at 2.25 V; 2450 V/V; Alternatively, with $W_1 = W_2 =$ 16.7 μm and $W_3 =$ 3.85 μm, and $I =$ 1.0 mA, the gain is – 1296 V/V.

6.48 a) ∞ Ω, 5 Ω, 80.9×10^3 V/V; b) 10.2 kΩ, 283 Ω, 4040 V/V.

6.49 Reduce R_4 to 1.15 kΩ, R_5 to 7.85 kΩ, R_6 to 0.75 kΩ, for which $A_3 =$ – 6.12 V/V, $A_4 =$ 0.998 V/V, $A =$ 8099 V/V, with $R_o =$ 71.4 Ω; With $R_L =$ 286 Ω, swing is from 9.8 V to – 4.1 V.

6.50 $R_0 =$ 18.6 kΩ, $R_1 =$ 12 kΩ, $R_2 =$ 12 kΩ, $R_3 =$ 2 kΩ, $R_4 =$ 1.3 kΩ, $R_5 =$ 10.7 kΩ, $R_6 =$ 2 kΩ; $R_{i1} =$ 10.2 kΩ, $R_{i2} =$ 2.55 kΩ, $R_{i3} =$ 67.6 kΩ, $R_{i4} =$ 102.3 kΩ; 3070.

Chapter 7
FREQUENCY RESPONSE

7.1 $T(s) = (1+R_1C_1s)/(1+R_1(C_1+C_2)s)$; a) Yes, low pass; b) See page 225: Pole at 182 rad/s, 0 at 2000 rad/s.

7.2 a) $T(s) = (10^2 s(1+s/10))/((1+s)(1+s/100)(1+s/10^5)(1+s/10^6))$; b) 0 V/V, 90°; 0 V/V, – 180°; c) Poles at $s = -1, -100, -10^5, -10^6$ rad/s; Zeros at $s = 0, -10, \infty, \infty$ rad/s; d) Gain is 10^3 V/V and phase 0° from about 10^3 to 10^4 rad/s; e) 60 dB, 57 dB; Better to prepare plots earlier, certainly by the end of part a), easiest after c), most useful before d).

7.3 57 dB, 39.8°; 52.8 dB, – 74.7°.

7.4 $A_m = 10^3$, $F_L(s) = s(s+10)/((s+1)(s+100))$, $F_H(s) = 1/((1+s/10^5)(1+s/10^6))$, $A_L(s) = 10^3 s(s+10)/((s+1)(s+100))$, $A_H(s) = 10^3/((1+s/10^5)(1+s/10^6))$.

7.5 $F_L(s) = s/(s+100)$, $F_H(s) = 1/(1+s/10^5)$, $A(s) = 10^3 s/((s+100)(1+s/10^5))$.

7.6 a) 100 rad/s; b) 99 rad/s; c) 99.00 rad/s.

7.7 a) 10^5 rad/s; b) 0.995×10^5 rads/s; c) 0.990×10^5 rad/s.

7.8 $T(s) = 10^3 s(s+10)/((s+1)(s+100)(1+s/10^6)(1+s/2\times10^6))$; 0.894×10^6 rad/s, 0.856×10^6 rad/s.

7.9 a) 0.333×10^6 rad/s, 0.447×10^6 rad/s, 0.374×10^6 rads/s; b) 0.476×10^6 rad/s, 0.673×10^6 rad/s, 0.601×10^6 rad/s; c) 0.833×10^6 rad/s, 0.905×10^6 rad/s, 0.895×10^6 rad/s.

7.10 a) 300 rad/s, 224 rad/s; b) 210 rad/s, 147 rad/s; c) 1200 rad/s, 1054 rad/s.

7.11 – 13.14 V/V, 2.28 Hz, 53.1 Hz, 334.3 Hz; 15.9 Hz.

7.12 $C_S = 30\ \mu F$, $C_C = 0.02\ \mu F$, $C_{C2} = 5\ \mu F$; 11.1 Hz, 1.14 Hz, 1.08 Hz; 0.53 Hz.

7.13 $\omega_z = 1/(C_S(R_S+r_S))$, $\omega_p = 1/(C_S(R_S+r_S(1+g_mR_S))/(1+g_mR_S))$; The equivalent transconductance is $g_m(R_S+r_S)/(g_mR_Sr_S+R_S+r_S)$; $r_S = 526\ \Omega$; $f_{ps} = 159$ Hz, $f_{zs} = 15.1$ Hz.

7.14 6.24 Hz, 8.4 Hz, 74.4 Hz, 1.94 Hz; – 18 V/V; 75.1 Hz.

7.15 $C_E = 37.2\ \mu F$, $C_{C1} = 12.0\ \mu F$, $C_{C2} = 4.2\ \mu F$; Alternatively, $C_{C2} = 16.2\ \mu F$, $C_{C1} = 3.12\ \mu F$.

7.16 – 6.58 V/V; 4.37 Hz, 8.42 Hz, 28.2 Hz, 1.86 Hz; 29.6 Hz.

7.17 a) 144.7 MHz; b) 155.3 MHz; c) 9.65 GHz.

7.18 62.1 fF, 8.1 fF, 2.04 GHz.

7.19 240 fF, 2220 fF, 140 fF, 120.2 fF.

7.20 4.76 pF, 0.577 pF, 0.368 MHz, 42.3 MHz, 0.37 MHz, 42.3 MHz, 387 Ω.

7.21 0.365 MHz, 234 MHz, 318 MHz, $f_H = 0.365$ MHz; 19.8 MHz, 392 MHz, $f_H = 19.8$ MHz.

7.22 0.477 pF, 500 MHz.

7.23 760 kΩ.

7.24 2.54 GHz, 4.51 pF, 13.8 MHz, 22.6 pF; $I_C \geq 0.22$ mA.

7.25 – 18 V/V, 3.0 MHz.

7.26 – 6.79 V/V, 6.6 MHz.

7.27 A positive pulse of 2.5 V amplitude, of 50 μs duration, with transitions of 7 ns, and sag of 1.6%.

7.28 20.7 V/V, 113 MHz.

7.29 – 13.6 V/V, 25.6 MHz.

7.30 0.963 V/V, 50.0 MHz, 15.6 Hz.

7.31 0.909 V/V, 1.46 MHz.

7.32 – 24.4 V/V, 3.96 MHz, 86 Hz.

7.33 a) – 18.7 V/V, 5.26×10^6 Hz, 68.5 Hz; b) – 15.0 V/V, 1.24 MHz, 51.2 Hz.

7.34 11.05 V/V, 7.75 MHz.

7.35 For R_L connected to the collector of the input transistor: – 7.12 V/V, 6 MHz; For R_L connected to the collector of the grounded-base amplifier: + 7.12 V/V, 4.54 MHz.

7.36 7.12 V/V, 29.3 MHz.

7.37 6.02 V/V, 13.3 MHz.

7.38 14.9 mV, 20 MHz, 88 MHz.

7.39 6.71 V/V, 30.5 MHz.

7.40 ($r_E = 167\Omega$); 3.66 V/V, 37.8 MHz.

7.41 723 MHz; 2.5 V/V, 119 MHz.

7.42 6.71 V/V, 30.61 MHz.

7.43 6.63 V/V, 22.1 MHz.

Chapter 8
FEEDBACK

8.1 0.01 V, 300 V/V, 0.33 V/V, 300 V/V, 3.0 V/V, 300 V; The output limits.

8.2 7.696, 0.115, 22 dB, 1 V, 0.115 V, 0.01 V, 4.12%.

8.3 As designed, 11; as fabricated, 6; 16.7%, 83.3.

8.4 47.9%.

8.5 10^6 Hz, 10^8 Hz, 10^8 Hz.

8.6 2×10^5 Hz, 0.707×10^4 V/V; The desensitivity factor maintains the gain for frequencies above cutoff.

8.7 100 V/V, 37 dB.

8.8 90.9 V/V, 1.1 mV, 0.1 V, 50 V/V, 20 mV, 1.0 V, 0 V/V.

8.9 a) Shunt-Shunt, $-1/R_2$; b) Shunt-Series, $-r/R_2$ or $-r/(+R_2)$; c) Series-Shunt, $R_1/(R+R_2)$; d) Series-Series, $rR_1/(R_1+R_2)$ or $rR_1/(r+R_1+R_2)$.

8.10 a) Series-Series, r; b) Shunt-Shunt, $-1/R_F$; c) Series-Shunt, 1; d) Series-Series, r.

8.11 2.22 S, 50 mΩ.

8.12 10.6 V/V, 11.2 kΩ, 1.06 Ω, 0.963 Vrms.

8.13 227 V/V, 0.05 V/V, 9.5 kΩ, 200 kΩ, 18.4 V/V, 478 kΩ, 47.1 Ω, 18.4 V/V.

8.14 10.9 V/V, 26 MΩ, 96.3 Ω, 26 MΩ, 96.3 Ω.

8.15 0.951 V/V, ∞ Ω, 166 Ω, 0.816 V/V, 29.8 mV.

8.16 0.974 V/V, 249 Ω, 0.998 V/V, 13.8 Ω.

8.17 26.0 V/V, 2.5 kΩ, 10 kΩ, 1.856 V/V, 362 Ω.

8.18 204.5 mA/V, 18.3 mA/V, 438 kΩ, 23.6 kΩ.

8.19 20.7 A/V, 11.9 Ω, 83.7 mA/V, 3.38 MΩ, 183 kΩ.

8.20 10 Ω, 10 Ω, 10 Ω, 260 mA/V, 72.2 mA/V, 244 kΩ, 36.8 kΩ.

8.21 18 MΩ, -10^{-5} A/V, 100 kΩ, 100 kΩ, – 2.8 V/μA, – 96.6 V/mA, 221 Ω, 17.5 Ω, – 93.3 V/mA, 9.65 kΩ.

8.22 Series-shunt at low frequencies; Shunt-series at high frequencies; – 0.167 A/A, 6 kΩ, 0.83 kΩ, – 183 A/A, 0.58 mA/V, 50 Ω, ∞ Ω.

8.23 – 1.0, 10 kΩ, ∞ Ω, – 0.98 V/V, 168 Hz, 3.98 Hz.

8.24 – 0.298 A/A, 4.45 kΩ, 1.24 kΩ, – 115.4 A/A, 0.335 mA/V.

8.25 1.77 μF, 1.14 μF.

8.26 $2\times(10/11+R_L)$ mA/V, 12.9 kΩ, 31 kΩ; For loads from 0 kΩ to 12.9 kΩ, the transconductance varies by only 3 dB, and by another 3 dB for R_L up to 31 kΩ.

8.27 0.65, 61.3 Ω, 26 Hz; 15.5 μF, 15.5%.

8.28 See page 293: 63.5 V/V, 1550 V/V, 5 Hz, 244 Hz, 3100 V/V, 48 V/V, 650 Ω.

8.29 1550 V/V, 63.3 V/V, 32.5 μF, 0.10 Hz, 2.45 Hz, 6.8 kΩ, 150 kΩ.

8.30 120 V/V, 1.21×10^4 V/V.

8.31 100.5 V/V, 0.0995 V/V.

8.32 a) 19.3 V/V, 0.951 V/V; b) 4.36 V/V, 0.813 V/V.

8.33 10^5 rad/s, 5 V/V, $\beta =$ 0.02 or less.

8.34 See page 294.

8.35 174.3 °, Oscillation does not occur; 454 Ω.

8.36 7.98 V/V, 2.5 MHz, 20 MHz; The pole is shifted by the amount of the feedback factor, namely 500×.

8.37 200, 199, 0.05, 19.9 V/V.

8.38 1.11, 1.11×10^4 V/V, with poles at 10^5 Hz and 9×10^5 Hz, 1.78×10^{-4}, 4×10^3 V/V.

8.39 $10.5\times10^6(1\pm j)$ Hz, 0.707, 14.9 MHz, 14.9 MHz, 0.043 V/V, 22.7 V/V.

8.40 See page 297: $\beta = 0.024$; poles at -1.41×10^5 rad/s and at $(-0.298 \pm j\ 0.298)\times10^5$ rad/s, 143°.

8.41 60°, 29°, 5.7°.

8.42 0.01, 0.001, 0.0032, 909 V/V, 306.5 V/V, – 15 °.

8.43 $-\tan^{-1}f/10^6-2\tan^{-1}f/10^8$; Margins are zero at $f = 1.01\times10^8$ Hz; Margins are 45° at $f = 4.3\text{x}^7$ Hz, with a gain of 196.2 V/V and $\beta = 0.0051$; Margins are 78 ° at $f = 1.4\times10^7$ Hz with a gain of 698 V/V and $\beta = 0.00143$.

8.44 $K = 1.11$, 900 V/V, 0.9 MHz, 0.9 MHz; $K = 0.10$, 10 V/V, 10 MHz, 10 MHz.

8.45 10^3 Hz, 10^2 Hz; 10^6 Hz, 10^6 Hz.

8.46 4.14×10^4 Hz, 4.14×10^3 Hz; 41 MHz, 41 MHz.

8.47 1.59 pF, 0.146 pF, 77.6 MHz, 10 MHz, 10 MHz; Factor of 2.4, 0.066 pF; Poles are at 24 kHz, 10 MHz, 39.8 MHz; 10 MHz, 4.1 MHz.

Chapter 9
OUTPUT STAGES AND POWER AMPLIFIERS

9.1 For a 1 kΩ load, peak load currents are 1.4 mA, 14 mA, 141 mA, with operation in modes A, A, AB, respectively; For a 0.25 kΩ load, peak load currents are 5.7 mA, 57 mA, 566 mA, with operation in modes A, AB, AB, respectively; Class B operation is possible, for large signals and low bias currents.

9.2 1.53 Vp, 2.7 Vp, ≥ 1.76 kΩ, ≥ 0.88 kΩ.

9.3 4.1 Vp, 4.56 mA.

9.4 a) 43.5 mW, 180 mW, 24.2%; b) 21.8 mW, 180 mW, 12.1%; c) 43.5 mA, 270 mW, 16.1%; d) 21.75 mW, 270 mW, 8.06%.

9.5 0 V, – 7 V for – 7.9 V input; 6.43 V for +7 V input; For v_O = 6.43 V: 41.3 mW, 238 mW, 17.4%; For v_O = – 7.0 V: 49 mW, 117 mW, 41.9%; For largest sine wave output of 6.43 Vp: 20.7 mW, 198.5 mW, 10.4%.

9.6 v_O is 9 Vp or 18 Vpp, with input of 20 Vpp for no load and 22 Vpp for a 10 kΩ load. Gain is 0.9 V/V or 0.82 V/V. Supply power is 0 mW or 6.36 mW. Load power is 0 mW or 4.05 mW. Efficiency is 100% or 63.7%.

9.7 +10.2 mV, +102 mV, +1.02 V.

9.8 For 6 Vp or 4.2 Vrms; 1.125 W, 1.43 W, 78.7%, 0.305 W. For 4 Vp: 0.5 W, 0.95 W, 52.6%, 0.45 W. For a +14.5 V supply and 6 Vp: 1.125 W, 1.73 W, 65%, 0.6 W.

9.9 2.5 mA, 1.35 V, 5.07 V, 0.982 V/V, 0.910 V/V, 0.995 V/V.

9.10 4.48 V.

9.11 21.9 mA, 1.00 V.

9.12 R_1 = 10 kΩ, R_2 = 8.7 kΩ; At the peak, 0.55 V/V; At 0 V, 0.84 V/V.

9.13 $r_{eq} = (r_e(R_1+R_2)+R_1R_2/(\beta+1))/(R_1+r_e) = (r_e(kR_1)+R_1^2(k-1)/(\beta+1))/(R_1+r_e)$, 114 Ω.

9.14 123.8 °C, 2.08 °C/W, 16l mA, 400 mV.

9.15 86.4 W, 102.5 °C, 0.29 °C/W, 1.68 °C/W, 10.4 cm, 1.79 cm, 52.3 cm.

9.16 42.3 W, 43.1 W; The problem lies in the transistor itself with its dominating thermal resistance.

9.17 0.595 Ω, 0.80 Ω.

9.18 2.76 mA, 15 Ω, 140 Ω.

9.19 a) 1.93 V; Current increases by more than 5 times; b) 2.04 V; Current increases by a few tens of %.

9.20 27 Ω, 100 mA.

9.21 For β from 50 to 150, g_m ranges from 204 mA/V to 207 mA/V, gain ranges from – 977 V/V to 981 V/V, R_n ranges from 946 Ω to 1010 Ω, R_{out} ranges from 4.75 kΩ to 4.79 kΩ; Rhat is, there is little effect.

9.22 95 kΩ, 1 μA, 5.05 kΩ, 0.0405 μA, 10.34 V, 10.56 V.

9.23 Raise R_1 to 115 kΩ with 57.5 kΩ in each half. Raise R_2 and R_3 to 2.3 kΩ and 57.5 kΩ, respectively. I_{12} reduces by 2.3 times while I_9 reduces by less than 2.3 times.

9.24 – 1818 V/V, 87.5 Hz.

9.25 28 Ω, Factor of 2.76, $R_5 = R_6$ = 35 Ω.

9.26 Use R_4 = 10 MΩ, R_3 = 100 kΩ, R_2 = 1 MΩ, R_1 = 101 kΩ, for an input resistance of 100 kΩ.

9.27 a) Drive A_1 as shown, but with R_3 connected to the output of A_1: R_1 = 10 kΩ, R_2 = 90 kΩ, $R_3 = R_4$ = 100 kΩ; b) Merge R_3 and R_1 into R_{13} = 10 kΩ with R_2 = 90 kΩ, R_4 = 100 kΩ, using 3 resistors overall.

9.28 7 V, 7.5 A, 6.0675 A, 18 A, 0.9 A/V, 3 A/V, 1.5 A/V.

9.29 7.12 V, 863 Ω; 0.429 of V_{14} must appear across Q_6; a) 0.927 V/V; b) 0.988 V/V; On average, the gain is 0.960 V/V.

9.30 0.1 mA; a) 0.899 V/V, 0 V, 1.22 V, – 1.22 V, 0 V; b) 0.952 V/V, 10.78 V, 12 V, 9.40 V, 10 V; On average, the gain is 0.928 V/V.

Chapter 10

ANALOG INTEGRATED CIRCUITS

10.1 0.733 mA, 0.2205 mA, 12 kΩ.

10.2 18.3 μA, 13.9 μA, 3.40 kΩ.

10.3 Use a transistor Q_{25} whose collector is connected to the emitter Q_{16}, emitter to a resistor R_{12} connected to $-V_{EE}$, and base to either the base of Q_{11} or the emitter of Q_7; 7.4 kΩ, 1 kΩ; The latter is best.

10.4 a) OK; b) From 8.4 V above the positive supply to – 20.7 V below the negative supply.

10.5 $V_T \ln(kI_{REF}/I_{C10}) = I_{C10}R_4$; $I_{C10} =$ 15.7 μA, 4 kΩ.

10.6 – 1.034 V.

10.7 $I_{C6}/I_{C3} = 1/(1+1/(2\beta^2)+1.19/\beta)$, 0.994.

10.8 For R_1 shorted, 0.727 A/A; for R_2 shorted, 1.375 A/A.

10.9 ± 16 mV.

10.10 38.1 kΩ, 56 kΩ.

10.11 For $R_6 = R_7 =$ 0 Ω, $R_{10} =$ 13.46 kΩ; For $R_6 = R_7 =$ 27 Ω, a design is not possible.

10.12 5.03 μA, 0.1 mA/V.

10.13 2.65 kΩ, 5.25 MΩ; Now 1000 V/V versus 815 V/V then.

10.14 0.497 mV, 67% larger.

10.15 ± 3.4%.

10.16 Add resistors R_E in series with the emitters of both Q_8 and Q_9, 18.2 kΩ, 15.6 MΩ, 0.0088 μA/V, 86.7 dB.

10.17 5.81 MΩ versus 4.0 MΩ previously.

10.18 55.4 Ω, 10.0 mA/V, 0.59 MΩ, 78.8 kΩ, – 788 V/V versus – 526.5 V/V previously, a gain increase of 50%.

10.19 4.38, – 421 V/V, 330 Ω.

10.20 21.7 mA, 21.6 mA, 17.9 mA.

10.21 111.2 dB, 2.77 Hz, 1.00 MHz, the same as before.

10.22 a) For 45°: 0.123 pF, 244 kHz; 0.0123 pF, 253 kHz; b) For 60°: 0.212 pF, 142 kHz; 0.0212 pF, 146 kHz.

10.23 154 V/μs, 1540 V/μs; 2.45 MHz, 24.5 MHz.

10.24 Class AB, 2.5 mA, 11.8×10^3 V/V, 13.5 pF.

10.25 For Q_1 through Q_8: I_D are 12.5, 12.5, 12.5, 12.5, 25, 25, 25, 25 μA; $|V_{gs}|$ are 1.35, 1.35, 1.35, 1.35, 1.50, 1.35, 1.50, 1.50 V; g_m are 70.8, 70.8, 70.8, 70.8, 100, 141.6, 100, 100 μA/V; r_o are 2, 2, 2, 2, 1, 1, 1, 1 MΩ; 35.4 V/V, 70.8 V/V, 2506 V/V; from 3.15 V to – 4.64 V on the input; from 4.5 V to – 4.65 V on the output.

10.26 For Q_1 through Q_8: I_D are 6, 6, 6, 6, 12, 12, 12, 12 μA; $|V_{gs}|$ are 1.28, 1.28, 1.35, 1.35, 1.40, 1.35, 1.40, 1.40 V; g_m are 42.5, 42.5, 34.6, 34.6, 60, 69.2, 60, 60 μA/V; r_o are 4.2, 4.2, 4.2, 4.2, 2.1, 2.1 2.1, 2.1 MΩ; – 89.3 V/V, – 72.7 V/V, 6488 V/V, from 3.32 V to – 6.5 V on the input; from 4.60 V to – 4.65 V on the output.

10.27 1.707 V, 2212 V/V, 3.3 mV.

10.28 6.76 pF, 14.5 kΩ, 1.10 MHz, 42.3°, 58.3 pF, 0.116 MHz; 1.78 V/μs, 0.206 V/μs.

10.29 a) 29.3 kΩ, 28.3 μA/V, 12.7 μA/V; b) 55.3 kΩ, 44.7 μA/V, 12.7 μA/V; $L = (2 - 1/m^{1/2})$; $m > 1$, to overcome component variations, but small enough to minimize $R_B \propto (1 - 1/m^{1/2})$.

10.30 Minimum voltage is $2V_{GS} - V_t$, $R_O = g_m r_o^2$; Thus, as measured by the output resistance and output-voltage overhead, the cascode and (modified) Wilson are the same.

10.31 484 MΩ, 18.7×10³ V/V.

10.32 $R_O = (g_{m4CC}\, r_{o4CC}\, g_{m4C}\, r_{o4C}\, r_{o3}) \,||\, (g_{m2C}\, r_{o2C}\, r_{o2})$, 243 MΩ, 14.4×10³ V/V, 3.23 V.

10.33 3.75 V, – 2.75 V, – 3 V, 242 MΩ, 9.36×10³ V/V.

10.34 62.8 μA/V, 6170 V/V, 162 Hz, 2 V/μs.

10.35 55.15 MHz, 550 MHz.

10.36 See page 319: 50 kHz, 21.7 Ω.

10.37 7 bits, 9.77 mV.

10.38 128 kΩ or 256 kΩ, 3.91 Ω or 7.81 Ω, 1.95 Ω or 3.91 Ω, 0.2%.

10.39 10 kΩ, 78.4 Ω, 78 Ω to 156 Ω.

10.40 5 bits.

10.41 See page 320: Use 3 comparators with references at – ½ V, 0 V, +½ V.

10.42 a) 0 V; b) 0 V; c) Saturated at ± 10 V; d) Stays saturated; Specifically, for i) $V_A > V_{REF}$, $V_O = 0$ V, 0 V, – 10 V, – 10 V, for a), b), c), d), respectively; for ii) $V_A < V_{REF}$, $V_O = 0$ V, 0 V, +10 V, +10 V for a), b), c), d), respectively.

Chapter 11

FILTERS AND TUNED AMPLIFIERS

11.1 See Table, page 323.

11.2 0 dB, – 1 dB, – 50 dB.

11.3 0.915 dB, 26 dB, 1.25.

11.4 2.86×10^3 rad/s, 100.5 rad/s, 28.5; 0.455×10^{-3} s, 0.35 kHz, 11.2 dB.

11.5 $T(s) = (s^3+0.1s^2)/(s^3+2s^2+1.89s+0.89)$.

11.6 See page 324.

11.7 For $N =$ 13, 43.9 dB, 0.21 dB, 44.8 dB; twelfth order would suffice.

11.8 Use $N =$ 8, $\omega_o =$ 136.7 krad/s; poles at $\omega_o(-0.1951\pm j0.9808)$, $\omega_o(-0.5556\pm j0.8315)$, $\omega_o(-0.8315\pm j0.5556)$, $\omega_o(-0.9808\pm j0.1951)$; $T(s) = \omega_o^8/[(s^2+0.3902\omega_o s+\omega_o^2)(s^2+1.111\omega_o s+\omega_o^2)(s^2+1.663\omega_o s+\omega_o^2)(s^2+1.962\omega_o s+\omega_o^2)]$, 10.1 dB, 42.3 dB.

11.9 For Butterworth, 12.45 dB; For 4 Chebyshev, 22.5 dB.

11.10 $N =$ 7, 48.5 dB, 0.074 dB, 43.8 dB.

11.11 For Butterworth, $N =$ 13, $\omega_o = 1.084\times10^3$ rad/s; Poles at $10^3(-0.131\pm j1.076)$ rad/s, $10^3(0.384\pm j1.013)$ rad/s, $10^3(-0.616\pm j0.892)$ rad/s, $10^3(-0.811\pm j0.719)$ rad/s, $10^3(-0.969\pm j0.505)$ rad/s, $10^3(-1.052\pm j0.259)$ rad/s, 1.08×10^3 rad/s; For Chebyshev $N =$ 7, $\omega_p = 10^3$ rad/s; Poles at $10^3(-0.057\pm j1.006)$ rad/s, $10^3(-0.160\pm j0.807)$ rad/s, $10^3(-0.231\pm j0.448)$ rad/s, $10^3(-0.256)$ rad/s.

11.12 10 kΩ, 100 kΩ, 159.2 pF, 100 kHz, 1 V/V; More precisely: $f_z =$ 100 kHz, 100 V/V, with $R_1 =$ 10 kΩ and $R_2 =$ 990 kΩ.

11.13 0.1 μF, 0.01 μF, 15.9 kΩ, 15.9 kΩ, with input resistance of 15.9 kΩ.

11.14 $T(s) = (R_{2a}/R_{1a})((s+1/(R_{1b}C_1))/(s+1/(R_{2b}C_2)))(s+1/(R_{2a} \| R_{2b}C_2))/(s+1/(R_{1a} \| R_{1b}C_1))$; $C_1 =$ 0.2 μF with $R_{1b} = R_{2b} =$ 79.58 kΩ and $R_{1a} = R_{2a} =$ 8.84 kΩ, $C_2 =$ 0.002 μF, for poles at 100 Hz and 1 kHz, and zeros at 10 Hz and 10 kHz.

11.15 $C_1 =$ 0.001 μF, $R_{1a} =$ 17.7 kΩ, $R_{1b} =$ 159 kΩ, $R_{2a} =$ 177 kΩ, $R_{2b} =$ 1.591 MΩ, $C_2 =$ 10 nF, with zeros at 100 Hz and 1 kHz, and poles at 10 Hz and 10 kHz.

11.16 $T(s) = (s-1/RC)/(s+1/RC)$, – 90 °; 524 Ω, 1051 Ω, 2.68 kΩ, 5.77 kΩ, 10 kΩ, 17.3 kΩ, 37.3 kΩ, 95.1 kΩ, 190.8 kΩ.

11.17 $T(s) = 200s/(s^2+200s+10^6)$, 414 rad/s, and 2414 rad/s.

11.18 a) $T(s) = s^2/(s^2+1.414s+1)$, with $Q =$ 0.707 and $\omega_o = \omega_{3db} =$ 1 rad/s; $A_{\min} =$ 12.3 dB; b) $Q =$ 1.707 with $\omega_o =$ 1.287; $T(s) = 0.707s^2/(s^2+0.754s+1.657)$ where $\omega_{3db} =$ 1 rad/s and $A_{\min} =$ 18.3 dB, a 6 dB improvement.

11.19 $T(s) = (s^2+377^2)/(s^2+s\,377/Q+377^2)$, with $Q =$ 6.136 and a 3 dB bandwidth of 9.78 Hz; 3dB frequencies at 55.3 Hz and 65.1 Hz; 1 dB frequencies at 49.0 Hz and 70.4 Hz; 1% frequencies at 25.9 Hz and 94.1 Hz.

11.20 796 pF, 264 μH, 10 Vrms.

11.21 530.6 pF, 0.0048 μH, 26 dB.

11.22 112.5 pF, 22.5 mH, $r_{series} =$ 283 Ω, $R_{parallel} =$ 707 kΩ.

11.23 $R_1 = R_2 = R_3 = R_5 =$ 10 kΩ, $C_4 =$ 0.1 μF for $L =$ 10 H and 1 nF for $L =$ 0.1 H; $R_1 = R_2 = R_3 =$ 10 kΩ, $C_4 =$ 10 nF, $R_5 =$ 100 kΩ or 1 kΩ.

11.24 a) $R_1 = R_2 = R_3 = 10^4\Omega$, $R_5 = 10^3\Omega$, $C_4 =$ 10 nF, $C =$ 254 nF, for $Q =$ 31.9; b) $R_1 = R_2 = R_3 = 10^4\Omega$, $R_5 = 10^5\Omega$, $C_4 =$ 10 nF, $C =$ 254 nF, for $Q =$ 0.32; c) $R_1 = R_2 = R_3 = 10^4\Omega$, $C_4 = C =$ 0.05 μF, $R_5 =$ 1.013 kΩ, for $Q =$ 6.28; with $C_4 = C =$ 0.1 μF and $R_5 =$ 253 Ω, $Q =$ 12.6.

11.25 For the first-order section: $\omega_o = 1/(2\pi 10^4)$, $Q =$ 0.5, $R_1 =$ 33 kΩ ∥ 33 kΩ, $R_2 =$ 33 kΩ, $C =$ 400 pF ∥ 80 pF; For one second-order section: $\omega_o = 1/(2\pi 10^4)$, $Q =$ 3.953, $C_4 = C_6 = C =$ 400 pF ∥ 80 pF, $R_1 = R_2 = R_3 = R_5 =$ 33 kΩ, $R_6 =$ 120 kΩ; For the other second-order section: $\omega_o = 1/(2\pi 10^4)$, $Q =$ 0.7198, $C_4 = C_6 = C =$ 400 pF ∥ 80 pF, $R_1 = R_2 = R_3 = R_5 =$ 33 kΩ, $R_6 =$ 24 kΩ.

11.26 Use $C_4 = C_6 = C$, with R_6 to control Q (and ω_o), and $R_1 = R_2 = R_3 = R$, with R_5 to control ω_o; $\omega_o = 1/(CR^{1/2}R_5^{1/2})$, $Q = R_6/(RR_5)^{1/2}$; R_6 must vary by a factor of 200.

11.27 $R_x =$ 2.7 kΩ, $R_1 = R_2 = R_{3a} = R_{3b} = R_f =$ 100 kΩ, – 39 V/V.

11.28 $R_L =$ 10 kΩ, $R_H =$ 22.5 kΩ, $R_B = \infty$ Ω, $R_F =$ 15.79 kΩ, $C =$ 1 nF, $R =$ 31.83 kΩ, $R_1 = R_2 = R_f =$ 10 kΩ, $R_3 =$ 190 kΩ.

11.29 $R_1 =$ 102.6 kΩ, $R_2 = R_3 = \infty$ Ω, $r =$ 100 kΩ, $R =$ 200 kΩ, $QR =$ 4 MΩ, $C_1 =$ 0, $C =$ 769 pF, $R_2 = R_3 = \infty$ Ω.

11.30 $C_1 = C_2 =$ 20 pF, $R_3 =$ 0.707 MΩ, $R_4 =$ 0.354 MΩ; Alternatively, use $C_1 =$ 10 pF, $C_2 =$ 20 pF, $R_3 =$ 1 MΩ and 62 kΩ in series, $R_4 =$ 470 kΩ; Alternatively, use $C_1 =$ 10 pF, $C_2 =$ 50 pF, $R_3 =$ 0.849 MΩ, $R_4 =$ 0.236 MΩ; Of the 3 solutions, the first with equal 20 pF capacitors is the most straightforward.

11.31 $R_3 =$ 637 Ω, $R_4 =$ 398 Ω, $R_4/\alpha =$ 318 kΩ, $R_4/(1-\alpha) =$ 398.5 Ω; $R_{in} =$ 318 kΩ at both low and high frequencies.

11.32 $Z_i = (R_4/\alpha)(s^2+s(1/R_3C_1+1/R_3C_2)+1/(C_1C_2R_3R_4))/(s^2+s(1/R_3C_1+1/R_3C_2)+(1-\alpha)/(C_1C_2R_3R_4))$ $= (R_4/\alpha)(s^2+\omega_o s/Q+\omega_o^2)/(s^2+\omega_o s/Q+(1-\alpha)\omega_o^2)$; $Z_i = R_4/(\alpha(1-\alpha))$, as s approaches 0; $Z_i = R_4/\alpha$ as s approaches ∞; at the centre frequency, $Z_i = R_4(1+j\alpha Q)/(\alpha(1+\alpha^2Q^2))$.

11.33 $C =$ 3.3 nF; For the first-order section, $R =$ 0.965 kΩ; For the lowest-Q Sallen and Key section, $R_1 =$ 0.998 kΩ, $R_2 =$ 0.933 kΩ; For the second Sallen and Key section, $R_1 =$ 1.281 kΩ, $R_2 =$ 0.727 kΩ; For the third Sallen and Key section, $R_1 =$ 2.68 kΩ, $R_2 =$ 0.348 kΩ.

11.34 $S_{C_{3a}}^{\omega_o} = -1/(2(1+k_3))$, $S_{C_{3b}}^{\omega_o} = -k_3/(2(1+k_3))$, $S_{C_{4a}}^{\omega_o} = -1/(2(1+k_4))$, $S_{C_{4b}}^{\omega_o} = -k_4/(2(1+k_4))$; All become – ¼; $S_{C_{3a}}^{Q} = -1/(2(1+k_3)$, $S_{C_{3b}}^{Q} = -k_3/(2(1+k_3))$; Both become – ¼; $S_{C_{4a}}^{Q} = 1/(2(1+k_4))$, $S_{C_{4b}}^{Q} = k_4/(2(1+k_4))$; Both become ¼.

11.35 $S_k^{\omega_o} =$ 0, $S_T^{\omega_o} = aT_0$ around $T = T_0$; $S_k^Q = -k^2$, $S_T^Q =$ 0.

11.36 0.1 pC, 0.1 μA, 10 MΩ, 50 mV in the negative direction, 400 cycles, – 50,000 V/s, 5,000 V/s.

11.37 $C_1 = C_2 =$ 2.000 pF, $C_3 = C_4 =$ 1.257 pF, $C_5 = C_6 =$ 1.777 pF, bandpass output, 10^5 Hz, 1.41×10^5 Hz, 1 V/V.

11.38 a) 57.2×10^6 rad/s, 87.7, 652 krad/s, – 51.1 V/V; b) 67.1×10^6 rad/s, 124, 540 krad/s, – 16.7 V/V.

11.39 a) 62.9×10^6 rad/s, 106, 592 krad/s, – 66.8 V/V; b) 70.3×10^6 rad/s, 139.2, 505 krad/s, – 19.6 V/V; Thus the tapped coil gives the best results in general; see page 341 for a comparison.

11.40 0.63 Ω, 25.1 kΩ, 126.7 pF, 8.38 kΩ.

11.41 10; Use three stages: For one stage, bandwidth is 3.16 MHz and selectivity is 31.6; For the cascade, bandwidth is 0.294 MHz and selectivity is 5.88.

11.42 $C_1 =$ 73.27 pF, $C_2 =$ 74.23 pF, $R_1 =$ 30.616 kΩ, $R_2 =$ 30.416 kΩ; Relative gain is 1.007.

Chapter 12

SIGNAL GENERATORS AND WAVEFORM-SHAPING CIRCUITS

12.1 $\omega_0 = 1/(LC)^{1/2}$, $G_m = n/R_o + nRC_{ls}/L + n/R_{cp} + 1/(nR_{in})$.

12.2 See pages 343 and 344: Note that amplifier polarity change can be accounted for by coil-tapped connection reversal.

12.3 $G_m = 1/(nR_o) + n/R_{cp} + n/R_{lp} + n/R_{in})$.

12.4 $R_1 = 10$ kΩ, $R_2 = 31.0$ kΩ, $R_3 = 5.6$ kΩ, $R_f = 50$ kΩ, 5.83 kHz.

12.5 For zeners which are normal {or better}, gain: for $\upsilon_O \approx 0$ is {– 1.33 V/V} (– 1.33 V/V), For $\upsilon_O >$ 0.7 V is – 1.18 V/V {– 1.32 V/V}, For $\upsilon_O >$ 7.5 V is – 0.67 V/V {– 0.67 V/V}, with the input threshold at 6.36 V {5.68 V}.

12.6 $C = 10$ nF, $R = 1.59$ kΩ, $R_1 = 3.3$ kΩ, $R_2 = 8.2$ kΩ.

12.7 0.280 MHz; The closed-loop op-amp gain must be 3.02/(1 + 0.755j).

12.8 $L(\omega) = (1+R_2/R_1)/\left[1+j\,\omega/(\omega_t(1+R_2/R_1)) - j/(\omega RC) + 1/(RC\,\omega_t(1+R_2/R_1))\right]$, $\omega = \left[\omega_t(1+R_2/R_1)/(RC)\right]^{1/2}$, $R_2/R_1 = 1/(RC\,\omega_t(1+R_2/R_1))$, $R_2 = 0.0625R_1$; Excess phase shift implies operation at a frequency lower than in the simple case; see page 347.

12.9 $L(s) = -R_f\,Cs/\left[1+(R_1+2R_2+3R_3)Cs+(R_1R_2+2R_2R_3+2R_1R_3)C^2s^2+R_1R_2R_3C^3s^3\right]$; $\omega_o = 1/(C(R_1R_2+2R_2R_3+2R_1R_3)^{1/2})$, which for $R_1 = R_2 = R_3 = R$ is $\omega_o = 1/(CR\sqrt{5})$ with $R_f = 6.26R$; $S^{\omega}_{R_1,R_2} = -0.3$, $S^{\omega}_{R_3} = -0.5$, $S^{R_f}_{R_1} = 0.174$, $S^{R_f}_{R_2} = 0.335$, $S^{R_f}_{R_3} = 0.497$.

12.10 $L(s) = -(R_f/R)\left[3+s(4RC+3R_f\,C)+s^2(R^2C^2+4RR_f\,C^2)+R^2R_f\,C^3s^3\right]^{-1}$.

12.11 Five sections needed, $\omega_o = 0.191/(RC)$, $|L| = 5804$ V/V.

12.12 $L(s) = -R_f\,C\left[1/(R^3C^3s^4)+6/(R^2C^2s^3)+10/(RCs^2)+4/s\right]^{-1}$, $\omega = 1.224/(RC)$, $R_f = 6.22$ R.

12.13 $R = 1.59$ kΩ, $C = 10$ nF, $Q = 12.5$, 1.78 Vpeak, $R_1 = 1.1$ kΩ, $R_6 = 19.9$ kΩ.

12.14 $C_2/C_1 = 10.83$, $C_1 = 2.767$ nF, $C_2 = 29.97$ nF; For high supply voltages, the peak voltage can be estimated as 6.66 V, 7.1 V or 8.1 V; For 3 volt supplies a peak signal of about 1 V results.

12.15 0.523 H, 0.0119 pF, 1325 Ω; $C_2/C_1 = 6.4$ for which $C_1 = 1.56$ pF with $R_1 = 0$, or $C_1 = 1$ pF with $R_1 = 4.1$ kΩ.

12.16 $R_1 = 1.07$ kΩ, $R_2 = 5.13$ kΩ; V_{IL} ranges from 1.04 V to 2.94 V, and V_{IH} ranges from 2.06 V to 3.96 V; Estimates of some critical delay times include 0.4 ns, 0.64 ns, 10 ns, 75.3 ns, 3 μs, (see pages 353 and 354).

12.17 $R_1 = 2.87$ kΩ with $R_2 = 7.8$ kΩ, or $R_1 = \infty$ Ω with $R_2 = 10$ kΩ and $R_3 = 5.6$ kΩ; $V_{TL} = +0.7$ V, $V_{TH} = +1.4$ V.

12.18 See page 355: $V_{IH} = 3.83$ V, $V_{IL} = 1.17$ V.

12.19 $R = 200$ kΩ, $R_1 = 33$ kΩ, $R_2 = 240$ kΩ, $R_3 = 3.9$ kΩ; The slope is 0.036 V/μs at $t = 0$, 0.028 V/μs at $t = 50$ μs, and 0.032 V/μs on average, $R = 424$ kΩ; Use $R_1' = 10$ kΩ and R_2' as 10 kΩ in series with 680 Ω.

12.20 $R_A = 1$ kΩ, $R_B = 7.2$ kΩ, wih thresholds of ± 1 V; $R_1 = 10$ kΩ, $R_2 = 82$ kΩ; $C = 1000$ pF, $R = 410$ kΩ, $R_3 = 1.5$ kΩ.

12.21 – 0.7 V, – 1.67 V, – 0.833 V, – 10 V, 1.07 V step, 0.25 ms pulse, > 118 V/s; R_4 controls recovery; Retriggering is possible 10 μs (plus the amplifier recovery time) after output falls.

12.22 54.2 μs; For very long input pulses, the loop is open, and the output fall time is an amplified version of that at node A; 124 ns.

12.23 $T = 0.11$ ms; Input less than 0.11 ms; For longer inputs, set and reset are applied together.

12.24 – 1.8%.

12.25 4.83 kHz, 66.7%; 12.1 kHz, 91.7%; 14.5 kΩ; There is no combination of resistors which will produce 10% duty cycle: Use 90% duty cycle and an inverter!

12.26 400 Ω; See the Table on page 359: The maximum error is about 8% at about 45°.

12.27 Progressively: $R_1 = 5$ kΩ, $V_2 = 2.4$ V, $V_3 = 6.4$ V, $R_2 = 1.15$ kΩ, $R_3 = 1.15$ kΩ; Errors: – 0.3 mA at 3 V, + 0.1 mA at 5 V, – 0.3 mA at 7 V, 0 mA at 10 V; With 1 mA diodes: $V_2 = 2.4$ V, $V_3 = 6.4$ V, $R_1 = 5$ kΩ, $R_2 = 1.139$ kΩ, $R_3 = 1.659$ kΩ; Errors: – 0.25 mA at 3 V, + 0.14 mA at 5 V, – 0.13 mA at 7 V, – 0.28 mA at 10 V.

12.28 Requires 7 amplifiers and 4 diodes in total, with extra inverters from input v_3 to the lower amplifier, and on the output.

12.29 This is an absolute-value circuit.

12.30 See page 362: $R_3 = R_1 || R_2 = R/2$ with $R \gg R_0$.

12.31 See page 361: R_1 is 56 kΩ in series with 620 Ω, $R_2 = 10$ kΩ, R_5 is to 12 kΩ resistors in parallel.

12.32 + 10 V, 0 V, + 10 V; Input resistance is infinite; The circuit is called a full-wave doubler rectifier.

12.33 Up to $v_I = 14$ mV, $v_O/v_I = 101$ V/V, with v_O reading 1.414 V; From $v_I = 14$ mV to $v_I = 150$ mV, $v_O/v_I = 51$ V/V, with v_O reading 8.35 V; Above $v_I = 150$ mV, $v_O/v_I = 1$ V/V.

12.34 The output is a dc level of 200 mV which remains when the input is lowered; $R_1 = 3.3 / (fC)$ to ground at the output; Output rises to represent the maximum peak-to-peak value of the combined signal at $f/100$; add $R_2 = 0.48 / (fC)$ from node B to ground.

Chapter 13
MOS DIGITAL CIRCUITS

13.1 See page 365; 0.0 V, 3.3 V, 1.2 V, 1.8 V; 1.5 V, 1.5 V; 1.2 V, 1.5 V.

13.2 See page 365; 2, 5, 5, 10; 1 ns; 1.091 ns, 0.909 ns.

13.3 See page 366; 198 μW, 0.182 pF, 0.198 pJ.

13.4 5 MHz; See page 366.

13.5 (3.0 μm/0.8 μm); 1.65 V, 1.3875 V, 1.9125, 1.3875, 1.3875.

13.6 $V_{th} = [V_{DD} - |V_{tp}| + \sqrt{k_n/k_p}\,V_{tn}]/[1 + \sqrt{k_n/k_p}]$.

13.7 Slope is: $2\,V_{Ap}\,V_{An}(V_{DD} - V_M - |V_{tp}|)(\sqrt{k_p/k_n} + 1)/(V_{Ap} + V_{An})$; – 38.1 V/V.

13.8 3.125 kΩ, 3.125 kΩ.

13.9 Both 640 μA; both 550 μA; both 150 μA.

13.10 V_{IL} from 1.59 V to 2.66 V, V_{IH} from 2.16 V to 3.59 V, V_{OL} always 0 V, V_{OH} from 3.75 V to 6.25 V; NM_H from 4.09 V to 0.16 V, NH_L from 2.66 V to 1.59 V.

13.11 22.5 fF; 77.3 ps, 77.3 ps, 77.3 ps.

13.12 $t_{PHL} = 1.6C/(k_n'(W/L)_n V_{DD})$; From constant current, t_{PHL} = 3232 C s; From Eq. 5.101, t_{PHL} = 3111 C s; From Eq. 13.18, t_{PHL} = 3434 C s; The constant-current approach, here, is the simplest.

13.13 $t_{PHL} = 1.37C/(k_n'(W/L)_n V_{DD})$; From constant current, $t_{PHL} = 1.23C/(k_n'(W/L)_n V_{DD})$; From Eq. 5.101, $1.31C/(K_n'(W/L)_n V_{DD})$; The constant current approach is good for rapid analysis.

13.14 90 μA, 13.5 μA, 50 μA, 63.5 μA total, 317.5 μW, 67.5 μW with no capacitor.

13.15 $t_{PHL} = t_{PLH}$ = 2.12 ns, 47.2 MHz, 1.25 pJ.

13.16 a) 0.8 to 1.3 V, 0 to 0.5 V, neither conducts; b) 0 to 1.3 V, 0.0 V, 1.3 V; c) 0.8 V, 0.5 V, no current flows; d) See page 370; e) 0.39 μs, 0.64 μs; f) 156 kHz.

13.17 V_{OH} = 5 V, V_{OL} = 0 V, V_{IL} = 1.275 V, V_{IH} = 2.02 V, V_{th} = 1.75V, t_{PLH} = 5.57 ns, t_{PHL} = 2.89 ns.

13.18 See page 71; not identical; variation in proximity to power-supply rail; 2 PUN, 2 PDN, 4 combinations.

13.19 See pages 371 and 372; 4 possible XOR units; 24 combinations.

13.20 See page 372; 42.2 μm^2, 12.6 × the area of a single inverter.

13.21 25.0 μm^2; 7.4 × the area of a single inverter.

13.22 17.8 μm^2; 5.3 × the area of a single inverter.

13.23 1.36 k; threshold is 2.10 V.

13.24 See page 374.

13.25 Inverter area is 39 units, NOR area is 72 units; inverter capacitance is 3 units, NOR capacitance is 9 units; For equivalent NOR, area is 111 units, and capacitance is 3 units.

13.26 r = 2; 3.3 V, 0.79 V, 2.80 V, 1.70 V, 2.16 V; High V_{OL} leads to additional leakage current in connected gate; noise margins are OK.

13.27 Mismatch of n and p drive capabilities; $V_{OL} \approx$ 1.5 V, $V_M \approx$ 2.30 V.

13.28 a) r = 4: $(W/L)_n$ = 1/1, $(W/L)_p$ = 1/2, $A = 3units^2$; r = 10: $(W/L)_n$ = 1/1, $(W/L)_p$ = 1/5, $A = 6units^2$, b) r = 4: $(W/L)_n$ = 2/1, $(W/L)_p$ = 1/1, $A = 3units^2$; r = 10: $(W/L)_n$ = 5/1, $(W/L)_p$ = 1/1, $A = 6units^2$; Current drive is $r/2$ × greater; c) r = 4: $(W/L)_n$ = 1.414/1, $(W/L)_p$ = 1/1.414, $A = 2.83units^2$; r = 10: $(W/L)_n$ = 2.236/1, $(W/L)_p$ = 1/2.236, $A = 4.47units^2$; Current drive is $\sqrt{r/2}\times$ better.

13.29 $r =$ 2.954; 3.3 V, 0.504 V, 2.414 V, 1.390 V, 1.958 V, 0.886 V, 0.886 V; 185 μA.

13.30 $r = (V_{DD} - V_t)^2 / [\alpha V_t (2V_{DD} - (\alpha + 2)V_t)]$; $r = 20.25 / [\alpha(9 - \alpha)]$, $r =$ 4.76.

13.31 a) $r =$ 5.61, $V_{OL} =$ 0.252 V; b) $r =$ 2.53; c) $r =$ 13.75, $V_{IL} =$ 0.789 V;
d) $r =$ 135.2, $V_{IL} =$ 0.62 V; e) $r =$ 4.02.

13.32 a) $t_{PLH} = 1.7C/(k_p V_{DD})$, $t_{PHL} = 1.7C/(k_p (r - 0.46) V_{DD})$; b) $t_{PLH} = 1.7C/(k_n V_{DD}/r)$, $t_{PHL} = 1.7C/(k_n V_{DD}/r)$, $t_{PHL} = 1.7C/(k_n (1 - 0.46/r) V_{DD})$; $t_{PLH}/t_{PHL} = (r - 0.46)$; $r =$ 1.46; 2.05 V, 1.18 V, $NM_H =$ 0.120 V, $NM_L =$ – 0.089 V; 1.46 × longer.

13.33 One input active: $r =$ 4: $t_{PLH} =$ 477 ps, $t_{PHL} =$ 135 ps; $r =$ 10: $t_{PLH} =$ 1.19 ns, $t_{PHL} =$ 125 ps; Two inputs active: $r =$ 4: $t_{PLH} =$ 485 ps, $t_{PHL} =$ 64.5 ps; $r =$ 10: $t_{PLH} =$ 1.21 ns; $t_{PHL} =$ 62 ps.

13.34 Area ratio is 18.2 to 1; Pseudo NMOS is 5.5% of the area of the CMOS.

13.35 $V_{OL} =$ 0.0 V, $V_{OH} =$ 2.24 V; 2.24 V; 15.9 μA, 0.064 V; 12.04 pF; $t_{PLH} =$ 68.3 ps, $t_{PHL} =$ 39.3 ps.

13.36 1.896 V; Q_R conducting reduces t_{PLH} and speeds up the upper end of the transition; $V_{OH} =$ 3.3 V; 56.25 ps.

13.37 0.0 V, 3.3 V; 18.9 fF; $t_{PLH} =$ 62.3 ps, $t_{PHL} =$ 50.1 ps.

13.38 See page 381.

13.39 $Y = A\ B\ C + A\ \overline{B}\ \overline{C} + \overline{A}\ \overline{B}\ C + \overline{A}\ B\ \overline{C}$; See pages 381 and 382.

13.40 See page 383. $F = C\ X\ Y\ \overline{Z}$; Use an additional p-channel MOSFET at the left, connected to the supply and $\overline{X}$; $(W/L)_4 =$ 10/5, $(W/L)_{1,2,3} =$ 60/5.

13.41 See page 383.

13.42 For the NAND: $t_{PLH} =$ 2.22 ns, $t_{PHL} =$ 1.31 ns; For the NOR: $t_{PHL} =$ 0.65 ns.

13.43 Internal node capacitances, $C_2 = C_3 =$ 7.73 fF; $C_{eq} =$ 30.4 fF; $C_{OL} =$ 42.5 fF; See page 385.

13.44 0.883 V; 0.727 V.

13.45 247 ps; 273 μA; 1.35 V.

13.46 $C_{X1} =$ 10.7 fF; $C_{Y1} =$ 10.1 fF; $t_{PLH} =$ 194 ps; Φ high for 1.94 ns.

13.47 1.3875 V, 1.9125 V; 1.65 V; – 38.0 V/V; 1444 V/V.

13.48 See page 386: Equal at (0 V, 0 V), (5 V, 5 V), (2.64 V, 2.64 V); gain is 74 V/V in the middle, and 0 V/V at the ends of the range.

13.49 1.41 V; 2.47 V.

13.50 See page 387.

13.51 $C =$ 26.4 fF; t_P to Q falling is 304 ps, t_P to $\overline{Q}$ rising is 474 ps.

13.52 $W_6 =$ 2.65 μm; $W_D =$ 3.04 μm.

13.53 $V_{th} =$ 1.414 V; For D high: At Q, $t_{PLH} =$ 725 ps; For D low: At Q, $t_{PHL} =$ 667 ps; Φ must be high for > 725 ps; $f_{max} =$ 657 MHz.

13.54 For D high, $t_P =$ 630 ps; For D low, $t_P =$ 653 ps; 769 MHz; faster, lower power, more reliable.

13.55 16 transistors; 19.2 μm; 33.6 μm; increases by 75%; See page 390.

13.56 See page 390; [3t, 3t] for both delays installed; [3t, 1t], for one delay installed; [1t, 1t], for no delays installed; Adding two more inverters increases the gap from 3t to 5t.

13.57 See page 391. Total width is 101 μm.

13.58 14.14 kΩ; 30 ns; regenerative turnoff is inhibited, and fall time increases to 28 ns; 215 ns; 15 ns.

13.59 30.2 μs, 4.96 V, 2.7 V, 0.024 V, 0.225 mA, 12.8 mA.

13.60 3.67 pF.

13.61 795 ns; C_i = 140 fF; C_o = 143 fF; For G_1, t_{THL1} = 84.5 ps,; t_{PHL1} = 42.2 ps; For G_2, $t_{PLH2} = t_{PHL2}$ = 51.0 ps, $t_T \approx$ 150 ps; Minimum trigger pulse is 92 ps; Maximum trigger pulse is 795 ns; 0 μA, 241 μA; capacitor voltage is 0 V; Recovery time is 1.83 μs, but can be much reduced with an added circuit. (See page 393).

13.62 22 kΩ with 220 kΩ at the gate.

13.63 This is an astable multivibrator; t_{high} = 5.5 μs; t_{low} = 10.6 μs. See page 394.

13.64 C_{eq} = 18.7 fF; t_p = 56.4 ps; f_{OSC} = 1.77 GHz; a) 1.34 GHz; b) 340 MHz.

13.65 See page 395.

13.66 9 bits; The 9 bits on the right are 0 0 1 0 0 1 1 0 0 ≡ 76_{10}.

13.67 Total gate area = 12.48μm^2; Square-cell dimensions: 5 μm × 5 μm; 1.39 mA pulling down; 3.7 μA pulling up; bit-line capacitance is 1.15 pF; 165 ps.

13.68 Net drive currents at V_M: 0.926 mA down, but essentially 0 up; switching is possible, but with only a one-sided drive; C_{eq} = 29.5 fF; Average switching drive is 2.10 mA; t_{reg} = 14.8 ps.

13.69 2.16 V, 62.3 μA to the cell, 31.3 μA from the cell; 3.6 μm; i_5 = 27.6 μA, i_1 = 105.1 μA: Writing cannot occur by Q_5 overpowering Q_1; i_6 = 240 μA, i_4 = 53.4 μA: Writing can occur by Q_6 overpowering Q_4, and the design is viable.

13.70 1.22 pF, minimum bit-line signals are + 23.5 mV and – 19.0 mV.

13.71 $1.12 \times 10^{12}\Omega$.

13.72 0.027 fF.

13.73 0.3%; 16384 amplifiers; 4.9%.

13.74 22.2 ns; 26.6 μm.

13.75 W_n is 3.17 × minimum width; 6.55 ns; 7.93 ns.

13.76 Word-line decoder has 10 input bits with 1024 outputs; Bit-line decoder has 8 input bits and 256 output lines; 10 bits; 10240 decoder NMOS, 1024 PMOS, 10 inverters.

13.77 256-line decoder has 8 layers with 510 transistors; 1024-line decoder has 10 layers with 2046 transistors; 2.47 kΩ; 1 layer!; increase device width and/or add buffers.

13.78 See page 399: 10 transistors.

13.79 1.56 μm × 1.04 μm, or perhaps 1.6 μm × 1 μm; $2^{19} = 5.2 \times 10^5$; 2^{14} = 16384 ≡ 16K words.

13.80 See page 399; 25 transistors in the array and 10 in the inverters; without the inverters, need 65 transistors; decoder needs 80 transistors.

13.81 See page 400 and 401.

13.82 a) high 1, b) low 1, c) low 1, d) high (threshold for) 1.

Chapter 14

BIPOLAR DIGITAL CIRCUITS

14.1 0.990, 0.200; $6.98 \times 10^{-13}A$, $3.46 \times 10^{-12}A$, $6.91 \times 10^{-13}A$; 617 mV, 88.2 mV; 6.68 mA, 0.368 mA; 6.68.

14.2 1.77 ns, 109 ns (or 14.9 ns), 56.3 ns (or 9.2 ns); 5.3 ns, 223 ns (or 30.6 ns, or other estimates), 117 ns (or 20.6 ns); 223 ns (or 5.3 ns).

14.3 380 ns, 18.1 ns, 1.89 ns.

14.4 15.2 ns, 35.8, 12.8 pF, 3 ns, 99.9 ns.

14.5 0.643 V, 0.625 V.

14.6 0.820 V, 0.972 V.

14.7 a) 16.7 μA; b) 1.376 V; c) 0.409 mA; d) 1.64 mA; e) Add another diode in series with D_3, D_4; $R_1 =$ 1.26 kΩ, 2.60 mA, fan out of 2.

14.8 0.41 mA, 2.31 mA.

14.9 1.46 mA, 64, 95.6 mV.

14.10 42.7 ns, 15.5 ns, 47.0 ns, 26.2 ns.

14.11 505 Ω, 15.5 ns, 5.16 ns.

14.12 $V =$ 2.0 V, $R =$ 600 Ω; For $V =$ 2.5 V, use $R =$ 1.0 kΩ for which $N \leq 21$; For $V =$ 2.0 V, use $R =$ 0.6 kΩ for which $N \leq 27$.

14.13 See page 409.

14.14 $V_{B3} =$ 0.7 V, $I_{1K} =$ 0.7 mA, $V_{C2} =$ 0.9 V, $I_{C2} =$ 2.56 mA, $V_{B2} =$ 1.4 V, $V_{B1} =$ 2.1 V, $I_{B1} =$ 0.725 mA, $I_{E1} =$ 35 μA, $I_{E2} =$ 3.33 mA, $I_{B3} =$ 2.59 mA, $V_{CEsat2} =$ 0.124 V, $V_{CEsat3} =$ 0.107 V, $N \leq 23$.

14.15 $V_{E1} =$ 0.3 V, $V_{B1} =$ 1.0 V, $V_{C1} =$ 0.5 V, $I_{B4} =$ 0.3 mA, $V_{B4} =$ 4.5 V, $V_{E4} =$ 3.8 V, $V_{out} =$ 3.1 V, $I_{B4} =$ 0.13 mA, $V_{B4} =$ 4.5 V, $V_{C4} =$ 4.64 V.

14.16 $V_{B3} =$ 0.7 V, $V_{B2} =$ 1.4 V, $V_{B1} =$ 2.1 V, $V_{E1} =$ 1.4 V, $V_{E4} =$ 1.96 V, $V_{B4} =$ 2.66 V, $V_O =$ 1.29V with $I_{C3} =$ 4.4 mA, $I_{B3} =$ 0.49 mA, $I_{E3} =$ 4.9 mA, $I_{E2} =$ 1.19 mA, $I_{E1} =$ 0.61 mA; $V_{CEsat1} =$ 72 mV.

14.17 a) $I_{C3} =$ 1.1 mA, $I_{E4} =$ 7.0 mA, $I_{B4} =$ 0.70 mA, $I_{C2} =$ 0.74 mA, $I_{E1} =$ 0.82 mA, $I_{B3} =$ 0.12 mA, $I_{C3} =$ 1.1 mA, with $V_O =$ 1.29 V and $I_L =$ 6.5 mA; b) $I_{C3} =$ 18 mA, $I_{B3} =$ 2 mA, $I_{E2} =$ 2.7 mA, $I_{C2} =$ 2.43 mA, $V_{C2} =$ 1.1 V, $I_{B2} =$ 0.27 mA, $I_{E1} =$ 0.46 mA, $V_O =$ 1.3 V, with Q_4 cut off.

14.18 $R_2 =$ 1.37 kΩ; Gross slope = – 1 V/V, $R_1 =$ 1.17 kΩ; Increase is 0.94 mA or 36%.

14.19 For source at – 55 °C and load at 125 °C: $NM_L =$ 0.70 V, $NM_H =$ 1.16 V; For source at 125 °C and load at – 55 °C: $NM_L =$ 1.42 V, $NM_H =$ 1.74 V.

14.20 4.76 mA, 17.8 ns.

14.21 1.34 V.

14.22 1.305 mA.

14.23 a) 0.75 V, 0.25 V, 19.6 μA, 0.98 mA; b) 0.767 V, 0.266 V, 0.04 mA, 0.96 mA; c) 0.810 V, 0.316 V, 0.216 mA, 0.784 mA; d) 0.825 V, 0.268 V, 0.392 mA, 9.61 mA; e) 0.810 V, 0.256 V, 0.216 mA, 9.78 mA.

14.24 a) Approximate values: 0.75 V, 0.25 V; 1.5 V, 1.00 V; 2.00 V, 1.5 V; 0.75 V, 0.25 V. b) More precise values: 776 mV, 244 mV; 1563 mV, 1064 mV, 1566 mV, 1064 mV, 761 mV, 321 mV.

14.25 157 Ω.

14.26 a) 6.02 Ω; b) 2.04 Ω.

14.27 a) 0.675 mA, 0.675 mA; 3.73 mW, 3.73 mW; b) 0.213 mA, 33.5 mA; 1.07 mW, 167.5 mW; DP = 54.9 pJ.

14.28 a) V_{OH} = – 0.75 V, V_{OL} = – (RI + 0.75)V; b) $V_{th}A$ = – (RI/2 + 0.75)V, V_{BE} = 682.7 mV; c) 748.7 mV; d) 675.1 mV, 73.6 mV; e) V_{IL} = – (RI/2 + 0.824)V, V_{IH} = – (RI/2 + 0.676)V; f) NM_H = (RI/2 – 0.074)V, NM_L = (RI/2 – 0.074)V; g) IR = 0.442 V; h) V_{OH} = – 0.750 V, V_{OL} = – 1.192 V, V_{IH} = – 0.897 V, V_{IL} = – 1.045 V, V_R = – 0.971 V.

14.29 285 Ω, – 0.750 V, – 1.89 V, – 1.493 V, – 1.147 V, 0.397 V, 0.397 V.

14.30 285 Ω; Use 2 diodes for V_{OH} = 2.75 V; Use R_1 = 570 Ω for V_{OL} = 0.47 V.

14.31 R_2 = 148 Ω, I_{R2} = 0.95 mA, V_{OH} = – 0.75 V or slightly more, and V_{OL} = – 1.04 V to – 1.14 V.

14.32 0.205 V.

14.33 9.44 V/V.

14.34 *Fanout of 1:* t_{PHL} = 173 ps; t_{PLH} = 132 ps, t_P = 152 ps; *Fanout of 10:* t_{PHL} = 1.11 ns, t_{PLH} = 174 ps, t_P = 642 ps.

14.35 2 cm.

14.36 Use sources: 4 mA, 0.66 mA, 0.66 mA to replace R_E, R_2, R_1, respectively; Raise R_{C1} to 245Ω = R_{C2}; For V_R: 2 mV/°C; For V_{OL}: 1.8 mV/°C; For V_{OH}: 2 mV/°C. See page 418.

14.37 a) 2.24 V, 5.0 V, 0.0 V; b) 2.24 V, 4.17 V, 0.68 V.

14.38 See page 419; 1.25 kΩ with output high; 1.25 kΩ with output low.

14.39 On the short term, V_{OH} = 2.6 V, V_{OL} = 0.7 V; On the longer term, V_{OH} = 2.8 V, V_{OL} = 0.5 V; For a workable design, Q_N must be made 6 times wider than the minimum. For this choice, V_{IL} = 1.61 V, V_{IH} = 1.91 V; Peak currents are 27.9 mA, pulling up, and 76.0 mA, pulling down; Corresponding currents at $V_{DD}/2$ are 16.2 mA and 58.9 mA; t_{PLH} = 0.52 ns, t_{PHL} = 0.17 ns; With the addition: On the short term, V_{OH} = 2.6 V, V_{OL} = 0.7 V; On the long term, V_{OH} = 3.3 V, V_{OL} = 0.0 V; $t_{TLH} \approx t_{THL} \approx$ 125 ns.

14.40 On the short term, V_{OH} = 2.6 V, V_{OL} = 0.0 V; On the long term, V_{OH} = 3.3 V; t_{PLH} = 0.64 ns, t_{PHL} = 0.65 ns; Supply current is 5.6 mA.

14.41 See page 422. $(W/L)_p$ = 2W/L, $(W/L)_n$ = W/L, V_{th} = 2.06 V.

14.42 V_{OH} = 0.7 V, V_{OL} = 0.114 V, V_{IL} = 0.51 V, V_{IH} = 0.497 V, NM_H = 0.203 V, NM_L = 0.396 V.

14.43 0.544 mA, 0.816 mW, 13.4 ps, 35.2 μW, 0.011 pJ.

14.44 V_{OH} = 0.7 V, V_{OL} = – 1.27 V, V_{IL} = – 0.26 V, V_{IH} = – 0.16 V, NM_H = 0.86 V, NM_L = 1 V; a) V_{OH} = 0.7 V, V_{OL} = – 1.29 V, V_{IH} = – 0.234 V, V_{IH} = – 0.146 V, NM_H = 0.85 V, NM_L = 1.06 V; b) V_{OH} = 0.7 V, V_{OL} = – 1.09 V, V_{IL} = – 0.293 V, V_{IH} = – 0.184 V, NM_H = 0.88 V, NM_L = 0.80 V; c) V_{OH} = 0.7 V, V_{OL} = 0.032 V, V_{IL} = – 0.582 V, V_{OL} = – 1.37 V, V_{IH} = – 0.532 V, NH_H = 1.23 V, NH_L = 0.79 V.

14.45 W_S = W_L = 20 μm, W_{PD} = 10 μm, V_{OH} = 2.1 V, V_{OL} = 0.13 V, V_{IL} = 1.14 V, V_{IH} = 1.24 V, NM_H = 0.86 V, NM_L = 1.01 V; For a fanout of 4, reduce W_{PP} to 2.5 μm.

STANDARD COMPONENT VALUES

• Standard 5% Values with Marked $\overline{10\%}$ and $\underline{20\%}$ Values

$\underline{\overline{1.0}}$, 1.1, $\overline{1.2}$, 1.3, $\underline{\overline{1.5}}$, 1.6, $\overline{1.8}$, 2.0, $\underline{\overline{2.2}}$, 2.4, $\overline{2.7}$, 3.0, $\underline{\overline{3.3}}$,

3.6, $\overline{3.9}$, 4.3, $\underline{\overline{4.7}}$, 5.1, $\overline{5.6}$, 6.2, $\underline{\overline{6.8}}$, 7.5, $\overline{8.2}$, 9.1, $\underline{\overline{(10)}}$.

• Standard 1% Values with Marked "$\underline{\textit{Unit}}$" Values

$\underline{1.00}$, 1.02, 1.05, 1.07, 1.10, 1.13, 1.15, 1.18, 1.21, 1.24, 1.27, 1.30, 1.33, 1.37,

1.40, 1.43, 1.47, 1.50, 1.54, 1.58, 1.62, 1.65, 1.69, 1.74, 1.78, 1.82, 1.87, 1.91,

1.96, $\underline{2.00}$, 2.05, 2.10, 2.15, 2.21, 2.26, 2.32, 2.37, 2.43, 2.49, 2.55, 2.61, 2.67,

2.74, 2.80, 2.87, 2.94, $\underline{3.01}$, 3.09, 3.16, 3.24, 3.32, 3.40, 3.48, 3.57, 3.65, 3.74,

3.83, 3.92, $\underline{4.02}$, 4.12, 4.22, 4.32, 4.42, 4.53, 4.64, 4.75, 4.87, $\underline{4.99}$, 5.11, 5.23,

5.36, 5.49, 5.62, 5.76, 5.90, $\underline{6.04}$, 6.19, 6.34, 6.49, 6.65, 6.81, $\underline{6.98}$, 7.15, 7.32,

7.50, 7.68, 7.87, $\underline{8.06}$, 8.25, 8.45, 8.66, 8.87, $\underline{9.09}$, 9.31, 9.53, 9.76, $\underline{(10.0)}$.

• "Unit" Values

1, 2, 3, 4, 5, 6, 7, 8, 9, (10).

• Tens Values (with {Geometric} and [Arithmetic] "Means")

1, {3.16}, [5], (10).